Anaerobic Bacteria
in Human Disease

Anaerobic Bacteria in Human Disease

SYDNEY M. FINEGOLD, M.D.

Infectious Disease Section
Wadsworth Hospital Center
Veterans Administration
and Department of Medicine
University of California
School of Medicine
Los Angeles, California

ACADEMIC PRESS   New York   San Francisco   London   1977
A Subsidiary of Harcourt Brace Jovanovich, Publishers

ACADEMIC PRESS, INC.
111 Fifth Avenue, New York, New York 10003

United Kingdom Edition published by
ACADEMIC PRESS, INC. (LONDON) LTD.
24/28 Oval Road, London NW1

Library of Congress Cataloging in Publication Data

Finegold, Sydney M
    Anaerobic bacteria in human disease.

    Includes bibliographical references and index.
        1.   Bacterial diseases.   2.   Bacteria, Anaerobic.
I.   Title.   [DNLM: 1.   Bacterial infections.
2.   Bacteria—Pathogenicity.   WC200 F495a]
RC115.F5   616.9'2   75-44766
ISBN 0—12—256750—1

TO MY FAMILY

Contents

List of Tables

xiii

Preface

This book is concerned with disease in humans caused by anaerobic bacteria. While the approach is primarily a clinical one, it is hoped that microbiologists will also benefit from it, just as clinicians find it valuable to read treatises on microbiology and other basic sciences. There has never been a clinically oriented monograph in this field. Some books (Beerens and Tahon-Castel, "Infections á Bactéries Anaérobies," and Prévot, "Biologie des Maladies dues aux Anaérobies") present a certain amount of clinical data but primarily in the form of listings of various types of clinical specimens from which anaerobes have been isolated. "Anaerobes in Clinical Medicine " by Kozakai and Suzuki and colleagues does have quite a bit of clinical information in addition to a detailed bacteriologic treatise. Unfortunately, it is available only in Japanese.

Much of the literature on anaerobes suffers, to some extent, from too narrow an approach to the subject. Many clinicians have failed to give adequate bacteriologic data and many microbiologists have not given proper clinical information. At times, information presented is confusing or even misleading. When a specimen is indicated as "sinus drainage," is the writer referring to the paranasal sinuses, to a venous channel, to an abnormal channel created in the course of an infection, or to one of the hollow spaces or recesses in the body so designated? Indication that a specimen has come from an "ulcer" or a "wound" is similarly inadequate. Does "empyema" refer to pleural infection or empyema of the gallbladder? If the specimen is urine, was it obtained during voiding (midstream or other?), by urethral or ureteral (if the latter, which side?) catheterization, by percutaneous bladder aspiration from the renal pelvis at surgery, or in some other manner? Was the organism recovered really a *Bacteroides fragilis* or even "*Bacteroides* species," or was it simply a nonpleomorphic, obligately anaerobic, gram-negative bacillus not further studied? Or were appropriate tests done and simply not reported? Is it the coccobacillary *Bacteroides melaninogenicus* or a destained anaerobic gram-positive coccus whose colonies also develop black pigment? Was the *Bacteroides fragilis* ss. *fragilis* recovered from the peritoneal cavity involved in peritonitis or was it normal colonic flora escaping through a fecal fistula? Was the growth of *Escherichia coli* from the wound so light that it may have represented either chance contamination or a very minor secondary organism in the infection? Unless the laboratory gives some type of quantitation, one can be misled. It is remarkable that some clinicians will write an excellent clinical treatise with all the details of historical and physical findings, radiologic and other studies, clinical course, and response to therapy and will either completely ignore bacteriologic findings or treat them in a cursory fashion. At times clinicians do not seem to regard it unusual that a significant percentage of cultures of frankly purulent material grow nothing, even when bacteria are seen on direct smear. Often direct smears will not even have been done. Similarly, some bacteriologists will present a superb microbiologic treatise with extensive bacteriologic and biochemical studies, but will ignore the fundamentals of proper specimen collection and transport (especially important with anaerobes), to say nothing of details of the clinical picture. These narrow approaches and lack of precision have been common in the literature on anaerobic infections and have certainly interfered with our understanding of these common and important problems. Let us have effective collaboration between clinicians and bacteriologists; both groups will benefit enormously, as will the victims of the diseases to be discussed. Of course, there are also clinicians and microbiologists who are not even thorough and accurate in their own disciplines. One must be very careful and critical in evaluating the literature.

Anaerobic infections are quite common but undoubtedly have been the most commonly overlooked of bacterial infections. There are a number of reasons for this. The primary one, of course, has to be lack of awareness or interest on the part of clinicians and laboratory workers. In the past, anaerobic culture and identification procedures were cumbersome, and classification schemes were much more complicated than they are now. A major problem is that anaerobic infections are frequently mixed infections. The mixtures may include six or eight organisms or more and there may be aerobic or facultative forms along with anaerobes. In the latter case, anaerobes can be overlooked entirely unless culturing is done carefully with the use of direct examination of the specimen for quality control and the use of selective as well as nonselective media. There has certainly been a great deal more recognition of these infections, both in the clinic and in the laboratory, in recent years. It is hoped that the present work will stimulate further interest on the part of medical students, clinicians, dentists, and bacteriologists. With regard to physicians, anaerobic infections really cross all lines. Any tissue or organ in the body may be affected by these organisms, and all clinicians, regardless of specialty or subspecialty interest, must deal with anaerobic infections.

Beyond the frequency with which they occur, anaerobic infections and intoxications are important because they may be fulminating and destructive. The course of illness is frequently prolonged and there is significant mortality with some of these conditions.

Since anaerobic infections are common and significant and since good treatment depends on early, specific diagnosis, it is very important for clinicians to be aware of these infections and of the proper way of obtaining and submitting specimens to the laboratory. Clinical microbiologists, also, must be able to culture, isolate, and identify the organisms involved.

The approach in this book will be to discuss, after certain general considerations, clinical aspects of anaerobic infections as they present themselves to the physician or dentist. Thus, there will be a discussion of specific infections presented in terms of the region of the body involved. To some extent this method of grouping will also coincide with areas of interest to particular specialists. This approach is used since the physician sees the patient first as a clinical problem of pneumonia or brain abscess, rather than as a *Clostridium ramosum* or anaerobic streptococcal infection. Next will follow a discussion of pediatric infections, miscellaneous anaerobic infections, then the intoxications, and a section on therapy and prophylaxis of these conditions. Finally, there is consideration of the role of anaerobes in miscellaneous pathologic conditions.

Inasmuch as anaerobic infections are frequently very much like similar infections due to aerobic or facultative bacteria except for a few distinctive features (such as foul odor or unusual morphology of organisms on direct gram stain of exudates), the presentation will usually not dwell on details of the clinical picture, general aspects of diagnosis, etc., but will concentrate instead on the distinctive features which may be noted. Such unique diseases as botulism, tetanus, and anaerobic cellulitis will be treated more extensively. *"This is not the end. It is not even the beginning of the end. But it is, perhaps, the end of the beginning"* (*Winston Churchill*).

SYDNEY M. FINEGOLD, M.D.

Acknowledgments

This monograph would not have been possible without the encouragement and assistance of many people. My parents have my deepest gratitude and respect for their many years of self-sacrifice which permitted me to complete my training and embark on an academic career. My wife, Mary, and children, Joe, Pat and Mike, have been most understanding and helpful over the years, especially during the recent period of concentrated effort on this book. Each has also helped in concrete ways.

I owe a great deal to my teachers: my bacteriology professors at UCLA (particularly Dr. Meridian Ball, Dr. T. S. Beckwith, and Dr. A. J. Salle); my clinical teachers at the University of Texas Medical Branch, Galveston (especially Dr. Raymond Gregory, Dr. Edgar Poth, and Dr. C. T. Stone); my microbiology professors at Galveston (especially Dr. Ludwig Anigstein and Dr. Morris Pollard); and the infectious disease clinicians who provided the final phase of my formal training (Dr. Wesley Spink and Dr. Wendell Hall at the University of Minnesota Medical School and Dr. William Hewitt of the UCLA Medical Center). All of these people were gifted teachers who had the capacity for making their special disciplines appear very exciting to a young student.

In Dr. Poth's laboratory, I first studied intestinal microflora; this work ultimately led me into the field of anaerobic bacteriology. Dr. Spink's enthusiasm and capacity for hard work were inspiring. Dr. Hall was most impressive as a precise clinician and microbiologist. Dr. Hewitt's clinical judgment and ability as a teacher provided me with important goals.

While working with Dr. Hewitt, I was studying the effect of neomycin on fecal flora. I was puzzled to note that material from serial tenfold dilutions of feces in tubes of fluid thioglycolate medium produced turbidity to the 10^{-10} dilution whereas few or no organisms could be detected on (aerobic) plate media. Smears of the broth tubes revealed bizarre forms which I finally appreciated were bacteria (since they grew on transfer to other tubes of broth). At that time textbooks of bacteriology were not very helpful (some are still not). Ultimately I was able to get a little surface growth of a few strains on plate media with the use of crude setups for anaerobiosis. At that point, Dr. Hewitt said, "Fine. Now get 50 strains growing well in pure culture and do antibiotic susceptibility tests." To my great surprise, I was eventually able to do this. It soon became apparent to me that these same anaerobic bacteria were involved in various infectious processes in humans. Thus was launched a series of clinical and bacteriologic investigations which were to get me more and more involved over the next 20 years. I was chagrined, after digging more deeply into the literature, to find that these organisms and their role in infections were well appreciated by French workers over 75 years ago.

After I finally became convinced of the role of anaerobes in infectious processes (aside from such well-known entities as gas gangrene), I recalled that one of my fellow residents, Gordon Cader, in medicine in Minneapolis had frequently mentioned the possibility of anaerobes in various clinical situations. Somehow it did not make much of an impression at that time.

Dr. Cader's mentors at Johns Hopkins, where he received his early training, had been impressed with the importance of these organisms.

The collaboration of many colleagues, too numerous to mention, has been of the utmost importance in the studies originating from our clinic and laboratory, which are referred to throughout the monograph, in giving me proper perspective and in stimulating the production of this monograph. Indeed, to paraphrase Dr. Spink's comment in his monograph, "The Nature of Brucellosis," this book represents something of an autobiography of a clinic and a laboratory devoted to anaerobic bacteria and their role in human disease. Vera L. Sutter, Ph.D., has directed the Wadsworth Anaerobic Bacteriology Research Laboratory for the past several years and has provided invaluable help. Important contributions were made by many workers in the Wadsworth Anaerobic Bacteriology Research Laboratory, by infectious disease fellows, and by colleagues from other sections and services at the Wadsworth Veterans Hospital, UCLA, and other institutions.

House officers and students rotating through the Infectious Disease Section are a constant source of inspiration; their questions are helpful in formulating ideas and approaches.

Esther Alperin, Tomoko Sugano, and Mili Weiman have been helpful in acquiring the necessary books and articles for the literature review. Lorrayne Adams and Kimi Ishii provided excellent typing and clerical skills

I am grateful to Dr. Martin McHenry for his generous contribution of unpublished illustrations and to all of the others who kindly permitted reproduction of figures or other material.

Historical Aspects and
General Considerations

Early History

The striking clinical features of tetanus and its mortality captured the attention of the earliest medical writers. Accounts of the disease are found in the writings of Hippocrates (460–355 BC) and Aretaeus (about 200 AD). Acute necrotizing ulcerative gingivitis has been known since at least the fourth century BC when Xenophon described Greek soldiers suffering from foul-smelling breath and sore mouths. John Hunter gave a precise description of this disease, clearly differentiating it from other diseases of the gums including scurvy. The disease occurred in troops of the French army of the nineteenth century in epidemic proportions.

In 1680, Leeuwenhoek communicated to Thomas Gale his observation that "animalcules" could exist in the absence of air. However, Louis Pasteur is generally conceded to have discovered anaerobiosis when he noted that butyric fermentation occurred in the absence of oxygen. He noted that this was due to bacilli that he called *Vibrion butyrique* which were noted, in a wet preparation under the microscope, to lose their motility rapidly when they chanced to get near the edge of the preparation and thereby were exposed to the air. This nonpathogenic anaerobic spore-forming bacillus is now called *Clostridium butyricum* and is the type species of the genus *Clostridium*. Most of the pathogenic clostridia had been discovered by the end of the nineteenth century, although it is likely that none was obtained in pure culture. Many of the anomalous bacteriologic results obtained in the early days of anaerobic work (and occasionally even today) are due to the special difficulties in obtaining pure cultures of many anaerobic bacteria. *Clostridium chauvoei* was discovered by Bollinger in 1875 and was documented as the cause of blackquarter in sheep and cattle by Feser in 1876. This organism was grown in liquid media in 1880, and on solid media by Kitasato in 1889. *Vibrion septique* (*Clostridium septicum*) was cultured in liquid media in 1877 by Pasteur and Joubert.

Carle and Rattone, in 1884, showed that material from the wound of a patient who died from tetanus produced similar disease when inoculated into rabbits. In the same year, Nicolaier was able to produce a tetanuslike disease in animals by injecting samples of earth. This worker noted the presence of long slender bacilli in the inflammatory exudate produced in the local lesions in these animals. Rosenbach noted bacilli resembling a pin or a drumstick (*Clostridium tetani*) in material from human cases of tetanus in 1886 and 1887. This organism was obtained in pure culture in 1889 by Kitasato. The role of the exotoxin of *Clostridium tetani* in the causation of tetanus was established by Faber in 1890, and in the same year the classic paper on immunization against tetanus was published by Behring and Kitasato. *Clostridium perfringens* was described by Welch and Nuttall in 1892 as *Bacillus aerogenes capsulatus*. Toward the end of the nineteenth century, van Ermengem isolated *Clostridium botulinum* and demonstrated its relationship to botulism.

W. D. Miller described *Fusobacterium* in patients with ulcerative stomatitis in 1883. This same worker published an extensive monograph on microorganisms of the mouth in 1890. In 1894, Plaut reported finding fusiform bacteria and spirochetes in patients with sore throat resembling diphtheria but in whom *Corynebacterium diphtheriae* was not present (1). The fusobacteria and spirochetes were found in smaller numbers in normal oral cavities. H. Vincent, in 1896, found the same organisms microscopically in a number of cases of hospital gangrene (2). Probably the first successful cultivation of fusobacteria was by Lewkowicz in 1903. In 1891, E. Levy described a patient with a gas-containing abscess of the upper right thigh continuous with a mass in the right parametrium (3). This was a postpartum infection. He anticipated an anaerobic infection and was properly set up to grow this type of organism. His cultures did yield, in addition to a small number of *Streptococcus pyogenes*, a strict anaerobic bacillus arranged in long chains and resembling, in colonial morphology, the anthrax bacillus. This organism lived for only short periods on artificial media. Professor Hoppe-Seyler, working with Levy, collected some of the gas from the wound under mercury immediately after the abscess was punctured. The gas was entirely free of oxygen; it consisted of carbon dioxide, nitrogen, and hydrogen. The pus from this wound was malodorous.

In 1893, M. A. Veillon reported recovery in pure culture of an obligately anaerobic micrococcus from a fetid suppurative bartholinitis (4). He noted that he had previously recovered the same micrococcus from three patients, but always in mixed culture with *Streptococcus pyogenes*; these earlier three cases were a fatal Ludwig's angina, a perinephric abscess, and another case of bartholinitis. He was able to demonstrate that the foul odor of the pus recovered from these latter three patients was due to the anaerobic micrococcus alone. In 1894, Lubinski (5) published a bacteriological study of 60 cases of various types of suppuration in which he used aerobic and anaerobic plate culture procedures in parallel. Anaerobic bacteria were observed in five of these cases but could be cultured in only two.

In a brief but classic paper published in 1897, Veillon and Zuber, working in the laboratory of Professor Grancher, described recovery of obligately anaerobic bacteria from 25 cases of gangrenous or fetid suppuration (6). The types of lesions studied included purulent arthritis, appendicitis, brain abscess, pulmonary gangrene, bartholinitis, and pelvic suppuration. These authors pointed out that when aerobes were also present in these lesions, they were present in small numbers. A large number of different types of anaerobic bacteria were recovered. These authors pointed out, furthermore, that the anaerobic bacteria were responsible for the fetid odor of the discharges and for the tissue necrosis. Further details on the role of anaerobic bacteria in infectious processes, together with a good deal of bacteriological information, were presented by Veillon and Zuber in a paper published in 1898 (7). In this paper is mentioned an 11-year-old girl who died of meningitis secondary to chronic otitis media and mastoiditis. At autopsy, in addition, a brain abscess, septic thrombophlebitis of the sphenoidal sinus, and gangrenous foci in both lungs were noted. Two anaerobes were isolated in the absence of aerobes. They further discussed 22 cases of appendicitis, 21 of which yielded a variety of anaerobic bacteria, usually accompanied by small numbers of facultative organisms. They also referred to a role of anaerobic bacteria in peritonitis, in infections about the uterus, and in periurethral suppuration.

This was the beginning of a series of remarkable investigations under the direction of Veillon at the Faculty of Medicine of Paris into the role of anaerobic bacteria in various human infections. Included among these workers was Rist, who studied suppurative disease of the middle ear and mastoid (8), septicemia of otitic origin, and pleural empyema; Guillemot, who studied gangrene of the lung (9); Hallé, who studied female genital tract infections (10); Jeannin, who studied putrid puerperal infections (11); and Cottet, who investigated urinary tract infections in the male and the female (12). These remarkable investigations, together with the earlier work of Veillon and Zuber, established effectively the significant role of anaerobic bacteria in a wide variety of human infections. A large number of different anaerobic bacteria were isolated, described, and characterized during these studies. Tissier carried out extensive investigations of the normal intestinal flora. Subsequently, many workers throughout the world contributed significantly to our knowledge of anaerobes and their role in human disease. However, particular notice should be given to the workers at the Pasteur Institute in Paris where so many important contributions were forthcoming in the early and mid-twentieth century. First, there were Jungano and Distaso, working in the laboratory of Metchnikoff. Following these workers were Weinberg, Seguin, Ginsbourg, Nativelle, and Prévot.

Actinomycosis was first described in the human in 1845 by von Langenbeck, but Israel first recognized and described the organism causing human disease in 1878.

What Is an Anaerobe?

It is surprisingly difficult to provide an acceptable definition of the term "anaerobe." This is remarkable considering how long anaerobic bacteria have been known and considering that all microbiologists use the term. There is considerable disagreement among microbiologists, even among those working actively with anaerobic bacteria, as to a proper definition for the term. One cannot say that anaerobic bacteria are those which die on exposure to atmospheric oxygen nor can one say that these organisms grow better in the absence of air than they do in its presence or that they require a low oxidation–reduction potential. However, as will be discussed subsequently, it may be entirely satisfactory, at least from a practical standpoint, to define anaerobic bacteria on the basis of the quantity of oxygen that they can tolerate and still grow satisfactorily.

While we cannot say that oxygen has direct toxicity against all anaerobes, it may well affect them in an indirect manner. Early workers felt that hydrogen peroxide formed by anaerobes in the presence of oxygen was the major factor in killing anaerobes, since most of these organisms lack catalase to break down hydrogen peroxide. This is clearly not the case because (i) a number of anaerobes are known to produce catalase, (ii) addition of purified catalase to media does not necessarily protect anaerobic bacteria, and (iii) some organisms may produce both hydrogen peroxide and catalase. Organic peroxides formed by the exposure of culture media to air may, however, be very detrimental to anaerobes. *Clostridium haemolyticum* will not grow on the surface of blood agar plates that have been allowed to stand for 3 or 4 hours before streaking unless these plates have been kept under anaerobic conditions during that interval (13).

An interesting new theory concerning anaerobiosis has been proposed by McCord, Keele, and Fridovich (14). These workers noted that obligate anaerobes generally did not possess catalase activity and never possessed the enzyme superoxide dismutase. Aerobic or facultative forms (those forms capable of growth under either aerobic or anaerobic conditions) containing cytochrome systems were found to contain both superoxide dismutase and catalase. Less fastidious anaerobes, which can survive exposure to air and metabolize oxygen to a limited extent but which do not contain cytochrome systems, do not have catalase activity but do have superoxide dismutase activity (at a lower level than was noted with the aerobic or facultative forms). McCord *et al.* propose that the major function of superoxide dismutase is to protect organisms that metabolize oxygen from the detrimental effects of the superoxide free radical, this radical being an intermediate that results from the univalent reduction of molecular oxygen. Smith (14a) points out that while the production of superoxide in the presence of air and its inactivation by superoxide dismutase may be important for many anaerobes, it is not for all anaerobes. *Eubacterium limosum* and *Clostridium oroticum* form significant quantities of superoxide dismutase but cannot grow aerobically. There may be strain variation in the amount of enzyme produced. Thus Hewitt and Morris (14b) found moderate activity in a strain of *Clostridium perfringens*, but Tally and co-workers (14c, 14d) found only low levels in other strains of this species. The latter workers noted that there was general, but not universal, correlation between the amount of aerotolerance and the amount of superoxide dismutase produced. Extremely oxygen sensitive anaerobes did not possess the enzyme, but all pathogenic anaerobes isolated from wounds did.

Another important factor in the growth of anaerobic bacteria is the oxidation–reduction potential or E_h. The importance of this factor, relative to oxygen, is said to be demonstrated by studies in which certain anaerobic bacteria (*Clostridium sporogenes*, *Clostridium perfringens*, and *Bacteroides fragilis* ss. *vulgatus*) were grown in broth through which streams of air were passed, while holding the oxidation–reduction potential at definite levels by electrical means [discussed by Smith (13)]. The ability of these anaerobes to grow appeared to be directly related to the oxidation–reduction potential and was not affected by the presence or absence of oxygen nor of peroxides that must have been formed in the medium. Unfortunately, similar studies have not been done with more fastidious anaerobes. There are upper limits of E_h beyond which specific anaerobes will not grow, and it appears that there are specific lower limits as well (W. J. Loesche, personal communication). The upper limit for *Clostridium sporogenes* is about $+150$ mV, for *Bacteroides fragilis* ss. *vulgatus*, about $+140$ mV, and for *Clostridium histolyticum*, about $+90$ mV. These results are affected by pH. For example, *Clostridium perfringens*, which is not a very fastidious anaerobe, has a limiting oxidation–reduction potential as low as $+30$ mV at pH 7.8 but as high as $+250$ mV at pH 6.0. Media with high oxidation–reduction potentials may result in failure of anaerobic growth or a considerable delay in such growth. Unfortunately, simple addition of reducing agents to media is not sufficient to guarantee good growth of all anaerobic bacteria. Recently the significance of the oxidation–reduction potential has been questioned by Walden and Hentges (14e). By individually regulating both oxygen concentration and oxidation–reduction potential, they found that three intestinal anaerobes important in infection (*C. perfringens*, *B. fragilis*, and *P. magnus*) were inhibited by oxygen even at an E_h of -50 mV. On the other hand, in the absence of oxygen there was no inhibition even at an E_h of $+325$ mV.

It has become apparent that the anaerobic bacteria that we know vary tremendously in their sensitivity to oxygen or air. Methane bacteria are unable to grow if the atmosphere contains as little as 0.03% oxygen, whereas *Clostridium perfringens*, as was indicated, is quite aerotolerant. Exposure of a thin layer of broth containing *Butyrivibrio* to air resulted in the death of 99.99% of the organisms in 6 minutes, whereas *Clostridium perfringens* can withstand similar exposure for many hours (13). In 1930, MacLeod (15) found that *Clostridium tetani* would not grow well unless the maximal oxygen tension was less than 2 mm Hg and that oxygen tensions of 4–5 mm Hg completely inhibited growth of this organism. *Clostridium perfringens* exhibited good growth at oxygen tensions of 10–30 mm Hg and limited growth as high as 70–80 mm Hg. Anaerobic streptococci grew well at 2 mm Hg but were partially or completely inhibited at concentrations of 3 mm Hg or higher. MacLeod also showed that *Clostridium perfringens* surface cultures could be exposed to air for periods of up to at least 4 days and still be successfully subcultured, whereas similar cultures of *Clostridium tetani* usually survived only a few hours, and at times did not survive as long as 90 minutes.

Rosebury (16) studied a number of strains of anaerobes and found that they varied considerably in terms of the limiting concentration of oxygen for surface growth. One strain of *Bacteroides melaninogenicus* grew well in the presence of 0.1% oxygen but not at 1.0%, whereas another strain could grow at 2% oxygen concentration but not at 4%. *Fusobacterium necrophorum* sometimes tolerated 2% oxygen but not 4%; the same was true for *Clostridium novyi* and *Clostridium tetani*. One strain each of *Bacteroides fragilis* and *Fusobacterium nucleatum* were able to tolerate as much as 4% oxygen when cultured on plates that were preincubated anaerobically and which were then incubated in an anaerobic glove-box; these strains tolerated only 2% oxygen with conventional media and anaerobic incubation. In general, however, differences between anaerobically and aerobically preincubated plates were unexpectedly small. *Clostridium haemolyticum* grew in 1% oxygen but not in 2%. Most strains of oral spirochetes studied did not tolerate as much as 0.02% oxygen, although occasional strains tolerated 0.1% and one strain 0.5%.

Loesche (17) studied the oxygen sensitivity of a number of strains of anaerobic bacteria and classified them into two groups. Strict anaerobes were those species not capable of growing on the surface of agar at oxygen levels greater than 0.5%. Included in this group were 3 species of *Treponema*, *Clostridium haemolyticum*, *Selenomonas*, and *Butyrivibrio*. The other group tolerated oxygen levels as high as 2–8% and were classified as moderate anaerobes. Included in this group were *Bacteroides fragilis*, *Bacteroides melaninogenicus*, *Bacteroides oralis*, *Fusobacterium nucleatum*, *Clostridium novyi* type A, and *Peptostreptococcus elsdenii*. Both *Vibrio foetus* and *Vibrio sputorum* were unique in that growth was greater at oxygen concentrations of 0.5% or higher than at lower oxygen concentrations. In the case of *Vibrio sputorum*, growth again diminished when a level of 12% oxygen was used. This growth pattern was described as typical for true microaerophilic bacteria. In this same study, strict anaerobes showed a significant decrease in numbers after exposure to air for 20 minutes but could still be recovered in small numbers after 1 hour's exposure. Moderate anaerobes, on the other hand, showed very little decrease in numbers in the first 100 minutes of exposure to air. Between 100 and 300 minutes exposure, there was a drop to about one-third of the original inoculum in the case of *Bacteroides oralis* and *Fusobacterium nucleatum*, and these organisms were no longer viable after 480 minutes of exposure. On the other hand, *Bacteroides fragilis* could be exposed for 360 minutes without any decrease in numbers, and 85% of the original inoculum survived 480 minutes of exposure.

Fredette and colleagues (18) determined the size of the inhibition zone produced by oxygen under various pressures when anaerobic bacteria were grown on a solid medium. They suggested that our present crude terminology relating to anaerobiosis might be replaced by precise values corresponding to these inhibition zone sizes. They pointed out that the size of the inoculum used in such studies would certainly influence results. Other factors that might well be important would include the age of the inoculum, whether the organism was studied soon after isolation or after being maintained in stock culture for some period, the reducing capacity of the medium used, and the presence or absence of peroxides in the medium.

Recently in our laboratory, Dr. Tally and colleagues (19) used relatively fresh clinical isolates of anaerobic bacteria for determination of oxygen sensitivity, in contrast to the studies referred to earlier which utilized stock cultures (not necessarily from clinical material). The organisms in this study were never stocked prior to testing and had undergone a maximum of two transfers between isolation and determination of oxygen sensitivity. Seven isolates failed to grow in greater than 0.4% oxygen; included among these were strains of *Peptostreptococcus*, *Bacteroides* other than *B. fragilis*, and nonsporulating gram-positive bacilli. Eleven isolates that were more aerotolerant included strains of *Bacteroides fragilis*, other *Bacteroides* species, *Clostridium* species, and nonsporulating gram-positive bacilli. Of particular interest is the fact that many of the strains in this study tolerated prolonged exposure to air without dying completely. Nine of the 11 more aerotolerant anaerobes survived for 72 hours, and the other 2 for 48 hours. Two of the strict anaerobes survived for 24 hours, and 2 others for 72 hours.

It must be appreciated that the studies referred to above dealt with anaerobes in pure culture in the laboratory. This artificial situation does not parallel nature where pure cultures do not exist. For the most part, the anaerobes in which we are interested are part of the indigenous flora of the body, and here may be effectively protected against oxygen in a variety of ways. Not the least of these is the presence of aerobic and facultative bacteria that facilitate the survival and growth of anaerobes by eliminating peroxides, providing reduced conditions, etc. Even on the surface of the skin and in the mouth, where one would anticipate that anaerobes might have difficulty in growing because of exposure to oxygen, these organisms may survive because of microscopic areas that are quite anaerobic. Anaerobic bacteria actually significantly outnumber aerobic or facultative forms, both in the mouth and on the skin.

In actual laboratory practice, definition of an anaerobe need not be as difficult as suggested by the previous discussion. A practical definition for operational purposes is that an *anaerobe* is a bacterium that requires a reduced oxygen tension for growth and fails to grow on the surface of solid media in 10% CO_2 in air (18% oxygen). *Facultative organisms* are those which can grow both in the presence or absence of air. Strictly speaking, *microaerophilic bacteria* would be those preferring reduced oxygen tension, such as that provided by 10% CO_2 in air, over either aerobic or anaerobic conditions. In actual practice, the term microaerophilic is used commonly for organisms that grow poorly or not at all in air but which grow distinctly better under 10% CO_2 in air (reduced O_2 content) or anaerobically. The role of the CO_2 per se and of the humidity present in jars, as compared to the reduced oxygen tension, has not ordinarily been investigated with these organisms. *Aerotolerant organisms* are anaerobes which tolerate oxygen just enough to grow on the surface of freshly prepared solid media.

Problems in Classification and Characterization

Classification and characterization has been a particular problem with anaerobic bacteria from the beginning of work in this field. While great progress has been made in recent years, many problems are still with us. One major problem is that anaerobic bacteria are always found in mixed culture as normal flora and are frequently found in mixed culture in clinical infections. The associations in these mixtures are so intimate that it may be extremely difficult to isolate each component of the mixture in pure culture. To some extent, this reflects the dependence of anaerobes on other anaerobes or facultative forms for provision of growth factors and optimum conditions for anaerobic growth. Some of the foremost workers in anaerobic bacteriology have been guilty of describing a new organism, only to discover years later that the organism only appeared to be different and was actually a mixture of two well-known organisms, the mixture showing characteristics distinct from either in pure culture.

Another major problem in anaerobic bacteriology is the confusion that has existed because of many different classification schemes and because of many synonyms that have been used for these organisms. Considerable progress has recently been made in terms of simplifying and standardizing classification schemes. Much of this improvement has resulted from the efforts of subcommittees of the International Committee on Systematic Bacteriology. The group in the Anaerobe Laboratory of Virginia Polytechnic Institute deserves particular credit for its efforts in this area. They were able to obtain a very large number of anaerobes of all types from the collection of Prévot of the Pasteur Institute of Paris and to compare these organisms with organisms from many other collections, using a large battery of tests, including such important tests as gas chromatographic analysis of end products of metabolism, cell wall analysis, DNA base ratios, and DNA homology. The Virginia Polytechnic Institute Anaerobic Laboratory Manual (20) is very helpful. A list of synonyms of most of the anaerobic bacteria commonly encountered in humans is given in Table 1.1.

The literature on anaerobic bacteriology and the reports of anaerobes in various disease processes would be improved enormously if each investigator would study anaerobic isolates carefully and indicate the details of such studies in his reports. Many people have used "Bacteroides" to mean a gram-negative obligately anaerobic bacillus. Even when species names are provided, these are not always reliable; for example, some workers use *Bacteroides fragilis* to indicate a gram-negative anaerobic bacillus that is nonpleomorphic and *Fusobacterium necrophorum* to indicate a gram-negative anaerobic bacillus that is very pleomorphic and has large round bodies in culture. These criteria are far from definitive or reliable.

Gram-Negative Anaerobic Bacilli
Bacteroides biacutus

Fusiformis biacutus
Fusobacterium biacutum
Ristella biacutus, Ristella biacuta

Bacteroides capillosus

Bacillus capillosus
Pseudobacterium capillosum
Ristella capillosa

Bacteroides clostridiiformis

Subspecies	Synonyms of Subspecies
B. clostridiiformis ss. clostridiiformis	*Bacterium clostridiiforme*
	Bacterium clostridiiformis
	Eggerthella clostridiiformis
	Ristella clostridiiformis
	Sphaerophorus clostridiiformis
B. clostridiiformis ss. girans	*Bacterium clostridieformis*
	Bacterium clostridiiformis mobilis
	Fusobacterium girans
	Fusocillus girans
	Zuberella clostridiiformis
	Zuberella clostridiiformis mobilis
	Zuberella girans

Bacteroides coagulans

Pasteurella coagulans
Pseudobacterium coagulans

Bacteroides constellatus

Zuberella constellata

Bacteroides corrodens

[Not same as facultative form with same name (the name *Eikenella* has been proposed for the facultative form)]
Ristella corrodens

Bacteroides fragilis

Subspecies	Synonyms of Subspecies
B. fragilis ss. distasonis	*Bacteroides distasonis*
	Pseudobacterium distasonis
	Ristella distasonis
B. fragilis ss. fragilis	*Bacteroides convexus*
	Bacteroides fragilis
	Eggerthella convexa
	Pasteurella convexa
	Pseudobacterium convexum
	Ristella pseudoinsolita
B. fragilis ss. ovatus	*Bacteroides gulosus* (probable)
	Bacteroides ovatus
	Pasteurella ovata
	Pseudobacterium ovatum

TABLE 1.1 (*continued*)

B. fragilis ss. *thetaiotaomicron*	*Bacillus thetaiotaomicron, B. variabilis*
	Bacteroides thetaiotaomicron, B. variabilis
	Capsularis variabilis
	Pseudobacterium thetaiotaomicron, P. variabilis
	Sphaerocillus thetaiotaomicron
B. fragilis ss. *vulgatus*	*Bacteroides vulgatus*
	Pasteurella vulgata
	Pseudobacterium vulgatum

Synonyms of Species
   Bacillus fragilis
   Bacteroides inaequalis, Bacteroides incommunis, Bacteroides uncatus
   B. sassmannshausen(?)
   Eggerthella incommunis, E. uncata
   Fusiformis fragilis
   Pseudobacterium fragilis, P. incommunis, P. uncatum
   Ristella fragilis, R. incommunis, R. uncata
   Sphaerophorus inaequalis, S. intermedius
Probable Synonyms
   Bacteroides exiguus, B. freundii Hauduroy *et al.*(?), *B. gulosus, B. insolitus, B. pyogenes* Hauduroy *et al., B. tumidus, B. uniformis*
   Pasteurella serophila
   Pseudobacterium gulosum
   Ristella freundii(?), *R. gulosa, R. glycolytica, R. insolita, R. pseudoinsolita, R. thermophila, R. tumida*
   Sphaerophorus glycolyticus, S. gulosus
   Spherophorus gulosus
Possible Synonyms
   Bacillus pyogenes anaerobius
   Buday's bacillus
   B. anaerobius pyogenes
   Spherophorus pyogenes

Bacteroides furcosus

Bacillus furcosus
Fusiformis furcosus
Pseudobacterium furcosum
Ristella furcosa

Bacteroides melaninogenicus

Subspecies
   B. melaninogenicus ss. *asaccharolyticus*
   B. melaninogenicus ss. *intermedius*
   B. melaninogenicus ss. *melaninogenicus*
Synonyms of Species
   Bacterium melaninogenicum
   Bacteroides nigrescens
   Fusiformis nigrescens
   Hemophilus melaninogenicum
   Ristella melaninogenica

Bacteroides ochraceus

B. oralis ss. *elongatus*

(*continued*)

11

TABLE 1.1 (*continued*)

Bacteroides oralis

Subspecies
    Bacteroides oralis ss. *elongatus* (see *Bacteroides ochraceus*)
    Bacteroides oralis ss. *oralis*

Bacteroides pneumosintes

Subspecies
    B. pneumosintes ss. *septicemiae*
    B. pneumosintes ss. *septicus*
Synonyms of Species
    Bacillus pneumosintes
    Bacterium pneumosintes
    Dialister pneumosintes
    Dialister pneumosintes var. *septicemiae* and *D. p.* var. *septicus*

Bacteroides praeacutus

Coccobacillus praeacutus
Fusobacterium praeacutum
Zuberella praeacuta and *Z. p.* var. *anaerogenes*

Bacteroides putredinis

Bacillus A. Heyde
Bacillus putredinis Weinberg *et al.* (not *Bacillus putredinis* Trevisan)
Pseudobacterium putredinis
Ristella putredinis

Bacteroides ruminicola

Subspecies
    B. ruminicola ss. *ruminicola*
    B. ruminicola ss. *brevis*
Synonyms of Species
    Ruminobacter ruminicola

Bacteroides serpens

Bacillus radiiformis
Bacillus serpens
Zuberella serpens

Bacteroides splanchnicus

No synonym

Bacteroides trichoides

See *Clostridium ramosum*

Fusobacterium abscedens

Ristella abscedens
Sphaerophorus abscedens
Spherophorus abscedens

Fusobacterium aquatilum

Fusobacterium novum
Zuberella nova

TABLE 1.1 (*continued*)

Fusobacterium bullosum

Bacillus bullosus
Bacterium bullosum
Bacteroides bullosus
Sphaerocillus bullosus
Sphaerophorus bullosus
Spherocillus bullosis, Spherocillus bullosus

Fusobacterium glutinosum

Bacillus glutinosus
Bacteroides glutinosus
Ristella glutinosa

Fusobacterium gonidiaformans

Actinomyces gonadiformis
Actinomyces gonidiaformans
Bacillus gonidiaformans
Bacteroides gonidiaformans
Pseudobacterium gonidiaformans
Sphaerophorus gonidiaformans
Spherophorus gonidiaformans

Fusobacterium mortiferum

Bacillus mortiferus, B. necroticus
Fusobacterium ridiculosum
Pseudobacterium mortiferum, P. necroticum
Sphaerophorus mortiferus, S. necroticus ridiculosus
Spherophorus mortiferus, S. ridiculosus, S. necroticus
Probably synonymous
  Sphaerophorus freundii Hauduroy *et al.* (*Bacteroides freundii, Sphaerophorus freundi,*
  Bacterium de Freund)

Fusobacterium naviforme

Bacillus naviformis
Pseudobacterium naviformis
Ristella naviformis

Fusobacterium necrophorum

Subspecies
  F. necrophorum ss. *funduliformis*
  F. necrophorum ss. *necrophorum*
Synonyms of Species
  Actinomyces cuniculi, A. necrophorus, A. pseudonecrophorus
  Bacille de Schmorl
  Bacillus diphteriae vitulorum, B. diphtheriae vitulorum, B. filiformis, B. fundibuliformis,
    B. funduliforme, B. funduliformis, B. necrophorus, B. necroseos, B. necrosus, B. thetoides
  Bacterium fundibuliformis, B. funduliforme, B. necrophorum
  Bacteroides fundibuliformis, B. funduliformis, B. necrophorus
  B. des Kälbernoma
  B. filiformis
  Cladothrix cuniculi

(*continued*)

TABLE 1.1 (*continued*)

Fusobacterium necrophorum (*continued*)

Corynebacterium de la nécrose
Corynebacterium necrophorum, C. necrophorus
Cohnistreptothrix cuniculi
Fusiformis hemolyticus, F. necrophorus
Necrobacterium necrophorus
Nekrosebacillus
Necrosis bacillus (of Bang)
Proactinomyces necrophorus
Pseudobacterium funduliformis
Schmorl's bacillus
Species C. of Veillon and Zuber
Sphaerophorus funduliformis, S. necrophorus, S. pseudonecrophorus
Spherophorus funduliformis
Streptothrix cuniculi, S. necrophora, S. necrophorus, S. necupthora
Probable Synonyms
    Actinomyces pseudonecrophorus
    Bacillus pyogenes anaerobius, B. anaerobius pyogenes
    Bacillus symbiophiles
    Buday's bacillus
    Leptothrix anaerobius tenuis, Leptothrix asteroide
    Sphaerophorus pseudonecrophorus, S. pyogenes

Fusobacterium novum

See *Fusobacterium aquatilum*

Fusobacterium nucleatum

Bacillus fusiforme, B. fusiformis
Corynebacterium fusiforme
Fusiformis fusiformis, F. nucleatus
Fusobacterium fusiforme
Group III, Baird-Parker
Probable Synonyms
    Fusobacterium polymorphum Baird-Parker
    Sphaerophorus fusiformis
Not Synonyms
    Fusobacterium plauti-vincenti Knorr
    F. fusiforme (Hoffman—Bergey's 7th edition)

Fusobacterium perfoetens

Subspecies	Synonyms of Subspecies
F. perfoetens ss. *lacticum*	*Sphaerophorus perfoetens* var. *lacticus*
	Ristella perfoetens var. *lacticus*
F. perfoetens ss. *perfoetens*	*Bacterium perfoetens*
	Bacteroides perfoetens
	Coccobacillus anaerobius perfoetens
	Coccobacillus perfoetans
	Ristella perfoetens
	Sphaerophorus perfoetens

TABLE 1.1 (*continued*)

Fusobacterium plauti

Bacille de Plaut
Bacillus plauti
Fusocillus plauti
Zuberella plauti

Fusobacterium prausnitzii

Bacillus mucosus anaerobius
Bacterium zoogleiformans
Bacteroides prausnitzi, B. prausnitzii
Capsularis zoogleiformans
Probable Synonym
Ristella abscedens

Fusobacterium russii

Bacillus influenzaeformis
Bacteroides russii
Sphaerophorus influenzaeformis
Spherophorus influenzaeformis

Fusobacterium stabilum

Capsularis stabilis
Capsularis stabilus

Fusobacterium symbiosum (Stevens) Moore and Holdeman

Bacteroides symbiosus (Stevens)
Z. pedipedis (Prévot) Sebald
Not Fusocillus pedipedis Prévot

Fusobacterium varium

Bacteroides varius
Pseudobacterium varium
Sphaerophorus varius, S. varius var. *sulfitoreductans*
Spherophorus varius

Gram-Positive Non-Spore-Forming Anaerobic Bacilli
Actinomyces bovis
(not a human pathogen)

Actinocladothrix bovis
Actinomyces bovis sulphureus
Bacterium actinocladothrix
Cladothrix bovis
Discomyces bovis
Nocardia actinomyces, N. bovis
Oospora bovis
Proactinomyces bovis
Sarcomyces bovis
Sphaerotilus bovis
Streptothrix actinomycotica, Streptothrix bovis, S. b. communis
Streptotrix [sic] *actinomyces*

(continued)

TABLE 1.1 (*continued*)

Actinomyces israelii

Actinobacterium israeli, A. israelii
Actinomyces hominis, A. israeli, A. wolff-israel
Brevistreptothrix israeli
Cohnistreptothrix israeli, C. wolff-israel
Corynebacterium israeli
Discomyces israeli
Nocardia israeli
Oospora israeli
Proactinomyces israeli
Streptothrix israeli

Actinomyces naeslundii

Actinomyces naeslundi

Actinomyces odontolyticus

No synonyms

Actinomyces viscosus

A. discofoliatus
Odontomyces viscosus

Actinomyces

Former *Actinomyces* species now in *Rothia* (*R. dentocariosus*) (Aerobic): *A. dentocariosus* (*Nocardia dentocariosus, N. salivae*)
Former *Actinomyces* species now in *Bifidobacterium*: *A. eriksonii, A. parabifidus*
Former *Actinomyces* species now in *Arachnia* (*A. propionica*): *A. propionicus* (*Propionibacterium propionicum*)

Bifidobacterium

Synonyms for Genus
   Actinomyces (in part)
   Bacillus (in part)
   Bifidibacterium (in part)
   Lactobacillus (in part)
   Tissieria (in part)

Eubacterium aerofaciens

Bacteroides aerofaciens
Pseudobacterium aerofaciens

Eubacterium alactolyticum

Ramibacterium alactolyticum, R. dentium, R. pleuriticum

Eubacterium budayi

Bacillus cadaveris butyricus
Bacterium budayi
Eubacterium cadaveris
Pseudobacterium cadaveris

Eubacterium cellulosolvens

Cillobacterium cellulosolvens

16

TABLE 1.1 (*continued*)

Eubacterium combesii

Cillobacterium combesi, Cillobacterium combesii

Eubacterium contortum

Catenabacterium contortum

Eubacterium cylindroides

Bacterium cylindroides
Bacteroides cylindroides
Pseudobacterium cylindroides
Ristella cylindroides

Eubacterium lentum

Bacteroides lentus
B. minutum Hauduroy
Corynebacterium diphtheroides (CDC Manual), *Corynebacterium* Group 3
Pseudobacterium lentum
Probable Synonym
 Coccobacillus oviformis, Bacteroides oviformis

Eubacterium limosum

Bacteroides limosus
Butyribacterium limosum, B. rettgeri
Mycobacterium limosum

Eubacterium moniliforme

Bacillus moniliformis, B. repazii
Cillobacterium moniliforme

Eubacterium multiforme

Bacillus multiformis Distaso (not *Bacillus multiformis* van Senus)
Bacteroides multiformis?
Cillobacterium multiforme

Eubacterium nitritogenes

No synonym

Eubacterium rectale

Bacteroides rectalis
Pseudobacterium rectale

Eubacterium saburreum

Catenabacterium saburreum
Leptotrichia aerogenes

Eubacterium tenue

Bacillus spatuliformis, B. tenuis spatuliformis
Bacteroides tenuis
Cillobacterium tenue, C. spatuliforme

(*continued*)

TABLE 1.1 (*continued*)

Eubacterium tortuosum

Bacillus tortuosus
Bacteroides tortuosus
Mycobacterium flavum var. *tortuosum*

Eubacterium ventriosum

Bacteroides ventriosus
Pseudobacterium ventriosum

Former *Eubacterium* species now in genus *Lactobacillus*
  E. (*Catenabacterium*) *catenaforme*
  E. crispatum
  E. disciformans
  E. minutum
Former *Eubacterium* species now in *Clostridium* (see separate list of synonyms)
  E. ramosum (*E. filamentosum, Catenabacterium filamentosum*)
Former *Eubacterium* species now in *Bifidobacterium*
  E. thermophilum

Propionibacterium

Synonym for Genus
Corynebacterium (in part)

Propionibacterium acnes

Bacillus acnes, B. anaerobius diphtheroides, B. parvus liquefaciens
Corynebacterium acnes, C. adamsoni, C. anaerobium Prévot (not all strains),
  C. liquefaciens, C. parvum infectiosum, C. parvum

Anaerobic Cocci
Acidaminococcus fermentans

No synonym

Megasphaera elsdenii

Peptostreptococcus elsdenii

Peptococcus aerogenes

Micrococcus aerogenes

Peptococcus activus

Staphylococcus activus

Peptococcus asaccharolyticus

Micrococcus asaccharolyticus, Micrococcus indolicus
Peptostreptococcus CDC Group 1
Staphylococcus asaccharolyticus
S. asaccharolyticus var. *indolicus*

Peptococcus constellatus

Diplococcus constellatus

Peptococcus glycinophilus

Diplococcus glycinophilus

TABLE 1.1 (*continued*)

Peptococcus magnus
Diplococcus magnus, Diplococcus magnus anaerobius
Peptostreptococcus magnus
Staphylococcus anaerobius Hamm
Micrococcus anaerobius
"Staphylocoque anaérobie"

Peptococcus morbillorum
Diplococcus morbillorum, Diplococcus rubeolae
Peptostreptococcus morbillorum

Peptococcus niger
Micrococcus niger

Peptococcus prevotii
Micrococcus prevoti, Micrococcus prevotii
Peptostreptococcus CDC Group 2

Peptococcus saccharolyticus
Micrococcus saccharolyticus

Peptococcus variabilis
Micrococcus variabilis

Peptostreptococcus anaerobius
Micrococcus foetidus Veillon and M. f. Flügge (not M. f. Klamann or M. f. Eisenberg)
Peptostreptococcus foetidus
Peptostreptococcus putridus
Streptococcus anaerobius
Streptococcus foetidus Prévot (not S. f. Migula)
Streptococcus putridus
Streptococcus putrificus
Streptococcus anaerob (sic)
"Stinkcoccus"

Peptostreptococcus intermedius
(Officially should be classified in genus *Streptococcus* as it is not always obligately anaerobic)
Streptococcus intermedius
Peptostreptococcus evolutus?

Peptostreptococcus lanceolatus
Coccus lanceolatus anaerobius
Streptococcus lanceolatus Prévot (not S. l. Gamaleia)

Peptostreptococcus micros
Streptococcus anaerobius micros, Streptococcus micros

Peptostreptococcus parvulus
Streptococcus parvulus non liquefaciens
Streptococcus parvulus Weinberg, Nativelle and Prévot
Not *Streptococcus parvulus* Levinthal

(continued)

19

TABLE 1.1 *(continued)*

Peptostreptococcus productus
Streptococcus productus

Sarcina ventriculi
No synonym

Veillonella alcalescens
Micrococcus alcalescens, Micrococcus gazogenes, Micrococcus gazogenes alcalescens, Micrococcus gazogenes alcalescens anaerobius, Micrococcus gingavalis, Micrococcus lactilyticus, Micrococcus minutissima, M. minutissimus, Micrococcus syzygios, M. syzygios scarlatinae
Syzygiococcus scarlatinae
V. alcalescens var. *gingivalis, V. alcalescens* var. *minutissimum, V. alcalescens* var. *syzygios*
Veillonella gazogenes

Veillonella parvula
Micrococcus branhamii, Micrococcus minimus, Micrococcus parvulus, Micrococcus thomsoni
Staphylococcus minimus, Staphylococcus parvulus
V. parvula var. *branhami, V. parvula* var. *minima, V. parvula* var. *thomsoni*

Gram-Positive Anaerobic Spore-Forming Bacilli

Clostridium
[See "Bergey's Manual of Determinative Bacteriology" (R. S. Breed, E. G. D. Murray, and A. D. Hitchens, eds.), 6th ed., 1948, for many more synonyms according to Spray.]

Clostridium bifermentans
Bacillus bifermentans, Bacillus bifermentans sporogenes, Bacillus centrosporogenes, Bacillus liquefaciens magnus
Clostridium foetidum, Clostridium foetidum carnis
Martellillus bifermentans

Clostridium botulinum
Bacillus botulinus
Botulobacillus botulinus
Clostridium luciliae (type C)
Clostridium parabotulinus bovis (type D)
Clostridium parabotulinum
Clostridium parabotulyinum equi (type E)
Ermengemillus botulinus

Clostridium butyricum
Amylobacter navicula
Bacillus amylobacter, Bacillus butyricus, Bacillus navicula
Bacterium navicula
Clostridium amylobacter, Clostridium multifermentans, Clostridium naviculum
Metallacter amylobacter
Vibrion butyrique Pasteur

Clostridium cadaveris
Bacillus capitovalis
Clostridium capitovale, C. capitovalis
Plectridium capitovalis

TABLE 1.1 (*continued*)

Clostridium carnis

Bacillus carnis, Bacillus VI Hibler
Clostridium sextum
Plectridium carnis
Bacillus sextus (probable)
B. lactiparcus (probable)

Clostridium chauvoei

Bacillus anthracis symptomatici, Bacillus carboni, Bacillus chauvoei
Bacterium chauvoei
Clostridium chauvaei, C. chauvei, C. chauvoeii, C. feseri

Clostridium cochlearium

Bacillus cochlearius, Bacillus type IIIc McIntosh
Flemengillus cochlearium
Plectridium cochlearium, Plectridium incertum

Clostridium difficile

Bacillus difficilis
Clostridium difficilis

Clostridium fallax

Bacille A Weinberg and Seguin
Bacillus fallax
Clostridium pseudofallax, C. pseudo-fallax
Vallorillus fallax

Clostridium haemolyticum

Bacillus hemolyticus
Clostridium haemolyticum bovis, Clostridium hemolyticus bovis, C. hemolyticum, C. novyi type D
  (British)

Clostridium hastiforme

Bacillus 4a Cunningham

Clostridium histolyticum

Bacillus histolyticus
Weinbergillus histolyticus

Clostridium innocuum

No synonym

Clostridium lentoputrescens

Plectridium lentoputrescens

Clostridium limosum

Clostridium species CDC Group P-1, *C. bubalorum* Prévot

Clostridium novyi

Bacillus gigas, Bacillus novyi, Bacillus oedematiens, Bacillus oedematis maligni II, *Bacillus oedematis thermophilus, Bacillus thermophilus*

(*continued*)

TABLE 1.1 (*continued*)

Clostridium novyi (*continued*)

Bacterium oedematis thermophilus
C. bellonensis, C. bubalorum Kraneveld, *C. hemolyticum, C. oedematis, C. oedematis – benigni,*
C. gigas, C. oedematiens, C. sarcoemphysematodes, C. toxinogenes, C. ukilii
Novillus maligni

Clostridium paraperfringens

C. barati

Clostridium paraputrificum

Bacillus diaphthirus, Bacillus paraputrificus, Bacillus paraputrificus coli
Plectridium paraputrificum
Tissierillus paraputrificus

Clostridium perfringens

Bacillus aerogenes capsulatus (Welch and Nuttall), *Bacillus emphysematosus, Bacillus welchii,*
B. perfringens, Bacillus phlegmones emphysematosae (Fraenkel)
Clostridium welchii
Gas bacillus
Welchia agni (and varieties)
Welchia perfringens (and varieties)

Clostridium putrificum

Bacillus putrificus
Clostridium lentoputrescens
Plectridium lentoputrescens, Plectridium putrificum

Clostridium ramosum

Actinomyces ramosus
Bacillus ramosus, B. ramosus liquefaciens
Bacillus trichoides, B. terebrans
Bacteroides ramosus, B. terebrans, B. trichoides
Catenabacterium filamentosum
Eubacterium filamentosum, Eubacterium ramosum
Fusiformis ramosus
Nocardia ramosa
Pseudobacterium trichoides, P. terebrans
Ramibacterium ramosum
Ristella trichoides, R. terebrans
"Wisp bacillus"

Clostridium septicum

Bacillus oedematis maligni, Bacillus septicus
B. edematis
Clostridium oedematis maligni, C. septique
Vibrio pasteurii
Vibrion septique Pasteur

TABLE 1.1 (*continued*)

Clostridium sordellii

Bacillus oedematis sporogenes, Bacillus sordellii
Clostridium oedematoides, C. tonkinensis

Clostridium sphenoides

Bacillus sphenoides
Douglasillus sphenoides
Plectridium sphenoides

Clostridium sporogenes

Bacillus enteritidis sporogenes Klein, *Bacillus sporogenes, Bacillus sporogenes* var. A, Metchnikoff
B. putrificus verrucosus B. saprotoxicus, B. tyrosinogenes
Clostridium flabelliferum, Clostridium parasporogenes, C. saprotoxicum, Clostridium tyro-
*   sinogenes*
Metchnikovillus sporogenes
Reading bacillus

Clostridium subterminale

Bacillus subterminalis

Clostridium tertium

Bacillus sporogenes non liquefaciens Jungano (probable)
Bacillus tertius
Henrillus tertius
Plectridium tertium

Clostridium tetani

Bacillus tetani
Nicolaierillus tetani
Pacinia nicolaieri
Plectridium tetani

Clostridium

Synonym for Genus
Inflabilis

---

[a] Species not listed are either uncommonly isolated or are of uncertain status in that the original isolates are not available for study and there have been no subsequent isolations of organisms which seem to be the same. It must be appreciated that, typically, in the latter category are organisms isolated many years ago and so inadequately studied (by modern standards) that it would often be impossible to determine whether recent isolates are identical or not. Reference to the 8th edition of Bergey's Manual of Determinative Bacteriology should ordinarily clear up questions on such points. Names listed are not always legitimate earlier names. There are many additional obscure synonyms for some of the listed species which are not given. The synonyms, and even the accepted names, have sometimes been misapplied to other organisms in the literature. Unfortunately, many times articles fail to include mention of the techniques used for bacteriological identification or even the types of tests employed. Thus the criteria used for identification may be difficult or impossible to judge. In other cases, the criteria are grossly inappropriate (viz., "non-pleomorphic anaerobic gram-negative bacilli are *Bacteroides fragilis* and pleomorphic anaerobic gram-negative bacilli are *Bacteroides funduliformis*") but the names may be perpetuated in subsequent references to the article.

Another important problem with regard to classification of anaerobic bacteria is the area of the so-called microaerophilic organisms. These tend to be overlooked or shunned by workers not familiar with anaerobic techniques. It is difficult or impossible to set up meaningful and sensible boundary lines for these organisms. A number of anaerobic bacteriologists have worked with aerotolerant or microaerophilic forms, such as *Actinobacillus actinomycetemcomitans* and *Campylobacter* (*Vibrio*) *foetus,* and yet may not consider *Haemophilus aphrophilus* and *Erysipelothrix* in their sphere. Furthermore, *Bacteroides ochraceus* will grow on the surface of solid media in an atmosphere of 5% CO_2 in air (but not aerobically). Since this organism is obviously very closely related to other organisms that are obligate anaerobes, it is reasonable to consider it with the anaerobes. Organisms such as *Brucella abortus,* which also grows in the candle jar, can be excluded from consideration on the basis of the fact that they will not grow anaerobically. Actually, *Haemophilus aphrophilus* can be excluded from direct consideration as an anaerobe because, although it grows in a candle jar, it is actually the moisture in the jar which is important, and this organism can be grown aerobically if a moist atmosphere is provided. This may well be true for many other "microaerophilic" organisms. Basically, the important considerations are that clinical laboratories should be able to grow all organisms that may be important in human disease and that, in describing an organism by name or in any other way, everyone else should know exactly what is being talked about.

Confusion exists also with regard to anaerobic cocci. There are anaerobic cocci that possess certain key characteristics of facultative forms, such as staphylococci and pneumococci. In the case of the staphylococci, the anaerobic forms may produce coagulase and be typable by bacteriophage active against the usual facultative forms. The anaerobic pneumococci may be bile soluble and react specifically with antisera prepared against the facultative forms. Furthermore "anaerobic pneumococci" may produce irreversible aerobic mutants [Roemer, cited by Prévot (21)]. A recent study by Austrian and Collins (22) questions the existence of true anaerobic pneumococci. This excellent study points out that 8% of pneumococci have a requirement for carbon dioxide if detectable growth is to occur on the surface of solid media. They demonstrated that these particular pneumococci did not grow either in air or anaerobically in the absence of CO_2. The CO_2-requiring strains of pneumococci were distributed among 18 pneumococcal capsular types, but two-thirds of them were found among five types (I, III, XVI, XXVIII, and XXXIII).

Perhaps the most confusing area of all is that of the so-called microaerophilic cocci and streptococci. Organisms in this category are certainly very important pathogens, as will be brought out later, and they will frequently be overlooked unless anaerobic conditions are provided for culture of clinical materials. Roemer [as quoted by Prévot (21)] indicated that many anaerobic streptococci, after evolving towards aerobiosis, are related to the Lancefield A and D groups. Roemer (23), in a later study, indicated that not many of the anaerobic streptococci are stable and that among those evolving toward aerobic forms are some from groups C, D, and G. On the other hand, Stone (24) indicated that only 4 of 40 anaerobic cocci that he studied over an extended period of time adapted to aerobic growth and then only after cultivation for 6 to 8 months. He noted further that antisera prepared against Lancefield groups A, B, and C reacted with 14 of 24 strains of anaerobic streptococci. There were more cross-reactions with C than with B and fewest with A. For the most part the reactions were relatively weak. Two strains reacted to both A and B antisera. Stone concluded that there was at least one antigen in hemolytic streptococci which also seems to be present in some anaerobic streptococci and that one might be able to utilize this information in a classification scheme, together with other characteristics. Anders *et al.* (25) noted that both *Streptococcus lactis* and *S. cremoris* (Lancefield group N) form hydrogen peroxide under aerobic conditions and that therefore there is decreased growth in the presence of high oxygen tension; 10% carbon dioxide in the atmosphere significantly enhances the growth of *Streptococcus anginosus* (Lancefield group F and type I, group G) [Breed *et al.* (26) and Duma *et al.* (27)]. Other characteristics of this organism which are seen in microaerophilic cocci are the minute size of the cells and colonies and the occasional occurrence of strains that are relatively resistant to penicillin (requiring 1 to 2 units/ml for bactericidal activity) [Duma *et al.* (27)]. A very important study on oral streptococci has been carried out by Carlsson (27a). This numerical taxonomic study embraced 89 strains of streptococci representing various distinct varieties from a total of 243 isolates from the oral cavity of four subjects. A number of reference strains were used as well. These organisms could be arranged in five major groups, some with subgroups. Group I consisted of organisms the majority of which were microaerophilic. Group IA bore some resemblance to *Streptococcus mitis* and IB conformed closely to *Streptococcus sanguis* (Lancefield group H, *Streptococcus* s.b.e.). Group II all grew poorly without carbon dioxide in the atmosphere, although none of these was microaerophilic. These organisms conformed to *S. mutans*. Group V resembled *S. mitis*, with the majority of strains in V A being microaerophilic and a few in V B similarly being microaerophilic. *Streptococcus salivarius* (Lancefield group K) was not microaerophilic.

Finally, there are two less well-known groups of cocci which remain to be considered. *Pediococcus* is microaerophilic and gives poor surface growth [Breed *et al.* (26)]. *Aerococcus viridans* is a true microaerophile [Evans and

Kerbaugh (28)] growing as a discrete band of many pinpoint colonies a few millimeters below the surface of semisolid fluid thioglycolate medium. This organism tends to form tetrads and has been confused with *Micrococcus tetragenus*. It is widely distributed in the hospital environment and is apparently the causative organism in a variety of infections, particularly endocarditis and urinary tract infections. Some of the microaerophilic cocci and streptococci will fit nicely into described species of both anaerobic and facultative classification schemes and therefore might be called one or the other, depending on which tests were used for identification, particularly if end products of carbohydrate fermentation were not examined. Studies by E. J. Harder, V. L. Sutter, and S. M. Finegold (unpublished) showed that all microaerophilic streptococci studied (about 60 clinically significant isolates) produced lactic acid as the major end product of metabolism and fit criteria for speciation within the genus *Streptococcus*.

This monograph will cover, in addition to obligately anaerobic bacteria, closely related more aerotolerant forms and the microaerophilic cocci and streptococci.

Classification of Anaerobes

I favor the use of subspecies and similar groupings unless it is established that these have no significance. In the case of *Bacteroides fragilis*, Werner and colleagues (29–31) have developed considerable evidence to indicate that it is clinically significant to use subspecies for these organisms (or to consider them as separate species).

The International Committee on Systematic Bacteriology's Subcommittee on Gram-Negative Anaerobic Bacilli recommended that *Bacteroides melaninogenicus* ss. *asaccharolyticus* should be placed in a separate species, *B. asaccharolyticus*. *Bacteroides melaninogenicus* ss. *melaninogenicus* and ss. *intermedius* will remain in *B. melaninogenicus; Bacteroides oralis* closely resembles *Bacteroides melaninogenicus* ss. *melaninogenicus*.

The genera of anaerobic bacteria encountered in humans are listed in Table 1.2. The species noted most frequently in infections will be detailed in Chapter 3.

Anaerobes as Normal Flora

Anaerobic bacteria are prevalent throughout the body as indigenous flora (Table 1.3). They are numerous on all mucosal surfaces and certain of them populate the skin as well. Certain anaerobes have demonstrated the ability to

TABLE 1.2

Genera of Anaerobic Bacteria Encountered in Humans[a]

Bacilli
    Spore-formers (spores may be difficult to demonstrate)
        Clostridium
    Non-spore-formers
      Gram-positive
        Actinomyces
        Arachnia
        Bifidobacterium[b]
        Eubacterium
        Lactobacillus[b]
        Propionibacterium[b]
      Gram-negative
        Nonmotile or motile with peritrichous flagella
        Bacteroides
        Fusobacterium
        Leptotrichia[c]
        Motile by other means
        Borrelia[d]
        Butyrivibrio[b]
        Campylobacter (*Vibrio*)[c]
        Selenomonas[c]
        Treponema[b]

Cocci
    Gram-positive
      Peptococcus
      Peptostreptococcus
      Ruminococcus[c]
    Gram-negative
      Acidaminococcus[b]
      Megasphaera[b]
      Veillonella[b]

[a] Footnotes refer to anaerobic species in genera which include microaerophiles or facultatives.

[b] Infrequent or rare cause of serious human infection (or only one or two species in genus cause such infection).

[c] Not known to be pathogenic.

[d] Anaerobic species may cause relapsing fever in man (not covered here).

adhere to mucosal epithelial cells (32, 32a, 32b). This property is important in determining the ecology of at least the oral anaerobes and may be important in pathogenicity as well.

Knowledge of the presence of specific anaerobes as normal flora at certain sites is useful in several ways. Most anaerobic infections arise in proximity to mucosal surfaces where the organisms reside. The clinician

TABLE 1.3
Anaerobes as Normal Flora in Humans[a]

Body area	Clostridium	Nonsporulating bacilli									
		Gram-positive					Gram-negative			Cocci	
		Actinomyces	Bifido-bacterium	Eubac-terium	Lacto-bacillus[c]	Propioni-bacterium	Bacteroides	Fuso-bacterium	Vibrio	Gram-positive	Gram-negative
Skin	0	0	0	±	0	2	0	0	0	1	0
Upper respiratory tract[b]	0	1	0	±	0	1	1	1	1	1	1
Mouth	±	1	1	1	1	±	2	2	1	2	2
Intestine	2	±	2	2	1	±	2	1	±	2	1
External genitalia	0	0	0	U	0	U	1	1	0	1	0
Urethra	±	0	0	U	±	0	1	1	±	±	U
Vagina	±	0	1	±	2	1	1	±	1	1	1

[a] Key to symbols: U, unknown; 0, not found or rare; ±, irregular; 1, usually present; 2, usually present in large numbers.
[b] Includes nasal passages, nasopharynx, oropharynx, and tonsils.
[c] Includes anaerobic, microaerophilic, and facultative strains.

will, therefore, be able to predict which organisms may be involved in a given infection (e.g., *Fusobacterium nucleatum*, *Bacteroides melaninogenicus*, and anaerobic or microaerophilic cocci in the case of aspiration pneumonia) and to initiate therapy in a rational manner pending definitive bacteriologic data. The microbiologist may use this type of information to choose selective and other media that may be helpful. Knowledge of the normal flora permits one to judge more readily whether or not a given isolate is significant. For example, *Propionibacterium acnes* in a blood culture is usually a contaminant from the patient's skin. In bacteremia of uncertain source, the presence of a particular organism may suggest the portal of entry (e.g., *Bacteroides fragilis* in a male patient would suggest the gastrointestinal tract).

Aside from *Propionibacterium* (32c), the only anaerobes usually present on the skin are the gram-positive anaerobic cocci. *Eubacterium* is found occasionally. On areas of the skin near the anus, however, there may be variable numbers of anaerobes from the bowel flora as transient flora. These organisms may assume significance in patients with surgical or other wounds. Thus, *Clostridium perfringens* from the fecal flora may invade and produce serious disease following hip surgery.

Anaerobes, and other organisms, vary in their distribution on various oral surfaces. *Bacteroides melaninogenicus* constitutes 4 to 8% of the cultivable flora of the gingival crevice but less than 1% of the flora of coronal tooth surfaces, of the tongue, and of the cheek. *Veillonella* make up 10–15% of the tongue flora, 5–15% of the gingival crevice flora, 1–3% of the coronal tooth surface flora, and less than 1% of the cheek flora. The microaerophilic streptococci are also prominent in all of the above locations, but particularly on the tongue and cheek (32).

The mean total anaerobic count in saliva is 1.1×10^8/ml and the aerobic count, 4×10^7/ml. Aside from the organisms already mentioned, other anaerobes prevalent in the mouth are *Fusobacterium nucleatum* and anaerobic streptococci and cocci. Present in smaller numbers are *Vibrio*, *Lactobacillus*, *Actinomyces* (*A. israelii*, *A. naeslundii*, *A. viscosus, and A. odontolyticus*), *Propionibacterium*, *Leptotrichia buccalis*, *Arachnia*, *Bifidobacterium*, *Eubacterium*, and *Treponema*. Plaque, a naturally occurring film on the surface of teeth, contains many anaerobes among the cultivable flora—anaerobic diphtheroids, 18%; *Peptostreptococcus*, 13%; *Veillonella*, 6%; *Bacteroides*, 4%; and *Fusobacterium*, 4% [Gibbons *et al.* (33)].

The stomach normally has $<10^3$ organisms per milliliter and, as a rule, no obligate anaerobes. The small bowel flora is also relatively simple, with total counts usually 10^4 to 10^5 organisms per milliliter or less, except for the distal ileum where the counts are about 10^6/ml. High in the small bowel, the flora is made up almost entirely of gram-positive facultative forms. The flora becomes more diversified as one progresses down the small bowel. In the terminal ileum, there are approximately equal numbers of aerobes and anaerobes; *Bacteroides* and *Bifidobacterium* are the major anaerobes encountered. In ileostomy patients, however, mean anaerobic counts were $10^{5.72}$ compared to aerobic counts of $10^{8.29}$ in one study (34). In this study, transverse colostomy effluent had a ratio of anaerobes to aerobes of $10:1$, whereas the ratio in feces is $100:1$ to $1000:1$. Fecal anaerobic counts in 18 subjects on a Western diet [Finegold *et al.* (35)] were as follows: *Bacteroides*, 10^{11}; anaerobic cocci and streptococci (including *Peptococcus, Peptostreptococcus, Veillonella, Acidaminococcus, Megasphaera, Sarcina,* and *Ruminococcus*), $10^{10.6}$; *Eubacterium*, $10^{10.1}$; *Clostridium*, $10^{9.5}$; *Bifidobacterium*, $10^{9.5}$; *Fusobacterium*, $10^{5.5}$; and *Lactobacillus* (including facultatives), $10^{4.0}$. *Bacteroides fragilis* was the dominant species encountered. An enormous number of species and groups, including many previously unrecognized species, was present (over 160 in this group of subjects and about 300 when another group, on a Japanese diet, is included). Figure 1.1 shows the counts of major elements of the fecal flora in 25 United States subjects (483). *Bacteroides fragilis* ss. *fragilis* was only the third most common subspecies of *B. fragilis* isolated from feces (35a, 483) behind ss. *vulgatus* and ss. *thetaiotaomicron*, whereas ss. *fragilis* is the subspecies most commonly involved in infection.

Although the intestinal flora undoubtedly plays an extremely important role in the body's nutrition and physiology, comparatively little is known about the subject. It is known that *Escherichia coli* and *Bacteroides fragilis* are capable of synthesizing vitamin K. Under certain circumstances, vitamin K production by intestinal bacteria may be important to the host. Since

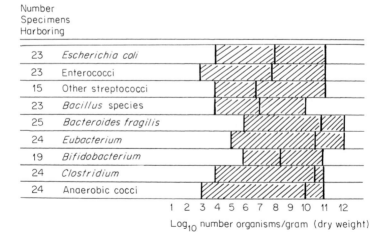

Fig. 1.1. Predominant fecal flora from 25 normal adult subjects. The length of the block represents the range; the vertical lines within the blocks represent the median counts.

B. fragilis greatly outnumbers *E. coli* in the normal intestinal flora, it is reasonable to assume that *B. fragilis* is much more important in this regard (36).

Bile acids play an essential role in fat absorption, bile formation, and regulation of cholesterol metabolism. Intestinal bacteria modify bile acids in various ways, including deconjugation of the attached glycine and taurine, dehydroxylation of the primary bile acids (cholic and chenodeoxycholic acids) to form the secondary bile acids (deoxycholic and lithocholic acids), and a number of other changes not all of which have been well characterized. The significance of some of the transformation products remains to be determined as well. In any case, it is known that the action of intestinal bacteria on bile acids is important for their conservation by means of enterohepatic circulation. Deconjugation of bile salts may be carried out by a number of bacteria, those of importance in the intestinal tract being *Bacteroides fragilis*, various species of *Fusobacterium*, *Bifidobacterium*, other gram-positive anaerobic bacilli, and, among the aerobic organisms, *Streptococcus faecalis* (37, 38). Dehydroxylation of bile acids is carried out by various gram-positive anaerobic bacilli and, to a limited extent, by some strains of *B. fragilis*, *Veillonella*, and some aerobes (38).

The normal bacterial flora of the intestinal tract is certainly important in relation to a variety of infections that may arise secondary to bowel perforation, strangulation, obstruction by carcinoma or other mechanisms, or following bowel surgery. *Bacteroides fragilis* is one of the major organisms likely to be found in such infections, along with coliforms and enterococci; other anaerobes are likely to be involved as well.

The normal flora of the bowel constitutes one of our most important defense mechanisms, a fact that is not generally appreciated. In this connection, the excellent studies of Bohnhoff *et al.* (39) are of interest. These workers showed that mice are rendered much more susceptible to experimental *Salmonella* infection after treatment with oral streptomycin, which reduces the counts of *Bacteroides* in the mouse intestine. In other animal studies, Hentges (39a) has shown that the normal bowel flora protects against *Shigella flexneri* infection. Levison (39b) demonstrated *in vitro* inhibition of *Pseudomonas aeruginosa* by normal mouse colonic contents, by certain organisms (including *Bacteroides fragilis*) from human colonic flora, and by acetic and butyric acids (common end products of metabolism of anaerobic bacteria).

Although much has been done regarding the effect of antimicrobial agents on the intestinal flora, it has not been generally appreciated that the reverse (effect of intestinal flora on antimicrobial agents) may be very important. Two examples of this type of activity are the hydrolysis of succinyl and phthalyl sulfathiazole to the active drug, sulfathiazole, and the effect of the bowel flora on sulfadimethoxine, which is normally excreted in the biliary tract as the glucuronide. The bowel flora deconjugates the glucuronide, thus liberating free drug that is then reabsorbed. This enterohepatic circulation may account for the long duration of activity of this drug. Drugs of other classes may also be affected by normal flora. The specific organisms involved are usually not known, but presumably anaerobes are important because of their numerical prevalence.

Mead and Louria, in 1969 (40), describe the composition of the normal vaginal flora as compiled from eight papers. *Bacteroides* were commonly isolated. Lactobacilli were found in from 49 to 73% of subjects, diphtheroids (not necessarily anaerobic) in 44–74%, *Bifidobacterium* in 26–72%, anaerobic streptococci in 12–59%, and *C. perfringens* in 0–9%. Suzuki and Ueno, in 1971 (41), studied the normal vaginal flora of 90 subjects. In this study, *Peptococcus* was commonly isolated, and *Peptostreptococcus* was found in 12–59%. *Veillonella, Bacteroides, Fusobacterium,* and *Propionibacterium* were usually found. Occasionally isolated were *Eubacterium* and *Bifidobacterium*. In a study of 280 pregnant women, de Louvois *et al.* (42) found lactobacilli and corynebacteria (not necessarily anaerobic) in 82 and 83%, respectively. Microaerophilic and anaerobic streptococci were recovered from 22% of the subjects and *Bacteroides* from 5%.

A study of the normal cervical flora of 30 healthy women was done by Gorbach *et al.* (43). *Bacteroides* (*B. oralis, B. fragilis, B. capillosus, B. clostridiiformis,* and *Bacteroides* sp.) was isolated from 57% of the women. *Peptostreptococcus* was found in 33%, and *Veillonella* in 27%. *Clostridium* (*C. bifermentans, C. perfringens, C. ramosum,* and *C. difficile*) was recovered from 17%. *Bifidobacterium* was noted in 10%, *Peptococcus* in 7%, and *Eubacterium* in 3%. Lactobacilli were found in 73% of women. Ohm and Galask (43a) did cervical cultures on 100 women who were to have hysterectomy (primarily for nonmalignant disease). Anaerobes were isolated (along with aerobes) from 86% of the women; 69% had two or more species/culture. Most commonly encountered were gram-positive cocci (154 isolates in 74 cultures); the most prevalent of these were *Peptococcus asaccharolyticus* and *Peptostreptococcus anaerobius*. There were 41 anaerobic gram-positive bacilli; *Lactobacillus* and *Eubacterium* were the most prevalent, with *Clostridium* (chiefly *C. perfringens*) next. There were 32 anaerobic gram-negative rods (including 4 *B. fragilis*). Sixteen gram-negative cocci were recovered. Sanders *et al.* (43b) studied 26 healthy women and noted that 92% yielded anaerobes on endocervical culture. In their study, *Bacteroides* were most common (65% of women). Peptococci were recovered from 57%, peptostreptococci from 50%, *Bifidobacterium* from 35%, and *Eubacterium lentum* from 15%. A study of products of conception of 47 patients with spontaneous abortion (43c) provides additional information in this area; 66% of cultures yielded anaerobes. The most prevalent isolates were *Peptococcus* (34%), *Peptostreptococcus* (17%), and Bacteroidaceae (15%). The uterine cavity is normally sterile.

Studies by Bran *et al.* (44), in a small number of subjects, noted the following anaerobes among the indigenous urethral flora of females: *Fusobacterium gonidiaformans*, *Bacteroides* sp., anaerobic diphtheroids, and another type of anaerobic gram-positive bacillus. The normal flora of the male urethra and of prostatic fluid from 46 subjects without known genitourinary tract disease was evaluated by Ambrose *et al.* (45). No anaerobes were recovered from the urethra and only 2 *Bacteroides* were found in the prostatic fluid. On the other hand, Finegold *et al.* (46) studied "urethral" urine (the first 10 to 20 ml of voided urine) in 17 subjects and recovered anaerobes (along with aerobes) from 8 of these (bladder urine, obtained immediately previously by percutaneous bladder aspiration, was negative for anaerobes in all cases). In most cases, counts of the anaerobes were between 10^2 and 10^4/ml. Organisms recovered included *Bacteroides fragilis*, *B. melaninogenicus*, *Bacteroides* sp., *Fusobacterium*, anaerobic gram-positive cocci and streptococci, anaerobic catalase-negative gram-positive bacilli, *Corynebacterium*, and *Lactobacillus*.

Brams *et al.* (47) found fusiform bacilli and spirochetes in the preputial secretions of 51 of 100 men studied, in the absence of disease. Davis and Pilot (48) also found these organisms about the female clitoris. *Bacteroides melaninogenicus* was also found under the prepuce and in the coronary sulcus, but less often on the glans penis or the meatus, and only occasionally on the skin of the scrotum and pubic area (49). Burdon (49) also found *B. melaninogenicus* regularly about the clitoris, frequently in the vagina, and occasionally in the cervix.

The physiologic role of the normal flora in locations other than the intestinal tract is even less well known than that of the bowel flora. One recent report (49a), however, showed *in vitro* inhibition of *N. gonorrhoeae* by some strains of *Bifidobacterium*, *Clostridium*, and *Fusobacterium*.

The normal flora of the body varies with age and is modified under certain abnormal conditions. In the first year of life, streptococci are the only organisms cultured consistently from the mouth, although anaerobes are found in the edentulous mouth of infants (50). Diet is undoubtedly important. Total restriction of dietary carbohydrate for several days leads to a reduction in bacteria that synthesize intracellular polysaccharide (*Bacteroides* and *Fusobacterium* store such material) [Geddes and Jenkins (50a)]. Hospitalized individuals who are quite ill acquire nosocomial pathogens such as *S. aureus*, *Klebsiella*, and *Pseudomonas* in their oral and intestinal flora, even in the absence of antimicrobial therapy. Acquisition of this abnormal flora implies the possibility of reduction of the normal anaerobic flora, but this has not been studied specifically except in connection with antimicrobial therapy.

In the case of the intestinal tract, there is a transient increase in small bowel flora following meals [Drasar *et al.* (51)]. In subjects with achlorhydria, small intestinal bacterial counts are higher. Intestinal motility is another important factor controlling small bowel flora, along with the pH of gastric contents. Thus the type of diet and frequency of feeding affect the intestinal flora, since they affect the rate of gastric emptying and the secretion of gastric acid. Bacteria, including anaerobes, increase in number in the stomach and small bowel after gastrectomy, vagotomy, ileostomy, and extensive small bowel resection (52, 52a, 52b). The small bowel microflora also varies with geographical location of subjects; whether this is related to diet and/or other factors is unknown. Jejunal contents of South Indians have 100 times as many bacteria as is true in English subjects, and *Bacteroides* is found commonly in the individuals from South India (53). In a group of 21 children in Jakarta with malnutrition and diarrhea (53a), anaerobes were recovered from gastric contents of 7 and from small bowel contents in 7 of 20 sampled (counts $> 10^5$/ml in 5). Controls had no gastric anaerobes and low counts of anaerobes in the small bowel. In hypogammaglobulinemia, increased numbers of anaerobes are found in the small intestine (54, 54a).

Antimicrobial compounds may effect significant changes in the normal anaerobic flora (55). Most studies relating to this are of the fecal flora. The changes may usually be inferred accurately from the antibacterial spectrum of the drug and the amount of excretion of active drug in the intestinal tract.

It is generally possible to manipulate the bowel flora to eliminate one or more of the normal components. Such manipulation could be used to determine the relative importance of various segments of the flora in various physiologic functions and in certain pathophysiologic states. For example, one can eliminate anaerobic bacteria with little or no change in aerobes by the use of oral lincomycin. One can eliminate aerobes with retention of much of the anaerobic flora by use of oral kanamycin or related agents. Oral polymyxin or colistin eliminates *E. coli* with no other changes except for increased streptococcal counts in some patients. Oral bacitracin may be used to decrease or eliminate streptococci and clostridia. Combinations, such as oral kanamycin plus either tetracycline, erythromycin, or lincomycin, would eliminate all or most bacteria, at least for brief periods [Finegold *et al.* (56)]. One must always, of course, do appropriate quantitative cultures to determine exactly what effect was produced in the particular patient studied.

Pathogenesis and Epidemiology of Anaerobic Infections

Virtually all anaerobic infections arise endogenously. As indicated previously, anaerobic bacteria are found as part of the normal flora of the skin and are prevalent as indigenous flora on all mucous membrane surfaces.

Under certain circumstances (surgical or other trauma, tumors arising at the mucosal surface, etc.) anaerobes from the indigenous flora have an opportunity to penetrate tissues and thus to set up infection. Occasionally, as in aspiration pneumonia, anaerobic bacteria from a site of normal carriage may be carried into another area normally free of bacteria and set up an infection. Poor blood supply and tissue necrosis lower the oxidation–reduction potential and favor the growth of anaerobic bacteria. Experimental abscesses in rats (56a) had a pO_2 of 0–75 mm Hg, pH of 6.2–7.2 and oxidation–reduction potential of -20 to $+40$ mV. Gorbach and Bartlett (56b) state that the E_h of the intestinal tract, necrotic tissue, and abscess cavities is -150 to -250 mV. Thus, vascular disease, epinephrine injection, cold, shock, edema, trauma, surgery, foreign bodies, malignancy, and gas production by microorganisms all may significantly predispose to anaerobic infection. Prior infection with aerobic or facultative bacteria may also make conditions more favorable for growth of anaerobic bacteria (56c). Kenney and Ash (56c) have measured the E_h of periodontal pockets and found a mean of -47.6 as compared to $+72.6$ mV for gingival sulci. As with infection due to aerobic or facultative bacteria, various conditions in which host defense mechanisms are impaired (56d–56i, 492), and prior antimicrobial therapy to which the infecting organism is resistant (especially aminoglycosides in the case of anaerobes) may pave the way for anaerobic infection (56j, 492). Gentamicin actually enhanced *B. fragilis* infection in rabbits (56k). There are conflicting reports regarding the effectiveness of polymorphonuclear leukocytes against anaerobic bacteria under conditions of anaerobiosis. Mandell (561) commented that oxygen is necessary for normal bactericidal activity of polymorphonuclear leukocytes (PMN's), yet found that anaerobic PMN's were able to kill *P. anaerobius, B. fragilis, C. perfringens*, and *P. magnus* normally. On the other hand, Keusch and Douglas (56m) noted normal phagocytosis but impaired killing of *C. perfringens* by leukocytes under anaerobic conditions. Casciato *et al.* (70) noted normal killing of *B. fragilis* by PMN's under aerobic conditions.

Extremely oxygen-sensitive (EOS) anaerobes, which may be found in large numbers in normal flora, have not been demonstrated to participate in a significant way in infectious processes. Thus, pathogenic anaerobes are more aerotolerant and some are surprisingly aerotolerant [Tally *et al.* (19)]. This feature may permit these organisms to survive, after the normally protective mucosal barrier is broken, until conditions are satisfactory for their multiplication and further invasion (prior growth of accompanying facultative forms lowering the E_h, etc.). Once proper conditions are achieved, anaerobes can multiply relatively rapidly and can maintain their own reduced environment by virtue of excretion of end products of fermentative metabolism such as short-chain fatty acids, organic acids, and alcohols.

Because anaerobes are part of our normal flora, they are considered by some to be innocuous. There is ample clinical and experimental data, however, to document an important infectious role for them under the right circumstances. This can be demonstrated by inoculating experimental animals subcutaneously with bacterial scrapings from the oral cavity or with human fecal suspensions (57–63). Necrotic infections, characterized by a mixed bacterial flora in which nonsporulating anaerobes are prominent, invariably develop. These infections may be of two types (62, 63). One type remains localized as an abscess, while the second is characterized by gangrenous spreading, usually accompanied by bacteremia. Exudate from such experimentally induced lesions contains a complex mixture of bacteria analogous to that present in the original inoculum. Such exudate may be used to transmit serially the infection to additional animals. Throughout serial transmission, the anaerobic segments of the flora persist.

Most investigators have found that pure cultures of indigenous bacteria lack infectious potential when inoculated subcutaneously in experimental animals. However, the entire bacterial complex is consistently pathogenic, and a number of investigators have successfully reproduced transmissible infections in experimental animals using defined mixtures of bacteria (62–67). Thus, these mixed anaerobic infections provide a clear-cut example of bacterial synergism in the production of disease.

In certain of these mixed infections, *Bacteroides melaninogenicus* was found to play a key role. Deleting it from the mixture led to a noninfective combination and adding it made the complex uniformly infective (63). A complex mixture of over 50 organisms could be reduced to a very small number, one of which had to be *B. melaninogenicus*. Certain other organisms were important only in terms of providing vitamin K, a necessary growth factor for some strains of *B. melaninogenicus* (62). *Bacteroides melaninogenicus* was also a key element in another experimental model (67a), along with a heparinase-producing *Bacteroides* and an anaerobic *Corynebacterium*.

Onderdonk *et al.* (68, 68a) and Bartlett *et al.* (68b) have also described an interesting experimental model for peritonitis in the rat. For subsequent intraabdominal abscess to develop, it was necessary to have both a facultative organism and an anaerobe in the original implanted mixture.

Thus two mechanisms for synergism have been demonstrated—production of a low oxidation–reduction potential by the facultative organism and supply of an essential growth factor. Other mechanisms undoubtedly exist. Certainly in some cases, the mucosal barrier may be penetrated initially by an infection involving only facultative bacteria.

Aside from mixed anaerobic–facultative infections that are common in all organ systems, there are examples of specific synergy. The most clear-cut of these is postoperative bacterial synergistic gangrene, which requires the presence of a microaerophilic (or anaerobic) streptococcus and *Staphylococcus aureus* (sometimes a gram-negative facultative bacillus will substitute for the latter) [Meleney (69)].

Involvement of anaerobic bacteria in infectious processes may also depend on resistance of certain of these organisms to normal body defense mechanisms. Thus, the fact that certain subspecies of *Bacteroides fragilis* (especially ss. *fragilis*) are more commonly involved in infection than others may relate, at least partially, to their resistance to normal bactericidal activity of serum [Casciato *et al.* (70)]. One strain of *B. melaninogenicus* has been shown to produce an antigenic capsule that appeared to relate to its infectivity (71, 71a). A thick capsule has been demonstrated about *B. fragilis* strains as well (71b); the author noted that this might protect the organism from phagocytosis and from serum bactericidal activity.

Certain toxins account for the virulence of some anaerobic infections or cause serious intoxication without infection (botulism). The α toxin of *Clostridium perfringens* is a potent lecithinase that is hemolytic and necrotizing. Tetanolysin, from *C. tetani*, causes lysis of human platelets and is lytic for other cellular or subcellular membranes (71c), although the significance of this toxin is still uncertain. Five clostridial species isolated from human infections were shown to produce a number of extracellular proteins in addition to lecithinase [hemolysin, gelatinase, a caseinolytic product, elastase, a staphylolytic enzyme, and deoxyribonuclease (71d)]. Strains of gram-negative anaerobic bacilli have been shown to contain endotoxins (72–77). However, while the structure and overall chemical composition of the cell wall of these bacilli is similar to that of the nonanaerobic gram-negative bacilli, the cell wall lipopolysaccharide of *Bacteroides* is devoid of heptose and 2-keto-3-deoxyoctonate (KDO) and that of *Fusobacterium* is apparently relatively low in KDO content (72a, 72b). Some gram-negative anaerobic bacilli also produce a neuraminidase, a fibrinolysin, a hemolysin, deoxyribonuclease, hydrogen sulfide, indole, ammonia, and β-glucuronidase (35a, 62, 78, 79, 79a). In experimental *Bacteroides* peritonitis (79b), elevation of serum lysozyme levels paralleled the course and severity of the illness; these levels dropped in response to therapy. They were considered a good prognostic indicator.

Most strains of *Bacteroides melaninogenicus* and some other anaerobic strains are highly proteolytic, exhibiting activity on a variety of proteins including casein, fibrin, and native collagen (79–82). The collagenolytic activity of *B. melaninogenicus* is of special interest, since it has been suspected of playing a role in the tissue destruction resulting from anaerobic infection. Strains of *B. melaninogenicus* hydrolyze native collagen from human gingiva and human dentine (83–85). Strains with high collagenolytic activity produce more acute infections in combination with anaerobic vibrios than strains with weak collagenolytic activity (86). Recently, Kaufman *et al.* (87) demonstrated that cell-free extracts of *B. melaninogenicus* possessing collagenolytic activity, when given with a live *Fusobacterium* species, produced nontransmissible lesions in rabbits which were more severe than those resulting from injection of either the extract or the organism separately.

The elaboration of heparinase by *Bacteroides* (88) and the acceleration of coagulation by *Bacteroides* sp. and fusobacteria demonstrated experimentally (89) may contribute to septic thrombophlebitis (see Fig. 1.2) (89a). This lesion may lead to metastatic abscesses and helps account for the difficulty in eradicating anaerobic infections (Figs. 1.3 and 1.4) (89a).

Although the vast majority of anaerobic infections are endogenous, some may be of exogenous origin. This is particularly true in the case of *Clostridium*, an organism widely distributed in soil in nature. Nonetheless, most workers feel that the majority of cases of clostridial myonecrosis ("gas gangrene") following war injury were of endogenous origin and related to difficulties in maintaining good personal hygiene, among other things, under battle conditions. Similarly, there is evidence to indicate that clostridial myonecrosis following surgery is traceable to the patient's own fecal flora rather than to spores of clostridia in the operating suite (though the presence of these spores in that setting has been noted). An unusual external source of *C. perfringens* leading to wound infection with myonecrosis following

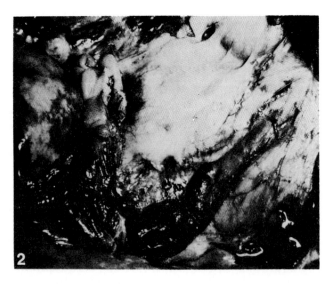

Fig. 1.2. Large thrombi filling pelvic veins. From Schwarz (89a), reproduced with permission.

Fig. 1.3. Large saddle embolus in main pulmonary artery extending into the major branches bilaterally. From Schwarz (89a), reproduced with permission.

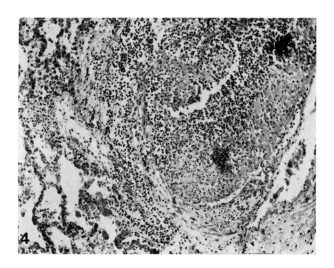

Fig. 1.4. Infected embolus in pulmonary artery. Notice the two dark-staining clumps of bacteria within the blood clot. Hematoxylin-cosin. × 120. From Schwarz (89a), reproduced with permission.

surgery is tanks of compressed air used to drive bone saws (P. Charache, personal communication).

Nosocomial infection as such does not involve anaerobes, but it is clear that anaerobic infection may be facilitated by diagnostic and therapeutic modalities utilized in (or out of) the hospital setting. Thus, radiation, cytotoxic drugs, and corticosteroids generally predispose to infection. Diagnostic and therapeutic maneuvers may lead to impaired blood supply, tissue necrosis, breakdown of the mucosal barrier, etc. Antibiotic therapy may select out resistant anaerobes such as *Bacteroides fragilis*, which may then cause infection under the appropriate conditions. Transmission of anaerobic infection from animal to man has, on rare occasion, been documented. Human to human transmission is even more rare, with the exception of certain venereal diseases. Infection following bites, of course, frequently involves anaerobic bacteria. Laboratory accidents may, rarely, lead to anaerobic infection. The epidemiology of anaerobic intoxications is discussed in Chapter 17.

Addendum

Recently it has been proposed [Cato and Johnson, *Int. J. Systematic Bacteriol.* **26**, 230 (1976)], that what are presently known as subspecies of *Bacteroides fragilis* be reinstated to full species rank. Thus there would be *Bacteroides fragilis*, *Bacteroides thetaiotaomicron*, etc., and the five organisms previously referred to as subspecies might be considered as the *Bacteroides fragilis* group.

2

Diagnostic Considerations

Clues to Presence of Anaerobic Infection

Tables 2.1 and 2.2 list certain clinical and bacteriologic clues to anaerobic infection. The foul odor of discharges or lesions is the most important and definitive clue to anaerobic infection. This odor, falsely attributed to *E. coli*, is not always found with anaerobic infections (certain anaerobes do not produce foul-smelling products and there may be no communication between the site of infection and the outside of the body) but is a definite indication that anaerobes are involved in an infection when it is present. It is interesting to note (89b) that when Altemeier first submitted a manuscript, in 1942, on his conclusion that the so-called "B. coli odor" was actually due to anaerobes, the *Annals of Surgery* rejected the paper on the basis of its representing an incorrect observation! Certain of the clinical clues, of course, are nonspecific, but combinations of two or more clues (tissue necrosis and gas in tissues, for example) would be very suggestive of the possibility of anaerobic infection.

Certain infections are so likely to involve anaerobes as significant pathogens that they should be regarded as anaerobic or mixed anaerobic–facultative until proved otherwise. Included in this category are such infections as brain abscess, cellulitis of the jaw or trismus following dental extraction, dental infections, aspiration pneumonia, lung abscess, bronchiectasis, peritonitis, intraabdominal abscess, wound infection following bowel surgery

TABLE 2.1

Clinical Suggestions of Possible Infection with Anaerobes

Foul-smelling discharge
Location of infection in proximity to a mucosal surface
Necrotic tissue, gangrene, pseudomembrane formation
Gas in tissues or discharges
Endocarditis with negative routine blood cultures
Infection associated with malignancy or other process resulting in tissue destruction
Infection related to the use of aminoglycosides (oral, parenteral, or topical)
Septic thrombophlebitis
Infection following human or other bites
Black discoloration of blood-containing exudates; these exudates may fluoresce red under
   ultraviolet light (*Bacteroides melaninogenicus* infections)
Presence of sulfur granules in discharges (actinomycosis)
Clinical setting suggestive of anaerobic infection (e.g., septic abortion, infection following
   gastrointestinal surgery)
Classical clinical picture (actinomycosis, clostridial myonecrosis)

---

or trauma, endometritis, tuboovarian abscess, perirectal or ischiorectal abscess, gas-forming and/or necrosing soft tissue infections, myonecrosis, bacteremia with major intravascular hemolysis, infection following human bite, and any other infection with putrid discharges. This does not mean, of course, that one can ignore other possible or concurrent causes. This could be disastrous, particularly in very sick patients. It simply means that statistically the probability of anaerobes being involved is very high and should not be overlooked. Aerobic or facultative bacteria may also be present, of course. Gray-black discoloration of tissues or discharges may be a clue to *B. melaninogenicus* infection (89b). Increased platelet count, apparently related to the infection, was described in one case of actinomycosis (89d).

TABLE 2.2

Bacteriologic Suggestions of Possible Infection with Anaerobes

---

Unique morphology on gram stain of exudate (or subsequent culture growth)
No growth on routine culture, particularly with purulent specimens
Failure of organisms seen on gram stain of original exudate to grow aerobically (failure to
   obtain growth in fluid thioglycolate medium is not adequate assurance that anaerobes are
   not present)
Growth in anaerobic zone of fluid media or of agar deeps
Growth anaerobically on media containing 75–100 μg of kanamycin, neomycin, or paromo-
   mycin per milliliter (or a medium also containing 7.5 μg of vancomycin per milliliter in the
   case of gram-negative anaerobic bacilli) or on other media highly selective for anaerobes
Gas, foul odor in specimen or culture
Characteristic colonies grown anaerobically on agar plates
Young colonies of *Bacteroides melaninogenicus* may fluoresce red under ultraviolet light

---

Specimens to Culture for Anaerobes

Since anaerobes are commonly involved in a variety of infections and since all types of infections may involve anaerobes, all types of specimens should be cultured for these organisms. Furthermore, a number of non-anaerobic organisms (notably streptococci) will grow better under anaerobic conditions than under aerobic. Certain organisms, viz., microaerophilic streptococci, may be missed if anaerobic cultures are not made. One unique report (89c) described a strain of *Salmonella typhi* requiring anaerobic conditions for primary isolation. However, since anaerobes are so prevalent as normal flora, it is also important to keep in mind that no specimen which is contaminated with normal flora should be cultured for anaerobic organisms; otherwise the results will be meaningless and the laboratory will be burdened unnecessarily. Since anaerobes characteristically produce abscesses, it is important that all abscesses be cultured anaerobically. When clinical or bacteriologic clues suggesting anaerobic infection are present, it is especially important to do anaerobic cultures.

The following types of specimens should *not* routinely be set up for anaerobic culture—throat culture, gingival swab, gastric contents, small bowel contents, feces, coughed sputum, voided urine, and vaginal swabs. Any other specimen which may be "contaminated" with normal flora should also not be submitted for anaerobic culture. Exceptions will have to be made in certain instances, For example, in suspected "blind loop syndrome" quantitation of aerobic or facultative and anaerobic flora of the small bowel or afferent gastric loop contents may be important diagnostically and in guiding therapy. When sulfur granules are present in sputum, they may be used successfully for diagnosis of actinomycosis; the granules are washed thoroughly before being cultured.

Collection and Transport of Specimens

Proper specimen collection and transport are absolutely essential if reliable and meaningful results are to be obtained on anaerobic culture. Ordinarily, these two important considerations will be the responsibility of the clinician. However, it may be necessary for the microbiologist to instruct the clinician in the importance of proper collection and transport and also in specific techniques that are appropriate.

In suspected anaerobic pulmonary infection, when empyema fluid is not available, percutaneous transtracheal aspiration is ordinarily the procedure of choice in adults (assuming there are no contraindications to this procedure) and direct lung puncture, in young children. Cultures of coughed sputum are routinely contaminated with anaerobes from the normal upper respiratory tract flora and are unsuitable for anaerobic culture (see Fig. 2.1). Specimens obtained at bronchoscopy or by bronchial brushing are also contaminated with normal oronasopharyngeal anaerobic flora and should not be cultured. It has not been satisfactorily demonstrated that this contamination problem does not exist with fiberoptic bronchoscopy. Specimens aspirated from patients with tracheostomies are probably suitable. Since the trachea at the level of the cricothyroid membrane is sterile (except in some patients with chronic obstructive pulmonary disease, and these are not

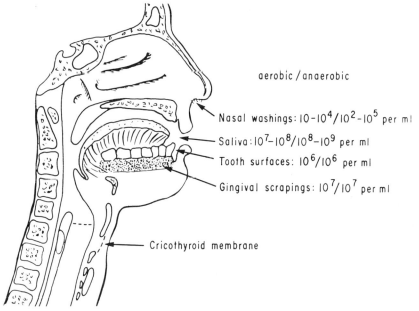

Fig. 2.1. Sagittal view of head and neck showing counts of indigenous flora at various sites. Dotted horizontal line indicates level below which tracheobronchial tree normally is sterile. [From P. D. Hoeprich (89e) with kind permission of author and publisher.]

known to have anaerobes present), transtracheal aspiration (Fig. 2.2) at this site is a reliable method of obtaining suitable culture material in pulmonary infection [Bartlett *et al.* (90)]. Complications have been very infrequent with transtracheal aspiration done by experienced operators on cooperative patients with no contraindications. Patients with severe hypoxia, hemorrhagic diathesis, or severe, intractable cough should generally be excluded. Additionally, patients unwilling to remain quietly in bed for 8 to 12 hours should also be excluded. Rarely, bleeding, hypoxia, subcutaneous emphysema, or arrhythmia complicate transtracheal aspiration. Cellulitis or

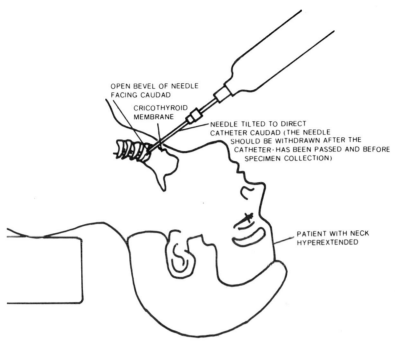

OPEN BEVEL OF NEEDLE
FACING CAUDAD

CRICOTHYROID
MEMBRANE

NEEDLE TILTED TO DIRECT
CATHETER CAUDAD (THE NEEDLE
SHOULD BE WITHDRAWN AFTER THE
CATHETER·HAS BEEN PASSED AND BEFORE
SPECIMEN COLLECTION)

PATIENT WITH NECK
HYPEREXTENDED

Fig. 2.2.   Diagrammatic sketch of transtracheal aspiration procedure (courtesy of Dr. John G. Bartlett). See procedure in Chapter 9.

abscess at the site of aspiration occurs in about 0.35% of procedures and is usually caused by local inoculation of the etiologic agent(s) from the primary pulmonary infection. Direct lung puncture in adults is associated with a relatively high incidence of pneumothorax.

Abscesses in contiguity with mucosal surfaces or skin should be cultured by aspiration with a syringe after the surface has been decontaminated with an appropriate germicide. If surface decontamination is not done, the normal anaerobic flora of the vicinity may contaminate the culture. Swabbing of open wound surfaces may yield secondary colonizers of wounds which have little clinical significance. Studies in which isolation of bacteria was achieved from such swabs may not convey an accurate picture of the infecting flora.

Midstream urine may be contaminated by anaerobic organisms present in the distal urethra or the adjacent perineal or vulvar areas (46). Thus midstream or catheterized urine cultures may contain anaerobic organisms unrelated to urinary tract infection. Percutaneous suprapubic bladder aspiration is required to confirm the presence of anaerobic bacteriuria. Of course, when a nephrostomy tube or suprapubic catheter is present this provides a source of urine free of indigenous flora.

Collection of specimens in suspected uterine infections requires aspiration of material through a syringe or plastic cannula after decontamination of the cervical os. Careful positioning of the patient and good exposure with a vaginal speculum are important. Even with good technique, normal vaginal or endocervical anaerobic flora will still contaminate the specimen to some extent; therefore, quantitation of growth may be useful. Aspiration of material through a tube that has been advanced in a protective sleeve and then exposed only after entering the uterine cavity is desirable (Figs. 2.3 and 2.4) (90a). Percutaneous transfundal aspiration may be suitable in some cases of postpartum endometritis (91). When pus has accumulated in the cul-de-sac, culdocentesis provides an excellent specimen (Fig. 2.5).

The preferred technique of collecting most specimens is by aspiration with needle and syringe. In doing so, it is important to protect the specimen as much as possible from contact with air. The plunger should be fully inserted into the barrel at the start and, when the volume is adequate, a drop of specimen should be expelled on an alcohol–cotton pledget; the syringe and needle should be entirely cleared of air. If the site to be sampled will yield very little material, about 1 ml of a sterile prereduced solution or broth (see below) is aspirated into the syringe before collection. It is also desirable to prerinse the syringe with this material with small specimens. The sample is then aspirated into this solution in the syringe.

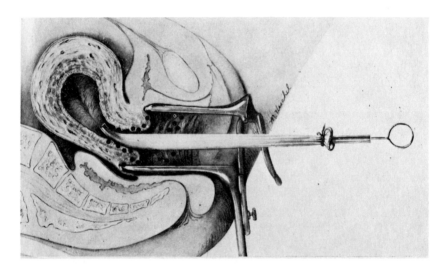

Fig. 2.3. Diagrammatic cross section showing culture tube through cervix beyond the internal os. Finger cot has been drawn taut and the stylet partially inserted preparatory to perforating the stretched rubber protecting the end of the metal tube. Vaginal and cervical flora indicated by dots. [From Guilbeau and Schaub (90a), reproduced with permission.]

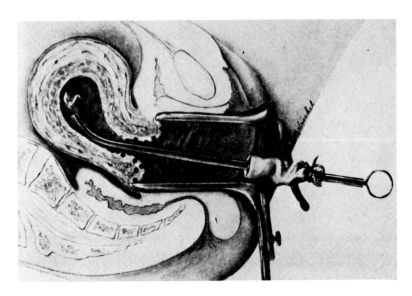

Fig. 2.4. Diagrammatic cross section revealing wire loop in place following perforation of the finger cot which is shown retracted into the vagina. Contamination with vaginal and cervical flora is avoided, and a culture representing the true uterine flora is thus obtained. [From Guilbeau and Schaub (90a), reproduced with permission.]

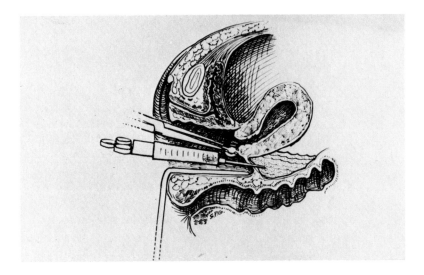

Fig. 2.5. Cul-de-sac aspiration, lateral view. Note ventral and caudal traction on the posterior cervical lip. From Schwarz (89a), reproduced with permission.

Whenever possible, it is desirable to obtain specimens of tissue (rather than just exudate) for anaerobic culture. Tissue specimens may more accurately indicate the bacteriology of the infectious process, and anaerobic bacteria survive more readily in this type of specimen. Collection procedures for specimens in cases of suspected anaerobic intoxications are given in Chapter 17.

The second major consideration relates to the possibility of anaerobes dying as a result of exposure to oxygen during the interval between obtaining the specimen and culturing it in the laboratory. Since some anaerobes are very tolerant to oxygen contact and can survive for a number of days if kept moist, every laboratory will have some degree of success, no matter what transport system is used. Accordingly, a laboratory that frequently isolates such relatively nonfastidious organisms as *Bacteroides fragilis* and *Clostridium perfringens* may not realize that it is overlooking many other more demanding anaerobes. Many anaerobes responsible for clinical infections do not tolerate oxygen exposure well, particularly in small specimens of thin fluid, so that special transport methods are needed to ensure survival from time of collection to the start of the analysis. Obligate aerobes and facultative organisms remain viable in an anaerobic system; therefore, the universal carrier for bacteriological specimens, where anaerobic bacteria as well as other forms are sought, should be an anaerobic container. Anaerobic syringe techniques, such as are used in analysis of blood gases, may prove very useful for transport (Fig. 2.6). If a sterile stopper is not available, one may use a nonsterile stopper, as from a Vacutainer tube. The laboratory should then be instructed to put a sterile needle on in place of the original one prior to setting up the specimen for culture. With the (plastic) syringe technique, it is important that the specimen be delivered to the laboratory and set up in culture within 20 to 30 minutes.

When longer periods are required for transport, a gassed-out tube with a butyl rubber stopper is convenient to use [Attebery and Finegold (92)] (Fig. 2.7). The tube is gassed out with oxygen-free gas and contains an indicator so that one may be certain that the atmosphere is anaerobic. The indicator may be incorporated either in agar or in a fluid diluent. Only a small amount of diluent (of low E_h and preferably containing cysteine HCl to facilitate reduction) should be used to minimize dilution of the specimen. Resazurin, a sensitive oxidation–reduction indicator, which turns pink on exposure to air, is a good one to use. A pink color may be noted briefly after addition of a specimen. Once the specimen has been injected into the tube, even a 2-hour delay between collection and setting up the specimen in culture is satisfactory. One can hold specimens in such tubes for extended periods (over 24 hours, if necessary) if the tube is refrigerated (not frozen) to prevent overgrowth of aerobic and facultative forms. Do not refrigerate anaerobic specimens that are not in a gas-tight container.

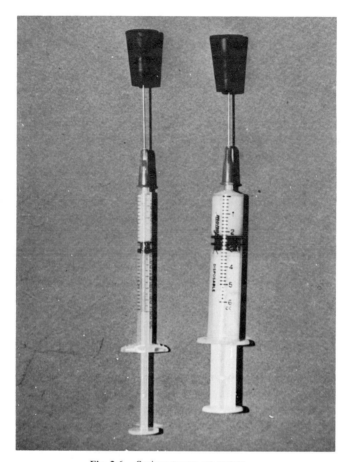

Fig. 2.6. Syringe transport system.

If it is necessary to use a swab to obtain a specimen, the swab itself should have been kept in an anaerobic atmosphere. After the specimen has been collected, the swab is placed into a second tube that contains a deep column of prereduced semi-solid transport medium such as Cary and Blair medium in an anaerobic atmosphere (Fig. 2.8).

Large amounts of fluid material may be transported by filling regular screw-topped culture tubes to the brim (thus displacing all air) and then capping. Tissue, or swabs, may be transported in a metal film cannister in which anaerobiosis is achieved by the reaction of acidified copper sulfate on steel wool (93) (Fig. 2.9). Specimens should always be transported and processed as rapidly as possible after collection to avoid loss of fastidious, oxygen-sensitive anaerobes and the overgrowth of facultative bacteria. The

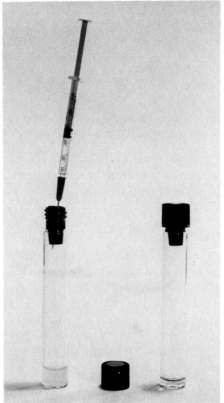

Fig. 2.7.   Gassed-out tube transport system.

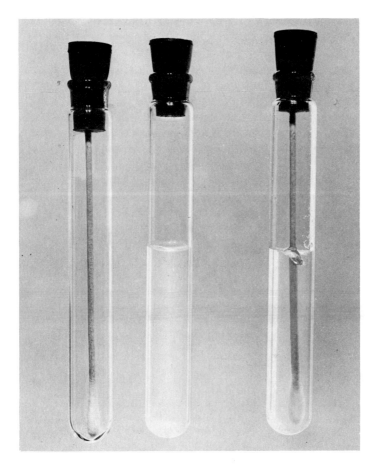

Fig. 2.8. Swab transport setup.

Fig. 2.9. Setup for transport of tissue specimen.

advantages of syringe collection and syringe or tube transport over the swab technique were emphasized in a recent bacteriologic study of peritonsillar abscesses (93b). A portable gassing-out device for making a tube anaerobic is useful for any type of specimen.

Direct Examination

Macroscopic examination of specimens includes notation of odor, color (89b), fluorescence under ultraviolet light (739), purulence, and presence of necrotic tissue fragments, gas, and sulfur granules.

Microscopic examination should include, in addition to gram stain, if possible, dark field or phase contrast microscopy. The latter is useful for demonstration of spirochetes or other motile forms and for spore detection. At times, Wright's stain will be useful for delineation of cellular elements in discharges although gram stain is ordinarily satisfactory for this purpose. Fluorescent antibody staining, while promising for the future, is not yet generally practical for clinical use. It is valuable in the case of actinomycosis, however, in specialized laboratories (93c, 233). Investigational work is also in process regarding the utilization of fluorescent antibody technique for rapid identification of gram-negative anaerobic bacilli (93d, 93e) and gram-positive anaerobic cocci (93f). Commercial reagents have proved valuable in fluorescent antibody identification of various clostridia from animal infections (93g).

Direct gas-liquid chromatography of clinical specimens has also been proposed as an aid in the rapid diagnosis of *Bacteroides fragilis* infection (93h). Large amounts of butyric acid in the absence of isoacids would suggest presence of *Fusobacterium*.

Direct examination is important because it gives immediate presumptive evidence of the presence of anaerobes. For example, small, faintly stained gram-negative bacilli, fusiform-shaped bacilli, spheroids, or other unique forms often will alert the microbiologist or clinician to the possibility of infection with anaerobes, and at the same time is a valuable guide to the selection of special media and procedures. Figures 2.10–2.19 show typical morphology of many clinically significant anaerobes. Direct examination also allows one to quantitate roughly the various types of organisms that are present, for the purpose of choosing initial therapy and for use as a check on the adequacy of the techniques employed. One should be able to recover, on culture, in proper relative proportions, all types of organisms seen on direct smear. Since it takes a considerable amount of time to grow anaerobes and to isolate and identify them, the gram stain may provide important preliminary information to the clinician in the case of fulminant infection.

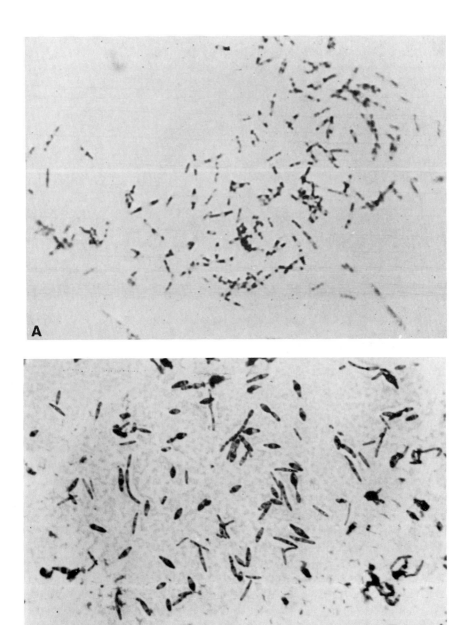

Fig. 2.10. (A) and (B) *Bacteroides fragilis*. Moderately pleomorphic, pale-staining, gram-negative bacillus. Irregular staining and bipolar staining, in particular, are common.

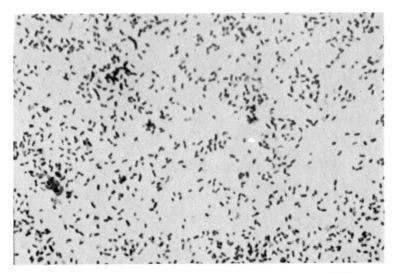

Fig. 2.11.   *Bacteroides melaninogenicus.* Pale gram-negative coccobacillus.

Anaerobic Culture Techniques

There are three major culture systems that can be used for the isolation of anaerobic bacteria: (a) The roll-streak tube system which uses oxygen-free tubes with prereduced anaerobically sterilized media. The oxygen-free atmosphere of the tubes is maintained by inoculation under a stream of oxygen-free gas. (b) The anaerobic chamber, in which all manipulations are performed in an oxygen-free atmosphere. (c) A jar system in which an anaerobic atmosphere can be achieved either by evacuation and replacement with oxygen-free gas or by the use of hydrogen to combine with and eliminate oxygen from the system. The roll tube and anaerobic chamber methods are sophisticated procedures and are capable of isolating extremely oxygen-sensitive organisms that cannot be recovered by jar techniques; however, they do not appear to be necessary for the isolation of clinically important anaerobic bacteria, since the presence of extremely oxygen-sensitive organisms has not been demonstrated in properly collected clinical material. Two studies (94, 95) have shown that, when clinical specimens are properly collected and transported, recovery of clinically significant anaerobes is as good with anaerobic jars [including the simple GasPak jar (95a)] as with other more complex methods. Availability of the simple GasPak jar with its disposable gas generating envelope has given a tremendous impetus to anaerobic bacteriology in the past decade. Unpublished studies (J. G. Finegold, personal communication) indicate that 98.8% of the oxygen is removed from the atmosphere within 1 hour of setting up a GasPak jar; the final concentration of oxygen is 0.17%. Laboratories which have anaerobic chambers or the equipment necessary for the Hungate roll tube technique or modifications of it may still find it convenient to use these procedures at various points in the culture technique. If specimen transport has been less than optimal, so that organisms in the specimen are reduced in numbers and perhaps are in poor condition, use of the chamber or Hungate techniques may result in a better yield than would be possible with anaerobic jar techniques. However, this is not established.

A number of different types of anaerobic jars are available. The use of a gas mixture containing hydrogen is followed by catalytic conversion of the hydrogen and oxygen in the jar to water, leading to anaerobic conditions. Carbon dioxide is always used as part of the atmosphere, since many anaerobes either require it or grow better with it. A "cold" catalyst, composed of palladium-coated alumina pellets, is the preferred type. This type of catalyst should be reactivated after each use by heating the container of pellets to 160°C in a drying oven for $1\frac{1}{2}$ to 2 hours. It is convenient to have a few extra containers of catalyst on hand for this purpose. Reactivated catalyst can be stored for extended periods in a dry place. With the GasPak technique, a commercially available envelope generates hydrogen and carbon dioxide after addition of water. An alternative procedure utilizes evacuation and replacement. Air is removed from the sealed jar containing the culture plates by drawing a vacuum of 700 mm Hg; then, the jars are filled with an oxygen-free gas such as nitrogen. This process is repeated 5 to 7 times. The final fill of the jar is made with a gas mixture containing 80% nitrogen, 10% hydrogen, and 10% carbon dioxide. This procedure achieves anaerobiosis more quickly than does catalytic conversion without any evacuation and replacement. This may be an advantage in the case of clinical specimens that have not been transported properly. With the GasPak system, it is important to make the jar anaerobic as soon as the inoculated plates have been put in. This can be done by activating the envelope, but if one does not have a full jar one could gas the jar with flowing oxygen-free gas (96) until additional specimens allow one to set up a full jar with the GasPak envelope.

It is beyond the scope of this work to give explicit details on processing and examination of clinical specimens or on identification of anaerobic bacteria. The interested reader is referred to other publications for such details. [*Wadsworth Anaerobic Bacteriology Manual* (97); *V.P.I. Anaerobe Laboratory Manual* (20); the *CDC Manual, Laboratory Methods in Anaerobic Bacteriology* (98); and the *ASM Manual of Clinical Microbiology* (99).]

The most important basic medium is the blood agar plate. *Brucella* agar base (Difco or BBL) yields much better growth of anaerobes than the usual

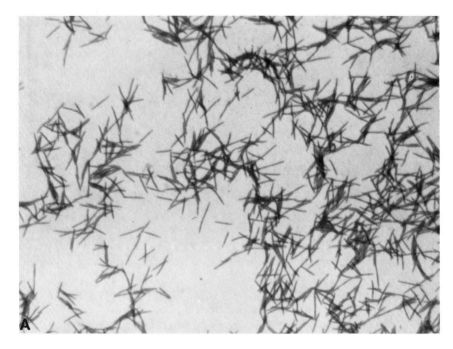

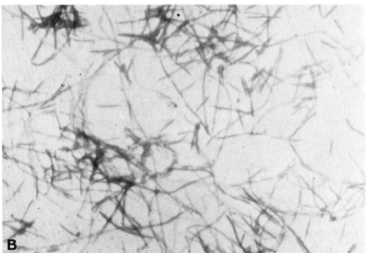

Fig. 2.12.   (A)–(D) *Fusobacterium nucleatum*. Pale gram-negative rod with tapered ends. (D) Dark field photomicrograph showing tapered ends, courtesy of Dr. A. C. Sonnenwirth.

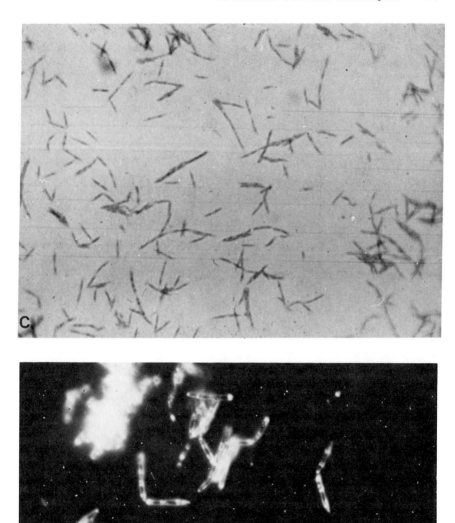

Fig. 2.12. (C) and (D).

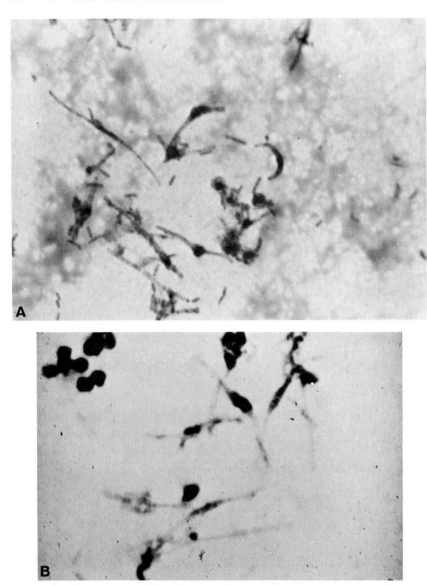

Fig. 2.13.   (A) *Fusobacterium necrophorum.* (B) *Fusobacterium mortiferum.* Both organisms are pale, irregularly staining, extremely pleomorphic gram-negative bacilli with filaments, swollen areas, and large round bodies.

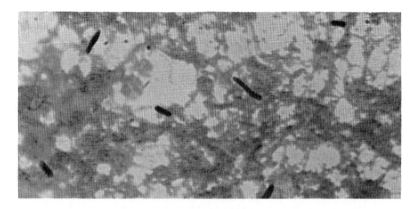

Fig. 2.14.   *Clostridium perfringens* in smear of exudate from patient with gas gangrene. Note large, broad gram-positive bacilli and damaged polymorphonuclear leukocytes. (The latter is related to the α toxin of the organism.)

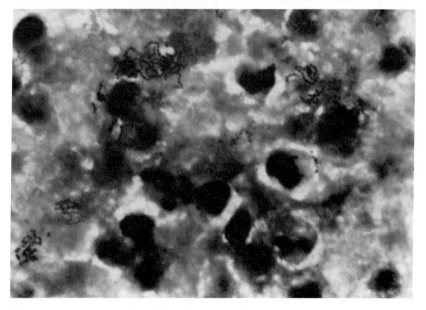

Fig. 2.15.   Direct smear from liver abscess showing microaerophilic streptococci. These organisms and anaerobic cocci may be quite small to tiny.

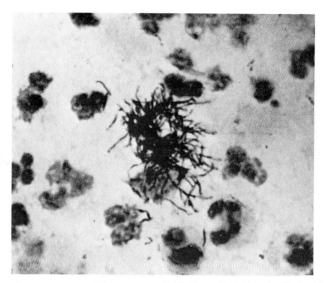

Fig. 2.16. *Actinomyces israelii* in empyema fluid. From A. I. Braude (93a) with the kind permission of the author and publisher. The organisms are thin, gram-positive filamentous, branching rods.

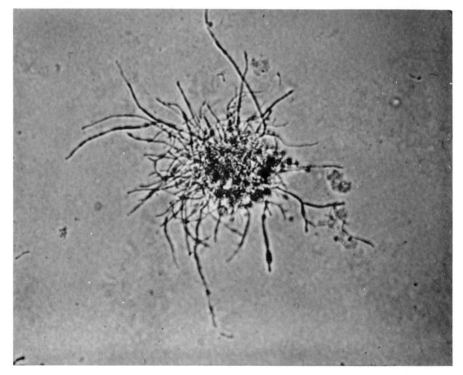

Fig. 2.17. Young colony of *Actinomyces naeslundii* showing branching filaments. Courtesy of Dr. L. Georg.

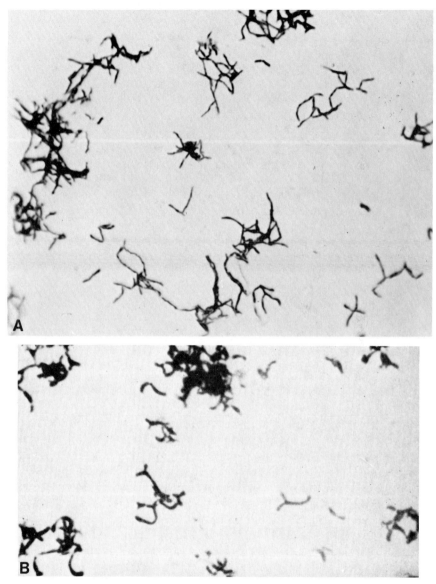

Fig. 2.18. (A) and (B) *Actinomyces israelii* showing morphology described in Fig. 2.16. (A) Courtesy of Dr. L. Georg, (B) courtesy of Dr. V. R. Dowell, Jr., both of Center for Disease Control, Atlanta.

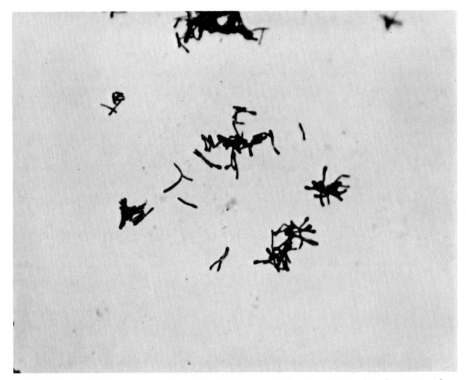

Fig. 2.19. *Actinomyces naeslundii.* Morphology similar to *A. israelii.* Courtesy of Dr. L. Georg.

heart infusion base. Vitamin K_1 is incorporated in a final concentration of 10 μg/ml. A variety of selective media is available. The one that is probably most useful is kanamycin–vancomycin laked blood agar. This incorporates 75 μg/ml base activity of kanamycin and 7.5 μg/ml vancomycin in *Brucella* blood agar plates with lysed or laked blood, rather than whole blood. Growth on this medium is presumptive evidence of the presence of an obligately anaerobic gram-negative bacillus (usually *Bacteroides*), and this group accounts for approximately 40% of all anaerobic isolates from clinical material. An additional advantage of the inclusion of a primary plating medium that contains laked blood is the earlier detection of isolates of *B. melaninogenicus*. A liquid medium may also be used, usually as a backup medium. Thioglycolate with glucose and without indicator, enriched with hemin 5 μg/ml, peptic digest of sheep blood (5%), sodium bicarbonate 1 mg/ml, and vitamin K_1 0.1 μg/ml, is a good medium for this purpose. Additional media for special purposes are noted in the references cited above.

All media used should be fresh. Commercial plate media are often not as satisfactory for growth of anaerobes as media made in the laboratory.

Plates are incubated in jars for 48 hours prior to initial examination and then are reincubated for 1 week or longer before making a final examination. Under emergency conditions, plates can be examined sooner. If this can be anticipated, it is desirable to set up duplicate cultures so that one set can be held for 48 hours, or to use clear anaerobic bags for incubation so that plates can be seen without exposing them to air (99a). Plates should be examined with the aid of both a hand lens ($8 \times$) and a stereoscopic microscope ($25 \times$). Each colony type should be described and picked for gram stain and further identification and should be streaked to blood agar for purity, with plates incubated anaerobically. Pure cultures should be tested for their ability to grow in 5 to 10% CO_2 and in air.

Blood cultures should be set up in one of the commercially available liquid media, such as Tryptic Soy Broth or Difco Thiol (99b, 403) prepared under vacuum with carbon dioxide. The medium should be inoculated with blood in a ratio of $1:10$ to $1:20$ (v/v) and left unvented. The addition of sodium polyanethol sulfonate (0.03 to 0.05%) may enhance the recovery of anaerobic bacteria, but certain anaerobic streptococci are inhibited by this compound (100, 101). Sodium amylosulfate apparently is not inhibitory and may prove to be superior (102). Incorporation of 0.1% agar in the medium is probably desirable.

Anaerobic bacteria are usually isolated in mixed culture, either as multiple anaerobes or, more frequently, mixed with facultative bacteria. Isolation and definitive identification is often tedious and rarely accomplished in time to be of practical value in the management of infection. Therapy, then, will often have to be based upon preliminary findings suggesting the presence of anaerobic bacteria, e.g., unique morphology on initial gram stain or characteristic colonies on anaerobically incubated plates, or, beyond this, gram stain results from pure culture plates of anaerobic isolates. Rapid identification procedures for anaerobes are being developed.

One may question whether definitive identification of anaerobes is necessary for the clinical laboratory. For most clinical purposes, it is not necessary to have final identification for management of the patient's infection. It is my feeling that all laboratories should be able to identify *Bacteroides fragilis* and *Clostridium perfringens* relatively rapidly. *Bacteroides fragilis* is the most common anaerobe encountered, accounting for about 50% of the gram-negative anaerobic bacilli in clinical infections. It is also important in that it is the most resistant of all anaerobes to antimicrobial agents. *Clostridium perfringens*, a much less commonly encountered pathogen, nevertheless is important inasmuch as it may produce very devastating illness. Beyond this, the immediate needs of the patient may be served adequately by informing the clinician as to the general nature of the organism (anaerobic gram-negative bacillus other than *B. fragilis*, anaerobic streptococcus, anaerobic gram-positive non-spore-forming catalase-negative bacillus, etc.) and, at times, to provide antimicrobial susceptibility testing of the isolate. However, *Fusobacterium nucleatum* and *Bacteroides melaninogenicus* can be identified rather readily. These two organisms, together with *B. fragilis*, *C. perfringens*, and the anaerobic cocci and streptococci account for about two-thirds of the anaerobic isolates from clinically significant infections.

Despite what has been said above, it is desirable for as many laboratories as possible to provide reliable identification of anaerobic isolates. There are a number of reasons for this. First of all, the broad general categories of bacteria may be very difficult to define in the case of anaerobes. For example, gram-positive anaerobes often destain readily and may appear gram-negative. Certain bacilli, such as *Bacteroides melaninogenicus*, are quite small and may appear to be cocci. It may be very difficult to demonstrate the presence of spores. Shortcuts in identification may not uncommonly lead to errors in the case of anaerobes.

Definitive identification of an isolate may provide valuable information to the clinician as to a likely portal of entry or source in the case of bacteremia of uncertain origin. The identity of a given isolate may influence the patient's prognosis. For example, the mortality in subacute bacterial endocarditis due to *Bacteroides fragilis* is distinctly higher than in endocarditis caused by other gram-negative anaerobic bacilli or anaerobic cocci. The strong association between *Clostridium septicum* infection and malignancy (103, 103a) underlines the desirability of knowing the identity of this isolate when it is encountered in infection. In certain cases, accurate identification of species may permit differentiation between relapse of an old infection and a new infection in the same patient. Finally, of course, careful bacteriologic (and clinical) study of anaerobic infection will permit us to gain much needed information concerning involvement of various anaerobes in different infectious processes and the prognosis associated with each.

Our laboratory has found it useful to utilize a number of simple tests and observations for preliminary grouping of anaerobic bacteria. Included are observations concerning colony morphology, pigment production, hemolysis, pitting of the agar, fluorescence under ultraviolet light, gram reaction, and microscopic morphology. All of these observations, as well as the spot indole test can be done using colonies on blood agar plates.

Young cultures in enriched thioglycolate broth should be examined for gram reaction, morphology, and motility. Colony morphology in broth may also be of interest at times. From the enriched thioglycolate broth, subcultures may be made to blood agar plates for antibiotic disc identification and to egg yolk–Nagler plates. The antibiotic disc identification procedure utilizes up to six antibiotic impregnated discs, most of which are not the standard concentrations used for the usual susceptibility testing. Details on the procedure for performance of these tests are given in the "Wadsworth Anaerobic Bacteriology Manual" (97). The egg yolk–Nagler plate permits identification of α-toxin producing clostridia. An additional simple test useful for characterizing anaerobes is the bile test.

Bacteroides fragilis can be identified by the fact that it is on obligately anaerobic gram-negative bacillus with rounded ends which is resistant (zone of inhibition less than 10 mm) to the discs containing colistin, kanamycin, penicillin, and vancomycin in the special disc identification procedure referred to above. It is sensitive to the erythromycin and rifampin discs. It is not inhibited, and is usually stimulated, by 20% bile. Subspeciation of *Bacteroides fragilis* can be carried out by the addition of three carbohydrate fermentation tests (mannitol, rhamnose, and trehalose) and the indole test.

Clostridium perfringens is identified rather readily. Its characteristic microscopic morphology, the double zone of hemolysis which may be seen, and the positive Nagler reaction all suggest this particular organism. However, there are three other α-toxin producers that may give a positive Nagler reaction that must be distinguished from *C. perfringens*. The lack of motility of *C. perfringens* distinguishes it from *C. bifermentans* and *C. sordellii* while its ability to hydrolyze gelatin distinguishes it from *C. paraperfringens*. Ultimately, identification of a number of anaerobes does require analysis of end products of metabolism by gas chromatography. This will not be practical for all laboratories, but should be available at least in teaching hospitals.

Toxin demonstration, important in the case of certain clostridia, is best left to specialized laboratories. Serologic tests are not yet practical for diagnosis of anaerobic infections except in the case of actinomycosis. These are also available through special laboratories such as the one at the Center for Disease Control, Atlanta.

The following is a list of common errors in anaerobic bacteriology.

1. Gram stain not prepared directly from clinical specimen. The gram stain alerts one to the possible presence of organisms requiring special media or conditions of incubation. It also helps one to realize that his techniques are defective if he fails to grow out organisms seen on the smear.

2. Failure to set up anaerobic cultures promptly from clinical specimens— or to keep these under anaerobic conditions pending culture.

3. Use of fluid thioglycolate medium as the only system for growing anaerobes. A number of anaerobes will not grow in this medium, even when it is enriched, and solid media are required to separate the various organisms

present in mixed culture. Errors are made at times even when anaerobes have grown in fluid thioglycolate medium. Some workers do not check this medium if growth appears on aerobic plates, assuming that the same organism is growing in both. A gram stain of the broth will often alert one to the additional presence of anaerobes.

4. Use of inadequate commercial media. Failure to use fresh media.

5. Failure to use supplements in media. Vitamin K_1 is required by some strains of *Bacteroides melaninogenicus* and often stimulates *B. fragilis*.

6. Failure to use selective media. Some anaerobes may be overgrown by facultative anaerobes and overlooked if selective media are not used.

7. Failure to use a good anaerobic jar. Brewer, McIntosh-Fildes, Torbal or Baird-Tatlock, and GasPak jars have been found to perform well.

8. Failure to check jars carefully for leaks after they are set up.

9. Catalysts not in good working order when using hydrogen in jars.

10. Using too few flushes when not using hydrogen in jars.

11. Using toxic gas in displacement procedure.

12. Failure to include CO_2 in jars. Carbon dioxide is essential for some anaerobes.

13. Failure to hold cultures for extended periods. Occasionally, fastidious organisms present in small numbers may require 2 to 3 weeks to grow.

14. Failure to minimize air exposure during processing.

15. Failure to use redox indicator or known fastidious anaerobe in jar.

16. Inaccurate identification and speciation.

17. Use of disc susceptibility technique without standards (or use of Kirby-Bauer aerobic standards).

18. Failure to determine whether or not organism is a true anaerobe.

Serology

Serologic tests have been used in a small number of laboratories, on a research basis, for the diagnosis and subsequent follow-up of patients with actinomycosis. Crossed immunoelectrophoresis appears to offer great promise in diagnostic serology in this disease (93c, 103b).

Scattered work on antibody response to gram-negative anaerobic bacilli has been done in the past, but interest in this area has picked up recently. By means of the sensitive passive hemagglutination test, Quick *et al.* (103c) showed that sera from 43–80% of normal adults reacted with various *Bacteroidaceae* antigens; 94% reacted with at least one such antigen. Antibody against various *Bacteroidaceae* was also noted in three patients with gram-negative anaerobic infections. Most of the antibody detected by these investigators was IgM. Similar findings were noted by Hofstad (103d). A radioactive antigen-binding assay for detection of antibody to *Bacteroides fragilis* ss. *fragilis* capsular polysaccharide (in rabbits) was developed by Kasper (71c). Immune responses to *B. fragilis* infections in humans have been detected by gel diffusion technique (103e, 103f) and also by agglutination and immunofluorescence procedures (103f).

3

General Aspects of Anaerobic Infection

Incidence of Anaerobic Infections

Most of the data available on the incidence of anaerobic infections is not entirely reliable. Bacteriologic data without any clinical correlation is not adequate, since the organisms are not necessarily significant. Similarly, clinical data with fragmentary bacteriological information is not ideal. In both types of papers, one often finds specimens cultured for anaerobes which clearly must be contaminated with normal flora (for example, coughed sputum and voided urine). The exact specimen type and source is not always recognizable. For example, some specimens are labeled "rectal" or "abdominal and rectal" or are labeled "sinus drainage." Fortunately, a number of reports have appeared in recent years in which both good clinical and bacteriologic data are presented.

Another problem that has received very little attention has to do with the suitability of discharges as compared to tissue per se. A study comparing tissue biopsy homogenates from surgical wounds with cultures of pyogenic exudate from wounds was carried out by Robson and Heggers in 1969 (104). These workers noted that the tissue homogenates yielded a single bacterial species in 87% of the cases, while the exudate yielded a single species in only 48% of the cases ($P < 0.01$). Although the study was not specifically directed at the role of the anaerobes, some anaerobes were recovered among the organisms encountered. One strain of *Peptostreptococcus* and three of *Bacteroides* were recovered in pure culture from pyogenic exudates, whereas in the tissue biopsy homogenates there were no peptostreptococci encountered in pure culture but *Bacteroides* was encountered in pure culture on six occasions. The authors postulated that the organism present in tissue was the significant organism in the wound and that additional species present in pyogenic exudates were there only as opportunists. As noted elsewhere, tissue specimens offer an additional advantage in that anaerobic bacteria remain viable during transport much more readily when they are present in tissue. A striking example of the importance of clinical correlation is noted in the paper by Wilson *et al.* (105) dealing with anaerobic bacteremia. In this study, anaerobes were isolated from blood cultures from 264 patients; however, only 25% had clinically significant bacteremia.

Dack (106) quotes Thompson who, as recently as 1939, stated the following: "In clinical bacteriology, anaerobes play a minor role. The finding of anaerobes is analogous to the occurrence of red-letter days on a calendar—when they occur, they are usually worthy of consideration." Studies in subsequent years have shown, with increasing effectiveness, that this statement was very inappropriate. In 1940, Dack noted that non-spore-forming anaerobes were recovered from just under 4% of 5180 specimens from the Department of Surgery at the University of Chicago Hospital. Lödenkamper and Stienen (107) found anaerobic bacteria to be the causative organisms in more than 700 cases of infection during the period 1947 to 1955. Stokes, in 1958 (108), reported a study of 4737 clinical specimens that yielded growth. From these, 496 yielded anaerobes, and, in 139 of them, the anaerobes were present in pure culture. In this study almost one-third of specimens of abdominal or genital pus yielded anaerobes. Werner and Pulverer (30) isolated 1244 anaerobic gram-negative bacilli in a period of somewhat less than 11 years. During 1970, Martin (96) recovered anaerobic bacteria from approximately 35% of specimens received in the Mayo Clinic Clinical Laboratory. Approximately 5000 isolations of anaerobic bacteria were made in this 1 year period; these accounted for 49.3% of all bacteria isolated. In the subsequent 2 year period Martin (109) isolated anaerobic bacteria from 49% of specimens that were culturally positive for any bacteria. A total of 10,998 anaerobic isolates was recovered. Hoffmann and Gierhake (110) did a very careful study of postoperative wound infection obtaining material for culture in all cases by needle puncture of wounds that were still closed but which showed impaired healing. In the first 2 years of their study they recovered anaerobes from 5 and 25% of the wounds examined; with improvements in technique in the subsequent 2 years of the study, they recovered anaerobes from one-third of all wounds studied. Altogether there were 120 anaerobic isolates, 103 of which were either gram-negative anaerobic bacilli or anaerobic gram-positive cocci. In

31% of the infections in which anaerobes were recovered, these organisms were present in pure culture. In the other cases, when aerobes were found in addition to anaerobes, the aerobes were usually present in much smaller numbers. Mitchell (110a) recovered 1067 strains of gram-negative anaerobic bacilli from clinical material in a 2 year period. Leigh (110b) noted an increase in isolations of *Bacteroides* from wound infections following intestinal surgery from 14% in 1971 to 81% in 1973; there was no increased incidence of wound infection. Clark *et al.* (110c) recovered *Bacteroides* from over 400 patients in a 19 month period.

Table 3.1 summarizes data from ten studies in terms of the frequency with which specific types of anaerobic bacteria are recovered from the clinical specimens. These studies from laboratories scattered widely over the world are remarkably similar in results. The two most common groups of organisms in clinical specimens, by far, are the gram-negative anaerobic bacilli and the gram-positive anaerobic cocci and streptococci. Together these two groups account for roughly two-thirds to three-quarters of all anaerobic isolates. The next most prevalent group is the Gram-positive non-spore-forming bacilli. In some of the reported studies, the majority of these isolates were *Propionibacterium acnes*, which is almost always a contaminant. Clostridia usually accounted for 5 to 10% of the isolates from clinical specimens, whereas gram-negative anaerobic cocci were found in only about 2% of speciments.

It is interesting to note that in 1955 Prévot (21) commented that, whereas in 1933 he collected over 100 anaerobic coccal isolates, that between 1940 and 1954 he and his colleagues observed only 130 cases of infection due to these organisms, or less than 10 cases per year. This experience has not been borne out in other laboratories, as noted in the data presented in Table 3.1. Furthermore, Mattman *et al.* (111) recovered anaerobic micrococci from 45% of 437 cultures that showed any microbial growth. Sandusky *et al.* (112) recovered nonhemolytic anaerobic streptococci from 170 lesions on a general surgical service between 1939 and 1941. They were present in pure culture 29 times and in association with other anaerobes 18 times.

While the frequency of clostridial infection is relatively low, as noted in Table 3.1, Gorbach *et al.* (113) recovered 136 strains of clostridia belonging to 25 different species from 112 patients over a 14 month period. The most common species encountered in this particular study were *Clostridium perfringens, C. ramosum, C. bifermentans, C. sphenoides, C. sporogenes, C. difficile, C. innocuum, C. butyricum,* and *C. sordellii. Clostridium perfringens* alone accounted for 37 isolates among a total of 57 clostridia recovered from blood culture.

At the University of Iowa Hospitals, with an annual admission rate of 17,000 to 53,000 patients, only 57 well-documented cases of actinomycosis

TABLE 3.1

Incidence of Anaerobic Bacteria in Clinical Specimens

	Anaerobic isolates (No.)	Gram-negative bacilli[k] (%)	Gram-positive cocci (%)	Gram-negative cocci (%)	Gram-positive non-spore-forming bacilli (%)	Clostridia (%)
University College Hospital[a]	643	41	47	7		5
Pasteur Institute, Lille[b]	464	37 (17%)	37	3	15	8
Juntendo University[c]	599	37	34	5	20	4
VPI[d]	144	39 (17%)	21	2	20	18
Mayo Clinic[e]	482	54 (31%)	29	0	4	13
Mayo Clinic[f]	5,029	44 (21%)	25	3	20	8
Gifu University[g]	1,181	31	28	1	35	5
Temple University[h]	750	47 (35%)	27	3	18	5
Mayo Clinic[i]	10,998	39 (23%)	26	1	27	7
Wadsworth VA Hospital[j]	890	44 (15%)	22	7	19	8

[a] From Stokes (108).
[b] From Beerens and Tahon-Castel (115).
[c] From Kozakai (360).
[d] From Moore et al. (1106).
[e] From Zabransky (1107).
[f] From Martin (96).
[g] From K. Ueno and S. Suzuki, personal communication (1972).
[h] From V. Vargo, unpublished data (1972).
[i] From Martin (109).
[j] From S. M. Finegold and V. L. Sutter, unpublished data (1972–1975).
[k] Figures in parentheses, *Bacteroides fragilis*.

were found in a 40 year period (113a). Pulverer (113b), in a reference laboratory serving all of Germany, has diagnosed 2008 cases of this disease in a
30 year period. Of these, 32 involved the thorax, 14 the abdomen, and the
others were cervicofacial in location. This worker estimates the annual
incidence of the disease in Cologne as 1:83,000 inhabitants. Pulverer noted
that a variety of other anaerobic and microaerophilic organisms were associated with the *Actinomyces* and *Arachnia*; most common were microaerophilic streptococci, *Bacteroides melaninogenicus*, and *Fusobacterium*.

With selected types of specimens, the incidence of anaerobic bacteria recovered may be higher than the overall average. This has already been mentioned in connection with pus from the abdominal cavity and from female
genital tract infections. Other studies to be cited in connection with Table 3.3
later in this chapter indicated a much higher incidence of recovery of anaerobes from intraabdominal and obstetrical and gynecological infections.
Nakamura *et al.* (114) recovered anaerobes from 35% of 55 dental specimens
which were positive for bacteria of any kind.

Significance of Anaerobic Isolates and of Specific Anaerobes

Clinical evaluation is obviously very important in determining the significance of anaerobic bacteria recovered on culture from clinical specimens.
Bacteriological features that help indicate that a given isolate is significant
include repeated isolation of the same organism from a patient over a period
of time, recovery of an organism in pure culture, recovery of an organism in
higher count or in larger numbers than is true for other organisms concurrently present, and presence of an organism with distinctive morphology
as the dominant organism on direct gram stain of clinical material.

With regard to the significance of specific organisms, *Bacteroides fragilis*
and *Clostridium perfringens* are organisms of the greatest importance. *Bacteroides fragilis* is the most commonly encountered and the most resistant to
antimicrobial agents of all anaerobes. Note in Table 3.1 that *Bacteroides
fragilis* itself accounts for roughly one-fourth of all anaerobic bacteria
isolated from clinical specimens. Beerens and Tahon-Castel (115) isolated
Bacteroides fragilis in pure culture more often than was true for any other
organism in their large series. Werner and Pulverer (30) recovered a total of
75 strains of *Bacteroides fragilis* ss. *fragilis*, of which 22 were present in pure
culture. Among 19 strains of *Bacteroides fragilis* ss. *thetaiotaomicron* recovered, 6 were found in pure culture. In a later study Werner (35a) noted
that 54 of 133 *B. fragilis* ss. *fragilis* isolates were recovered in pure culture
as compared to 11 of 28 isolates of *B. fragilis* ss. *thetaiotaomicron*. Beerens
and Tahon-Castel (115), isolated *Clostridium perfringens* in pure culture 2

times among a total of 18 isolates. Although *Clostridium perfringens* may be isolated from relatively benign types of infection and even in absence of disease (115a, 115b, 486), it is also capable of producing devastating illness with high mortality.

Stokes (108) recovered anaerobic gram-positive cocci in pure culture approximately one-third of the time (total of 308 isolates). Gram-negative anaerobic bacilli, by contrast, were recovered in pure culture 59 times from a total of 265 isolates of this type, and five clostridia were recovered in pure culture from a total of 34 isolates. Overall, Beerens and Tahon-Castel (115) recovered anaerobes in pure culture 59 times among a total of 471 isolates. The recovery of *Bacteroides fragilis* in pure culture in this series has already been commented upon. The organism isolated in pure culture by these workers next most often was the anaerobic streptococcus (14 of 124 isolates). These workers also recovered *Fusobacterium nucleatum* in pure culture nine times and *Fusobacterium necrophorum* in pure culture six times, among a total of 69 *Fusobacterium* isolates. On a percentage basis, these organisms were isolated in pure culture more frequently than were any other anaerobes. Werner and Pulverer (30) also isolated *Fusobacterium necrophorum* in pure culture on three occasions out of four total isolations.

Fusobacterium necrophorum is clearly a virulent anaerobe. As noted elsewhere, it is encountered less frequently in the antimicrobial era. However, when it is seen, despite its exquisite susceptibility to many antimicrobial agents, it often produces overwhelming sepsis, and not uncommonly metastatic disease.

Bacteroides melaninogenicus is rarely found in pure culture. Indeed, Heinrich and Pulverer (116) described only four cases in which it was recovered in pure culture among a total of 621 cases of infection yielding this organism. With this particular organism, however, this does not necessarily mean lack of virulence, but rather it likely represents the dependence of this organism on other organisms to supply growth factors that it requires. Not all of these growth factors have been identified, but even when known factors such as hemin, blood, serum or ascitic fluid, vitamin K_1, and CO_2 in the atmosphere are supplied, this organism may be found to satellite about colonies of other bacteria that are obviously producing some nutrient necessary for its growth. *Bacteroides oralis* is almost never found in pure culture in clinical infection. There is evidence to indicate that it is closely related to or identical with one of the subspecies of *Bacteroides melaninogenicus* (ss. *melaninogenicus*).

We have isolated *Bacteroides ruminicola* (mostly ss. *brevis*) from 16 clinical specimens during the past 2 years, always in mixed culture. A newly described species of *Bacteroides*, *B. splanchnicus* [(Werner) (117)] has been isolated from only a small number of infections but never in pure culture. We have recovered this organism from 3 clinical infections in the past 2 years.

The frequency with which anaerobic cocci are found in pure culture has already been commented upon. It should also be noted that microaerophilic streptococci (which officially belong to the genus *Streptococcus*, but are considered here along with anaerobes because they are not ordinarily recovered except by anaerobic techniques) are found in pure culture in a variety of serious infections with some frequency, probably more often than is true for the obligately anaerobic cocci.

Fusobacterium varium is not encountered very commonly. It is also relatively resistant to antimicrobial agents. Occasional strains are resistant to penicillin; one-fourth of strains require >8.0 $\mu g/ml$ or more of clindamycin for inhibition; and some strains require 50 $\mu g/ml$ or more.

The importance of *Clostridium perfringens* has already been stressed. *Clostridium ramosum* is isolated with almost the same frequency as *Clostridium perfringens*. Furthermore, it is important in that it is, next to *Bacteroides fragilis*, the most resistant of the anaerobes to antimicrobial agents (118). As much as 6.2 units/ml of penicillin G is required to inhibit the more resistant strains. About 15% of strains are highly resistant to clindamycin and even more so to lincomycin. Many strains are resistant to tetracycline and to erythromycin. This organism has been isolated from infections throughout the body but has undoubtedly been overlooked or misclassified (as a non-spore-former) frequently. While a number of reports indicate that infections with *Clostridium ramosum* are often benign, even when bacteremia is present, the report of Armfield *et al.* (119) noted that recovery of this organism usually indicated serious infection; there were 13 fatalities among 42 patients who yielded this organism.

Among the gram-positive non-spore-forming bacilli, *Actinomyces* and *Arachnia* are clearly the best documented pathogens. Although they are often present in mixed culture, they are clearly pathogenic in their own right and may produce widespread devastating disease anywhere in the body. Only one species of *Bifidobacterium*, *B. eriksonii*, has been proved to be pathogenic (120). This organism has been isolated primarily from pulmonary infection, including a case with granuloma formation.

Organisms in the genus *Eubacterium* seem to be quite benign on the whole. Anaerobic lactobacilli, with the exception of *L. catenaforme*, are nonpathogenic. *Lactobacillus catenaforme* seems to be on a par with *Eubacterium* in terms of pathogenicity. These organisms are almost invariably isolated as part of a mixed flora.

Propionibacterium is not a pathogen ordinarily and is found virtually only in association with implanted prostheses or as a cause of endocarditis on previously damaged valves. Many nonpathogenic organisms may be involved in infection under these circumstances.

Spirochetes seem to be essentially nonpathogenic, with the exception of *Treponema pallidum* and related forms that are not the subject of this book.

Spirochetes may be found in association with other anaerobic bacteria, particularly anaerobic gram-negative bacilli, but they behave like commensals.

While the gram-positive anaerobic cocci are very important pathogens, the gram-negative anaerobic cocci seem to be quite unimportant by comparison. Note in Table 3.1 that these are seldom encountered in clinical infection. They are invariably found as part of a mixed flora, and there is no evidence to indicate that they are important components of the infecting flora.

Table 3.2 lists the major anaerobes encountered clinically. Taken into consideration in devising this table were the frequency with which the organisms were encountered, evidence for their pathogenicity, and resistance to antimicrobial agents.

TABLE 3.2

Major Anaerobes Encountered Clinically

Gram-negative bacilli
    Bacteroides fragilis[a]
    B. melaninogenicus[a]
    Fusobacterium nucleatum[a]
    F. necrophorum
    F. varium
    F. mortiferum

Gram-positive cocci[a]
    Peptococcus (especially *P. magnus, P. asaccharolyticus, P. prevotii*)
    Peptostreptococcus (especially *P. anaerobius, P. intermedius,*[b] *P. micros*)
    Microaerophilic cocci and streptococci

Gram-positive spore-forming bacilli
    Clostridium perfringens[a]
    C. ramosum
    C. septicum
    C. novyi
    C. histolyticum
    C. sporogenes
    C. sordellii

Gram-positive non-spore-forming bacilli
    Actinomyces
    Arachnia
    Eubacterium (especially *E. lentum, E. limosum, E. alactolyticum*)
    Bifidobacterium eriksonii

   [a] These five organisms or groups of organisms account for two-thirds of all clinically significant anaerobic isolates.

   [b] *P. intermedius* is actually microaerophilic and belongs in the genus *Streptococcus.*

TABLE 3.3

Infections Commonly Involving Anaerobes

Infection	Incidence (%)	Proportion of cultures positive for anaerobes yielding only anaerobes	Reference
Bacteremia	20	4/5	(1108)
Central nervous system			
Brain abscess	89	2/3	(352)
Extradural or subdural empyema	10		(1109)
ENT-dental			
Chronic otitis media			
Chronic sinusitis	52	4/5[a]	(1110)
Dental and oral infections			(1111)
Thoracic			
Aspiration pneumonia	93	1/2[b]	(468)
Lung abscess	93	2/3	(470)
Bronchiectasis			
Empyema (nonsurgical)	76	1/2	(469)
Intraabdominal			
Intraabdominal infection (general)	86	1/10	c
Liver abscess (pyogenic)	50–100	2/3	(387, 1112)
Appendicitis with peritonitis	96	1/100?	(546)
Other intraabdominal infection	93		
(postsurgery)		1/6	(484)
Obstetrical-gynecological			(401)
Vulvovaginal abscess	74	1/2 ⎫	
Salpingitis and pelvic peritonitis	56	1/5 ⎪	
Tubo-ovarian and pelvic abscess	92	1/2 ⎬ (477, 1113, 1114)	
Septic abortion and endometritis	73	1/5 ⎪	
Postoperative wound infection	67	1/4 ⎪	
Total	72	1/3 ⎭	
Soft tissue and miscellaneous			
Gas gangrene (anaerobic myonecrosis)			(69)
Gas-forming cellulitis			(747)
Perirectal abscess			(790)
Breast abscess			(1012)

[a] 23/28 cultures (82%) yielding heavy growth of one or more organisms had only anaerobes present.

[b] Aspiration pneumonia occurring in the community, rather than in the hospital, involves anaerobes to the exclusion of aerobic or facultative forms 2/3 of the time.

[c] D. J. Flora, P. Wideman, V. L. Sutter, and S. M. Finegold, unpublished data.

Infections in Which Anaerobes Are a Major Cause

Table 3.3 is a compilation of infections that commonly involve anaerobic bacteria. Specific data on the incidence of anaerobes in most of these types of infections and of the frequency with which anaerobes are isolated to the exclusion of other types of bacteria are noted, along with references. Note that among commonly encountered infections (thoracic, intraabdominal, and obstetrical–gynecological) anaerobes are found in 70 to 95% of most of these infections and that they are found in pure culture in one-half to two-thirds of selected thoracic infections and one-third of obstetrical-gynecological infections.

4

Oral and Dental Infections

Anaerobic infections of oral and dental structures may extend locally or disseminate widely to produce serious infection elsewhere. In particular, the reader is referred to Chapter 6, concerning infections of the jaw, face and neck, and sinusitis, to Chapter 7, concerning anaerobic infections of the brain and meninges, and to Chapter 8, regarding bacteremia and bacterial endocarditis.

Pulp Infections

Normally the pulp of the tooth is protected from infection by oral micro-organisms by the enamel and dentin. However, this barrier may be breached in a variety of ways to permit entrance of bacteria into the pulp or periapical areas: (a) through a cavity caused by trauma, dental caries, or operative dental procedures; (b) through the tubules of cut or carious dentin; (c) in periodontal disease, by way of the gingival crevice and by invasion along the periodontal membrane; (d) by extension of periapical infection from adjacent teeth that are infected; or (e) by way of the bloodstream during bacteremia. Pulp infection is typically chronic and causes changes so slowly and gradually that symptoms may not appear for a considerable time. Eventually there will be occasional aching and then dull throbbing or sharp pain. A striking finding may be exquisite sensitivity to drinking hot liquids. This apparently causes pain by means of expanding the gases that have been produced in the root canal through bacterial action. With acute suppurative pulpitis, there may be excruciating pain, soreness of the tooth, and swelling of the face, as well as low grade fever.

Table 4.1 is a summary of a number of large studies of the bacteriology of root canals in which anaerobes have been detected. The quality of these studies varies considerably. In general, the older studies are deficient in not avoiding any possible contamination of the root canal by microorganisms of the oral flora, and in terms of inadequate anaerobic techniques. These criticisms apply to some later studies as well. The interested reader is referred to the superb monograph by Möller (121) in which these matters are discussed in great detail, along with specific information on techniques that overcome the problems. Obviously the data from the initial culture of previously closed necrotic pulps will be the most reliable. In general, anaerobic streptococci and other anaerobic cocci, along with aerobic streptococci, are important causes of root canal infection, along with anaerobic gram-positive bacilli and, to a lesser extent, anaerobic gram-negative bacilli. It is interesting to note that *Bacteroides fragilis* was isolated from root canals in two recent studies (1123a, 1123b). Möller noted that, as his technique improved, he obtained more anaerobes and fewer organisms other than streptococci and anaerobes. He was able to recover *Borrelia* and *Treponema* from a small number of specimens. Melville and Birch (122) found that the use of aseptic technique resulted in higher percentages of isolation of anaerobic diphtheroids, lactobacilli, and *Fusobacterium* as compared to other series where salivary contamination was more likely. However, Crawford and Shankle (123) found that lactobacilli, diphtheroids, and *Veillonella*, as well as α-hemolytic streptococci, were all detected less frequently in closed canal than in open canal specimens. Most investigations seem to support the latter group.

As long ago as 1908, Baumgartner (124) recovered a fusiform bacillus from the root pulp of a molar tooth and saw fusiforms and spirochetes on smear of the foul pulp of another tooth, which grew only aerobic streptococci. He noted anaerobic cocci and gram-positive thread forms from the gangrenous pulp of other cases. In 1927, Hadley and Rickert (125) recovered anaerobes from 8 previously unopened pulpless teeth. The organisms recovered included fusiform bacilli, *B. bifidus*, *Micrococcus gazogenes*, and thread and other rod forms. In a study specifically designed to detect spirochetes, Hampp (126) studied unexposed canals of 38 pulp-involved teeth. Microscopically, he noted spirochetes in 21 of the 38 specimens. In 10 cases he was able to grow these organisms in pure culture and in 4 others in mixed culture. Two were definite gas producers. All were small type treponemes.

TABLE 4.1

Bacteriologic Studies of Root Canals: Involvement of Anaerobes

Number of teeth cultured	Number with growth	Number organisms isolated	Number anaerobes[a] isolated	Number of initial cultures of closed, necrotic pulps	Number with growth	Number with anaerobes[a]	Number of anaerobic strains	Number of Actinomyces
12			109[b]	12			109[b]	
10	10		17					
39			3	39				
40			19[b]	40				
263			9	?				
50	45	90	23	50	45	90	23	
				70		17		
46	38	71	25	46	38	71	25	
154			134					33
4186 cults.,	1141 cults.		1417				195	20
> 1000 teeth								(17 pure)
				218	111 (132 strains)		15	2 (2 pure)
				36	21		32	
57			55	31			22	6
60[b]	45		93					1
				35	29	28	64	
563	198	242	12	?	47	?		
100	12		0	100?				
125	119		49	?				2
20	19	34	(41 teeth) 27	20	19	17	27	1
24	16	?	19	24	16	12	19	5

[a] Includes lactobacilli, even if they are microaerophilic or facultative. Other microaerophilic organisms counted as anaerobes.
[b] Many more cases were studied, but these were patients with necrotic pulp studied with best technique.

He also recovered *Spirillum sputigenum* in pure culture on one occasion and saw it in other mixed cultures.

Microscopic control of bacteriologic studies is very important. In the study by Brown and Rudolph (127) 23.9% of 70 specimens from unexposed canals of pulp-involved teeth grew anaerobic bacteria. However, these investigators saw much more microscopically than could be cultured. The noncultivable organisms were primarily filamentous forms, fusiforms, and vibrios. Crawford and Shankle (123) also found a number of organisms microscopically which were not recovered on culture. These were chiefly gram-positive cocci and gram-positive rods, as well as some spirochetes. Conventional stained smears are not adequate. Thus, Mazzarella et al. (128) in a study of 50 previously unexposed pulp canals with dead pulp noted that whereas direct gram stains showed that 72% contained organisms, dark

Number anaerobic streptococci	Number of other anaerobic cocci	Number of fusiforms	Number other gram-negative anaerobic rods	Number lactobacilli	Number other gram-positive anaerobic rods	Number other anaerobes	Reference	Year
		67			42		Sommer (1115)	1915
	1				16		Kamer (1116)	1929
		· 1			2		Andre (1117)	1932
Occas.			2	4	10		Helm (1118)	1934
9							Gruchalla and Hamann (1119)	1947
							Mazzarella et al. (128)	1955
							Brown and Rudolph (127)	1957
2	12	2	6		3		MacDonald et al. (1120)	1957
25	11	1	8	48	8		Leavitt et al. (1121)	1958
69		8		98			Winkler and Van Amerongen (131)	1959
4 in pure cult.)		(5 pure)		(57 pure)				
3		2		8				
(3 pure)		(1 pure)		(3 pure)				
8	3	11	1	3	6		Engstrom and Frostell (129)	1960
5	11		1	17	6		Crawford and Shankle (123)	1961
←——26——→		4	11	7	44		Möller (121)	1966
←——19——→		6	11	3	25		Möller (121)	1966
	7	1		4			Goldman and Pearson (1122)	1969
							Fourestier et al. (133)	1970
3	13		26		5		Mizuno and Tamai (1123)	1972
5	2	3	8	1	7		Berg and Nord (1123a)	1973
0	3	5	2		3	1	Kante and Henry (1123b)	1974

field and phase contrast microscopic examinations revealed that 96% of the materials contained organisms.

The incidence of positive cultures is highest in teeth with necrosis or gangrene of the pulp and with X-ray evidence of periapical destruction (129, 130). Previously root-filled teeth without periradicular rarefaction had the lowest number of positive cultures.

Follow-up cultures are important in terms of endodontic therapy. Persistence of infection at the time of the root filling leads to an adverse effect on the prognosis (130). Teeth with necrosis or gangrene of the pulp require a larger number of antibacterial dressings than previously root-filled teeth and may cause treatment failure. According to Winkler and van Amerongen (131) anaerobes and most aerobic streptococci and perhaps lactobacilli tend to persist longest in root canals on repeated cultures. Among 14 cases in which anaerobic streptococci were still present at the time of root filling (130), there were 2 cases in which the long-term results were uncertain, and 3 cases in which there was failure (anaerobic streptococci were present in pure culture in all of these cases).

Dentoalveolar Infections

Dentoalveolar abscesses occur following pulp infection. These occur at the normal and abnormal foramina of the roots or at a perforation made artificially by instrumentation. The most common location of such abscesses is at the apex of the tooth, but they may occur lateral to the root of the tooth or between roots in multirooted teeth, at the site of accessory root canals or perforations. Although bacteria enter the periapical tissues primarily by extension from an infected dental pulp, they may occasionally reach such tissues by way of lymphatic drainage of the alveolodental periosteum or by way of the bloodstream or lymphatics from more remote parts. The extension of infection from a pulp into the periapical area causes necrosis of bone and tissue in the affected area and the accumulation of pus. Such alveolar abscesses may be either acute or chronic. The abscess may progress to involve more and more tissue, including adjacent teeth, and the pressure of accumulated pus may produce a fistula to the outside or to the oral or nasal cavity (see Figs. 4.1 and 4.2). There may also be widespread dissemination to produce serious and fatal systemic illness.

Chronic periapical infection may be referred to either as an abscess, a dental granuloma, or a cyst. Pyogenic "granulomas" are not true granulomas. The term reflects the fact that granulation tissue is often an integral component of these lesions. This vascular tissue has potential for destroying microorganisms and for initiating tissue repair. Granulomas tend to remain intact and may come away with the tooth during the course of an extraction, but infection may extend from this type of process as well (Fig. 4.3). The common presence of epithelial remnants in periapical areas may give rise to chronic cystic formation. At times the host response to degeneration and infection of the pulp can result in a chronic apical lesion with bone hyperplasia rather than resorption. This type of response is known as chronic apical condensing osteitis. This stimulative, rather than destructive, periapical response results in a thickening of the periodontal membrane with diffuse condensation of bone in the apical and ascending regions of the root (132).

Melville and Birch (122) point out that although the periapical area may be sterile even when the pulp canal is not, typically the periapical flora is identical with or forms part of the root canal flora. In about 50% of cases, the labial bone plate flora is consistent with the apical and root canal flora

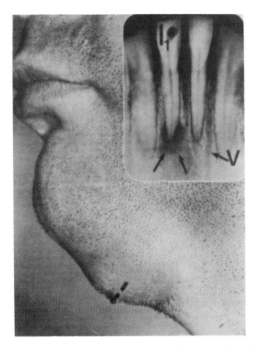

Fig. 4.1. Submental abscess, recurred four or five times in the previous $1\frac{1}{2}$ years, arising from the lower right central incisor. I_1 = 1st incisor. Arrows point to loss of alveolar bone. Treatment: incision along dotted line, followed by apicectomy of the tooth. V, vascular channel. From Boering (208), reproduced with permission.

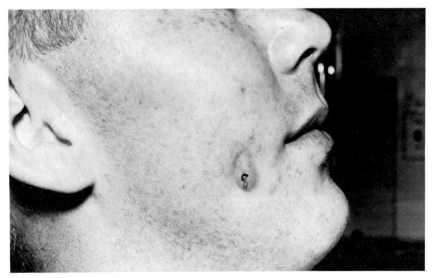

Fig. 4.2. Periapical abscess leading to draining fistula of the face (AFIP 56–12889. Reproduced with permission of the Armed Forces Institute of Pathology.) From Burnett and Scherp "Oral Microbiology and Infectious Disease," 3rd ed. copyright 1968. The Williams & Wilkins Co., Baltimore.

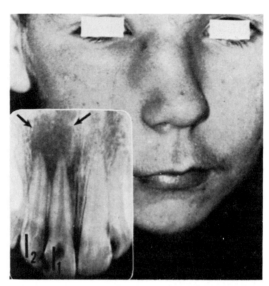

Fig. 4.3. Small odontogenic abscess adjacent to the nose originating from a granuloma at the upper right central incisor. After rupturing, a fistula may remain. From Boering (208), reproduced with permission.

and therefore reflects the widespread nature of the periapical lesion. In a study of 100 extracted teeth, Fourestier *et al.* (133) obtained positive cultures from the apex of only 19. Of these, only one was an anaerobe (a fusiform bacillus). In contrast to this, Heinrich and Pulverer (116) isolated *Bacteroides melaninogenicus* from 332 infections related to the teeth and jaw; no specific information is given. Meleney (69) indicated that he had isolated *Bacteroides fragilis* from dental infections, the number not specified. Other workers have noted this also (see Table 4.2). This is significant in view of the resistance of this organism to antimicrobial agents.

There are several reports of involvement of anaerobes in dental infection without specific information given as to the site of involvement. Prévot (134) described recovery of *Ramibacterium ramosum* from a buccal abscess of dental origin. The same worker in 1955 (21) reported isolation of *Catenabacterium catenaforme* from a dental infection; from 4 other "dental infections", he isolated *Diplococcus morbillorum*, *Streptococcus lanceolatus*, *Staphylococcus aerogenes*, and *Micrococcus grigoroffi*. Dietrich *et al.* (135) recovered anaerobic gram-negative cocci in pure culture from an infected tooth, and in 5 other infected teeth he noted mixed anaerobes and aerobes.

Table 4.2 is a summary of reports of the involvement of anaerobes in dental abscesses and phlegmons (the distinction between phlegmon and abscess is not always clear-cut, although in general phlegmon denotes more induration of tissues). Most of these are single reports, and some larger series show a relatively low incidence of anaerobes. However, studies such as that of Möller (121) on root canals suggests strongly that, with appropriate technique, anaerobes would be found to be far and away the most important causes of dental abscesses. A variety of anaerobes is involved, with anaerobic cocci and anaerobic gram-negative bacilli being isolated most commonly, and with a significant number of anaerobic gram-positive bacilli as well. In culturing periapical abscesses, as in other cultures from the mouth, it is important to use special precautions to avoid contamination with normal flora and to make control smears and cultures from adjacent areas, as Davis and Moorehead suggested in 1923 (136). Seitz in 1899 reported recovering *Bacillus hastilis* from 12 of 28 cases of dental abscess (137). However, he also found this organism in all 5 cases of diphtheria and in 6 of 10 cases of pulmonary tuberculosis, indicating that it must have been prevalent as normal flora. Gilmer and Moody (138) cultured acute and chronic alveolar abscesses but did not give specific data. Aerobic streptococci predominated, but fusiform bacilli were found in some abscesses, sometimes almost in pure culture. In three abscesses, these workers found a slow-growing black pigment-producing anaerobic organism.

There are a number of studies of the bacteriology of pyogenic granulomas. Bulleid (139) cultured 80 and isolated anaerobes from only 11, never in pure culture. Anaerobes isolated were *Bacillus fusiformis* and *Leptotrichia buccalis*. Pesch and Ruland (140) cultured 47 granulomata. Fifteen yielded aerobic (facultative) streptococci. The other 32 had, in addition to facultative streptococci, anaerobic streptococci—a total of 36 strains. Eleven of the 36 strains proved to be stable obligate anaerobes, even after numerous subcultures over an extended period of time. The other 25 ultimately adapted to growing in air (after 3 to 20 subcultures). Pulverer (141) cultured three tooth granulomas. He found *Bacteroides melaninogenicus* in all three, plus "aerobic and anaerobic mouth flora" in two and *Clostridium perfringens*, and viridans streptococci in the third. Grumbach and Hotz (142) recovered *Bacillus anaerobius diphtheroides* from a granuloma of the first molar tooth and *Actinomyces* was isolated from a periapical granuloma by Browne and O'Riordan (143). *Actinomyces israelii* was isolated from a case of chronic granulomatous inflammation (143a).

Twelve dental cysts were cultured by Bulleid (144) and two of these yielded fusiform bacilli. This worker also cultured two dentigerous cysts and recovered a fusiform bacillus from both the fluid and wall of one. Beerens and Tahon-Castel, 1965, cultured two infected dental cysts; one yielded *Streptococcus anaerobius, Diplococcus morbillorum, Streptococcus micros, Corynebacterium liquefaciens*, and aerobic nonhemolytic streptococci (115). The other yielded *Diplococcus morbillorum, Streptococcus anaerobius* plus

TABLE 4.2   Anaerobes in Dental Abscesses and Phlegmons

Number of cases	Number with anaerobes	Type and location of lesion(s)	Anaerobes isolated	Aerobes isolated	Comments	Reference	Year
1	1	Submaxillary phlegmon	Fusiforms, spirochetes	*Streptococcus*	Putrid, related to wisdom tooth	Vincent (146)	1905
17	7	Abscesses	Fusospirochetes	?		Vincent, cited by Babes (200)	1906–1907
1	1	Abscess, jaw	*Bacteroides funduliformis*	*Staphylococcus, Streptococcus*	Gas, foul odor, pulmonary infection, empyema, neck abscess	Heyde (1124)	1910
1	1	Periostitis, subgingival abscess	Anaerobic streptococci, *Bacillus ramosus,* and unidentified anaerobe	*Streptococcus, Bacillus*		Idman (1125)	1913
1	1	Periostitis, subgingival abscess	*Bacillus ramosus*	*Streptococcus, S. albus, M. tetragenus,* unidentified aerobe			
1	1	Periostitis, subgingival abscess	Anaerobic streptococci	*Staphylococcus, Streptococcus*			
1	1	Periostitis, subgingival abscess	*Bacillus ramosus*	*Staphylococcus*			
1	1	Periostitis, subperiosteal abscess	Anaerobic streptococci, *Bacillus ramosus, Bacillus thetoides, Clostridium perfringens*	Diphtheroid			

1		Periostitis, subperiosteal abscess	Anaerobic streptococci, Staphylococcus parvulus	Streptococcus, unidentified aerobe		
1		Periostitis, subperiosteal abscess	Bacillus ramosus, B. thetoides, Clostridium perfringens, B. bifidus communis, unidentified anaerobe	0		
1		Periostitis, subperiosteal abscess	Bacillus ramosus	Unidentified aerobe		
1		Periostitis, subperiosteal abscess	2 anaerobic staphylococci, Bacillus thetoides, B. ramosus, B.fusiformis	0		
1		Apical abscesses	Bacillus fusiformis	Streptococcus, S. albus	Thoma (1126)	1916
130 specimens, 100 cases	90?	Apical abscesses	Gram-negative coccobacillus-90; Bacillus bifidus, B. acidophilus, and related organisms-45	124 streptococci (mostly viridans streptococci)	Head and Roos (1127)	1919
107	1	"Abscessed teeth"	Fusospirochetes	?	8 sterile, 23 contaminants	Smith and Ludwick (1128)	1919
20	20	Chronic alveolar abscess	20 fusiforms, 16 spirochetes (smear and/or culture)	20 viridans streptococci	Davis and Moorehead (136)	1923

(continued)

TABLE 4.2 (*continued*)

Number of cases	Number with anaerobes	Type and location of lesion(s)	Anaerobes isolated	Aerobes isolated	Comments	Reference	Year
3	3	"Tooth abcess"	*Bacteroides melaninogenicus*	?		Burdon (799)	1927
1	1	Abscess next to first molar	Fusiforms	*Staphylococcus aureus*	External sinuses	Balyeat (1129)	1930
16	3	Acute alveolar abscesses	Gram-negative bacilli	Yes, not specified	1 abscess contained *B. acidophilus*	Bulleid (144)	1931
1	1	Abscess	*B. melaninogenicus*	?	Following extraction	Shevky *et al.* (281)	1934
1	1	Abscess upper jaw	*Bacillus anaerobius diphtheroides*	0	Following extraction; bacteremia, lung abscesses and empyema	Grumbach and Hotz (142)	1939
1	1	Phlegmon	Buday bacillus	?0	Following extraction; sepsis, liver abscess, peritonitis, empyema	Hornung (1130)	1940
1	1	Abscess near second molar	*Sphaerophorus varius*	*Streptococcus*	Extended to orbit and maxillary sinus	Ingelrans (308)	1947
1	1	"Gingivodental abscess"	*Ristella glutinosa, R. fragilis, Veillonella parvula, Ramibacterium ramosum*	*Streptococcus pyogenes*		Prévot (171)	1948
1	1	Phlegmon	*Eubacterium crispatum, Streptococcus micros*	0		Brygoo and Aladame (1131)	1953

		Diagnosis	Organism(s)		Comments	Reference	Year
1	1	Phlegmon, floor of mouth	Fusospirochetes	?	Following extraction of second molar; indurated, fetid	Durante (1132)	1953
1	1	Phlegmon	Ramibacterium pleuriticum	0		Beerens (332)	1953–1954
1	1	Phlegmon, floor of mouth	Fusobacterium nucleatum	?		Prévot (21)	1955
1	1	Dental abscess	Ristella insolita	?			
1	1	Phlegmon	Eubacterium crispatum	?			
1	1	Phlegmon	Ristella insolita dentium	?			
1	1	Dental abscess	Veillonella alcalescens	?			
1	1	Dental abscess	Streptococcus micros	?			
1	1	Phlegmon near masseter	Fusobacterium nucleatum	?	Following trauma to wisdom tooth	Caroli et al. (172)	1956
1	1	Abscess	Bifidibacterium bifidum	20		Beerens el al. (1133)	1957
1	1	Phlegmon, floor of mouth	Bacteroides melaninogenicus, fusiform	?		Pulverer (141)	1958
17	17	Phlegmon, floor of mouth	B. melaninogenicus	?		Heinrich and Pulverer (116)	1960
1	1	Dental abscess	Anaerobic streptococci and few unclassified microaerophilic gram-negative rods and streptococci	0	Led to brain abscess	Heineman and Braude (352)	1963
2	2	"Dentoalveolar" abscess	Bacteroides	?		Bornstein et al. (275)	1964

(continued)

TABLE 4.2 (*continued*)

Number of cases	Number with anaerobes	Type and location of lesion(s)	Anaerobes isolated	Aerobes isolated	Comments	Reference	Year
1	1	"Dentoalveolar" abscess	Anaerobic streptococci	?	Drained to submaxillary node		
1	1	Dental abscess	*F. fusiformis, Staphylococcus anaerobius, Streptococcus intermedius*	0		Beerens and Tahon-Castel (115)	1965
1	1	Dental abscess	*Eubacterium convexa, E. lentum*	0			
1	1	Dental abscess	*Streptococcus evolutus, Staphylococcus anaerobius, Diplococcus, Streptococcus micros, Corynebacterium liquefaciens,* unidentified gram-negative rod	0			
1	1	Periapical abscesses → chin and below jaw	Heavy *Bacteroides fragilis,* anaerobic streptococci	0	Trauma to chin, abscesses by lower central incisors	Goldsand and Braude (244)	1966
1	1	Abscess extending to palate	*Actinomyces*	Nonhemolytic streptococci	Complicated by brain abscess	Stevenson and Gossman (150)	1968
1	1	"Pus from molar extraction"	*Actinomyces israelii*	?		Coleman and Georg (319)	1969

			Anaerobic organisms	Other organisms / findings	Notes	Reference	Year
1	1	Cellulitis, abscess	Anaerobic streptococci	1 culture—*S. faecalis*, 1 culture—0	Pt. on steroids; repeated relapses; orbital cellulitis	Mason (158)	1970
259 strains, 110 patients	1	Gingival abscess	*Peptococcus*	0		Baba *et al.* (272)	1972
111 strains, 29 patients		Gingival abscess	*Bacteroides, Veillonella, Peptostreptococcus, Peptococcus, Actinomyces, Lactobacillus, Clostridium, Corynebacterium, Bacteroides corrodens*	148 strains		Fukuda and Tamai (1134)	1972
7	7	"Infected dental root"	*Actinomyces* + ?	?		(1134a)	1973
8	At least 7	Acute alveolar abscess	*Fusobacterium nucleatum* (7), *Fusobacterium* sp. (1). *Bacteroides* sp. (4), *Peptostreptococcus* sp.. (2) *Lactobacillus* sp. (2), *Actinomyces* sp. (1)	Facultative streptococcus (6), gram-positive facultative rod (1), *S. epidermidis* (1)		Sabiston and Gold (1134c)	1974
66	7?	"Acute soft tissue abscesses secondary to caries"	Microaerophilic streptococci (3), *Actinomyces* sp. (2), *Peptostreptococcus* (1). Gram-negative anaerobic rod (1)	?		Turner *et al.* (1134b)	1975

91

Haemophilus influenzae and aerobic nonhemolytic streptococci. *Sphaerophorus* and *Ramibacterium* were isolated from an infected paradental cyst by Morin and Potvin (145).

Vincent (146) cultured 17 cases of what he called dental periostitis that had fetid pus; fusiforms were recovered from seven of these, with or without spirochetes. One patient with maxillary osteitis and purulent gingival fistula yielded *Eggerthella convexa, Streptococcus anaerobius,* and aerobic nonhemolytic streptococci [Beerens and Tahon-Castel (115)].

The postextraction syndrome commonly referred to as dry socket has many other names—painful socket, alveolalgia, necrotic alveolar socket, alveolar osteitis, degeneration of the blood clot, and delayed repair of the extraction site. Regardless of the terminology, it is generally recognized that the process is one of a localized osteomyelitis or osteitis. The major manifestations of this syndrome are pain, the absence of an organized blood clot, and a foul taste or odor emanating from the socket. There may be inflammatory response in the soft tissue, leading to cellulitis and trismus. There may also be local infection or even bacteremia (147). While this type of infection is not as extensive as apical infection, it is very painful and responds rather slowly. Schroff and Bartels (148) were impressed with the frequency with which they saw fusiforms and spirochetes in this picture. The pain and the odor were also very reminiscent of Vincent's angina or gingivitis. They found that sodium perborate therapy locally was of significant benefit. Francis and de Vries (149) recommend gently removing debris remaining in the socket and then irrigating with warm saline. They use iodoform locally even when pus is present and recommend lincomycin therapy. They point out that it may take 5 to 10 days to control the infection.

There are two basic presentations of dental actinomycosis [Bramley and Orton, 1960, quoted by Stevenson and Gossman (150)]: (a) low grade fever and painless swelling (typically at the point where the facial vessels cross the lower border of the mandible) and (b) high fever, toxicity, and an acute, painful abscess at variable sites. Stevenson and Gossman point out that oral actinomycosis commonly involves mandibular teeth, particularly the third molar (because the posterior gum flap allows accumulation of materials and anaerobic conditions). Kapsimalis and Garrington (151) described a case of actinomycotic infection intimately associated with the apices of six teeth and leading to pulp death in these six anterior mandibular teeth. Endodontic therapy plus prolonged penicillin therapy led to a cure. *Actinomyces naeslundii* was isolated from a draining sinus following tooth extraction by Coleman *et al.* (152). Bezjak and Arya (153) also isolated *Actinomyces naeslundii,* along with *Corynebacterium acnes* and *Bacteroides,* from a case of low grade chronic actinomycosis of the cheek secondary to an infected tooth. As indicated earlier, Browne and O'Riordan (143) isolated *Actinomyces* from a periapical granuloma. Prévot *et al.* (154) described a case of cervical actinomycosis of buccodental origin. From this lesion they recovered *Fusobacterium nucleatum*, *Actinomyces bovis*, and *Clostridium sporogenes*. Several cases of actinomycosis of dental origin were presented by Schlegel and Hieber (143a).

There are many examples of serious infection arising from dental foci. A few will be cited here; others are cited elsewhere in the book, particularly in Chapters 6, 7, and 8. Veszprémi (155) described a case of purulent periostitis of the upper jaw due to fusiforms and spirochetes which spread to form a phlegmon near the temporal muscles and ultimately produced meningitis and metastatic lung abscesses. In 1929, Thompson (156) described a patient with Vincent's gingivitis who, following a dental extraction, developed swelling of the face, trismus and abscess in the sphenopalatine fossa, and then orbital abscesses and a brain abscess. Ikemoto *et al.* (157) discussed a patient who developed pain throughout the face and otalgia following tooth extraction. The patient was treated intermittently with antibiotics. Eventually he developed fever and meningeal signs. The cerebrospinal fluid was found to be purulent and foul smelling. The patient developed paraplegia and very likely had an epidural or other abscess in the lumbar area. *Actinomyces* and *Bacteroides melaninogenicus* were recovered from the spinal fluid. An interesting case is described by Mason (158). The patient was an asthmatic on maintenance corticosteroid therapy. She developed submandibular cellulitis following a tooth extraction; foul pus from this area yielded anaerobic streptococci. The infection extended to the face and orbital cellulitis developed. The patient's infection was finally controlled but three additional readmissions were required for exacerbations, at least two of which were related to the patient's increasing her dose of corticosteroids because of "chest infection." All in all, the infection persisted for a period of over $4\frac{1}{2}$ months. A fulminating cellulitis of the face following dental extraction due to *Bacteroides necrophorus* was described by Meleney (69). Weiss (80) described a case of cellulitis of the face following dental infection; he recovered from the lesion *Bacteroides melaninogenicus*, fusiform bacilli, viridans streptococci, diphtheroids, and *Staphylococcus albus*. Schulz (159), described a case of endocarditis due to the Buday bacillus secondary to dental disease with fistula formation. Crystal *et al.* (160) described five cases of Ludwig's angina, two of which followed extractions and the others other dental manipulation and/or disease. A *Fusobacterium* was cultured from one of these. Additional cases of Ludwig's angina related to dental disease or manipulation are noted in Chapter 6.

Pulmonary infection is a well-recognized complication of dental disease or manipulation (see Chapter 9 also). Four cases described by Bartlett and Finegold (161) are illustrative. One was a patient with a right lower lobe

pneumonia and empyema due to *Fusobacterium*, microaerophilic streptococci, and *Bacteroides oralis*; the patient was an alcoholic with a dental
abscess. A second alcoholic patient with pyorrhea and a recent dental
extraction presented with necrotizing pneumonia of the right upper and
lower lobes; transtracheal aspiration yielded *Fusobacterium*, microaerophilic streptococci, and *Bifidobacterium*. A third patient had pyorrhea and
a fractured jaw. He developed necrotizing pneumonia of the right upper
and lower lobes; transtracheal aspiration yielded *Bacteroides melaninogenicus* and *Bacteroides oralis*. The last patient had a periapical abscess as
well as bronchiectasis and epilepsy. He had a left lower lobe pulmonary
abscess. Transtracheal aspiration yielded *Bacteroides fragilis, Fusobacterium,
Peptococcus, Bacteroides oralis*, and *Propionibacterium*. Prévot *et al*. (154)
described the bacteriology of three pulmonary infections of buccal or bucco-
pharyngeal origin. One of these yielded *Fusobacterium nucleatum* in pure
culture, another yielded *Fusobacterium nucleatum* plus *Veillonella parvula*
and *Corynebacterium anaerobium* (this patient also had tuberculosis). The
third was a putrid abscess which yielded *F. nucleatum* plus *Actinobacterium
meyeri* and *Streptococcus intermedius*. Witebsky and Miller (162) discussed a
diabetic with putrid empyema due to *Bacteroides* and a β-hemolytic *Streptococcus*; this was thought to be related to extraction of an abscessed tooth
3 weeks prior to onset.

Larson and Barron (163) discussed a fatal case of *Fusobacterium* bacteremia in a patient with severe pyorrhea. At autopsy, the soft palate was
found to be necrotic and the bone of the superior maxilla was blackened and
necrotic. The bone destruction extended as far as the antrum of Highmore
and into the nasal cavities. Gröschel (164) reported a case of hematogenous
osteomyelitis due to *Sphaerophorus necrophorus* (however, the organism was
described as being gram-positive), following a tooth extraction. Bacteremia
due to *Fusobacterium biacutum* in a 15-year-old boy with a dental infection
was described by Laporte *et al*. (165); the patient had a small cyst in the
area of the first premolar tooth and a larger cyst by the first molar. An
infected paradental cyst led to maxillary sinusitis and bacteremia due to
Bacillus ramosus in a patient described by Boez *et al*. in 1928 (166). The
patient had an underlying tuberculous meningitis but despite this the
bacteremia proved to be a mild illness. Prévot (21) reported bacteremia
due to *Streptococcus intermedius* following a dental accident and later (167)
reported *Sphaerophorus pseudonecrophorus* septicemia of dental origin.

A patient described by Felner and Dowell (168) had pyorrhea, gingivitis,
and periodontal abscess. Following dental extraction, he developed bacteremia and endocarditis due to *Bacteroides fragilis* and microaerophilic
streptococci; he had underlying arteriosclerotic heart disease. Felner and
Dowell (168) also reported three additional cases of bacteremia and endocarditis related to dental problems. One of these patients had periodontosis

and had extractions that led to bacteremia and endocarditis due to both *Bacteroides melaninogenicus* and anaerobic streptococci. A second patient (with ventricular septal defect) had extractions, following which he developed endocarditis with *Fusobacterium fusiforme* and a brain abscess following emboli (the culture of the abscess yielded *F. fusiforme*, *Bacteroides corrodens*, and *Streptococcus faecalis*). A third patient reported by these workers had had a ventricular septal defect repaired surgically. Postoperatively, he had dental extractions following which he developed endocarditis due to *Fusobacterium necrophorum*. Vic-Dupont *et al.*, in 1963, reported a case of anaerobic streptococcal endocarditis of dental origin (169). Prévot, in 1955, reported a case of *Corynebacterium anaerobium* endocarditis of dental origin, as well as a septicemia due to *Sphaerophorus freundii*, again of dental origin (21).

Mutermilch and Séguin (170) described a case of fatal fusospirochetal infection secondary to radium therapy of an epithelioma of the floor of the mouth.

Infections of the Jaw

Infection of the periosteum or marrow leads to inflammatory and necrotic processes in the maxilla and mandible. Trauma is a common underlying cause. Dental extractions, particularly if they are difficult, may serve as the precipitating event. Osteomyelitis also follows fractures or injection of local anesthetics. Infection of the jaw may follow periapical or even periodontal infection (Fig. 4.4). Other predisposing factors include diabetes, radiation or other therapy for malignancy, serious systemic illness, and vitamin deficiency. Infections of the jaw are also discussed in Chapter 6.

Prévot (171) discussed the bacteriology of four cases of cellulitis of the jaw. All of these contained anaerobes and one each contained *Staphylococcus aureus* and *S. albus*. The anaerobes isolated included *Staphylococcus anaerobius* [3], *Ramibacterium alactolyticum* [2], *Actinomyces israelii* [1], *Streptococcus micros* [1], and *Ristella melaninogenica* [1]. He also mentioned a case of osteophlegmon of the jaw from which was cultured *Staphylococcus anaerobius*, *Ramibacterium alactolyticum*, and *Staphylococcus albus*. Finally there was an abscess of the jaw secondary to a dental problem that yielded *Ramibacterium alactolyticum* and *Streptococcus pyogenes*. Six cases of infection of the jaw involving anaerobes are discussed by Beerens and Tahon-Castel (115). A firm cellulitis of the chin of dental origin yielded *Eubacterium tortuosum*, *Streptococcus anaerobius*, *Dialister*, an anaerobic diplococcus, and aerobic nonhemolytic streptococci. A second patient had a woody cellulitis of dental origin in the perimaxillary area. From this was cultured *Ramibacterium pleuriticum*, *Eubacterium tortuosum*, *Fusobacterium fusiformis*

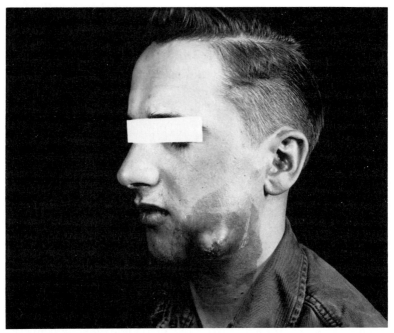

Fig. 4.4.   Cellulitis of the jaw arising from an abscessed pulp and periapical infection (AFIP 0–7744. Reproduced with permission of the Armed Forces Institute of Pathology.) From Burnett and Scherp, "Oral Microbiology and Infectious Disease," 3rd ed. Copyright 1965. The Williams & Wilkins Co., Baltimore.

and *Dialister*. A cellulitis of the chin following a dental extraction yielded *Streptococcus evolutus* in pure culture. *Streptococcus anaerobius* was isolated in pure culture from an abscess under the angle of the jaw secondary to an infected wisdom tooth. Another patient had an abscess with a fistula at the angle of the jaw, secondary to a fractured jaw and injury to a wisdom tooth. On culture this yielded *Eggerthella convexa*, *Veillonella alcalescens* and *Dialister*. The last patient had a voluminous abscess at the angle of the jaw secondary to an apico-dental cyst. From this last patient was isolated *Diplococcus plagarumbelli*, *Streptococcus foetidus*, *Streptococcus anaerobius*, *Catenabacterium contortum*, and *Corynebacterium liquefaciens*. Heinrich and Pulverer (116) isolated *Bacteroides melaninogenicus* from eight cases of jaw abscess. Pulverer (141) isolated *Bacteroides melaninogenicus* and "anaerobic mouth flora" from an abscess of the jaw and *Bacteroides melaninogenicus* plus Leptotrichia and anaerobic streptococci from a second such abscess.

Prévot *et al.* (154) recovered *Corynebacterium granulosum* and *Fusobacterium nucleatum* from an abscess of the maxilla following an open fracture. A second case reported by these authors was of a purulent collection near the masseter muscle in the course of dental phlegmon; this yielded *Fusobacterium nucleatum* in pure culture. Caroli *et al.* (172) reported an abscess of the maxilla following fracture which yielded *Fusobacterium nucleatum* and *Corynebacterium granulosum.* A second case, a paraparotid abscess of buccodental origin, yielded *Fusobacterium nucleatum* in pure culture. Prévot *et al.* (173) isolated *Bifidobacterium appendicitis* from a jaw abscess of dental origin. *Ramibacterium dentium* was isolated from a phlegmon of the maxilla of dental origin (174). Degnan *et al.* (175) reported a case of osteomyelitis of the mandible in the area of the second and third molars. The patient's course was prolonged, with several explorations and extended antibiotic therapy. Anaerobic streptococci were isolated on culture. After removal of a sequestrum and extraction of the teeth the patient was ultimately cured. A patient with osteomyelitis of the mandible, source not specified, yielded *Peptostreptococcus* (176).

The pathogenesis of actinomycosis of the jaw is not different from that of other anaerobic infections at this site. The potentially pathogenic *Actinomyces* are present normally on the mucosal surfaces of the area. Caron and Sarkany (177) reported five cases of cervicofacial actinomycosis related to periodontal disease or trauma. Heinrich and Pulverer (178) reported five cases of actinomycosis of the jaw and one of actinomycosis of the chin, all of which yielded, in addition to *Actinomyces israelii, Actinobacillus actinomycetemcomitans.* One of these cases also yielded anaerobic streptococci and two, aerobic hemolytic streptococci. Beerens and Tahon-Castel (115) reported three cases of cervicofacial actinomycosis related to dental extractions. One of these yielded *Actinomyces israelii* in pure culture, a second had this organism plus *Eubacterium lentum,* unknown gram-positive anaerobic rods, and *Staphylococcus epidermidis.* The third case grew *Corynebacterium liquefaciens, Fusobacterium fusiforme,* and *Staphylococcus epidermidis.* Figure 4.5 is from a case of actinomycosis of the jaw in association with malignancy.

Prévot *et al.* (154) isolated *Actinobacterium abscessum* and *Fusobacterium nucleatum* from a case of cellulitis of the chin and jaw following trauma to a wisdom tooth. These workers also reported a case of retromaxillary actinomycosis of buccodental origin which yielded *Actinomyces israelii* and *Fusobacterium nucleatum.* Linhard (178a) reported a case of cellulitis of the jaw of dental origin which yielded *Actinobacterium, Staphylococcus anaerobius,* fusiform bacilli, and rare spirochetes. Hamner and Schaefer (179) reported a case of maxillary actinomycosis secondary to trauma to the teeth. *Actinobacterium abscessum* and *Fusobacterium nucleatum* were isolated from a cellulitis of the chin secondary to a tooth injury, reported by Caroli *et al.* (172). Pulverer (141) recovered *Bacteroides melaninogenicus* as well as

Fig. 4.5.   Actinomycosis of the jaw. Note the thin, branching filamentous organisms.

Actinomyces israelii from a lower jaw abscess. Degnan *et al.* (175) reported a case of alveolitis and cellulitis following extraction of a second molar. Repeated incisions for drainage were required, both internally and externally. Organisms recovered were *Actinomyces bovis* and anaerobic streptococci. Cure followed prolonged, intensive antibiotic therapy as well as the drainage procedures. *Actinomyces* was recovered from two dentigerous cysts with no apparent oral communication by Sprague and Shafer (180). Additional cases of actinomycosis with involvement of the jaw have been reported by Schlegel and Hieber (143a).

Infections of the Gingiva

Acute necrotizing ulcerative gingivitis (ANUG) (Fig. 4.6) is known by a variety of names, including Vincent's infection, Plaut–Vincent's infection, trench mouth, and fusospirochetosis. The disease was probably known as early as 401 BC judging from the account of Xenophon who, describing the retreat from Persia, indicated that many of his soldiers suffered from tender

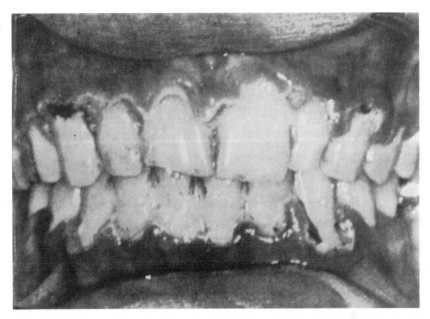

Fig. 4.6. Serious acute necrotizing gingivitis. From Boering (208), reproduced with permission.

mouths and fetid breaths. The Roman armies were also troubled with this disease [Thoma and Goldman (181)]. Bergeron, 1859, gave a classical description of the disease in French troops and pointed out that it may occur in chronic as well as in acute form. During World War I epidemics occurred, especially among troops on duty in the trenches, hence the name trench mouth.

ANUG is seen primarily in people between adolescence and age 30. There is some preponderance of the disease in males. There is a seasonal prevalence during fall and winter. The anterior areas of the mouth are most commonly involved. While ANUG is not always associated with poor oral hygiene, there is a general relationship between local environmental factors and the disease. Third molar flaps, faulty contacts between teeth with resultant food impaction, and calculus formation may all predispose to it. Pellagra and scurvy are frequently complicated by Vincent's infection. Psychosomatic factors and stress may also predispose to this infection.

The prime diagnostic feature of ANUG in the fully developed state is extreme pain. The disease often has a rather sudden onset, beginning around one or two teeth. The major diagnostic criteria according to Wentz and Pollack (182) are pain, hemorrhagic tendency, distinctive foul odor, and destruction of interdental papillae with formation of pseudomembrane.

In hopes of settling some areas of disagreement concerning ANUG and to more firmly establish accurate diagnostic criteria, Barnes *et al.* (183) undertook a controlled survey in a military population and dependents. The results of this study are depicted in Table 4.3. Bleeding of the gums, blunting and cratering of the interdental papillae, fetid breath, pain, and numbness were the major features noted. In this survey, only 0.19% of the 113,000 patients examined at Fort Knox, Kentucky, exhibited symptoms of ANUG. Most other surveys have revealed an incidence of 1 to 2% in military and university populations. In a university health service, Giddon *et al.* (184) noted that approximately 0.9% of university students developed ANUG during a 1-year period. Of the 20% of students who actually use the dental clinic, 4.2% were treated for ANUG.

Unusual large spirochetes (possessing up to 20 or more axial fibrils) appear to be specifically associated with ANUG. They have been observed in the gingival tissues only in this disease (184a, and S. Socransky and M. Newman, personal communication).

TABLE 4.3

Signs and Symptoms in Patients with Acute Necrotizing Ulcerative Gingivitis and in Control Patients[a,b]

Signs and symptoms	ANUG subjects		Control patients	
	Number	Percentage	Number	Percentage
Gingival bleeding	208	95.5	51	47.2
Profuse	20	9.2	0	0.0
Moderate	100	45.9	9	8.3
Slight	88	40.4	42	38.9
No bleeding	10	4.5	57	52.8
Blunting of papilla	205	94.0	46	42.3
Pain	188	86.2	22	20.4
Severe	31	14.2	0	0.0
Moderate	69	31.6	0	0.0
Mild	88	40.4	22	20.4
No pain	30	13.8	86	79.6
Fetid odor	184	84.4	17	15.3
Cratering of papilla	174	79.8	13	11.9
Pseudomembrane	160	73.4	2	1.7
Wooden or wedge-like sensation of the teeth	88	40.4	0	0.0
Bad taste in mouth	87	39.9	0	0.0
Total cases observed	218	100.0	108	100.0

[a] From Barnes *et al.* (183), with the kind permission of the authors and the publisher.
[b] Based upon a sample of 108 military and dependent subjects of equivalent age, sex, and race who did not have ANUG.

Therapeutically, the various drugs that are active against anaerobes in general, including penicillin G, are effective in the management of ANUG. Davies *et al.* (185) obtained good results in 15 patients with Vincent's gingivitis treated with metronidazole. They noted that good concentrations of the drug were achieved in the saliva. Good results were also noted by Shinn *et al.* (186), Wade *et al.* (187), Duckworth *et al.* (188), and Fletcher and Plant (189). Lozdan *et al.* (190) noted that the related compound, nitrimidazine, is as effective as metronidazole. Topical antimicrobial therapy with a number of agents, including bacitracin (191) and vancomycin (192), is effective. Wade and co-workers (193) noted that sodium peroxyborate was approximately 75% as effective as penicillin V in treating ANUG. Wade and Mirza (194) found that hydrogen peroxide was almost as effective as sodium peroxyborate. Francis and de Vries (149) recommended that, if the patient is febrile or the infection extensive, the patient be treated with systemic antimicrobial compounds and nascent oxygen (in the form of hydrogen peroxide or urea peroxide in glycerine, which releases oxygen slowly). After the acute infection has subsided, calculus should be removed, gingival pockets cleaned out and packed, and gingivectomy done as required.

Periodontal disease may be complicated by purulent gingival pockets or gingival abscesses. Prévot (21) isolated *Veillonella alcalescens* from purulent gingivitis. Beerens and Tahon-Castel (115) recovered *Diplococcus morbillorum* and nonhemolytic aerobic streptococci from gingival pus in a case of pyorrhea. Felner and Dowell (168) recovered *Fusobacterium fusiforme* from gingival abscesses and from the bloodstream in a case complicated by bacteremia. Pulmonary infection may also complicate gingival disease as noted by Bartlett and Finegold (161). In the latter report, one patient with gingivitis and coma due to head trauma developed a necrotizing pneumonia involving the right lower and upper lobes, with *Bacteroides melaninogenicus* recovered in pure culture on transtracheal aspiration. Another patient, an alcoholic and diabetic with gingivitis, had a necrotizing pneumonia of the left lower lobe; transtracheal aspiration in this patient yielded *Bacteroides melaninogenicus* and *Fusobacterium nucleatum*. In the overall series, 22 cases of 47 with anaerobic pleuropulmonary infection who had dental examinations had periodontal disease, gingivitis, and/or periapical abscess as compared to only 7 with these dental diseases among matched controls.

Pericoronitis is inflammation of the gingiva about the crown of a partially erupted tooth, usually a third molar. According to Francis and de Vries (149) organisms get under the gum flap and may get distal to the tooth and then set up cellulitis. Antibiotics should be used until the infection is controlled, at which time the third molar should be extracted.

In 1894, Lubinski (5) recovered anaerobic bacilli (not further described) from the green, foul pus of a parulis (gum boil). Rentsch (195) recovered anaerobes from two cases of parulis. In one case, the organism recovered was *Bacteroides funduliformis*, and in the other case it was a *Corynebacterium*. Hara *et al.* (176) recovered a *Peptococcus* from a closed abscess of the gums, and Weiss and Mercado (196) recovered a fusiform bacillus from a gum abscess. Prévot (197) recovered *Sphaerophorus ridiculosus* from a secondary infection of an epithelioma of the gum.

A review of the literature concerning gingival and other oral infections in patients with white blood cell deficiencies was given by Kyle and Linman (198). Gingival inflammation and/or ulceration and periodontal problems have been noted in patients with neutropenia, and particularly with acute agranulocytosis.

Stomatitis

The stomatitis to be discussed here is infectious stomatitis—an acute inflammation of the buccal mucosa caused by bacteria. Noma is discussed in depth in Chapter 13. This infectious stomatitis is the equivalent of acute necrotizing ulcerative gingivitis. The most severe form of this is cancrum oris or noma. Most of the reports are from the older literature. Buday (199) studied several cases of gangrenous infections of the mouth and throat. Several types of organisms were noted superficially in the necrotic areas—cocci, diphtheroid-like organisms, *E. coli* and smaller numbers of spirilla, fusiforms, and threads. Deep in the tissues, however, were masses of thread forms, *Lepthothrix*, spindle-form bacilli, comma forms, and spirilla. Babes (200) noted that fusiform bacilli occurred in all types of putrid and gangrenous processes of the oral mucosa and that they secondarily invaded mercurial stomatitis, luetic lesions, and scurvy. In a boy with extensive gangrenous stomatitis of the right cheek with a foul odor, Campbell and Shaw (201) noted thread forms with beading, along with conventional bacteria in direct smear of the tissue. The thread forms were grown in mixed culture, which had the peculiar odor of *Bacillus necrophorus*. Carnot and Blamoutier (202) described a case of extensive stomatitis involving the whole mouth and throat and extending to the nose and conjunctivae. The disease was ulceromembranous in character and numerous fusiforms and spirochetes were noted. They also described two additional cases, one secondary to bismuth stomatitis and the other to mercurial stomatitis. Lahelle (203) described a case of fusospirochetal stomatitis manifested by a 2 × 3.5 cm ulceration on the inside of the cheek. There was pronounced fetid odor.

The disease may be relatively benign and self-limited. Lichtenberg *et al.* (204) described 16 cases of ulcerative stomatitis in patients with concurrent throat infections. All cases healed in 4 to 7 days without therapy. Vincent's organisms were seen on smear of 14 cases that were examined. Prévot (171) isolated *Ristella fragilis* from a case of ulcerative stomatitis; along with it was *Staphylococcus pyogenes* and *S. albus*. Francis and de Vries (149) commented that they have seen the equivalent of Vincent's gingivitis in edentulous people wearing dentures. The process was manifested by tissue necrosis primarily under the dentures, but elsewhere at times as well, and a foul odor. Stephens (205) also described a case of fusospirochetal stomatitis under a partial denture. Jelliffe (206,207) noted that even severe infections such as cancrum oris responded well to small doses of penicillin (150,000 units of procaine penicillin daily), along with a high protein diet and debridement of necrotic tissue.

Other Oral Infections

Fetor oris or foul-smelling breath may be caused by a variety of things, including disease of the mouth, nose or throat, metabolic disease, digestive tract disturbances, and disease of the lower respiratory tract. Lesions in the mouth which may cause this condition are periodontal disease (especially when there are deep gingival pockets), acute or chronic ulcerative gingivitis, poor dental hygiene, alveolitis, and oroantral fistula with maxillary sinusitis. Good oral hygiene is of major importance in eliminating this problem when its origin is in the mouth. Thorough cleaning of the teeth and treatment of cavities and of teeth with gangrenous pulps may eliminate undesirable odor (208). That these odors are often related to the presence of anaerobic infection is supported by the report of Roth (209) to the effect that chlortetracycline generally helped a variety of oral conditions that caused bad odors.

Cope (210) cited some interesting work by Soderlund and Naeslund concerning the nature of salivary calculi and tartar. When these materials are carefully decalcified, there is left a coherent organized stroma that can be seen microscopically to be made up of interwoven filaments of *Actinomyces* and other thread forms. *Actinomyces* can be cultured from such material. Furthermore, *Actinomyces* grown in lime-containing sterilized saliva becomes calcified unless it has been killed by heat before being immersed in such fluid.

Barker and Miller (211) described a patient with marked oral sepsis and gingivitis, particularly about the posterior molars. The patient had a very fetid breath. A punched out ulcer, the size of a dime, was noted on the hard palate. Studies for syphilis were negative. Smears of the creamy exudate overlying the ulcer showed enormous numbers of fusiforms and many spirillae.

A number of studies point out the role of anaerobes in infections of the cheek. Heinrich and Pulverer (116) recovered *Bacteroides melaninogenicus*

from 95 cases of cheek abscess. Pulverer (141) recovered *Bacteroides melaninogenicus* from five cheek abscesses. Four of these also yielded *Actinomyces
israelii*, two yielded anaerobic streptococci, and one a *Leptotrichia* and a
fusiform bacillus. Three of the lesions also had *Staphylococcus albus* in addition to the anaerobes. The same author described an infiltrated cheek lesion
from which was recovered *Bacteroides melaninogenicus*, *A. israelii*, *Actinobacillus actinomycetemcomitans*, and *S. albus*. Heinrich and Pulverer (178)
recovered *A. israelii* and *Actinobacillus actinomycetemcomitans* from two
cheek abscesses and one fistula of the cheek. One of these lesions also yielded
S. albus. Schlegel and Hieber (143a) reported a cheek abscess involving *Actinomyces*. *Bacteroides* was recovered from a cheek abscess by Mitchell (110a).
Fritsche (211a) studied two cases of cheek infection, one an abscess. The
abscess yielded *Actinomyces israelii*, *Bacteroides melaninogenicus*, and *Bacteroides* sp. The other yielded *A. israelii*, *B. melaninogenicus*, *Fusobacterium*,
and *Actinobacillus actinomycetemcomitans*. Werner and Pulverer (30) recovered *Sphaerophorus necrophorus* from a cheek abscess. Prévot (21) recovered
Ramibacterium ramosum from a buccal abscess.

Harvey *et al.* (212) described a case of actinomycosis that originated in the
sublingual tissues. Actinomycosis of the tongue, although uncommon, is
certainly not rare, as about 3% of all cases of the disease occur in the tongue
(210). Actinomycosis of the tongue may present as an acute abscess, as a
subacute or chronic circumscribed inflammatory nodule, or as an infiltrating
indurated mass spreading from the tongue. Cecchi (213) reported two cases
of actinomycosis of the tongue, and Sodagar and Kohout (214) reported two
other cases of actinomycosis of the tongue presenting as pseudotumors.
Three cases of necrotic ulcer of the tongue due to fusospirochetes were
reported by Quérangal in 1933 (215).

Patients with leukemia or other diseases treated with cytotoxic agents may
suffer damage to the oral mucosa. Inflammation and necrosis may ensue and
serious local or systemic infection may follow, particularly in patients who
are neutropenic. Storring and Gerken (215a) have obtained improvement in
such patients with drugs primarily or exclusively active against anaerobes
(clindamycin and metronidazole).

5

Eye Infections

The role of anaerobic bacteria in eye infections seems to be a minor one, but it is clear that there are many defects in the studies done to date in this area. It is also clear that on occasion anaerobic infections of the eye may be devastating. The studies of Matsuura (216) are a step in the right direction; this worker has compared the anaerobic flora isolated from various infectious processes in the eye with that obtained from normal eyes. This type of study is extremely important in most eye infections inasmuch as it is difficult or impossible to obtain material for examination without contamination by elements of the indigenous flora that may be present. Many of the anaerobic infections of the eye described in the literature must be regarded as presumptive since the diagnosis is based on observation of organisms, such as fusiform bacilli and spirochetes, in the exudate which morphologically resemble anaerobes. In many of these cases, anaerobic cultures were not performed or were unsuccessful in recovering the suspected anaerobes observed on direct smear.

Conjunctivitis

Scholtz (217) described a case of gangrene of the conjunctiva in a person who had had an injury to the eye 10 years previously. Most of the eye had been destroyed by the original injury, and a large foreign body remained at the time of the conjunctival infection. The pus was very foul-smelling and showed, on direct smear, gram-negative spindle-form bacteria in addition to streptococci, staphylococci, and pneumococci. Culture was not successful. Removal of the foreign body and local treatment resulted in cure. Dunnington and Khorazo (218) described a case of purulent conjunctivitis persisting for over 2 months in a 20-year old girl. There was frankly purulent secretion, congestion and thickening of the palpebral conjunctivae, and a few enlarged follicles in the lower cul-de-sac. The upper lid was slightly swollen. Direct smear of the exudate revealed a number of fusiform bacilli and occasional spirochetes. The fusiform bacillus was recovered in almost pure culture anaerobically, whereas *Staphylococcus aureus* and a diphtheroid were grown in aerobic cultures of both the infected and the normal eye. The fusiform bacillus was not recovered from the normal eye. There was no evidence of disease due to this organism elsewhere in the body. A poorly documented case of fusospirochetal conjunctivitis is presented by Rocha (219).

Carnot and Blamoutier (202) described a case of fusospirochetal tonsillitis that ascended to produce a rhinitis and conjunctivitis. The conjunctivitis was bilateral, with edematous eyelids, chemosis and considerable injection, some granularity of the palpebral conjunctivae, and, to a lesser extent, the bulbar conjunctivae. There were abundant purulent secretions at the inner angle of each eye, and there was a marked photophobia. On direct smear of the mucopurulent conjunctival secretion, there were noted numerous cocci, some in chains, and a great abundance of spiral and fusiform bacilli. A case of Vincent's infection involving the eyes as well as the mouth and penis was reported by Bowman (220). This case was diagnosed by smear only. There was a very severe bilateral conjunctivitis with photophobia. Herrenschwand (221) described conjunctivitis in himself as a result of accidental contamination of his right eye by sputum from a patient whom he was examining. The patient had undergone surgery the previous day for a dental fistula that discharged purulent material. The conjunctiva of the lower lid was red, and there was chemosis of the bulbar conjunctiva. The infection lasted for 8 days. Direct smear of the exudate revealed sparse leukocytes, fusiform bacilli, and numerous spirochetes. Culture attempts were unsuccessful.

The role of fusiform bacilli and spirochetes must be considered uncertain in the 14-year-old girl who had purulent conjunctivitis that proceeded to extensive deep ulceration of the cornea in both eyes, with turbidity of the aqueous humor and hypopyon [Goudie and Sutherland (222)]. *Staphylococcus aureus* was isolated in culture from this case, whereas the fusiform bacilli and spirochetes could not be, although they were present in large numbers as these smears were described. Dejean and Temple (223) described a chronic fusospirillary conjunctivitis of one eye which lasted for many months in an elderly man. The organisms were not cultured. A case of unilateral follicular conjunctivitis and canaliculitis of 6 years' duration, and showing fusiform bacilli and spirochetes on direct smear, has been described by Burns *et al.* [quoted by Gutierrez (224)]. Burns and co-workers also described 2 other patients with chronic conjunctivitis of shorter duration due to fusiform bacilli. Both of these latter cases also had involvement of the canaliculi; one had severe papillary purulent conjunctivitis and the other had follicular conjunctivitis. Fusiform bacilli were cultivated in both cases. One of the patients improved rapidly after dilation, curetting, and irrigation of the canaliculi. From 6 cases of focal bilateral conjunctivitis and 1 of unilateral phlyctenular conjunctivitis, Matsuura (216) was able to recover only anaerobes that he regarded as elements of the normal flora. Included were various species of anaerobic *Corynebacterium*, *Peptococcus variabilis*, and *Clostridium nigrificans*.

Henkind and Fedukowicz (225) described two cases of conjunctivitis thought to be caused by *Clostridium perfringens*. The picture was unique in that the conjunctivitis was localized primarily in the lower fornices and adjacent conjunctival tissue where folds of conjunctiva might create a more favorable environment for anaerobes. Another unique feature, in addition to the localized involvement, was the presence of multiple small hemorrhages in the involved area of conjunctiva. The organism was recovered on more than one occasion from both of these patients. Gorbach and Thadepalli (486) isolated *Clostridium perfringens* in pure culture from pus of monocular conjunctivitis. The isolation of *Bacteroides fragilis* from the eye (and cervix) of a patient with conjunctivitis and arthritis thought possibly to have Reiter's disease (226) must be regarded as of questionable significance. The organism was recovered in pure culture from the eye. It could not be isolated from other sources in this patient. *Ristella pseudoinsolita* (*Bacteroides fragilis* ss. *fragilis*) plus *Corynebacterium anaerobium* were recovered from a case of conjunctivitis by Prévot (226a).

Perkins *et al.* (226b) cultured 273 inflamed conjunctivae with comparative cultures of uninvolved eyes in the same individuals. Anaerobic organisms were recovered from 172 (63.0%). The major pathogen was *Peptostreptococcus* sp., isolated from 29.3% of infected eyes and only 6.3% of normal eyes.

Mikuni *et al.* (227) reported two cases of phlegmon of the conjunctival sac that yielded anaerobic bacteria. One of these grew *Bacteroides* and *Peptostreptococcus* (neither speciated) in the absence of aerobes, and the other case grew both a *Peptococcus* and a pneumococcus.

Actinomyces israelii has also been isolated from the canaliculus of a patient who had moderate injection of the bulbar and palpebral conjunctivae in the adjacent area (228).

Our laboratory has isolated an anaerobic gram-negative bacillus distinct from *Bacteroides fragilis*, but not identified, in pure culture from an elderly man with a low-grade conjunctivitis with purulent exudate.

Infections of Lacrimal Apparatus

Anaerobic bacteria seem to be involved in lacrimal canaliculitis fairly frequently. However, the distinction between true *Actinomyces* and the morphologically similar *Nocardia* has frequently not been made in the earlier literature. When cultures have been made, the organism isolated was more often anaerobic than aerobic (210). It is clear that actinomycosis of the lacrimal canals is clearly distinct from all other types of actinomycosis. It is purely a local process and never extends outside the socket into surrounding tissues, never causes serious symptoms, and always responds readily to appropriate treatment. *Actinomyces* attacks only the lacrimal canals and has not been known to extend to the lacrimal sac. The lower lacrimal canal is involved twice as frequently as the upper canal. *Actinomyces* infections of the lacrimal canal are much more common in females than in males. The organism forms masses of filaments that may become grossly visible as a small granule and ultimately may attain the size of a pea. There may be low-grade inflammation of the adjacent tissue of the lacrimal canal and, in one unusual case recorded by Wegner [quoted by Cope (210)], the granulation tissue formed a polypoid mass that prolapsed through the upper lacrimal orifice. Symptomatically, there is irritation at the inner angle of the conjunctival sac, and there may be local congestion of the conjunctiva, tearing of the eye, stickiness of the lids in the morning, and swelling in the region of the lacrimal canal. A small amount of mucopurulent discharge may be noted at the lacrimal opening, and pressure may discharge more of this type of material. Although canalicular signs usually predominate, with minimal or no inflammation of the conjunctiva, marked conjunctival inflammation may occur and there may be considerable purulent exudate. Acute corneal ulceration may also occur {Elliott [quoted by Pine *et al.* (228)]}.

Although actinomycotic canaliculitis is almost invariably limited to the lacrimal canal, on rare occasions there may be such complications as erosion through the wall of the canaliculus onto the conjunctival surface {Moore [quoted by Pine *et al.* (228)]} and suppuration of the preauricular lymph node {as reported by Ridley and Smith [quoted by Pine *et al.* (228)]}.

Several species of *Actinomyces* have been implicated in lacrimal canaliculitis. Thus, Pine and co-workers (228, 229) reported two cases of lacrimal canaliculitis due to *Actinomyces israelii*. Subsequently, Buchanan and Pine (230) reported that one of the two afore-mentioned *Actinomyces* strains was actually *Actinomyces propionicus*. Still later, *Actinomyces propionicus* has been reclassified and is now known as *Arachnia propionica* (231). Gerencser and Slack (232) recovered both *Actinomyces propionicus* (*Arachnia propionica*) and *Actinomyces odontolyticus* from material from a case of lacrimal canaliculitis. Coleman and co-workers [quoted by Georg (233)] recovered *Actinomyces naeslundii* from a case of lacrimal canaliculitis. (Among 67 cases of actinomycosis reported in Great Britain in 1971 and 1972, 7 were of nasolacrimal canaliculitis (1134a).

Therapy of actinomycotic canaliculitis consists of simply removing the concrements from the lacrimal canal. However, in the presence of significant conjunctivitis, antimicrobial therapy is also important.

As noted earlier, fusiform bacilli, with or without spirochetes, have been implicated in lacrimal canaliculitis and conjunctivitis {Burns *et al.* [as quoted by Gutierrez (224)]}. Nectoux and Suchet [noted by Prévot (21)] isolated *Veillonella parvula* on two occasions from a man with persistent suppurative lacrimal canaliculitis. Bonnet *et al.* (234) noted recovery of a catalase-negative gram-positive anaerobic rod, identified as a *Corynebacterium* by the Institut Pasteur of Lille, from a patient with lacrimal canaliculitis. Gifford (235) described a woman with a small ulcer surrounding her left lower punctum. This ulcer was covered with a thin whitish membrane, a smear of which revealed gram-positive diplococci and a fairly large number of gram-negative rods, some of which were said to be typical of fusiforms. These were not recovered in culture. This same author noted somewhat similar fusiform bacilli in another patient which turned out to be an aerobic, spore-forming, fusiform-shaped bacillus.

Dacryocystitis may also involve anaerobic bacteria. Oishi *et al.* (235a) described a case of chronic dacryocystitis, purulent discharge from which grew *Peptococcus* and an aerobic gram-negative rod. In a 1 year period, they recovered anaerobes (primarily cocci and gram-negative bacilli) from 14 of 29 cases of chronic dacryocystitis. Wakisaka (236) recorded finding fusiform bacilli on two occasions in such cases, once in pure culture and once mixed with influenza bacillus. Löhlein (237) noted masses of spirochetes and fusiform bacilli and only sparse staphylococci from the fetid discharge of a patient with dacryocystitis. He was unable to culture the organisms. Other gram-negative anaerobic bacilli recovered from cases of dacryocystitis include a *Bacteroides* [recovered together with coagulase-negative staphylococci by Mikuni *et al.* (227)] and *Bacteroides* (*Zuberella*) *constellatus* isolated by Martres *et al.* (238). Mikuni *et al.* (227) isolated anaerobic corynebacteria from 3 additional cases of dacryocystitis, one of which also showed *Staphylococcus epidermidis*, and *Peptococcus* in pure culture from still another case of dacryocystitis. An anaerobic *Corynebacterium* was also isolated in pure culture and in large numbers from a case of purulent dacryocystitis by

Matsuura (216). Beneditti [referenced by Henkind and Fedukowicz (225)] described a case of chronic unilateral dacryocystitis caused by *Clostridium perfringens*. Subsequent to this documentation, the patient involved suffered a perforating ocular injury that ultimately led to gas gangrene panophthalmitis. Beneditti also isolated several anaerobic organisms from various other ocular infections, in a brief report.

According to Brons (239), Morax and Veillon found, in a case of gangrenous lacrimal sac phlegmon, *Fusobacterium necrophorum*, another anaerobe identified as a "coccobacillus," and *Streptococcus pyogenes*. Prévot (134) referred to another report by Veillon and Morax—a case of gangrenous peridacryocystitis from which they isolated an anaerobe that was subsequently identified as *Ramibacterium ramosoides* (a gram-positive anaerobic bacillus whose current classification is uncertain). It appears that Morax and Veillon had two similar cases with different bacteriological findings.

Lid Infections

Heinrich and Pulverer (116) recovered *Bacteroides melaninogenicus* from a patient with an abscess of the eyelid. Mikuni *et al.* (227) recovered *Bacteroides* and *Peptococcus*, along with *Staphylococcus epidermidis*, from an abscess of the margin of an eyelid. These same workers recovered a *Peptococcus*, along with *Staphylococcus aureus*, from an external stye and an anaerobic *Corynebacterium* plus *Peptostreptococcus*, along with *Staphylococcus aureus*, from a case of blepharitis. *Actinomyces* has also been recovered from 2 patients with an abscess or cyst on an eyelid (1134a).

Corneal Ulcer

Matsuura (216) studied several patients with unilateral corneal ulcer or corneal abscess with both aerobic and anaerobic culture technique and using the normal eye as a control. This excellent technique revealed no important difference in the two eyes, permitting the conclusion that the anaerobes that were present in small numbers (mostly *Corynebacterium*, but also including two strains of *Peptococcus* and one of *Bacteroides capillosus*) were simply present as part of the normal flora.

Slansky and co-workers (240) detected collagenase activity in the corneal epithelium of 10 cases with either acute or chronic corneal ulceration and could not detect this enzyme in 16 other cases without corneal ulcers (some of these latter cases did have disease of the cornea, but without ulceration). Since certain anaerobic bacteria produce collagenase and since some of these organisms, such as *Bacteroides melaninogenicus*, may be difficult to grow, it is desirable to do additional careful anaerobic cultures on patients with corneal ulcers, using the normal eye as a control. We have already referred to a patient reported by Goudie and Sutherland (222) in whom there was conjunctivitis and extensive ulceration of the cornea of both eyes. Mikuni *et al.* (227) reported isolation of *Bacteroides* along with *Staphylococcus epidermidis* from a patient with corneal ulcer. Gingrich and Pinkerton (241) reported a case of corneal ulcer due to *Actinomyces* (which they identified as *Actinomyces bovis*) following injury by a fragment of oyster shell and the use of an antibiotic-steroid preparation following injury. This case responded to treatment, but ultimately required corneal transplant. These authors also referred to two other cases of corneal ulcer due to anaerobes without previous perforating trauma. These two cases were both due to *Clostridium tetani* and were originally reported by Quentin and Tsutsui. Henkind and Fedukowicz (225) referred to Pringle's case of ulcerative keratitis associated with a lid wound and apparently caused by *Clostridium perfringens*.

Endophthalmitis and Panophthalmitis

The cases already referred to of Goudie and Sutherland (222) and Gingrich and Pinkerton (241) both had involvement of the anterior chamber with hypopyon, in addition to the corneal ulceration (and, in the first instance, conjunctivitis as well). The case of Bertozzi (242) was of metastatic Vincent's infection of the eye in the course of measles in an infant. In this case, there was intense conjunctivitis with photophobia, slight ciliary injection, and an intense inflammatory response in the aqueous humor such that it was impossible to view the fundi. Brons (239) referred to cases from the literature of vitreous body infection with *Clostridium*; all of these were caused by penetration of splinters. Weiss (243) reported the case of a young boy who developed widespread infection, including involvement of the ethmoid and maxillary sinuses, the periorbital area, and the eye itself. The eye involvement was an endophthalmitis and involved *Fusobacterium*, which was recovered on culture along with *Staphylococcus aureus* and viridans streptococci. Weiss stated that endophthalmitis following trauma also involved *Clostridium perfringens* and *Clostridium tetani*. Endophthalmitis, secondary to an injury, due to *Butyrivibrio fibrisolvens* was described by Wahl (243a). Endophthalmitis may also appear spontaneously, as noted by Goldsand and Braude (244). In an 8-year-old boy reported by these authors, the endophthalmitis was caused by microaerophilic streptococci that were recovered in culture from a portion of the vitreous obtained by needle aspiration. The boy was successfully treated with penicillin and streptomycin, but with some reduction in visual acuity. The interesting report of Golden *et al.* (245) concerning observation of two spirochetes, which were definitely not *Treponema pallidum*, among 55 specimens of aqueous humor taken at the time of cataract surgery is of uncertain significance. Both of these specimens were from one patient and there was no evidence that the presence of the spirochetes was associated with any pathology. The authors speculated that the spirochetes were of oral origin.

Brons (239) referred to a report of Mayweg in which the author collected a total of 16 cases of tetanus associated with injuries to the eyeball. In 13 of these cases, the injuries were perforating, and 9 of these resulted in panophthalmitis.

Leavelle (246) reviewed the literature on gas gangrene panophthalmitis up-to the time of his report (1955). There were a total of 53 cases, to which the author added 3 cases. Certain characteristics were common to all of these cases. Infection followed a perforating wound of the globe. Despite all therapeutic measures undertaken, the infection progressed and destroyed all visual function. The infected eye had to be either eviscerated (two-thirds of the cases) or enucleated, or the entire orbit had to be exenterated (two cases). The postoperative course was essentially uneventful, and recovery was complete. The injury frequently resulted from hammering at or chipping metal or stone. Pain and loss of vision occurred within 12 hours following the injury. By 18 hours, there was evidence of a fulminating panophthalmitis with chemosis and brawny swelling of the lids, proptosis, hypopyon, ring abscess of the cornea, coffee-colored discharge, increased intraocular tension, gas bubbles in the anterior chamber, immobilization of the globe, and necrosis of the wound margins. Systemic toxicity was noted later in the course, but was not severe compared to the local reaction. The diagnosis was suspected clinically within 24 hours in a number of cases but was not established by laboratory means in any case until after 48 hours following the injury. When the foreign body was found by X ray or at surgery, it was invariably in the vitreous. *Clostridium perfringens* was always recovered, either in pure culture or mixed with other anaerobic and/or aerobic cocci and bacilli. In the author's three cases, *Clostridium perfringens* was isolated in pure culture. Experimental work has shown that panophthalmitis does not result unless the anaerobes are inoculated directly into the lens or the vitreous. The author reported a fourth case of a steel chip foreign body that led to immediate loss of vision in the injured eye. The patient received penicillin and streptomycin $2\frac{1}{2}$ hours after injury, and the foreign body was removed some hours later. Postoperatively, the patient again received penicillin and streptomycin and when there was no light perception present in the destroyed eye, it was enucleated on the third day despite no evidence of inflammation. Cultures of the foreign body removed earlier yielded *Clostridium perfringens* in pure growth. Even more encouraging is a later report of *Clostridium perfringens* panophthalmitis [Levitt and Stam (247)]. The patient in this report was a 7-year-old boy who poked a stick into his left eye. He was first seen 18 hours later, at which time a small nub of uveal tissue extruded at the periphery of the cornea and there was marked hypopyon and a turbid aqueous humor. Smear of the left eye showed many pus cells and gram-positive rods which on culture proved to be *Clostridium perfringens*. Surgery was deferred because of the severity of the infection. The patient was treated with ampicillin and chloramphenicol. Little change was noted for the subsequent 4 days, following which under local anesthesia the extruding tissue was cut off and a thin splinter of wood was extracted from the eye. The hypopyon vanished and the aqueous cleared within 5 days, following which the patient was noted to have a completely opaque lens. The patient was left with light perception in the eye, and enucleation or evisceration was not required. Frantz *et al.* (247a) noted that 8 additional clostridial infections of the eye have been reported since Leavelle's review; they also added a case of their own. The latter case is interesting in that it was secondary to a bacteremia associated with a perforated gangrenous gallbladder; ophthalmoscopic examination revealed a gas bubble in the anterior chamber.

Orbital Cellulitis or Phlegmon

As already indicated, the case report by Weiss (243) due to *Fusobacterium, Staphylococcus aureus*, and viridans streptococci involved periorbital abscess in addition to endophthalmitis. Seecof (248) reported a similar case in a young girl in which there was also extensive sinusitis and meningitis in addition to orbital cellulitis; there was no injury involved in this latter case, however. An incision made into the left upper eyelid yielded a large amount of thick malodorous pus, which on smear showed gram-positive streptococci, large gram-negative rods, innumerable spirochetes, and fusiform bacilli. The anaerobes were not recovered on culture. The sinusitis was apparently the primary infection in this girl, as it was in 3 additional cases of orbital phlegmon following perforation or empyema of the ethmoid sinus reported by Kompanejetz [cited by Herrenschwand (221)]. Brons (239) cited the report of Stanculeanu and Baupp of fetid orbital phlegmon secondary to maxillary sinusitis. From the sinus infection, the following anaerobic bacteria were recovered: *Clostridium ramosum, Clostridium perfringens, Bacteroides fragilis, Bacteroides serpens*, and *Fusobacterium necrophorum*. Sevel *et al.* (248a) reported a case of paranasal sinusitis and orbital cellulitis with gas and foul pus in the orbit. Culture grew streptococci in broth only (microaerophilic?), but smears of the pus and of the broth also showed pleomorphic gram-positive rods and fusiform bacilli. Oishi *et al.* (235a) also reported a case of orbital cellulitis in association with sinusitis (maxillary). The foul pus from this case grew only *Peptostreptococcus*. These workers also referred to two other cases of anaerobic orbital cellulitis which they saw in the preceding year. Harvey *et al.* (212) also reported a case of orbital cellulitis secondary to maxillary sinusitis; the infecting organism in this case was *Actinomyces*. Cope (210) stated that involvement of the orbit in the course of actinomycosis is rare and occurs mostly from extension of the infection from soft parts of the face deeply through the temporal fossa and the sphenomaxillary fissure. Infection by way of the ethmoid or other sinuses is still less common. There may also be direct extension to the orbit from a lesion on the exterior of the upper jaw. Exophthalmos results when infection gains access to the orbit by the posterior route. Infiltration of the ocular muscles by the disease may interfere with movements of the eye. From the posterior orbit, the infection may proceed anteriorly by the formation of draining sinuses or posteriorly through the sphenoidal fissure into the cranium and thence to the brain.

Miscellaneous

Hartl (249) reported septic metastasis in the fundus oculi of a patient with disseminated infection due to *Fusobacterium necrophorum* originating as a tonsillitis. No further details are given concerning the problem in the eye. Klein *et al.* (249a) described a case of gas gangrene of the eye in a patient with bowel cancer and widespread gas gangrene. *Clostridium perfringens* was isolated from muscle. Of note in this case was the presence of gas in retinal vessels, seen on ophthalmoscopy.

6

Ear, Nose, Throat, and
Head and Neck Infections

In the series of Lodenkämper and Stienen (250), covering 722 anaerobic infections, 194 of the specimens that yielded anaerobes, plus an unspecified number of cases of tonsillogenic sepsis, originated in the ear, nose, and throat system.

Infections of the Ear and Mastoid

There seems to be little discussion of the role of anaerobes in external otitis. Barenberg and Lewis (251) described two cases of membranous external otitis with a fetid odor and much edema of the canal. Fusiforms and spirochetes were seen on smear. Saito *et al.* (252) recovered *Peptostreptococcus* from two subcutaneous abscesses of the external ear. The external ear may be involved in the course of facial actinomycosis (210), but is rarely the primary site of the disease.

Otitis media and/or mastoiditis, usually chronic, is the most common underlying source of anaerobic infection of the central nervous system, particularly brain abscess and meningitis. This type of infection may also serve as the precursor for subdural and extradural empyema. This is discussed in detail in Chapter 7. In considering the role of anaerobes in otitis media and mastoiditis, those cases that subsequently were complicated by central nervous system infection are discussed in Chapter 7, rather than here.

It seems to be generally conceded that anaerobic bacteria are not a common or important cause of *acute otitis media or mastoiditis*. However, the evidence for this is really not good and additional studies are much needed. For example, in a report by Howie *et al.* (253) 858 cases of otitis media in children were studied. Fluid was obtained from the middle ear by puncture of the eardrum but the material was cultured only aerobically; 29.5% of the specimens were either sterile or contained nonpathogens. Similarly, Mortimer and Watterson (254) cultured fluid from the middle ear of 68 children with otitis media, again not doing anaerobic cultures; 25 of these cases yielded sterile cultures. Stickler and McBean (255) reviewed six series in the literature concerning the etiology of acute otitis media. In three of these, the incidence of sterile cultures varied from 11 to 23%. Rist, in 1901, commented that anaerobes are found occasionally in acute otitis media but are usually not present (256).

Felner and Dowell (168) reported four cases of bacteremia with anaerobic, gram-negative bacilli secondary to otitis media; it is not specified that the otitis media was acute in these cases but this seems likely. The organisms involved were *Bacteroides oralis*, *Bacteroides* NCDC F3, and *Fusobacterium necrophorum* (2). One of these patients manifested shock, and another, disseminated intravascular coagulation. In 1927–1928, Wirth (257) described two cases of virulent acute otitis media and mastoiditis with involvement of a nearby venous sinus. Both cases proved fatal. One patient had meningitis and the other bacteremia, and other foci of metastatic infection were noted in both cases. The organisms were thought to be *Bacteroides fundibuliformis*.

There are a number of individual case reports of anaerobes involved in acute otitis media, with or without mastoiditis. In 1910, Schottmüller (258) described a case of acute otitis media with lateral sinus thrombosis, pulmonary gangrene and bacteremia due to an anaerobic streptococcus. In 1913, Massini (259) described a patient with acute bilateral otitis media with mastoiditis on the left side. This patient's illness was complicated by sepsis and pleuropulmonary infection, and the patient died. The exudate was foul. No aerobes were recovered, the culture yielding only *Bacillus diphtheroides anaerobius*. In 1930, Barenberg and Lewis (251) described a patient with chronic gingivitis who developed acute otitis media that showed fusiforms and spirochetes on smear. The discharge from the ear was fetid. Jame and Jaulmes (260) described a fatal case of acute bilateral otitis media and mastoiditis, complicated by bacteremia due to *Sphaerophorus funduliformis*. A fatal case of acute otitis media and mastoiditis following mild rhinopharyngitis in an 8 year old was described by Pham (261). The illness was complicated by bacteremia due to *B. funduliformis*, and the patient died. The following year, Grumbach *et al.* (262) described a case of otitis and mastoiditis, presumably acute, complicated by bacteremia and purulent arthritis of the knee. This involved *Bacteroides funduliformis* and was fatal. Prévot and Senez (263) described a case of acute antromastoiditis that flared following surgery; *Micrococcus grigoroffi* was the infecting organism. Fisher and McKusick (264) described a case of mastoiditis (acute?) due to *B. funduliformis* and anaerobic streptococci. A case of acute otitis media following measles and due to *Eggerthella convexa* plus *Staphylococcus aureus* was described by Beerens and Tahon-Castel (115). Oda [cited by Baba, in Kozakai and Suzuki (265)] described a case of *Clostridium perfringens* mastoiditis following acute otitis media. Lin and Arcala (266) reported a case of *Fusobacterium necrophorum* septicemia complicating acute otitis media and mastoiditis in a 3-month-old girl. The discharge from the ear was fetid. Fass *et al.* (266a) recovered a *Peptococcus* in pure culture from a case of acute mastoiditis.

It is clear that the majority of acute cases that are reported are those with severe or unusual complications. Since anaerobes are so commonly involved in chronic otitis media and mastoiditis, it seems likely that they may be more frequently involved in acute processes than we have recognized. We have seen one young girl with acute purulent otitis media and mastoiditis complicated by bacteremia due to an unidentified, anaerobic, gram-negative bacillus, probably *Fusobacterium necrophorum*.

Chronic otitis media and mastoiditis are commonly due to anaerobic bacteria. Rist, in 1901 (256), stated that chronic otitis media was always fetid and that the pus contained numerous anaerobes, with few aerobes. He pointed out that the anaerobes often appeared early—as early as 10 days following the beginning of an acute otitis media. He indicated that the anaerobes that were commonly involved included *Clostridium perfringens*, *Bacillus ramosus*, *Micrococcus foetidus*, and *Staphylococcus parvulus*. Aerobes which might be encountered, according to Rist, included *Streptococcus pyogenes*, *Staphylococcus tenuis*, *Staphylococcus aureus*, diphtheroids, and *Proteus vulgaris*. Rist also indicated that fetid mastoiditis was much more common and serious than the nonfetid variety. Anaerobes most commonly found in this type of mastoiditis, according to Rist, were *Bacillus ramosus*, *Bacteroides serpens*, *Clostridium perfringens*, *Bacteroides funduliformis*, *Spirillum nigrum*, *Micrococcus foetidus*, and *Staphylococcus parvulus*. A number of specific cases of chronic otitis media and chronic mastoiditis, or mastoiditis secondary to fetid otitis, are cited in the report by Rist (8) published in 1898. Anaerobes recovered from these cases, in addition to those mentioned above, included unidentified anaerobic cocci, an anaerobic gram-positive bacillus, an anaerobic *Streptobacillus*, *Bacillus radiiformis*, and *Bacillus thetoides*.

In 1899, Guillemot (9) reported eight cases of pulmonary gangrene secondary to otitis media, in at least seven of which the otitis was chronic. Some of these had previously been reported by Rist. Organisms recovered included *Bacteroides fragilis*, *B. funduliformis*, fusiform bacilli, *Bacillus thetoides*, *Bacteroides serpens*, *Bacillus ramosus*, *Clostridium perfringens*, a spirillum, *Micrococcus foetidus*, a black colony-forming anaerobic coccus, and others. Aerobes were recovered from most of these cases, but there was usually only one aerobe per case and they were greatly outnumbered by the anaerobes present. Pilot and Pearlman (267), found fusiform bacilli and spirochetes in 15 patients with chronic otitis media—with the greatest numbers of anaerobes in the most fetid secretions. These organisms were not found in 3 additional patients with foul discharges and in 12 patients with chronic otitis media whose discharges were nonfetid. Brisottq (268) recovered anaerobes from 10 of 25 cases of chronic fetid otorrhea. Included were *Spirillum nigrum* (isolated three times), *Bacillus ramosus* [4], *Micrococcus foetidus* [2], and *Bacillus radiiformis* [1]. At least one of these cases also had aerobic organisms present. He also studied four cases of mastoiditis secondary to fetid otitis and found that all had anaerobes without aerobes; organisms recovered included *Bacillus ramosus*, *Spirillum nigrum*, and *Micrococcus foetidus*. This author also had three cases of chronic otitis media with endocranial complications, all involving anaerobes (see Chapter 7). Busacca (269) was impressed with the role of fusiform bacilli and spirochetes in 10 cases of chronic otitis media.

The incidence of anaerobes is lower in certain other published series. For example, Palva and Hallstrom (270), in 1965, cultured 100 patients with chronic otitis media and recovered anaerobic organisms (streptococci) from only 5. However, 20 cultures were either negative or were not obtainable due to dry ears and at least 10 of the positives yielded only organisms that would ordinarily be regarded as contaminants. Palva *et al.* (271) reported on 480 cases of chronic otitis media from which anaerobes were recovered only 10 times (3 micrococci, 6 streptococci, and 1 *Bacteroides*); 83 cultures yielded only contaminants. There were 40 negative cultures and 39 ears that could not be cultured because of lack of discharge. Baba *et al.* (272) found anaerobes in 2 of 9 cases of chronic otitis media. One of these was a *Peptococcus* isolated together with *Staphylococcus epidermidis* and the other was *Corynebacterium acnes* isolated in pure culture. Mitchell (110a) cultured 422 patients with "infected ears" and recovered *Bacteroides* on 26 occasions. Eighteen of these cases gave a history of chronic otitis media, many with foul-smelling drainage.

There are several reports in which anaerobes were recovered from small numbers of patients with chronic otitis media, with or without mastoiditis. Dietrich *et al.* (135) recovered *Bacteroides* in pure culture from three cases of otitis, anaerobic staphylococci in pure culture from two cases of otitis, three to four anaerobes without any aerobes from another case of chronic otitis, and an aerobe plus an anaerobe from still another. These authors also found mixtures of various anaerobes without any aerobic organisms present in two cases of mastoiditis. Krumwiede and Pratt (273) recovered fusiform bacilli from three cases of otitis, presumably chronic. Hansen (274), recovered *Bacteroides funduliformis* plus anaerobic streptococci, with no aerobes, from three cases of chronic otitis media. Bornstein *et al.* (275) recovered *Bacteroides* from a case of chronic otitis media complicated by osteomyelitis of the temporal bone. They also reported anaerobes recovered from two cases of chronic otitis media that were complicated by cerebellar abscess. The report of Morin and Potvin (145) concerned two cases of otitis media of uncertain age. One yielded only anaerobes (*Fusobacterium*, anaerobic streptococci, and *Veillonella discoides*). The second case yielded fusiform bacilli as well as staphylococci of unspecified type.

A very interesting report by Black and Atkins (276) described two cases of tetanus originating from chronic otitis media. At least one of these patients also had mastoiditis. *Clostridium tetani* was cultured from the foul discharge from the ears of both patients. Smith and Ropes (226) recovered *Bacteroides* from 2 patients with chronic otitis media. Boez *et al.* (277) reported two cases of *Bacteroides fragilis* septicemia in patients with chronic otitis media complicated by acute mastoiditis. In one of these patients enterococci were also recovered from the bloodstream, and the patient had suppurative pulmonary infarcts. The second patient had gangrenous changes of the auricle; and *Bacillus ramosus* and α-hemolytic streptococci were recovered from this site. The same authors (166) reported a case of mastoiditis, lateral sinus thrombophlebitis, and bacteremia due to *Bacillus ramosus* in a patient with underlying chronic otitis media. De Vos *et al.* (278) reported a case in a 14-year-old boy of *Bacteroides fragilis* septicemia with shock secondary to chronic otitis media with cholesteatoma. The boy also had bronchopneumonia with abscess formation. Two cases of actinomycosis of the tympanomastoid were reported by Leek (278a).

There are several additional reports of single cases of chronic middle ear disease or mastoiditis involving anaerobic bacteria. Rist and Ribadau-Dumas (279) reported a case of chronic otitis media complicated by thoracic empyema from which they recovered *Bacillus thetoides*, *Bacteroides serpens* and *Clostridium perfringens*. Plaut (280) noted numerous spirochetes and fusiforms, as well as occasional gram-positive cocci and gram-negative rods, from a case of chronic otitis media. Culture yielded anaerobic streptococci plus a probable *Proteus*. Shevky *et al.* (281) recovered *Bacteroides melaninogenicus* from a case of mastoiditis, type not specified. Beigelman and Rantz (282) recovered *Bacteroides*, hemolytic streptococci (aerobic), *Staphylococcus albus*, and diphtheroids from a case of chronic otitis media. Beerens and Tahon-Castel (115) recovered *Clostridium perfringens* plus *Klebsiella aerogenes*, *Streptococcus faecalis*, and *Staphylococcus aureus* from a case of chronic otitis media. We have recovered *Bacteroides fragilis* and viridans streptococci from a case of chronic otitis media and mastoiditis. We saw two additional cases with chronic otitis media with secondary central nervous system infection. One of these patients had infection due to *Bacteroides fragilis, Sphaerophorus*, enterococcus, and *Staphylococcus aureus*. He experienced an exacerbation of chronic otitis media with foul drainage and was found subsequently to have an extensive epidural abscess. The second patient's infection was due to *Bacteroides fragilis, Escherichia coli*, and *Staphylococcus aureus*. His involvement included osteomyelitis of the petrous bone and meningitis.

The seriousness of chronic mastoiditis is underlined by the potential complications, outlined in Fig. 6.1.

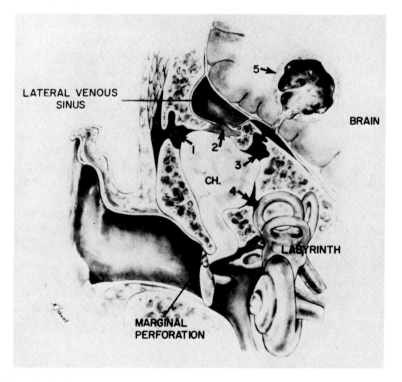

Fig. 6.1.   Cholesteatoma (CH.) may erode the bony walls of the mastoid in several locations. (1) Subperiosteal abscess; (2) perisinus abscess which may lead to septic thrombosis of the lateral venous sinus and jugular vein; (3) epidural abscess which may penetrate the dura to produce subdural abscess or meningitis; (4) labyrinthitis results from erosion of the horizontal semicircular canal; (5) brain abscess may occur in the cerebellum if the posterior wall breaks down or in the temporal lobe if the infection extends through the roof of the mastoid. [From Miglets and Harrington (1196) reproduced with permission].

Sinusitis

As in the case of otitis, sinusitis is a not uncommon underlying source of central nervous system infection. Cases of brain abscess, meningitis, and subdural and extradural empyema involving anaerobes originating in infections of the sinus are noted in Chapter 7. These cases will not be summarized again here. Table 6.1 lists a number of cases of anaerobic infection of the paranasal sinuses collected from the literature. Cases with central nervous system complications are specifically omitted (see Chapter 7).

Clearly, anaerobic infection of the sinuses is not a rare event. Anaerobes of all types may be involved and not infrequently are present in the absence of aerobic or facultative forms. Baup and Stanculeanu [cited by Jungano and Distaso (283)] frequently found anaerobes such as *Bacillus ramosus* and *Bacteroides fragilis* in sinusitis of dental origin. Babes (200) cited Vincent as finding fusospirochetal organisms in 11 of 19 specimens obtained from sinuses and the appendix (number of each is not specified). In 1949, Urdal and Berdal (284) reported on a study of 81 patients, 45 of whom had acute sinusitis and 36 chronic. A total of 8 yielded anaerobes (see Table 6.1). Twenty cases yielded no growth, although 3 of these showed organisms on smear. Urdal and Berdal also quote Sparrevohn and Buch, who found that 34% of their cultures were sterile. Fredette *et al.* (285) studied maxillary sinus contents from patients with chronic sinusitis. The material was obtained by having the patient blow the contents of each nostril into a petri dish. Fifty-seven patients were studied, and 47 of these were cultured anaerobically. Anaerobic bacteria were recovered from 18, none in pure culture. Eleven of the anaerobes were gram-positive rods, and one would wonder whether some of these might not be *Corynebacterium acnes* which is common in the normal flora of the anterior nose. There were five anaerobic diplococci and one anaerobic *Neisseria* recovered, and one patient yielded anaerobic gram-negative cocci and anaerobic rods. Bornstein *et al.* (275) noted that in a period of $1\frac{1}{2}$ years, they isolated *Bacteroides* 30 times from cases of chronic sinusitis and orbital cellulitis.

Baba (265, 286) reported on a large study of chronic sinusitis with specimens obtained from the maxillary sinus by puncture via the nose. There were a total of 388 specimens obtained from 208 patients; 55.4% of specimens were sterile, 31.4% yielded aerobes only, 5.4% anaerobes only, and 10.3% mixtures of aerobes and anaerobes. The greater the volume of fluid obtained and the more purulent the fluid obtained, the higher the incidence of anaerobic bacteria. In all, 64 strains of anaerobes were isolated from 61 specimens representing 43 cases; 27 strains were *Peptostreptococcus*, 24 *Peptococcus*, 7 *Bacteroides*, 2 *Veillonella*, 2 anaerobic corynebacteria, and 2 clostridia. Pure cultures of *Bacteroides* were obtained from 6 cases of sinus infection

TABLE 6.1

Anaerobic Infection of Paranasal Sinuses

Sinus(es) involved	Acute or chronic infection	Anaerobes recovered	Aerobes recovered	Comments	Reference	Year
?	?	*B. melaninogenicus*	?		Burdon (799)	1927
?	Acute	*C. perfringens*	?	Gas-containing phlegmon, face; fatal	Hausmann (1135)	1928
Ethmoid	Chronic	*Clostridium*	*Staphylococcus aureus*	Pseudomembranous rhinitis; fatal		
Ethmoid	Chronic	"Anaerobic bacilli"	*Staphylococcus*			
Maxillary	?	Fusiform bacilli and spirochetes	?	Fusiform bacillus bacteremia	Larson and Barron cited by Smith (290)	1932
Ethmoid	?	Fusospirochetes	?	2 cases	Pilot and Lederer cited by Smith (290)	1932
Ethmoid	?	Fusospirochetes	?		Kompanejetz cited by Smith (290)	1932
Ethmoid	?	Fusospirochetes	?	3 cases; all eroded into orbit	Brandt cited by Smith (290)	1932
Maxillary	?	Fusospirochetes	?		Silberschmidt cited by Smith (290)	1932
Maxillary	?	Fusospirochetes	?		Jay cited by Smith (290)	1932
Maxillary	?	Fusospirochetes	?		Broughton-Alcock cited by Smith (290)	1932
Maxillary	?	Fusospirochetes	?		Lichwitz and Sabrazes cited by Smith (290)	1932
Maxillary, frontal	Chronic	Anaerobic streptococci	0	Complicating osteomyelitis, frontal, and nasal bones	Williams and Nichols (291)	1943
Maxillary	Chronic	*Bacteroides necrophorum*	0	Antral irrigation, foul pus	Greenblatt and Greenblatt (1136)	1945
?Maxillary	Acute	*Sphaerophorus varius*	0	Of dental origin	Prévot et al. (1137)	1947
Maxillary	Acute	*Sphaerophorus varius*	*Streptococcus*	Following dental extraction, "horribly fetid" pus, cellulitis, abscess face	Ingelrans (308)	1947

Sinus	Chronicity	Organisms		Complications	Reference	Year
Maxillary	Chronic	*Bacteroides*	0		Beigelman and Rantz (282)	1949
Maxillary	Acute	Anaerobic streptococcus	0	Fetid secretions	Urdal and Berdal (284)	1949
Maxillary	Chronic	Anaerobic GNR[a]	0	Fetid secretions		
Maxillary	Chronic	Anaerobic GNR	0	Fetid secretions		
Maxillary	Chronic	Anaerobic GNR, anaerobic streptococcus	0	Fetid secretions		
Maxillary	Chronic	Anaerobic GNR, anaerobic streptococcus	0	Fetid secretions		
Maxillary	Chronic	Anaerobic GNR, anaerobic streptococcus	Nonhemolytic streptococcus	Fetid secretions		
Maxillary	Chronic	Anaerobic GNR	Pneumococcus	Fetid secretions		
Maxillary	Chronic	Anaerobic GNR	?	Fetid secretions		
?	?	15/17 fetid sinus washings yielded anaerobes	?	Fetid secretions	Björkwall [cited by Frederick and Braude (1110)]	1950
?	?	*Ramibacterium pleuriticum*	?		Beerens (332)	1953–1954
Frontal	?	Anaerobic streptococcus	0?		Thomas and Hare (811)	1954
Maxillary	?	*Bacteroides*, *Veillonella parvula*	?		Lodenkämper and Stienen (287)	1955
Maxillary	Chronic	*Actinomyces*	0	Spread to orbit, fatal	Harvey et al. (212)	1957
?	?	*Diplococcus constellatus*, *Veillonella alcalescens*, *Eubacterium tortuosum*, *Corynebacterium liquefaciens*	Nonhemolytic streptococcus		Beerens and Castel (845)	1958–1959
?	?	*Veillonella alcalescens*, *Ristella melaninogenica*, *Fusiformis fusiformis*	0			
Maxillary, ethmoid, frontal	Acute	*Actinomyces*	0	Oro-antral and antrofacial fistulae; destruction of floor of orbit; followed tooth extraction	Nathan et al. (1138)	1962

(continued)

123

TABLE 6.1 (continued)

Sinus(es) involved	Acute or chronic infection	Anaerobes recovered	Aerobes recovered	Comments	Reference	Year
Maxillary	Acute	Anaerobic streptococcus	0	Probably of dental origin	Ameriso and Caffarena (1139)	1963
?	?	*Veillonella alcalescens*	0		Rentsch (195)	1963
?	?	*Actinomyces naeslundii*	?		Howell, cited by Georg et al. (1140)	1964
?	?	*Bacteroides*	0	Three cases	Dietrich et al. (135)	1965
?	?	Anaerobic gram-negative cocci	0			
?	?	*Bacteroides*, anaerobic or microaerophilic *Corynebacterium*	0			
?	?	Anaerobic streptococci and anaerobic or microaerophilic *Corynebacterium*	0			
?	?	Mixture of at least 3 anaerobes, types not specified	0			
?	?	Mixed anaerobic-aerobic infection, organisms not specified		Two cases		
Maxillary	?	*Eubacterium tortuosum, Streptococcus evolutus*	0	Foul discharge	Beerens and Tahon-Castel (115)	1965
Pansinusitis Frontal, ethmoid	Chronic	*Streptococcus anaerobius*	0			
?	?	*Streptococcus evolutus*	0			

Sinus	Type	Organisms		Comments	Reference	Year
Maxillary	?	*Eubacterium tortuosum, Diplococcus constellatus, Corynebacterium liquefaciens, Veillonella alcalescens*	0		Beerens and Tahon-Castel (115) (continued)	1965
Maxillary	?	Anaerobic GNR, *Eubacterium lentum*	0	Dental origin		
Maxillary	?	*Eubacterium lentum, Eubacterium tortuosum, Streptococcus lanceolatus*	0			
Maxillary	?	*Dialister, Eubacterium lentum*	0			
Maxillary	?	*Fusiformis fusiformis*	*Staphylococcus aureus*			
Frontal, ethmoid	Chronic	Anaerobic diphtheroids, 2 different occasions	0	Smear showed GPR[a]	Goldsand and Braude (244)	1966
Frontal, ethmoid, sphenoid	Acute	*Bacteroides fragilis, Veillonella*	0			
Maxillary	Chronic	Anaerobic streptococci	Rare colonies *S. albus*, viridans streptococcus			
Pansinusitis	Acute?	*Bacteroides fundiliformis*	0		Himalstein (1521)	1967
Maxillary	?	Anaerobic streptococci, 2 patients	0	1 moderate growth, 1 heavy growth	Gullers *et al.* (288)	1969
Maxillary	Chronic	*Actinomyces israelii*	*Haemophilus* 0?	Antral lavage, foul pus	Stanton (1141)	1969
Maxillary 6 cases	2 chronic, 4 acute	*Actinomyces*		From literature review, 3 cases followed dental extraction		
?	?	Penicillin-sensitive *Bacteroides* (blood culture)	?		Bodner *et al.* (460)	1970
Sphenoid	Acute	*C. perfringens*	Streptococci	Gas gangrene sepsis, fatal	Gujer *et al.* (1142)	1970
Pansinusitis	?	*B. fragilis* from blood culture	?	Leukemia, bacteremia	Felner and Dowell (168)	1971

(continued)

125

TABLE 6.1 (*continued*)

Sinus(es) involved	Acute or chronic infection	Anaerobes recovered	Aerobes recovered	Comments	Reference	Year
Maxillary	Acute	*F. fusiforme* from blood culture	?	Bacteremia, lateral sinus thrombosis	Felner and Dowell (168) (continued)	1971
Frontal	Acute	*Bacteroides*	0	Foul odor	Baba *et al.* (272)	1972
Maxillary	Acute	Anaerobic gram-negative rod, 9 cases	4/9 had aerobes, type not specified	These 14 anaerobic infections were among 243 sinuses cultured from 171 patients; 22% of cultures had no growth and 9% *Staphylococcus albus*	Axelsson and Brorson (1142a)	1972
Maxillary	Acute	Anaerobic streptococci, 4 cases	3/4 were mixed, other organisms not specified			
Maxillary	Acute	Anaerobic diphtheroids (1 case)	0			
Maxillary	Acute?	*Peptostreptococcus*	0	Culture was from associated orbital abscess	Oishi *et al.* (235a)	1973
Maxillary and ethmoid	Acute?	α-Hemolytic streptococcus (?microaerophilic); fusiforms and pleomorphic GPR seen on smear	0	Smears and cultures from associated orbital abscess	Sevel *et al.* (248a)	1973
Sphenoid	Chronic	*Actinomyces israelii*	0?	In 28 patients, heavy growth was obtained which correlated well with gram stain. In this group, 23 gave a pure, heavy growth of anaerobes and only 3 a pure heavy growth of aerobes (*H. influenzae,*	Per-Lee *et al.* (1142b)	1974
Frontal, ethmoid and maxillary	Chronic	43/83 sinuses yielded anaerobes (anaerobic streptococci from 28 specimens, corynebacteria from 19, *Bacteroides* from 19, and *Veillonella* from 14)	36/83 sinuses yielded aerobes; included were *Staphylococcus aureus*, *S. epidermidis*, viridans streptococci, *Haemophilus influenzae*, *Neisseria*, pneumococcus, *Escherichia coli* and miscellaneous others	*influenzae,*	Frederick and Braude (111)	1974

126

Site	Type	Anaerobes/Organisms	Aerobes	Comments	Reference	Year
				pneumococcus, and *Alcaligenes faecalis*). Anaerobes recovered (number of strains) included anaerobic streptococci (16). *Bacteroides funduliformis* (4), *B. fragilis* (4), *Veillonella* (3), and *B. melaninogenicum* (1); in 21 specimens, cultures were sterile; approximately half of these showed bacteria on gram stain	Frederick and Braude (111) (continued)	1974
				Total of 24 adults; 17 with acute sinusitis, 3 with subacute and 4 with chronic	Evans et al. (1142c)	1975
Maxillary	Acute?	*Fusobacterium necrophorum* ($>10^8$/ml)	0			
Maxillary	?	*Bacteroides ruminicola* (6×10^3/ml)	0			
Maxillary	Subacute	*Bacteroides ruminicola* plus two other anaerobes ($>10^8$/ml)	Three aerobes ($>10^8$/ml)			
Maxillary	Chronic	Aerotolerant *Streptococcus intermedius* ($>10^7$/ml)	0			
Maxillary	Chronic	Anaerobic *S. intermedius* ($>10^4$/ml)	0	< 1000 wbc/mm^3 in sinus secretion		

[a] GNR, gram-negative rod; GPR, gram-positive rod.

127

by Lodenkämper and Stienen (287); no other details were given. Gullers *et al.* (288) found anaerobic streptococci in two of 17 specimens from patients with maxillary sinusitis. One other specimen showed no growth. These authors noted furthermore that whereas doxycycline orally gives adequate concentrations in the sinuses, 7 of the 17 patients treated with oral penicillin V (1.6 gm per day) had unmeasureable levels (less than 0.2 μg/ml) in their sinus secretions.

Felner and Dowell (289) felt that sinusitis was the underlying cause of *Bacteroides fragilis* endocarditis in one patient in their series; however, this patient had a periodontal abscess as well. One of our patients with carcinoma of the nasopharynx extending to the maxillary sinus developed purulent sinusitis due to *Bacteroides fragilis*, anaerobic streptococci, and anaerobic diphtheroids. We have seen two additional cases of sinusitis involving anaerobes, with extension to nearby bones to produce osteomyelitis. The first of these patients had had surgery on his frontal sinus, following which he developed a supraorbital fistula with foul drainage and osteomyelitis of the frontal bone. The organisms involved were *Bacteroides fragilis*, *Fusobacterium*, *Bacteroides melaninogenicus*, *Staphylococcus aureus*, Providence, and α-hemolytic streptococci. The second patient was a diabetic who had chronic maxillary sinusitis with involvement of the bone of the maxilla itself, as well as the wall of the orbit. Organisms involved in this patient were *Bacteroides melaninogenicus*, *Fusobacterium*, *Bifidobacterium*, *Pseudomonas aeruginosa*, Klebsiella–Enterobacter, *E. coli*, and α-hemolytic streptococci. We have also recovered from the foul antral pus of a patient with nasopharyngeal carcinoma and maxillary sinusitis an anaerobic streptococcus, a *Bacteroides*, and *Propionibacterium*.

Frederick and Braude (1110) pointed out that the relatively frequent involvement of anaerobes in chronic sinusitis is related to poor drainage and increased intrasinal pressure related to inflammation. Mucosal blood flow is compromised, leading to reduced intracavitary oxygen tension. Vasoconstricting nasal sprays would aggravate mucosal ischemia. Other predisposing factors include viscous secretions and impaired ciliary action. All of these factors lead to lower partial pressure of oxygen and pH and thus an optimum oxidation–reduction potential for anaerobes.

Nasal Infections

Nasal infections are uncommon and usually represent extension of infection from an adjacent site (such as the mouth, throat, or paranasal sinuses) or occur following surgery or trauma. Carnot and Blamoutier (202) described fusospirochetal rhinitis secondary to processes in the mouth and throat.

Smith (290) reviewed a few cases of nasal noma or gangrene from the literature. A case of osteomyelitis of the nasal bones secondary to sinusitis and
due to anaerobic streptococci was reported by Williams and Nichols (291)
in 1943. Schwabacher *et al.* (78) recovered *Bacteroides melaninogenicus*
along with *Staphylococcus aureus* and a diphtheroid from an abscess of the
nasal septum. Heinrich and Pulverer (116) recovered *Bacteroides melaninogenicus* from one case each of rhinitis and nasal abscess (no further details
given).

Von Gusnar and Globig (292) described an extensive infection due to the
Buday bacillus in a patient who sustained an injury resulting in fracture of
a nasal bone. The fracture site became infected, producing a small abscess
locally with subsequent spread to produce abscesses in the liver and along
the psoas muscle, as well as pulmonary gangrene and purulent arthritis of
the hip. Osteomyelitis of the nasal bones due to the Buday bacillus secondary
to an injury with a broken nose was described by Nathan (293) in 1931.
Three cases of delayed healing following submucous septal resections due
to Vincent's infection were described by Hollender (294) in 1929. All 3
patients had a fetid discharge and typical smears, and all ended up with a
septal perforation. Thomas *et al.* (295) described actinomycosis of the nasal
bones following implantation of Surgibone for atrophic rhinitis.

Tonsillar and Pharyngeal Infections

A problem with this type of infection is that anaerobes are normally
prevalent on the surface of the tonsil and the pharynx so that cultures taken
directly from these areas are difficult to interpret. Therefore, in order to
document the role of anaerobes in this type of infection, a table has been
constructed (Table 6.2) listing reports of anaerobes in a variety of serious
infections complicating tonsillar infection. In other words, the portal of
entry for the more serious infection was the tonsil. Material for culture in
these cases was obtained from various sources where there would not be
confusion with normal flora (blood culture, pus from thoracic empyema,
etc.). Again in this table, infections involving the central nervous system
have been omitted; these are included in tables in Chapter 7. There are
clearly many serious infections with the tonsils serving as the portal of
entry. It is interesting to note, however, that the vast majority of these
reports antedate the availability of antimicrobial agents. In the review of
the literature on *Bacteroides* septicemia, published by Gunn in 1956 (296),
representing 148 cases, nasopharyngeal infection was by far the most common site of the primary lesion in these cases of sepsis. There were 54 such
cases, with the second most common underlying primary lesion (traumatic

TABLE 6.2

Anaerobes in Infections Complicating Tonsillar Infection

Type of complication	Tonsillar infection	Anaerobes isolated[a]	Aerobes isolated	Comments	Reference	Year
Abscesses, lungs, and soft tissues	Exudative tonsillitis	? Anaerobic streptococcus	0	Thrombosis, jugular vein	Fraenkel (1143)	1925
Lung abscesses, purulent arthritis	Tonsillar abscess	Anaerobic streptococcus	0	Jugular vein thrombophlebitis		
Empyema, thoracic	Tonsillitis, peritonsillar abscess	Anaerobic streptococcus	0	Jugular vein thrombophlebitis		
Lung abscesses, localized meningitis	Tonsillar abscess	Anaerobic streptococcus (blood)	Hemolytic streptococcus	Jugular vein thrombophlebitis		
Lung abscesses, pleural empyema	Tonsillopalatine abscess	Anaerobic rods, no further identification	0			
Bacteremia, thoracic empyema, pyarthrosis	Tonsillitis	Bacillus anaerobius pyogenes	0		Hegler and Jacobsthal (1144)	1925
24 cases. Bacteremia, 18; empyema, 6; metastatic abscesses, few	Tonsillitis; 2/3 of cases had retrotonsillar abscesses	Streptococcus putrificus in pure culture 9 times, together with anaerobic GNR 2 times, and 1 with hemolytic streptococci; anaerobic GNR in pure culture 12 times	Hemolytic streptococcus, 1 case	32 cases in series altogether; 4 aerobic and 4 probably anaerobic (not proved); venous thrombosis common	Kissling (1145)	1929
Bacteremia, osteomyelitis sternum, presternal abscess	Tonsillitis	Bacillus fragilis	0		Lemierre et al. (1146)	1929
Bacteremia, retroperitoneal abscess, arthritis, lung abscesses	Tonsillitis, ? pharyngitis	Actinomyces necrophorus	0		Cunningham (323)	1930
Retropharyngeal abscess, pneumonia, empyema, pericarditis, abscesses neck	Tonsillitis	A. necrophorus, anaerobic streptococcus	Pneumococcus from neck only			

Clinical manifestations	Throat	Organism	No.	Reference	Year
Otitis, mastoiditis, endocranial and other abscesses, bacteremia	Tonsillitis	Gram-negative anaerobic rods	0	Franklin (482)	1933
Bacteremia, lung abscesses, peritonitis	Tonsillitis	*Bacillus funduliformis*, anaerobic streptococcus	0	Lemierre *et al.* (1147)	1933
Bacteremia, empyema	Acute tonsillitis	*Bacillus funduliformis*	0	Cathala *et al.* (1148, 1149)	1933
Purulent arthritis	Tonsillar phlegmon	*Bacillus fragilis*	0	Richon *et al.* (1150)	1934
Bacteremia, pleuropulmonary infection	Tonsillitis	*Bacillus funduliformis*	0	de Font-Réaulx (1151)	1935
Bacteremia, thoracic empyema, pyarthrosis	Exudative tonsillitis	*Bacillus funduliformis*	0	Lemierre (1152)	1935
Bacteremia, pleural empyema, lung abscess, soft tissue abscesses	Ulcerative exudative tonsillitis	*Bacillus funduliformis*	0	Lemierre and Meyer (1153)	1935
Bacteremia, pleurisy, arthralgia	Exudative tonsillitis	*Bacillus funduliformis*	0		
Bacteremia, thoracic empyema, mediastinitis	Ulcerative exudative tonsillitis	*Bacillus funduliformis*	0	Pham (261)	1935
Bacteremia, thoracic empyema, purulent arthritis, lung abscesses	Ulcerative tonsillitis	*Bacillus funduliformis*	0		
Bacteremia, thoracic empyema, pneumonia, purulent arthritis	Tonsillitis	*Bacillus funduliformis*	0		
Bacteremia, pneumonia, purulent pleurisy	Tonsillitis	*Bacillus funduliformis*	0		
Bacteremia, lung abscesses, empyema, purulent arthritis	Exudative tonsillitis	*Bacillus funduliformis*	0		

(continued)

131

TABLE 6.2 (*continued*)

Type of complication	Tonsillar infection	Anaerobes isolated[a]	Aerobes isolated	Comments	Reference	Year
Bacteremia, pneumonia, pyopneumothorax, "arthritis"	Tonsillitis, lateral pharyngeal abscess	*Bacillus funduliformis*	0		Pham (261) (continued)	1935
Bacteremia, lung abscesses, suppurative arthritis, psoas abscess	Exudative tonsillitis	*Bacillus funduliformis*	0			
Pyarthrosis several joints	Exudative tonsillitis	*Bacillus funduliformis*	0			
Pyarthrosis, lung abscesses	Tonsillitis	*Bacillus funduliformis*	0			
Lung and liver abscesses, suppurant arthritis, psoas abscess	Ulcerative exudative tonsillitis	*Bacillus funduliformis*	0			
Bacteremia, pulmonary abscesses	Tonsillitis	*Bacillus funduliformis*	0		Donzelot et al. (1154)	1936
Bacteremia, lung abscesses, pleurisy, arthritis	Tonsillitis	*Bacillus funduliformis*	0	Jugular vein thrombophlebitis		
Bacteremia, lung abscesses, empyema, pyarthrosis	Tonsillitis	*Bacillus funduliformis*	0		Grumbach et al. (262)	1936
Bacteremia, lung abscesses	Abscess	*Fusobacterium nucleatum*	0	Phlegmon both piriform sinuses and base of tongue	Grumbach and Verdan (350)	1936
Gaseous phlegmon face, neck and chest, pre-sternal abscess	Peritonsillar phlegmon	*Bacillus funduliformis*	Hemolytic streptococci		Lemierre et al. (1155)	1936
Bacteremia, lung abscesses, arthralgia, supraclavicular abscess	Ulcerative tonsillitis	*Bacillus funduliformis* (blood), anaerobic streptococcus (abscess)	0		Lemierre and Moreau (1156)	1936
Bacteremia, lung abscesses, empyema, suppurative arthritis	Exudative tonsillitis with microabscesses	*Bacillus funduliformis*	0		Lemierre et al. (1157)	1936

132

Bacteremia	Tonsillitis	*Bacillus pyogenes anaerobium*	?0		Weyrich (452)	1936
Bacteremia + misc. complications	Tonsillitis	*Streptococcus putrificus*	?0	12 cases		
Bacteremia + misc. complications	Tonsillitis	*Streptococcus putrificus* and anaerobic GNR[b]	?0	10 cases		
Bacteremia + misc. complications	Tonsillitis	Anaerobic GNR	?0	15 cases		
Bacteremia	Tonsillitis	*Bacillus fragilis*, anaerobic streptococcus	0		Ternois (1158)	1938
Bacteremia, lung abscess	Tonsillar phlegmon	*Bacillus funduliformis*	0		Lemierre (1159)	1939
Bacteremia	Exudative tonsillitis	*Bacillus funduliformis*	0			
Bacteremia, arthralgia	Tonsillitis	*Bacillus funduliformis*	0			
Bacteremia, pleural empyema	Tonsillitis	*Bacillus funduliformis*	0			
Bacteremia, pleurisy	Exudative tonsillitis	*Bacillus funduliformis*	0			
Bacteremia, subcutaneous abscesses	Exudative tonsillitis	*Bacillus funduliformis*	0			
Pleural empyema, lung and splenic abscesses, bacteremia	Retrotonsillar abscess	Buday bacillus	0		Hornung (1130)	1940
Bacteremia, pleurisy, arthralgia	Tonsillitis	*Bacillus funduliformis*	0		Lemierre *et al.* (1160)	1940
Bacteremia, lung abscess, arthralgia	Phlegmonous tonsillitis	*Bacillus funduliformis*	0			
Bacteremia, lung abscess	Phlegmonous tonsillitis	*Bacillus funduliformis*	0			
Bacteremia, thoracic empyema	Tonsillitis	*Bacillus funduliformis*	0			

(continued)

133

TABLE 6.2 (continued)

Type of complication	Tonsillar infection	Anaerobes isolated[a]	Aerobes isolated	Comments	Reference	Year
Bacteremia, pleurisy	Exudative tonsillitis	*Bacillus funduliformis*	0		Lemierre *et al.* (1160) (continued)	1940
Bacteremia, severe arthralgia	Tonsillitis	*Bacillus funduliformis*	0			
Bacteremia, purulent arthritis, buttock abscess	Tonsillitis	*Bacillus funduliformis*	0			
Bacteremia, buttock and thigh abscesses	Tonsillitis	*Bacillus funduliformis*	0			
Bacteremia	Tonsillitis	*Bacillus funduliformis*	0			
Bacteremia, pleurisy, arthralgia	Tonsillitis	*Bacillus funduliformis*	0			
Pleural empyema, purulent bursitis, subcutaneous abscesses	Ulcerative exudative tonsillitis	*Bacterium necrophorum*	0		Martin (532)	1940
Bacteremia, lung abscesses, empyema	Tonsillitis	*Bacillus funduliformis*	0		Naville *et al.* (1161)	1940
Purulent arthritis, buttock abscess	Tonsillitis	*Bacillus funduliformis*	0		Siegler (1162)	1940
Sternoclavicular abscess, ?lung abscess	Phlegmonous tonsillitis	*Bacillus funduliformis*	0			
Bacteremia, empyema, pneumonia, ?liver abscess	Tonsillitis with necrotic ulcer	*Bacteroides funduliformis*	0	Thrombophlebitis jugular vein with adjacent cellulitis	Brown *et al.* (959)	1941
Bacteremia	Exudative tonsillitis, abscess	*Bacillus funduliformis*	0		Delbove and Reynes (351)	1941
Bacteremia, pleurisy	Tonsillar abscess	*Bacillus funduliformis*	0		Ramadier and Mollaret (1163)	1941

Bacteremia, pyopneumothorax, buttock abscess	Tonsillitis	Bacillus fundiliformis	Streptococcus		Ravault et al. (1164)	1941
Thoracic empyema, soft tissue abscesses	Tonsillar abscess	Bacillus fundiliformis	0		Kemkes (1165)	1943
Bacteremia	Tonsillar abscess	Bacteroides	0		Smith (1166)	1943
Bacteremia, pneumonia, pleuritis	Tonsillitis	Bacteroides fundiliformis	0		Waring et al. (1167)	1943
Gas-containing abscess, neck; pleurisy	Tonsillar phlegmon	Bacillus fundiliformis	0		Aussannaire (1168)	1943
Bacteremia, pleurisy	Tonsillitis	Bacillus fundiliformis	0		Smit (1169)	1944
Empyema, thoracic; bacteremia	Tonsillitis	Bacillus fundiliformis (?) (from blood and empyema)	0			
Abscess, neck; lung abscesses	Tonsillitis	? Bacillus fundiliformis	Foul pus, viridans streptococcus 0		Rüedi (1170)	1944
Purulent thrombophlebitis jugular and retrotonsillar veins	Tonsillitis	Bacillus fundiliformis				
Soft tissue phlegmons	Peritonsillar abscess	Bacillus fundiliformis	0			
Bacteremia, pleuropulmonary infection	Tonsillitis, abscess	Bacillus fundiliformis	0	Jugular vein thrombophlebitis	Desbuquois and Iselin (1171)	1945
Bacteremia, pyarthrosis	Exudative tonsillitis, abscess	Bacteroides fundiliformis	0		Smith and Ropes (226)	1945
Bacteremia	Exudative tonsillitis	Bacillus fundiliformis	0		Lemierre et al. (1172)	1945
Bacteremia, prevesical abscess	Exudative tonsillitis	Bacillus fundiliformis	0			
Bacteremia, abscess of the neck	Exudative tonsillitis	Bacillus fundiliformis	0			

(continued)

135

TABLE 6.2 (*continued*)

Type of complication	Tonsillar infection	Anaerobes isolated[a]	Aerobes isolated	Comments	Reference	Year
Bacteremia, pulmonary infection	Tonsillitis	*Bacteroides gonidiaformans*	0		Reid *et al.* (1173)	1945
Septic arthritis, neck abscess, ?retropharyngeal abscess	Tonsillitis, probable abscess	*Bacteroides funduliformis*	0			
Bacteremia, pneumonia	Probable tonsillitis	*Bacteroides funduliformis*	0			
Bacteremia	Tonsillar phlegmon	*Bacillus funduliformis*	0		Lemierre *et al.* (1174)	1946
Bacteremia	Tonsillar phlegmon	*Bacillus funduliformis*	0			
Bacteremia	Tonsillar phlegmon	*Bacillus funduliformis*	0			
Voluminous phlegmon of neck	Tonsillar phlegmon	*Bacillus funduliformis*	0			
Bacteremia	Tonsillar phlegmon	*Bacillus funduliformis*	0			
Bacteremia, pleural empyema	Tonsillitis	*Bacteroides funduliformis*	0		Ruys (1175)	1947
Bacteremia	Peritonsillar abscess	*Bacteroides funduliformis*	0			
Bacteremia, pleurisy	Tonsillitis	*Sphaerophorus pseudonecrophorus*	0		Tardieux and Nabonne (1176)	1949
Bacteremia, thrombophlebitis	Tonsillitis	*Bacillus funduliformis*	Hemolytic streptococcus		Hartl (249)	1950
Sepsis, pleuropulmonary infection	Tonsillitis	*Bacillus funduliformis*, anaerobic cocci	0		Ernst (347)	1951

Bacteremia	Tonsillar abscess	*Bacillus funduliformis*	0	Thrombosis jugular vein	Ernst and Hartl (1177)	1951
Bacteremia, lung abscess, arthralgia, subcutaneous abscess	Exudative tonsillitis	*Bacillus funduliformis*	0		Le Sueur (1178)	1951
Bacteremia	Tonsillectomy	*Bacillus pyogenes anaerobium*	0		Willich (961)	1951
Abscess chest wall, pyarthroses, liver infection	Tonsillar abscesses	*Bacteroides funduliformis*, anaerobic streptococci	0		Fisher and McKusick (264)	1952
Bacteremia, lung abscesses	Probable tonsillitis	*Bacteroides funduliformis*	0		McVay and Sprunt (621)	1952
Purulent pleurisy	Acute tonsillitis	*Bacillus funduliformis*	0		Brocard *et al.* (463)	1954
Thoracic empyema	Membranous tonsillitis	*Bacillus necrophorus*	0		Alston (302)	1955
Bacteremia	Tonsillitis	*Bacillus funduliformis*	?		Prévot (21)	1955
Bacteremia	Exudative tonsillitis	*Bacteroides*	0	Hemolytic streptococcus	Tynes and Frommeyer (489)	1962
Bacteremia, shock, jaundice	Pharyngitis, ?tonsillitis	*Bacteroides funduliformis*	0			
Bacteremia, arthralgia, jaundice	Tonsillitis	*Bacteroides funduliformis*	0			
Bacteremia, osteomyelitis, jaundice	?Tonsillitis	*Bacteroides fragilis*	0	S-C hemoglobin		
Osteomyelitis, mandible	Infected tonsillar carcinoma	*Bacteroides*	0		Bornstein *et al.* (275)	1964
Bacteremia (?), pericarditis	Peritonsillar phlegmon	*Fusobacterium gonidiaformans*	0	Complication of tonsillectomy	Rubenstein *et al.* (1178a)	1974

a From site of complicating infection.
b GNR, gram-negative rod.

wounds) accounting for only 24 cases. Gunn pointed out that tonsillitis or tonsillar or peritonsillar abscess was particularly common. This has not been true in most recent series of anaerobic bacteremia, and localized tonsillar infection due to anaerobes is not commonly recognized at the present time. The advent of the antimicrobial era, then, has made a tremendous impact on this type of infection. While local tonsillar infection undoubtedly still occurs, it is seldom recognized as being due to anaerobes, and the use of antimicrobial therapy has resulted in the prompt response of the infection without later development of the serious complications that were so notable in the past.

Some of the earliest medical writers, such as Hippocrates, Aretaeus, and Celsus, described diseases of the tonsils and throat which must have been what subsequently became known as fusospirochetal angina or Vincent's (297) or Plaut-Vincent's angina or anaerobic tonsillitis. The description by Aretaeus (298) is vivid·

Ulcers occur on the tonsils; some, indeed of an ordinary nature, mild and innocuous; but others of an unusual kind, pestilential and fatal. Such as are clean, small, superficial, without inflammation and without pain, are mild; but such as are broad, hollow, foul, and covered with a white, livid, or black concretion, are pestilential. Aphtha is the name given to those ulcers. But if the concretion has depth it is an Eschar and is so called: but around the eschar there is formed a great redness, inflammation, and pain of the veins, as in carbuncle; and small pustules form, at first few in number, but others coming out, they coalesce, and a broad ulcer is produced. And if the disease spread outwardly to the mouth, and reach the columella (uvula) and divide it asunder, and if it extend to the tongue, the gums, and the alveoli, the teeth also become loosened and black; and the inflammation seizes the neck; and these die within a few days from the inflammation, fever, foetid smell, and want of food. But, if it spread to the thorax by the windpipe, it occasions death by suffocation within the space of a day. For the lungs and heart can neither endure such smells, nor ulcerations, nor ichorous discharges, but coughs and dyspnoea supervene. The manner of death is most piteous; pain sharp and hot as from carbuncle; respiration bad, for their breath smells strongly of putrefaction, as they constantly inhale the same again into their chest; they are in so loathesome a state that they cannot endure the smell of themselves; countenance pale or livid; fever acute, thirst as if from fire, and yet they do not desire drink for fear of the pains it would occasion; for they become sick if it compress the tonsils, or if it return by the nostrils; and if they lie down they rise up again as not being able to endure the recumbent position, and if they rise up, they are forced in their distress to lie down again; they mostly walk about erect, for in their inability to obtain relief they flee from rest, as if wishing to dispel one pain by another. Inspiration large, as desiring cold air for the purpose of refrigeration, but expiration small, for the ulceration, as if produced by burning, is inflamed by the heat of the respiration. Hoarseness, loss of speech supervene; and these symptoms hurry on from bad to worse, until suddenly falling to the ground they expire.

The emphasis on ulceration and foul discharges clearly point to this disease rather than to diphtheria. A good clinical description of this disease was presented by Vincent in 1898 (297). He indicated that this diphtheria-like angina is found on one or both tonsils and sometimes on the neighboring tonsillar pillar. At the beginning of the infection, the tonsil is covered with a thin white or gray film that can be detached to leave a surface that bleeds rather easily. This irregular area increases gradually to cover a larger area and forms a stronger and more adherent membrane. There may be a super-ficial ulcer underneath the membrane. By the third or fourth day, the pseudo-membrane is thick and caseous in appearance and contributes a foul smell to the breath. It can now be detached more easily. The mucosa nearby is edematous and erythematous. The cases that Vincent saw were all relatively mild and self-limited, and he did not note suppuration locally, nor metastatic infection. There was relatively little involvement of neighboring lymph glands in the cases that he described. Finally, Vincent pointed out the presence of characteristic organisms on smear of the exudate or membrane—a long (10 to 12 μm) bacillus (*Bacillus fusiformis*) with tapered ends and a spirillum or spirochete that did not stain as well. He was unable to cultivate the organisms.

Other workers have subsequently shown that with anaerobic tonsillitis one may often have enlarged submandibular lymph node enlargement, and there may be significant periadenitis, edema, and even trismus. It is also clear from Table 6.2 that serious complications of this disease may occur and that many different anaerobic bacteria may be involved, with *Bacteroides funduliformis* (*Fusobacterium necrophorum*) one of the major causes. With better anaerobic techniques, it is likely that *Bacteroides melaninogenicus* would be found to be more commonly involved in this condition than has been documented in the past. Although most infections of this type are endogenous in origin, infection may be exogenous as well. Reports cited by Smith (290) showed that there may be transmission from person to person through intimate contact, such as kissing or even by means of common towels or toothbrushes.

The differential diagnosis would include diphtheria, hemolytic strepto-coccal infection, nonstreptococcal exudative pharyngitis (viral), and in-fectious mononucleosis. The most unique features of anaerobic tonsillitis or tonsillopharyngitis are the fetid or foul odor and the presence of fusiform bacilli, spirochetes, and other organisms which have the unique morphology of anaerobes on direct smear of the membrane. It must be remembered that anaerobic tonsillopharyngitis may coexist with diphtheria, hemolytic strep-tococcal infection, etc.

Haymann (299) noted that in a large series (size not given) of patients with tonsillogenous sepsis, only 5% were due to anaerobic bacilli and 2% to *Streptococcus putridus*. Christ (300) described a case of *Sphaerophorus pseudonecrophorus* sepsis following pharyngitis. Ware (301) described a very severe case of so-called Vincent's angina in which early there was inflammation of the pharynx and a small ulcer on one tonsil. The ulcer increased in size and eventually led to extensive destruction of the right side of the soft palate and profuse hemorrhage. A typical odor and typical organisms on smear were noted. External carotid ligation was carried out and the patient eventually recovered, although he was left with an extensive defect. Heinrich and Pulverer (116) noted recovery of *Bacteroides melaninogenicus* from 13 cases of tonsillitis; no further details are given. However, the question of whether some or all of these strains represented normal flora must be raised.

There are several reports of recovery of anaerobes from *tonsillar or peritonsillar abscesses*. The most striking study is that of Hansen (274) who examined 153 samples of pus from peritonsillar abscesses. He recovered a total of 151 strains of anaerobic gram-negative bacteria, 48 of which were anaerobic gram-negative cocci, and the remainder anaerobic gram-negative bacilli. Among the bacilli, there were 29 strains of *Bacteroides funduliformis* (14 in pure culture), 25 fusiform bacilli (one in pure culture), and 5 *Bacteroides fragilis* (none in pure culture). The gram-negative cocci were never in pure culture. Hallander *et al.* (93b) reported studies on 30 patients with peritonsillar abscess; 26 yielded anaerobic bacteria. Included were *Bacteroides*, *Fusobacterium*, *Peptostreptococcus*, *Peptococcus*, microaerophilic cocci, *Veillonella* and *Bifidobacterium*. Sprinkle *et al.* (301a) did quantitative bacteriology on 6 individuals with peritonsillar abscess, 4 of which yielded anaerobes. One yielded a pure growth of *Bacteroides fragilis* (8×10^4/ml) and another yielded *B. fragilis* (10^4/ml) and higher counts of *Staphylococcus aureus* and *Klebsiella pneumoniae*. A third case had *Bacteroides necrophorus* (9×10^5/ml) plus 10^5/ml *Streptococcus pyogenes*. The last had *Bacteroides melaninogenicus* (10^4/ml) plus β-hemolytic non-group-A streptococci (10^7/ml) and small numbers of pneumococci. Lodenkämper and Stienen (287) recovered *Bacteroides* in pure culture from 6 cases of retrotonsillar abscess and *Bacteroides* along with *Veillonella parvula* from one case of tonsillar abscess. Baba *et al.* (272) studied 4 cases of peritonsillar abscess. One yielded *Peptococcus*, one *Peptococcus* plus *Bacteroides*, one *Peptostreptococcus*, and the fourth *Peptostreptococcus* plus *Neisseria*.

There are also some reports of single cases. Prévot (21) recovered *Ramibacterium pseudoramosum* from a case of peritonsillar abscess, and Alston (302) recovered *Bacteroides necrophorus* plus many other organisms from a case of peritonsillar abscess. *Sphaerophorus* (*freundi?*) and hemolytic streptococci were recovered from a tonsillar phlegmon by Morin and Potvin (145). A tonsillar abscess studied by Beerens and Tahon-Castel (115) yielded *Bacteroides funduliformis*, *Fusiformis fusiformis*, plus nonhemolytic streptococci. *Actinomyces* was recovered along with aerobes from a peritonsillar abscess (143a).

Grüner (303) found *Actinomyces* 17 times in a routine survey of surgically removed tonsils. He stated that a review of the literature shows that tonsils have been described as the portal of entry for actinomycosis, and, in a few cases, the disease has started following tonsillectomy. Harvey *et al.* (212) biopsied a case of chronic tonsillitis and found *Actinomyces*. Excision of the tonsil led to cure of the condition. Brock *et al.* (304) described an interesting case. The patient had a sore throat for 2 weeks, following which he developed marked swelling of the hypopharynx with exudate. There was no response to therapy with demethylchlortetracycline or penicillin. Biopsy showed only nonspecific inflammatory changes. Six weeks later, the patient developed a draining abscess from which was cultured *Arachnia propionica*. The lesion healed after drainage and excision of a sinus tract.

Infections of the Head and Face

Anaerobes are not uncommonly involved in infections about the face. These vary in severity from mild furuncle-like lesions to rapidly spreading life-threatening infections. Actinomycosis must always be given serious consideration in the differential diagnosis of lesions of all types about the head and face, as well as the neck. However, a variety of other anaerobic organisms aside from *Actinomyces* may be involved in such infections. Mutermilch and Séguin (170) described a gaseous phlegmon with fetid pus involving the face and neck. This infection arose secondary to an epithelioma of the floor of the mouth which had been treated with radium. Pulmonary gangrene complicated the infection of the head and neck. This infection was fusospirochetal in nature. Sandusky *et al.* (112) recovered anaerobic streptococci as well as other organisms from a furuncle of the face. *Fusobacterium necrophorum* plus anaerobic streptococci were isolated from a fulminating cellulitis of the face following dental extraction [Meleney (69)]. Thjotta and Jonsen (305) also described a case of cellulitis of the face, in this case yielding *Bacteroides funduliformis*, anaerobic gram-positive cocci, plus two other organisms. Prévot *et al.* (306) pointed out that suppurative lesions of the face may be caused by *Ramibacterium* and *Corynebacterium*, as well as by *Actinomyces*.

Beeuwkes *et al.* (307) recovered *Bacteroides* after incision and drainage of a chronic infection of the face near the parotid gland. Caroli *et al.* (172) recovered *Fusobacterium nucleatum* from a paraparotid abscess in pure culture; this infection was of dental origin. Rentsch (195) recovered an anaerobic *Corynebacterium* from a plasmacytoma on the cheek. Beerens and Tahon-Castel (115) described a patient with chronic swelling and fistula formation of the cheek from whose lesion they recovered *Ramibacterium pleuriticum* plus *Staphylococcus aureus*. Ingelrans (308) described a case of cellulitis and abscess of the face as well as maxillary sinusitis following dental extraction. The pus from this patient was horribly fetid and yielded on culture, *Sphaerophorus varius* and aerobic streptococci. Fass *et al.* (266a) recovered *Peptococcus* and a microaerophilic streptococcus, along with *Neisseria* and viridans streptococci, from a facial abscess. One of our patients with an infected facial wound, secondary to carcinoma of the parotid gland, cultured *Bacteroides melaninogenicus*, a gram-positive anaerobic coccus, and a gram-negative anaerobic coccus, as well as Klebsiella–Enterobacter and other coliform organisms. Schaller (309) reported a case with retroauricular abscesses due to *Actinomyces*. Smear of the material from these abscesses also revealed long gram-negative bacilli and fine gram-positive cocci, but these were not recovered in culture.

There are numerous reports of *anaerobic infection of the jaw and chin*, many or most of which are related to dental disease or manipulation. Prévot (197) described a patient with acute necrosis of the maxilla, secondary to an epithelioma of the gum. The organism recovered from this lesion was *Sphaerophorus ridiculosus*. Gins (310) described three infections of the lower jaw involving anaerobic bacteria. The first was a case with multiple abscesses which yielded *Spirillum sputigenum*, *Bacillus fusiformis*, *Leptothrix*, and *Bacillus maximus buccalis*. The second infection was a putrid purulent process secondary to paradentosis and following infiltration anesthesia. The organisms recovered from this patient were *Leptothrix maxima*, *Leptothrix lanceolata*, *Bacteroides melaninogenicus*, *Streptococcus lactis*, and nonhemolytic staphylococci. The third case was one of purulent infection of the ascending ramus; this infection yielded *Bacteroides melaninogenicus*, *Bacterium pneumosintes*, *Bacillus fusiformis*, and *Leptothrix*, with no aerobes present.

A patient with tuberculous meningitis who developed *Bacillus ramosus* bacteremia secondary to a suppurative infection of a maxillary cyst, was described by Modjallal (311). Sandusky *et al.* (112) recovered anaerobic streptococci as well as other organisms from an abscess of the upper jaw. *Ramibacterium ramosoides* was recovered from an abscess of the chin by Prévot *et al.* (306). Prévot and Thouvenot (312) recovered *Staphylococcus asaccharolyticus* and *Sphaerophorus ridiculosus* from a maxillary abscess. Prévot and Taffanel (313) described recovery of *Ramibacterium alactolyticum* from a case of osteomyelitis of the maxilla, and the same organism was recovered from a case of cellulitis of the jaw (21). Prévot (21) also described recovery of *Fusiformis fusiformis* from a jaw abscess and *Fusiformis nucleatus* from a case of cervicofacial cellulitis. Caroli (172) recovered *Fusobacterium nucleatum* plus *Corynebacterium granulosum* from an abscess of the jaw following fracture and *Fusobacterium nucleatum*, *Veillonella parvula*, and *Micrococcus grigoriffi* from a case of cervicofacial cellulitis secondary to tonsillitis. Prévot *et al.* (314) recovered *Ramibacterium pleuriticum*, *Clostridium perfringens*, and *Staphylococcus albus* from a case of osteophlegmon of the maxilla and *Ramibacterium pleuriticum* and *Fusiformis fusiformis* from a case of pseudoactinomycosis of unspecified site. Pulverer and Heinrich (81) recovered *Bacteroides melaninogenicus* from 4 patients with abscess of the chin or jaw; no details are given. Rentsch (195) recovered *Bacteroides fragilis* from a case of maxillary osteomyelitis and *Fusobacterium fusiforme* from another.

Beerens and Tahon-Castel (115) described several anaerobic infections of the chin or jaw. *Corynebacterium liquefaciens* and *Staphylococcus epidermidis* were recovered from one patient with cellulitis of the chin. Another patient with infection of the chin of dental origin yielded *Eubacterium tortuosum*, *Staphylococcus anaerobius*, *Dialister*, and an anaerobic diplococcus, as well as an aerobic nonhemolytic streptococcus. A similar patient, who also had a fistula, yielded *Corynebacterium liquefaciens*, *Diplococcus paleopneumoniae*, *Streptococcus micros*, as well as *Staphylococcus epidermidis*. A case of perimaxillary cellulitis of dental origin yielded *Ramibacterium pleuriticum*, *Eubacterium tortuosum*, *Fusiformis fusiformis* and *Dialister*. Three patients with osteitis of the maxilla following fracture were studied. Pus from these lesions yielded, in one case, a spirochete, *Veillonella alcalescens*, *Staphylococcus anaerobius*, and an aerobic nonhemolytic streptococcus; from a second case, *Fusiformis polymorphus* plus *Staphylococcus epidermidis* and nonhemolytic aerobic streptococci; from the third, *Diplococcus morbillorum*, *Dialister*, and nonhemolytic aerobic streptococci; and, from the fourth case, (unassociated with fracture) pus yielded *Corynebacterium granulosum*. Saito *et al.* (252) recovered a *Peptostreptococcus* from a patient with osteomyelitis of the maxilla. A case of osteomyelitis of the mandible secondary to trauma, with a compound comminuted fracture of the jaw, was reported by Leake (315). The infecting organisms were *Bacteroides fragilis* and anaerobic streptococci. Prévot *et al.* (154) recovered *Fusobacterium nucleatum* in pure culture from an abscess near the masseter muscle; this infection was of dental origin. Monaldo *et al.* (315a) reported a case of a draining submandibular ulcer extending to the neck and originating in a sinus tract from the mandibular right second and third molars; the foul discharge grew anaerobic streptococci, *Bacteroides*, and *Staphylococcus epidermidis*. Fritsche (211a) reported a lower jaw abscess which grew *Bacteroides melaninogenicus*, *Candida*, and *Citrobacter*; an acute submandibular abscess which yielded *Bacteroides* sp. and microaerophilic streptococci; and a paramandibular abscess with *B. melaninogenicus*, *Fusobacterium*, and *Leptotrichia*.

Graham *et al.* (316) described a case of facial actinomycosis misdiagnosed as tetanus because of the presence of significant trismus. These authors stress the importance of keeping the possibility of actinomycosis in mind in the case of obscure conditions about the face and neck, particularly if there is a mass lesion or an infiltrative process. Tyldesley (317) described a patient with a diffuse, almost painless, swelling of the right side of the face with slight induration which proved to be actinomycosis. Schubert and Tauchnitz (318) described a patient with a furuncle-like lesion of the upper lip which also proved to be actinomycosis. Coleman and Georg (319) recovered *Actinomyces israelii* from one case of facial infection and another of cervicofacial infection with abscess formation. *Actinomyces naeslundii* was recovered in pure culture from a patient with facial actinomycosis by Bezjak and Ayra (153).

Actinomycosis commonly involves the jaw as well. Schwabacher *et al.* (78) recovered *Bacteroides melaninogenicus* as well as *Actinomyces* from a patient with an abscess of the jaw. Caroli *et al.* (172) recovered *Fusobacterium nucleatum* as well as *Actinomyces* from a case of cervicofacial actinomycosis. Prévot *et al.* (154) recovered *Fusobacterium nucleatum*, *Actinobacterium abscessum* and an aerobic *Diplococcus* from a patient with cellulitis of the jaw following an accident involving damage to a tooth. They also recovered *Fusobacterium nucleatum* and *Actinomyces* from a case of retromaxillary actinomycosis.

Another mixed infection is described by Leake (315) in a patient with osteomyelitis of the mandible secondary to a tooth extraction. Material from this lesion yielded *Actinomyces israelii*, *Bacteroides melaninogenicus*, and a few diphtheroids. One of our patients with actinomycosis of the jaw, secondary to carcinoma of the floor of the mouth, yielded on culture, in addition to *Actinomyces*, *Bacteroides fragilis*, an unidentified anaerobic gram-negative bacillus, *Fusobacterium nucleatum*, anaerobic streptococci, and an anaerobic and a microaerophilic catalase-negative gram-positive bacillus.

Infections of the Neck

Abscesses of the neck may involve various anaerobic bacteria in addition to *Actinomyces*. The lesions may represent extension of infection from the tonsils or the teeth, may reflect metastatic infection carried by the bloodstream or may be associated with malignancy or trauma. Koch and Rinsche (320) recovered *Bacterium pneumosintes* in pure culture from a neck abscess. Anaerobic streptococci, along with other organisms, were recovered from a neck abscess by Sandusky *et al.* (112) and *Ramibacterium alactolyticum* was recovered by Prévot and Taffanel (313). Schwabacher *et al.* (78) recovered *Bacteroides melaninogenicus*, another anaerobic gram-negative bacillus, anacrobic streptococci, and an aerobic gram-negative bacillus from a neck abscess. Prévot *et al.* (306) pointed out that suppurations of the neck may be caused by *Ramibacterium* species (*R. ramosum*, *R. alactolyticum*, *R. ramosoides* and *R. dentium*) and by various species of anaerobic *Corynebacterium*, particularly *Corynebacterium granulosum*, as well as by *Actinomyces*. Beeuwkes *et al.* (307) recovered *Bacteroides* in pure culture from a walnut-sized abscess of the neck in a young boy. *Bacteroides funduliformis* and anaerobic streptococci were recovered from a neck abscess by Fisher and McKusick (264). Alston (302) recovered *Bacteroides necrophorus* and nonhemolytic streptococci from a neck abscess and, in another patient with a neck abscess secondary to septic tonsillitis, *B. necrophorus* was recovered from the bloodstream. Morin and Potvin (145) recovered *Staphylococcus aerogenes* and anaerobic streptococci and no aerobes from a deep cervical abscess. Heinrich and Pulverer (116) recovered *Bacteroides melaninogenicus* from 12 cases of abscess and phlegmon of the neck. A *Bacteroides* was recovered from a neck abscess by Keusch and O' Connell (321). *Bacteroides fragilis* was recovered in pure culture from a neck abscess by Sprinkle *et al.* (301a). A case of phlegmon of the neck (534) yielded anaerobic streptococci, *Fusiformis dentium*, *Borrelia vincenti*, and *Staphylococcus aureus*. Bøe (322) recovered a *Fusobacterium* from a submandibular abscess. Pulverer and Heinrich (81) recovered *Bacteroides melaninogenicus* from a submandibular abscess, and Rentsch (195) recovered anaerobic corynebacteria from a submandibular abscess. Fritsche (211a) recovered *A. israelii*, *Bacteroides corrodens*, *B. melaninogenicus*, *Fusobacterium*, and microaerophilic streptococci from an acute submental abscess and *A. israelii*, *B. melaninogenicus*, and *Fusobacterium* from a case of submandibular actinomycosis. Fourteen cases of mandibular abscess due to *Actinomyces* were noted in a 2 year period in Great Britain (1134a). We have recovered a pure culture of *Fusobacterium nucleatum* from a lemon-sized submandibular abscess in a patient with gingivitis. We have seen anaerobic neck abscesses in two patients with malignancy. One of these was secondary to a silk suture, following excision of a carcinoma of the floor of the mouth the previous year; this abscess yielded fusiform bacilli, anaerobic streptococci, and *E. coli*. The second patient had a carcinoma of the epiglottis. His neck abscess yielded microaerophilic cocci, anaerobic catalase-negative gram-positive bacilli, and α-hemolytic streptococci.

Anaerobic bacteria may also be involved in more extensive and severe infections of the neck. Cunningham (323) described a patient with multiple abscesses of the neck, with much gas in the soft tissues, secondary to tonsillitis. There was extensive necrosis of muscle fibers along with cellulitis.

Organisms recovered included *Actinomyces necrophorus* and anaerobic streptococci, as well as pneumococci. Gram-positive rods seen on smear were not recovered. *Ramibacterium alactolyticum* was recovered from a case of cellulitis of the neck by Prévot and Taffanel (313). Prévot (167) recovered *Sphaerophorus pseudonecrophorus* from a phlegmon of the neck. Morin and Potvin (145) recovered *Streptococcus foetidus* in pure culture from a phlegmon of the neck. Wey (324) described a fatal case of gas phlegmon of the right side of the neck which spread very rapidly. This was preceded by a sore throat. Incision and drainage of the neck lesion yielded thin hemorrhagic pus from the peritonsillar and parapharyngeal spaces. Culture of this pus yielded *Bacteroides melaninogenicus* and smaller numbers of *Escherichia coli*, *Staphylococcus aureus*, and nonhemolytic streptococci.

Anaerobic neck infection may also follow dental extraction or surgery. Richardson *et al.* (324a) described necrotizing fasciitis complicating dental extraction. Only β-hemolytic streptococci were recovered from several related abscesses; however, the retrosternal space contained foul-smelling fluid and gas. Baird (324b) recovered *Bacteroides melaninogenicus* and other *Bacteroides* from three mild neck infections, one involving burns and the other two following carotid endarterectomy. Infection following a radical neck dissection and grafting led to clostridial bacteremia in a patient described by Wynne and Armstrong (56e). In unpublished studies done in collaboration with G. Becker, J. Parell, and D. Busch, we have noted anaerobic involvement in 6 of 7 postoperative infections following major head and neck surgery for malignant disease. In this group, there was a mean of 7.1 bacterial isolates per patient (4.1 anaerobes and 3 facultatives). All 6 infections involving anaerobes were mixed, but in all cases but one the anaerobes predominated, both in terms of the number of species isolated and quantitation of the growth. Anaerobes recovered included *Bacteroides melaninogenicus* ss. *intermedius* (2), *B. melaninogenicus* ss. *melaninogenicus* (1), *B. oralis* (2), *B. ruminicola* ss. *brevis* (1), *Bacteroides* sp. (1), *Fusobacterium nucleatum* (2), *F. necrophorum* (1), *Peptococcus prevotii* (2), *P. magnus* (1), *P. asaccharolyticus* (1), *Peptostreptococcus anaerobius* (2), *P. micros* (2), *Propionibacterium* sp. (3), *Actinomyces* sp. (1), *Lactobacillus delbrueckii* (1), *L. minutus* (1), *Lactobacillus* sp. (2), *Bifidobacterium* sp. (1), *Veillonella alcalescens* (1), and *V. parvula* (1).

An excellent discussion of the compartments or spaces of the neck and infections thereof is given by Levitt (325). The submandibular space may be further subdivided into the sublingual space (superiorly) and the submaxillary space (inferiorly). The entire space is well known because of its involvement in Ludwig's angina, a very serious infection (Fig. 6.2). Ludwig originally described this infection as a "gangrenous induration of the neck"

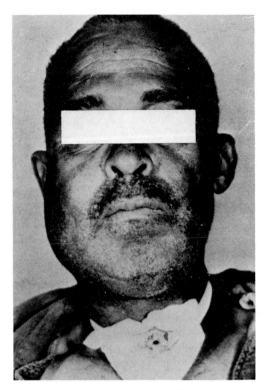

Fig. 6.2. Ludwig's angina. Massive edema, now subsided, necessitated tracheotomy to provide airway.

(290). Ludwig's angina is a cellulitis, rather than an abscess, of the submandibular space. The majority of infections in this area are of dental or periodontal origin. Although it has been said that streptococci and staphylococci are the commonest infecting organisms, this seems unlikely, not only because of the source of infections as indicated above but also because a putrid or foul odor is very characteristic of these infections. The first evidence of infection is usually noted in the oral tissues adjacent to the initial focus. As the infection proceeds, the tongue becomes greatly swollen and is pushed superiorly and posteriorly against the palate causing respiratory embarrassment. There is bulging of the submaxillary space associated with severe pain, trismus, inability to swallow, and marked respiratory distress. After the infection penetrates the myohyoid muscle, symptoms progress at a very rapid rate. Ultimately, there is boardlike firmness of the tissues of the floor of the mouth due to the severe cellulitis. Ludwig [as cited by Levitt (325)] stressed the following points in his original description: the comparatively slight inflammation of the throat itself, the "woody" hardness of the cellular tissue, the hard sublingual swelling and the swelling of the floor of the mouth, the well-defined border of the hard edema of the neck, and the absence of infection in regional lymph nodes. Grodinsky [also cited by Levitt (325)] reviewed the literature on the subject and stated that the term Ludwig's angina should be reserved for those cases in which the infection starts in the floor of the mouth, usually from a lower second or third molar tooth and spreads to the submental and submaxillary triangles through fascial planes (not by way of lymphatics). This would essentially eliminate the throat, tonsils, and pharynx as the site of the primary lesion, since infections in these sites usually spread by lymphatics and extend into the retropharyngeal and parapharyngeal spaces. Grodinsky's criteria for true Ludwig's angina are as follows: cellulitis (rather than abscess) of the submandibular space; bilateral involvement, usually; gangrene with serosanguinous putrid infiltration but little or no frank pus; involvement of the connective tissue, fascia, and muscles but not glandular structures; and spread of the cellulitis by continuity rather than by lymphatics.

In what must be one of the earliest documented anaerobic infections Veillon (4) recovered an anaerobic micrococcus from a fatal case of Ludwig's angina. Melchoir [cited by Smith (326)] found fusiform bacilli and cocci in 4 cases of Ludwig's angina and fusiform bacilli and spirochetes in an additional case. Hansen [also cited by Smith (326)] found fusiform bacilli and spirochetes in large numbers in a case of Ludwig's angina. Smith (326) was able to produce an infection similar to Ludwig's angina in guinea pigs by inoculating mixed anaerobic flora into the inner margin of the gum of the lower posterior teeth. Crystal et al. (160) recovered Fusobacterium from a case of Ludwig's angina which followed extraction of several teeth. Mason (158) also described a case of Ludwig's angina following dental extraction. Anaerobic streptococci and Streptococcus faecalis were recovered from this patient. Marks et al. (326a) reported a case of Ludwig's angina involving Bacteroides and α-hemolytic streptococci.

Cervicofacial actinomycosis accounts for over one-half of all cases of actinomycosis (Fig. 6.3). This condition frequently produces a different type of lesion than is found with other anaerobes causing infection in the same area. Actinomycosis tends to spread along connective tissue planes with little tendency toward ulceration or lymphatic involvement. As part of the inflammatory reaction, there is characteristically produced a very hard ("woody") fibrosis, and draining sinuses or fistulae are found relatively frequently. Chronic suppuration is also very common. Discrete colonies of the organism are formed in tissues with some frequency. These

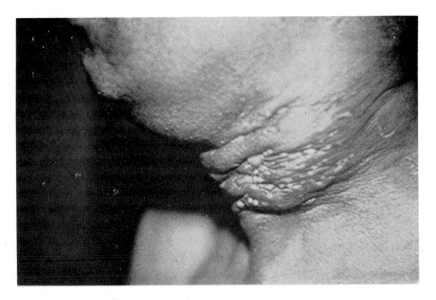

Fig. 6.3.   Actinomycosis of the neck with multiple draining sinuses.

"sulfur granules" are discharged by way of sinuses along with purulent material. Any of a number of species of *Actinomyces* and of certain related organisms, such as *Arachnia propionica*, may produce this disease. Other anaerobic bacteria, *Actinobacillus actinomycetemcomitans* and certain aerobic organisms, may all be found in conjunction with *Actinomyces*. Although actinomycotic infections may be acute, more often they are subacute or chronic and not infrequently they are insidious. In the course of a long-term infection, particularly in the neck, puckered scars that are quite characteristic of the disease are formed. *Actinomyces* is part of the normal flora of the mouth, and these infections are endogenous in origin. Infection with *Actinomyces* may be precipitated by trauma, malignancy, dental disease or manipulation, or other factors producing tissue destruction.

   Linhard (178a) described a case of cellulitis of the jaw of dental origin due to *Actinobacterium cellulitis*. McVay *et al.* (327) described 2 cases of actinomycosis secondary to extraction of molars. Lane *et al.* (328) described 7 cases, at least 4 of which were related to dental disease or extraction. Prévot *et al.* (154) described a case of cervical actinomycosis of buccodental origin from which they cultured, in addition to *Actinomyces*, *Fusobacterium nucleatum* and *Clostridium sporogenes*. Bramley and Orton (329) described 11 cases of cervicofacial actinomycosis, 6 of which were acutely painful and which clinically could not be distinguished from pyogenic abscesses. All were of dental origin. Accompanying organisms included *Bacteroides* and various aerobic cocci. Holmes (330) discussed 12 cases of cervicofacial actinomycosis, most of which were related to dental infection or manipulation. Herrell *et al.* (331) treated 3 cases of cervicofacial actinomycosis with erythromycin. All did well, but one with bony involvement required penicillin therapy in addition. In the series of 37 cases of actinomycosis reported by Harvey *et al.* (212) 24% were primary in the cervicofacial area, and 24% were secondary infections in this area. One of the latter cases arose in the sublingual tissues. Caron and Sarkany (177) reported 5 patients with cervicofacial actinomycosis related to periodontal disease or trauma. Cultures of 12 patients with actinomycosis, 10 of which involved the head and neck, yielded *Actinobacillus actinomycetemcomitans* in addition to *Actinomyces* in all 12 cases [Heinrich and Pulverer (178)]; these workers also isolated one anaerobic streptococcus, one *Bacteroides melaninogenicus*, and two aerobic hemolytic streptococci. Beerens and Tahon-Castel (115) reported six cases, three of which were related to dental extraction. In three cases, *Actinomyces israelii* was obtained in pure culture. One case yielded *Actinomyces israelii* plus *Actinobacillus actinomycetemcomitans*, one case yielded *A. israelii* plus *Eubacterium lentum* and *Staphylococcus epidermidis*. The last case yielded only *Fusiformis fusiformis*, *Corynebacterium liquefaciens*, and *Staphylococcus epidermidis*. Along the same lines, Beerens (332) reported 36 cases of infection, clinically like cervicofacial actinomycosis, 11 of which yielded *Ramibacterium pleuriticum*. On two occasions this organism was isolated in pure culture, and in the rest of the cases was mixed with various aerobes and anaerobes. A patient with submental abscesses due to *Actinomyces* was reported by Schaller (309). An unusual case was reported by Graybill and Silverman (333). Their patient developed a supraclavicular abscess that obstructed the left subclavian vein. This patient developed multiple abscesses over the chest and extremities. *Actinomyces israelii* was obtained in pure culture.

The very firm fibrosis that may be found in this disease may lead the clinician to seriously entertain a diagnosis of carcinoma. Macoul and Souliotis (334) reported a patient with a solitary nodule in the neck attached to the sternocleidomastoid muscle which resembled metastic carcinoma. It turned out to be an actinomycotic infection. Sugano (335) also reported a case of actinomycosis of the neck simulating malignancy. Harvey *et al.* (212) reported 3 cases of actinomycosis which were primary in the neck and then disseminated widely; one to the anterior chest wall and axilla; one to the paranasal sinuses, the chest wall, the scapula, sacrum, mastoid, lung and pleura; and the third to the mediastinum and lung. Garrod (336) isolated a ·fusiform bacillus in addition to *Actinomyces* from a case of cervical actinomycosis. Brock *et al.* (304) reported a typical case of actinomycosis (no other details are given) due to *Arachnia propionica*.

From a patient with actinomycosis of the neck of uncertain origin we isolated, in addition to *Actinomyces israelii*, another anaerobic, catalase-negative, gram-positive bacillus and *Fusobacterium nucleatum*. Another of our patients had actinomycosis of the mandible and jaw and surrounding soft tissues. In addition to *Actinomyces israelii*, we isolated from him *Actinobacillus actinomycetemcomitans*. The patient improved on lincomycin therapy but the *Actinobacillus* persisted, and it was necessary eventually to treat with tetracycline in order to eliminate the other organism and to eventually achieve cure.

There are several other miscellaneous infections of the neck involving anaerobic bacteria which have been reported. Prévot *et al.* (337) reported a case of "ulceration of the neck" from which *Staphylococcus anaerobius* and *Micrococcus niger* were recovered. Morin and Potvin (145) recovered *Sphaerophorus* in pure culture from a case of cervical adenitis. Anaerobic streptococci were involved in an infection that spread from a dentoalveolar abscess to a submaxillary node (275). Sandusky *et al.* (112) isolated anaerobic streptococci in pure culture from an infected branchial cyst. Meleney (69) isolated *Bacillus necrophorus* in pure culture from an infected thyroglossal cyst. Several reports describe anaerobic anterior cervical infections as a complication of transtracheal aspiration. All responded readily to appropriate therapy. Deresinski and Stevens (337a) recovered *Bacteroides fragilis* ss. *fragilis*, *Peptostreptococcus anaerobius* and α-hemolytic streptococci from a subcutaneous abscess; all 3 organisms had been recovered (along with 2 other aerobes) from the transtracheal aspirate. Lourie *et al.* (337b) recovered *B. fragilis* and an enterococcus from a similar abscess; again, the same organisms plus other aerobes had been present in the initial transtracheal aspirate. Yoshikawa *et al.* (337c) saw gram-positive cocci on smear of a paratracheal abscess and on the original transtracheal aspirate; all cultures were negative. We have seen two patients with similar problems.

Salivary Gland Infections

Shevky *et al.* (281) recovered *Bacteroides melaninogenicus* from a parotid abscess. Beigelman and Rantz (282) recovered *Bacteroides* and anaerobic streptococci in small numbers along with two other organisms (not specified) from a case of parotitis in a diabetic. Heck and McNaught (338) recovered *Bacteroides* from a parotid abscess that subsequently extended to the subtemporal area. Heinrich and Pulverer (116) recovered *Bacteroides melaninogenicus* from two parotid abscesses (no information is given on other organisms that may have been present).

Bock (339) described a patient with inflammation of the sublingual gland associated with a bad taste in his mouth. Bock noted numerous spirochetes and a few fusiform bacilli. He also cultured numerous pneumococci and a few hemolytic streptococci.

Baba et al. (272) isolated a Peptococcus in pure culture from a purulent submaxillary gland infection. There are three other reports of submaxillary abscess, but it is not clear whether these refer to abscess of the submaxillary gland per se or of the submaxillary area. Shevky et al. (281) recovered Bacteroides melaninogenicus from one of these. Sandusky et al. (112) recovered anaerobic streptococci plus other unspecified organisms from another. Prévot et al. (337) recovered Micrococcus niger and Staphylococcus anaerobius from the third.

Söderlund [cited by Cope (210)] has carried out some very interesting investigations on the role of actinomycotic infection in the formation of salivary calculi. He points out that Actinomyces commonly infects the salivary ducts, producing diffuse inflammation of the ducts at first, followed by general swelling of the gland and usually abscess formation, and, finally, inflammation spreading through the capsule of the gland to the adjacent parts. In the submaxillary and sublingual glands, infection seldom or never spreads outside the salivary gland, but the surrounding area is sometimes affected in the case of the parotid. It seems to be well established that salivary duct calculi are formed by a deposition of insoluble salts about Actinomyces filaments. Since removal of the calculi does not necessarily clear the infection, additional therapy may be required.

Sazama (340) reported 5 cases of actinomycosis of the parotid gland, one of which also involved the cervicofacial area and another involved osteomyelitis of the ramus. Coleman and Georg (319) reported recovery of Actinomyces israelii from two parotid gland abscesses.

Thyroid Infections

Lemierre (341) stated that he had seen thyroiditis in the course of Bacteroides funduliformis septicemia. Beigelman and Rantz (282) described a patient with a known thyroid adenoma who developed fever and enlargement and tenderness of the thyroid gland. Two cultures of the gland yielded Bacteroides without any other organisms, and histologically thyroiditis was demonstrated. The patient was cured by thyroidectomy. Warren and Mason (342) described a patient who had a carcinomatous ulcer of the colon who developed a gas-forming suppurative thyroiditis and bacteremia due to Clostridium septicum. Hawbaker (343) recovered Bacteroides fragilis in pure culture from the foul pus of an abscess of the left lobe of the thyroid. The patient originally had had a hysterectomy and perineal repair. This was complicated by a large infected hematoma in the lower abdomen and by peritonitis, subphrenic abscess, and septic shock. *Bacteroides fragilis* was also recovered from the blood, from the peritoneal fluid, and from the subphrenic abscess. Sharma and Rapkin (345) recovered *Bacteroides melaninogenicus* in pure culture from foul-smelling pus in a case of acute suppurative thyroiditis in a 2-year-old boy.

According to Cope (210), actinomycosis of the thyroid gland is quite rare. He was able to find a total of 8 cases of involvement of the thyroid gland in the course of actinomycosis. Leers *et al.* (344) described a case of suppurative thyroiditis from which *Actinomyces naeslundii* was recovered in pure culture; sulfur granules were found in the pus. Coleman *et al.* (152) described a postoperative thyroidectomy wound abscess due to *Actinomyces naeslundii*. Gorbach and Thadepalli (344a) recovered *Actinomyces israelii, Peptostreptococcus, Bacteroides melaninogenicus, Fusobacterium necrophorum,* and *Lactobacillus catenaforme* from a thyroid abscess. Blanc and Jenny (344b) described a case of suppurative thyroiditis due to *Actinomyces.*

Miscellaneous Infections

There are several cases of anaerobic *retropharyngeal abscess* in the literature. Cunningham (323) described a case complicating tonsillitis. The entire retropharyngeal space down to the arch of the aorta was involved. There was gelatinous edema of the palate, epiglottis, and aryepiglottic folds. Myerson (346) described two cases of anaerobic retropharyngeal abscess, secondary to swallowing small bones, characterized by tissue sloughing, foul odor, and gas. One case yielded an anaerobic gram-negative bacillus and hemolytic streptococci. The second case was not cultured anaerobically. Aerobes recovered were viridans streptococci and *Staphylococcus albus.* Prévot (21) recovered *Sphaerophorus gonidiaformans* from a retropharyngeal abscess. Ernst (347) recovered *Bacteroides funduliformis* among other organisms (not specified) from a retropharyngeal abscess. Janecka and Rankow (348) reported a case of retropharyngeal gas-forming abscess that began as a sore throat and proceeded to mediastinitis and pleural infection. Pus obtained at the time of tracheotomy yielded *Bacteroides* and anaerobic streptococci, and pus from the pleural space yielded *Bacteroides* and *Klebsiella.* The pus had a putrid odor. This case was ultimately fatal. Baba *et al.* (272) recovered a pure culture of *Peptostreptococcus* from a closed retropharyngeal abscess. Bryan *et al.* (348a) recovered 200–250 ml of foul-smelling pus from a retropharyngeal space infection; cultures were negative. Heinrich and Pulverer (116) recovered *Bacteroides melaninogenicus* from three cases of *parapharyngeal phlegmon.*

Peters (349), described a case of tertiary lues involving the *larynx* with secondary fusospirochetal infection. Grumbach and Verdan (350) described a patient with bilateral tonsillar infection and abscess with bacteremia who subsequently developed a phlegmon of both piriform sinuses and at the base of the tongue. *Fusobacterium nucleatum* was the infecting organism. Delbove and Reynes (351) reported a case of suppurative perichondritis of the larynx due to *B. funduliformis*.

Felner and Dowell (168) reported a case of *Bacteroides fragilis* bacteremia in a patient with hypopharyngeal carcinoma receiving cytotoxic therapy and a case of *Bacteroides variabilis* bacteremia in a patient with nasopharyngeal carcinoma, also receiving cytotoxic therapy.

7

Central Nervous System Infections

It is now clear, particularly since the paper by Heineman and Braude (352), that anaerobic bacteria are the major cause of brain abscess. The role of these organisms in other types of central nervous system infection is not generally recognized. Table 7.1 presents the distinguishing characteristics of various intracranial inflammatory diseases.

Brain Abscess

Figs. 7.1 and 7.2 show brain abscesses demonstrated by radioisotope brain scan and angiography, respectively. The incidence of anaerobic bacteria in various reported series of brain abscess is noted in Table 7.2. The incidence obviously varies considerably from series to series. It is likely that the low recovery rates of anaerobes in a number of these series reflects failure to do anaerobic culture at all or inadequate transport or cultivation techniques for recovery of fastidious anaerobes.

A literature survey of brain abscess involving anaerobic bacteria is summarized in Tables 7.3 and 7.4. Among 171 cases of brain abscess involving anaerobes reported in some detail, only 42 had aerobic or facultative organisms present as well, with nine cases uncertain on this point. Eight of the aerobic or facultative organisms that were recovered were present in small

TABLE 7.1

Differential Diagnosis of Intracranial Inflammatory Diseases[a,b]

	Localized headache	Hemiplegia	CSF			Frontal sinusitis	Convulsions	Rapid mortality
			1500 Lymphocytes	Normal sugar	WBC 20,000			
Purulent meningitis	+	–	–	rare	++	rare	±	++
Partly treated	+	–	±	++	++	rare	–	–
Viral meningoencephalitis	±	rare	+	++	–	–	±	–
Tuberculous meningitis	–	+	±	1/2	rare	–	+	±
Brain abscess	++	±	±	+	±	±	+	±
Subdural abscess	++	++	+	++	++	++	++	++

[a] Reprinted from McLaurin (1179). "Cranial and Intercranial Suppuration" (E. S. Gurdjian, ed.), 1969. Courtesy of Charles C. Thomas, Publisher, Springfield, Illinois.

[b] –, not seen; ±, occasionally seen; +, frequently seen; ++, frequently and characteristically seen.

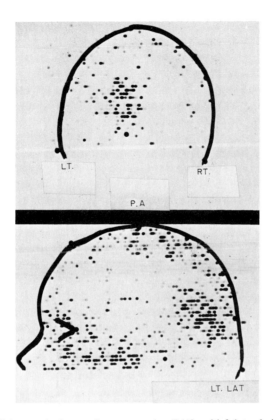

Fig. 7.1. Radioisotope brain scan [posteroanterior (PA)] and left lateral views showing an occipital lobe abscess.

numbers or were unlikely pathogens. Specific information on the amount of growth frequently was not given. Concurrent central nervous system infection of several types is not unusual, as is evidenced by the 47 cases of meningitis and 11 of extradural or subdural abscess noted among these 141 brain abscess patients. Simultaneous presence of two or more anaerobes in a single specimen is not uncommon; this was noted in 48 patients among the overall group of 321. Details on the specific anaerobes isolated from cases of brain abscess (and other central nervous system infections) are presented in Table 7.10 also, see Fig. 7.3.

The etiologic agents of anaerobic brain abscess are arranged in major groupings according to type and/or source of underlying condition in Table 7.4. *Bacteroides fragilis* (over 1% of all isolates) and *Clostridium ramosum* are singled out because of their relative resistance to antimicrobial agents, and *Clostridium perfringens* is also considered separately because infections

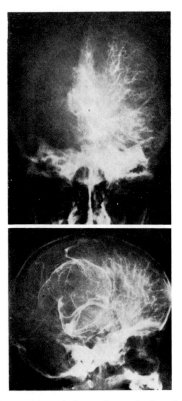

Fig. 7.2. Anteroposterior and lateral views of a cerebral angiogram. Early venous phase. The lateral view shows a distinct demarcation of the abscess capsule in the left parietal region, while the anteroposterior view shows a nonspecific shift and only a vague outline of the vascular marking. From Chou *et al.* (1207), reproduced with permission.

in which it is involved may be devastating. Far and away the most important underlying condition was otitis media, with or without mastoiditis. The ear infections were usually chronic, although not always; cholesteatoma formation was not unusual. Brain abscess following chronic otitis media and mastoiditis is most often in the temporal lobe and less frequently in the cerebellum or other intracranial sites. *Bacteroides* species were the most common isolates in this setting, with *Fusobacterium* and anaerobic streptococci or cocci common causes also. Of particular interest and importance is the fact that 14 cases of brain abscess secondary to ear or mastoid infection involved *Bacteroides fragilis*. *Clostridium ramosum* was isolated on three occasions from this source.

TABLE 7.2

Incidence of Anaerobes and of Sterile Cultures or Insignificant Isolates in Several Reported Series of Brain Abscesses

No. cases cultured	No. from which anaerobes recovered	No. with sterile cultures or nonpathogen[a] recovered	Reference
44	12	9	Ballantine and Shealy (1180)
25[b]	6	7	Baltus and Noterman (1181)
37	11	2	Beerens and Tahon-Castel (115)
74	2	18	Beller et al. (1200b)
21	2	8	Buchheit et al. (1182)
89	2	24	Fog (1183)
14[c]	5	1	Gregory et al. (1184)
63	9	20	Gurdjian and Thomas (367)
18	16	2	Heineman and Braude (352)
11	3	2	Heinz and Cooper (1185)
16	1	7	Langie et al. (1186)
88	?12	26	Loeser and Scheinberg (1187)
47	22/62[d]	2	McFarlan (1188)
30	4	14	McGreal (1189)
607	8[e]	143	Schiefer and Kunze (1190)
62	2	12	Sperl et al. (1191)
60	10		Swartz (357)
70	16	6	Swartz and Karchmer (1109)
372	9	58	Tamalet et al. (1192)

[a] Coagulase-negative staphylococci or micrococci, aerobic diphtheroid, or *Bacillus* species.
[b] Eleven specified as not cultured anaerobically.
[c] Eight not cultured anaerobically.
[d] Of the 62 organisms recovered, 22 were anaerobes.
[e] All *Actinomyces*.

Infection of the lung or pleural space was the second most common underlying condition for anaerobic brain abscess. Here also, the underlying condition tended to be subacute or chronic with lung abscess, bronchiectasis, empyema, and necrotizing pneumonia the more common specific types of infection involved. Brain abscesses that are metastatic from distant sites of infection, such as the pleuropulmonary area, tend to be multiple and are most likely to occur at the junction of the gray and white matter. These abscesses secondary to pleuropulmonary disease affect the frontal, parietal, and occipital lobes of the brain with equal frequency but only rarely involve the temporal lobe or the cerebellum. Brain abscess, secondary to congenital heart disease, is the one exception to the rule that metastatic brain abscess

TABLE 7.3

Anaerobic Bacteria in Brain Abscesses (Literature Survey)

Number of cases involving anaerobes which are reported in some detail (see Table 7.4 for references)	171
Number in which aerobes or facultatives are also present	42
Number in which presence of aerobes or facultatives is uncertain	9
Number with light growth of nonanaerobes or growth of unlikely pathogen[a]	8
Entamoeba histolytica also present	1
Number with concurrent meningitis	47
Number with concurrent extradural or subdural abscess	11
Number with concurrent bacteremia (exclusive of endocarditis cases)	6
Number of cases involving anaerobes but reported in inadequate detail	150[b]
Aerobes or facultatives known to be present also	3
Aerobes or facultatives known not to be present	21
Number known to have concurrent meningitis	13
Number known to have concurrent bacteremia	2
Total group of brain abscesses involving anaerobes	321
Number with two or more anaerobes	48

[a] Coagulase-negative micrococcus or staphylococcus, aerobic diphtheroid or *Bacillus* species.
[b] Data from (21, 30, 115, 116, 167, 210, 212, 226a, 264, 287, 290, 312, 367, 487, 531, 709, 811, 1106, 1109, 1134a, 1140, 1182, 1187, 1193–1200, 1200a–1200j).

tends to be multiple. Various members of the genus *Fusobacterium* were the most common cause of brain abscess related to the lung or pleural space. *Clostridium ramosum* was seen on one occasion in a brain abscess from such a source. *Bacteroides fragilis* was not seen at all from brain abscesses secondary to pulmonary or pleural disease in this series. However, since we know that *Bacteroides fragilis* may be recovered from as many as one-fifth of patients with anaerobic pleuropulmonary infection (353) one should be alert to the possibility that this organism may be involved in brain abscess from this origin.

Sinusitis was the third most common underlying site for anaerobic brain abscess. Most often the frontal and sphenoidal sinuses are involved, leading to brain abscess in the frontal and temporal lobes, respectively. It is important to note here again that *Bacteroides fragilis* is a relatively common cause of anaerobic sinusitis as judged from this particular series. Additional conditions underlying anaerobic brain abscess include dental or oral infection, infection in the tonsillar or pharyngeal area, congenital heart disease, and trauma. Infection following dental, tonsillar, or oropharyngeal disease is typically caused by anaerobic bacteria that are susceptible to penicillin G *in vitro*, but in the present literature survey one infection in these categories was due to *Bacteroides fragilis*. Congenital cardiac defects with a right to

TABLE 7.4

Summary of Source and Etiology of Reported Cases of Brain Abscess due to Anaerobic Bacteria (142 Patients) (Literature Survey)[a,b]

Type and/or source of underlying condition	Anaerobic streptococci	Microaerophilic streptococci	Anaerobic gram-positive cocci	Veillonella and unspecified gram-negative cocci	Bacteroides fragilis	Other Bacteroides or unspecified	Fusobacterium	Other anaerobic gram-negative bacilli or unspecified	Clostridium perfringens	Clostridium ramosum	Other Clostridium or unspecified	Actinomyces	Eubacterium	Propionibacterium	Other NSF[c] gram-positive bacilli or unspecified	Miscellaneous anaerobes	Total
Sinusitis	5			1	7	3	3								1		20
Otitis media and/or mastoiditis	16	1	3	4	21	13	14	5	1	3		3	1	3	2	1	91
Dental or oral infection	3	2				1	1					3	1				11
Orbit	1						1					1					3
Tonsillar, pharyngeal area		1			1	1	5				1						9
Pulmonary and/or pleural infection	7	3			1	2	16	4		1		2		1			37
Congenital heart disease	3	5				4	1					1		1			15
Endocarditis						1	2								1		4
Intraabdominal infection			1	1			1										3
Genitourinary tract	2	2	2	1			1		2							1	11
Trauma	1		2						6		4	1			2		16
Surgery	1	1											1				3
Unknown	7	4					5	4			1	1		2			24
Total	46	19	8	7	30	25	50	13	9	4	7	12	3	7	6	2	248

[a] From (7, 8, 21, 112, 115, 134, 150, 156, 168, 210, 244, 265, 268, 274, 275, 300, 347, 352, 354, 355, 364, 370, 372, 479, 480, 482, 486, 560, 816, 950, 984, 1023, 1173, 1181, 1184–1186, 1189, 1200–1240, 1240a–1240o).

[b] Numbers refer to number of isolates, usually one per case.

[c] NSF, non-spore-forming.

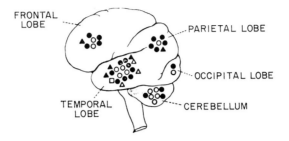

o   *Bacteroides*
●   Anaerobic streptococci
□   Anaerobic diphtheroid
△   *Actinomyces*
▲   *Veillonella*

Fig. 7.3.   Cerebral localization of anaerobic brain abscesses; compilation from Heineman and Braude and series at Massachusetts General Hospital. From Swartz (357), reproduced with permission.

left shunt, which permits short-circuiting of the pulmonary circulation, may be associated with brain abscess. Brain abscess rarely complicates the course of subacute bacterial endocarditis, but may be seen, particularly in the form of miliary abscesses, in acute bacterial endocarditis on occasion. Note the relative frequency with which *Clostridium perfringens* and other clostridia are encountered in brain abscess secondary to trauma.

Newton [cited by Gorbach and Bartlett (56b)] reviewed the literature on brain abscess complicating congenital heart disease; anaerobes were isolated from 10 of the 32 cases which had positive cultures (5 were *Actinomyces*). Anguiano *et al.* (353a) encountered both cerebral abscess and meningitis among 82 anaerobic infections (further details not given).

Typically, brain abscess secondary to disease of the middle ear and mastoid or sinuses results from direct extension of the infectious process. It has been generally felt that brain abscess secondary to pulmonary or pleural disease involves transport of infected material via the valve-free spinal venous system with the help of elevated intrathoracic pressure, as during coughing. However, Prolo and Hanbery (354) present serial arteriographic studies to indicate that in their case the organism first invaded the parenchyma of the brain in the distribution of the right middle cerebral artery. They noted intense arterial spasm, suggesting either local cerebritis, vasculitis, or both. Heineman and Braude (352) raise the interesting possibility that previous cerebrovascular accidents may predispose to brain abscess if there is an opportunity for infarcted areas to become secondarily infected. Anaerobic growth would be favored by the low oxygen tension in the infarct.

Cope (210) noted the curious fact that in six cases of isolated actinomycosis of the brain the lesions were situated in the third ventricle in or near the septum lucidum and the anterior commissure. He speculated that *Actinomyces* reached this unusual site from the nose or throat along the olfactory nerve sheaths. In contrast to actinomycosis elsewhere in the body, the lesions in the brain do not invade the surrounding tissue, but instead behave like a tumor mass surrounded by a gelatinous capsule. *Actinomyces* may involve the central nervous system more commonly either by extension or metastasis, just as other anaerobic bacteria may.

Our laboratory has recovered anaerobic bacteria from 15 patients with brain abscess. In 9 of these, anaerobes were present in the absence of any aerobic or facultative forms, whereas in the other 6 cases there were a total of 7 strains of facultative organisms, 2 of which were present only as a few colonies. The aerobic forms isolated included three β-hemolytic streptococci, one gram-positive coccus that was not further identified, and one gram-negative bacillus. There was a total of 23 strains of anaerobic bacteria isolated from these patients. Included were 5 microaerophilic streptococci; 1 microaerophilic coccus; 4 anaerobic cocci (one identified as *Peptococcus asaccharolyticus*); 1 anaerobic streptococcus (*Peptostreptococcus micros*); 3 anaerobic non-spore-forming, catalase-negative gram-positive bacilli; 2 *Propionibacterium* (in mixed culture); 1 *Bacteroides fragilis*, 1 *Bacteroides melaninogenicus*; 1 unidentified *Bacteroides*; 2 *Fusobacterium nucleatum*; 2 *Fusobacterium* species; and 1 *Actinomyces*. The underlying conditions were similar to those recounted in the literature survey, except for one patient whose brain abscess was secondary to a pulmonary arteriovenous fistula. One of these cases of brain abscess has been reported previously (161).

During the stage of acute suppurative encephalitis, intensive antimicrobial therapy and control of increased intracranial pressure constitute the regimen of choice. With early diagnosis and appropriate antimicrobial therapy, it will be possible occasionally to avoid surgery [Heineman *et al.* (355)]. When the abscess has become localized, surgical treatment is mandatory. There is some difference of opinion regarding optimal surgical therapy. While some abscesses may be cured by repeated aspiration and intensive antimicrobial therapy, most workers feel that excision is preferable. Whenever possible, excision should be carried out immediately after the initial aspiration at the time of exploration. However, no matter how well encapsulated an abscess may be, infection is typically present in the surrounding tissue. Accordingly, intensive antimicrobial therapy is still indicated in order to minimize the possibility of satellite abscesses or relapse. If the abscess is deep, poorly encapsulated, or located in a strategic area, repeated aspiration is usually preferable to excision. Brain abscess cavities require many weeks for closure. This relates to the fact that the brain is relatively rigid, tending to hold cavities open, and to the relative avascularity of the brain (abscesses often originate in poorly vascularized areas, as already discussed, and increased intracranial pressure further compromises vascularity).

In antimicrobial therapy of brain abscess, the blood–brain barrier must be taken into consideration. This means that certain antimicrobial compounds, otherwise suitable for many anaerobic infections, such as clindamycin and lincomycin, should not be used because the evidence to date indicates that they do not achieve satisfactory concentrations in the central nervous system. High dose therapy is important in order to penetrate the blood–brain barrier and relatively avascular areas. Treatment should be started early and maintained for prolonged periods of time. The choice of antimicrobial agents to be used depends also on the infecting organisms and their susceptibility pattern. Penicillin G is very useful for anaerobes other than *Bacteroides fragilis*, which is usually so resistant as to preclude effective therapy with this agent, particularly in the case of central nervous system infections. Occasional strains of *Fusobacterium* are also resistant to penicillin G. Many strains of *Clostridium ramosum* and occasional strains of anaerobic cocci require as much as 6.2 U/ml of penicillin G for inhibition. The dosage of penicillin G should be 30,000,000 to 40,000,000 U/day intravenously. Chloramphenicol is, without any question, the drug of choice for anaerobic infections of the central nervous system. It penetrates the blood–brain barrier well and is active against virtually all anaerobic bacteria; only rare strains are resistant. The initial dosage of chloramphenicol should be at least 50–60 mg/kg of body weight intravenously. Tetracycline is effective for central nervous system infections, but many strains of anaerobic bacteria are resistant to this agent. Accordingly, *in vitro* susceptibility studies are imperative. Erythromycin may also be useful on occasion but, again, *in vitro* susceptibility determinations would be required prior to its use. Metronidazole is still experimental and at the present time is only available for use by the oral route. It is, however, very active against most anaerobic bacteria and there is preliminary evidence to indicate that it may cross the blood–brain barrier well.

Meningitis

Anaerobic bacteria are an uncommon cause of meningitis, but they undoubtedly occur more commonly in this condition than has been appreciated. Statements to the effect that anaerobic bacteria are essentially never involved in meningitis are quite inappropriate. Swartz and Dodge (356) found only anaerobes in just one case of anaerobic meningitis among 207 patients with this disease (19 spinal fluids were sterile in this series), whereas Swartz (357) found that 16% of isolates from 60 cases of brain abscess were anaerobes. Lutz *et al.* (358) found only three anaerobes (streptococci) among 323 organisms isolated from 309 patients with meningitis (17 cases had sterile spinal fluid). Oguri and Kozakai (359) found only one anaerobe among 77 spinal fluids that yielded growth of some type on culture. On the other hand, a few small series suggest anaerobes may be somewhat more common. Kozakai (360) found anaerobes in four of 45 cases among which there were 10 others with no growth or unlikely pathogens recovered from spinal fluid. K. Ueno (personal communication) found anaerobes in 7.8% of spinal fluids studied. Nakamura *et al.* (361) found anaerobes in 3 of 29 positive spinal fluids, and Stokes (108) found anaerobes in 3 of 33 spinal fluids yielding growth of some type. Randall (361a) recovered 9 strains of anaerobes among 218 spinal fluids cultured; no case details are given. Morrison *et al.* (1267f) noted 110 cases of actinomycotic central nervous system infection in a literature survey. Adler and von Graevenitz (1267e) found 17 clinically detailed cases of clostridial meningitis in the literature. A case of questionable meningitis, in which lactobacilli were cultured from brain tissue at autopsy and direct smear revealed many gram-positive bacilli, was reported by Sharpe *et al.* (361b); there were no postmortem signs of meningitis.

A survey of the literature revealed many more cases of meningitis than might have been anticipated from some of the incidence data cited above. Data from this survey is presented in Tables 7.5 and 7.6. There were 125 well-documented cases presented in sufficient detail for meaningful analysis. In addition, there were 73 other cases reported with inadequate details given. In all then, there were 198 cases, 16 of which yielded two or more anaerobes. Among the 125 cases reported in some detail, there were only 13 in which aerobic or facultative organisms were also present and 14 in which the presence of such organisms was uncertain. Thus, as compared with the brain abscess data presented earlier, anaerobic meningitis is more apt to be a monomicrobial infection and less likely to be a mixed anaerobic–aerobic infection. Note that anaerobic meningitis is not uncommonly part of a more extensive intracranial infection. Thus there were 43 patients with concurrent brain abscess, and 12 with concurrent extradural or subdural abscess. There were 5 cases in which the meningitis was localized and therefore not clinically significant as meningitis. The very fact that meningitis is associated with brain abscess as often as was noted should indicate that anaerobes would be involved in meningitis with a certain frequency.

The data in Table 7.6, correlating type and/or source of underlying condition and anaerobic bacteria involved in meningitis, is based on the 78

TABLE 7.5

Anaerobic Bacteria in Meningitis (Literature Survey)

Number of cases involving anaerobes which are reported in some detail (see Table 7.6 for references)	125
Number in which aerobes or facultatives are also present[a]	13
Number in which presence of aerobes or facultatives is uncertain[a]	14
Number with light growth of nonanaerobe or growth of unlikely pathogen[a,b]	2
Number with concurrent brain abscess	43
Number with concurrent extradural or subdural abscess	12[c]
Number with concurrent bacteremia	22[d]
Number with only localized meningitis	5
Number of cases involving anaerobes but reported in inadequate detail	73[e]
Aerobes or facultatives known to be present also[a]	4
Aerobes or facultatives known not to be present[a]	14
Number known to have concurrent brain abscess	15
Number known to have concurrent bacteremia	6
Cryptococcus also present	1
Total group of cases of meningitis involving anaerobes	198
Number with two or more anaerobes[a]	16

[a] Excluding cases with concurrent brain abscess.
[b] Coagulase-negative *Micrococcus* or *Staphylococcus*, aerobic diphtheroid or *Bacillus* species.
[c] Six of these also had brain abscess.
[d] Including one case with endocarditis.
[e] From (21, 134, 145, 210, 226a, 283, 287, 290, 356, 360, 531, 709, 1109, 1145, 1195, 1196, 1199, 1200, 1200h, 1201, 1219, 1241–1244, 1244a–1244e, 1267e).

cases of well-described meningitis in which brain abscess was not concurrently present. The most common underlying condition by far was otitis media and/or mastoiditis, both usually chronic. The other underlying causes were all proportionately much less common than was true in the case of brain abscess, with the exceptions of trauma and surgery, both of which formed a relatively greater proportion of the underlying conditions than was true in brain abscess. In the case of underlying ear or mastoid disease, *Fusobacterium* species were the most common isolates. Again, as in the case of brain abscess, *Bacteroides fragilis* was isolated with some frequency, an important consideration in terms of therapy. There was also one isolate of *Clostridium ramosum* from this source, again important in terms of therapy. Note the overwhelming importance of *Clostridium perfringens* and other clostridia in cases of meningitis subsequent to trauma or surgery. Almost all clostridial isolates were in these categories of patients.

We have seen four patients with meningitis involving anaerobic bacteria. One of these originated in the lung in a patient who had carcinoma of the lung; this case was associated with brain abscess as well and was caused by

TABLE 7.6

Summary of Source and Etiology of Reported Cases of Meningitis Due to Anaerobic Bacteria (125 Patients) (Literature Survey)[a,b]

Type and/or source of underlying condition	Anaerobic streptococci	Microaerophilic streptococci	Bacteroides fragilis	Other Bacteroides or unspecified	Fusobacterium	Other anaerobic gram-negative bacilli or unspecified	Clostridium perfringens	Clostridium ramosum	Other Clostridium or unspecified	Actinomyces	Eubacterium	Propionibacterium	Other NSF[d] gram-positive bacilli or unspecified	Miscellaneous anaerobes	Total
Sinusitis	1[c]				1								1		3
Otitis media and/or mastoiditis	4	1	4	2	14	5		1	1				2	3	37
Dental or oral infection					2					2					4
Tonsillar, pharyngeal	2			1	4										7
Pulmonary and/or pleural	4		1	1	1					1					8
Genitourinary tract							1								1
Trauma	1		1	1			3		5	1					12
Surgery or manipulation[e]	1	1	3	3	1		5					2			16
Bacteremia									1						1
Unknown		1	1	2	1		1		1	3	1	2		2	15
Total	13	3	10	10	24	5	10	1	8	7	1	4	3	5	104

[a] Cases with concurrent brain abscess not included in this table.
[b] From (8, 108, 110a, 155, 168, 210, 248, 258, 262, 265, 268, 275, 282, 302, 350, 362, 363, 365, 372, 373, 411, 412, 460, 480, 490, 534, 558, 621, 659, 1143, 1144, 1202, 1219, 1233, 1240o, 1245–1267, 1267a–1267g).
[c] Numbers refer to number of isolates, usually one per case.
[d] NSF, non-spore-forming.
[e] Mostly on or near nervous system.

microaerophilic streptococci that were isolated in pure culture. Another case was associated with subdural empyema with gas formation in the subdural space. This patient also had a focal encephalitis at autopsy. His underlying disease was chronic otitis media and mastoiditis with an acute and chronic petrositis with osteomyelitis of the petrous bone. This case, which also involved septic shock, was a dual infection with *Bacteroides fragilis* and *Escherichia coli*. A third case was a postoperative complication in a patient who had an Ommaya valve inserted because of chronic meningitis due to *Coccidioides immitis*. This case involved acute purulent meningitis, bacteremia, and localized abscess formation about the Ommaya valve and was due to a microaerophilic streptococcus that was recovered in pure culture. The patient recovered with intensive penicillin therapy and removal of the Ommaya valve. A fourth case was an uncomplicated meningitis due to *Bacteroides fragilis*. A detailed listing of all anaerobic bacteria recovered in cases of meningitis, together with their frequency in pure culture, is noted in Table 7.10.

Unique pathologic findings were reported in a case of *Clostridium perfringens* meningitis by Conomy and Dalton (362). In addition to the usual meningeal reaction, these workers noted acute necrotic vasculitis in supraglenoid vessels of all sizes. Thrombosis was present in many of these vessels, leading to hemorrhagic cerebral infarction and supraglenoid hemorrhage. Antemortem gas bubbles were found to a limited extent in the occipital lobe where they formed microcysts. Extensive brain necrosis was also noted by Worster-Drought (363) in a case of meningitis due to a fusiform bacillus and an aerobic streptococcus. The association with brain abscess and subdural and extradural abscess has already been commented on. Frühwald (364) also noted a pituitary abscess in addition to other brain abscesses. Alexander (365) described acute hemolytic anemia in the course of meningitis due to *Clostridium perfringens* (presumably bacteremia was involved, but this was not specified).

Aside from the acute hemolytic anemia mentioned above, other distinctive features that might suggest the possibility of anaerobic infection included foul-smelling pus and even foul odor to the spinal fluid obtained on lumbar puncture, the occasional presence of gas, the site and type of predisposing or underlying condition, the unique morphology of certain of the anaerobes on direct gram stain, and the failure to obtain growth from a grossly purulent specimen. Bacteremia is a relatively common accompaniment of anaerobic meningitis (20 instances among 115 patients), in contrast to brain abscess where bacteremia was found to be quite uncommon.

Therapy is similar to that described for brain abscess except that surgery is not indicated other than for underlying conditions such as mastoiditis or accumulations of pus in the subdural or extradural space, etc. Therapy should be intensive but need not be as prolonged as in the case of brain abscess.

Subdural Empyema

Subdural empyema (Fig. 7.4) is a very serious condition. It is crucial that it be recognized early since the disease is curable if diagnosed early, drained surgically, and treated with antimicrobial agents. It may be distinguished from brain abscess by the very rapid evolution, the spinal fluid findings (fluid under high pressure with several hundreds to thousands of neutrophils but sterile on culture), and angiography. Orbital swelling may also be noted in patients with acute subdural empyema.

Subdural empyema may be distinguished from meningitis by the presence of hemiparesis, hemiplegia, or aphasia; the frequently normal sugar content in the spinal fluid in the case of subdural empyema; and the frequency with which subdural empyema follows frontal sinusitis.

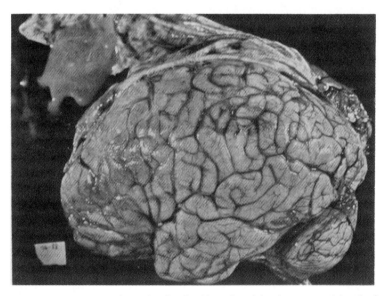

Fig. 7.4. Postmortem specimen showing liquid pus in subdural space overlying the frontal lobe. The subarachnoid reaction and perivascular exudate may be seen on the surface of the frontal lobe with flattening of the convolutions due to swelling. From McLaurin (1179). "Cranial and Intercranial Suppuration" (E. S. Gurdjian, ed.), 1969. Courtesy of Charles C. Thomas, Publisher, Springfield, Illinois.

TABLE 7.7
Subdural Empyema Involving Anaerobes (Summation from Literature)

Comments on infection	Source of infection	Anaerobes isolated	Aerobes isolated	Reference
32 cases—no individual case detail	Variable	1 anaerobic streptococcus and 1 fusiform bacillus isolated (1 case each); 16 cases had no growth	12 aerobes in all	Anagnostopoulos and Gortvai (1268a)
Foul discharge	Cholesteatoma, attic perforation	Anaerobic streptococcus	None	Beeden et al. (1268b)
	Frontal and maxillary sinusitis	Streptococcus anaerobius	None	Beerens and Tahon-Castel (115)
37 cases—no individual case detail	Variable	2 cases yielded anaerobic streptococcus; 13 sterile	24 aerobes in all	Bhandari and Sarkari (1268c)
? Infected hematoma	Trauma, surgery	Clostridium hastiforme, Clostridium sporogenes	Nonhemolytic streptococcus	Cairns et al. (372)
Foul odor, subgaleal abscess also	Pansinusitis	?	Haemophilus influenzae, hemolytic streptococcus (not group A or O)	Coonrod and Dans (1268)
	?	Anaerobic streptococcus	None	Crocker et al. (1240n)
	?	Micrococcus foetidus	None	Dereymaeker et al. [in Wiesmann (370)]
Hematoma present	?	Bacteroides melaninogenicus	?	Felner and Dowell (168)
Also had brain abscess, meningitis, extradural abscess and pituitary abscess	Foreign body led to retropharyngeal and prevertebral abscesses	Bacillus fusiformis	None	Frühwald (364)
Also had subarachnoid abscess, brain abscess and meningitis	? Lung abscess, ? appendicitis	Bacillus fusiformis (by smear only)	None	Ghon and Mucha (1211)
6 cases in all. One anaerobic case, with associated brain abscess	?	Microaerophilic streptococcus	None	Gurdjian and Webster (1240j)

19 cases—no individual case detail; 3 anaerobic cases (including above case by Gurdjian and Webster); 5 cases—no growth 5 cases, 2 anaerobic	Variable	One case—microaerophilic streptococcus. One case—anaerobic micrococcus plus *Bacteroides*	11 aerobes in all	Gurdjian and Thomas (367)
	Dental, paranasal sinuses, associated epidural abscess, meningitis	None, but pus foul-smelling	*Staphylococcus aureus* and nonhemolytic streptococcus	Hollin *et al.* (1268d)
	Dental, inframandibular abscess; several small subcortical abscesses also present	*Micrococcus foetidus*	None	
7 cases, 2 anaerobic	Lung abscess? bronchiectasis; associated brain abscess	Microaerophilic streptococcus	None	Keith (1268e)
	Maxillary sinusitis—foul pus	Microaerophilic streptococcus	None	
Subdural empyema in lumbar area; also had brain abscess and meningitis	Chronic otitis media, bacteremia	*Bacteroides fragilis*, anaerobic streptococcus	None	King *et al.* (1220)
Also had meningitis, fetid pus	Otitis media, mastoiditis	*Bacteroides funduliformis* (blood culture)	None	Lamy and de Font-Réaulx (1258)
	Sinusitis, ? frontal bone osteomyelitis	Anaerobic streptococcus	None	Murphy and Wilkes (1269)
Brain abscess also	Lung	*Actinomyces* (new species)	None	Prolo and Hanbery (354)
3 cases, 1 anaerobic	Frontal sinusitis, associated bacteremia	Fusiform bacillus	None	Ray and Parsons (1268f)
Also had brain abscess and meningitis	Posttonsillectomy	*Bacteroides funduliformis*	None	Salibi (1227)

(continued)

TABLE 7.7 (*continued*)

Comments on infection	Source of infection	Anaerobes isolated	Aerobes isolated	Reference
Concurrent brain abscess and meningitis	Chronic otitis media	*Eggerthella convexa* (from brain abscess)	None	Schaffner (1240a)
	Frontal sinusitis	*Eggerthella convexa*	Nonhemolytic streptococcus	
	?	*Eggerthella convexa*	None	
16 cases—no individual case detail	Variable	15% of isolates were anaerobic streptococci	?	Swartz and Karchmer (1109)
Also had brain abcess	Sinusitis, osteomyelitis	*Bacteroides*	Enterococcus	Wiesmann (370)
	Osteomyelitis	Anaerobic streptococcus	None	
	Frontal, ethmoid, and maxillary sinusitis	Anaerobic streptococcus (from frontal sinus)	*Staphylococcus epidermidis*	Wilkins and Goree (1270)
"Combined extradural–intradural sepsis"	Postoperative (intracranial surgery)	*Clostridium*	None	Wright (1029)
	None identifiable	Microaerophilic streptococcus	None	Yamada *et al.* (1271)
	Frontal sinusitis	*Peptostreptococcus anaerobius*	None	Yoshikawa *et al.* (368)
Associated epidural abscess, brain abscess, bacteremia	Chronic otitis media, mastoiditis	*Bacteroides fragilis* ss *thetaiotaomicron*	None (*Staphylococcus aureus* isolated, in addition, from brain abscess)	
	Frontal and maxillary sinusitis	*Peptococcus variabilis*	Streptococcus, not group A or D	
	Frontal sinusitis (with air-fluid level)	*Peptostreptococcus micros*, *Eubacterium lentum*, *Propionibacterium acnes* (gram-negative pleomorphic bacilli seen on smear, not recovered)	None	
Associated bacteremia	Bilateral sinusitis (frontal?)	*Peptostreptococcus intermedius*	None	

172

Anaerobic bacteria probably play a bigger role in subdural abscess than has been appreciated. It is certainly established that anaerobic bacteria are important in sinusitis, a frequent forerunner of subdural empyema. Hitchcock and Andreadis (366) found that anaerobic streptococci were the causative agents of 6 cases among a total of 27 cases of subdural empyema. Seven of these 27 patients either showed no growth on culture of the subdural pus or grew only unlikely pathogens; 2 of the 4 cultures that were sterile showed gram-positive cocci by smear. Gurdjian and Thomas (367) found anaerobes in 4 of 19 cases of subdural empyema, 5 of which were sterile or showed only nonpathogens. Yoshikawa *et al.* (368) described five cases of subdural empyema due to anaerobes. Bacteria recovered included four strains of anaerobic cocci, *Bacteroides fragilis*, *Eubacterium lentum*, and *Propionibacterium acnes*; only one aerobe, a streptococcus, was recovered. Schiller *et al.* (369) discussed 10 cases of subdural empyema due to anaerobic streptococci; a β-hemolytic aerobic streptococcus was also recovered in one case. Wiesmann (370) found anaerobes in 2 of 22 cases of subdural empyema, 8 of which were sterile or showed nonpathogens. We have seen two cases of subdural abscess involving anaerobes, one referred to in the section on meningitis and the second caused by microaerophilic streptococci.

Table 7.7 is a summary of cases of subdural empyema involving anaerobes culled from the literature. Among the 43 cases listed, there were 30 in which the source or probable source of infection was indicated, and among these the source was sinusitis (usually frontal) in 14 cases. Trauma, foreign body, and/or surgery accounted for 5 cases, as did otitis media and mastoiditis. Anaerobic or microaerophilic cocci were involved in 25 patients, as were 14 gram-negative anaerobic bacilli of the genera *Bacteroides* and *Fusobacterium*. *Bacteroides fragilis* was found in 5 patients. Again the association between *Clostridium* and trauma or surgery as a background for central nervous system infection is noted. Specific organisms found in both subdural and extradural empyema in this review are noted in Table 7.10. The frequent association with other types of serious central nervous system infection in this group of patients is important to note; at least 13 patients also had brain abscess and/or meningitis.

Therapy of subdural empyema is similar to that for brain abscess, with surgery playing a key role. Early surgical drainage is imperative. Exploration is indicated in any patient with meningitis in whom there is frontal sinusitis and hemiparesis, hemiplegia or aphasia. Membrane and granulation tissue do not form about the empyema readily. The extension of the pus depends on the primary site of infection and the anatomical peculiarities of the subdural space. Collections beneath the hemispheres are uncommon, whereas dorsolateral and interhemispherical collections of pus are common.

The recumbent position of the patients aids spread posteriorly as well. Neurosurgical exploration will be aided by knowledge that the extension of pus from the various primary sites appears to follow fairly well-defined paths (366).

Extradural Empyema

Extradural empyema is a much more benign condition than subdural empyema. It rarely causes increased intracranial pressure or focal neurological signs. The major manifestations are irregular fever, localized headache, and tenderness. There may be minor to moderate leukocytic pleocytosis of the spinal fluid.

Table 7.8 summarizes 20 cases of extradural empyema from the literature aside from those patients who had both subdural and extradural empyema listed in Table 7.7. Eight of these 20 patients had, as an associated condition, meningitis and/or brain abscess, and the extradural empyema was a minor part of the picture. In 4 patients, the epidural abscess was in the spinal canal, and here, of course, this type of infection may be serious because of local pressure effects that may lead to paralysis. Otitis media and/or mastoiditis was the most common underlying source of infection, with gram-negative anaerobic bacilli (including one isolate of *Bacteroides fragilis*) present in each of the seven patients with this type of underlying disease. Sinusitis was the underlying process in three patients, and two patients with *Clostridium perfringens* infection developed extradural empyema following craniotomy. One of our own cases of extradural abscess is mentioned in Chapter 6.

In the case of extradural empyema, antimicrobial therapy is often curative without surgical intervention, provided that there is not another focus, such as brain abscess, which would be an indication for surgery.

Spinal Epidural Abscess

This condition is important to diagnose early in order to prevent compression of the spinal cord and permanent neurologic damage. A case of spinal epidural abscess was noted in the above section. Parker [cited by Cope (210)] also described a case of actinomycosis presenting as an extradural mass causing compression of the spinal cord. See Table 7.8 for additional cases.

Miscellaneous Central Nervous System Infections

The six cases of third ventricular mass as a manifestation of isolated intracranial actinomycosis which were culled from the literature by Cope (210) have already been mentioned. Cope also cited a description by Zajewloschin of actinomycotic infection of the pituitary gland.

Other miscellaneous central nervous system infections involving anaerobes are listed in Table 7.9. Some of these have already been touched on because of the concomitant presence of meningitis and/or brain abscess. One of these latter patients also had ependymitis and one had extensive necrosis of both frontal lobes along with meningitis.

Parker and Collins (371), from a review of the literature, found a total of 49 cases of intramedullary abscess of the spinal cord, two of which yielded anaerobes on culture and 25 of which were either sterile or had insignificant organisms recovered on culture.

Gurdjian and Thomas (367) failed to find any anaerobes in a series of 24 patients who developed intracranial suppuration following penetrating wounds; 8 of these cases were either sterile or yielded nonpathogenic organisms on culture. On the other hand, Cairns *et al.* (372) described 16 cases of head wound infection (2 with meningitis and 2 with brain abscess) following trauma in which anaerobes were involved. These workers were impressed with the fact that the skull and head wounds involving clostridia were relatively benign for the most part and that, as in wounds of the limb, one might have clostridia present without any evidence of infection. The intensity of the infection appeared to be directly correlated with the size of the wound, rather than with the species of organism recovered. Cairns *et al.* (372) commented further that Small *et al.* found clinical evidence of toxemia in clostridial infections of head wounds only when the temporal muscle was involved. They further cited studies of Grashchenkov of Russian soldiers during the second World War. Several hundred patients with brain wounds were studied, and *Clostridium perfringens* was recovered from 24%, *Clostridium novyi* from 4%, *Clostridium sordelii* from 5%, and *Clostridium butyricum* from 10%. There was a relatively low incidence of gas infection, the rest of the patients showing either a relatively low grade and chronic type of infection or no evidence of infection whatever. As shown in Table 7.9 there was a total of 24 anaerobes, 21 of them clostridia, isolated from the 16 cases described by Cairns *et al.*, along with 21 facultative or aerobic organisms, at least 2 of which were probably contaminants. Bornstein *et al.* (275) described two cases of skull and head wound infection from which anaerobic streptococci were recovered. Wright (373) described a subgaleal abscess and an osteomyelitis of the skull, each due to anaerobic streptococci,

TABLE 7.8

Extradural Empyema Involving Anaerobes[a] (Summation from Literature)

Comments on infection	Source of infection	Anaerobes isolated	Aerobes isolated	References
39 cases in all, no individual case detail for most. 5 cases yielded no growth on culture	?	Anaerobic streptococcus	None	Baker et al. (1271a)
	?	Propionibacterium	α-Hemolytic streptococcus	
	Dental extraction	Bacteroides melaninogenicus, Veillonella, and anaerobic streptococcus	None	Beerens and Tahon-Castel (115)
Meningitis, probable extradural empyema	Surgery for tumor 4 months previously	Clostridium perfringens	None	
	Otitis and ?mastoiditis	Eggerthella convexa	None	
Meningitis also	Otitis, mastoiditis	Bacteroides funduliformis	None	Chandler and Breaks (1247)
Brain abscess and abscess over frontal lobe also	Otitis, mastoiditis	Fusiformis fusiformis group (from blood culture)	None	Franklin (482)
	Otitis, mastoiditis	Fusiformis fusiformis group (from blood culture)	None	
Spinal epidural abscess; foul pus	Ischiorectal and psoas abscesses	Gram-positive cocci on smear, not cultured	None	Frumin and Fine (542)
Meningitis also	Otitis, mastoiditis	Fusobacterium nucleatum	None	Grumbach and Verdan (350)
Associated subdural abscess and meningitis	Maxillary and frontal sinusitis	None, but pus from sinuses was foul-smelling	Staphylococcus aureus, nonhemolytic streptococcus	Hollin et al. (1268d)
	Ethmoid sinus, orbit	Fusiforms and spirilla seen, not cultured	Staphylococcus aureus	Kompanejetz (1272)

Spinal epidural abscess, vertebral osteomyelitis, liver abscess, subphrenic abscess	Pneumonia, lung abscesses, pleural empyema	*Actinomyces* seen in sections at autopsy, no culture	None?	Krumdieck and Stevenson (1271b)
Infected cephalhematoma, newborn	Trauma of birth, ?vagina of mother	*Bacteroides*	None	Lee and Berg (1273)
Spinal epidural abscess; associated meningitis	Cervical myelography and corticosteroid injection	Microaerophilic streptococcus, group **F**	None	Lerner (1240o)
Also had brain abscess, meningitis and necrotic temporal bone	Otitis, mastoiditis	Fusiform	None	Maresch (1224)
Meningitis also	Sinusitis, frontal bone osteomyelitis, orbital cellulitis	Fusiforms and spirilla seen, not cultured, *Lactobacillus*	*E.coli Streptococcus faecalis*	Seecof (248)
Also had brain abscess and meningitis	Otitis, mastoiditis	*Bacteroides funduliformis*	None	Smith *et al.* (1233)
Spinal epidural abscess	?	*Actinomyces*	?	Wilson and Kammer (cited by Krumdieck and Stevenson, 1271b)
Also had meningitis and subgaleal abscess	Postcraniotomy	*Clostridium perfringens*	None	Wright (373)

[a] Those with subdural empyema also not listed—see Table 7.7.

TABLE 7.9 Miscellaneous Central Nervous System Infections Involving Anaerobes (Literature Summary)[a]

Type of infection	Source of infection	Anaerobes isolated	Aerobes isolated	References
Skull and head wound infection	?	Anaerobic streptococcus	?	Bornstein et al. (275)
Skull and head wound infection	?	Anaerobic streptococcus	?	
16 cases of head wound infection (2 with meningitis and 2 with brain abscess)	Trauma	9 Clostridium sporogenes 8 Clostridium welchii 3 Clostridium bifermentans 1 Clostridium capitovale 1 anaerobic NSF GPR[b] 1 anaerobic streptococcus 1 fusiform	5 coliforms 3 E. coli 2 Proteus 2 Streptococcus 1 viridans streptococcus 2 S. faecalis 1 S. aureus 1 S. albus 2 gram-positive cocci 1 diphtheroid 1 gram-negative coccobacillus 4 sterile	Cairns et al. (372)
6 cases of mass in third ventricle	?	Actinomyces	?	Cope (210)
Ependymitis; also brain abscess and meningitis; gas cysts in brain	Postoperative excision of urethroperineal fistula	Bacillus aerogenes capsulatus	None	Howard (1218)
Extradural mass causing compression of spinal cord	?	Actinomyces	?	Parker [cited in Cope (210)]
Probable septic cerebral emboli	Probable bacterial endocarditis	"Actinomyces bovis" (spinal fluid)	None	Schwartz and Christoff (1273a)
Infected epidermoid cystic tumor of pons	?	Catenabacterium filamentosum	None	Stacy et al. (1274)
Bilateral frontal lobe necrosis plus meningitis; foul pus	Pyorrhea	Fusiform, smear only	Streptococcus	Worster-Drought (363)
Subgaleal abscess	Postoperative craniotomy	Anaerobic streptococcus	Staphylococcus albus	Wright (373)
Osteomyelitis of skull	Postoperative craniotomy	Anaerobic streptococcus	Diphtheroid	Zajewloschin [In cope (210)]
Infection in pituitary	?	Actinomyces	?	

[a] See Tables 7.7 and 7.8 for cases of subgaleal abscess, pituitary abscess, and subarachnoid abscess (plus subdural or extradural empyema).
[b] NSF GPR, non-spore-forming gram-positive rod.

178

TABLE 7.10

Specific Anaerobes Isolated from Central Nervous System Infections[a]

Organism	Brain abscess[b]		Meningitis		Extradural and subdural empyema		Other CNS infections		Total isolations all CNS infections[d]
	No. cases isolated from	Isolated in pure culture[c]	No. cases isolated from	Isolated in pure culture	No. cases isolated from	Isolated in pure culture	No. cases isolated from	Isolated in pure culture	
Anaerobic cocci[e]									
Microaerophilic *Streptococcus*	19	12	2	2	5	5			30
Peptococcus	8	2	13	8	3	2			16
Peptostreptococcus	48	18			9	7	3	0	148
Veillonella	5				1	1			7
Other anaerobic cocci (+ unspecified)	3				1	1			4
Gram-negative anaerobic bacilli									
Bacteroides corrodens	1	0							1
B. fragilis	25	9	10	5	3	2			47
B. melaninogenicus	6	0			1				13
B. oralis	1		1	1					1
B. pneumosintes	2	1							4
B. serpens	3	1	1	1					3
Bacteroides, unknown or unspecified	9	2			1	1			48
Fusobacterium gonidiaformans	1	1							1
F. glutinosum									1
F. mortiferum									1
F. necrophorum	15	8	6	6					33
F. nucleatum	15	6	4	1					25
F. russii									1
Fusobacterium-unknown or unspecified	2	1	4	4	1	1			11
Spirillum	2	0							2
Vibrio	1	0	1	0					2
Anaerobic gram-negative rods, unknown or unspecified	8	5	5	2					17

(continued)

179

TABLE 7.10 (*continued*)

Organism	Brain abscess[b] No. cases isolated from	Brain abscess Isolated in pure culture[c]	Meningitis No. cases isolated from	Meningitis Isolated in pure culture	Extradural and subdural empyema No. cases isolated from	Extradural and subdural empyema Isolated in pure culture	Other CNS infections No. cases isolated from	Other CNS infections Isolated in pure culture	Total isolations all CNS infections[d]
Anaerobic non-spore-forming gram-positive bacilli									
Actinomyces	12	9	4	4			1	1	37
Eubacterium alactolyticum	1	1							3
Eubacterium lentum	1	0			1				3
Eubacterium tortuosum			1	0					1
Lactobacillus crispatum	1	1	1	1					2
Propionibacterium	7	3			1				25
Other non-spore-forming gram-positive rod	2	1	1	1			1	0	8
Clostridia									
C. bifermentans	1	0	2	0			2	0	5
C. capitovale							1	0	1
C. cochlearium			1	0					1
C. hastiforme	1	0			1	0			2
C. perfringens	9	4	7	5	1	1	5	0	34
C. ramosum	4	0					1	1	8
C. sordellii			1						1
C. sporogenes	1	0	3	0	1	0	7	0	12
C. tertium	1								1
Clostridium, unknown or unspecified	3	1	1	0	1	1			6
Anaerobe, not further specified	3	0	2	0					11
Microaerophilic diphtheroid			1	1					1

[a] Cases with inadequate case detail or isolated only from blood not included. Organisms seen but not cultured not counted.

[b] If have multiple types of intracranial infection, coded under brain abscess if that was present; if not, under meningitis if that was present.

[c] Only one anaerobic species present.

[d] This column also includes definite isolates from patients in whom data regarding possible mixed culture is inadequate, plus isolates from indirect sources (such as blood). Organisms from cases with inadequate clinical detail are included.

[e] Very few designated as to species.

following craniotomy. We have also seen two postcraniotomy infections with anaerobes, one the case of infection around an Ommaya valve described earlier, and the other a persistent scalp infection in a middle-aged alcoholic female who required surgery for a subdural hematoma. This latter infection was caused by an unidentified *Bacteroides*.

Suppurative cerebral venous thrombophlebitis is rarely manifest today. Prior to the antimicrobial era, anaerobic infections of the ear, nose, and throat commonly spread to these intracranial venous sinuses causing thrombosis and leading to many local and distant complications. Prime local complications, of course, were meningitis and brain abscess.

An overall summary of specific anaerobes isolated from various categories of central nervous system infection is provided in Table 7.10. Anaerobic and microaerophilic cocci and gram-negative bacilli were the major organisms encountered.

8

Cardiovascular Infections

Bacteremia

Transient bacteremia may occur spontaneously or may follow manipulation of various types in areas of the body where there is a normal flora or where an infection exists. Since anaerobes comprise a significant part of the normal flora of the body, it is natural that they would commonly be involved in such transient bacteremias and in the consequences of these bacteremias. Nevertheless, anaerobic bacteremia of this type is not as well documented as is bacteremia due to other organisms, reflecting less knowledge of and interest in anaerobes in the past as well as less efficient techniques for recovering anaerobes from the bloodstream.

The best documentation of such bacteremia is in the bacteremia of dental origin. Such transient bacteremia may follow vigorous dental prophylaxis, endodontic therapy, and other minor dental procedures as well as extractions. Even the rocking motion of teeth during the course of chewing, particularly of firm materials, may lead to bacteremia. One of the most impressive studies of transient bacteremia following dental manipulation is that of Rogosa *et al.* (374). These workers demonstrated bacteremia within 5 minutes of maximal trauma in 28 of 34 patients (82%) requiring extraction of teeth and in 29 of 33 (88%) after curettage, gingivectomy, and other periodontal procedures. In all, 117 strains of bacteria were recovered on blood culture; of these, 72 were anaerobes. There were also 36 streptococci that were not further classified and a number of these must surely have been either anaerobic or microaerophilic. Anaerobes (or microaerophilic organisms) recovered included diphtheroids, *Vibrio*, *Spirillum*, *Bacteroides*, *Veillonella*, *Fusobacterium*, *Actinomyces*, *Leptotrichia*, and unclassified anaerobes. In the study reported by Francis and de Vries (149) 18 of 50 patients undergoing dental extraction had bacteremia 10 minutes later. A total of 38 strains of bacteria were recovered from the bloodstream (1 to 4 organisms per patient), of which 32 were anaerobes. Included among the anaerobes were *Bacteroides melaninogenicus*, *Fusobacterium fusiforme*, *Peptostreptococcus anaerobius*, *Peptostreptococcus evolutus*, *Veillonella alcalescens*, and *Corynebacterium acnes*. In this study it was noted that penicillin given prophylactically worked well when the repository forms were avoided and large doses were given. They gave 3 million units in 3 doses for 2 days, then 1 million units 1 hour prior to extraction. Erythromycin and tetracycline were less effective. Lincomycin was very good and was the second choice of the authors, after penicillin.

In 1937, Marseille (375) studied 100 patients with root abscess or granuloma who underwent tooth extraction. Forty-two of these developed bacteremia, eight of them mixed. Among the organisms recovered were 13 anaerobic streptococci, 9 *B. fusiformis*, and an unspecified number of anaerobic gram-negative cocci. Khairat, in 1966 (376), studied 100 patients undergoing dental extraction. Sixty-four of these had bacteremia, yielding a total of 155 strains of bacteria. Of these, 46 were obligate anaerobes, 5 were true microaerophiles, and 37 grew either anaerobically or in 5% carbon dioxide. Sixty-seven grew aerobically and in other atmospheres.

Other studies have shown a lower incidence of anaerobic bacteremia following dental manipulation. Thus, McEntegart and Porterfield (377) studied 200 patients undergoing extraction. One hundred and eight of these had bacteremia, included among which were six bacteremias due to *Veillonella gazogenes*. Among 84 patients with dental extraction studied by Müller (378), 40 had bacteremia; in 4 instances anaerobes were involved. Conner *et al.* (379) studied the incidence of bacteremia following periodontal scaling. Thioglycolate broth was used to recover anaerobes. Among 109 patients, there were 38 with positive blood cultures, including four with anaerobes (*Actinomyces odontolyticus*, *Fusobacterium*, and *Veillonella*). The use of antimicrobial prophylaxis, of course, modified the results. Schirger *et al.* (380) studied 127 patients undergoing oral surgical procedures, including dental extraction; 77 of these received oral penicillin V for 24 hours prior to the procedure, and the other 50 were untreated. The total number of positive blood cultures at the time of surgery was 27 in the untreated group and 13 in those receiving penicillin V. Eighteen hours later, 5 and 7, respectively, had positive blood cultures. The only anaerobe isolated was one strain of *Peptococcus* which was recovered 18 hours post-surgery in the untreated group. Fifty patients undergoing oral surgery and receiving ampicillin prophylaxis were studied by Tolman *et al.* (381). Three of these patients developed bacteremia. Of these, one had three anaerobes in his blood culture (*Peptococcus, Bacteroides funduliformis,* and a *Bacteroides* species). There are several recent studies on this subject. Berry *et al.* (381a) studied 34 children requiring general anesthesia (and nasotracheal intubation) for extensive dental repair. Blood cultures were drawn before and after intubation, and after various types of dental manipulation, but data regarding specific organisms recovered are not broken down as to time drawn. Among 26 organisms recovered in all, there were 5 anaerobic diphtheroids and one *Bacteroides melaninogenicus.* A study in children undergoing oral prophylaxis (381b) revealed bacteremia in 28% following the manipulation. Among the 28 strains recovered, 21 were anaerobes (9 diphtheroids, 5 *Veillonella alcalescens,* 3 *Bacteroides melaninogenicus,* 2 *Peptostreptococcus anaerobius,* 1 *Eubacterium,* and 1 *Fusobacterium*). Crawford *et al.* (381c) detected bacteremia postoperatively in 23 of 25 patients undergoing tooth extraction. Included were 22 *Bacteroides* (14 *B. melaninogenicus* and 1 *B. fragilis* among these), 11 *Fusobacterium* (10 *F. nucleatum*), 8 *Vibrio,* and a number of strains of the following genera: *Peptostreptococcus, Peptococcus, Veillonella, Actinomyces, Arachnia, Leptotrichia,* and *Propionibacterium.* Berger *et al.* (381d) noted 8 instances of bacteremia among 30 persons using an oral irrigation device; anaerobes encountered were *Bacteroides melaninogenicus* and *Peptostreptococcus intermedius.* Although most of these bacteremias of dental origin were transient and not clinically significant, it is clear that this type of bacteremia, as with any bacteremia, may lead to serious consequences. Thus, Gröschel (164), described a case of hematogenous osteomyelitis due to *Sphaerophorus necrophorus* (but gram-positive!) following tooth extraction. There are a number of cases of anaerobic endocarditis following dental extraction (see Table 8.3). A very interesting case was described by Puklin *et al.* (382). A patient developed pulmonary infection secondary to extraction of an infected tooth. In the course of an accompanying sepsis, he developed Osler's nodes from which were recovered on culture *Actinomyces israelii* and *Fusobacterium. Fusobacterium* sepsis has been described following a human bite [Murphy *et al.* (383)].

While most of the studies of bacteremia relate to dental manipulation, it is clear that transient or more serious bacteremia may follow manipulation of other parts of the body or may be associated with infection in various parts of the body. W. R. Winn, M. Kraut, and S. M. Finegold (unpublished observations) have noted anaerobic bacteremia following nasotracheal intubation. Burman and Hill (383a) reported one instance of anaerobic streptococcal bacteremia following bronchoscopy. A case of transient bacteremia due to *Bacteroides fragilis* associated with percutaneous liver biopsy was reported by LeFrock *et al.* (383b); the patient had an appendiceal abscess. Also reported by LeFrock and co-workers (383c) was a study of transient bacteremia associated with barium enema; among 20 episodes of bacteremia (175 patients studied in all), there were 3 due to *Bacteroides* and one due to *Veillonella*. In some of these cases, bacteremia persisted for 15 minutes, but not for 30 minutes. LeFrock *et al.* (384) studied 200 patients undergoing sigmoidoscopy. Nine of these had bacteremia, of which one had a *Bacteroides* bacteremia. *Bacteroides* was recovered at 1, 5, 10, and 15 minutes following the procedure, but not at 30 minutes. Anaerobic bacteremia occurs with some frequency following gastrointestinal surgery (56d, 56g, 384a, 1573) and in relation to disease, particularly malignancy, of the gastrointestinal tract (56d, 56e, 56g, 384b, 384c, 1573). *Fusobacterium* septicemia has been described in a newborn with diarrhea (385). Bacteremia due to a *Clostridium* species was described in a cirrhotic who developed spontaneous peritonitis [Ansari (386)]. Fifty-four percent of a group of 25 patients with anaerobic liver abscess had positive blood cultures for anaerobes [Sabbaj *et al.* (387)]. These authors noted that positive blood cultures for anaerobes, without an obvious portal of entry, may be an important clue to the presence of liver abscess. Female genital tract disease or manipulation is a relatively common background for anaerobic bacteremia (56b, 56d, 56g, 384a, 387a, 1114, 1573). Sullivan *et al.* (388) recovered 11 anaerobic organisms following genitourinary tract manipulation, primarily urethral dilatation. The organisms recovered included 3 strains of *Peptococcus*, 2 *Peptostreptococcus*, 1 *Veillonella*, 1 microaerophilic streptococcus, 1 *Bacteroides melaninogenicus*, 1 *Bacteroides* species, 1 *Bifidobacterium adolescentis*, and 1 *Propionibacterium avidum*. Biorn *et al.*, in 1950 (389); described a case of anaerobic streptococcal bacteremia following transurethral resection of the prostate. Pien *et al.* (390) described a case of bacteremia due to *Peptococcus* following dilatation of a urethral stricture. Okubadejo *et al.* (390a) noted *Bacteroides* bacteremia in a patient following prostatectomy. Mackenzie and Litton (384a) encountered *Bacteroides* bacteremia in a patient with cystitis who underwent cystoscopy and in another patient who had carcinoma of the bladder who underwent transurethral resection. Vinke and Borghans (390b) reported *Bacteroides* bacteremia in 2 patients with urinary retention who were catheterized. Chow and Guze (1573) reported *Bacteroides fragilis* bacteremia in two patients with obstructive uropathy requiring surgical manipulation. A third patient of theirs, with bladder metastasis of carcinoma of the rectum and obstructive uropathy, had bacteremia with 4 anaerobes—Bacteroidaceae sp., *Clostridium* sp., *Eubacterium* sp. and an anaerobic streptococcus. It is not clear whether or not this was precipitated by cystoscopy. Obstructive urinary tract disease was also noted by Meyer (56g) as an underlying disorder in 3 patients with

Bacteroides fragilis bacteremia. A case of *Clostridium perfringens* bacteremia following nephrectomy for calculous pyonephrosis was described by Mencher and Leiter in 1938 (391). Fishbach and Finegold (392) noted bacteremia due to *Fusobacterium gonidiaformans* in a patient with a prostatic abscess.

Anaerobic bacteremia may also be seen in relation to intravenous therapy. This presumably implies contamination of the skin of the patient or the attending physician or nurse with the anaerobes that are involved. Altemeier *et al.* (393) described "third day surgical fever"—sepsis related to continuous intravenous infusion devices. They described this situation in 54 patients, 13% of whom had infections due to *Bacteroides* (*Serratia* accounted for 30% of these infections; *Klebsiella* 16%; and *Staphylococcus* 11%). Beazley *et al.* (394) noted 3 cases of *Bacteroides* sepsis in association with intravenous catheters. Hirsch *et al.* (395) described 1 patient with *Bacteroides* sepsis secondary to a subclavian catheter. A venous cutdown led to *C. perfringens* wound infection and bacteremia (396). One strain of *Clostridium* sp. and one of *C. perfringens* were recovered among 57 positive routine cultures of intravenous catheters at the time of removal (no associated sepsis) by Crossley and Matsen (396a). A study (396b) of unused intravenous cannulae and other medical devices yielded anaerobic streptococci and *Propionibacterium acnes*, among other organisms.

Hermans and Washington (397) pointed out that *Bacteroides*, in particular, and other anaerobes as well, are not uncommonly involved in polymicrobial bacteremia and that, therefore, they may be easily overlooked. This was also stressed by von Graevenitz and Sabella (398) who described 3 cases of mixed bacteremia with *Bacteroides* plus facultative organisms in which the *Bacteroides* could only be identified by subculture to a selective medium (kanamycin blood agar). McHenry *et al.* (398a) noted 8 cases of polymicrobial bacteremia involving anaerobes. Sanford (398b) noted that 57% of 23 patients with polymicrobic bacteremia had anaerobes as part of the infecting flora. Cole *et al.* (399) recommended culturing lymph from the thoracic duct in patients with suspected bacteremia and negative blood cultures. However, the anaerobes isolated in their series (which were not recovered from blood but only from lymph) were probably all contaminants. Included were anaerobic streptococci and clostridia.

The portal of entry for anaerobic bacteremia in a number of published series is noted in Table 8.1. Note that the upper and lower respiratory tract represented the portals of entry much more commonly in the preantimicrobial era than following the introduction of antimicrobial agents. "Angina" (acute bacterial tonsillopharyngitis) was a particularly important source of anaerobic bacteremia prior to the advent of antimicrobial drugs. Presently, the gastrointestinal tract, and to a lesser extent the female genital

TABLE 8.1

Portals of Entry, Anaerobic Bacteremia (Number of Cases)

Gastrointestinal tract	Female genital tract	Lower respiratory infection	Ear, sinus, oropharyngeal area	Skin, soft tissue	Miscellaneous or unknown	Total	Reference	Year
26	17	3	72	24	6	148	Gunn (296)[a]	1956
213	116	36	33	29	31	458	Love (649)[b]	1972
8		7	4		1	20	Tynes and Utz (1275)[a]	1960
6	3	2				11	McHenry et al. (530)	1961
8	5	3	5	3	1	25	Tynes and Frommeyer (489)	1962
18	8			7	2	35	DuPont and Spink (647)	1969
24	7		1		7	39	Bodner et al. (460)	1970
5	18	1	1	4	6	35	Gelb and Seligman (490)	1970
89	11	1		11	11	123	Marcoux et al. (420)	1970
16	2	1	2		5	26	Sinkovics and Smith (411)	1970
139	41	30	26	14	—	250[c]	Felner and Dowell (168)	1971
15					1	16	Kuklinca and Gavan (491)[d]	1971
46	1			13	7	67	Wilson et al. (105)	1972
48	40	3	3	12	6	112[c]	Chow and Guze (1573)	1974

[a] From a literature review, most cases from the pre-antibiotic era.
[b] From a literature review, 1956–1971.
[c] Only Bacteroidaceae considered.
[d] Postmortem blood cultures.

187

tract, represent the major portals of entry for bacteremia due to anaerobic organisms. In the study reported by Martin and McHenry (400), the portal of entry was the gastrointestinal tract in 70% of their patients and the genito-urinary tract in 20%.

While female genital tract infections do not account for the majority of cases of anaerobic bacteremia, anaerobic bacteremia does occur with considerable frequency in subjects with such infections. Thadepalli et al. (1114) noted that at Cook County Hospital in Chicago the Gynecology Ward had the highest incidence of positive blood cultures for anaerobes, one per 50 admissions, as compared to 1 per 300 admissions for Adult Medicine and Surgery Wards. Rotheram and Schick (401) noted that 34 of 56 patients with septic abortion had anaerobic bacteremia. Smith et al. (416) reported on 76 cases of bacteremia in patients who had septic abortions; 63% of these involved anaerobes. Now that criminal abortion is seldom practiced and most abortions are done under good circumstances in hospitals, the incidence of infections of all types, including anaerobic bacteremia, following abortion is relatively low, and the very serious anaerobic bacteremias that sometimes were seen in such patients are now seldom seen.

Douglas and Davis (402) performed 524 blood cultures on 295 patients with puerperal infection. There were 90 positive cultures with a total of 98 organisms recovered. Included among the bacteria isolated were 24 strains of anaerobic streptococci, one anaerobic gram-negative bacillus, and three anaerobic diphtheroids. Unusual portals of entry for *Bacteroides fragilis* bacteremia include the middle ear and the lung (278).

The incidence of anaerobic bacteremia, like the incidence of anaerobic infections, in general, is difficult to determine since it is not clear that optimum anaerobic techniques are always utilized and it is not certain that anaerobes which have been recovered are clinically significant. Data on incidence from the literature are summarized in Table 8.2. In general, it appears that 5 to 15% of cases of bacteremia involved anaerobic organisms. However, the series by Washington and Martin (403) had an incidence of 27% positive for anaerobes, all of which were clinically significant. Cabrera et al. (404) described 19 cases of acute clostridial sepsis in patients with malignancy out of a total of 4145 autopsies, an incidence that they regarded as relatively high. Kagnoff et al. (405) encountered 55 cases of *Bacteroides* bacteremia in a 5-year period in a hospital dealing primarily with malignant disease. On the other hand, Bodey et al. (406) in a similar hospital, found only one *Bacteroides* and one *Clostridium* out of 126 organisms involved in polymicrobial bacteremia. Four of 28 episodes of bacteremia in surgical patients with intestinal diseases involved anaerobes (406a). Anaerobic bacteria accounted for 28.8% of positive blood cultures in a 6 month study involving 188 patients with significant bacteremia (398b). *Bacteroides* accounted for

TABLE 8.2

Incidence and Mortality, Anaerobic Bacteremia

Number of years covered	Number of blood cultures	Number positive	Number positive, anaerobes	% Positive, anaerobes[a]	Number of patients positive, anaerobes	Mortality	Reference	Year
3		144	6				Rosenow and Brown (1276)	1938
4	634		4[b]	0.6			Dack (106)	1940
	8,620	2,362	94	3.8			Reynes (418)	1947
1	5,000	665			5 (3 significant)		Kotin (1277)	1952
3	2,773	133 (½ significant)	8				Munroe and Cockcroft (1278)	1955
15		gram-negative rod only 137 patients	4				Spittel et al. (488)	1956
		85	9	10.5			Stokes (108)	1958
						15/20[c]	Tynes and Utz[c] (1275)	1960
11		1,026 patients		2.2	23	3/11	McHenry et al. (530)	1961
6						5/25	Tynes and Frommeyer (489)	1962
2		108 patients			5[d]		Ames and Fischer (1279)	1966
5		97	19	19.6			Kozakai (360)	1966
12		398 patients		4.5[d]	18[d]	6/15	Altemeier et al. (431)	1967

(Continued)

189

TABLE 8.2 (*continued*)

Number of years covered	Number of blood cultures	Number positive	Number positive, anaerobes	% Positive, anaerobes[a]	Number of patients positive, anaerobes	Mortality	Reference	Year
15	15,543	2,410		~2			Dalton and Allison (1280)	1967
1		106		5.0 (Bacteroides)			McCutchan and Pagano (1281)	1967
3	2,180 patients	314 patients			21		Rosner (626)	1968
3	3,453	287	4	1.4			Brumfitt and Leigh (1282)	1969
9	81,000	273	10	5.5[d]	43	14/43	DuPont and Spink (647)	1969
5				3.7			Ridley (1283)	1969
7	(2190 patients)	(248 patients)				15/39	Bodner et al. (460)	1970
	4,642	403			8		Crowley (1284)	1970
5	25,000		50[d]	0.2	40	13/35	Gelb and Seligman (490)	1970
10			296[d]		134	35/123	Marcoux et al. (420)	1970
					9		Rosner (1285)	1970
?	2,352 patients	259 patients		2.2 (Bacteroides)	26	18/26	Sinkovics and Smith (411)	1970
9 months	237[e]	103	16	15.5	250	79/250	Felner and Dowell (168)	1971
					16	—	Kuklinca and Gavan (491)	1971
1.5	447	161	2	1.2			Myerowitz et al. (1286)	1971
		48	6	12.5			Nakamura et al. (114)	1971

2.5	38,847	3,795 (1,718 patients)	447	11.8	213		Washington (417)	1971
	4,816 (1,000 patients)	261	20 (11 *Bacteroides*, 9 anaerobic streptococci)	7.7			Rosner (1287)	1972
9 months	13,162	1,116 (566 patients)	125	11.2	63		Washington (1288)	1972
15 months	1,368	360		27.0[f]	264[g]		Wilson *et al.* (105)	1972
2	8,472	250		3	22/67		Shanson (Survey of 59 labs) (1289)	1973
	103	28 (all clinically significant)		27			Washington and Martin (403)	1973
2	453			9.5	Total unknown (27 *Bacteroides fragilis*)	67% (*B. fragilis* patients)	Schoutens *et al.* (1289a)	1973
6 (surgical unit only)	149			15	23 (19 *B. fragilis*)	42% (*B. fragilis* patients)	Mackenzie and Litton (384a)	1974
5	285 (gram-negative bacilli only)			16	46 (7 mixed with aerobes)	36% (of 39 cases due to Bacteroidaceae)	McHenry *et al.* (398a)	1975

[a] Percent positive for anaerobes of all positive cultures.
[b] Only non-spore-forming anaerobes considered.
[c] From a literature review, most cases from the pre-antibiotic era.
[d] Only Bacteroidaceae considered.
[e] Postmortem blood cultures only.
[f] 10% if *Propionibacterium* isolates excluded.
[g] 67 clinically significant.

7.6% of all bacteremias at the Peter Bent Brigham Hospital in 1971, occurring as frequently as *Pseudomonas, Proteus,* or *Serratia* (406b). Nobles (56d) reported 43 patients with bacteremia among 112 patients with *Bacteroides* infections seen in an 18 month period; in a 6 month period, approximately 10% of all positive blood cultures yielded this organism. In a study involving 2318 positive blood cultures over a 3 year period (1573), 216 (9.3%) were positive for Bacteroidaceae (129 patients involved). At the Yale–New Haven Hospital *Bacteroides fragilis* accounted for 6.5% of all gram-negative bacteremias in the first half of 1973 (406c).

Factors predisposing to anaerobic bacteremia include malignancy (56e, 56g, 103, 404, 405, 407–413, 413a), other hematologic disease (103), transplantation of organs (56g, 407, 414, 415, 415a), dialysis (56g), recent surgery (103, 105, 384a, 410, 412, 1573), intestinal obstruction (168), other intraabdominal conditions (56d, 56g, 409, 1573), diabetes mellitus (412), the newborn age group (409), postsplenectomy (407), the use of cytotoxic agents or corticosteroids (105, 412), and/or the use of preoperative "bowel preparation" or other antimicrobial agents (56g, 105, 412, 1289a).

Fig. 8.1.   Blood culture bottle showing rounded colonies of *Fusobacterium necrophorum.*

The specific organisms involved in anaerobic bacteremia depend to a large extent on the portal of entry and the underlying disease. Where the gastro-intestinal and urinary tracts are the portals of entry, *Bacteroides* is the most common isolate. Overall, as was already indicated, *Bacteroides* (and espe-cially *B. fragilis*) is the most common anaerobe in bacteremia (56d, 406b, 406c, 1573). From 1970 through 1972, Bacteroidaceae represented the second most frequent gram-negative bacillary isolate from blood cultures at the Mayo Clinic and affiliated hospitals (415b). *Fusobacterium* is the most likely cause of bacteremia, however, when the oropharynx is the portal of entry (Figs. 8.1 and 8.2). In infections involving the female genital tract, bacteremia is most likely to be due to anaerobic streptococci, with *Bacteroides* and clostridia less common anaerobic isolates. Thus, in the 76 cases of bacteremia associated with septic abortion reported by Smith *et al.* (416), among the 48 strains that were anaerobic there were 31 anaerobic streptococci, seven *Bacteroides*, seven with both anaerobic streptococci and *Bacteroides*, and three *Clostridium perfringens*. Anaerobic gram-positive cocci (*Peptostrepto-coccus* and *Peptococcus*) and clostridia account for 1–3% of all bacteremias (56b). In the study of Gorbach and Thadepalli (486) clostridia represented 2.6% of all positive blood cultures at Cook County Hospital, Chicago, over

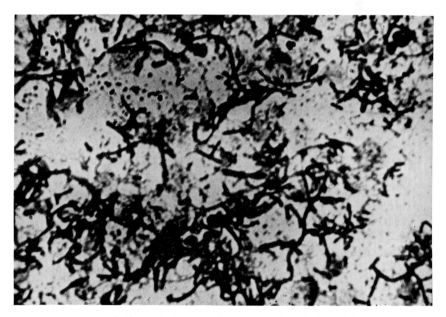

Fig. 8.2. Gram stain from blood culture bottle shown in Fig. 8.1. Note pleomorphism, filaments, and round bodies.

a 14 month period. *Clostridium perfringens* accounted for 37 of the 65 clostridial isolates, with *C. ramosum* the next most prevalent (3 strains recovered). *Clostridium* bacteremia is most likely to originate in the gastrointestinal tract. *Clostridium septicum* septicemia is commonly associated with malignancy (103) so that one should always consider that an underlying malignancy is a likely possibility when this organism has been isolated.

Data from the Mayo Clinic give us some idea as to the relative frequency with which the various anaerobes may be recovered from bacteremia in a general hospital population. In the paper by Washington (417), the number of patients yielding various anaerobes on blood culture was as follows: Bacteroidaceae 145, *Clostridium* 30, *Peptostreptococcus* 16, *Peptococcus* 12, *Bifidobacterium eriksonii* 4, *Eubacterium lentum* 2, and *Veillonella* 1. Among the 67 patients with clinically significant anaerobic bacteremia described by Wilson *et al.* (105) 52 had bacteremia due to Bacteroidaceae, 6 had gram-positive spore-formers, 5 gram-positive cocci, 3 gram-positive non-spore-forming bacilli, and 1 a gram-negative anaerobic coccus. Isolation of *Actinomyces* from the blood has been quite unusual, but Schain *et al.* (408) reported recovery of "*Actinomyces bovis*" from the blood and/or bone marrow of four patients with Hodgkin's disease and three patients with miscellaneous conditions. The cultures were said to have been confirmed by Conant and by the Army Medical School. Little clinical data was given. Additional cases of *Actinomyces* bacteremia have been reported (1134a, 1267g). Bacteremia due to *Succinivibrio*, an anaerobic vibrio and primarily a rumen organism, has been recorded by Southern (417a). Sharpe *et al.* (361b) described several cases of bacteremia due to *Lactobacillus*. Included was one case in association with pneumonia and due to an atypical strain of *L. brevis*.

Reynes (418) did quantitative blood cultures in some cases of anaerobic bacteremia with the following results: 1 to 20 colonies per milliliter in 3 patients, 21 to 50 colonies in 2 patients, 51 to 100 colonies in 3 patients, 101 to 500 colonies in 2 patients, and over 500 colonies in 2 patients. In certain other quantitative studies, counts of over 100 to several hundred colonies per milliliter were not rare. On occasion, over 1000 anaerobic organisms per milliliter were recovered.

In general, the clinical features of anaerobic bacteremia are not distinctly different from those encountered with other types of bacteremia. The one type of anaerobic bacteremia that is distinctive is that of severe sepsis with *Clostridium perfringens* in which there may be a most dramatic clinical picture consisting of hemolytic anemia, hemoglobinemia, hemoglobinuria, disseminated intravascular coagulation, bleeding tendency, bronze-colored skin, hyperbilirubinemia, shock, and oliguria or anuria. This picture is described in more detail in Chapter 12. It must be appreciated, however, that the severe intravascular hemolysis, so typical of *C. perfringens* sepsis, is not an invariable accompaniment (56e, 115b). For example, Rathbun (419) reported 20 cases of sepsis due to *Clostridium* (mostly *C. perfringens*), none of which showed intravascular hemolysis.

Certain serious complications of sepsis may be seen in anaerobic bacteremia as well as in bacteremia due to other organisms. Table 8.3 shows the incidence of certain selected complications in patients with anaerobic bacteremia. Despite the lack of 2-keto-3-deoxyoctanate and heptose in the endotoxin of *Bacteroides* (72a, 420a), septic shock is seen with some frequency in *Bacteroides* bacteremia. It is also seen in the course of bacteremia due to other anaerobes. Jaundice was frequent in some of the series noted in Table 8.3, but the incidence is probably not different from that in other types of bacteremia (647). Septic thrombophlebitis was common in one of the series. Various hematologic problems, including hemolytic anemia, bleeding diathesis, and disseminated intravascular coagulation, were seen in small numbers of patients. The incidence of these various complications was not always stated, and it is not clear that they were looked for consistently. Metastatic infection which was common in the preantimicrobial era, is much less common now. Nevertheless, it was noted in 28 cases of 123 instances of *Bacteroides* septicemia by Marcoux *et al.* (420). In the series reported by Felner and Dowell (168) there were two instances of septic arthritis and one of meningitis.

In addition to the series noted in Table 8.3, there are a number of individual reports bearing on the incidence of these complications. Wilson *et al.* (421) described 132 patients with septic shock. Included in this series were 17 with clostridial sepsis and 11 with *Bacteroides* bacteremia. In a paper by Litton (422) in which 34 patients with septic shock were discussed, there were two cases involving *Bacteroides*. Hara *et al.* (176) described shock in a case of bacteremia involving three different anaerobes—*C. perfringens*, *Bacteroides*, and *Peptostreptococcus*. Dilworth and Ward, (423) described a fatal case of *Bacteroides fragilis* sepsis following criminal abortion; this patient had gross intravascular hemolysis and shock. Nathan (424) described a fatal case of *Sphaerophorus* sepsis with shock. Rapin *et al.* (427) described a patient with bacteremia, due to *Ristella pseudoinsolita*, with shock and jaundice. In addition to the cases of shock indicated in Table 8.3, there are several others related to *Bacteroides* sepsis (278, 390b, 419a, 419b, 419c, 419d). In the report on 14 cases of clostridial septicemia by Wynne and Armstrong (56e), it was noted that 13 of the patients had hypotension. In the paper by Attar *et al.* (429), 4 patients with anaerobic or mixed bacteremia had shock (4 patient episodes?); 2 were due to *Bacteroides* and 2 to *Clostridium perfringens* plus *Escherichia coli*. Winslow *et al.* (419e) described 3 cases of bacteremic shock due to anaerobes, one each to *Clostridium*

TABLE 8.3

Incidence of Selected Complications in Patients with Anaerobic Bacteremia

Number of patients	Anaerobes involved	Number of patients with						Reference	Year
		Shock or hypotension	Jaundice	DIC[a]	Hemolytic anemia	Bleeding diathesis	Septic phlebitis		
11	Bacteroides	2	2					McHenry et al. (530)	1961
25	Bacteroides	7	10				3	Tynes and Frommeyer (489)	1962
19	Clostridium	"Common"			1			Cabrera et al. (404)	1965
15	Bacteroides melaninogenicus	5						Altemeier et al. (431)	1967
21	Clostridium septicum	4			0			Alpern and Dowell (103)	1969
35	Bacteroides	Present (number not specified)	11			1		Gelb and Seligman (490)	1970
123	Bacteroides	31	12	1			6	Marcoux et al. (420)	1970
26	Bacteroides	0						Sinkovics and Smith (411)	1970
48	Anaerobic streptococci, Bacteroides, Clostridium	8 (6 anaerobic streptococcus, 1 anaerobic streptococcus plus Bacteroides, 1 C. perfringens)	1		3			Smith et al. (416)	1970

86	Nonhistotoxic clostridia	"Common"		1		Alpern and Dowell (409)	1971
9	Anaerobic gram negative rod	4				Farquet *et al.* (723)	1971
250	Many different types	27	2		48	Felner and Dowell (68)	1971
39	*Bacteroides*	8				Goodman (412)	1971
16	13 clostridia, 3 *Bacteroides*	3 (2 clostridia and 1 *B. fragilis*)	3	"Several"	2	Kuklinca and Gavan (491)	1971
55	*Bacteroides*	19				Kagnoff *et al.* (405)	1972
264	Several types	22	6	3	5	Wilson *et al.* (105)	1972
43	"*Bacteroides*"	7	5	2	4 (+4 probable)	Nobles (56d)	1973
112	Bacteroidaceae	31 (6 died)	8 (5 with other possible causes)	Serum bilirubin > 1.2 mg/100 ml in 59/82 patients with this measurement	6	Chow and Guze (1573)	1974
49	47 *Bacteroides fragilis*, 2 other *Bacteroides*	5	0	1		Meyer (56g)	1975

[a] Disseminated intravascular coagulation.

perfringens, an anaerobic streptococcus, and a microaerophilic streptococcus. Rubenberg *et al.* (425) described a case of *Clostridium perfringens* sepsis with disseminated intravascular coagulation and severe intravascular hemolysis. Vermillion *et al.* (426) described two patients with anaerobic streptococcal bacteremia who had shock. One of these patients also had a distinct leukopenia (1900 wbc/mm³). Smith *et al.* (416) noted liver function abnormality in two patients with anaerobic streptococcal bacteremia and in another with double bacteremia due to anaerobic streptococci and *Bacteroides*. In the report of Felner and Dowell (168) 11 of the 48 patients with thrombophlebitis had recurrences despite heparin therapy. Some patients in this series required very large doses of heparin. Emboli were noted in 74 patients.

The fever is usually high in anaerobic septicemia, and it may be prolonged. Cabrera *et al.* (404) reported that a dissociation of pulse and temperature was common among their 19 cases of acute clostridial sepsis. Wynne and Armstrong (56e) also saw this in 4 patients. Leukocytosis is the rule but leukopenia is seen on occasion, as noted above. A leukemoid reaction was noted in five patients among 123 with *Bacteroides* sepsis by Marcoux *et al.* (420). This group of workers also noted among their 123 cases, 27 patients with significant central nervous system symptoms related to bacteremia including mental clouding, coma, marked apprehension, lethargy, dizziness, restlessness, hallucinations, euphoria, delirium, toxic psychosis, headache, and hemiplegia. In addition, they noted inappropriate antidiuretic hormone (ADH) secretion in one patient and supraventricular tachycardia in two. Martin and McHenry (400) emphasized that diarrhea occurred in one-third of their patients with *Bacteroides* bacteremia. While acknowledging that this occurs in other bacteremias as well, they felt that the presence of diarrhea in a patient with suspected sepsis should lead one to consider strongly the possibility of *Bacteroides* bacteremia, as well as *Salmonella* bacteremia. McCabe (428) noted a fall in serum complement (C3) in a patient with *Bacteroides* bacteremia.

The α-toxin of *Clostridium perfringens* is a lecithinase. Its action on the red blood cells not only leads to lysis, but results in the formation of lysolecithin (Fig. 8.3) which also contributes to the hemolytic process. In addition to this, hemolysis may be related to the microangiopathic hemolytic anemia. The disseminated intravascular coagulation may be maintained by release of thromboplastin from toxin-damaged red blood cells and platelets. The marked microspherocytosis is probably due to a direct effect of the α-toxin. Severe hemoglobinemia and hemoglobinuria result. Renal failure is generally secondary to the hemolysis as well as to hypotension and fibrin deposition (Fig. 8.4). The toxin also causes hydrolysis of platelet phospholipid producing lysis without aggregation and release of thromboplastin and thrombocytopenia. Boggs *et al.* (413) pointed out that the adrenal cortex, with its

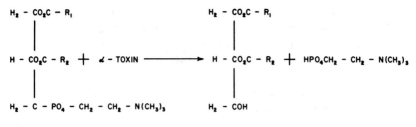

LECITHIN + LECITHINASE ⟶ LYSOLECITHIN + PHOSPHORYLCHOLINE

Fig. 8.3. Action of *C. perfringens* α-toxin on lecithin. From Ref. (644) L. P. Smith, A. P. H. McLean, G. B. Maughan, *Clostridium welchii* septicotoxemia, *Am. J. Obstet. Gynecol.* **110**, 135–149 (1971).

high phospholipid content, is theoretically vulnerable to destruction and thus one should look for adrenal insufficiency in patients with clostridial sepsis.

Attar *et al.* (429) reported a hypercoagulable state (decreased clotting time and increased fibrinogen) initially, followed by a hypocoagulable state (increased clotting time and decreased fibrinogen) in sepsis due to *Bacteroides* and other gram-negative bacilli. Using the *Limulus* assay, Sonnenwirth *et al.* (430) found endotoxin in all 29 strains of gram-negative anaerobic bacilli studied, at concentrations of 10^{-2} to 10^{-6} μg/ml. Included in the strains studied were *Bacteroides fragilis*, *Bacteroides variabilis*, *Fusobacterium varium*, *Bacteroides oralis*, *Bacteroides melaninogenicus*, *Sphaerophorus necrophorus*, and *Bacteroides putredinis*. Ketodeoxyoctonate, 0.033 to

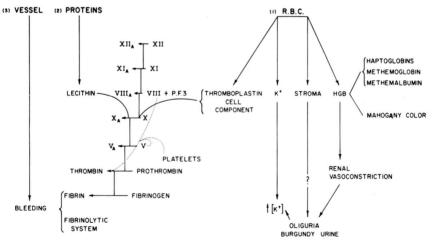

Fig. 8.4. Sites of action and effects of *C. perfringens* α-toxin. From Ref. (664) L. P. Smith, A. P. H. McLean, and G. B. Maughan, *Clostridium* welchii septicotoxemia. *Am. J. Obstet. Gynecol.* **110**, 135–149 (1971).

0.041 μmoles/mg, and traces of heptose were found in four strains examined. Furthermore, these authors noted circulating endotoxin in five patients with *Bacteroides fragilis* sepsis. The endotoxin of *Bacteroides oralis* was the least potent. Gram-positive organisms were always negative in the *Limulus* assay.

The mortality in anaerobic bacteremia is quite variable, since it depends on so many factors, for example, the age and underlying disease of the infected person, the nature of the organism, the intensity of the bacteremia, the speed with which the diagnosis is made and appropriate therapy [surgical (56g) and medical (1573)] undertaken. Table 8.2 gives data on mortality from several series in the literature. It is clear that, even in the antimicrobial era, roughly one-third of the patients with anaerobic bacteremia die. The mortality in obstetrical and gynecologic patients tends to be quite low, reflecting the good resistance of young female patients. Thus, in the series of 76 cases of bacteremia related to septic abortion, two-thirds of which involved anaerobes [Smith *et al.* (416)], there was only one death, that in a patient with *C. perfringens* sepsis. There are several reports which give data on mortality which have not been included in Table 8.2. In the paper by Altemeier *et al.* (431), 5 of 15 patients with *Bacteroides melaninogenicus* sepsis had shock, and 3 of these died. In the series of Goodman (412) among 39 cases of *Bacteroides* sepsis, there were 15 fatalities. The mortality in the 112 cases of Bacteroidaceae bacteremia reported by Chow and Guze (1573) was 43%. Eleven of 15 *Bacteroides* bacteremias occurring on a surgical service [Beazley *et al.* (394)] were fatal. In the report of Alpern and Dowell (103) on 86 cases of bacteremia due to nonhistotoxic clostridia, there were 55 survivors. Among the 21 cases of *Clostridium septicum* septicemia, nine died [Alpern and Dowell (103)].

In patients with underlying malignancy, of course, the mortality is frequently higher. Thus, in the study of Sinkovics and Smith (411) of 26 patients with *Bacteroides* septicemia, 18 ultimately died. Death was associated with extensive injury to the intestines or the nasopharynx as a result of the tumor and/or therapy. In the report by Kagnoff *et al.* (405) 71% of the 55 patients with *Bacteroides* bacteremia, 50 of which were associated with malignancy, died ultimately; the mortality directly related to infection was 35%.

Altemeier *et al.* (432) recovered L forms or atypical bacterial forms from the blood or surgically removed thrombi of 50 patients with thrombophlebitis or thromboembolic disease. These forms, particularly of *Bacteroides*, were felt to have probably played an important etiologic role in the thrombophlebitis. There was a particular association of such forms with pregnancy and administration of birth control pills in female patients, suggesting the possibility of a hormonal effect that (a) enhances the growth of these forms (which was demonstrated in one case), (b) interferes with the activity of heparin, or (c) causes chemical degradation of heparin.

In terms of diagnosis, only the unique overwhelming type of bacteremia with marked intravascular hemolysis caused by *Clostridium perfringens* is distinctive, as noted earlier. Nevertheless, anaerobic bacteremia may be suspected on the basis of portal of entry (commonly the gastrointestinal tract, the female genital tract, to a lesser extent the upper respiratory tract, and decubitus ulcers of the thighs and buttocks). Early invasion of the regional veins, thrombophlebitis, and septic emboli are also suggestive of anaerobic bacteremia. Certain other clues suggesting the possibility of anaerobic infection elsewhere in the body may serve to suggest the likelihood of a bacteremia being anaerobic in etiology (see Chapter 2). In the study of Martin and McHenry (400), of *Bacteroides* bacteremia, the organisms were recovered from one-third of the patients within 48 hours. In the others, it took 5 to 20 days for blood cultures to become positive. Accordingly, they recommend incubating blood cultures for 3 weeks. Therapy is considered in Chapter 19.

Endocarditis

Endocarditis due to anaerobic bacteria is not a rare phenomenon, as documented by Table 8.4. It will be noted that quite a wide variety of anaerobic bacteria have been reported from endocarditis; however, the most common infecting organisms are anaerobic and microaerophilic streptococci. Dental and oral disease or manipulation was a not uncommon precursor of endocarditis, as is true with this disease involving aerobic or facultative organisms. Narcotic addiction was the underlying problem in a small number of cases. It is interesting to note (432a) that *Clostridium perfringens* was isolated from 11 of 100 samples of street heroin and from 31/100 samples of injection paraphernalia, although this organism has not yet been recorded as a cause of endocarditis in addicts. The balance of the cases had a variety of underlying anaerobic infections. Most patients did have previous valvular disease, usually rheumatic, but there were a number of cases in which the disease was engrafted on previously normal valves. The disease involved any of the valves, but particularly the aortic or mitral valves, septal defects, or the endocardium of the atrium. There was a significant mortality, even in the antimicrobial era.

The exact incidence of anaerobes in endocarditis remains to be determined. Our own experience over a number of years (see below) is that some 10% of cases of endocarditis involved either anaerobic or microaerophilic organisms.

TABLE 8.4

Anaerobic Bacterial Endocarditis

Probable source	Other underlying condition	Prior heart disease[a]	Endocardial location	Organism(s)	Outcome	Reference	Year
i.v. heroin	—	None	Tricuspid valve	*Bacteroides fragilis*	Lived	Child *et al.* (972)	1969
Gastroenteritis	—	None	?	*Bacteroides fragilis*	Lived	Felner and Dowell (289)	1970
Septic abortion, groin abscess	—	None	?	*Bacteroides fragilis*	Lived		
Appendix, pelvic abscess	—	None	Mitral valve	*Bacteroides fragilis*	Died		
Cholecystitis	—	ASHD	?	*Bacteroides fragilis*	Lived		
Decubitus ulcer	Diabetes	RHD	?	*Bacteroides fragilis*	Lived		
Periodontal abscess, sinusitis	—	ASHD	?	*Bacteroides fragilis*	Lived		
Diverticulosis	Diabetes	None	?	*Bacteroides fragilis*	Lived		
Infected aortofemoral graft	Aortoduodenal fistula	RHD	"Systolic and diastolic murmurs"	*Bacteroides fragilis*	Died		
Peritonitis, pelvic abscess	—	RHD	?	*Bacteroides fragilis*	Died		
Decubitus ulcer	—	ASHD	"Two systolic murmurs"	*Bacteroides fragilis*	Died		
Pneumonia, empyema	—	ASHD	?	*Bacteroides fragilis*	Died		
Appendicitis	—	RHD	Left atrium	*Bacteroides fragilis* ss. *fragilis*	Died	Nastro and Finegold (433)	1973
Hemicolectomy, pelvic abscess	—	RHD, prosthetic aortic and mitral valves	?	*Bacteroides fragilis*	Died	McHenry and Hawk (1289b)	1974
?	—	Prosthetic valve	?	*Bacteroides fragilis*	?	Wilson *et al.* (1289c)	1975

"Dental caries"	Duodenal ulcer	ASHD	?	*Bacteroides melaninogenicus*	Lived	Felner and Dowell (289)	1970
"Dental caries"	Gastric ulcer	None	?	*Bacteroides oralis*	Died	Ueno and Suzuki (1257)	1967
? Tonsil	—	CHD	Ventricular septal defect	*Bacteroides serpens* and *S. aureus*	Died		
?	—	None?	Mitral valve	*Bacteroides symbiophilus*	Died	Arjona et al. (1290)	1946
Unknown	—	None	Aortic valve	*Bacteroides*	Lived	Cressy et al. (1291)	1948
?	?	?	Mitral valve	Gram-negative anaerobic rod	Lived	Tumulty and Harvey (1292)	1948
?	?	?	Mitral and tricuspid valves and wall of left auricle	Gram-negative anaerobic rod	Died		
Tooth extraction	None	Yes	Mitral valve	*Bacteroides*	Lived	Wallach and Pomerantz (1293)	1949
Dental extraction	—	Probable RHD	?mitral valve	"*Bacteroides*", but there was slight growth under 10% CO_2	Lived	deHay et al. (1294)	1950
Unknown	—	Probable RHD	Mitral valve	*Bacteroides*	Relapse, then cure	Rantz (1295)	1951
Unknown	Pregnancy	RHD	Mitral valve	*Bacteroides*	Lived		
Unknown	Pregnancy	RHD	Unknown	*Bacteroides*	Lived		
?[c]	Pregnancy	?	Aortic and mitral valves	*Bacteroides*	Lived	Hermans et al. (1296)	1966
?	?	?	Mitral valve?	*Bacteroides*	Lived		
?	?	Prosthetic valve	Mitral valve prosthesis	*Bacteroides*	Lived	Ellner and Wasilauskas (740)	1971
?Dental	Alcoholism	RHD	Aortic and mitral valves	*Bacteroides* R to PC[b] ?*B. fragilis*	Died	Masri and Grieco (441)	1972
Diarrhea at onset	—	"Floppy" mitral valve	Mitral valve	*Bacteroides*	Died	(1296a)	1974
Bowel perforation	—	?	Tricuspid, mitral and aortic valves	Bacteroidaceae plus *Candida tropicalis*	Died	Chow and Guze (1573)	1974

(continued)

TABLE 8.4 (*continued*)

Probable source	Other underlying condition	Prior heart disease[a]	Endocardial location	Organism(s)	Outcome	Reference	Year
Dental	?	?	Aortic valve	Bacillus Buday	Died	Schulz (159)	1935
Dental fistula	—	None	Aortic valve	Fusobacterium necrophorum (Sphaerophorus funduliformis)	Died		
Tonsillar abscesses	—	None	Mitral valve	Fusobacterium necrophorum (Sphaerophorus funduliformis)	Died	Lupu et al. (1297)	1937
Unknown	—	Probable RHD	Mitral valve	Fusobacterium necrophorum (Sphaerophorus funduliformis)	Died	Arjona et al. (1290)	1946
?	?	?	?	Fusobacterium necrophorum (Sphaerophorus funduliformis)	?	Ales et al. (1298)	1952
Suppurative tonsillitis	Previous SBE	Aortic insufficiency, etiology?	Aortic valve	Fusobacterium necrophorum (Sphaerophorus funduliformis)	Died	Caselitz et al. (1299)	1968
Tonsil?	—	Yes (type?)	Aortic valve	Sphaerophorus funduliformis	Died	Felner and Dowell (289)	1970
Pyorrhea	—	CHD	VSD	Fusobacterium necrophorum (Sphaerophorus funduliformis)	Lived		
Pneumonia, empyema, subphrenic abscess	—	None	?	Fusobacterium necrophorum (Sphaerophorus funduliformis)	Lived		
"Dental caries"	—	CHD	VSD	Fusobacterium nucleatum	Lived	Felner and Dowell (289)	1970
Pneumonia	Diabetes	None	"Two systolic murmurs"	Fusobacterium nucleatum	Lived		

?Duodenal ulcer	—	ASHD	?	*Fusobacterium nucleatum*	Lived	Felner and Dowell (289)	1970
?	Widespread infection	?	Mitral valve?	*B. fusiformis*	Lived	Dormer (435)	1958
Rectum		?	Mitral and tricuspid	Fusiform bacillus	Died	Ghon and Roman (543)	1916
Rectum?	CA rectum Liver abscess GB disease	?	Mitral and aortic	Fusiform bacillus	Died		
?	None	Prob. RHD	Mitral valve	*?Dialister granuliformans*	Died	Magrassi (1300)	1944
?	?	?	?	*Dialister granuliformans*	Died	Prévot (21)	1955
Dental abscesses, extraction	—	RHD	Mitral and aortic valves	Unidentified anaerobic GNR[a]	Lived; erosion of mitral valve	Tumulty and McGehee [cited in Fisher and McKusick (1301)]	1953
Unknown	—	RHD	Mitral and tricuspid valves and left atrium	Unidentified anaerobic GNR	Ruptured chordae tendinae, died		
Unknown	—	Probable RHD	Tricuspid or mitral valve	Unidentified anaerobic GNR	Lived	Fisher and McKusick (1301)	1953
Unknown	?	?	?	Anaerobic GNR	?	Werner *et al.* (438)	1967
Unknown	—	None	Tricuspid valve	Anaerobic GNR	Died	Bodner *et al.* (460)	1970
?	?	?	?	Bacteroidaceae	Died	Gelb and Seligman (490)	1970
?	?	?	?	*Treponema pallidum*	?	Finland and Barnes (1302)	1970
?	?	?	?	2 cases due to *C. welchii*	Died?	Janbon [cited in More (1303)]	1943
?	?	?	Systolic murmur at apex and base	*C. perfringens* (*C. velchii*)	Died	Labraque-Bordenave (1304)	1935

(continued)

TABLE 8.4 (*continued*)

Probable source	Other underlying condition	Prior heart disease[a]	Endocardial location	Organism(s)	Outcome	Reference	Year
Dental	?	?RHD	Mitral and aortic murmurs	*C. perfringens* (*C. welchii*)	Died	Labraque-Bordenave (1304)	1935
Appendicitis	?	None?	Mitral murmur	*C. perfringens* (*C. welchii*)	Died	More (1303)	1943
Cancer cervix	Local radium reaction	RHD	Aortic valve	*C. perfringens* (*C. welchii*)	Died	More (1303)	1943
?	Chronic renal disease	?	Systolic and diastolic murmurs	*C. perfringens* (*C. welchii*)	Lived	Felner and Dowell (289)	1970
?	Hiatal hernia	ASHD	?	*C. perfringens* (*C. welchii*)	Lived		
Diverticulitis	—	?	?	*C. perfringens*	Lived; shock		
?	?	None	Mitral valve	*C. perfringens* (*C. welchii*)	Died	Robinson *et al.* (1305)	1972
Pelvic inflammatory disease	—	RHD	?	*Clostridium sporogenes*	Lived	Felner and Dowell (289)	1970
?	?	?	"Two systolic murmurs"	*Clostridium sordellii*	Lived		
?	Bleeding duodenal ulcer	?	?	*Clostridium sordellii*	Lived		
Pyelonephritis	?	?	?	*Clostridium cochlearium*	Lived; hypotension		
?	?	RHD	Systolic and diastolic murmurs	*Clostridium* species NCDC Group P-1	Lived		
?	?	?	?	*Cillobacterium*	Died	Prévot (197)	1948

?	?	Congenital ventricular septal defect	Septal defect	*Eubacterium ventriosum*	Lived	Watanabe and Ueno (1306)	1968
?	?	?	?	*Eubacterium crispatum*	Died	Kurimoto *et al.* (1219)	1961
Dental extraction	—	RHD	Aortic valve	*Eubacterium aerofaciens*	Lived	Sans and Crowder (1306a)	1973
Postoperative valvular surgery	?	Prosthetic aortic valve	?	*Propionibacterium acnes* (*Corynebacterium acnes*)	Lived	Levin [cited by Johnson and Kaye (1202)]	1970
?	?	?	?	Transitional forms of *Corynebacterium acnes*	Died	Zierdt and Wertlake (1307)	1969
Rectal abscess?	Regional enteritis	?	"2 systolic murmurs"	*Propionibacterium acnes* (*Corynebacterium acnes*)	Shock; lived	Felner and Dowell (289)	1970
?	?	CHD	Systolic and diastolic murmurs	*Propionibacterium acnes* (*Corynebacterium acnes*)	Lived		
Skin?	Advanced psoriasis	RHD	?	*Propionibacterium acnes* (*Corynebacterium acnes*)	Lived		
?	?	?	?	*Propionibacterium acnes* (*Corynebacterium acnes*)	Lived; coronary embolus		
Decubitus ulcer	Diabetes	?	?	*Propionibacterium acnes* (*Corynebacterium acnes*)	Ruptured chordae tendinae, died	Felner and Dowell (289)	1970
?	?	RHD, prosthetic valve	Aortic valve	*Propionibacterium acnes* (*Corynebacterium acnes*)	Died (dehiscence of valve)	Wilson *et al.* (105)	1972
?	?	?	?	3 cases *Corynebacterium anaerobium*	?	Prévot (167)	1953
Dental	?	CHD	?	*Corynebacterium anaerobium*	Lived	Campeau *et al.* (436)	1960
?	?	?	?	*Corynebacterium anaerobium*	Lived	Prévot (1199)	1963

(continued)

207

TABLE 8.4 (continued)

Probable source	Other underlying condition	Prior heart disease[a]	Endocardial location	Organism(s)	Outcome	Reference	Year
?	?	?	?	3 cases C. avidum	?	Prévot (167)	1953
?	?	?	?	1 case C. granulosum	?		
?	?	?	?	C. granulosum	Lived	Prévot (1199)	1963
?	?	CHD	Patent ductus	Corynebacterium parvum	Lived		
?Dental	—	RHD and prosthetic valve	?Aortic prosthesis	Anaerobic diphtheroid	Lived	Kaplan and Weinstein (558)	1969
?Postoperative, valvular surgery	?	Prosthetic aortic and mitral valves	?	Anaerobic diphtheroid	Lived	Cobbs [cited by Johnson and Kaye (1202)]	1970
?	?	RHD	Mitral valve	Anaerobic diphtheroid	Died	Griffin et al. (1308)	1972
Unknown	—	CHD	VSD	Microaerophilic diphtheroid	Lived	Wittler et al. (1309)	1960
Postoperative, valvular surgery	—	RHD	Aortic valve	Microaerophilic diphtheroid	Cusp defect, valve surgery, died	Davis et al. (1310)	1964
Postoperative, valvular surgery	—	RHD	Aortic valve	Microaerophilic diphtheroid	Cusp defect, valve surgery, died		
Postoperative, valvular surgery	—	Dilated aortic ring	Aortic valve	Microaerophilic diphtheroid	Cusp defect, valve surgery, died		
Postoperative, valvular surgery	—	Traumatic valvular damage	Aortic valve	Microaerophilic diphtheroid	Lived		

Source of infection	Underlying disease	Heart disease	Organism	Valve	Outcome	Reference	Year
	None	RHD					
Dental	None	None	*Lactobacillus*	?Mitral and aortic valves	Died	Biocca (1312)	1944
?	None?	RHD	"Doederlein Bacillus vaginalis"	Mitral valve	Died	Marschall (1313)	1938
Dental	—	RHD	*Lactobacillus casei*	Mitral valve	Lived	Tenenbaum and Warner (1313a)	1975
?Throat	None	Yes?	*Actinomyces israelii*	Mitral valve	Died	Dutton and Inclan (444)	1968
?	Pulmonary tuberculosis	?	*Peptostreptococcus intermedius* (*Streptococcus intermedius*)	?	Died	Kurimoto et al. (1219)	1961
Dental	None	None	*Peptostreptococcus intermedius* (*Streptococcus intermedius*)	Aortic valve	Died	Mathieu and Lefebvre (1314)	1964
?	None?	None?	*Veillonella gazogenes*	?Mitral valve	Lived	Loewe et al. (1315)	1946
Postabortal	None	None?	Anaerobic staphylococcus	Mitral	Died	Langeron et al. (661)	1948
?	?	?	2 cases—anaerobic staphylococci	?	?	Dietrich et al. (135)	1965
?	?	?	Microaerophilic staphylococcus + *B. bronchiseptica*	?	?	Dale and Geraci [cited by Lerner and Weinstein (434)]	1966
?	None	CHD	CO_2-dependent staphylococcus	Aortic valve	Died	Spink et al. (1316)	1962
?	None	None?	"Anaerobic pneumococcus"	?Mitral valve	Died	Hollander and Landsberg (1317)	1940
?	None	None	*Peptostreptococcus evolutus* (*Streptococcus evolutus*)	Mitral valve	Died	Lemierre et al. (1318)	1935
?	?	?	3 cases *Streptococcus evolutus*	?	2 Lived, 1 died	Campeau et al. (436)	1960

(continued)

TABLE 8.4 (*continued*)

Probable source	Other underlying condition	Prior heart disease[a]	Endocardial location	Organism(s)	Outcome	Reference	Year
Dental	None	RHD	Aortic valve	*Peptostreptococcus evolutus* (*Streptococcus evolutus*)	Lived	Lefebvre and Côté (1319)	1958
Dental	None	CHD	Ventricular septal defect	*Peptostreptococcus evolutus* (*Streptococcus evolutus*)	Died	Mathieu and Lefebvre (1314)	1964
Dental	None	Yes—RHD?	Aortic valve	*Peptostreptococcus evolutus* (*Streptococcus evolutus*)	Lived		
Furuncles	None	None	Mitral valve	*Peptostreptococcus evolutus* (*Streptococcus evolutus*)	Lived		
?	None	Yes—RHD?	Aortic and mitral valves	*Peptostreptococcus evolutus* (*Streptococcus evolutus*)	Lived		
?	?	CHD	Aortic valve?	*Peptostreptococcus evolutus* (*Streptococcus evolutus*)	Lived	Ikemoto [cited in Kozakai and Suzuki (265)]	1968
?	?	?	?	Microaerophilic streptococcus	?	Harvey *et al.* (1320)	1949
?	?	?	?	Microaerophilic streptococcus	?		
?	?	?	?	6 cases due to microaerophilic streptococcus	?	Cates and Christie (437)	1951
?	?	?	?	Microaerophilic streptococcus	?	Wedgwood (439)	1955
Aural sepsis	?	?	?	Microaerophilic streptococcus	Lived	Dormer (435)	1958
?	?	?	?	Microaerophilic streptococcus	Died		

Dental	?	?	?	Microaerophilic streptococcus	?	Lived; popliteal mycotic aneurysm	Dormer (435)	1958
Dental	?	?	?	Microaerophilic streptococcus	?	Lived		
?	?	?	?	Microaerophilic streptococcus	?	Died		
?	?	?	?	Microaerophilic streptococcus	?	Lived; mycotic aneurysms, interior iliac and inferior gluteal arteries		
?	?	?	?	Microaerophilic streptococcus	?	Lived		
?	?	?	?	Microaerophilic streptococcus	?	Lived		
?	?	?	?	Microaerophilic streptococcus	?	Lived	Vogler et al. (1321)	1962
?	?	?	?	Microaerophilic streptococcus	?	?	Tompsett (440)	1964
Dental 3/13	—	7 cases (4 RHD) None in 6 cases	?	13 cases due to microaerophilic streptococcus	?	?	Lerner and Weinstein (434)	1966
i.v. heroin	—	None known	Aortic valve	Microaerophilic streptococcus	?	Lived (aortic valve replaced)	Carey and Hughes (1322)	1967
?	?	?	?	Microaerophilic streptococcus	?	?	Laxdal et al. (1323)	1968

(continued)

TABLE 8.4 (*continued*)

Probable source	Other underlying condition	Prior heart disease[a]	Endocardial location	Organism(s)	Outcome	Reference	Year
?	None	?	?	2 cases due to microaerophilic streptococcus	Lived	Rahal et al. (1324)	1968
?	?	?	?	10 cases due to microaerophilic streptococcus	?	Hayward et al. (1325)	1969
?	—	?	?	Microaerophilic streptococcus	Lived	Jawetz (1326)	1970
?	?	CHD	?	Microaerophilic streptococcus	Lived	Levison et al. (1327)	1970
?	?	None	?	Microaerophilic streptococcus	Lived		
?	?	None	?	Microaerophilic streptococcus	Lived	Griffin et al. (1308)	1972
?	?	?	?	"Diphtheroid streptococcus" (microaerophilic)	Died	Lamanna (1328)	1944
Dental	None	Yes, RHD?	Mitral valve?	Streptococcus lactis (microaerophilic)	Lived	Wood et al. (1329)	1955
?	?	?	?	Anaerobic streptococcus	Died	Vaucher and Woringer (449)	1925
Female genital tract	—	None	Tricuspid valve	Anaerobic streptococcus	Died	Lehmann (660)	1926
Female genital tract	—	None	Aortic and mitral valves	Anaerobic streptococcus	Died		
Pelvic abscess	Pleural empyema	?	?	Anaerobic streptococcus	Died	Colebrook (1330)	1930

Source	Predisposing condition	Heart disease	Valve	Organism	Outcome	Reference	Year
?Appendix	—	RHD	Tricuspid valve	Anaerobic streptococcus	Died	Bingold (442)	1932
Postabortal	Thrombophlebitis, splenic abscess	?	Mitral valve	Anaerobic streptococcus	Died		
	?	Yes?	Tricuspid valve	Anaerobic streptococcus	Died		
Cholangitis	?	?	Pulmonic valve	Anaerobic streptococcus and E. coli	Died		
Lung	—	None	Mitral valve	Anaerobic streptococcus	Died	Fisher and Abernethy (1331)	1934
	?	CHD	?	Anaerobic streptococcus	Lived	Geraci (1332)	1955
	?	RHD	Mitral valve	Anaerobic streptococcus	Died	White and Varga (965)	1960
	?	?	?	Anaerobic streptococcus	Lived	Vogler et al. (1321)	1962
Dental	None	Yes, RHD?	Aortic valve	Anaerobic streptococcus	Lived	Vic-Dupont et al. (169)	1963
	?	?	?	4 cases, anaerobic streptococci	?	Dietrich et al. (135)	1965
	—	Congenital subaortic stenosis	Aortic valve	Anaerobic streptococcus	Lived	Goldsand and Braude (244)	1966
	1/3 cancer of breast	1/3 cases	?	3 cases due to anaerobic streptococcus	?	Lerner and Weinstein (434)	1966
	?	?	?	Anaerobic streptococcus	Lived	Rahal et al. (1324)	1968
	?	RHD	Aortic and? mitral valve	Anaerobic streptococcus	Lived	Stason et al. (1333)	1968
	?	?	?	Anaerobic streptococcus	?	Hayward et al. (1325)	1969
	?	RHD	Mitral valve?	Anaerobic streptococcus	Lived	Griffin et al. (1308)	1972
	Receiving hemodialysis	None?	Aortic and mitral valves	Anaerobic streptococcus	Lived	Leonard et al. (1334)	1973
Female genital tract	—	None	Papillary muscle and chordae tendinae	*Streptococcus putrificus*	Died	Lehmann (660)	1926

(continued)

TABLE 8.4 (*continued*)

Probable source	Other underlying condition	Prior heart disease[a]	Endocardial location	Organism(s)	Outcome	Reference	Year
Uterus	?	?	Tricuspid valve	*Streptococcus putrificus*	Died	Bingold (442)	1932
?	?	?RHD	Mitral valve	*Peptostreptococcus micros*	Lived	Ko et al. [cited by Nakamura et al. (1335)]	1972
?	?	CHD	?	*Peptostreptococcus micros*	Lived	Ikemoto et al. (1336)	1971
?Pharynx	—	RHD	Mitral and aortic valves	*Peptostreptococcus parvulus* (*Streptococcus parvulus*) and viridans streptococcus	Died	Arjona et al. (1290)	1946
Urinary tract	—	?	Mitral valve	*Peptostreptococcus parvulus* (*Streptococcus parvulus*)	Lived	de la Barreda and Centenera (1337)	1949
?	?	CHD	VSD	*Peptostreptococcus* NCDC Group 3 and facultative organism	Lived	Felner and Dowell (289)	1970
Cystitis	?	RHD	?	*Peptostreptococcus* NCDC Group 3 and facultative organism	Lived		1970

?[c]	?	?	Peptostreptococcus (2 kinds)	Died	Ikemoto et al. [cited by Nakamura et al. (1335)]	1972
Dental	—	Aortic valve	Peptostreptococcus	Lived	Milovanov et al. (1337a)	1969
?	CHD	Patent ductus arteriosus	Peptostreptococcus	Lived	Sawada et al. [cited by Nakamura et al. (1335)]	1972
?	?	?	Peptostreptococcus	Lived	Nakamura et al. (1335)	1972
?	?	?	Peptostreptococcus	Lived	Wilson et al. (105)	1972
?	?	?	6 Cases due to microaerophilic or anaerobic streptococcus	?	Cherubin and Neu (1338)	1971
?	?	?	One or more cases each with Bacteroides, microaerophilic streptococci and anaerobic streptococci	?	Kaye et al. (1339)	1962
?	?	?	Anaerobes (type not specified) (3 cases)	?	Stokes (108)	1958

[a] RHD, rheumatic heart disease; CHD, congenital heart disease; ASHD, arteriosclerotic heart disease; VSD, ventricular septal defect.

[b] R to PC, resistant to penicillin.

[c] Bacteremia, questionable endocarditis.

[d] GNR, gram-negative rod.

One of our cases was due to *Bacteroides fragilis* (433); the rest were caused by anaerobic or microaerophilic streptococci or cocci. Lerner and Weinstein (434) found 13 cases due to microaerophilic streptococci, and 3 due to anaerobic streptococci among 100 cases of endocarditis. Dormer (435) studied 82 cases of endocarditis over a 12-year period. One was due to *B. fusiformis* and 8 to microaerophilic streptococci. Two of the latter organisms were "very resistant" to penicillin. Two patients died, one with an embolus of the coronary artery and two had mycotic aneurysms of peripheral arteries. Among 35 cases of endocarditis studied over a 6-year period, Campeau *et al.* (436) found four anaerobes (three *Streptococcus evolutus* and one *Corynebacterium anaerobium*). Other series, however, suggest a much lower incidence of anaerobes in endocarditis. In their review of 408 cases of endocarditis from 14 centers, Cates and Christie (437) found only six cases due to microaerophilic streptococci. Werner *et al.* (438) studied 206 cases of endocarditis in 197 patients over a 16-year period. One was due to a *Peptostreptococcus* and at least one other to *Bacteroides*. There were seven due to lactic streptococci, which may be microaerophilic. Wedgwood (439) studied 65 cases of subacute bacterial endocarditis over an 18-year period. Fifty-six of these had blood cultures done; of these, only one yielded a microaerophilic streptococcus. Tompsett (440) studied 46 cases of endocarditis in a 6-year period. Only one of these involved a microaerophilic streptococcus.

Streptococcus mutans has recently been recognized as an important cause of endocarditis by Harder *et al.* (440a). These authors reported 9 cases of *S. mutans* endocarditis and another was recorded by Lockwood *et al.* (440b). *Streptococcus mutans* accounted for 6% of cases of streptococcal endocarditis seen at the Mayo Clinic in a 7 year period. This organism is considered to be one of the "viridans streptococci," but usually requires microaerophilic conditions or anaerobiosis for surface growth; it may also be confused with group D streptococci. It is typically very sensitive to penicillin G, but some strains have a minimal bactericidal concentration of >5 μg/ml (440c). In a 12 year period at Wadsworth Veterans Hospital (S. Finegold, E. Harder, and R. Rothman, unpublished data), we noted that 10 of 109 cases of bacterial endocarditis were caused by anaerobic [1] or microaerophilic [9] streptococci. Six of the latter 9 strains were fully characterized; 3 were *S. mutans* and 1 each *S. bovis*, *S. sanguis* I and *S. salivarius*. Most of these patients had underlying heart disease, usually rheumatic, but 2 had no evidence of prior valvular disease. At least 2 of the patients had prior dental disease or manipulation and one was a narcotic addict. Eight of the cases had 6 positive blood cultures each and the others 3. Although the clinical features of the illness in these patients were not unlike those of "viridans streptococcus" endocarditis in general, 5 of the patients had significant valvular destruction or distortion as a result of the disease. All patients were cured bacteriologically, but 4 patients died within 4 months of completion of therapy due to cardiac failure or a major embolus. The fifth patient had an aortic valve prosthesis inserted and was well after a one year followup. We (L. J. Nastro and S. M. Finegold) have also seen endocarditis due to a microaerophilic streptococcus which keyed out as a group M streptococcus.

Masri and Grieco (441) reviewed 27 cases of *Bacteroides* endocarditis and presented one new case. The portal of entry was found to be the oropharynx in 3 cases, the gastrointestinal tract in 5, genitourinary tract 3, skin 2, lung 2, and unknown 7. Preexisting heart disease was noted in 15 of the cases (rheumatic heart disease 7, arteriosclerotic heart disease 6, and ventricular septal defect 2). There were three diabetics, two patients with decubitus ulcers, and one heroin user. There were two reports of involvement of the tricuspid valve. There was a relatively high incidence of thrombophlebitis (6 patients) and of embolization (15). The mortality was 36% in the total group. Felner and Dowell (289) noted septic shock in three cases of anaerobic bacterial endocarditis (*B. oralis*, *P. acnes*, and *C. perfringens*) and hypotension in another (*Clostridium cochlearium*). Sharpe *et al.* (361b) mentioned recovery of lactobacilli from 3 cases or suspected cases of bacterial endocarditis; few case details were given. An early myocardial abscess was noted in the course of one anaerobic streptococcal endocarditis case by Bingold in 1932. (442). Prévot (442a) noted that he had studied 26 cases of endocarditis due to anaerobic *Corynebacterium*.

A requirement for certain reducing substances was noted in streptococci from three cases of endocarditis reported by Cayeux *et al.* (443). One of these organisms belonged to group N. These organisms were characterized by satellite growth about another streptococcus and requirement for either thioglycolate or dithiothreitol, in addition to L-cysteine or reduced glutathione, for optimum growth. Not all reducing substances were suitable, and anaerobic conditions alone were unsatisfactory. Media designed to grow L forms were also unsatisfactory. Two cases of endocarditis due to similar organisms, classed as *Streptococcus mitior* (*mitis*), were noted by Carey *et al.* (443a). George (443b) also discussed this type of organism and cited its isolation from two cases of endocarditis by Burdon and from an additional case of endocarditis by Frenkel and Hirsch.

Dutton and Inclin (444) reviewed from the literature 86 cases of actinomycosis involving the heart. Thirty-five percent of these had endocardial involvement. Of these, 43% had valvular involvement, most often the mitral and/or aortic valves. Another case of probable endocarditis due to *Actinomyces* has been reported (1273a). Marked thrombocytopenia was noted in a case of *Bacteroides* endocarditis (1296a).

Pericarditis

The incidence of anaerobes in purulent pericarditis is apparently very low. In an extensive review, Boyle *et al.* (445) found one case due to *Clostridium* plus β-hemolytic streptococci; this was a case of purulent pericarditis following an acute myocardial infarction. They also noted that pericardial involvement occurred in 2% of cases of actinomycosis, with clinical features in only one-third of these. Large amounts of pericardial fluid may be formed, and eventually there may be complete obliteration of the pericardial sac. Dutton and Inclin (444) found that 71% of 86 cases of actinomycosis of the heart had pericardial involvement. Five of these cases had a friction rub, and five had constrictive pericarditis. Two had pericardial effusion. There are several single case reports of involvement of the pericardium by *Actinomyces*. Coodley (446) reported a case of pericarditis, presumably serious, in a patient with actinomycosis involving the pleura and chest wall. Verme and Contu (447) reported constrictive pericarditis due to *Actinomyces*. Pericarditis with effusion due to *Actinomyces* was reported in a single patient by Goldsand and Braude in 1966 (244). Mohan *et al.* (448) reported a case of purulent pericarditis due to *Actinomyces*. The organism was seen in but not cultured from microabscesses in a pericardial biopsy. The patient had tamponade and, later, fibrosis of the pericardium. On one occasion 600 ml of purulent fluid was removed, and on another, 788 ml. There were no pulmonary lesions but pleural effusion occurred later in the course. Datta and Raff (448a) reported a case of pleuropericarditis showing *Actinomyces* in sulfur granules involving the pericardium, as well as the lung and mediastinum; *Fusobacterium fusiforme* was cultured from the pericardial fluid in addition.

Involvement of other anaerobes in pericarditis is also noted in several case reports. Prévot (197) reported two cases of fatal purulent pericarditis without clinical details. One of these was due to *Ristella trichoides* and one to *Sphaerophorus floccosus*. Vaucher and Woringer (449) reported a case of benign pericarditis with spontaneous cure secondary to anaerobic streptococcal bacteremia. Leys (450) reported a case of pneumopyopericardium secondary to esophageal traction diverticulum related to a caseous tuberculous paratracheal node. The pus was foul and from it was recovered a *Klebsiella* plus a "gas-forming anaerobe." Avierinos and Turries (451) also reported a case of pyopneumopericardium. The pus and the gas contained in the pericardial sac were both fetid. There was partial spontaneous drainage through a pericardiobronchial fistula. No bacteriology was reported on this case. These authors also cited another case of putrid pericarditis published by Dufour and Baruk. In 1930, Cunningham (323) reported a case of *Actinomyces necrophorus* sepsis of tonsillar origin. At autopsy, fibrinous purulent pericarditis was noted. From the pericardial fluid *A. necrophorus* and occasional anaerobic streptococci were cultured. Weyrich (452) reported a case, without clinical details, of *Streptococcus putrificus* infection with metastatic spread to the pericardium. A case of suppurative pericarditis was reported by Ross in 1939 (453). The pus, which was not foul-smelling, yielded on culture a short, oval gram-negative rod that grew in brain broth but not on a blood agar plate. This organism was said to have the appearance of *B. pertussis*. This might or might not represent an anaerobe. Smith *et al.* (416) recorded a case of pericarditis due to anaerobic streptococci in a patient with septic abortion and bacteremia. Rubin and Moellering (453a) noted involvement of *Bacteroides* sp. in 1 of 26 cases of purulent pericarditis they studied; there was an antecedent bacteremia. It was felt that the original process was probably a viral pericarditis. Guneratne (453b) reported purulent pericarditis due to ruptured myocardial abscess involving *Clostridium perfringens* (see below). Three cases of purulent pericarditis involving anaerobic cocci or streptococci (once in pure culture) were reported by Gould *et al.* (452a).

Myocardial Infection

We have already cited the case of myocardial abscess reported by Bingold (442). Tennant and Parks (454) reported a most unusual case of a diabetic with *C. perfringens* empyema of the gallbladder who developed multiple myocardial abscesses, one of which ruptured and caused death due to pericardial tamponade. This abscess was localized in a recent myocardial infarct. Guneratne (453b) reported a patient with multiple myocardial abscesses due to *C. perfringens* developing in a recent myocardial infarct; perforation led to purulent pericarditis. Korns [cited by Pittman *et al.* (455)] also reported a case of myocardial abscess due to *Clostridium perfringens*. A myocardial abscess due to *Bacteroides* is noted in the Case Records of the Massachusetts General Hospital (Case 27–1970); this abscess appeared in an area of infarction and led to rupture (456). Lewis (455a) reported a similar case involving *Bacteroides fragilis* ss *fragilis*. Chow and Guze (1573) noted microabscesses in the heart of a patient with endocarditis apparently involving both Bacteroidaceae and *Candida tropicalis*. Harvey *et al.* (212) reported myocardial abscesses due to *Actinomyces* in one patient.

Weyrich (452) discussed metastatic spread of anaerobic infections to the myocardium without giving clinical details. One of his cases involved *Streptococcus putrificus* and two additional cases, *Streptococcus putrificus* plus gram-negative bacilli. A very questionable case is reported by Schenken and Heibner (457). The patient had a patchy but diffuse myocarditis. There were a moderate number of inflammatory cells present, primarily polymorphonuclear leukocytes. No organisms were seen on gram stain, but microaerophilic streptococci were isolated in pure culture. Postmortem blood culture was negative. There was no evidence of endocarditis. The authors felt that the organism isolated was responsible for the myocarditis. In the review of 86 cases of actinomycosis involving the heart by Dutton and Inclin in 1968 (444), these authors noted that 50% of the patients had myocardial involvement. Myocardial granulomata due to *Actinomyces* was reported by Verme and Contu (447).

Involvement of the heart muscle during the course of widespread clostridial sepsis has been reported on a number of occasions. Necrosis of the heart muscle in a case of clostridial sepsis was reported by Schottmüller in

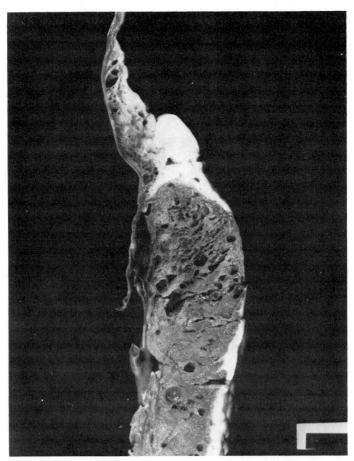

Fig. 8.5. Gas cysts in myocardial wall. Section is oriented with left atrium at top, posterior leaflet of mitral valve at left, and left ventricle occupying most of the photograph. From Ref. (485) C. W. Roberts and C. W. Berard, *Am. Heart. J.* **74**, 428–488 (1967).

1910 (258). Roberts and Berard (458) noted that 9 of 17 patients with widespread systemic clostridial infection had invasion of the cardiac muscle with foci of myonecrosis containing numerous organisms and gas cysts without any inflammatory response (Fig. 8.5). Clumps of organisms were noted within the lumina of cardiac vascular and lymphatic channels. Warkel and Doyle (459) noted gas-filled cysts with many organisms in the myocardium (and many other organs) in the course of clostridial septicemia in a patient with lymphoma.

Miscellaneous Cardiovascular Infections

Bodner *et al.* (460) reported an interesting patient who had an infected abdominal aortic prosthesis with abscess formation in a large infected hematoma. The patient had *Bacteroides* bacteremia. He previously had had a colectomy for recurrent diverticulitis. Another infected vascular graft is reported by Szilagyi *et al.* (461). This patient had an aortofemoral graft to bypass an aortoiliac occlusion. The distal anastomosis became infected. Recovery ensued after segmental replacement of the limb of the graft. Infected aortic grafts were reported in two papers by Gorbach and Thadepalli—*Bacteroides fragilis, B. melaninogenicus* and *Escherichia coli* in one case (344a) and *Clostridium perfringens* plus *Staphylococcus epidermidis* in the second case, who developed an inguinal fistula (486). A very unusual case of systemic-to-pulmonary artery fistula (multiple connections between the intercostal arteries and the left pulmonary artery) in a case of pleuropulmonary actinomycosis was reported (461a). Three cases of mycotic aneurysm involving anaerobes were reported by Anderson *et al.* (461b). One of these, in an addict, involved the common femoral artery in association with a gas-forming thigh abscess; it was due to a *Peptostreptococcus*. The other two cases were fatal. One involved the abdominal aorta (*Bacteroides*), and the other, the common iliac (*Clostridium perfringens, Bacteroides, Enterobacter,* and *Streptococcus*).

Waitzkin (462) reported microabscesses in the bone marrow in one patient. Several cultures were positive for *Corynebacterium acnes* and then became negative following a course of erythromycin therapy.

Venous involvement has been noted earlier (infected cut-downs, etc.). Brocard *et al.* (463) noted phlebitis in the lower leg in the course of *B. funduliformis* sepsis; this was considered an unusual location. Fass *et al.* (464) reported a case with multiple subcutaneous abscesses of the legs from which were recovered *Bacteroides, E. coli, Proteus mirabilis, Klebsiella,* and enterococcus. These occurred in a patient with venous insufficiency in the legs who had previously had vein ligation and stripping. The interesting association of L forms and other atypical forms of gram-negative anaerobes with thrombophlebitis and thromboembolism as reported by Altemeier *et al.* (432) has already been mentioned. Portal vein bacteremia due to *Bacteroides* was demonstrated in one case of chronic ulcerative colitis undergoing total colectomy—both before and after mobilization of the colon (465). There is a strong association between septic thrombophlebitis and anaerobic infection in various parts of the body. This is a very distinctive, although not unique, feature of anaerobic infection.

9

Respiratory Tract and Other Thoracic Infections

General

Intrathoracic infections, especially those of the lung and pleural space, caused by anaerobic bacteria are relatively common. However, they are frequently overlooked in the usual clinical practice and, because of difficulties inherent in obtaining appropriate specimens for diagnosis, these infections are often not documented even when suspected. Anaerobic pleuropulmonary infections have been the subject of intensive investigation by my colleagues (particularly John G. Bartlett and Sherwood L. Gorbach) and me, and most of the data presented in this chapter will be from our studies (90, 161, 353, 466–472). Material derived from a review of the literature is also included for comparative purposes.

The first major report of involvement of anaerobes in infections of the lung and pleural space was that of Guillemot et al. (473) in 1904. In this study of putrid pleurisy, the authors noted that the empyema fluid from 13 patients showed multiple bacteria on smear, but that aerobic cultures were either sterile or yielded unlikely pathogens in small numbers. On the other hand, anaerobic culture yielded multiple anaerobic species in every case. Because the anaerobic isolates resembled those previously identified as normal oral flora, the investigators hypothesized that aspiration and pneumonitis preceded the empyema. Studies over the next several decades, before any antimicrobial therapy became available, provide us with important information on the natural course of untreated anaerobic pleuropulmonary infections of various types. The role of anaerobic bacteria in the processes was readily documented because of availability of autopsy and surgical specimens. Such specimens are less often available today.

Since the introduction of antimicrobial agents, there have been relatively few reports of bacteriologically documented anaerobic pleuropulmonary infections. The more recent studies have emphasized the emergence of such pulmonary pathogens as staphylococci, Enterobacteriaceae, and *Pseudomonas aeruginosa*. The literature of this recent postantibiotic period, then, suggested that such entities as aspiration pneumonia, lung abscess, necrotizing pneumonia and empyema were now being caused by various aerobic and facultative bacteria rather than by anaerobes. However, the use of special techniques, and in particular percutaneous transtracheal aspiration, to bypass normal oropharyngeal flora, and the use of optimal techniques for transporting the specimens to the laboratory under anaerobic conditions, soon made it clear that anaerobic bacteria were still very common and important causes of pulmonary infection. Still another factor has operated to make diagnosis of anaerobic pleuropulmonary infection more difficult in the antimicrobial era. Pharyngeal and tonsillar infections due to *Fusobacterium necrophorum* were relatively common before the advent of antimicrobial drugs. These infections were not infrequently complicated by septic thrombophlebitis and bacteremia, and, in the absence of effective therapy, metastatic infection occurred with some frequency. The lung and pleural space were often involved in such dissemination of disease. This type of infection seldom occurs today, probably because of widespread use of antimicrobial agents for upper respiratory infections. The incidence of bacteremia with anaerobic pleuropulmonary infection is extremely low at the present time, and *Fusobacterium necrophorum* is seldom encountered in such infections.

Types of Anaerobic Pleuropulmonary Infection

The various types of anaerobic infection involving the lung and pleural space are indicated in Table 9.1. The upper part of the table summarizes our own experience with these infections in two different time periods, 1958–1968 and 1969–1974. During the latter period, we were more aware of the character of anaerobic infections and certain clues suggesting the presence of anaerobes in pulmonary infection; we were more actively looking for such infections; and we were using percutaneous transtracheal aspiration much more commonly when other appropriate sources of mate-

TABLE 9.1

Types of Anaerobic Pleuropulmonary Infection

	No. of cases	
Infection (Wadsworth and Sepulveda Veterans Hospitals)	1958–1968	1969–1974
Pulmonary abscess without empyema	6	24
Pulmonary abscess with empyema	4	11
Necrotizing pneumonitis without empyema	7	11
Necrotizing pneumonitis with empyema	6	4
Pneumonitis with empyema	20	4
Pneumonitis without empyema or abscess	0	44
Infected bronchogenic carcinoma cavity	1	0
Empyema without evident parenchymal infection	0	1
Total	44	99
Total with empyema	30	20

	No. of cases	
Infection (literature review)	1899–1944	1945–1970
Pulmonary abscess without empyema	8	21
Pulmonary abscess with empyema	15	25
Necrotizing pneumonia without empyema	28	5
Necrotizing pneumonia with empyema	17	9
Multiple discrete abscesses without empyema following septic emboli	26	9
Multiple discrete abscesses with empyema following septic emboli	10	5
Empyema without abscess[a]	65	93
Pneumonitis without empyema or abscess(es)	4	10
Miscellaneous pulmonary lesion[b]	0	8
Total	173	185
Total with empyema	107	132

[a] Many cases were associated with pneumonitis.

[b] Infected cysts (2), pneumonitis with possible cavitation (1), infected tuberculous cavity (2), empyema with possible parenchymal abscess (3).

rial for culture were not available. For these various reasons, we not only diagnosed significantly more cases in a much shorter time period, but there was a difference in the type of case that we diagnosed during this later period. Many more cases without empyema or abscess formation were diagnosed in the later period. The lower portion of Table 9.1 shows the types of infections encountered in a literature review, also divided into two time periods, the period prior to introduction of antimicrobial agents and the antimicrobial era. In the latter period, the incidence of cases with empyema was significantly higher. In addition, there were many fewer cases with multiple discrete abscesses following septic emboli and fewer cases with necrotizing pneumonia. On the other hand, the incidence of solitary lung abscess increased.

Predisposing or Underlying Conditions

Predisposing and underlying conditions are reviewed in Table 9.2, the upper half of the table again showing our own data, and the lower half, data from the literature review. In our own series, by far the most important background factor for anaerobic pleuropulmonary infection was documented or suspected aspiration, usually related to altered consciousness. Accordingly, dependent pulmonary segments (Fig. 9.1) are most often involved. The second most common factor was periodontal disease. Similar

TABLE 9.2

Predisposing and Underlying Conditions, Anaerobic Pleuropulmonary Infection

Predisposing conditions in 143 cases of anaerobic pleuropulmonary infections[a]	No. of cases	% of cases
Dental status (87 patients evaluated)		
Periodontal disease	53	
Gingivitis	1	
Suspected aspiration		
Altered consciousness	104	73
Alcoholism	43	
Cerebrovascular accident	12	
Drug overdose or addiction	11	
General anesthesia	9	
Seizure disorder	9	
Miscellaneous causes	16	
Dysphagia		
Esophageal disease	5	
Neurologic cause	10	
Intestinal obstruction	6	
Pulmonary conditions	29	20
Bronchogenic carcinoma	12	
Bronchiectasis	9	
Suspected pulmonary embolism with infarction	7	
Pulmonary tuberculosis	1	
Intraabdominal suppuration	8	6
Subphrenic abscess	8	
No established underlying condition	15	10

TABLE 9.2 (*continued*)

Underlying or predisposing conditions (literature survey)	No. of cases	
	1899–1944	1945–1971
Aspiration		
Altered consciousness	31	53
Alcoholism	3	13
General anesthesia	19	25
Other	1	7
Esophageal dysfunction or intestinal obstruction	3	7
Tonsillectomy or tooth extraction	7	4
Preceding extrapulmonary anaerobic infection[b]		
Periodontal disease	12	6
Pharyngitis	43	8
Otitis–mastoiditis	11	3
Female genital tract infection	16	10
Intraabdominal infection (other)	11	13
Bacterial endocarditis	4	1
Other	4	4
Postthoracotomy or penetrating chest wound	15	17
Local underlying conditions		
Bland pulmonary infarction	1	3
Bronchogenic carcinoma	3	15
Bronchiectasis	8	7
Foreign body	1	
None apparent	33	53

[a] Patients at Wadsworth and Sepulveda Veterans Hospitals. Some patients had two or more underlying conditions.

[b] 67 patients had documented anaerobic sepsis accompanying the preceding extrapulmonary infection.

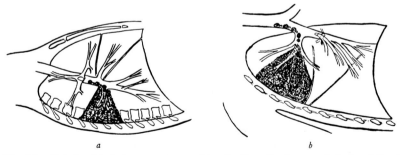

a   *b*

Fig. 9.1. Relationship between posture and localization of pulmonary infection following aspiration. (a) With the patient on his back, the apical segment of the lower lobe is vulnerable. (b) With the patient on his side, the lateral and posterior portion of the upper lobe is affected. From Ref. (1588) R.C. Brock, "Lung abscess." 1952 Courtesy of Charles C. Thomas Publisher, Springfield, Illinois.

predisposing factors were noted in the cases reviewed from the literature. Note, however, that in the period 1899–1944, there were 43 cases with pharyngeal infection as a predisposing cause but only 8 such cases in the period 1945–1971, when antimicrobial therapy was available. Five cases of pulmonary infection involving anaerobes in patients who had undergone cardiac transplantation have been described (473a, 473b, 473c).

Diagnostic Considerations

When empyema or bacteremia complicate anaerobic pulmonary infection, obtaining a reliable specimen for culture, free of normal flora, is no problem. However, it is desirable to diagnose anaerobic infections as early as possible, before complications such as empyema ensue. As indicated earlier, bacteremia is no longer frequently associated with anaerobic infections of the lung. Coughed sputum is not suitable for diagnosis of infections of the lung parenchyma because of the prevalence of large numbers of anaerobes as indigenous flora in the mouth and upper respiratory tract. Even quantitative culture will not distinguish between anaerobes as infecting flora and anaerobes as normal flora. One can do a percutaneous transthoracic aspiration (direct lung puncture), but the incidence of complications is relatively high in adults, and one samples only a very small portion of the lung and obtains a small specimen. This type of needle biopsy was used successfully in diagnosing 2 of the 5 anaerobic pulmonary infections in cardiac transplant patients cited previously (473b). Beerens and Tahon-Castel (115) have used direct lung puncture successfully to establish the role of anaerobes in 21 cases of lung abscess and 4 cases of superinfected lung cysts and cavities. Percutaneous transtracheal aspiration (Fig. 9.2) is preferred if there are no contraindications to its use. Details on the use of this procedure in the diagnosis of anaerobic pulmonary infection are presented in a paper by Bartlett *et al.* (90). Localized anaerobic neck infections as a complication of transtracheal aspiration in patients with anaerobic pulmonary infection are discussed in Chapter 6. Preliminary studies do suggest that a washing and liquefaction procedure might permit use of coughed sputum (467).

Table 9.3 lists the culture sources for material used to diagnose anaerobic pleuropulmonary infection in our own studies and in the material reviewed from the literature. Note in the literature review portion of the table the frequency with which diagnosis was made from material obtained at autopsy or from sites of infection other than the lung (either primary or secondary). Also note the very large number of cases diagnosed by pleural fluid culture in the literature review series, and the virtual lack of percutaneous transtracheal aspirates for culture.

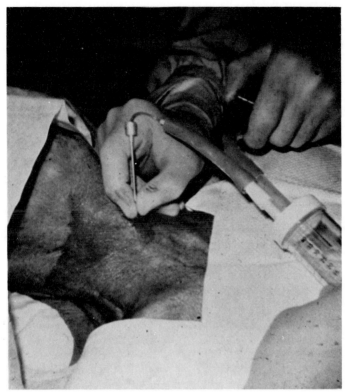

Fig. 9.2. Patient undergoing percutaneous transtracheal aspiration. (Courtesy of Dr. J. G. Bartlett.)

TABLE 9.3

Culture Sources, Anaerobic Pleuropulmonary Infection

Culture sources: Wadsworth and Sepulveda Veterans Hospitals (143 cases)	No. of cases[a]
Percutaneous transtracheal aspirate	98
Empyema fluid	43
Thoracotomy specimen	9
Blood culture	2
Autopsy material	2
Percutaneous transthoracic aspirate	1
Metastatic site	1

TABLE 9.3 (*continued*)

Culture sources: Literature review (358 cases[b])	No. of cases[a]
Pleural fluid	221 (24)[c]
Transtracheal aspiration	1
Percutaneous transthoracic aspiration	30 (2)
Resected pulmonary tissue	10 (1)
Blood culture[d]	77 (36)
Distant metastatic site[d]	20 (13)
Nonpulmonary primary site[d]	12 (6)
Autopsy material	64 (29)

[a] Two or more sources yielded anaerobes in some cases.

[b] Includes 32 cases which were documented only by smear from these sources.

[c] Numbers in parentheses indicate cases where two or more of the sources listed yielded anaerobes.

[d] Extrapulmonary infection was clearly related to the pleuropulmonary disease.

Anaerobic Pneumonitis without Necrosis or Abscess Formation

This type of anaerobic pulmonary infection is by far the most easily overlooked of all because three very important clues to anaerobic pulmonary infection (putrid discharge, tissue necrosis, and subacute or indolent course) are generally absent. One would almost always have to depend on percutaneous transtracheal aspiration or a similar procedure to obtain material for culture. One major clue that is present in this group is the suspicion of aspiration as the underlying mechanism. Thirty-one of 44 patients (70%) with anaerobic pneumonitis that we have studied had associated conditions resulting in compromised consciousness or dysphagia.

Anaerobic pneumonitis is usually characterized by a relatively acute course and a rapid response to appropriate therapy. This is in contrast to other forms of anaerobic pulmonary infection in which patients usually exhibit more chronic symptomatology. It is likely that most cases of anaerobic pneumonitis represent the early stage of an infection which would ultimately be complicated by abscess formation or empyema if allowed to continue without therapy. This view is supported by experimental studies done before chemotherapy was available. On the other hand, 2 of our 44 patients had anaerobic pneumonitis for several months with no evidence of tissue necrosis.

The bacteriology in our 44 cases of anaerobic pneumonitis without necrotizing features or abscess formation is detailed in Table 9.4. Over one-third of the cases yielded anaerobes to the exclusion of other types of bacteria. There was an average of 3.7 bacterial isolates per case, 2.8 of

TABLE 9.4

Bacteriology of Anaerobic Pneumonitis without Necrotizing Features or Abscess Formation (44 Cases)[a]

Bacteriological results	No. of cases	% of cases
Only anaerobes recovered	16	36
Anaerobes and aerobes or facultatives recovered concurrently	28	64
Anaerobic isolates		
Fusobacterium nucleatum	17 (2)[b]	
F. varium	1	
F. necrophorum	1	
Bacteroides melaninogenicus	19	
B. fragilis	8	
B. oralis	7	
B. corrodens	1	
B. pneumosintes	1	
Unidentified gram-negative bacilli	1	
Peptostreptococcus	22	
Peptococcus	11	
Microaerophilic streptococcus	4	
Veillonella	6 (3)	
Eubacterium lentum	5	
Eubacterium sp.	3	
Catalase-negative, non-spore-forming gram-positive bacilli	3 (1)	
Propionibacterium acnes	4	
Clostridium[c]	7 (1)	
Aerobic and facultative isolates		
Streptococcus pneumoniae	6	
Staphylococcus aureus	8	
Streptococcus faecalis	3	
Klebsiella sp.	6	
Enterobacteriaceae (other)	8	
Pseudomonas aeruginosa	2	
Haemophilus influenzae	5	

[a] Average number of anaerobes per case, 2.8; average number of aerobes per case, 0.9. Data from Wadsworth and Sepulveda Veterans Hospitals.

[b] Numbers in parentheses are numbers isolated in pure culture.

[c] Includes *C. perfringens* (3), *C. sordellii* (1), *C. sporogenes* (1), *C. bifermentans* (1), and *Clostridium* sp. (1).

which were anaerobic. By far the most common anaerobic isolates were *Bacteroides melaninogenicus*, *Fusobacterium nucleatum*, and various species of *Peptostreptococcus* and *Peptococcus*. Of interest is the fact that there were eight isolates of *Bacteroides fragilis*. The fact that only 36% of cases were associated with a flora that was exclusively anaerobic probably relates to the frequency of hospital-acquired aspiration pneumonia in this group

(22 of the 44 patients). As will be noted below in our discussion of aspiration pneumonia, when aspiration takes place in a hospital setting, aerobic and facultative bacteria are found with considerably greater frequency than when aspiration takes place outside of the hospital. More than 70% of cases in each of the other categories of anaerobic pulmonary disease to be discussed below represented infections acquired in the community rather than in the hospital.

Response to therapy in these patients was generally what one would expect with other forms of bacterial pneumonia. Antimicrobial agents were usually given for relatively brief periods, the mean duration of therapy in our series being just 10 days. Fever typically cleared within 3 days, and chest X rays were usually clear at 3 weeks. This contrasts with cases of lung abscess and necrotizing pneumonia in which these evidences of disease persisted for longer periods after therapy was instituted. The anaerobic pneumonitis was a contributing factor to death in 6 patients, all of whom had severe underlying conditions complicated by aspiration pneumonia.

Recent studies have been directed at assessing the role of anaerobes in the overall picture of pneumonia (473d) and of hospital-acquired pneumonia (473e). Ries et al. (473d) studied 76 patients with bacterial pneumonia using specimens obtained by transtracheal aspiration. Anaerobes were found in pure culture in 5 patients and mixed with aerobic organisms in 25. O'Keefe et al. (473e) studied 159 cases over a 40 month period using cultures of transtracheal aspirates, empyema fluid, and blood. Anaerobes were found alone in 11 cases (7%) and together with aerobes in 45 patients (28%). In all there were 149 anaerobic isolates.

Anaerobic Necrotizing Pneumonia

This condition is a suppurative pneumonitis characterized by multiple areas of necrosis and cavity formation (usually small cavities), primarily within one pulmonary segment or lobe (Fig. 9.3). The role of anaerobes in this type of infection was first noted by Rona (474) in 1905. This type of pneumonia may spread rapidly, producing destruction characterized by ragged greenish discoloration of pulmonary parenchyma and large putrid sloughs of tissue, leading to the designation "pulmonary gangrene." This process is quite distinct from lung abscess on the basis of pathology, X-ray findings, and prognosis.

In our studies, we have encountered 28 patients with necrotizing pneumonia involving anaerobic bacteria. Of these, 18 had the characteristic X-ray appearance when first seen, and the other 10 had no evidence of cavitation initially but did on subsequent roentgenograms. Ten of the

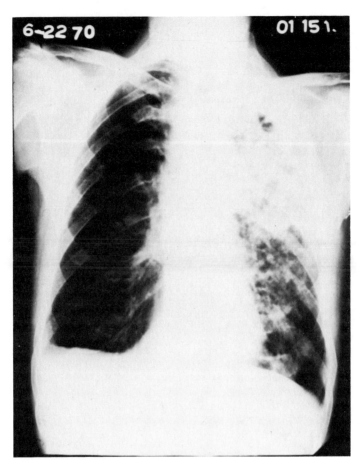

Fig. 9.3.   Extensive necrotizing pneumonia, left lung; empyema, right base. Six anaerobic species were recovered on transtracheal aspiration.

patients had an associated empyema. These patients were typically quite ill, with a mean temperature of 102.4°F and a mean peripheral white blood cell count of 24,200/mm³; 50% of the patients had more than one lobe involved. Putrid sputum or empyema fluid was noted in 17 of the cases (61%).

While necrotizing pneumonia tends to be a fulminating infection with a relatively high mortality, the course is quite variable. Three of the patients in our series were asymptomatic, the infection being discovered on a routine chest X ray taken at the time of hospitalization for another condition.

The bacteriology of anaerobic necrotizing pneumonia is noted in Table 9.5. In 20 of the cases, anaerobes were the exclusive isolates (71%). Overall,

TABLE 9.5

Bacteriology of Anaerobic Necrotizing Pneumonia (28 Cases)[a]

Bacteriologic results	No. of cases	% of cases
Only anaerobes recovered	20	71
Anaerobes and aerobes or facultatives recovered concurrently	8	29
Anaerobic isolates		
Fusobacterium nucleatum	11 (2)[b]	
Bacteroides melaninogenicus	11	
B. fragilis	3	
B. oralis	3	
B. pneumosintes	1	
B. corrodens	1	
Unidentified gram-negative bacilli	5	
Peptostreptococcus	6	
Peptococcus	5 (1)	
Microaerophilic streptococcus	8 (5)	
Veillonella	1	
Eubacterium lentum	2	
Catalase positive, non-spore-forming gram-positive bacilli	2	
Bifidobacterium	1	
Propionibacterium acnes	2	
Aerobic and facultative isolates		
Streptococcus pneumoniae	2	
Staphylococcus aureus	2	
Streptococcus pyogenes	1	
Streptococcus faecalis	1	
Escherichia coli	1	
Pseudomonas aeruginosa	1	

[a] Data from Wadsworth and Sepulveda Veterans Hospitals. Average number of anaerobic species per case, 2.3; average number of aerobic species per case, 0.4.

[b] Numbers in parentheses are numbers recovered in pure culture.

there was an average of 2.7 strains of bacteria isolated per case, 2.3 of which were anaerobic. The specific organisms recovered were not notably different from those recovered in cases without the necrotizing features. The most common isolates again were *Bacteroides melaninogenicus, Fusobacterium nucleatum,* and anaerobic or microaerophilic cocci or streptococci.

Anaerobic necrotizing pneumonia was the principal cause of death in 2 patients in our group, and contributed to death in 3 others. In each of these fatal cases, the antimicrobial therapy utilized was appropriate, although in 1 case the dose of penicillin utilized was relatively low.

Anaerobic Lung Abscess

Our series of anaerobic pulmonary infections included 45 patients with lung abscess, 15 of which were complicated by empyema (Figs. 9.4–9.6). In 29 of the patients, the lung abscess was apparent on the admission chest X ray, whereas in the others there was only pneumonitis initially. We arbitrarily defined lung abscess as a cavity at least 2 cm in diameter. The mean cavity diameter in our series of 45 cases was 4.5 cm (range 2 to 9 cm). Several of the patients had multilocular abscesses. All but one had a solitary lung abscess; the one exception had two distinct abscess cavities in different pulmonary segments.

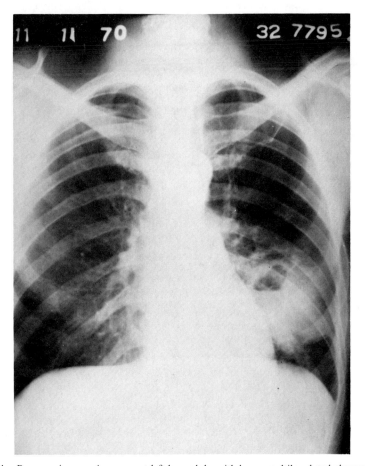

Fig. 9.4. Pneumonia, superior segment left lower lobe with large, multiloculated abscess cavity.

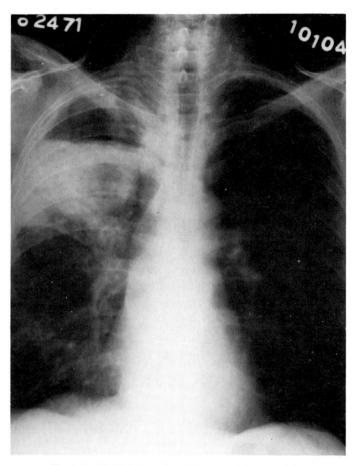

Fig. 9.5. Multiple large lung abscesses, right upper lobe.

Aspiration was the usual underlying mechanism. Approximately two-thirds of the patients had associated conditions leading to altered consciousness or dysphagia. Thirty-three of the 45 patients had dental evaluations; periodontal disease was noted in 22 of these, and gingivitis in 1. Four patients were edentulous. There is an old clinical adage that lung abscess never occurs in edentulous patients; however, the underlying processes in these particular cases were not related to aspiration. Three had carcinoma, and 1 a subphrenic abscess. Although the sputum is classically putrid in lung abscess, such an odor to sputum or empyema fluid was noted in only 47% of our cases.

The location of the lesion on chest X ray supported the notion that

Fig. 9.6. Autopsy specimen, patient in Fig. 9.5 (who died of other causes). Note extensive necrosis in sectioned right upper lobe.

aspiration was the primary underlying mechanism, since dependent segments were the primary ones involved (Fig. 9.7). Twelve cases involved the posterior segment of the right upper lobe, 7 the posteroapical segment of the left upper lobe, 7 the superior segment of one of the lower lobes, and 10 the basilar segments of lower lobes.

The average duration of symptoms when the patient initially presented was 12 days. Weight loss and anemia were often observed on admission, indicating that the infections were indeed indolent. The mean weight loss was 19 lbs among patients who reported weight loss of any degree. Temperature was usually low grade, averaging 101.8°F.

It was possible to determine the time required for cavitation to appear in 11 patients whose time of aspiration could be accurately determined. Serial X rays showed pneumonitis first and subsequently excavation. The mean time for appearance of an air–fluid level or cavity was 12 days; the earliest appearance of a cavity was 7 days after the time of aspiration. Similar data was obtained prior to the availability of antimicrobial agents when it was noted that lung abscess usually appeared 1–2 weeks after aspiration during general anesthesia or tonsillectomy [Smith (475)].

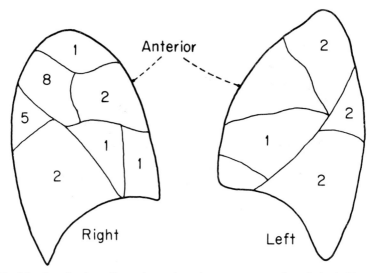

Fig. 9.7.   Localization of lung abscess by pulmonary segment (lateral view). 26 cases.

Bacteriologic data on the cases of lung abscess are noted in Table 9.6. Fifty-eight percent of these cases yielded only anaerobic bacteria. An average of 3.0 organisms per case was recovered, with 2.4 of these being anaerobic. Again, *Fusobacterium nucleatum*, *Bacteroides melaninogenicus*, and anaerobic cocci and streptococci were the predominant organisms recovered; *Bacteroides fragilis* was found in a moderate number of cases as well.

Five patients early in the study had relapse after discontinuation of antimicrobial therapy before X ray clearing or stabilization was achieved. For this reason, prolonged therapy was used in all subsequent cases, the average duration being 2 months. All relapses responded to a second, more extended course of chemotherapy. Surgery was required only in connection with drainage of accompanying empyemas, and, in two cases, a diagnostic thoracotomy was carried out to rule out an underlying neoplasm. Neoplasm was not encountered, and no resection was performed. In several recent studies of lung abscess, resection of the involved portion of lung has been performed in as many as 20–30% of cases, with the primary indication for surgery in these reports being "delayed closure" (persistence of the cavity after 6 weeks of medical therapy). There were 10 cases in our series that would fit these criteria; in each instance a satisfactory response was eventually achieved when therapy with an appropriate agent was continued for a long enough time. Surgical therapy is actually contraindicated in lung abscess, unless there is an accompanying lesion such as carcinoma, since

TABLE 9.6
Bacteriology of Anaerobic Lung Abscess (45 Cases)[a]

Bacteriologic results	No. of cases	% of cases
Only anaerobes recovered	26	58
Anaerobes and aerobes recovered concurrently	19	42
Anaerobic isolates		
Fusobacterium nucleatum	19 (3)[b]	
F. necrophorum	1	
Bacteroides melaninogenicus	18 (1)	
B. fragilis	9	
B. oralis	5	
B. corrodens	2	
Unidentified gram-negative bacilli	4	
Peptostreptococcus	14 (4)	
Peptococcus	14 (1)	
Microaerophilic streptococcus	6 (3)	
Veillonella	4	
Eubacterium lentum	4·	
Eubacterium sp.	3	
Catalase-negative, non-spore-forming, gram-positive bacilli	3	
Propionibacterium acnes	6	
Clostridium perfringens	2 (1)	
Aerobic and facultative isolates		
Staphylococcus aureus	6	
Streptococcus faecalis	3	
Streptococcus pneumoniae	2	
Streptococcus pyogenes	1	
Klebsiella pneumoniae	3	
Escherichia coli	6	
Proteus mirabilis	2	
Haemophilus influenzae	1	
Pseudomonas aeruginosa	3	

[a] Data from Wadsworth and Sepulveda Veterans Hospitals. Average number of anaerobic species per case, 2.4; average number of aerobic species per case, 0.6.
[b] Numbers in parentheses are numbers recovered in pure culture.

there is a hazard of intrabronchial spread of infection during surgery and even of "drowning" secondary to spillage of contents from large abscesses.

In all cases of lung abscess, it is important to look for such associated problems as obstructing foreign body, obstructing tumor (Fig. 9.8) and associated tuberculosis.

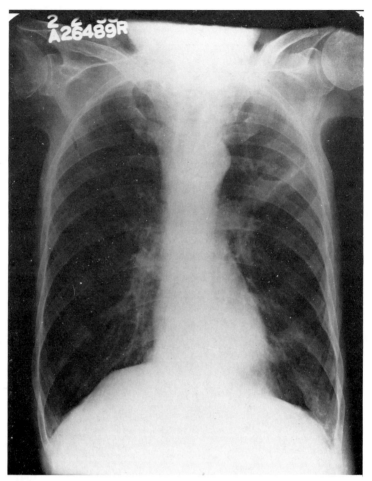

Fig. 9.8.   Left upper lobe lung abscess with bronchogenic carcinoma, left hilum.

Aspiration Pneumonia

Aspiration pneumonia obviously overlaps with other categories, since aspiration is a common background factor for most anaerobic pulmonary infections, and the clinical expression varies from pneumonitis without necrotizing features to necrotizing pneumonia and lung abscess with or without empyema. We have carried out a prospective study of patients with aspiration pneumonia, all of whom had to fulfill the following criteria: observed aspiration or predisposition to aspiration, subsequent development of infection in a dependent pulmonary segment, and availability of a reliable specimen for bacteriologic study prior to institution of antimicrobial therapy. Seventy patients met the criteria for inclusion in the study. There were 67 instances of compromised consciousness and 9 of dysphagia, with some overlap. The principal causes of compromised consciousness were alcoholism (23), cerebrovascular accident (13), seizure disorder (9), general anesthesia (9), and drug ingestion (8). Seven of the 9 patients with dysphagia had a neurological disorder and the other 2 had esophageal stricture. Thirty-eight of the 70 patients had pneumonitis, 5 with complicating empyema; 20 had lung abscess, 2 of which also had empyema; and 12 had necrotizing pneumonia, 1 with empyema in addition. As to source of culture material, there were 63 transtracheal aspirates, 8 empyema fluids, and 4 positive blood cultures.

Duration of symptoms at the time the condition was diagnosed was 3 days or less in approximately half of the patients. This reflects both acute illness and aspiration in the hospital setting in many cases. Putrid sputum was produced by only about one-fourth of patients. Most patients were fairly sick, with mean peak temperatures of 102.5°F and mean peripheral white blood cell counts of $17,000/mm^3$.

The bacteriology of all of our cases of aspiration pneumonia is indicated in Table 9.7. Overall, 46% of the cases yielded only anaerobes from appropriate specimens. However, the incidence of anaerobes in pure culture was distinctly higher among the group of 38 patients who aspirated in the community, as compared to the 32 patients aspirating in the hospital setting (25 versus 7 cases). Overall, anaerobes (with or without accompanying aerobes or facultative organisms) were found in 35 of the 38 patients aspirating in the community setting, against 26 of 32 aspirating in the hospital setting. The anaerobic bacteria recovered were similar to those already described for the types of anaerobic infections previously discussed. Of particular note is the fact that patients who aspirate in the hospital setting had a significant incidence of potential nosocomial nonanaerobic pathogens from the hospital environment. Included were *Staphylococcus aureus* and various gram-negative aerobic or facultative bacilli, such as *Pseudomonas, Klebsiella,* and *Enterobacter.* Lorber and Swenson (476) found similar results in a prospective study of aspiration pneumonia. They recovered anaerobic bacteria from 21 of 24 patients who aspirated in the community, and in 13, these were the only isolates recovered. In hospital-acquired aspiration pneumonia, these authors found anaerobes in 8 of 23 cases, and noted that anaerobes were the only isolates in just 2 cases. They also noted, in the hospital-acquired group, various aerobic and facultative gram-negative bacilli and *Staphylococcus aureus.*

In these patients, many of whom are acutely ill, therapy should be intensive and possibly prolonged. In general, in patients who aspirate in the hospital

TABLE 9.7

Bacteriology of Aspiration Pneumonia (70 Cases)[a]

Bacteriologic results	Hospital acquired	Community acquired	% of total cases
Only anaerobes recovered	7	25	46
Anaerobes and aerobes or facultatives recovered concurrently	19	10	41
Only aerobes or facultatives recovered	6	3	13
Total cases	32	38	

Bacteriologic results	No. of cases
Anaerobic isolates	
Bacteroides melaninogenicus	27 (1)[b]
B. fragilis	10 (1)
B. oralis	9
B. corrodens	2
B. pneumosintes	1
Fusobacterium nucleatum	19
F. necrophorum	1
Unidentified	4 (1)
Peptostreptococcus	23 (5)
Peptococcus	11 (1)
Microaerophilic streptococcus	9 (1)
Veillonella	4
Eubacterium	6
Propionibacterium	6
Bifidobacterium	2
Clostridium	2
Aerobic and facultative isolates	
Streptococcus pneumoniae	11 (2)
Staphylococcus aureus	11 (2)
Enterococcus	3
Group A β-hemolytic streptococcus	2
Klebsiella sp.	8 (2)
Pseudomonas aeruginosa	7
Escherichia coli	6
Enterobacter cloacae	4
Haemophilus influenzae	2
Citrobacter freundii	1
Pseudomonas maltophilia	1

[a] Data from Wadsworth and Sepulveda Veterans Hospitals.

[b] Numbers in parentheses are numbers recovered in pure culture.

setting, therapy should include drugs appropriate for the aerobic and facultative forms that may commonly be encountered in this situation. We did note, however, in 11 patients in our series who received therapy active only against the anaerobic isolates, despite the presence of enteric gram-negative bacilli or *Pseudomonas* in addition to the anaerobes, that the response was satisfactory in all cases. Nevertheless, particularly in patients who are quite ill, we strongly urge that therapy cover all organisms that may be present. Mortality in aspiration pneumonia is seen primarily in individuals with necrotizing pneumonia.

It should be pointed out that aspiration may lead to a variety of problems in the lung. Aspiration of acid gastric contents may lead to a chemical pneumonitis (Mendelson's syndrome). Hydrocarbons and other irritants may cause similar pathology. Aspiration of solid particles may produce acute bronchial obstruction. Although infection may complicate these events, it often does not, and infection certainly plays no role in the immediate events following these forms of aspiration. Bacterial pneumonia ultimately developed in only 8 of Mendelson's 66 patients (56b).

Empyema

The last of the major categories of anaerobic infection is empyema (Fig. 9.9). Our series includes 83 cases, virtually all of which had associated parenchymal disease. The predisposing conditions were similar to those found in other categories of anaerobic pleuropulmonary disease, with aspiration the primary background factor. There were several cases in which the empyema formed above a subphrenic abscess, suggesting trans-diaphragmatic spread. The majority of the patients had been sick for at least 1 week prior to presentation, and a number had been sick for several weeks. Significant weight loss was found in over half of the patients. Relatively high fever and peripheral white blood cell counts were common.

The bacteriology of our 83 cases of empyema is detailed in Table 9.8. It should be pointed out that we specifically excluded from our series cases of empyema arising postoperatively, and also cases of tuberculous or fungal empyema. Among our patients, 76% yielded anaerobes, and in 35% these were the exclusive isolates. The specific bacteriology was very much like that described for other types of anaerobic pleuropulmonary infection, with *Fusobacterium nucleatum*, *Bacteroides melaninogenicus*, and anaerobic or microaerophilic cocci and streptococci predominating. However, a larger number than usual of *Bacteroides fragilis* and of various species of *Clostridium* was recovered from these cases.

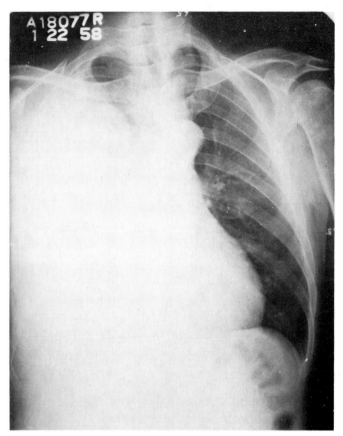

Fig. 9.9.   Massive anaerobic empyema right pleural space.

In virtually all cases, the empyema fluid was purulent or became purulent. Often, the empyema was loculated and proved difficult to aspirate, requiring multiple thoracenteses, sometimes at several sites. Surgical drainage is the procedure of choice for all but the earliest cases, and open thoracotomy with rib resection is the specific procedure of choice when the fluid is frankly purulent. Delay in achieving adequate drainage frequently resulted in extended periods of toxicity, and the mortality that occurs in this disease is always ascribable to failure to obtain adequate drainage or delay in performing an optimal surgical procedure. Antimicrobial therapy is only an adjuvant and is certainly inadequate in the absence of proper surgical drainage. It will be most useful for cases with associated parenchymal disease. Convalescence is slow, averaging nearly 6 months in cases followed to closure of the thoracostomy wound.

TABLE 9.8

Bacteriology of Empyema[a]

Total cases	83
Anaerobes only	29 (35%)
Aerobes or facultatives only	20 (24%)
Anaerobes plus aerobes or facultatives	34 (41%)
Total with anaerobes	63 (76%)

Specific organisms recovered	No. of cases
Anaerobic isolates	
Fusobacterium nucleatum	16 (3)[b]
Bacteroides melaninogenicus	13
B. fragilis	13 (1)
B. oralis	8
B. pneumosintes	2
Unidentified	3
Peptostreptococcus	12
Peptococcus	14 (1)
Microaerophilic streptococcus	15 (5)
Veillonella	6
Eubacterium sp.	5 (1)
Propionibacterium sp.	4
Lactobacillus sp.	5
Unidentified catalase-negative, nonsporulating gram-positive rod	9
Clostridium sp.[c]	13 (1)
Actinomyces israelii	1
A. naeslundii	1
Aerobic and facultative isolates	
Staphylococcus aureus	17 (6)
S. epidermidis	5
Streptococcus pneumoniae	5 (2)
S. faecalis	5
S. pyogenes	4
Streptococcus (other)	8
Escherichia coli	11
Klebsiella sp.	6 (1)
Proteus mirabilis	2
Pseudomonas aeruginosa	10 (2)
Haemophilus influenzae	1

[a] Excludes cases following thoracic surgery and cases due to mycobacteria and fungi. Data from Wadsworth and Sepulveda Veterans Hospitals.

[b] Number of cases isolated in pure culture.

[c] Includes *C. perfringens* (3); *C. ramosum* (1); *C. innocuum* (1); *C. subterminale* (1); *C. limosum* (1); *C. sporogenes* (1); and *Clostridium* sp. (5).

Sullivan *et al.* (476a) studied the role of anaerobes in 226 cases of culture-proven empyema. Anaerobes, chiefly *Bacteroides* and anaerobic streptococci, were recovered from 19% of cases. The authors noted that anaerobes are undoubtedly more important since 53% of the empyema cases they reviewed had negative cultures.

Additional Bacteriological Data

The bacteriology of 358 cases of various types of anaerobic pleuro-pulmonary infection culled from the literature is shown in Table 9.9. There are a number of distinct differences from the bacteriology as determined in our current series. Some of these differences, such as the significantly greater incidence of *Bacteroides melaninogenicus* in our current series, are attributable to improvements in anaerobic bacteriologic technique. Modern techniques would also lead to a smaller number of unidentified anaerobes of different types. Changing patterns of disease account for other changes. The most notable example of this is in the case of *Fusobacterium necrophorum* (*Sphaerophorus necrophorus*). In the literature review series, this organism was encountered in 88 cases (24% of the total anaerobic isolates identified) and was isolated in pure culture 65 times (primarily from blood cultures). As indicated previously, this organism was found primarily in cases of pharyngitis and tonsillitis in the preantimicrobial era. A number of other organisms equally sensitive to antimicrobial agents have persisted as important causes of anaerobic pulmonary infection during the antimicrobial era. The difference may relate to the site of infection (pharyngitis and tonsillitis) and the frequency with which antimicrobials may be used in these clinical situations. It may also relate to the high incidence of septic thrombophlebitis adjacent to the site of infection, which was seen so commonly prior to the advent of antimicrobial agents and which is much less common at this time, at least in those particular sites. Differences in classification of certain anaerobes may account for certain differences in the data also. For example, the organism referred to as *Bacteroides fragilis* in the literature review series undoubtedly included another organism, *Bacteroides oralis*, which is similar in many respects and which was described only recently as a new species. *Bacteroides oralis* is actually much more similar to *Bacteroides melaninogenicus* ss *melaninogenicus*. The relatively high incidence of *Clostridium perfringens* in pure culture in the older literature is surprising.

In Table 9.10, we see a correlation of the infecting organisms isolated from cases of anaerobic pulmonary infection and the conditions underlying these infections. The strong association between *Fusobacterium necrophorum* and tonsillar infections has already been commented on. *Bacteroides*

TABLE 9.9

Bacteriology of Anaerobic Pleuropulmonary Infection (Literature Review Series, 358 Cases)

Bacteriologic results	Number of cases isolated from	Isolated in pure culture	% of total anaerobic isolates identified
Anaerobic gram-negative bacilli			50
Bacteroides fragilis	32	10	9
B. viscosus	2	0	
B. melaninogenicus	1	0	
B. serpens	1	0	
B. glycolytica	1	0	
Fusobacterium sp.[a]	44	17	12
Sphaerophorus necrophorus	88	65	24
S. glutinosus	1	1	
S. freundii	1	1	
Dialister pneumosintes	6	1	2
Anaerobic spirillum	1	0	
Anaerobic vibrio	2	0	
Unidentified anaerobic gram-negative bacilli	54	27	
Anaerobic cocci			32
Veillonella	1	0	
Peptostreptococcus	71	23	19
Peptococcus	9	0	2
Anaerobic gram-positive coccus (not further characterized)	16	2	5
Microaerophilic streptococcus	20	9	6
Anaerobic, non-sporulating, gram-positive bacilli			7
Eubacterium lentum	7	1	2
E. tortuosum	2	0	
E. disciformans	2	0	
E. ventriosum	1	0	
Eubacterium (not speciated)	3	0	
Bifidobacterium eriksonii	1	1	
Ramibacterium pleuriticum	1	0	
Ramibacterium (not speciated)	1	0	
Corynebacterium pyogenes	4	1	
Corynebacterium granulosum	1	0	
Corynebacterium liquefaciens	2	0	
Unidentified anaerobic gram-positive bacilli	10	1	
Clostridia			13
Clostridium perfringens	26	21	7
Clostridium ramosum	11	3	3
Clostridium sporogenes	4	2	
Clostridium (not speciated)	4	0	
Concurrent pathogenic aerobes	46		

[a] In 35 additional cases, no anaerobic culture was performed but gram stain of appropriate specimens showed typical fusiform bacilli.

TABLE 9.10

Correlation of Infecting Organism and Conditions Underlying Anaerobic Pleuropulmonary Infection[a]

Bacteria	Aspiration	Tonsillitis, tonsillectomy	Gingivitis, dental extraction, pyorrhea	Otitis, mastoiditis	Bronchiectasis	Bronchogenic carcinoma	Chest trauma, thoracotomy	Peritoneal infection or source in bowel	Pelvic infection
Bacteroides fragilis	11	1	0	7	3	1	3	16	1
B. melaninogenicus	13	0	7	0	2	0	0	2	0
Fusobacterium	24	2	7	2	4	4	4	6	1
Sphaerophorus necrophorus	2	45	2	0	1	2	0	5	4
Peptostreptococcus and *Peptococcus*	27	5	9	4	4	2	10	6	8
Microaerophilic streptococcus	17	0	5	0	0	4	0	5	1
Anaerobic	6	1	0	3	1	2	1	3	0
non-spore-forming catalase-negative gram-positive rods									
Clostridium	6	3	1	0	0	0	15	7	1

[a] Data from Wadsworth Veterans Hospital and literature review.

fragilis, as expected, was seen most often when the underlying condition was intraabdominal in location. However, there were moderate numbers of isolates of *Bacteroides fragilis* from cases with otitis and mastoiditis and with aspiration. To some extent, this may relate to the problem in distinguishing between *Bacteroides fragilis* and *Bacteroides oralis* already referred to; but it is clear from our current data, in which *Bacteroides oralis* was distinguished, that *Bacteroides fragilis* may still be found with a certain frequency in anaerobic pulmonary infections related to aspiration. The predominant source of *Clostridium* was trauma to the chest or chest surgery.

There are a number of other reports concerning involvement of *Bacteroides fragilis* in pulmonary infection which were not cited in the literature review incorporated in the tables. Schaffner (1240a) recovered *B. fragilis* (*Eggerthella convexa*) from 4 patients with lung abscess (by transthoracic puncture), never in pure culture; one of these infections originated below the diaphragm. He also recovered it from 9 cases of empyema, 3 times in pure culture; 4 of these cases had an original source below the diaphragm. Prévot (226a) recovered *B. fragilis* (*Ristella pseudoinsolita*) in pure culture from 3 cases of purulent pleurisy and one of "pleuropulmonary abscess". A fifth case, one of lung abscess, yielded *B. fragilis* on blood culture. No information was given about possible sources in these cases. De Vos *et al.* (278) also reported on *Bacteroides fragilis* bacteremia in a man with multiple lung abscesses and with a history of probable aspiration; two aerobes were found, in addition, in the abscess fluid (method of collection not specified). Although *B. fragilis* is not recognized as part of the normal flora of the upper respiratory or gastrointestinal tracts, it may colonize these areas under certain circumstances and clearly may be involved in a number of types of infection, other than pleuropulmonary, above the diaphragm (see Chapters 4, 6, 7, and 10).

Clues to Anaerobic Pleuropulmonary Infection

The major clues to anaerobic etiology of pulmonary infection are noted in Table 9.11. Putrid discharge, when present, is a clear indication of the presence of anaerobes in an infectious process. At times the putrid odor to the breath is the only or first manifestation of anaerobic pulmonary infection (476b). There is one note of caution to observe, however. The sputum of patients with necrotizing gingivitis (an anaerobic process itself) may smell foul, even when the pulmonary process itself is not actually anaerobic, simply from contact of the sputum with the foul-smelling oral discharges. Material obtained from the tracheobronchial tree by percutaneous transtracheal aspiration in such situations will not have a foul odor. The absence of a foul odor does not exclude the possibility of anaerobic infection, because certain

TABLE 9.11

Clues to Anaerobic Etiology of Pulmonary Infection

---

1. Pleuropulmonary infection with tissue necrosis
   a. Pulmonary abscess(es)
   b. Empyema with or without bronchopleural fistula
   c. Necrotizing pneumonia
2. Subacute or chronic presentation[a]
3. Underlying conditions
   a. Aspiration or suspected aspiration
   b. Extrapulmonary anaerobic suppuration
      i. Dental
      ii. Abdominopelvic
   c. Bronchial obstruction or pulmonary necrosis (bronchogenic carcinoma, bronchiectasis, bland pulmonary infarct)
4. Fetid discharge (sputum or empyema fluid)
5. Suggestive morphology on gram stain of clinical specimen
6. Failure to recover a likely pathogen on aerobic culture

---

[a] Symptoms present more than 2 weeks prior to hospitalization or discovered on routine chest X ray.

organisms (particularly some of the cocci) do not produce the end products of metabolism responsible for such an odor, and in other cases there may not be communication between the site of the lesion and the tracheobronchial tree.

Subacute or chronic presentation, or course, of illness is relatively unique among bacterial infections of the lung. It is, of course, common in myco-bacterial infections. As has already been indicated on several occasions, aspiration or suspected aspiration and periodontal or gingival disease are very common in anaerobic pleural and pulmonary infection; accordingly, the presence of such conditions constitutes an important clue to the likely role of anaerobes in infection in the lung. Abscess formation or other evidence of tissue necrosis is nonspecific but is so commonly seen in anaerobic infection that one should consider the possibility of anaerobes when tissue necrosis is present. Blank et al. (473c) noted that rapid cavitation within a dense segmental consolidation suggests anaerobic infection and that rapidly enlarging multiple nodular lesions, with or without cavitation, may also indicate this among other possibilities.

A number of anaerobes, including *Fusobacterium nucleatum* and *Bacteroides melaninogenicus*, which are commonly involved in anaerobic pleuro-pulmonary infection, have morphology which is unique enough to suggest their specific presence. Pyopneumothorax, in the absence of bronchopleural fistula or prior thoracentesis, suggests the possibility of gas formation by bacteria involved in the infection. While this is not specific for anaerobic infection, it is more common with anaerobes than with other organisms.

Course and Outcome

The upper portion of Table 9.12 describes the course and outcome of 100 cases of anaerobic pleuropulmonary infection from our series. As indicated earlier, cases showing only pneumonitis typically have a short duration of symptoms prior to presentation to the physician. The other categories of illness all had symptoms for 2–4 weeks, on an average, prior

TABLE 9.12

Anaerobic Pleuropulmonary Infection, Course and Outcome

	Wadsworth and Sepulveda Veterans Hospitals (100 cases)				
	Abscess without empyema	Necrotizing pneumonia ± empyema	Empyema	Pneumonitis only	Series
Duration of symptoms prior to therapy (weeks)	4	2	4	0.5	3.5
Duration of fever after therapy (weeks)	1.0	1.8	4.0	0.3	2.5
Duration of infection after therapy (weeks)	8	15	20	3	13
Mortality (total cases)	1	6	5	3	14
Total number of cases	26	24	41	15	100

	Literature review					
	1899–1944			1945–1971		
Pulmonary condition	Cure	Incomplete[a]	Died	Cure	Incomplete[a]	Died
Abscess without empyema	3	1	4	16	3	2
Abscess with empyema	4	3	8	13	9	3
Necrotizing pneumonia without empyema	0	4	24	0	0	5
Necrotizing pneumonia with empyema	1	1	15	1	3	5
Infection following multiple septic infarctions without empyema	4	1	21	3	0	6
Infection following multiple septic infarctions with empyema	1	1	8	1	0	4
Pneumonitis without empyema or abscess	3	1	0	6	2	2
Empyema without abscess	25	15	25	43	35	15
Total with empyema	31	20	56	58	47	27
Total cases	41	27	105	81	52	40

[a] Infections were not followed to complete cure or patients died of other causes.

to institution of therapy. Note that fever may persist for 1–4 weeks despite appropriate therapy, except in the case of patients with pneumonitis alone, who typically respond very quickly. Empyema patients have the longest duration of fever after therapy, reflecting failure to effect optimal surgical drainage early. The duration of infection is quite prolonged in all cases other than pneumonitis alone. The overall mortality in this group of 100 cases was 14%, with the most significant mortality in the group of patients with necrotizing pneumonia (25%). Complications such as metastatic brain abscess (Fig. 9.10a–d) may be seen.

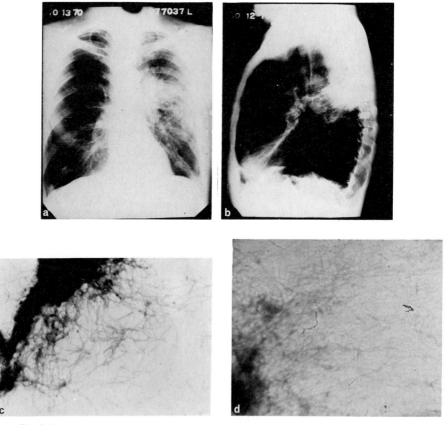

Fig. 9.10.   Anaerobic pneumonitis with metastatic brain abscess. (a) PA chest X ray, admission. (b) Lateral chest X ray, admission. Pneumonia is seen to be in the superior segment of the left lower lobe. (c) Patient subsequently developed brain abscess that was drained surgically. Gram stain of abscess contents showed these thin, filamentous, pale gram-negative bacilli. *Fusobacterium nucleatum* was recovered in pure culture. (d) Percutaneous transtracheal aspiration was done. Gram stain of this material showed same organism.

In the lower portion of Table 9.12, data on outcome is summarized from the literature. It is clear that prior to antimicrobial therapy, the mortality was very impressive in all types of anaerobic pulmonary infection except for pneumonitis that was uncomplicated by empyema or abscess formation. Note that necrotizing pneumonia, with or without empyema, was almost always fatal in the preantimicrobial era. Even with the advent of antibiotics, mortality has remained high in cases of necrotizing pneumonia and in cases of sepsis with multiple septic infarctions of the lung.

Therapy

Therapy has already been discussed under each of the individual major types of anaerobic pleuropulmonary infection. Specific comment should be made with regard to *Bacteroides fragilis*, since this organism may be found in 15—20% of anaerobic pleuropulmonary infections, and since it is more resistant to antimicrobial agents than all other anaerobes. It is typically resistant to penicillin G and to many other penicillins and cephalosporins. Bartlett (477) treated 7 patients with mixed anaerobic pulmonary infections involving *Bacteroides fragilis* as well as other anaerobes with penicillin G alone, and all patients did well despite the presence of *B. fragilis*. Thus, elimination of the majority of the infecting flora may permit the body to handle the residual *Bacteroides fragilis*. Pending additional studies, however, it would be my recommendation that an agent active against *Bacteroides fragilis* be incorporated in the regimen for all patients with severe pulmonary infection and in whom *Bacteroides fragilis* has been isolated, or in whom the *Bacteroides fragilis* has not yet been excluded as a possible infecting organism. This would be especially important if the primary source of the pleuropulmonary infection was in the lower bowel or even the female genital tract. I would certainly want to cover this organism in all cases of necrotizing pneumonia in which it might be present, and I would not rely on penicillin G in cases where *Bacteroides fragilis* was the sole infecting organism (this is rare). Other aspects of therapy are considered in Chapter 19.

Prevention

Since aspiration and periodontal disease and gingivitis are important underlying causes, one should direct attention to them in considering prevention of anaerobic pleuropulmonary infection. Avoiding aspiration is difficult in those predisposed. Proper positioning of a patient during anesthesia may certainly help. When tonsillectomy was done in the past on anesthetized patients in the semisitting position, oropharyngeal secretions, blood, and excised tissue commonly drained directly into the trachea of the patient. Anaerobic pulmonary infection was not an uncommon sequel. A similar sequence of events may still follow tooth extraction. Whether or not prophylactic antimicrobial therapy is indicated, and when, following observed aspiration of oropharyngeal or gastric contents, is unknown. Clearly, proper management of periodontal disease and of gingivitis will minimize the likelihood of anaerobic pleuropulmonary infection secondary to these conditions.

Miscellaneous Intrathoracic Anaerobic Infections

Many cases of so-called fusospirochetal bronchitis were reported in the early 1900's. These cases are quite unsatisfactory, however, as studies to exclude bronchiectasis and parenchymal disease were not performed as a rule, and, furthermore, the designation of etiology was based entirely on direct microscopic examination of expectorated sputum or bronchoscopically aspirated material. While anaerobes may play a role in certain types of bronchitis, there is no clear-cut evidence that this is the case. Specifically, in bronchitis associated with chronic obstructive pulmonary disease there is no evidence that anaerobes play a role, although studies performed to date are inadequate. Our group has studied about a dozen such patients with transtracheal aspiration, and we have failed to obtain anaerobes from any. A recent report by Wong et al. (477a) noted more anaerobes, with higher counts, in patients studied by fiberoptic bronchoscopy than when they were evaluated by transtracheal aspiration.

Bronchiectasis is undoubtedly a situation in which anaerobes play a very important role. The penetrating putrid odor of the sputum of patients with this disease and the fact that these lesions commonly precede or follow anaerobic parenchymal infections clearly establishes the importance of anaerobes in this condition, although they may be only secondary invaders at times. Greey, in 1932 (478), studied the bacteriology of 9 cases of bronchiectasis in whom lobectomy was carried out. Fusiform bacilli were seen in stained sections from 5 patients and were recovered on culture from 4 of these, always in association with facultative cocci. Spirochetes, usually in small numbers, were demonstrated in 4 by smear. In patients who came to autopsy in the preantimicrobial era, stained sections of bronchiectatic dilatations revealed fusiform bacilli and spirochetes. Metastatic anaerobic infection originating in an area of bronchiectasis has also been described (479, 480).

Multiple septic pulmonary emboli, secondary to septic thrombophlebitis associated with anaerobic infection elsewhere in the body, were formerly seen with some frequency but are not common at the present time. The most common primary lesions were tonsillar or pharyngeal infections and anaerobic pelvic infections.

Anaerobic infection following thoracotomy or penetrating chest wounds is also seen much less commonly than was true in the preantimicrobial era. This type of complication of chest or other surgery has now been nearly eliminated by good surgical technique, with frequent suctioning, use of light anesthesia, and proper positioning of the patient. There were a number of cases of clostridial empyema associated with traumatic hemothorax during World War I. Improved management of the war-wounded has decreased this complication significantly.

Extension of an anaerobic pulmonary and pleural process to the chest wall itself may occur, but is rare in the absence of actinomycosis or associated tuberculosis or tumor; one such case due to *Bacteroides funduliformis* and an anaerobic streptococcus was described by Fisher and McKusick (264). Moon *et al.* (481) described a diabetic patient with anaerobic streptococcal pneumonia and empyema who developed gangrene of the chest wall around the track of a tube inserted for drainage of the empyema.

Goldsand and Braude (244) described an unusual case of pulmonary disease in which two biopsies showed interstitial pneumonitis from which a heavy growth of anaerobic streptococci was obtained in pure culture. Noncaseating granuloma formation and miliary pattern on chest X ray said to be related to pulmonary infection with *Bifidobacterium eriksonii* has been described (481a).

Acute tracheobronchial lymphadenitis may occur during the course of anaerobic pulmonary infection (472, 480, 482). This may lead to considerations of carcinoma or fungus infection.

Empyema and mediastinitis following esophageal perforation may involve anaerobic bacteria (282). Anaerobic mediastinitis may also follow extension of anaerobic infection from other sites (324a). Cunningham (323) described a case of retropharyngeal abscess with gangrene and extension into the peritracheal and subcutaneous tissues as well as the upper anterior and posterior mediastinum. Foul greenish pus from these tissues yielded an anaerobic streptococcus, *Actinomyces necrophorus*, and a pneumococcus; a gram-positive bacillus seen on smear was not recovered. Two cases of mediastinal involvement in actinomycosis were noted by Weese and Smith (113a).

Actinomycosis involves the thorax not uncommonly. Although consolidation may occur in the lung, the primary process is usually an abscess, often in one of the lower lobes. The infection gradually extends to the pleural surface to produce empyema and then may involve the chest wall to produce external sinuses and/or periostitis of the ribs (482a). (Fig. 9.11).

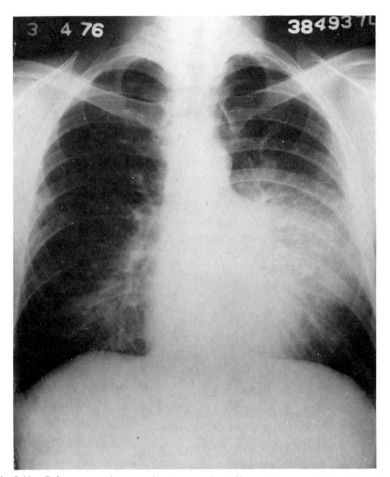

Fig. 9.11. Pulmonary actinomycosis. An extensive infiltrate is present in left midlung field. This extended through to the anterior chest wall to produce a soft tissue abscess. In addition, there is left hilar adenopathy and a small infiltrate overlying the right fourth anterior rib.

The mediastinum is invaded through the esophagus, and the infection spreads to the pleura and pericardium without involving the lung. This usually breaks through the chest wall in the paravertebral area. Thoracic lesions have a greater tendency than cervicofacial actinomycosis to disseminate to the brain and vertebrae. A case of actinomycosis involving only 2 separate sites of the thoracic wall with no pleuropulmonary or other site of infection was reported by Morris and Kilbourn (482b). Granuloma formation may be noted in actinomycosis.

10

Abdominal and Perineal Infections

General

Considering the fact that anaerobic bacteria account for well over 99% of the normal colonic flora (483), it is surprising that they have not been reported more frequently from intraabdominal infections. Clearly, as in the case of other types of infections in which anaerobes have been commonly overlooked, this relates to inadequate specimen collection and transport and use of less than adequate anaerobic culture techniques. Our group (D. Flora, V. L. Sutter, and S. M. Finegold, unpublished data, 1973–1975) has studied 73 intraabdominal infections occurring in 62 patients. Included were 16 cases of peritonitis, 36 of intraperitoneal abscess, 3 retroperitoneal abscesses, 1 liver abscess, 9 wound infections, and 8 biliary tract infections. The average number of isolates per case was 4.5, with an average of 2.5 anaerobes and 2 aerobic or facultative forms; the range was 1 to 12 organisms per specimen. Anaerobes recovered included *Bacteroides fragilis* (57), other *Bacteroides* (30), *Fusobacterium* (13), *Clostridium perfringens* (11), other clostridia (20), *Peptococcus* (17), *Peptostreptococcus* (10), *Eubacterium* (12), and others (11). The total number of anaerobes obtained was 181 and of aerobes or facultatives, 160. *Escherichia coli* was by far the predominant facultative organism recovered (150 isolates). There were 33 strains of group D streptococci. The frequency with which various organisms were recovered varied according to the site of the disease process or surgery. *Bacteroides fragilis* was recovered 9 times from processes associated with the upper gastrointestinal tract, 46 times from the lower GI tract, and 2 times from the biliary tract. The respective counts for clostridia were 5, 21, and 5 and for anaerobic cocci 16, 15, and 1. Thirteen positive blood cultures occurred in 11 patients in this series; there were 15 organisms recovered. Included were *E. coli* (4), *Bacteroides* (4), enterococcus (3), *Clostridium perfringens* (1), *Klebsiella* (1), and *Candida albicans* (1).

A recent study by Gorbach *et al.* (484) also shows that anaerobes can indeed be found to play a major role in abdominal infections when appropriate techniques are used. This study was carried out on 46 patients with intraabdominal sepsis. Thirty-two had intraabdominal abscess, 10 generalized peritonitis, and 4 miscellaneous infections. Predisposing conditions included trauma in 22 patients, carcinoma of the colon, pancreas or kidney in 7 patients, intestinal surgery in 7, perforated appendix in 4, cirrhosis with spontaneous peritonitis in 3, and infection following peritoneal dialysis in 3. Three patients had positive blood cultures without an abdominal site available for culture; in the other 43 cases, specimens were collected from the abdomen. Seven patients had an exclusively anaerobic flora, and 3 had only aerobic or facultative organisms. The remaining 33 specimens from the abdomen contained mixtures of aerobes and anaerobes. One to 13 different species of organisms were isolated from each sample, with an average of 5 isolates per specimen. This included an average of two aerobes and three anaerobes. There were a total of 88 strains of aerobic or facultative organisms recovered and 104 strains of anaerobic organisms. *Bacteroides fragilis* was the most common anaerobic isolate, recovered 28 times in all. There were 8 other isolates of *Bacteroides* species, 4 of which were *Bacteroides melaninogenicus*. There were 6 isolates of *Fusobacterium* (including one *Fusobacterium varium*), 11 *Peptostreptococcus*, 3 *Peptococcus*, 1 *Veillonella*, 31 clostridia (including 3 *C. perfringens* and 3 *C. ramosum*), 11 *Eubacterium* (including 6 *E. lentum*), 2 *Propionibacterium*, 2 *Lactobacillus catenaforme*, and 1 *Bifidobacterium*. The major facultative organism recovered was *Escherichia coli*, which was present 28 times. There were 15 isolates of *Staphylococcus epidermidis*, which undoubtedly was a contaminant. Positive blood cultures were found in 13 of the 46 patients; 11 of these had only anaerobes and 2 had both aerobes and anaerobes. In the case of bacteremia as well, *Bacteroides fragilis* was the predominant organism; it was found in 8 of the 13 positive cases.

Still another recent large study of intraabdominal infection was carried out at Temple University (384b, 484a). Anaerobic bacteria were recovered from 84% of the 76 cases and were the only isolates in 39%. There was an average of 3.9 isolates per case (2.6 anaerobic, 1.3 facultative or aerobic). *Bacteroides fragilis* accounted for 36% of all anaerobes, with *Peptococcus* and *Peptostreptococcus* also relatively common. *Clostridium perfringens* was isolated in 7 cases. Bacteremia occurred in 23 patients; 20 of the 30 isolates were anaerobes (15 *B. fragilis*).

In a study of 123 wound and intraabdominal infections complicating emergency celiotomy carried out because of disease or trauma, Stone *et al.* (384c) noted anaerobes in two-thirds of cases and aerobes in all but one patient. Patients were all treated with either cephalothin or clindamycin from the outset.

Of interest are the preliminary studies of Thadepalli and Gorbach (485) in which 12 patients with mixed aerobic and anaerobic abdominal and pelvic infections were treated with clindamycin alone. This antibiotic, despite lack of activity against such aerobic forms as *E. coli* and enterococcus, gave good clinical results. There was an average of two aerobic or facultative organisms in each of these 12 cases in addition to anaerobes; aerobic or facultative forms recovered included *E. coli*, *Proteus*, *Klebsiella*, and *Pseudomonas*.

In a study on the role of clostridia in infection, Gorbach and Thadepalli (486) recovered 152 clostridia from 144 patients; most of these were from intraabdominal sites. Armfield and co-workers (119) found that *Clostridium innocuum* and *Eubacterium filamentosum* were usually isolated from infections involving the gastrointestinal tract. Weiss, in 1943 (80), reported recovery of *Bacteroides melaninogenicus* along with other organisms from 10 miscellaneous abdominal infections related to the bowel or bowel surgery. Lodenkämper and Stienen (287) reported 12 pure *Bacteroides* infections from the abdominal cavity (Douglas' abscess, appendicitis, and peritonitis). The same workers in 1956 (107) noted that among 722 anaerobes isolated from various infectious processes, 137 were from the abdominal area.

Stokes (108) found that approximately one-third of 528 specimens of abdominal pus that yielded a positive culture (173) yielded anaerobes, 30 of them in pure culture. Saksena *et al.* (487) reported on the role of Bacteroidaceae in 112 surgical infections. Sixty of these were related to the gastrointestinal tract (32 to the large bowel, 10 to the small bowel and stomach, 11–14 to the appendix, and the remainder to the biliary tract).

Gunn, in his report in 1956 (296), reviewed 148 cases of *Bacteroides* bacteremia and noted that 26 were related to abdominal lesions. There are a number of other reports of bacteremia due to anaerobes in which the portal of entry was the abdomen. Spittel *et al.* (488) reported four cases of *Bacteroides* bacteremia with the gastrointestinal tract as the portal of entry. Tynes and Frommeyer (489) reported five cases of *Bacteroides* bacteremia with the GI tract serving as a portal of entry. In a series of 123 cases of

Bacteroides bacteremia, Marcoux *et al.* (420) found that the gastrointestinal tract was the most common portal of entry, accounting for 61.8% of the bacteremias. Bodner *et al.* (460) reported that 24 of their 39 cases of *Bacteroides* bacteremia originated in the gastrointestinal tract. Nine of these cases followed bowel surgery; most of the patients received a "bowel prep" prior to surgery. There were two cases of trauma to the bowel, one related to transrectal prostatic biopsy, one mesenteric artery thrombosis, one carcinoma of the bowel with perforation, and one a case of gastroenteritis of unknown cause. Five of the 35 cases of bacteremia due to Bacteroidaceae reported by Gelb and Seligman in 1970 (490) originated from the gastrointestinal tract. In 1971, Kuklinca and Gavan (491) reported on 237 postmortem blood cultures, 126 of which were sterile. There were a total of 16 anaerobic isolates, of which 13 were *Clostridium* (*C. perfringens, C. septicum*, and *C. histolyticum*) and 3 were *Bacteroides* (two *B. fragilis* and one *B. serpens*). Only 1 of the 16 cultures positive for anaerobes came from a patient without a gastrointestinal lesion, for an incidence of 0.6% among 1/6 patients studied. The other 15 came from patients with gastrointestinal lesions; the incidence of anaerobic bacteremia in these patients was 23%. Underlying causes included carcinoma of the bowel; leukemia producing ulceration in the bowel; hemorrhagic diathesis with ulcerative lesions in the bowel; perforation of the cecum, rectum, and a gastric ulcer; mesenteric artery occlusion; hemorrhagic colitis; and necrotizing enteropathy.

The association of anaerobic abdominal infection and carcinoma of the bowel is a common and important one. This type of infection may be the first manifestation of the malignancy. In 1971, Finegold *et al.* (492) reported 13 cases of anaerobic infection in association with carcinoma of the colon and mentioned an additional 58 cases from a review of the literature. Additional cases have been reported by a number of other workers. In the paper by Marcoux *et al.* referred to above (420), there were 21 cases of carcinoma of the colon; and in the report of Saksena *et al.* (487), also mentioned above, there were 18 cases of carcinoma of the colon. Bornstein *et al.* (275) reported 6 cases of *Bacteroides* infection as well as some intraabdominal abscesses due to anaerobic streptococci associated with carcinoma of the bowel. Jones *et al.* (493) reported a case of septicemia due to *Clostridium sporogenes* and microaerophilic streptococci secondary to carcinoma of the colon in a patient with leukemia. Isenberg (494) reported 3 cases of *Clostridium perfringens* septicemia, 2 of whom had perforation of the colon secondary to carcinoma. In the series of bacteremia patients reported by Wilson *et al.* (105), there were 4 cases of clostridial bacteremia (including bacteremia due to *C. perfringens* and *C. septicum*) in patients with obstructing or perforating colon lesions secondary to carcinoma. The same authors also noted that among 52 patients with septicemia due to anaerobic gram-negative bacilli, 23% had carcinoma of the colon. Keusch (56h) stated that isolation of C. *septicum* from a patient without a grossly contaminated deep traumatic wound is an indication of an underlying malignant process, most likely colorectal cancer.

The important role of anaerobes in infections related to penetrating abdominal trauma is emphasized by the studies of Thadepalli *et al.* (495) and Nichols *et al.* (496). These studies included 100 patients with penetrating injuries to their gastrointestinal tract. Patients were randomized between two drug regimens, the drugs ordinarily being started shortly before the patients were taken to surgery for repair of the injuries. The two drug regimens were cephalothin plus kanamycin and clindamycin plus kanamycin. The initial cultures at the time of surgery yielded a total of 31 strains of anaerobes in the cephalothin plus kanamycin group and 18 in the clindamycin plus kanamycin group. *Bacteroides fragilis* accounted for 22 of all of these anaerobic isolates and was by far the most commonly encountered of all anaerobes. During therapy, anaerobes were isolated from 39 patients in the cephalothin plus kanamycin group and only 2 in the clindamycin plus kanamycin group. With regard to infectious complications in these patients with abdominal trauma, the group treated with cephalothin plus kanamycin had three infections with aerobes alone, five with anaerobes alone and six mixed infections. In the clindamycin plus kanamycin group, however, there was only one anaerobic infection, no mixed infections, and four infections due to aerobes. The anaerobic infection in the clindamycin plus kanamycin group was a bacteremia due to *Clostridium perfringens* and *Clostridium tertium*; this patient was subsequently cured with chloramphenicol. Felner and Dowell (168) reported three cases of bacteremia due to *Bacteroides fragilis* and one due to *Bacteroides incommunis*, secondary to trauma to the bowel.

Other bowel lesions (vascular, obstructive, inflammatory, etc.) may also be the source of anaerobic infection. In the Case Records of the Massachusetts General Hospital (Case 41-1968) (497) a case report is presented of a diabetic with ischemic colitis with bacterial invasion, pneumatosis cystoides intestinalis, and peritonitis. This patient also had bacteremia with *Bacteroides* and anaerobic streptococci. Ishiyama *et al.* (498) described a patient with a gangrenous small bowel and peritonitis due to *Clostridium perfringens*, *Peptostreptococcus*, and *Lactobacillus*. In the series of anaerobic bacteremia reported by Felner and Dowell (168) there are two cases of *Bacteroides fragilis* bacteremia, one each secondary to a vascular problem in the bowel and intestinal obstruction. There are also four patients with bacteremia secondary to chronic ulcerative colitis, one each due to *Bacteroides fragilis*, *Bacteroides oralis*, *Bacteroides variabilis*, and *Fusobacterium necrophorum* plus anaerobic streptococci. Fifteen percent of the 52 patients with anaerobic, gram-negative rod bacteremia in the series by Wilson *et al.* (105) had small bowel perforation as the underlying source of the bacteremia.

It is well known that intraabdominal infections may present in the thigh as an abscess or cellulitis. Duncan and Samuel (499) reported a patient with three separate carcinomas of the colon and an intraabdominal abscess that extended to the thigh. Anaerobic streptococci were recovered. Mzabi *et al.* (499a) described two diabetics with gas gangrene of the pelvis and lower extremity as the presenting feature of perforating cecal carcinoma; *Clostridium septicum* was isolated from one of these.

Visceral gas gangrene is among the most overwhelming of all infections. It may be related to surgery, trauma, or various lesions of the bowel, notably malignant lesions. It may also be related to impaired host defense mechanisms. Govan (500) described visceral gas gangrene complicating sarcoma of the ilium. Boggs *et al.* (413) described four fatal cases of clostridial sepsis in leukemics, three of which were associated with visceral gas gangrene. The organisms isolated were *Clostridium perfringens* (twice), *C. septicum*, and another clostridium that was probably *C. septicum*. Fethers (501) described a case of visceral gas gangrene of endogenous origin which the author felt was related to severe hematemesis. Canipe and Hudspeth (502) described visceral gas gangrene and septicemia following blunt, mild abdominal trauma. Fifteen cases of clostridial infection, either sepsis or gas gangrene, or both, were described by Cabrera *et al.* (404); these patients all had malignancy and ulceration of the intestinal tract related either to the tumor, therapy for the tumor, or other cause. Visceral gas gangrene due to *Clostridium sphenoides* in a young girl with periodic neutropenia was described by Felitti in 1970 (503). Heyworth *et al.* (504) described visceral gas gangrene in a pregnant woman with Crohn's disease.

Wiot and Felson (505) described a diabetic with portal vein gas secondary to mesenteric artery thrombosis. Blood culture in this patient yielded *Clostridium paraputrificum*, *Proteus mirabilis*, and *P. aerogenoides*. Barrett in 1962 (506) described two cases of gas embolism of the portal vein branches within the liver, both associated with intestinal gangrene (one with mesenteric thrombosis and the other with enteritis necroticans). *Clostridium perfringens* was isolated from the portal venous blood in one case, and from liver slices in the second. Fred *et al.* (507), described 6 cases of hepatic portal vein gas, and an additional 46 cases encountered in a review of the literature (including the two cases of Barrett's described above). Cultural data were available in 25 of the 52 cases. Three showed no growth, and eight yielded *Clostridium* (7 *C. perfringens* and 1 *C. paraputrificum*). In four cases the clostridia were present in mixed culture, along with aerobic or facultative forms. An infected mesenteric cyst yielding 5 liters of fluid gave a pure culture of *Bacteroides* (275).

Actinomycosis may be found in the abdominal cavity, most commonly involving the ileocecal region. Cope (210) felt that most of these cases actually arise in the appendix. This disease results in firm, woody masses within the abdomen and often abscess formation as well as draining sinuses and fistulae. Lesions of the colon usually develop external to the gut, according to Cope and, because of the firm masses that are produced, are readily confused with carcinoma. Grechiari [cited by Cope (210)], however, described a case of actinomycosis originating in the mucosa of the large intestine. The small intestine may be involved in actinomycosis [Wheeler cited by Cope (210)], and the mesentery or mesocolon is not uncommonly involved [Gordon-Taylor and König cited by Cope (210), Harvey *et al.* (212), and Palmisano and Russin (508)].

Actinomycosis within the abdomen is much less common than other forms of this disease. Thus, Holm (509) found that among approximately 650 patients with actinomycosis there were only 36 in whom the disease was found within the abdomen or pelvis. The excellent paper on actinomycosis by Harvey *et al.* (212) described 10 cases of actinomycosis originating in the gastrointestinal tract (3 in perforated gastric ulcers, 4 in perforated colons, 1 in a perforated rectum, and 2 in the cecum). These authors also noted secondary involvement of the intestinal tract including the mesentery, the stomach wall (3 cases), or the bowel wall (10 cases). As in other sites of actinomycotic infection, anaerobic organisms other than *Actinomyces* may be found associated with *Actinomyces* in abdominal actinomycosis. Thus, Gins (310) recovered *Bacteroides melaninogenicus* and *Leptothrix* in addition to *Actinomyces* from a patient with an abdominal wall fistula due to this disease.

Postoperative and Other Wound Infections

When good transport and culture techniques are used, anaerobes are found to play a major role in postoperative infections related to the abdomen. Thus, Spaulding *et al.* (510) cultured 55 postoperative wounds, 45 of which were positive for one organism or another, and among these, 37 involved anaerobic organisms (only 1 in pure culture). All together, 82 strains of anaerobes, representing 38 different generic or specific groups, were recovered. Most numerous was *Bacteroides fragilis* (recovered 18 times), followed by *Bacteroides melaninogenicus* (10), *Peptococcus prevotii* (7), *Fusobacterium nucleatum* (6), and *F. ridiculosum* (5).

Among a group of 301 cases of postoperative wound infection seen in a period of somewhat over 2 years, Hoffmann and Gierhake (110) found anaerobes involved in 33.2%. In this study, cultures were obtained by needle puncture of wounds that were still closed. Members of the *Bacteroides* group and anaerobic streptococci were found with approximately equal frequency and greatly outnumbered all other types of anaerobes recovered. In 69% of the infections in which anaerobes were involved, aerobic or facultative bacteria were found in addition, although usually in much smaller numbers. Anaerobes were much more commonly recovered in the case of laparotomies involving opening of hollow organs (appendectomies and surgery on the colon or rectum, small intestine, stomach, or gallbladder and bile ducts). Anaerobes were also recovered in other types of laparotomies, but the incidence was much lower and the severity of the wound infection much less. In the study of Azar and Drapanas (511), *Bacteroides* species accounted for 29% of wound infections following elective surgery on the colon. In this connection, it is also interesting to note that cultures of the bowel at the time of surgery in 29 patients who were prepared with castor oil, a low residue diet, and 6 to 10 doses of oral neomycin plus oral sulfathalidine yielded *Bacteroides* species from 10 patients and *Clostridium* species from 2. Baird (324b) noted heavy growth of *Bacteroides* from the bowel lumen of 8 of 12 patients who had oral kanamycin or neomycin prior to surgery. Washington *et al.* (56j) noted some persistence of anaerobes in the bowel lumen of patients on oral neomycin (10/26 cultured) or placebo (15/32) and fewer in the case of patients on neomycin plus tetracycline (4/25). The incidence of anaerobes in postoperative wound infections paralleled their presence in the lumen, with by far the fewest wound infections and anaerobes in such infections in the double drug group. Further discussion of the question of desirability of preoperative bowel preparation will be found in Chapter 20.

Gillespie and Guy (512) studied 46 cases selected as suspicious of anaerobic infection postoperatively. From this group they recovered 31 strains of *Bacteroides*, 18 anaerobic streptococci, and 2 clostridia. Saksena *et al.* (487) reported 29 wound infections due to Bacteroidaceae following gastrointestinal surgery; in addition, they noted 3 abscesses and 2 septicemias with these organisms in such patients. Sandusky *et al.* (112) recovered anaerobic streptococci from nine wound infections following surgery on the gastrointestinal tract. One of these was present in pure culture and the other eight were mixed; in the latter group, in addition to aerobic or facultative organisms, one *Bacteroides* and two clostridia were recovered. Beazley *et al.* (394) recovered *Bacteroides* from 30 patients who had had intestinal surgery and developed infection subsequently; one-third of this group had the surgery performed for malignancy. Pyrtek and Bartus (513) reported *Clostridium perfringens* infections complicating gastrointestinal surgery in seven cases, and Eickhoff (514) reported two cases of postoperative gas gangrene of the abdominal wall. McNally *et al.* (515) reported three cases of gas gangrene of the abdominal wall following abdominal surgery. Single cases of anaerobic wound infection following abdominal surgery are reported by Witebsky and Miller (516) (*Clostridium perfringens* infection following surgery for incarcerated inguinal hernia), Nathan (424) (*Sphaerophorus* sepsis following leakage from the bowel subsequent to a Billroth II operation), Bartlett *et al.* (517) (wound infection following colostomy in a patient with perforated colon, due to *Bacteroides fragilis*, *B. melaninogenicus*, and *Clostridium septicum*), and Ishiyama *et al.* (498) (infection of the anastomotic site following gastric resection, due to *Bacteroides*, *Sphaerophorus*, *Catenabacterium*, and *E. coli*). Bernard and Cole (518) advocated delayed primary wound closure following potentially contaminated operations in order to minimize the risk of anaerobic infection. They noted that 5 of 33 patients who had primary wound closure developed serious wound infections, 3 of which involved anaerobes, whereas none of 40 patients having delayed primary wound closure developed infection.

Abdominal Wall Infections

Abdominal wall infections may follow a variety of infectious, traumatic, or other pathologic processes within the abdomen or the female genitalia.

Fromm and Silen (519) presented two cases of postoperative clostridial infection of the abdominal wall and reviewed the literature on this subject. Their review covers 111 cases of postoperative clostridial infection of the abdominal wall. The mortality was 54% in the absence of peritonitis and 86% in the presence of definite peritonitis. The mean overall mortality was 60%. This complication most frequently followed operation on the appendix, colon, small bowel, biliary tract, and upper gastrointestinal tract. The authors point out that *Clostridium perfringens* is not only a normal inhabitant of the gastrointestinal tract but also that it has been cultured from numerous other sites, including the abdominal skin of patients, 90% of appendices, 2 to 19% of gallbladders removed surgically, and the stomach (where it does not grow if the pH is less than 4.5). Diagnosis of clostridial infection following abdominal surgery is difficult prior to development of signs of wound infection. However, since the mortality is so high, it is crucial that a high index of suspicion be maintained in order to attempt to diagnose this condition before obvious signs of wound infection are present. Hypotension, fever, tachycardia out of proportion to the fever, jaundice, renal failure, or mental changes may all precede signs of wound infection. Wound crepitus is a late sign. If the clinical picture is compatible with the diagnosis, material should be obtained from the operative wound for smear and culture. If organisms resembling clostridia are found on gram stain, treatment must be started promptly.

Heinrich and Pulverer (116) reported isolation of *Bacteroides melanino-genicus* from 11 cases of abdominal wall abscess. Isenberg (494) described eight cases of gas gangrene of the abdominal wall, one of which followed perforation of the sigmoid colon due to cancer, and seven of which were postoperative (two following stomach surgery, four following bowel surgery, one cholecystectomy and one an above the knee amputation with resection of an arterial aneurysm). Three of the surgical cases in Isenberg's series had malignancy. Bittner *et al.* (520) reported a case of gangrene of the abdominal wall following colectomy; organisms recovered included *Clostridium bifermentans, E. coli,* and enterococcus. Gas gangrene of the abdominal wall following bowel surgery was reported by Arnar *et al.* (521).

Lodenkämper and Stienen (287) reported eight abdominal wall abscesses yielding pure cultures of *Bacteroides* plus two additional abscesses with mixed cultures involving *Bacteroides* and other organisms. Furuta and Tsuchiya (522) reported seven abscesses of the abdominal wall, five of which yielded *Bacteroides fragilis* in pure culture, one yielding *Bacteroides fragilis* and *Fusobacterium,* and the other *Bacteroides fragilis* and an α-hemolytic aerobic streptococcus. Six of the patients in this last series had had appendectomy prior to their infection. Bornstein *et al.* (275) reported three cases of abdominal wall abscess due to anaerobic streptococci, as well as three cases of synergistic gangrene of the abdominal wall involving the same organism. Sandusky *et al.* (112) described five cases of abdominal wall abscess due to anaerobic streptococci mixed with other organisms, including *Actinomyces, Bacteroides,* and other anaerobic gram-negative bacilli. Werner (523) found *Eggerthella* in three abdominal wall abscesses, twice in pure culture and once with a *Clostridium.* Morin and Potvin (145) reported a case of abdominal wall infection due to *Fusobacterium, Sphaerophorus,* and *Streptococcus lanceolatus.* Georg *et al.* (120) reported a case of intramuscular abscess in the rectus muscle at a drain site following gastric resection; the offending organism was *Bifidobacterium eriksonii.*

Single cases of abdominal wall abscess involving anaerobes were reported by Grumbach *et al.* (262) (*Bacteroides funduliformis* plus *Actinomyces* following appendectomy), Weens (524) (anaerobic streptococci), Ernst (347) (*B. funduliformis* plus anaerobic cocci), and Werner and Reichertz (525) (*Bacteroides splanchnicus*). Harvey *et al.* (212) reported 15 cases of actinomycosis with secondary spread to the abdominal wall.

Subphrenic and Other Intraperitoneal Abscesses

The origin of subdiaphragmatic (Figs. 10.1–10.3) and other intraperitoneal abscesses suggests that anaerobes would commonly be involved. A survey of the literature by Kazarian and Ariel (526) showed that among over 6000

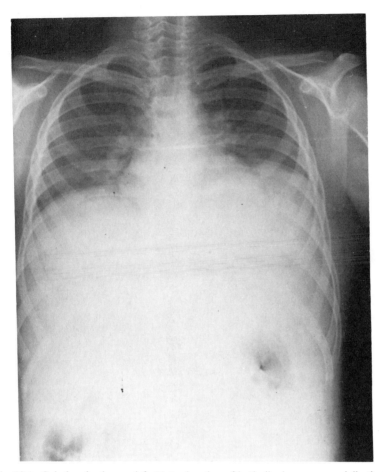

Fig. 10.1   Subphrenic abscess, left. Note elevation of both diaphragms, especially the left.

subdiaphragmatic abscesses, 26% had their origin in the appendix, 29% in the stomach and duodenum, and 15% in the liver and biliary tract. Similarly, in the paper by Altemeier *et al.* (527), which included 501 patients with 540 intraabdominal abscesses, most of which were intraperitoneal, 19% originated with appendicitis, 8% with lesions of the biliary tract, 7% with diverticulitis, 4% with actinomycosis, 2% with a leaking suture line after bowel anastomosis, and 0.6% with regional enteritis. In this last study, at least half of the bacteria found in intraabdominal abscesses were anaerobic. The authors noted that when careful technique was used, the frequency of anaerobic pathogens was even greater and approached 60–70% of cases. Infections were commonly mixed, with an average of two to three strains

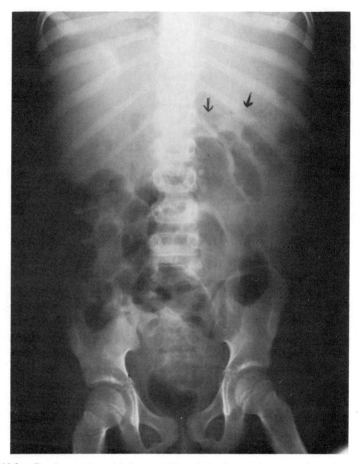

Fig. 10.2.   Gas (arrows) outside lumen of bowel (in abscess) in upper left quadrant. Same patient as in Fig. 10.1.

per specimen. The most commonly isolated bacteria were *E. coli, Bacteroides,* anaerobic streptococci, and the *Klebsiella–Enterobacter* group. Relating the bacteriologic findings to the primary disease, Altemeier *et al.* found that *Bacteroides* was recovered from 17% of cases of intraabdominal abscess secondary to appendicitis, 44% of those related to the female genitourinary tract, 27% to the male urinary tract, 20% to diverticulitis, and 24% to trauma. In the case of the anaerobic streptococcus, this organism was found in 25% of intraabdominal abscess where the primary disease was appendicitis, 38% related to female genitourinary tract disease, 23% to male

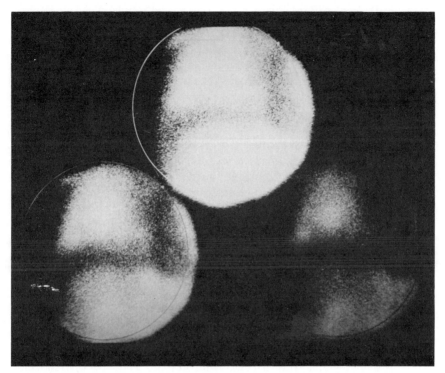

Fig. 10.3. Double isotope scan outlining both lung and liver. The abnormally large space between the two organs reflects the subphrenic abscess.

genital or urinary tract disease, 11% to diverticulitis, and 6% to trauma. Another paper from Altemeier's group (89b) discussed giant "horseshoe" intraabdominal abscesses in 12 patients. Ten of these yielded *Bacteroides*, 9 *Peptostreptococcus*, and 2 *C. perfringens*.

Magilligan (528) found an incidence of 5% *Clostridium perfringens* and 3% anaerobic streptococci among 82 cases of subphrenic abscess; however, 14% of cases were sterile on culture, and 10% yielded *Staphylococcus epidermidis*, an obvious contaminant. In the paper on anaerobic bacteremia by Wilson *et al.* (105), among 52 patients with anaerobic gram-negative bacillus bacteremia, 8% had intraabdominal abscess. Haldane and van Rooyen (529) reported one subphrenic abscess from which *Bacteroides* was isolated in pure culture and three other intraperitoneal abscesses yielding anaerobes. One of these, secondary to cholecystitis and carcinoma of the cervix, yielded *Bacteroides*; another, secondary to cholecystitis and small bowel obstruction, yielded *Bacteroides*, *Proteus mirabilis*, and *E. coli*; and the third, associated with carcinoma of the colon, yielded *Bacteroides* and anaerobic streptococci. Heinrich and Pulverer (116) isolated *Bacteroides melaninogenicus* from two subphrenic and two other intraabdominal abscesses. McHenry *et al.* (530) described two cases of subphrenic abscess with *Bacteroides* bacteremia, one of which originated with appendicitis and the other in relation to carcinoma of the colon. Bornstein *et al.* (275) reported eight intraabdominal abscesses due to anaerobic streptococci, all of which were related to either bowel or gallbladder surgery. Harvey *et al.* (212) reported four cases of subphrenic abscess due to *Actinomyces*. Shoemaker (531) reported 13 abdominal abscesses from which anaerobic gram-negative bacilli were cultured, four times in pure culture. Sandusky *et al.* (112) reported 12 intraabdominal abscesses yielding anaerobic streptococci, one in pure culture and the others mixed with other anaerobes (*Clostridium perfringens* and *Actinomyces*) and aerobes. Beerens and Tahon-Castel (115) reported three cases of intraperitoneal abscess. One of these yielded eight different anaerobic strains plus *E. coli*; another had *Eggerthella convexa* plus *S. micros*; and the third had *Bacteroides funduliformis* plus two aerobic organisms. Smith and Ropes (226) reported two cases of *Bacteroides* abdominal abscesses, one of which was associated with bacteremia. Felner and Dowell (168) also reported a case of bacteremia due to *Bacteroides fragilis* associated with an intraabdominal abscess. There are many other single case reports of subphrenic, subhepatic, or other intraabdominal abscess involving anaerobic bacteria [Martin (532), Hartl (249), Henderson (533), Morin and Potvin (145), McDonald *et al.* (534), Goldsand and Braude (244), Nathan (424), Saksena *et al.* (487), Jacobs *et al.* (535), and Beazley *et al.* (394)].

Retroperitoneal Abscesses

Retroperitoneal infections related to the kidney are considered in Chapter 11. Altemeier and colleagues (527) found in their large series of intra-abdominal abscesses 19 cases of anterior retroperitoneal abscess secondary to appendicitis and two others related to diverticulitis. They also noted 19 retroperitoneal abscesses secondary to pancreatitis and pancreatic tumor. While these cases were not analyzed separately in terms of their bacteriology, it is clear that anterior retroperitoneal abscesses would commonly involve anaerobic organisms. Müller (536) reported two cases of retroperitoneal abscess from which *Clostridium perfringens* was isolated in pure culture. These cases were secondary to retroperitoneal fibrosis. Harvey *et al.* (212) reported three cases of retroperitoneal abscess due to *Actinomyces* and Weese and Smith (113a) reported a single case.

Ortmayer (537) reported a bilateral psoas abscess, probably originating in the pleural space, which yielded on culture *Bacteroides necrophorus* and which showed gram-positive diplococci on smear which could not be recovered in culture. Pham (261) reported two cases of psoas abscess as part of metastatic infection during the course of *Bacteroides funduliformis* sepsis. Debré *et al.* (538) reported a patient with psoas abscess due to *Bacteroides fragilis*; this patient earlier had sepsis and pulmonary gangrene. Swenson *et al.* (384b) recovered multiple anaerobes from 2 retroperitoneal abscesses. Wyman (539) reported a case of endogenous gas gangrene complicating carcinoma of the colon which involved a psoas abscess presenting clinically at the level of the upper thigh. Christiaens *et al.* (540) reported a case of psoas abscess due to *Fusiformis fusiformis* originating from a perforated gangrenous appendix. Nettles *et al.* (541) reported a psoas abscess extending to the left sacroiliac joint due to *Bacteroides*. Prévot (226a) recovered a pure growth of *Ristella pseudoinsolita* from a voluminous retroperitoneal abscess. A case, related to trauma, described by Gorbach and Thadepalli (344a) yielded four anaerobes including *B. fragilis*. *Bacteroides melaninogenicus* was isolated from a psoas abscess by Shevky *et al.* (281). In a case reported by Frumin and Fine (542) a psoas abscess in a diabetic yielded foul pus and showed numerous gram-positive cocci on smear. No mention was made of culture, but this must have been an anaerobic infection.

There are a number of other individual case reports of retroperitoneal abscess with a variety of underlying lesions including appendicitis, retroperitoneal sarcoma, and septic abortion. The organisms involved include *Bacteroides*, *Actinomyces israelii*, *Sphaerophorus necrophorus*, fusiform bacilli, and anaerobic streptococci [(Schottmüller (258), Ghon and Roman (543), Cunningham (323), Beigelman and Rantz (282), Anderson (544), Saksena *et al.* (487), Coleman and Georg (319), Nettles *et al.* (541), Hara *et al.* (176), and Ishiyama *et al.* (498)].

Peritonitis

There are a number of causes of peritonitis, many of which predispose to infection with anaerobes (Fig. 10.4). Included in the latter category would be diseases and injuries of the gastrointestinal tract, certain lesions of the female genital system, lesions of the biliary tree, and abdominal surgery—particularly surgery relating to the gastrointestinal tract. Somewhat surprisingly, peritonitis secondary to perforation of the small bowel may carry a higher mortality than a similar lesion of the large bowel [Altemeier (67)]. Some reasons for this are the irritating effect of digestive

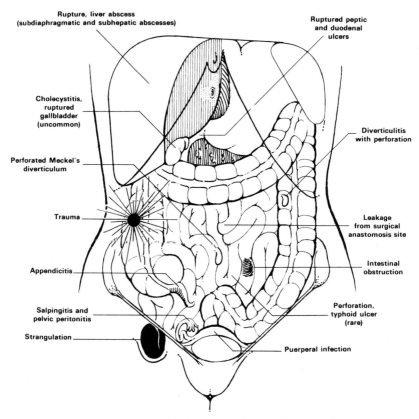

Fig. 10.4.   Various causes of peritonitis.

fluids from the small intestine, the greater fluid content of the small bowel, and the greater motility of the small bowel, with the latter two items favoring the spread of contamination.

There is considerable evidence that peritonitis is frequently a synergistic infection. Multiple strains of bacteria are commonly isolated from cases of peritonitis, and Meleney *et al.* (545) and Altemeier (546) have shown that the more organisms that are present in peritoneal exudate, the worse the prognosis. There is no correlation between severity of the infection and any one specific organism, including *Clostridium perfringens*. When there are secondary or metastatic abscesses in cases of peritonitis, essentially the same mixed flora can be isolated from the latter abscesses as from the peritoneal exudate of the original infection. This indicates that these are true mixed infections.

As long ago as 1902, Friedrich (547) noted that peritoneal exudate is rich in anaerobic bacteria, particularly *Clostridium septicum*, *Clostridium tetani*, and anaerobic streptococci. He noted that there were almost invariably anaerobes involved in peritonitis secondary to inflammation or perforation in or about the cecum. Altemeier, in a classic paper in 1938 (546) described the bacterial flora of peritonitis secondary to acute appendicitis with perforation. One hundred cases were studied and anaerobic bacteria were cultivated in 96 of these. At least 18 different species of anaerobes were recovered. *Bacteroides melaninogenicus* was recovered from 89 cases, anaerobic streptococci from 64, anaerobic cocci from 27, *Bacillus thetoides* from 11, and unidentified gram-negative anaerobic bacilli from 10. There was only one isolate of *Clostridium perfringens*, but there were 12 isolates of other *Clostridium* species. Multiple species of anaerobes were commonly recovered; 42 cases yielded two anaerobes, 45 cases yielded three, and 4 cases had four different species of anaerobes. The average number of different species of anaerobes per case exceeded that of aerobes. Counting both anaerobes and aerobes, the average number of species per case was slightly over four. Meleney *et al.* (548) reported on a series of 106 cases of peritonitis. Among the 71 cases that had positive cultures for any organism, 16% yielded anaerobic cocci from peritonitis following small bowel lesions. *Clostridium perfringens* was isolated from half of these cases. In six cases of peritonitis following large bowel lesions, the incidence of *C. perfringens* was 83%. Most of the cases in this series were secondary to appendicitis.

In the report of Wilson *et al.* (105) on anaerobic bacteremia, 13% of the 52 patients with anaerobic, gram-negative bacillary bacteremia had peritonitis as the underlying problem. Schatten (549) reported on 38 cases of peritonitis. Eight of these were negative by culture; 18 yielded anaerobes (*C. perfringens*, anaerobic streptococci, and *Bacteroides*), usually mixed with aerobes. Seven of these cases were secondary to appendicitis, three to perforated gastric or duodenal ulcers that were closed surgically, one was secondary to a tubo-ovarian abscess, and the rest were related to lesions of the bowel. Beazley *et al.* (394) isolated *Bacteroides* from 21 cases of peritonitis, 9 of which were fatal. Gillespie and Guy (512) recovered 9 *Bacteroides*, 5 anaerobic streptococci, and 2 clostridia from 18 cases of peritonitis other than that related to appendicitis. In a series of 16 cases of peritonitis due to ruptured gastroduodenal ulcers, Wright *et al.* (550) found that 5 had anaerobes, 3 of which were in pure culture (anaerobic streptococci). The anaerobes isolated from the other two cases were anaerobic streptococci, *Bacteroides melaninogenicus*, *Bacteroides funduliformis*, and *Clostridium tetanosporum*. One of the cases was not cultured, and four showed no growth on culture. Shibata *et al.* (551) cultured seven specimens of peritoneal fluid and found that all yielded anaerobes (nine strains). The anaerobes outnumbered the aerobes in this series. Sandusky *et al.* (112) recovered anaerobic streptococci, always in mixed culture, from 10 cases of peritonitis, 8 of which followed appendicitis. Other anaerobes recovered included one *Bacteroides* and one anaerobic gram-positive bacillus. Hartl (249) recovered *Bacteroides funduliformis* and streptococci from five cases of peritonitis.

Beerens and Tahon-Castel (115) cultured four cases of peritonitis secondary to bowel perforation or disease; all yielded mixtures of aerobes and anaerobes. Among the anaerobes were three strains of *Eggerthella convexa*, two of anaerobic streptococci, and one each of *Clostridium perfringens*, *Fusiformis biacutus*, and *Bifidobacterium bifidum*. Saito *et al.* (252) recovered anaerobes in pure culture from three cases of peritonitis. The organisms isolated were *C. perfringens*, *Bacteroides*, and *Bacteroides* plus *Peptococcus*. Hardy *et al.* (552) studied peritonitis secondary to intestinal perforation following exchange transfusion in the newborn. They reviewed 28 cases from the literature plus their own case. One of these 29 had *Bacteroides* present in pure culture. Snyder and Hall (553) recovered *Clostridium capitovalis* from four cases of peritonitis. Smith and King (554) reported two cases of peritonitis due to *Clostridium difficile*, one in pure culture, where the underlying lesions were bowel obstruction and mesenteric thrombosis. Haldane and van Rooyen (529) reported a case of *Bacteroides* peritonitis and septicemia secondary to small intestinal infarction. From a case of peritonitis secondary to small bowel obstruction, Goldsand and Braude (244) recovered *Clostridium perfringens*, anaerobic streptococci, and *Veillonella*. A case of spontaneous gas peritonitis with visceral gas gangrene due to *C. perfringens* and *Klebsiella* was described by Silverstein *et al.* (555).

A case of spontaneous or primary peritonitis was reported by Ansari (386) in a patient with Laennec's cirrhosis. From the ascitic fluid, *Clostridium perfringens* and α-hemolytic streptococci were recovered. *Clostridium perfringens* was also recovered on blood culture. As noted earlier, Gorbach *et al.* (43) had three cases of spontaneous peritonitis in cirrhotics in their series of 46 intraabdominal infections. Three additional cases involving anaerobes were reported by Correia and Conn (555a). The responsible organisms were *Bacteroides* in one case, anaerobic diphtheroids in a second, and *Peptococcus* and *Peptostreptococcus anaerobius* in the third. Randall (361a) noted recovery of anaerobes (mostly *Clostridium* and *Bacteroides*) from 2% of peritoneal dialysis fluids studied (presumably after use). A case of intestinal perforation due to peritoneal dialysis yielded *Bacteroides* and 3 aerobes (555b).

Infections of the Liver

Considerable evidence is accumulating to indicate that anaerobic bacteria are a major cause of liver abscess, undoubtedly the predominant cause. In the series of liver abscesses reported by Sabbaj et al. (387) anaerobes were present in 45% of the cases. However, only a small percentage of the specimens were cultured in the research laboratory, which had a much higher recovery rate for anaerobes than was true for the clinical laboratory. Indeed, it is not even certain that all specimens were cultured anaerobically. It seems likely that anaerobes were responsible for well over half of the cases of liver abscess in that series. These authors also reviewed several major series of liver abscess and found that overall among the total of 249 cases, in 12 different series, 22% of the cultures were sterile. The clinical features of anaerobic liver abscess, as noted in the paper by Sabbaj et al., are not different from those of pyogenic liver abscess in general, in keeping with the conclusion that most pyogenic liver abscesses are caused by anaerobic bacteria. One factor that helps account for the fact that anaerobes are not reported as the major cause of liver abscess in most series is that a number of authors include in their series cases in which abscesses were apparent only on microscopic examination of liver sections at autopsy. Thus, patients with sepsis or endocarditis due to various aerobic streptococci, staphylococci, and other aerobes are included. These microabscesses are not clinically significant as pyogenic liver abscesses.

Table 10.1 summarizes isolations of anaerobic organisms from pyogenic liver abscess. Included are a total of 379 strains isolated from 310 patients in a survey of the literature. Meleney (69) indicated that he isolated *Bacteroides* in pure and mixed culture from liver abscesses but did not give any numerical data. Altemeier (556) commented that in the previous 3 years, 8 of 12 liver abscesses that he saw yielded only anaerobic bacteria (anaerobic streptococci, *Bacteroides*, and *Actinomyces*). Sparberg et al. (557) reported two cases of liver abscess complicating regional enteritis, both of which were sterile on culture. One of these, which yielded 600 ml of frank pus, showed gram-positive cocci on smear. Pitt and Zuidema (1353a) noted that the percentage of anaerobes recovered in their series increased from 16% in the first half of the study to 36% in the second; 6 of the 10 patients seen during 1972 yielded anaerobes. Swenson et al. (384b) indicated that they recovered anaerobes from 13 hepatic abscesses, but no details were given. Sabbaj and Villanueva (557a) recovered anaerobes from 2 hepatic and subphrenic abscesses; the specific organisms recovered were not given.

Fulginiti et al. (415) recorded five infections of liver transplant patients due to *Bacteroides fragilis*. Each of these patients had bacteremia, and one each also had subphrenic and hip abscesses and a septic infarction of the

TABLE 10.1

Anaerobic Organisms in Pyogenic Liver Abscess (310 Patients)

Organisms	No. of isolates	References
Anaerobic streptococci	58	112, 168, 387, 534, 563, 564, 659, 948, 949, 1331, 1340–1353, 1353a–e
Microaerophilic streptococci	38	387, 948, 1244a, 1344, 1352, 1353a, 1353c–e, 1354–1357, 1358a, 1358c, 1358e, 1358g
Anaerobic gram-positive cocci	7	105, 387, 473, 591, 817, 1353c, 1357
Anaerobic gram-negative cocci	1	560
Bacteroides species	39	282, 307, 390a, 390b, 394, 489, 490, 534, 563, 914, 949, 1244a, 1353a–d, 1358, 1358a, 1358b, 1358d, 1358f
Bacteroides fragilis	33	115, 168, 226, 387, 394, 473, 1226, 1341, 1346, 1359–1361, 1361a
Bacteroides oralis	2	168
Bacteroides serpens	3	1358, 1362, 1363
Bacteroides melaninogenicus	7	116, 281, 387, 949
Bacteroidaceae	6	1353e
Fusobacterium species	34	168, 172, 387, 565, 949, 1209, 1211, 1224, 1250, 1340, 1346, 1354, 1364–1367, 1367a
Fusobacterium girans	3	168, 1353c
Fusobacterium necrophorum	47	168, 261, 293, 327, 347, 410, 621, 1130, 1228, 1297, 1301, 1341, 1343, 1347, 1354, 1359, 1365, 1368–1378
Unidentified anaerobic gram-negative rods	16	387, 543, 1331, 1340, 1379–1381
Clostridium	38	948, 1341, 1351, 1353a, 1353d, 1353e, 1354, 1358g, 1382–1391, 1391a
Actinomyces	32	113a, 210, 212, 293, 319, 327, 387, 447, 594, 948, 949, 1271b, 1351, 1353a, 1365, 1366, 1392–1396, 1396a, 1396b
Eubacterium species	2	259, 1250
Lactobacillus species	1	1358a
Bifidobacterium	1	1359
Propionibacterium	2	1353c, 1359
Bacterium halosepticum	1	1211
Unidentified anaerobic gram-positive rods	5	560, 591, 1211
"Anaerobe"	4	591, 1244a, 1397, 1398
Total	379 strains	

liver. Kaplan and Weinstein (558) reported a case of "bacterial hepatitis" due to anaerobic diphtheroids, the significance of which must be regarded as questionable. Kahn *et al.* (559) reported a case of *Clostridium septicum* infection of a metastasis in the liver from colon carcinoma. Anaerobes may be secondarily involved in amebic liver abscess (560, 1358d), and in hydatid cysts (561, 562).

Tests important in establishing the diagnosis of liver abscess include alkaline phosphatase and liver scan (Fig. 10.5). Flat plate X rays of the abdomen may also reveal gas in the area of the liver—either with gas-fluid levels or as scattered pockets (Figs. 10.6 and 10.7). Blood cultures, usually

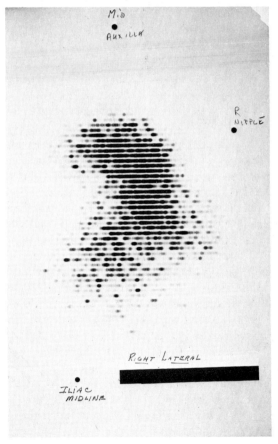

Fig. 10.5.   Radioisotope scan of liver (right lateral view). Large defect posteriorly is due to liver abscess.

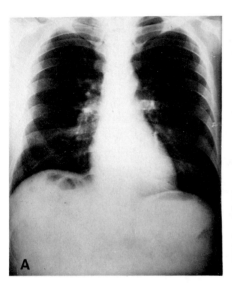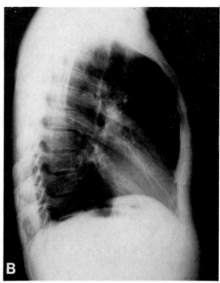

Fig. 10.6. Posteroanterior (A) and lateral (B) chest X rays showing elevated right hemidiaphragm and multiple pockets of gas beneath it. Patient had liver abscesses.

reported as negative in this condition, were positive in 54% of the cases described by Sabbaj *et al.* (387). Recovery of anaerobes on blood culture in the absence of an obvious portal of entry may be an important clue to liver abscess. Other helpful procedures include radionuclide scans, ultrasound, and arteriography. All five of the patients in the series of Sabbaj *et al.* who were diagnosed only at autopsy had anaerobic bacteremia.

Most clinicians favor surgical drainage plus antimicrobial therapy. Patterson (563), however, recommended aspiration under direct visualization, after exploration (rather than open surgical drainage); he claimed fewer complications than with open drainage. McFadzean *et al.* (564) had good results with the use of closed aspiration and antibiotic therapy in a group of 13 patients. Gilbert (565) has recommended treatment with antimicrobials alone, without any type of drainage procedure, and reports one case successfully treated in this manner. Perhaps with the availability of bactericidal agents that concentrate well in the liver, such as metronidazole, it may ultimately be feasible to treat pyogenic liver abscess with antimicrobial therapy alone; however, this must certainly be regarded as experimental at this time and should not be attempted except under protocol and where the patient may be kept under very close supervision. Multiple liver abscesses (Fig. 10.8) may require multiple drainage procedures and very prolonged

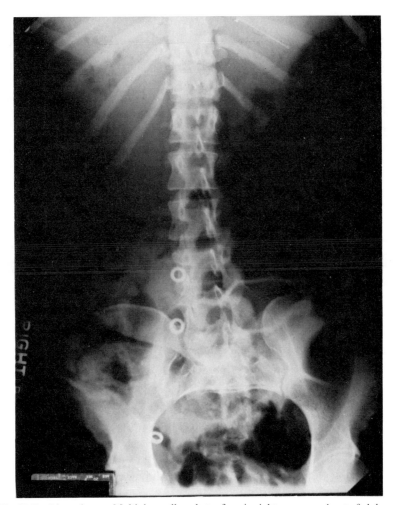

Fig. 10.7. Liver abscess. Multiple small pockets of gas in right upper quadrant of abdomen.

chemotherapy. Administration of antibiotics via the reopened umbilical vein has been suggested for the management of multiple liver abscesses where adequate surgical drainage may not be feasible (1353c). Both cases treated in this manner by Ranson *et al.* (1353c) recovered from the hepatic problem; one, unfortunately, died of pulmonary embolisms.

Pylephlebitis is rarely recognized as such in the antimicrobial era. It is certainly an important background for many cases of pyogenic liver abscess.

Fig. 10.8.   Multiple multiloculated liver abscesses. From Ref. (1360). C. Futch, B. A. Zakria, and H. C. Neu, *Bacteroides* liver abscess. *Surgery* **73**, 59–65 (1973).

Marson *et al.* (566) did document a case of pylephlebitis secondary to appendiceal abscess involving *Bacteroides* and a β-hemolytic aerobic streptococcus.

Infections of the Gallbladder and Biliary Tree

The concept of asymptomatic bactibilia is an important one to keep in mind. Although bacteria of any type, and particularly anaerobes, are uncommonly found in either the walls or contents of normal gallbladders [Andrews and Henry (567), Edlund *et al.* (568), and Flemma *et al.* (569)] in the presence of cholecystitis and/or cholelithiasis, and particularly with obstruction of the common duct, appreciable numbers of various bacteria, including anaerobes, may be found in both the contents and the wall of a significant percentage of gallbladders. As in asymptomatic pyelonephritis, there may be bacteria present in concentrations of 10^5/ml or more without any symptomatology. The incidence of bactibilia is higher in acute cholecystitis than in chronic and is particularly high when there is obstruction of the common bile duct [Andrews and Henry, (567) and Edlund *et al.* (568)]. The organisms found in the gallbladder are commonly the same as those in the normal intestinal flora—both aerobic and anaerobic. It is not clear how these organisms reach the gallbladder and bile ducts, but the best evidence is that spread is by an ascending mode. This is based partly on the fact that the number of organisms of the intestinal flora is higher in gallbladder bile than in the wall of the gallbladder [Edlund et al. (568)]. Counts in the gallbladder bile are higher than those in common duct bile, which may be explained by better conditions for growth in the relatively stagnant gallbladder bile. The proposal of ascending spread of organisms presumes a bacterial flora in the duodenum, and it is known that such a flora exists under certain circumstances and in relation to meals in all people. The types of bacteria found also negate the possibility that organisms arrive in the gallbladder from the general circulation. While it has been demonstrated that bacteria travel in the lymph from the duodenum wall to the gallbladder wall, if this were an important means of colonizing the gallbladder one would anticipate higher counts in the gallbladder wall than in the contents. It is postulated that asymptomatic bactibilia is converted to clinically manifest infection by trauma of one sort or another. Included would be surgery or percutaneous or operative cholangiography. Another important factor in converting asymptomatic bactibilia to symptomatic infection, either cholangitis or empyema of the gallbladder, is obstruction—particularly an acute change in the degree of obstruction. However, complete obstruction is less likely to lead to difficulties of these types than is partial obstruction. An acute change in partial obstruction leads to a sudden increase in intraductal pressures, which may rupture the biliary mucosa and allow organisms to enter the bloodstream to produce bacteremia. Table 10.2 summarizes a number of routine surveys of gallbladders and gallbladder contents in the absence of active infection. Engstrom et al. (569a) cultured duodenal contents and operative specimens from the liver and biliary tract of 49 patients with disease of the gallbladder or common bile duct. Anaerobes (anaerobic streptococci) were recovered from only 1 patient (operative specimen). Table 10.3 summarizes reports of infection of the gallbladder and biliary tree involving anaerobic bacteria. Paredes and Fernández (1404a) found that *Clostridium* constituted a greater percentage of recovered bacteria in patients with empyema of the gallbladder (21.6%) than in patients with acute or chronic cholecystitis (16.9%, 16.1%).

Acute gaseous cholecystitis is seen most often in elderly diabetic men. Systemic symptoms are more severe than in ordinary acute cholecystitis and gas formation is distinctive. The classic X-ray picture shows gas within the gallbladder lumen, frequently with a gas–fluid level, gas in a ring along the contour of the gallbladder wall, and gas in the tissues surrounding the

TABLE 10.2

Anaerobes in Routine Surveys of Gallbladders and Contents (No Specific Indication of Active Infection)

Material sampled	Number sampled	Number or % positive, any organism	Number positive, anaerobes	Number with anaerobes only	Specific anaerobes isolated	Comments	Reference	Year
Contents			5	3	*Bacillus ramosus, Micrococcus ovalis, B.funduliformis, C. perfringens, B. radiiformis,* anaerobic streptococci	Calculous cholecystitis with contents suppurative or not	Gilbert and Lippmann (1399)	1902
Contents			7		Same organisms as above plus *B.fragilis, B. nebulosus, Micrococcus foetidus,* and *Sarcina minuta*	Same as above	Gilbert and Lippmann (1400)	1902
Bile	29	16	2		Fusiforms		Rosenow (1401)	1916
Stones	62	29	7		*Clostridium perfringens*			
Walls	32	27	3		2 fusiforms and 1 *C. perfringens*			
					C. perfringens in 18% of gallbladders at surgery		Gordon-Taylor and Whitby [cited by Edinburgh and Geffen (1402)]	1930 1958
Bile			1		Buday's bacillus	"Cholecystitis"	Hegler and Nathan (1375)	1932
			4		*C. perfringens*		Posselt (1402a)	1934
Wall	91	49%	10	2	8 *Clostridium* (7 *C. perfringens*), 2 anaerobic streptococci		Andrews and Henry (567)	1935
Bile	91	33%	7		6 *Clostridium perfringens*, 1 anaerobic streptococcus			
Gallbladder walls	340		13		Nonsporulating anaerobes		Dack (106)	1940
Bile	371		3		Nonsporulating anaerobes			

282

Material					Anaerobes isolated	Remarks	Reference	Year
Duodenal bile A	247				of all types		Lodenkämper and Stienen (287)	1955
bile B	149	41	10		Wide variety of anaerobes of all types	"Cholecystitis"		
		4	4		3 Bacteroides, 1 anaerobic coccus		Gordon-Taylor and Whitby [cited by Edinburgh and Geffen (1402)]	1958
					C. perfringens in 13% of gallbladders at autopsy			
Bile	125	48	5		⎱ Anaerobic gram-positive bacilli, anaerobic streptococci, and lactobacilli (anaerobic?)		Edlund et al. (568)	1959
Wall	96	61	23		⎰			
Stones	57	30			30 Actinomyces		Rains et al. (1404)	1960
Bile	100	34	2		C. perfringens		Pyrtek and Bartus (513)	1962
Intrahepatic bile	75	32	2	2?	2 Microaerophilic streptococci (also 2 lactobacilli)	Bile obtained during percutaneous transhepatic cholangiography	Flemma et al. (569)	1967
Bile, gallbladder wall, stones	?	188 total bacterial strains isolated	35 anaerobic strains isolated	?	Clostridium, 28 strains Non-spore-formers, 7 strains		Paredes and Fernández (1404a)	1967
Wall, bile and stones	100	3		2	3 Actinomyces, 1 with Sarcina ventriculi	2 Positive from wall, 1 each from bile and stone	Czarnecki and Kolsut (1405)	1969
Bile	85	8	1	2	Bacteroides convexus		Gupta and Bhatia (1241)	1970
Bile	59	59	12	2	6 Veillonella, 3 Peptostreptococcus, 3 gram-positive rods, 2 gram-negative rods, 1 Clostridium	Probably included in data of Nakamura et al. (361)	Nakamura et al. (631)	1970

(continued)

TABLE 10.2 (*continued*)

Material sampled	Number sampled	Number or % positive, any organism	Number positive, anaerobes	Number with anaerobes only	Specific anaerobes isolated	Comments	Reference	Year
Bile	101	72	16	2	3 *Peptococcus* and *Peptostreptococcus*, 7 *Veillonella*, 4 *Clostridium*, 3 gram-positive rods, 2 gram-negative rods	Probably includes data from Nakamura et al. (631)	Nakamura et al. (361)	1971
Bile	16	13	4		3 *Clostridium*, 1 *Veillonella*		Nakamura et al. (114)	1971
Bile		15	9			Anaerobes always outnumbered aerobes	Shibata et al. (551)	1971
	34		7	4	Anaerobic streptococci (5 isolates), C. perfringens, Bacteroides		Keighley and Graham (1406)	1971
	31		11		Anaerobic gram-positive rods, C. perfringens, Peptococcus, Sphaerophorus, and Bacteroides		Ohnishi et al. (582)	1972
Bile, gallbladder mucosa	501	234	24	?	Clostridium perfringens, 22 Bacteroides, 2		Fukunaga (1420b)	1973
Gallbladder, bile	42	15	3	?	Bacteroides fragilis		Shimada et al. (1420c)	1973
Bile	231	166 total strains isolated	20 anaerobic strains isolated	4	Anaerobic streptococci, 7 (4 pure), Clostridium welchii, 11 Bacteroides sp., 2		Keighley et al. (1406a)	1975

Infection Due to Anaerobes Related to Gallbladder and Biliary Tree

Type of infection	Underlying problem	Anaerobes recovered	Aerobes recovered	Comment	Reference	Year
Perforated gallbladder, peritonitis	?	*Bacillus ramosus*, anaerobic cocci, *? Staphylococcus parvulus*, *? B. fragilis*	Aerobic streptococci, small numbers		Zuber and Lereboullet (1407)	1898
Gallbladder disease, empyema and liver abscess		*B. fragilis*, *Staphylococcus parvulus*	*E. coli*		Guillemot *et al.* (473)	1904
Abscess of gallbladder		Fusiform			Ghon and Roman (543)	1916
Suppurative cholecystitis		*B. trichoides*			Potez and Compagnon (1408)	1922
Emphysematous cholecystitis		*C. perfringens*			Kirchmayr [cited by Wilson (1409)]	1925
Emphysematous cholecystitis		*C. perfringens* (atypical)			Hegner [cited by Wilson (1409)]	1931
Emphysematous cholecystitis		"Sporing anaerobe"			Simon [cited by Wilson (1409)]	1932
Cholangitis, bacteremia		Anaerobic streptococcus	*E. coli*		Bingold (442)	1932
Actinomycosis of gallbladder		*Actinomyces*			Mayo-Robson [cited by Cope (210)]	1938
Emphysematous cholecystitis		*C. perfringens*			Del Campo and Otoro [cited by Wilson (1409)]	1940
Emphysematous cholecystitis		Anaerobe resembling *C. perfringens*	*E. coli*		McCorkle and Fong [cited by Wilson (1409)]	1942
Cholecystitis?, suppurative (3 cases)		2 Anaerobic streptococcus Microaerophilic streptococcus	0 "Mixed"		Sandusky *et al.* (112)	1942

(continued)

285

TABLE 10.3 (*continued*)

286

Type of infection	Underlying problem	Anaerobes recovered	Aerobes recovered	Comment	Reference	Year
Pericholecystic abscess		Anaerobic streptococcus	0		Sandusky et al. (112)	1942
Postoperative abscess (2 cases)	Cholecystectomy	Anaerobic streptococcus	"Mixed"			
Peritonitis		Bacteroides melaninogenicus	E. coli; Staphylococcus aureus, and hemolytic streptococcus		Weiss (80)	1943
Emphysematous cholecystitis		Anaerobic streptococcus	Aerobic streptococcus		Stevenson [cited by Wilson (1409)]	1944
Emphysematous cholecystitis		Clostridium perfringens			Hutchinson [cited by Wilson (1409)]	1946
Suppurative cholangitis	Carcinoma pancreas	Anaerobic streptococcus	E. coli		Cole (1410)	1947
Pneumocholecystitis Pneumocholecystitis		Clostridium filiforme Bacteroides	Streptococcus faecalis, paracolon		Heifetz and Senturia (1411)	1948
Emphysematous cholecystitis		Clostridium oedematiens			Jemerin [cited by Wilson (1409)]	1949
Gallbladder or biliary tract infection (21 cases)		20 C. perfringens 1 C. septicum		9/11 with abdominal X rays showed gas in gallbladder 7/17 operated on showed frank pus Mortality 50% —visceral gas gangrene	Schottenfeld (570)	1950

Clinical presentation	Predisposing factor	Aerobes	Anaerobes	Comments	Reference	Year
Widespread infection including liver abscess and bacteremia	?		*Bacteroides funduliformis*		Ernst (347)	1951
Recurrent cholangitis			*B. funduliformis*	Organism isolated from small bowel contents; assumed to be from bile	Rubin *et al.* (1412)	1951
Pyopneumocholecystitis	?	*E. coli,* numerous	"Some anaerobes" *Sphaerophorus funduliformis*		Tedesco (1413)	1951
"Biliary infection"		?	*Sphaerophorus funduliformis*		Beerens (1414)	1954
Bacteremia			*Sphaerophorus freundii*		Prévot (21)	1955
Emphysematous cholecystitis			Clostridia found in ~ 40% of cases, anaerobic streptococci 1 time, *C. perfringens* mostly in gallbladder wall and stones	Literature review, 50 cases; 11 not cultured anaerobically	Edinburgh and Geffen (1402)	1958
Wound infection	Cholecystectomy		*Bacteroides,* anaerobic staphylococcus		Witebsky and Miller (162)	1958
Empyema of gallbladder, bacteremia	Diabetes mellitus	0?	*Clostridium perfringens*		Tennant and Parks (454)	1959
Gaseous cholecystitis, bacteremia			*C. perfringens*		Bigler (1415)	1960
Bacteremia (2 cases)	Cholecystectomy	*E. coli*	*C. perfringens*	All 3 patients died	Pyrtek and Bartus (513)	1962
Gas gangrene abdominal wall	Cholecystectomy		*C. perfringens*			

(continued)

TABLE 10.3 (*continued*)

Type of infection	Underlying problem	Anaerobes recovered	Aerobes recovered	Comment	Reference	Year
Wound infection	Resection of choledochal cyst	*Bacteroides*	*E. coli, Aerobacter aerogenes,* 2 aerobic streptococci		Bernard and Cole (518)	1963
Wound infection	Cholecystectomy	*C. perfringens*	*E. coli, A. aerogenes,* intermediate coliform, diphtheroids		Brummelkamp *et al.* (1002)	1963
Intraabdominal gas gangrene	Cholecystectomy	*C. perfringens, C. fallax*	?			
Abdominal wall gas gangrene, ?intraperitoneal	Cholecystectomy	*C. perfringens*	?			
Gangrenous gallbladder, pneumocholecystitis (4 cases)	?	*C. perfringens* (2), with microaerophilic streptococcus (1), also 2 gram-positive rods on smear but no growth		Foul pus	Parsons (1416)	1963
Septicemia, acute hemolytic anemia	Cholecystectomy	*C. perfringens*		Spherocytic anemia, acute	Bennett and Healey (1417)	1963
Emphysematous cholecystitis	Diabetes	*C. perfringens*			Sawyer *et al.* (1418)	1963
Abdominal wall gas gangrene extending to back, chest and leg	Cholecystectomy and sphincterotomy, Diabetes	*Clostridium*	?		Bornstein *et al.* (275)	1964

Bacteremia	Acute biliary tract disease	*Clostridium*			Bornstein *et al.* (275)	1964
Bacteremia (2 cases)	Cholecystitis (1 pneumo-cholecystitis)	*C. perfringens*		All 3 died despite therapy surgery and antimicrobial	Plimpton (1391)	1964
Gas gangrene liver and abdominal muscles	Cholecystitis	*C. perfringens*				
Gas gangrene (5 cases) (bacteremia in 4)	Cholecystectomy (5) Choledochotomy (1) Gastrectomy (1) Appendectomy (2) Diabetes (1)	*C. perfringens* (5)	*E. coli* (1)	All 5 cases fatal	Aldrete and Judd (1419)	1965
Empyema gallbladder Bacteremia		*Bacteroides* *Bacteroides*, *Propionibacterium*	0 0		Dietrich *et al.* (135)	1965
"Biliary infection"	Cancer gallbladder	*Eggerthella convexa*, *C. liquefaciens*	*Staphylococcus aureus*, *E. coli*, *Streptococcus zymogenes*		Beerens and Tahon-Castel (115)	1965
"Biliary infection"	?Stricture common bile duct	*Clostridium*, *F. fusiformis*, *Streptococcus anaerobius*	*Pseudomonas aeruginosa*, *E. coli*, *Streptococcus faecalis*			
Peritonitis	Perforation gallbladder	*B. funduliformis*, anaerobic streptococcus	0			
Acute cholecystitis, bacteremia	?	*Ristella pseudoinsolita*	0		Prévot (226a)	1965
Emphysematous cholecystitis	—	*Clostridium welchii*	0	Foul-smelling pus and gas	Sarmiento (1419a)	1966
Local abscess, visceral gas gangrene	Common duct surgery	*C. perfringens*			Irvin *et al.* (595)	1967

(continued)

TABLE 10.3 (*continued*)

Type of infection	Underlying problem	Anaerobes recovered	Aerobes recovered	Comment	Reference	Year
Wound infection (6 cases)	Gallbladder surgery (3 with appendectomy also)	Bacteroidaceae			Saksena et al. (487)	1968
Empyema gallbladder	?	A. naeslundii	?		Coleman et al. (152)	1969
Infection, abdominal scar	Cholecystectomy several months previous	Actinomyces	?		Schubert and Tauchnitz (318)	1969
Empyema of gallbladder, sepsis	?	Bacteroides fragilis			Labowitz and Holloway (968)	1969
Cholecystitis, endocarditis	?	B. fragilis			Felner and Dowell (289)	1970
Bacteremia (3 cases)	Gallbladder or biliary tract surgery	Bacteroides			Bodner et al. (460)	1970
Bacteremia	Carcinoma of pancreas	Bacteroides			Sinkovics and Smith (411)	1970
Bacteremia Cholangitis		Bacteroides Bacteroidaceae		Fatal cholangitis documented at autopsy, number of patients uncertain	Gelb and Seligman (490)	1970
Emphysematous cholecystitis, visceral gas gangrene		C. perfringens			Boerema and McWilliam (1420)	1970
Bacteremia		B. fragilis (6 cases), B. oralis (1 case), Bacteroides CDC F1 (1 case), Fusobacterium ridiculosum (1 case)			Felner and Dowell (168)	1971

(continued)

Condition	Predisposing factor	Bacteroides	Other organisms	Comments	Reference	Year
Bacteremia		B. fragilis	Enterococcus		Kuklinea and Gavan (491)	1971
Peritonitis, septicemia	Cholecystectomy, diabetes mellitus	Ristella pseudoinsolita	?		Farquet et al. (723)	1971
Five cases of acute cholecystitis, bacteremia	?	Bacteroides			Nobles (56d)	1973
Probable cholecystitis, subphrenic abscess	None	Actinomyces	Streptococcus		Davies and Keddie (1420a)	1973
2 cases of wound infection following cholecystectomy	?	Bacteroides	0?		Baird (324b)	1973
3 cases of wound infection following cholecystectomy	?	C. perfringens (2 cases), Bacteroides	E. coli (2 cases), 0		Fukunaga (1420b)	1973
Gangrenous cholecystitis	—	B. fragilis (bile), C. perfringens (gallstone)	Several	Disseminated intravascular coagulation, death	Shimada et al. (1420c)	1973
Cholangitis, bacteremia	Common duct stone	B. fragilis (bile and blood)	E. coli, Enterobacter and enterococcus (bile only)			
?	Cancer stomach	B. fragilis (bile, ascitic fluid, sub-hepatic abscess)	E. coli (from same 3 sites)			
Five cases of empyema of gallbladder	?	Bacteroides	?	No case details	Okubadejo et al. (390a)	1973
Cholangitis, bacteremia	Clonorchis sinensis infestation	Bacteroides	Eikenella corrodens		Von Hoff and Kimball (1420d)	1974
Two cases of cholangitis and bacteremia	?	Not specified	?	No case details	Chow et al. (1420e)	1974
Cholangitis	?	Bacteroides fragilis	?		Leigh (1420f)	1974
Cholecystitis	?	Bacteroides fragilis	?			

TABLE 10.3 (*continued*)

Type of infection	Underlying problem	Anaerobes recovered	Aerobes recovered	Comment	Reference	Year
Perforated gangrenous gallbladder, bacteremia		*Clostridium perfringens*	0	Secondary panophthalmitis. Died	Frantz *et al.* (247a)	1974
Acute gangrenous cholecystitis, pericholecystic abscess, liver abscess	Acute myelogenous leukemia	*Bacteroides* (liver abscess)	0		Novy *et al.* (1353b)	1974
Chronic cholecystitis, cholelithiasis, liver abscess, bacteremia	Dysplastic granulocytic hyperplasia	*Bacteroides* (liver abscess, blood)	0			
Giant horseshoe intraabdominal abscess following cholecystectomy		*Bacteroides* sp., *Sphaerophorus* sp. *Peptostreptococcus* (all from abscess)	*S. aureus, E. coli, Micrococcus* (all from abscess)		Altemeier *et al.* (89b)	1975
Acute cholecystitis	Complication of retrograde cholangio-pancreatography	None, but bile was foul-smelling	*Proteus, E. coli*		Davis *et al.* (1420g)	1975

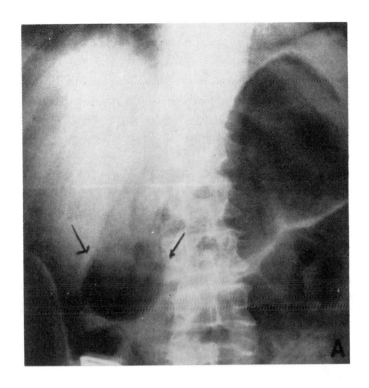

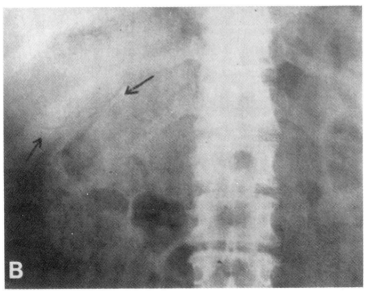

Fig. 10.9. Gas in pericholecystic tissues outlines the gallbladder wall. Some air is present within the gallbladder. From Sawyer *et al.* (1418). Reproduced with permission.

293

gallbladder (Fig. 10.9A and B). Barium swallow may be required to rule out a fistulous connection with the gastrointestinal tract. Thrombosis of large veins in the area is not uncommon, according to Schottenfeld (570), who also pointed out that the prognosis is worse when the common duct is involved. Sarmiento (1419a), in 1966, found 101 cases of emphysematous cholecystitis in the world literature since 1901. One-third of the patients were diabetics; 87 patients were above age 50. Clostridia were recovered from 25% of cases and accounted for almost half of the positive cultures; almost all were *C. welchii*.

Clostridium perfringens infection should be considered in a febrile patient with a rising pulse rate in the immediate postoperative period following gallbladder surgery. Such patients will also manifest toxemia and ultimately shock. Pyrtek and Bartus (513) recommend routine culturing of the bile at the time of surgery, and I would certainly endorse this. Many workers advocate prophylactic antimicrobial therapy for gallbladder surgery. It would seem wise in selected cases at least (perhaps based on gram stain of gallbladder contents) to use prophylaxis against the uncommon but devastating clostridial infections that may be seen in relation to gallbladder surgery. For this purpose penicillin could be used; this would result in minimal disruption of normal flora of the body, and, therefore, the possibility of superinfection would not be great.

Appendiceal Infections

A very good quantitative bacteriologic study of normal appendices was carried out by Werner and Seeliger (571) and reported in 1963. Eleven appendices were cultured; all grew *Escherichia coli*, aerobic streptococci, *Bifidobacterium*, and gram-negative anaerobic bacilli. There were four biotypes of *Bifidobacterium* and the counts ranged from 10^4 to 10^8 per gram. The gram-negative anaerobic bacilli were numerically dominant in seven of the specimens, and all but two of the specimens had counts of 10^7 to 10^{10} or greater per gram. Less commonly encountered gram-negative anaerobic bacilli included *Bacteroides melaninogenicus*, *Fusobacterium nucleatum*, *Fusobacterium necrophorum*, and *Bacteroides insolita*. *Lactobacillus* of five different species was present in 10 specimens, and *Veillonella* and anaerobic *Corynebacterium* were found in one each.

The importance of appendicitis (Fig. 10.10) in anaerobic infections as a whole is emphasized by the series of Bornstein *et al.* (275) in which one-fourth of the *Bacteroides* infections in a large series accumulated over a 45 month period involved the appendix. In the study of Felner and Dowell (168) involving 250 cases of bacteremia due to gram-negative anaerobic bacilli, over 10% of the cases originated with appendiceal lesions. Table 10.4 summarizes a number of reports of anaerobic bacteremia secondary

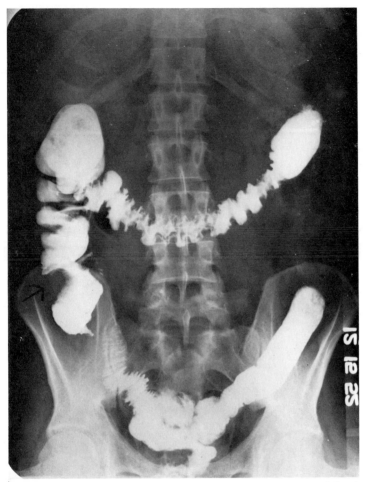

Fig. 10.10. Appendiceal abscess producing a constant extrinsic filling defect in cecum outlined by barium enema.

to appendicitis and its complications or to appendectomy. The frequency with which anaerobes may be recovered from lesions related to the appendix undoubtedly depends heavily on techniques of collection and transport of specimens and on anaerobic culture techniques in the laboratory. Results vary widely from series to series. The excellent study of Altemeier (546) on the bacterial flora of peritonitis associated with acute perforated appendicitis has already been described above. Anaerobes were recovered from 96 of the 100 cases of peritonitis in that study. On the other hand, in a more recent study, Altemeier (527) reported on 97 cases of appendicitis; 17%

TABLE 10.4

Anaerobic Bacteremia Secondary to Appendicitis and Its Complications or to Appendectomy

Number of cases	Anaerobes recovered	Comments	Reference	Year
1	Anaerobic streptococcus		McDonald et al. (534)	1937
1	Sphaerophorus gulosus, S. mortiferus		Patocka and Laplanche (1421)	1947
2	Bacteroides		Gunn (296)	1956
2	1 Bacteroides and anaerobic streptococcus		McHenry et al. (530)	1961
	1 Bacteroides			
3	Bacteroides		Tynes and Frommeyer (489)	1962
1	Bacteroidaceae		Saksena et al. (487)	1968
1	Bacteroides fragilis	Endocarditis	Felner and Dowell (289)	1970
3	Bacteroides		Bodner et al. (460)	1970
1	B. fragilis		Gelb and Seligman (490)	1970
8	Bacteroides		Marcoux et al. (420)	1970
16	B. fragilis		Felner and Dowell (168)	1971
1	B. oralis			
2	B. variabilis			
1	B. incommunis			
1	B. terebrans			
2	Bacteroides CDC group F1			
1	Bacteroides species			
2	Fusobacterium ridiculosum			
1	Sphaerophorus funduliformis	E. coli and enterococcus also	Farquet et al. (723)	1971
3	S. funduliformis			

yielded *Bacteroides*, and 25% anaerobic streptococci on culture. These later cases, however, were not necessarily complicated by peritonitis and perhaps antimicrobial therapy may have influenced the results to some extent as well.

The results in the study by Veillon and Zuber (7) in 1898 are impressive indeed. These workers studied 22 cases of appendicitis. One yielded only aerobic bacteria, two yielded only anaerobic bacteria, and the other 19 had anaerobes as the predominant flora, with only rare streptococci and *E. coli*. The anaerobes found, in the order of frequency, were *Bacillus fragilis*, *Bacillus ramosus*, *Bacillus perfringens*, *Bacillus fusiformis*, *Bacillus furcosus* and *Staphylococcus parvulus*. Lanz and Tavel in 1904 (572) recovered five strains of *Clostridium septicum* and five of *Bacillus pseudotetani* from eight normal appendices. They also studied 138 pathologic appendices from which they recovered 59 strains of *B. pseudotetani* and 49 of *C. septicum*. Heyde in 1911 (573) reported on 102 cases of appendicitis from which he recovered a total of 274 anaerobic strains and 153 aerobic strains. Most prevalent among the anaerobes were *Bacillus thetoides* (45 isolates), *Bacillus ramosoides* (27), *Bacillus ramosus* (24), anaerobic streptococci (23), and *Bacillus fusiformis* (25). Heyde commented that pure infections with anaerobes in purulent and gangrenous appendicitis was not uncommon. In 1923, Brütt (574) reported the recovery of 45 anaerobic streptococci from 107 cases of appendicitis.

Weinberg *et al.* (575), in 1926, stressed the importance of culturing pus and not simply the lumen of the appendix in appendicitis. In that study, they found that 30% of such specimens contained *Clostridium perfringens*; also recovered were *C. sporogenes*, *C. fallax*, *C. septicum*, and *C. histolyticum*, as well as some non-spore-forming anaerobes. In a later study Weinberg *et al.* (576) reported on 150 cases of appendicitis. From these they recovered *C. perfringens* 49 times along with 60 gram-negative anaerobes, 40 anaerobic cocci and streptococci, 16 *B. ramosus*, 5 *C. fallax*, 2 *B. bifidus*, 3 *C. bifermentans*, 2 *C. septicum*, 2 *C. histolyticum*, 2 *C. sporogenes*, and a total of 26 other anaerobic strains. From the same group, *E. coli* was recovered 128 times along with 41 enterococci, 14 *Proteus*, 12 *Staphylococcus*, 10 other streptococci, as well as small numbers of other aerobic or facultative forms. The series was expanded to 160 cases in a still later report by Weinberg *et al.* (577). The results were quite similar. The total number of isolates was 204 anaerobes and 264 aerobes. They recovered 1 to 7 organisms per specimen and noted that when four or more microbes were cultured from a single appendix the anaerobic species recovered outnumbered the aerobes. In all, 113 of the 160 cases yielded anaerobes.

In 1930, Schmitz (578) reported recovery of *Bacillus fusiformis* 25 times from 100 appendices. Löhr and Rabfeld in 1931 (579) reported that the normal flora of the appendix contained *C. perfringens* 79% of the time, *C. tertium* 37%, *C. butyricum* 35%, *C. bifermentans* 24%, and *C. sphenoides* 22%. In severe phlegmonous appendicitis these workers noted that fine anaerobic gram-positive rods were most numerous. Also in 1931 Meleney *et al.* (548) studied 31 cases of appendicitis with localized peritonitis, found no growth from 77% and *C. perfringens* in 6%. They also studied 23 cases of appendicitis with diffuse peritonitis or abscess formation and found that 58% yielded *C. perfringens*. Maccabe and Orr (580), in a study of 28 normal appendices, recovered *C. perfringens* twice, other clostridia three times, and anaerobic streptococci once. Among 172 diseased appendices they found that 11% yielded *C. perfringens*, 8% other clostridia, 4% anaerobic streptococci, and 10% *Bacteroides*. Holgersen and Stanley-Brown (581) cultured 100 cases of perforated appendicitis and recovered only two *C. perfringens*, one *Bacteroides*, and one microaerophilic streptococcus. Ohnishi *et al.* (582) cultured 75 cases of appendicitis and found that 30% of the organisms isolated were anaerobes, mostly *Peptococcus* and *Bacteroides*. Forty-three of the 75 cases had only aerobic organisms. They also cultured seven specimens from cases of peritonitis following appendicitis, and three of these yielded anaerobes.

Several much smaller series have been reported. In 1933, Cazzamali and Miglierina (583) found that among 14 cases of localized peritonitis secondary to appendicitis, 24% yielded anaerobes, whereas in 26 cases of diffuse peritonitis related to the appendix 33% yielded anaerobes, with *C. perfringens* and *Bacteroides fragilis* dominant. Jennings, in 1931 (584), recovered *C. perfringens* from the lumen of the appendix 90% of the time in appendicitis. He also found this organism on seven occasions out of seven attempts in the appendiceal wall, in 16 cases of localized collections of pus, and in 10 of 40 cases of free fluid in the pelvis or general peritoneal cavity. Wright *et al.* (550) studied 35 cases of peritonitis due to appendicitis. Twelve of these had anaerobes, four in pure culture. One specimen was not cultured, and eight yielded no growth. The anaerobes recovered included anaerobic streptococci, anaerobic cocci, *Bacteroides melaninogenicus*, *Bacteroides funduliformis*, *Clostridium perfringens*, and *Clostridium tetanosporum*. Gillespie and Guy (512) cultured the walls of 13 appendices from acute appendicitis and recovered 11 *Bacteroides*, 10 anaerobic streptococci, and 2 clostridia. They also cultured 34 cases of peritonitis with secondary abscess formation, the specimens not being handled as well in this particular case, and recovered 16 *Bacteroides*, 10 anaerobic streptococci, and 1 *Clostridium*. Perrone (585) studied 14 cases of appendicitis, found one sterile, seven with *Bacteroides fragilis*, six *Clostridium perfringens*, and one a fusiform. McDonald *et al.* (534) studied five cases of gangrenous appendicitis, four of which ruptured. All had anaerobic streptococci in the local infection, and one had a positive blood culture as well. There was also one *Bacteroides fragilis* and one other "anaerobe" as well as two aerobes isolated. Sandusky *et al.* (112) recovered anaerobic or microaerophilic streptococci from 11 appendiceal abscesses, once in pure culture. These organisms were also recovered from eight cases of peritonitis secondary to appendicitis. Along with the anaerobic cocci there was one *Bacteroides* and four clostridia recovered.

Excellent data is provided in a monograph by Beerens and Tahon-Castel (115). From 15 cases of peritonitis secondary to appendicitis, these workers recovered 12 *Eggerthella convexa*, 1 *Bacteroides funduliformis*, 2 *Fusiformis biacutus*, 3 *Clostridium perfringens*, 4 other clostridia, 2 *Eubacterium lentum*, 1 *Eubacterium limosum*, 6 anaerobic streptococci, and 3 anaerobic cocci. Three cases of appendicitis yielded 3, 4, and 6 anaerobes, respectively, as well as 1 to 2 aerobes per specimen. The anaerobes included *E. convexa* (3), *F. fusiformis* (1), *Dialister* (1), *E. lentum* (1), *E. limosum* (1), *Sarcina ventriculi* (1), other anaerobic cocci (3), and anaerobic streptococcus (1). There were two cases of intraabdominal abscess secondary to appendicitis. One of these yielded *Eggerthella convexa* in pure culture and the other yielded *Bacteroides funduliformis*, *Fusobacterium biacutus*, *Streptococcus foetidus*, as well as *Streptococcus faecalis*. Saksena *et al.* (487) recovered Bacteroidaceae from 11 infections related to the appendix, including five wound infections and one septicemia. In a study of 14 specimens from 12 cases of appendicitis, Shibata *et al.* (551) found anaerobes in 86%, a total of 22 strains. Saito *et al.* (252) recovered only anaerobes from six cases of infection related to appendicitis (four *Bacteroides*, one *Sphaerophorus*, and one *Peptostreptococcus*). Beazley *et al.* (394) recovered *Bacteroides* from 12 cases of appendicitis.

Two recent studies are of interest. Leigh *et al.* (585a) found *Bacteroides* in 78% of 322 swab specimens taken at the time of appendectomy; coliforms were found in 27–29% of cultures. *Bacteroides* was isolated twice as often in perforated appendicitis as in less affected appendices. Wound infection occurred in 19% of these surgeries, and *Bacteroides* was isolated from over 90% of wound infections. Werner *et al.* (585b) did quantitative cultures of the contents of 49 inflamed appendices. Anaerobic gram-negative bacilli were recovered from 43 appendices in counts of 10^3 to 10^9 per gram. *Bacteroides fragilis* was detected in 31 appendices and was the predominant organism in 18.

In addition to those mentioned previously, several workers have documented the presence of certain less commonly encountered anaerobes in appendicitis. Weinberg and Prévot (586) recovered *Fusobacterium biacutum* from six cases of appendicitis. Weiss (80), recovered *Bacteroides melaninogenicus* from nine cases of appendicitis. Heinrich and Pulverer (116) recovered *Bacteroides melaninogenicus* from eight cases of appendicitis. From a case of abscess in the pouch of Douglas following appendicitis, Prévot *et al.* (314) recovered *Ramibacterium pleuriticum* plus *Fusobacterium girans*. *Bacteroides putredinis* was isolated from two cases of appendicitis by Suzuki [cited by Werner (587)]. Armfield *et al.* (119) reported *Eubacterium fila-mentosum* from 16 cases of complicated appendicitis.

The appendix may also be involved in actinomycosis. Harvey *et al.* (212) in 1957 described nine cases of actinomycosis which were primary in the appendix. Five cases of actinomycosis originating in the appendix are described by Davies and Keddie (1420a). Single cases were also described by Nathan (293) and by Nash (588). Nash described a case involving the cecum as well as the appendix and, because of the firm mass that resulted, the surgeons originally thought that they were dealing with carcinoma. Brown and George [cited in Georg (233)] isolated *Actinomyces odontolyticus* from a case of gangrenous appendicitis.

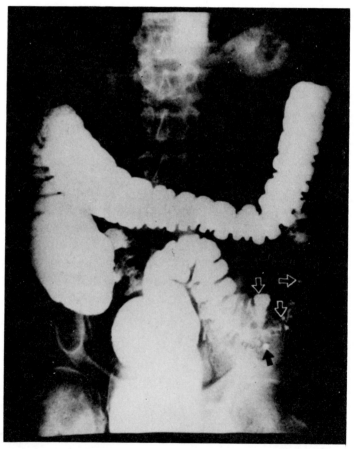

Fig. 10.11.   Barium enema. Arrows point to diverticula. Upper left arrow points to ruptured diverticulum and paracolic abscess cavity filled with barium.

Diverticulitis

It seems not to be generally appreciated that anaerobes are almost invariably involved in infections associated with diverticulitis (Figs. 10.11 and 10.12). In Table 10.5 reports of involvement of anaerobes in various complications of diverticulitis are summarized. The significance of such infections is underlined by the frequency with which they predispose to anaerobic

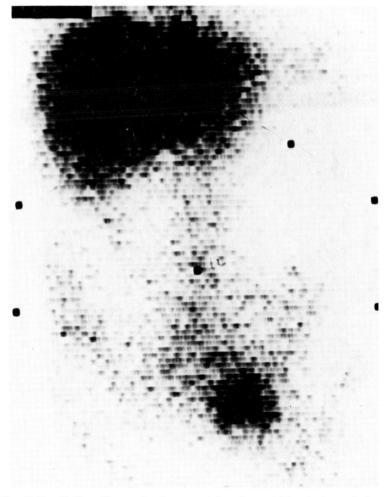

Fig. 10.12. Gallium-67 scan showing normal liver (upper left) and abscess in lower left quadrant. The abscess resulted from diverticulitis with rupture of a diverticulum.

TABLE 10.5 Involvement of Anaerobes in Complications of Diverticulitis

Number of cases	Complications	Anaerobes recovered	Aerobes recovered	Reference	Year
1	Abscess	Bacteroides melaninogenicus	E. coli, Pseudomonas aeruginosa	Weiss (80)	1943
1	Bacteremia	Bacteroides funduliformis		Ruys (1175)	1947
1	Bacteremia	Bacteroides		Herrell et al. (596)	1950
1	Liver abscess	Anaerobic gram-negative rod, filamentous		Knowles and Rinaldo (1381)	1960
1	Bacteremia	Fusobacterium		Tynes and Utz (1275)	1960
1	Bacteremia	Bacteroides, anaerobic streptococci		McHenry et al. (530)	1961
6	Various	Bacteroides	?	Bornstein et al. (275)	1964
1	Peritonitis	Clostridium perfringens, Eggerthella convexa	E. coli, Streptococcus faecalis	Beerens and Tahon-Castel (115)	1965
1	Perineal fistula	Actinomyces	E. coli, streptococcus (viridans)	McCarthy and Picazo (607)	1969
1	Endocarditis	C. perfringens		Felner and Dowell (289)	1970
1	Endocarditis	B. fragilis			
1	Bacteremia	Bacteroides		Bodner et al. (460)	1970
1	Abscess	B. fragilis, C. perfringens	E. coli, enterococcus	Sinkovics and Smith (411)	1970
1	Bacteremia	B. fragilis			
8	Bacteremia	Bacteroides		Marcoux et al. (420)	1970
14	Bacteremia	B. fragilis		Felner and Dowell (168)	1971
1	Bacteremia	B. variabilis			
2	Bacteremia	B. incommunis			
1	Bacteremia	Fusobacterium fusiforme			
1	Bacteremia	F. necrophorum			
1	Bacteremia	F. ridiculosum			
1	Peritonitis, abscess, bacteremia	Eubacterium lentum, Bifidobacterium sp.		Wilson et al. (105)	1972
3	Perforation, bacteremia	Anaerobic gram-negative rod			
1	Intraabdominal abscess	Bacteroides	Pseudomonas aeruginosa	Haldane and van Rooyen (529)	1972
1	Intraabdominal abscess	Bacteroides, anaerobic streptococcus, Clostridium			
1	Peritonitis, bacteremia	Bacteroides, C. perfringens			
1	Postoperative wound infection	Bacteroides, C. perfringens	Klebsiella, Citrobacter		
37	Intraabdominal abscess	20% Bacteroides, 11% anaerobic streptococcus	38% E. coli, 3% Klebsiella–Aerobacter	Altemeier et al. (527)	1973

bacteremia. Thus, Felner and Dowell (168) found that in their series of 250 cases of *Bacteroides* bacteremia 20 originated from diverticulitis. In the series of Wilson *et al.* (105) diverticulitis with perforation accounted for 6% of bacteremias due to anaerobic gram-negative bacilli. Marcoux *et al.* (420) found that 8 of 123 patients with *Bacteroides* bacteremia had diverticulitis as the portal of entry.

Pancreatic Infections

Pancreatitis does not usually result in necrosis of pancreatic tissue, but when this does occur there is risk of infection. The origin and route of bacterial contamination has not been established, but the fact that half of pancreatic abscesses grow multiple organisms with a heavy predominance of coliforms and other enteric species suggests that direct transmural penetration of the adjacent transverse colon may commonly occur [Warshaw (589)]. Similar organisms may also frequently be recovered by paracentesis during acute pancreatitis. It is surprising, therefore, that anaerobes have not been recovered more frequently from pancreatic abscesses. Again, this may reflect inadequate collection, transport, and anaerobic culture techniques. For example, Steedman *et al.* (590) found that 2 of 13 cases of pancreatic abscess showed no growth, even though frank pus was cultured. Felner and Dowell (168) reported two cases of *Bacteroides fragilis* bacteremia (one with microaerophilic streptococci also present) in patients with pancreatitis. Bodner *et al.* (460) felt that one of their cases of *Bacteroides* bacteremia was related to metastatic pancreatic carcinoma. *Bacteroides fragilis* bacteremia was noted in one acute pancreatitis patient each by Kodesch and DuPont (591a) and by Jones *et al.* (591b). At least one of these patients had a pancreatic abscess. Anaerobic streptococcal bacteremia was found by Davis *et al.* (1420g) in a patient who developed an infected pancreatic pseudocyst following retrograde cholangiopancreatography. Carey *et al.* (443a) recovered a microaerophilic *Streptococcus mitior* (*mitis*) from a pancreatic abscess. Okubadejo *et al.* (390a) recovered *Bacteroides* from a patient with acute hemorrhagic pancreatitis with abscess. Norris (591) reported a pancreatic abscess that yielded anaerobic cocci and pleomorphic anaerobic rods. Siler and Wulsin (592) recovered *Clostridium perfringens* from one case of hemorrhagic pancreatitis but found no growth from 6 of 14 specimens of suppurative pancreatitis. An interesting report by Altemeier and Alexander (593) concerns a chronic alcoholic with pancreatic abscess and a pancreatobronchopleural fistula and erosion into a large blood vessel. The abscess contained 500 ml of pus which, on culture, yielded anaerobic streptococci plus *Staphylococcus epidermidis*, the latter presumably a contaminant. Serrano-Rios *et al.* (594) recovered *Actinomyces israelii* from multiple abscesses in the head of the pancreas of a patient who also had liver abscesses that yielded the same organism. Actinomycosis involving the tail of the pancreas as a primary site was described by Weese and Smith (113a).

The question of whether prophylactic antimicrobial therapy would prevent pancreatic abscess following pancreatitis remains unsettled. Warshaw (589) pointed out that in the largest reported series of pancreatic abscesses all patients had been given tetracycline during the antecedent pancreatitis. Before surgery for pancreatic abscess, it is probably desirable to start antimicrobial therapy of a type that would cover both anaerobes and other enteric organisms.

Splenic Infections

The spleen may be involved with gas cysts in the course of visceral gas gangrene. The other type of anaerobic infection involving the spleen is abscess. A summary of reports of anaerobes in splenic abscess is given in Table 10.6.

Infections Related to the Stomach and Duodenum

A summary of the literature on involvement of anaerobic bacteria in various types of infections of the stomach is given in Table 10.7. Irvin *et al.* (595) reported a case of visceral gas gangrene in a diabetic following gastrectomy for carcinoma. These workers noted also that only 2 of 41 patients undergoing vagotomy and pyloroplasty or partial gastrectomy had *Clostridium perfringens* in the stomach postoperatively. They point out that the organism does not tolerate a pH less than 4.5. Herrell *et al.* (596) reported a case of *Bacteroides* septicemia secondary to carcinoma of the stomach, with perforation to the colon. Weyrich (452) described anaerobic staphylococcal sepsis in a patient with gastric carcinoma. Harvey *et al.* (212) reported three cases of primary actinomycosis originating in perforated gastric ulcers, as well as three cases of secondary involvement of the wall of the stomach in actinomycosis. Davies and Keddie (1420a) cited other reports of perigastric actinomycosis following a perforated gastric ulcer, gastric surgery, and a leaking duodenal stump. Okubadejo *et al.* (390a) described a *Bacteroides* wound infection following gastrectomy. Ishiyama *et al.* (498) found *Bacteroides*, *Sphaerophorus*, *Catenabacterium*, and *E. coli* in a purulent infection of an anastomosis following gastric resection. Gordon-Taylor [cited

TABLE 10.6

Anaerobes in Splenic Abscess

Anaerobes recovered	Aerobes recovered	Comments	Reference	Year
Fusiform		Multiple splenic abscesses and perisplenic abscess	Kaspar and Kern (1340)	1910
Identity uncertain	"Mixed"		Massini (259)	1913
Anaerobic streptococcus		Endocarditis	Bingold (442)	1932
Buday bacillus		Endocarditis	Schulz (159)	1935
Fusiform, anaerobic streptococci, anaerobic gram-negative rod			Wechsler (1422)	1935
Buday bacillus		Multiple bacteremia	Hornung (1130)	1940
Sphaerophorus necrophorus		Primary infection in genital tract	Alston (302)	1955
Gram-negative rod on smear, negative culture (no antibiotics)		500 ml foul pus	Podgorny (1423)	1971
Clostridium	0	Chronic myelogenous leukemia, splenic infarction	Rosenblum *et al.* (1423a)	1974
Anaerobe, not specified	?		Swenson *et al.* (384b)	1974
Bacteroides (most common isolate among those isolated from 16 patients with multiple splenic abscesses)	?		Gadacz *et al.* (1423b)	1974

by Cope (210)] reported a case of chronic peritoneal actinomycosis following a perforated gastric ulcer.

Felner and Dowell (168) reported three cases of bacteremia secondary to ruptured gastric or duodenal ulcers; two were due to *Bacteroides fragilis*, and one to *Bacteroides* CDC group F-1. Wright *et al.* (550) reported 16 cases of peritonitis due to ruptured gastroduodenal ulcers, five of which yielded anaerobes. There were three cases of anaerobic streptococci recovered in pure culture; the other two cases involved anaerobic streptococci, *Bacteroides melaninogenicus*, *Bacteroides funduliformis*, and *Clostridium tetanosporum*. One of the 16 cases was not cultured, and 4 yielded no growth. Schatten (549) reported three cases of anaerobic peritonitis following perforated gastric or duodenal ulcers with surgical repair. Baron *et al.* (596a) described *Bacteroides fragilis* bacteremia following gastroduodenectomy.

In 1966, Rose and Bukosky (597) reported a case of *Clostridium perfringens* peritonitis, sepsis, and visceral gas gangrene following perforation of a

TABLE 10.7 Involvement of Anaerobes in Infections of Stomach

Type of gastric involvement	Anaerobes recovered	Aerobes recovered	Comments	Reference	Year
Inflammatory thickening, mucosa	Fusiform	?	Purulent thrombi in gastric vessels	Kaspar and Kern (1340)	1910
Gastric ulcer	*Actinomyces*	?		Nathan (293)	1931
?	*Actinomyces*	?			
At site of gastric ulcer repair (perforation)	*Actinomyces*	?		Keynes [cited by Cope (210)]	1938
Phlegmonous, with gas formation	"Cultures negative at 48 hours"		No mention of bacteria in sections; presumptively anaerobic infection; "ruled out vascular problem and poisoning	Rosenthal and Tobias (1424)	1943
Emphysematous		*Proteus*	Fecal odor, gas bubbles, presumptively anaerobic infection	Weens (524)	1946
Ulcerative, necrotizing	Fusiform (spirochetes seen, not cultured)	*E. coli, Pseudomonas aeruginosa*, others	Stomatitis, glossitis, esophagitis as well; entire stomach involved, fusiform also cultured from peritoneal fluid, esophagus and mouth	Behrend *et al.* (1425)	1954
Acute necrosis	*Clostridium perfringens*			Fethers (501)	1959–1960
Emphysematous	*C. perfringens*			Morton and Stabins [cited by Gonzalez *et al.* (1426)]	1963
Emphysematous	*C. perfringens*		Secondary to carcinoma of stomach	Smith (1427)	1966
Intramural	*Actinomyces*	?	Authors found 7 similar cases in literature	Urdaneta *et al.* (1428)	1967
Emphysematous, perforation	Numerous clostridia (histologically)	?	Complication of phytobezoar	Lagios and Suydam (1429)	1968
Phlegmonous	*C. perfringens*		?Began with lesser curvature ulcer	Ringler and Tarbiat (1430)	1969
Subacute phlegmonous	*Bacteroides*	None	Dense fibrosis, patient on corticosteroids for ulcerative colitis	Smith (1430a)	1972

duodenal ulcer. Weiss (80) recovered *Bacteroides melaninogenicus* from two cases of abdominal infection secondary to perforated duodenal ulcer. Weens (598) reported a case of intraperitoneal abscess due to anaerobic strepto-coccus secondary to a perforated duodenal ulcer.

Enteric Infection

A number of workers have described intestinal spirochetosis or fuso-spirochetosis as a cause of diarrhea. The early literature on this subject is reviewed extensively by Smith in his book published in 1932 (290). This condition is said to occur primarily in patients with vitamin deficiency, especially vitamin C deficiency, and patients with other underlying debili-tating diseases, particularly those related to the gastrointestinal tract. Many of these patients were said to have had bloody diarrhea. The earlier works suggested that arsenicals and vitamin replacement constituted effective therapy. In some cases, distinctive proctoscopic findings were noted. Included were ulceromembranous changes, patches of pseudomembrane [de Lavergne and Florentin (599)] and a particular type of granular proctitis called the "strawberry lesion" [Shera (600)]. One of the problems in attempting to assess the significance of spirochetes in diarrheal diseases is the fact that these organisms are found fairly commonly in man and in various animals, even in the absence of intestinal symptomatology or disease [Macfie and Carter, (601), Smith (290)]. R. M. Smibert (personal communication) has found very fastidious anaerobic spirochetes normally in feces in counts of 10^6/gm of stool. To attach any clinical significance to intestinal spirochetes, one would need to demonstrate in a controlled fashion, in which the examiner knew nothing of the history of the patient from whom he received specimens, that there was a significantly greater incidence, or higher count, or both of spirochetes in patients with diarrhea than in a group of controls without any gastrointestinal symptoms or disease. These studies should include cultural recovery of the organism, as well as microscopic data, if possible. In terms of evaluating therapy, this should be done in a double-blind ran-domized type of study employing active drug and placebo. Leach *et al.* (601a) offered a possible explanation of intestinal spirochetosis. They noted in rats that mucosa-associated spiral organisms appeared in the stool in large numbers when diarrhea was induced with magnesium sulfate.

Another type of interesting condition of the large bowel related to spiro-chetes is described by Harland and Lee (602) and by Lee *et al.* (603). In this condition, the surface epithelium is infested by short spirochetes that are oriented in the long axis of the cells and are adherent to the surface of the epithelium. The spirochetes occasionally penetrate short distances into the epithelial cells, brushing the plasma membrane before them. The organisms are surrounded by a cluster of microvilli which seems to account for the consistent longitudinal orientation of the organisms. These spirochetes stain readily with hematoxylin and eosin and are periodic acid–Schiff-positive (PAS-positive), suggesting that they are *Borrelia* rather than *Treponema*. The organisms are best visualized by electron microscopy, but their presence can be detected by light microscopy. In one patient, the organisms were found both in the appendix and subsequently in a rectal biopsy. There is also evidence that the spirochetosis may persist for considerable periods of time as in one patient five biopsies taken over a period of 6 years all showed presence of the organisms. Of 790 appendices removed surgically, 7.8% showed these spirochetes with no particular difference in incidence between cases of acute appendicitis and appendices removed incidentally during the course of other surgery. Seven of 14 patients in whom these spirochetes were demonstrated on rectal biopsy had carcinoma, but the carcinomatous lesions themselves were not colonized with the spirochetes. Except for two patients with diarrhea, symptoms found in patients showing spirochetosis could readily be attributed to the presence of carcinoma or other underlying disease.

Two cases of actinomycosis involving the large bowel were reported by Olsson (603a). A case of sigmoid actinomycosis is noted by Davies and Keddie (1420a).

Fethers (501) described a patient with acute inflammatory necrosis of the jejunum, with a 12-inch segment showing gangrenous changes. There was no evidence of involvement of vessels to explain the condition. Large numbers of gram-positive rods were noted in the mucosa histologically, and the culture yielded *Clostridium perfringens*. Barrett (506) described another case of enteritis necroticans associated with gas embolism of the portal vein branches within the liver. This patient also had intestinal gangrene, and *Clostridium perfringens* was cultured from liver slices. Jarkowski and Wolf (604) described a patient with acute lymphatic leukemia treated with prednisone and 6-mercaptopurine. This patient ultimately died, and at autopsy, in addition to peritonitis and visceral gas gangrene, was found to have confluent ulceration of the mucosa of the ileum and cecum with hemorrhagic infarction. The bowel was loaded with *Clostridium perfringens*. Kuklinca and Gavan (491) described two patients with bacteremia. One of these had *Clostridium perfringens* in the blood culture and hemorrhagic colitis, with disseminated intravascular coagulation. The second patient's blood culture yielded *Bacteroides serpens*, and this patient had a necrotizing enteropathy. Amromin and Solomon (605) noted necrotizing enteropathy in 63 of 280 leukemia patients and in 6 of 128 lymphoma patients. They acknowledged that there are many causes for this, but speculated that intestinal anaerobes

TABLE 10.8

Anaerobes in Perirectal and Similar Infections

Process	Number of cases	Anaerobes recovered	Aerobes recovered	Comments	Reference	Year
Perirectal abscess	1	Anaerobic gram-negative rod			Russ (1431)	1905
Periproctitis		Fusiform			Ghon and Roman (543)	1916
Ischiorectal abscess	2	*Bacteroides melaninogenicus*			Shevky et al. (281)	1934
Perirectal abscess	9	Anaerobic streptococcus			Sandusky et al. (112)	1942
Perirectal abscess	28	Anaerobic streptococcus (27) *Clostridium* (3) Anaerobic gram-negative rod (5) Microaerophilic streptococcus (2)	Mixed			
Ischiorectal abscess	1	*B. melaninogenicus*, *Propionibacterium*			Weiss (80)	1943
Abscess adjacent to rectal carcinoma		*Bacteroides*, anaerobic streptococcus			Beigelman and Rantz (282)	1949
Multiple abscesses near sacrum		*B. funduliformis*, anaerobic streptococcus, anaerobic gram-negative coccus		Followed hemorrhoid surgery; local infection, fistula to bladder; widespread infection ultimately fatal	Lodenkämper (1432)	1949
2 perirectal abscesses		*B. funduliformis*	*Streptococcus*		Hartl (249)	1950
Perianal abscess		*Pasteurella serophila*			Bokkenheuser (1433)	1951
Six cases periproctitic abscess		*B. funduliformis*	Mixed		Ernst (347)	1951
Multiple isochiorectal abscesses		*Bacteroides*, anaerobic streptococcus		Postoperative, vaginal hysterectomy	Clark and Wiersma (654)	1952

(continued)

309

TABLE 10.8 (*continued*)

Process	Number of cases	Anaerobes recovered	Aerobes recovered	Comments	Reference	Year
Perirectal abscess		*Bacteroides*	Few *E. coli*	Complicating carcinoma of vulva	Carter *et al.* (653)	1953
Ischiorectal abscess		Anaerobic gram-negative rod			Henderson (533)	1953
Gangrenous phlegmon ischiorectal fossa		*Ramibacterium pleuriticum*			Beerens (332)	1953–1954
Perianal abscess		*Actinomyces israelii*			Herrell *et al.* (331)	1955
Perianal abscess		Anaerobic streptococcus, *Sphaerophorus*			Morin and Potvin (145)	1957
Perianal abscess		Anaerobic streptococcus, *Ristella trichoides*				
Perianal abscess		*Ristella* (?*putredinis*)				
Ischiorectal abscess	1	*Ramibacterium pleuriticum*, *Sphaerophorus gulosus*	*Streptococcus*		Prévot *et al.* (314)	1958
Perianal abscess	1	*Bacteroides*			Witebsky and Miller (162)	1958
"Rectal" abscess	1	Fusiform	Mixed		Pereira (1434)	1958
2 cases periproctitis		*B. melaninogenicus*	?		Heinrich and Pulverer (116)	1960
4 cases anal abscess		*B. melaninogenicus*	?			
Perirectal abscess	3	Anaerobic gram-negative rods			Shoemaker (531)	1960
Perirectal abscess	3	Anaerobic gram-negative rods	Mixed			
Pelvic gas gangrene		*C. perfringens*	?	Secondary to infected hemorrhoids	Brummelkamp *et al.* (1002)	1963
Pararectal abscess	1	*B. putidus*			Rentsch (195)	1963
	1	*S. funduliformis*				
	1	*S. putridus*				
	1	Anaerobic micrococcus				
Perianal abscess	1	*Eggerthella convexa*	*E. coli*, *S. faecalis*, paracolon	Small connection with rectum	Schaffner (1240a)	1963

abscess					
Periproctitic abscess	1	*Bacteroides*		Dietrich *et al.* (135)	1965
Perianal abscess	1	*E. convexa*		Beerens and Tahon-Castel (115)	1965
Perianal abscess	1	*Veillonella parvula*, anaerobic staphylococcus (2)			
Perianal abscess	1	*E. convexa*, anaerobic streptococcus, anaerobic diplococcus, *B. bifidum*			
Perianal abscess	4	*E. convexa* (4) / Anaerobic streptococcus or cocci (4) / *C. perfringens* (2) / *E. limosum* (1) / *B. bifidum* (1) — Mixed			
Perianal fistulae	12	*Eggerthella* (8) / Anaerobic streptococcus (7) / Anaerobic cocci (3) / *Eubacterium* (2) / Fusiform (1) / *Corynebacterium* (1) / *Clostridium* (6) — All mixed—Enterobacteriaceae and 2 enterococci			
Periproctitic abscess	1	*Eggerthella*		Werner (523)	1967
Large pararectal abscess		*Bacteroides*		Nettles *et al.* (541)	1969
Rectal abscess		*Bacteroides* (blood culture)	Bacteremia 2° to rectal abscess	Bodner *et al.* (460)	1970
Perianal abscess		*Bacteroides* (blood culture)	Complicating bacteremia	Sinkovics and Smith (411)	1970
Perianal abscess	1	*S. funduliformis*		Farquet *et al.* (723)	1971
Ischiorectal abscess	1	*S. funduliformis*, anaerobic streptococcus			
Perianal abscess	3	1 *Bacteroides* / 2 *Peptococcus* — 3 *E. coli*		Ohnishi *et al.* (582)	1972
Perirectal abscess	6	*Bacteroides* — ?		Beazley *et al.* (394)	1972

(continued)

TABLE 10.8 (*continued*)

Process	Number of cases	Anaerobes recovered	Aerobes recovered	Comments	Reference	Year
Perirectal abscess	1	*Bacteroides*			Haldane and van Rooyen (529)	1972
Perirectal abscess	1	*B. fragilis* ss. *fragilis*, *Clostridium difficile*			Danielsson et al. (1435)	1972
Sacral abscess	1	*Actinomyces*	None		Davies and Keddie (1420a)	1973
Fistula-in-ano	11	*Bacteroides* 1 case, *C. perfringens* 1 case	Several	Other cases probably had anaerobes as well (foul odor, organisms seen on gram stain not recovered on culture)	Marks et al. (1435a)	1973
Ischiorectal abscess	11	*Bacteroides*	?	No case details	Mitchell (110a)	1973
Perianal abscess	4	*Bacteroides*	?			
Perirectal abscess	8	*Bacteroides*	?	One had bacteremia	Nobles (56d)	1973
Ischiorectal abscess	2	*Bacteroides*	?		Okubadejo et al. (390a)	1973
Perianal abscess	3	*Bacteroides*	?			
Ischiorectal phlegmon	2	*C. perfringens*	?		Larcan et al. (1435b)	1974
Ischiorectal abscess	2	Anaerobes, type not specified	?		Sabbaj and Villanueva (557a)	1974
Perirectal abscess, myonecrosis, bacteremia	1	Anaerobes, type not specified	?		Chow et al. (1420e)	1974
Ischiorectal and rectal abscess	5	*Bacteroides fragilis* + ?	?		Leigh (1420f)	1974
Perianal abscess	4	*Bacteroides fragilis* + ?	?			
Perianal abscess	1	*Clostridium* + 2 other anaerobes	0		Gorbach and Thadepalli (486)	1975
Perianal abscess	1	*Actinomyces*	?		Weese and Smith (113a)	1975
Rectal abscess	1	*Peptococcus niger* + ?	?		Wilkins et al. (1435c)	1975

may contribute to the lesion and/or invade it secondarily. In this same connection one might wonder whether toxic megacolon, a dreaded complication of idiopathic ulcerative colitis, might not similarly involve anaerobes from the bowel flora. Chapter 17 contains a more detailed commentary on enteritis necroticans.

Perineal, Perirectal, and Similar Infections

Anaerobic bacteria are very commonly involved in these infections. Bornstein *et al.* (275) reported a perineal wound infection due to anaerobic streptococci. Henriksen (606) found an anaerobic gram-negative bacillus with unusual spreading colonies from a case of perineal abscess. Morin and Potvin (145) recovered *Ristella* from a perineal abscess.

Actinomycosis has been reported several times as a cause of perineal lesions. McCarthy and Picazo (607) reported recovery of *Actinomyces* along with *E. coli* and viridans streptococci from a case of diverticulitis with perineal fistula formation. Schubert and Tauchnitz (318) recovered *Actinomyces* from a patient with multiple perineal and buttock abscesses and fistulae of 30 years' duration. Fry *et al.* (608) described a case of primary actinomycosis of the rectum with a large mass in the rectal wall almost occluding the rectum. This patient had multiple perineal and perianal fistulae from which *Actinomyces israelii* was grown. Brewer *et al.* (608a) described 9 cases of primary anorectal actinomycosis.

Ohnishi *et al.* (582) recovered anaerobes (*Bacteroides* plus two *Peptococcus* strains) from three of 17 cases of perianal abscess. Table 10.8 is a summary of the literature concerning anaerobes involved in perirectal and similar infections. Our experience is that perirectal and perianal abscesses routinely yield multiple anaerobes and facultatives in mixed culture.

11

Urinary Tract Infections

Certain aspects of urinary tract infection involving anaerobic bacteria are considered in other chapters. Infections of the perineal area, such as Fournier's gangrene, are covered in Chapter 13. Diseases of the genitalia and infections of the scrotum, groin, and buttock are considered in Chapter 16.

Spirochetal Urinary Tract Infection

A number of authors have discussed urinary tract infection of various types due to spirochetes. This type of infection, however, is not well documented. Among the types of infection which have been ascribed to spirochetes are urethritis, abacterial pyuria, and chronic prostatitis (609–617). Some of these infections have been thought to respond to treatment with arsenicals and others have not.

On the other hand, Stoddard (618) found spirochetes in the urethras of 44 of 100 men and noted there was no correlation between the presence of these organisms and evidence of disease. He concluded therefore that the diagnostic significance of this finding was questionable. Kon and Watabiki (619) found spirochetes in hyaline casts and hyaline substances in the renal tubules, again without any relation to disease. Castellani (620) noted that in spirochetal urethritis, the discharge may be abundant and frankly purulent and contain enormous numbers of spirochetes. However, he also pointed out that spirochetes may be found in the normal urethra as well.

Anaerobes in Voided Urine

Many workers have reported recovery of anaerobic bacteria from voided urine. A number of these reports fail to comment on the lack of validity of such isolations. Often, little or no clinical information is given, but at times evaluation was made and there was no clinical evidence of infection in patients whose voided urine yielded anaerobes.

In 1952, McVay and Sprunt (621) reported seven cases with isolation of *Bacteroides* from urine. In four cases, the organisms were obtained in pure culture from catheterized urines. In only two (possibly three) of these cases were *Bacteroides* of possible clinical importance. The authors concluded that this group constituted their most ill-defined and clinically insignificant category of *Bacteroides* infections. Also, in 1952, Metzger *et al.* (622) cultured urine from patients with various types of acute and chronic urinary tract infection. *Bacteroides* were recovered from three of 113 cases, microaerophilic staphylococci from 2.6%, and anaerobic streptococci from 11.5%. There was no documentation of any possible significance of these anaerobic isolates. In a study of over 6000 positive urine cultures, Lutz *et al.* (623) found few anaerobes (a total of nine strains). Pyuria was noted in about half of these cases. In 1961, Ambrose *et al.* (45) found *Bacteroides* as normal flora in 1 urine specimen of 46 from males without known genitourinary infection. Dietrich *et al.* (135) noted a total of 74 isolates of anaerobes from urine. These were recovered in either pure or mixed culture; no clinical or other details were given. A survey of 15,250 consecutive urine cultures carried out by Headington and Beyerlein (624) yielded a total of 158 anaerobes from 147 patients. There was no good evidence for clinical infection in these cases. Slotnick and Mackey (625) performed 810 urine cultures on 600 individuals. From these, they recovered anaerobes from 41, five in pure culture. Takase [in Kozakai and Suzuki, (265)] studied 84 patients with cystitis. Anaerobes (26 strains) were isolated from 24 patients. An anaerobe was also isolated from a patient with pyelonephritis. In all cases, the specimen was voided urine. There were no clinical details and relatively little bacteriological information in this particular report.

A rather unusual study was reported by Rosner in 1968 (626). In this study, patients who had negative results after 48 hours of incubation of conventional urine cultures, had duplicate specimens incubated for a prolonged period (7 days). Each patient had two successive daily cultures and only those situations in which 1000 or more organisms per milliter of the same type were recovered on both cultures were considered positive. Only two types of organisms were recovered from these cultures with prolonged incubation—an obligately anaerobic streptococcus and a microaerophilic streptococcus. A total of 233 patients out of 1386 studied were positive for one or the other of these two organisms. Three different categories of patients, in approximately equal numbers, were studied. The first group consisted of patients who had been treated 2 weeks earlier for a urinary tract infection with aerobic or facultative organisms; 50 patients in this group were positive for anaerobes on prolonged incubation. The second group consisted of patients with no evidence of urinary tract infection; 20 of these showed anaerobes on prolonged incubation. The final group was a group of patients with a clinical diagnosis of chronic pyelonephritis; 153 of these yielded anaerobes on prolonged incubation. Subjects in the first two groups yielded only microaerophilic streptococci, whereas those in the latter group yielded either anaerobic streptococci or microaerophilic streptococci.

Lodenkämper (627) found anaerobes in 96 urine samples, 1.7% of all samples studied. In 51 cases, anaerobes were found to the exclusion of aerobes (0.9% of samples), whereas the rest were mixed with aerobic or facultative organisms. However, the latter were always present in counts of less than 10^5/ml. In those cases in which anaerobes were recovered to the exclusion of other forms, the counts ranged from several thousand to greater than 500,000/ml. Kuklinca and Gavan (628) studied 200 urine specimens at random. Ninety-three of these grew aerobic or facultative forms by standard techniques and were not studied further. Sixty-six others had growth in heavily inoculated thioglycolate broth (0.5 ml of urine plus 7.5 ml of broth). Fifty-six grew aerobes, and 10 grew anaerobes. Analysis of the clinical records of these 10 patients revealed that nine had no evidence of urinary tract infection. The tenth patient did but a second urine culture, 1 month later, in his case yielded a different anaerobic organism. In a study of 471 urine cultures, Bittner et al. (629) recovered a total of five strains of Clostridium (including 1 C. perfringens). There was no evidence that these organisms were involved in infection. Caselitz et al. (630) did urine cultures on 877 patients. Ten percent of males and 8.4% of females yielded anaerobes, always in mixed culture with aerobic or facultative forms.

Nakamura et al. (631) recovered anaerobes from 8.7% of 253 positive urine cultures; 8 of 22 were in pure culture. In 1971, Nakamura et al. (114) noted 9 anaerobic isolates among 50 positive urine cultures. Furuta et al. (632) recovered anaerobes from 47 of 275 persons. A 20% incidence of anaerobes from 238 urine samples from patients with urinary tract infections was noted by Kumazawa *et al.* (633). Among 110 urines from patients without urinary tract infection, only 10% yielded anaerobes. The anaerobes were isolated only from patients with chronic infections (cystitis and pyelonephritis). Mitchell (110a) recovered *Bacteroides* from 17 of 114 urines. All isolates were from females. In 7 cases, there were significant numbers of aerobes present also. In 6 of the remaining 10, there were very few leukocytes present. Shimizu *et al.* (633a) recovered anaerobes (98 strains) from 40 of 397 urine specimens. The anaerobes were present in counts of $> 10^5/$ml twelve times and in pure culture twice. The authors commented that the anaerobes recovered were similar to those found normally in the anterior urethra. Shimizu and Mo (1466b) recovered anaerobes from 42 of 423 urine specimens. Among 95 cases with urinary tract "inflammation," 27% yielded anaerobes on urine culture. Kumazawa *et al.* (633b) recovered anaerobes from 10% of urines of 110 individuals without urinary tract infection and from 20% of 228 patients with such infection. They also documented the presence of anaerobic bacteria of a number of different types in the urethra (by swab or suction culture).

Pien *et al.* (390) noted two isolations of anaerobic gram-positive cocci from the urine of women, both in counts of less than 1000/ml; they concluded that these probably represented contamination. A report by Alling *et al.* (634) concerned repeated urine cultures on hospitalized geriatric patients. There was a marked spontaneous variation in the number of anaerobes (and aerobic or facultative organisms) from week to week. Anaerobes were recovered from 6 of 13 patients without catheters and 25 of 31 patients with catheters. The incidence of anaerobes was higher in females (26 of 28) than in males (5 of 16). In many cases, counts of anaerobes recovered ranged from 10^5 to $5 \times 10^7/$ml. Nobles (56d) mentioned 3 urinary tract infections in his series of *Bacteroides* infections, but gave no details.

Certain data on the presence of anaerobes, including spirochetes, as part of the normal flora of the urethra has already been presented. Other studies may be cited concerning presence of anaerobes as normal urethral flora. Jungano and Distaso (283) studied the normal urethral flora of 17 people. One of these yielded no organisms. From the others there were 17 aerobes and 32 anaerobes recovered. Included among the anaerobes were *Micrococcus foetidus* (8 strains), *Clostridium perfringens* and *Bacillus neigeux* (5 strains each) *Clostridium ramosum* (3), other clostridia (2) and *Bacillus thetoides*, (1). Bran *et al.* (44) recovered a *Bacteroides* species and an anaerobic gram-positive bacillus from six cultures of normal urethras. Stamey (635) commented that *Bacteroides* is commonly recovered from "urethral urine" of normal females in counts of 100 to 1500 organisms per milliliter.

Prostatic Cultures

The study of prostatic secretions is subject to the same difficulties already cited in culturing voided urine. Thus, one cannot judge the significance of a report such as that of Ghormley *et al.* (636) in which the prostatic secretions of 105 patients were cultured; 7 anaerobes were recovered, 4 in pure culture. Ambrose *et al.* (45) found *Bacteroides* as normal flora in prostatic secretions on two occasions among 46 males without known genitourinary tract disease. There was a similar incidence on 45 males with urethroprostatitis. Anaerobes were recovered from prostatic tissue on two occasions among 34 males who underwent transurethral resection of the prostate or open simple prostatectomy. Three patients undergoing radical perineal prostatectomy for adenocarcinoma afforded the best opportunity for an uncontaminated sampling of the flora of the prostate. Two yielded *Bacteroides* and one an anaerobic *Corynebacterium*, all together with aerobic diphtheroids, streptococci, and/or micrococci. The seminal vesicles of one of these patients had a similar flora. The authors conclude that a varied and profuse flora may exist and that infections of the area may occur as the result of altered host resistance or organism virulence. Justesen *et al.* (636a) studied 23 patients with chronic prostatitis and 23 with chronic urethritis for the possibility of anaerobic infection. Anaerobes were found in the urine in 1 patient but anaerobes were not localized with certainty to the prostatic tissue or secretions or to the urethra. Mårdh and Colleen (636b) studied a variety of specimens from 79 patients with chronic prostatitis and 20 healthy controls and failed to find evidence for a role for anaerobes. Of possible significance, however, was the recovery of peptostreptococci from the ejaculates of 7 patients. Several cases of suspected anaerobic prostatic infection are noted in Table 11.1; two cases of prostatic abscess with accompanying anaerobic bacteremia are included.

Documented Infections

It is important to remember that anaerobes are found as normal flora in the urethra. Accordingly it is seldom satisfactory or reliable to obtain voided urine specimens for diagnosis of urinary tract infection due to anaerobic bacteria. In the study reported by Finegold *et al.* (46), anaerobes were recovered from 14 of 100 random urines (kept for several hours under conditions that might have been unfavorable for anaerobes). Relatively high counts of anaerobes were recovered in certain cases. In follow-up studies, these authors failed to recover any obligate anaerobes from 19 specimens of urine obtained by percutaneous bladder aspiration, whereas

TABLE 11.1

Anaerobic Infections of the Urinary Tract

No. of cases	Source of culture	Types of urinary tract infections	Anaerobic organism(s)	Aerobic or facultative organisms	Documentation (or comments)	Reference	Year
1	Pus	Periurethral abscess	Anaerobic coccus	*Streptococcus pyogenes*	Good	Veillon (4)	1893
1	?	Perinephric abscess	*Micrococcus foetidus*	?	?	Veillon [cited in Jungano and Distaso (283)]	1897
1	Pus	Pyonephrosis	Anaerobic coccus	*E. coli*, streptococci	Fetid pus, necrosis, miliary cortical abscesses	Albarran and Cottet (1436)	1898
1	Pus	Renal abscess secondary to renal tuberculosis	*Bacteroides*	*Mycobacterium tuberculosis*			
1	Urine	Cystitis–bladder neoplasm	Anaerobic coccus	Aerobes present	Fetid urine		
11	Pus	Periurethral cellulitis and abscess	*Bacteroides funduliformis* (3), *Bacteroides fragilis* (5), other *Bacteroides* (5), anaerobic staphylococci (3), other anaerobic cocci (7)	Present in 5 cases— in smaller numbers than anaerobes	Pus fetid in all but one case; in 4 cases, anaerobes in pure culture	Cottet (12)	1899

(continued)

319

TABLE 11.1 (*continued*)

No. of cases	Source of culture	Types of urinary tract infections	Anaerobic organism(s)	Aerobic or facultative organisms	Documentation (or comments)	Reference	Year
9	Pus	Periurethral cellulitis and abscess, some with gangrene and some with fistulae (multiple)	*B. funduliformis* (1), *B.fragilis* (1), other *Bacteroides* (2), anaerobic staphylococci (1), *Diplococcus reniformis* (4), other anaerobic cocci (4), anaerobic gram-positive bacilli (2)	Aerobes or facultatives present in 6, in equal numbers to anaerobes in 2	Anaerobes in pure culture (3), anaerobes predominant (4)	Albarran and Cottet (1437)	1900
2	Urine	Cystitis	*D. reniformis* (2), *Bacteroides* (1)	Rare aerobes in one case	One case, with no aerobes, also had no good evidence for cystitis		
1	Pus	Prostatic abscess, periprostatic phlegmon	*Clostridium perfringens*	*Staphylococcus* (type?)	Significance uncertain		
3	Pus	Pyonephrosis, calculi	*D. reniformis* (3), fusiforms (2), *B. fragilis* (1), *B. ramosus* (1)	Third case was mixed with aerobes	One pure; one had only few *Staphylococcus albus*		
1	Many organs other than kidney (abscesses)	Small metastatic renal abscess	*Bacterium halosepticum*	None	Leg injury with infection and then widespread metastases and death	Wyss (1438)	1904
1	Urine?	Cystitis, prostatitis	"*Bacillus nevosus*"	None	Fetid urine?	Jungano (1439)	1907
1	Pus?	Cowperitis	"*Bacillus nevosus*"	None			

No.	Specimen	Condition	Organisms	Aerobes	Results	Reference	Year
2	Pus?	Periurethral gangrene	"*Bacillus nevosus*" (1), anaerobic staphylococcus (1)	None		Jungano (1439)	1907
1	Pus?	Renal abscess	"*Bacillus nevosus*"	None			
1	Pus?	Pyonephrosis	"*Bacillus nevosus*"	None			
1	Pus?	Chronic urethritis	*M. foetidus*	None			
1	Urine?	Cystitis	Anaerobic staphylococcus	None	Fetid urine?		
1	Urine, left kidney (by ureteral catheter)	Pyelonephritis, calculi	Anaerobic staphylococcus; "*Bacillus albarrani*" (*Bacteroides*)	*E. coli*, gram-positive rod	Inadequate, urine from left ureter, mixed aerobes and anaerobes	Jungano (1440)	1907
11	Urethral secretion	Chronic urethritis	*C. perfringens* (4), *B. fragilis* (1), other *Bacteroides* (1), anaerobic coccus (3), *B. ramosus* (2), anaerobic "*Streptothrix*" (1)	Absent in some cases	Poor	Jungano (1441)	1908
3	Urethral secretion following gland massage	Chronic cowperitis	Anaerobic staphylococcus (2), anaerobic coccus (1), *B. ramosus* (2), *Bacteroides* (1)	None	Poor		
8	Pus	Periurethral suppuration, abscess—penile and perineal, some gangrenous, some gas-forming	Bifid bacillus (1), *D. reniformis* (1), anaerobic staphylococcus (3), anaerobic coccus (6), *Bacteroides* sp. (2), *C. perfringens* (6), *B. ramosus* (4)	Absent or in smaller numbers in most cases	Generally good, in most cases anaerobes were either predominant or in pure culture		

(continued)

321

TABLE 11.1 (*continued*)

No. of cases	Source of culture	Types of urinary tract infections	Anaerobic organism(s)	Aerobic or facultative organisms	Documentation (or comments)	Reference	Year
2	Secretion after prostatic massage	Chronic prostatitis	*M. foetidus* (2), *Clostridium* (1)	Yeast, *S. albus*	Poor	Jungano (1441)	1908
4	Urine	Chronic cystitis	Anaerobic staphylococcus (1), anaerobic coccus (6), *Bacteroides* (1), *C. perfringens* (1)	Other 3 cases all had *E. coli* and 2 had *Proteus*	Good in 1 case only— anaerobic cocci isolated from both urine and testicular abscess		
1	Urine, right ureter	Pyelonephritis	*C. perfringens*, anaerobic coccus, *B. fragilis*	*Staphylococcus albus*			
1	Pus	Pyonephrosis, perirenal abscess	*C. perfringens*	Several aerobes, including *E. coli*, also present	Foul-smelling pus		
2	Pus (at time of nephrectomy)	Tuberculous and mixed pyonephrosis	*C. perfringens* (2), *B. fragilis* (1), other *Bacteroides* (2), anaerobic staphylococcus (1), anaerobic coccus (1), *B. ramosus* (1)	Aerobes, other than *M. tuberculosis*, also present	Fetid pus		
1	Pus	Acute urethritis	Anaerobic staphylococci	*S. albus*	Poor	Jungano (1442)	1908
1	Kidney	Gas gangrene bladder, scrotum, penis, kidneys (with probable sepsis as well)	*Clostridium* sp.	See documentation	Excellent documentation, organisms seen in tissues chiefly in absence of other organisms	Babes and Babes (1443)	1909

No.	Source	Diagnosis	Organism	Other organisms	Notes	Reference	Year
1	Pus	Large abscess inferior pole left kidney	Fusiforms and spirilla (smears only)	*E. coli*, 2 types of cocci, *Bacillus*	Fatal widespread infection	Costà (1223)	1909
1	Urine Kidney Bladder Seminal colliculus	Cystitis and pyelonephritis	Bifid bacillus	Paracolon (all sites)	Good, organisms demonstrated in tissues histologically and by culture	Rach and von Reuss (1444)	1909
1	Urine	Cystopyelitis	Anaerobic streptococcus	None	Pyuria, fever; urine contained the anaerobe repeatedly, urine fetid when bloody.	Schottmüller (258)	1910
1	Blood?	Metastatic abscess(es?) kidney	Fusiform bacillus	?	Periproctitis, bacteremia and endocarditis with multiple abscesses at autopsy	Ghon and Roman (543)	1916
1	Pus	Renal abscess	Fusiform bacillus	None	Large numbers of similar organisms seen on direct smear	Mellon (1445)	1919
1	Urine, septic renal vein thrombus?	Septic renal vein thrombophlebitis, bilateral, involvement of renal cortical parenchyma	Anaerobic streptococci	None?	Clot had putrid odor, urine consistently grew anaerobic streptococci	Bingold (659)	1921
1	Renal abscess?	Renal abscesses	Anaerobic streptococcus	None?	Pelvic abscess blocking left ureter		

(continued)

TABLE 11.1 (*continued*)

No. of cases	Source of culture	Types of urinary tract infections	Anaerobic organism(s)	Aerobic or facultative organisms	Documentation (or comments)	Reference	Year
3	Urine	Vesicovaginal fistulae	Anaerobic streptococcus	None?	Underlying diseases—Douglas' abscess, parametritis, uterine abscesses, repeated urine cultures positive in 1 case	Bingold (659)	1921
1	Urine	"Suspected focal infection of kidney"	B. melaninogenicum	?	Poor—no further details	Oliver and Wherry (1446)	1921
1	Pus, blood	Small renal abscess	B. funduliformis	None	One of a number of abscesses in septic patient	Teissier et al. (1368)	1929
1	Catheterized urine	None?	Micrococcus niger, ? Bacteroides	S. faecalis, S. albus	Poor documentation, no pyuria	Hall (1447)	1930
1	Pus	Metastatic renal abscess	Buday bacillus	None	Good	Hegler and Nathan (1375)	1932
1	Blood	Bacteremia related to cancer of bladder and suprapubic drainage?	B. fragilis	None	Poor	Thompson and Beaver (1448)	1932
1	?	Cystitis emphysematosa	C. welchii	None		Antoine [cited in Lee (1449)]	1934
1	?	Cystitis emphysematosa	C. welchii	E. coli		Redewill [cited in Lee (1449)]	1934

324

(continued)

		infection					
						and Leiter (391)]	
5	?Kidneys at autopsy	Metastatic kidney infection during sepsis	1 Anaerobic gram-negative rod, 1 *S. putrificus*, 1 *S. putrificus* + anaerobic gram-negative rod, 2 other anaerobes	None	nephrotomy for stone, 1 following nephrectomy None?	Weyrich (452)	1936
1	Blood	Abscesses, left lobe of prostate (? metastatic)	*B. funduliformis*	None?	Carcinoma of rectum, bacteremia due to *B. funduliformis*. Multiple abscesses at autopsy; no mention of direct culture of abscesses	Dixon and Deuterman (410)	1937
1	Blood	Periurethral abscess	Anaerobic streptococcus	None	Secondary pelvic cellulitis, marked necrosis, many cocci on smear	McDonald *et al.* (534)	1937
1	Pus?	Perinephric abscess, pyonephrosis ?	"Anaerobic organisms"	?		Doering [cited in Mencher and Leiter (391)]	1938
1	Urinary calculus	?	*C. welchii*	None?	Organisms obtained from center of calculus after flaming, crushing	Ferrier and Bliss [cited in Mencher and Leiter (391)]	1938

TABLE 11.1 (*continued*)

No. of cases	Source of culture	Types of urinary tract infections	Anaerobic organism(s)	Aerobic or facultative organisms	Documentation (or comments)	Reference	Year
1	?	Cystitis emphysematosa	C. welchii	None		Levin [cited in Lee (1449)]	1938
1	Urine	Pyelonephritis, periprostatitis	C. welchii	?	Poor	Mencher and Leiter (391)	1938
1	Urine	Necrotizing cystitis	C. welchii	?	Followed laparotomy		
3	Urinary calculus	?	C. welchii	None?	Organisms obtained from center of calculus after flaming, crushing		
11	Wound (11), blood also (1)	Postoperative wound infection	C. welchii	None	Well documented, 4 cases nephrectomy, 2 prostatectomy (suprapubic), 2 ureterolithotomy, and 1 each pyelotomy and nephrotomy, nephroureterectomy, and drainage renal cysts and cystostomy		
3	None	Postoperative tetanus	C. tetani	None	All followed two stage prostatectomy		

No.	Source	Condition	Organism	Associated organisms	Documentation	Reference	Year
1	Urethra, bladder	Cystitis?	C. welchii	Staphylococcus, streptococcus	Poor	Weiser [cited in Mencher and Leiter (391)]	1938
1	Urine, right ureter; pus, post-nephrectomy wound infection	Pyelonephritis, right followed by postnephrectomy wound infection	B. fragilis, anaerobic streptococcus	None	Well documented, urine from left ureter and bladder negative	Schulte (1450)	1939
1	Pus	Pyonephrosis	Bacterium symbiophiles	None	Nephrectomy, organism seen in large numbers on direct smear	Schultz (1451)	1939
1	Pus	Perinephric abscess	C. welchii	None	Foul odor, gas, well documented	Madison (1452)	1940
1	?	Postoperative infection	C. welchii	?		Lazarus [cited in Willis (1453)]	1942
1	?	Cystitis emphysematosa	C. welchii	None		Burns [cited in Lee (1449)]	1943
1	Prostatic fluid	Prostatitis	C. welchii	?	Organism recovered consistently from prostatic fluid	Drummond [cited in Willis (1453)]	1943
1	Urine	Cystitis?	B. melaninogenicus	E. coli, nonhemolytic streptococcus, Proteus, Pseudomonas	No documentation, diverticuli of bladder	Weiss (80)	1943
1	Pus	Perirenal abscess	B. funduliformis	None	Good, 1000 ml foul pus	Smith and Ropes (226)	1945
2	Prostatic secretion?	Chronic prostatitis	Bacteroides	Yes—number and type not specified	Poor		

(continued)

TABLE 11.1 (*continued*)

No. of cases	Source of culture	Types of urinary tract infections	Anaerobic organism(s)	Aerobic or facultative organisms	Documentation (or comments)	Reference	Year
1	Kidneys-histology	Pyelitis, bilateral honeycombing of cortices, bladder thickened and massively ulcerated	*Clostridium*	None?	Patient with sarcoma, paraplegia, urinary retention, only other organ with honeycombing was liver	Govan (500)	1946
1	Pus, blood	Minute renal abscess	*B. necrophorum*	None	Not clear whether renal abscess per se was cultured: this was one of many metastatic abscesses in fatal case of bacteremia	Owen and Spink (1228)	1948
1	Urine	Pyuria	*Bacteroides*	Few aerobic streptococci	Large numbers of *Bacteroides*	Beigelman and Rantz (282)	1949
1	Urine	Pyelonephritis, right with calculi	*Bacteroides*		Pure culture twice; heavy growth		
1	Urine	Possible pyelonephritis	*Bacteroides*	*S. albus* (400/ml)	Few *Bacteroides*		
1	Urine	None known	*Bacteroides*		Few *Bacteroides*, no evidence of infection		
1	Pus	Perinephric abscess	*B. funduliformis*	?	No clinical details	Sevin and Beerens (1197)	1949
1	Pus	Periurethral abscess	*B. funduliformis*	None	Numerous gram-negative rods on direct	Thjötta and Jonsen (305)	1949

No.	Specimen	Condition	Organism	Type	Clinical details	Reference	Year
1	Pus	Perinephric abscess	*Ristella glutinosa*	None	Patient died of metastatic brain abscess	Vinzent and Linhard (1235)	1949
1	Pus—remote abscess	Perinephric abscess	Fusiform	None	Not documented, organism isolated from a remote abscess after perinephric abscess was drained and healed	Christiaens *et al.* (540)	1950
1	Pus?	Periurethral abscess	*B. funduliformis*	Type not specified	Mixed infection	Ernst (347)	1951
1	?	?	*Actinomyces*	?	No specific information given	Rasovic [cited in Galinovic-Weisglass, (1454)]	1952
1	?Pus	Paraurethral abscess	*Bacteroides*	None	No clinical details	Carter *et al.* (653)	1953
1	?	Cystitis, pyelonephritis	"Anaerobes"		Carcinoma of cervix, anaerobic pelvic abscesses with direct extension to bladder		
1	?Pus	"Urinary tract abscess"	*Sphaerophorus pseudonecrophorus*	None	?	Prévot (167)	1953
2	?Urine	Purulent cystitis	1 *S. ridiculosus*, 1 *S. gulosus*	None	?		
1	?Pus	Perinephric abscess	*S. funduliformis*	None	No clinical details	Beerens (1414)	1954
1	Pus	Perinephric abscess	*S. freundi*	None	Appendicitis followed by multiple metastatic abscesses	Prévot *et al.* (1455)	1954

(continued)

329

TABLE 11.1 (*continued*)

No. of cases	Source of culture	Types of urinary tract infections	Anaerobic organism(s)	Aerobic or facultative organisms	Documentation (or comments)	Reference	Year
6	Pus—abscess	Prostatic abscess	Gram-positive cocci seen on smear; culture negative for cocci	5 No growth, 1 E. coli	5 no growth, gram-positive cocci on smear of 4, gram-positive cocci also on smear of 6th case which grew only E. coli, pus obtained by perineal incision and drainage	Persky et al. (1456)	1955
1	?Pus	Perinephric abscess	R. glutinosa	None	?	Prévot (21)	1955
1	?	?	D. reniformis	?	None		
1	Urine?	Pyelonephritis	D. magnus	?	None		
1	Urine?	"Pyuria"	S. putridus	?	None		
1	Pus?	"Urinary tract abscess"	S. foetidus	?	None		
1	Urine?	Purulent cystitis	S. anaerobius	?	None		
1	Urine?	Purulent cystitis	Micrococcus prevoti	?	None		
3	Pus—or biopsy	Metastatic renal abscess	Actinomyces	None?	Good, either culture or histologic proof; several additional cases of spread to retroperitoneal and psoas areas	Harvey et al. (212)	1957

No.	Source	Clinical	Anaerobe	Aerobe	Comments	Reference	Year
1	Urine (ureteral?)	Ureteritis	Anaerobic Corynebacterium	None	Poor	Lodenkämper et al. (1457)	1957
1	Pus	Prostatitis	Anaerobic staphylococcus	None	Poorly documented; source of specimen?	Morin and Potvin (145)	1957
1	Bladder wall	Cystitis, gangrene of bladder with rupture	C. perfringens	Streptococci. E. coli	Foul urine, foul drainage and gas from gas gangrene (which extended to abdominal wall)	Amar and Ratliff (1458)	1958
1	Urine	Cystitis?	Bacteroides	None	No documentation	Witebsky and Miller (162)	1958
1	Urine?	Purulent cystitis	Micrococcus niger, S. abscedens	?	No clinical details	Prévot et al. (337)	1959
1	Urine?	Gas gangrene, bladder	Clostridium sp.	?	No clinical details	Christenson et al. (975)	1960
1	Pus, renal abscess	Bilateral pyelonephritis with papillary necrosis	Actinomyces	E. coli	Well documented, no actinomycosis elsewhere in body	Jutzler et al.[a] (1459)	1961
1	—	Pyonephrosis communicating with 800 ml retroperitoneal abscess	No cultures done	No culture	Malodorous pus, therefore presumably anaerobic, no smear or culture	Stahlgren and Thabit (1460)	1961
1	?	Pelvic gas gangrene	?	?	Gas gangrene developed after transvesical prostatectomy	Brummelkamp (1001)	1962

(continued)

331

TABLE 11.1 (*continued*)

No. of cases	Source of culture	Types of urinary tract infections	Anaerobic organism(s)	Aerobic or facultative organisms	Documentation (or comments)	Reference	Year
1	Blood	Hemorrhagic, gas-forming infection of kidneys as part of general sepsis	*C. perfringens*	None?	Well documented, typical organisms in kidney sections at autopsy	Jarkowski and Wolf (604)	1962
1	Urine, blood	Pyelonephritis	*Bacteroides*	None	Bullet wound traversed kidney, liver, stomach and intestines, calyceal-cutaneous fistula	Tynes and Frommeyer (489)	1962
1	Kidneys	Micronodular lesions cortex and medulla—both kidneys	*Actinomyces*	None?	Part of widespread dissemination in fatal case of actinomycosis	Verme and Contu (447)	1962
1	Wound?	Postoperative cellulitis	*C. welchii*	?		Dickinson and Edgar [in Willis (1453)]	1963
1	Urine from renal pelvis and blood × 2	Pyelonephritis and bacteremia (leukemic infiltration of kidneys and ureters)	*C. perfringens*	None	Well documented at autopsy	McHenry et al. (1461)	1963

1	Urine?	"Pyuria"	*Bifidobacterium appendicitis*	?	No details given	Prévot (1199)	1963
1	Blood culture	"Urinary infection"	*Ristella pseudoinsolita*	None	No clinical data	Prévot (226a)	1965
1	?	Acute epididymitis	*Ristella pseudoinsolita*	None	No clinical data		
1	Urine	Cystitis	*Eggerthella convexa, B. funduliformis, S. asaccharolyticus, E. lentum*	None	Purulent fetid urine, dilated inflamed bladder adherent to loop of small bowel	Beerens and Tahon-Castel (115)	1965
1	Pus	Acute prostatitis	*B. funduliformis*	*E. coli*			
1	Pus	Intrarenal abscess	*Eggerthella convexa, B. funduliformis*	*M. tuberculosis*	Coexistent renal tuberculosis		
1	Urine (voided)	?Prostatitis	*B. fragilis* (10^3/ml), *B. funduliformis* (8×10^2/ml), *B. melaninogenicus* (10^2/ml), anaerobic non-spore-forming catalase negative gram-positive rods (10^2/ml), capnophilic streptococci (10^2/ml), few *Propionibacterium* and *Lactobacillus*	None	6 quantitative aerobic urine cultures essentially negative, patient refused percutaneous bladder aspiration, mild urethral stricture	Finegold et al. (46)	1965

(continued)

333

TABLE 11.1 (*continued*)

No. of cases	Source of culture	Types of urinary tract infections	Anaerobic organism(s)	Aerobic or facultative organisms	Documentation (or comments)	Reference	Year
1	Ureteral urine, right	Periureteritis secondary to regional ileitis	*B. fragilis,* anaerobic catalase neg. non-spore-forming gram-positive rod	None		Finegold *et al.* (46)	1965
1	Urine (voided) + blood	?Pyelitis, pyelonephritis	*B. fragilis*	None	Simultaneous *B. fragilis* bacteremia, 10 day history of fever, mild right flank pain, microscopic hematuria, responded to tetracycline therapy		
1	Urine, blood (×4), nephrectomy site, renal homograft	Nephrectomy wound infection, renal homograft infection (+ rejection reaction)	*Bacteroides* species (?*B. fragilis*)	Nephrectomy site— none Blood cultures— none Rejected kidney— *E. coli* and *P. mirabilis*	Transplanted, on immunosuppressive therapy, ileoileal intussusception with strangulation. Bacteroides bacteremia; other infections probably secondary to this. Foul pus.		

			Bacteroides fragilis	Three facultative organisms	Foul pus	Finegold et al. (46)	1965
1	Pus	Urethrocutaneous fistula					
1	Blood	Urethral phlegmon, perineal abscess	_Peptostreptococcus_	_S. aureus_		Lufkin et al. (1462)	1966
1	?	Periurethral abscess	_Clostridium perfringens_	?	No clinical data	Paredes and Fernández (1404a)	1967
1	?	Perinephric abscess?	_Actinomyces_	?	"Renal actino-mycosis with secondary infection"	Salvatierra et al. (641)	1967
3	Pus	Perinephric abscess	1 Anaerobic streptococcus, 2 no anaerobes recovered	1 coliforms, 1 coliforms + enterococcus, 1 streptococci	All 3 cases had carcinoma of colon with invasion of perinephric fat; pus foul-smelling in all cases	Feldman et al. (1463)	1968
2	Pus?	Periurethral infection (abscess?)	_Bacteroides_	"Mixed"	Poor documentation	Smith et al. (651)	1968
1	Pus	Pyonephrosis, perinephric abscess	_B. melaninogenicus_	None	Good documentation	Ueno (in Kozakai and Suzuki) (265)	1968
1	Blood	Necrotic prostatitis, postoperative	_C. perfringens_	None	Incomplete transurethral resection of prostate, organisms seen on smear of prostatic tissue at autopsy	Conomy and Dalton (362)	1969

(continued)

335

TABLE 11.1 (*continued*)

No. of cases	Source of culture	Types of urinary tract infections	Anaerobic organism(s)	Aerobic or facultative organisms	Documentation (or comments)	Reference	Year
1	Purulent material from bladder, peritoneal fluid	Gangrene of bladder	Anaerobic streptococcus, *Bacteroides*	None	Good. Same organisms from bladder and from peritoneal fluid (after rupture of bladder), transmucosal migration of organisms into muscularis layer of bladder	Daines and Hodgson (639)	1969
1	Tissue	Circumscribed mass, right kidney	*Actinomyces*	None?	Diagnosis made histologically, no apparent disease elsewhere	Anhalt and Scott (643)	1970
1	Tissues?	?	*C. welchii*	?	Gas gangrene following exploratory ureterotomy	Bittner *et al.* [in Nielson and Laursen (637)]	1970
1	Urine, blood	Cystitis	*Peptostreptococcus* species CDC Group 3	None?	Bacterial endocarditis with GU infection portal of entry	Felner and Dowell (289)	1970

1	Urine, blood	Pyelonephritis	*Clostridium cochlearium*	None?	Bacterial endocarditis with genitourinary infection portal of entry	Felner and Dowell (289)	1970
3	Blood culture	?Cystitis	1 *Bacteroides fragilis* 2 Bacteroidaceae	?	"Resulted from complications of in-dwelling Foley catheters"	Gelb and Seligman (490)	1970
1	Urine, suprapubic bladder tap	Cystitis	*Sphaerophorus necrophorus*	None?		Nishiura et al. (1464)	1970
1	Pus?	Periurethral abscess	*Bacteroides*	?	No details given	Pearson and Anderson (1465)	1970
1	Pus	Testicular abscess	*Peptococcus variabilis, Peptococcus asaccharolyticus*	None	Sour-smelling pus, history of chronic prostatitis	Caselitz et al. (1466)	1971
1	Blood (3 times)	?	*S. funduliformis*	*E. coli* + nonhemolytic streptococci[1] (1 of 3)	Nature of urinary tract infection and reason for suspecting it as portal of entry not given	Farquet et al. (723)	1971
3	Blood culture	?	*Bacteroides fragilis*	None	Related to renal homotransplants and immuno-suppressive therapy	Felner and Dowell (168)	1971
1	Blood culture	Perinephric abscess	*Bacteroides fragilis*	None?	No clinical details		

(continued)

337

TABLE 11.1 (*continued*)

No. of cases	Source of culture	Types of urinary tract infections	Anaerobic organism(s)	Aerobic or facultative organisms	Documentation (or comments)	Reference	Year
1	Blood culture	"Urinary tract infection"	*Bacteroides oralis*	?	Poor, patient with cystocele and prolapsed uterus and urinary tract infection developed bacteremia	Felner and Dowell (168)	1971
1	Blood culture and urine	Pyelonephritis	*Bacteroides* CDC Group F-1	None	Ureteral calculi		
1	Pus from abscess	Perinephric abscess	*Clostridium, B. melaninogenicum*	*E. coli Enterobacter Klebsiella Citrobacter*	Cellulitis and necrotizing fasciitis of abdominal wall also, *E. coli* bacteremia	Altemeier *et al.* (1466a)	1971
1	Urine (twice)	Cystitis	*C. welchii* ($>10^5$/ml)	None (repeated cultures)	Good: clear evidence of infection, pyuria. Large gram-positive rods on direct smear of urine. Negative aerobic cultures, successfully treated with cotrimoxazole	Nielsen and Laursen (637)	1972

No.	Source	Condition	Anaerobe(s)	Other organism(s)	Probable etiology
1	Suprapubic bladder tap urine	?	B. fragilis	M. tuberculosis	Probable colocutaneous fistula, renal stones
1	Suprapubic bladder tap urine	?	Bacteroides sp. and Peptostreptococcus	Pseudomonas	Nonfunctioning left kidney (etiology?)
1	Suprapubic bladder tap urine	?	B. incommunis	S. aureus and group D streptococcus	Urethral diverticulum
1	Suprapubic bladder tap urine	Infected ureteral stump	B. fragilis and Peptococcus	Enterococcus and P. morganii	Atonic neurogenic bladder
1	Suprapubic bladder tap urine	?	Peptococcus	Citrobacter and E. coli	Chronic urethral stricture; repeated instrumentation
1	Suprapubic bladder tap urine	Chronic pyelonephritis, infected renal stones	Bacteroides CDC Group F-2	P. mirabilis and viridans streptococcus	
1	Suprapubic bladder tap urine	?	P. acnes	Corynebacterium	Carcinoma of prostate, acute urinary retention, ?contaminant
1	Suprapubic bladder tap urine	Chronic cystitis	B. fragilis and B. incommunis	Viridans streptococci and group D streptococcus	Previously had bladder leukoplakia, probably ureterovesical reflux

(continued)

339

TABLE 11.1 (*continued*)

No. of cases	Source of culture	Types of urinary tract infections	Anaerobic organism(s)	Aerobic or facultative organisms	Documentation (or comments)	Reference	Year
1	Suprapubic bladder tap urine	Chronic cystitis, chronic pyelonephritis	*B. melaninogenicus*	*E. coli*	Bilateral reflux, previous hemi-nephrectomy	Segura *et al.* (638)	1972
1	Suprapubic bladder tap urine	?	*Peptostreptococcus* and *Peptococcus*	Viridans streptococcus	Cicatricial urethritis, recurrent urinary tract infection		
1	Urine and kidney	Renal abscess	*Arachnia propionica*	None?	Organism still recovered from urine over 4-year period following nephrectomy; patient died of unrelated cause, no autopsy	Brock *et al.* (304)	1973
1	Prostatic fluid, blood	Prostatic abscess	*Sphaerophorus gonidiaformans* (2 blood cultures), *Peptococcus* (1 blood culture), *S. gonidiaformans* + *Bacteroides* species from prostatic fluid	None		Fishbach and Finegold (392)	1973

340

	Source	Clinical condition	Anaerobe(s)	Other organisms	Comments	Reference	Year
1	Urine	Cystitis, acute	Anaerobic cocci, 3 times (> 10⁵/ml, 8 × 10⁴/ml, 5 × 10³/ml)	None	Good documentation. Symptoms and findings of infection and response to therapy. Repeated cultures grew only anaerobes	Shimizu and Mo (1466b)	1973
1	Urine	Cystitis, chronic	Anaerobic cocci, 5 times. Anaerobic gram-negative rod, 2 times	None	Cancer of bladder. Aerobic culture negative 5 times		
2	Wound?	Wound infection following suprapubic prostatectomy	*Bacteroides*	?		Baird (324b)	1973
1	Pus?	Perinephric abscess	*Clostridium sporogenes*	*Proteus mirabilis, Staphylococcus aureus*		Werner et al. (1466c)	1973
1	Wound	Wound infection after nephrectomy	*Bacteroides*	?		Okubadejo et al. (390a)	1973
1	Wound	Wound infection after prostatectomy	*Bacteroides*	?			
4	?	Infection related to prostatectomy (2), carcinoma of bladder, vesico-colic fistula	*Bacteroides fragilis* + ?	?		Leigh (1420f)	1974

(continued)

341

TABLE 11.1 (*continued*)

No. of cases	Source of culture	Types of urinary tract infections	Anaerobic organism(s)	Aerobic or facultative organisms	Documentation (or comments)	Reference	Year
1		Periurethral abscess	Bacteroides fragilis	?		Leigh (1420f)	1974
1		Pyonephrosis	Bacteroides fragilis	?			1974
1	Pus?	Posttraumatic renal abscess, bacteremia	Bacteroides fragilis, Eubacterium lentum, Clostridium sphenoides, Clostridium perfringens, Propionibacterium	Enterococcus, Escherichia coli	B. fragilis in blood culture	Gorbach and Thadepalli (344a)	1974
1	Blood	Bladder infection, bacteremia	Bacteroidaceae sp., Clostridium sp., Eubacterium sp., anaerobic streptococcus	None (from blood)	Metastatic carcinoma of bladder. Pus in bladder	Chow and Guze (1573)	1974
1	Abscess?	Prostatic abscess	Clostridium + 5 other anaerobes	None?		Gorbach and Thadepalli (1486)	1975
1	?	"Kidney infection"	Actinomyces	?		Weese and Smith (113a)	1975
1	?	"Testis infection"	Actinomyces	?			1975
1	Urine, blood	Cystitis (?), bacteremia	Bacteroides fragilis (>10^5/ml, urine; and 4 blood cultures)	None?	Vesico-colic fistula	Ingham et al. (984)	1975
1	Wound, blood	Nephrectomy wound infection, bacteremia	B. fragilis	None?	Renal transplant, with rejection		1975

[a] Forty-two additional cases from literature, 12 of which are definitely documented, are also tabulated by authors.

29 specimens of "urethral" urine (the first 10 to 20 ml voided) and "midstream" urine yielded anaerobes in mixed culture in 13 instances. Thus 45% of these two types of urine specimens contained anaerobes, whereas suprapubic bladder puncture urine in the 19 patients studied failed to yield any obligate anaerobes. The anaerobes recovered from the voided specimens clearly represented normal urethral flora. In this study anaerobes were recovered in counts of 10^3 to 10^4/ml or greater on a number of occasions. Thus, even quantitative anaerobic culture would not likely be helpful in distinguishing between infection and the presence of anaerobes as normal flora.

One report in which involvement of anaerobes in urinary tract infection was documented in a satisfactory fashion despite use of only voided urine is that of Nielsen and Laursen (637). In this case, a patient with cystitis had repeated negative aerobic urine cultures despite pyuria and other clear evidence of infection. This patient had a count of *Clostridium welchii* of greater than 10^5/ml on anaerobic culture of voided urine, and large gram-positive rods characteristic of clostridia were seen on direct smear of the urine specimen. Two similar cases were reported by Shimizu and Mo (1466b).

The report of Segura *et al.* (638) confirms the validity of suprapubic bladder puncture for documentation of anaerobic urinary tract infection. This report will be discussed in greater detail later in this chapter.

Table 11.1 is a listing of anaerobic infections of the urinary tract culled from the literature. Unfortunately, there is not sufficient clinical or bacteriological detail given in all cases to judge the reliability of the data. In other cases, the data presented suggests that the infections are not well documented. However, there clearly are a good number of well-documented infections of all types as well. The types of infections of the urinary tract in which anaerobes have been involved include para- or periurethral cellulitis or abscess (some with gangrene and some with multiple fistulae), acute and chronic urethritis, cowperitis, cystitis (including necrotizing cystitis and cystitis emphysematosa), acute and chronic prostatitis (occasionally necrotizing prostatitis), prostatic abscess, periprostatic phlegmon, ureteritis, periureteritis, pyelitis, pyelonephritis (sometimes with papillary necrosis), renal abscess, metastatic renal infection (usually with abscess formation during sepsis), pyonephrosis, perinephric abscess, retroperitoneal abscess related to the kidney, nephrectomy wound infection, other postoperative wound infection, postoperative tetanus, renal homograft infection, bacteremia related to the urinary tract, septic renal vein thrombophlebitis, gangrene (nonclostridial) of the bladder, perineal abscess or gangrene, gas gangrene involving various parts of the urinary tract, and testicular abscess. Calculi are found in conjunction with certain of the infections due to anaerobic bacteria. Subsequent to publication of our own cases (see Table 11.1), we have recovered anaerobic gram-positive cocci from an infected nephrostomy wound.

In addition to the data presented in Table 11.1 Lodenkämper and Stienen (287) reported a number of infections involving anaerobes but gave very little detail. They recovered *Bacteroides* in pure culture from 12 cases of pyonephrosis and recovered *Bacteroides* as part of mixed infections from one additional case of pyonephrosis and from nine cases of epididymitis and urethritis. Microaerophilic streptococci were recovered from one renal abscess.

A very unusual case is reported by Daines and Hodgson (639). This was a fatal case of bladder rupture secondary to gangrene of the bladder caused by anaerobic streptococci and *Bacteroides*. The patient had a urethral stricture and prostatic hypertrophy. The rupture was probably related to long-standing obstruction and relative fixation of the bladder size by markedly trabeculated mucosa. Salvatierra *et al.* (640) reported on 71 perinephric abscesses. Although no anaerobes were recovered in this group, there were four instances of abscesses of this type which were sterile on culture.

Segura *et al.* (638) performed suprapubic bladder aspiration in two groups of patients, one group whose aerobic cultures did not reveal organisms that were present in significant numbers on gram stain and a second group who required suprapubic bladder aspiration for other reasons, such as an inability to void. The second group served as a control group. The study was carried out over a period of 6 months. There were a total of 5781 midstream urine cultures performed. Of these, 795 revealed significant numbers of organisms by gram stain. Aerobic culture results correlated with gram stain in all but 25 specimens. Suprapubic bladder aspiration was done on 17 of these and yielded anaerobes in 10. Accordingly, at least 1.3% of patients with significant bacteriuria had anaerobic organisms involved. Among the seven that were negative for anaerobes by suprapubic bladder aspiration, four were also negative aerobically. The control group of 36 patients all failed to yield anaerobes on culture, confirming the validity of suprapubic bladder aspiration for the diagnosis of suspected anaerobic urinary tract infection. Among the 10 suprapubic bladder taps that were positive for anaerobes, there was only one instance in which an anaerobe was recovered in pure culture; this was an isolate of *Bacteroides fragilis* in a patient with known renal tuberculosis. Five of the subjects had two anaerobes recovered. All ten patients had chronic recurrent urinary tract infections and had had multiple courses of antimicrobial therapy. All but one had significant urologic disease as well as surgical or pathological derangement of the normal urinary tract anatomy. Four of the ten had obstructive uropathy, and two had stones. The tenth patient had cicatricial urethritis. The suggestion that certain microaerophilic corynebacteria ("NSU corynebacteria") may play a role in nonspecific urethritis has no confirmation to date [Furness *et al.* (640a)]. *Clostridium difficile* has been isolated more frequently from the urogenital tract of males and females attending a venereal disease clinic than from others attending family planning and urology clinics [Hafiz *et al.* (640b)]. Whether this represents secondary colonization or infection or whether this organism may be a primary pathogen remains to be determined. Shapiro and Breschi (640c) suggested the possibility of anaerobes playing a role in epididymitis in view of the presence of inflammatory cells in urine and negative urine cultures.

Actinomycosis of the Urinary Tract

According to Cope (210) actinomycosis does not often involve the kidneys. He quoted von Lichtenberg's statement that only 9 of 128 fatal cases he studied showed renal lesions. In addition to cases involving direct spread or metastatic spread to the kidneys, there are cases involving only the kidney known as "isolated," "solitary," or "primary" (the last undoubtedly is not accurate). There were 20 solitary cases described in the literature up to 1938. Pathologically, actinomycosis takes three forms—circumscribed chronic suppurative lesions, pyonephrosis, and pyelonephritis. On occasion, there may be associated stones. The circumscribed chronic lesions extend gradually to the capsule of the kidney and eventually break through to produce a tender perinephric mass and flank pain [Goldsand (641)]. Anhalt and Scott (642) described a solitary mass of the right kidney due to actinomycosis in which the firm mass resembled tumor, even at surgery. Henthorne *et al.* (643) reported a possible *Actinomyces* recovered in mixed culture from urine; the patient had no bladder symptoms.

Anaerobic Bacteremia Related to the Urinary Tract

The presence of anaerobes in the urinary tract, either as normal flora or as infecting organisms, may predispose to bacteremia. This may be secondary to the urinary infection per se, or may result from manipulation of one type or another of the urinary tract. Table 11.2 summarizes cases of anaerobic bacteremia following urinary tract manipulation reported in the literature. In the report by Marcoux *et al.* (420) cited in Table 11.2, it is indicated that there were six patients with urologic problems as underlying factors in *Bacteroides* bacteremia. Some of these (number not specified) underwent manipulation or surgery that precipitated the bacteremia.

TABLE 11.2

Anaerobic Bacteremia following Urinary Tract Manipulation

Type of manipulation	No. of cases	Anaerobic organism(s)	Comments	Reference	Year
Catheterization	?	C. perfringens		Dunham [cited in Conomy and Dalton (362)]	1897
Cureting of fistula, urethral dilatation, catheterization	1	Bacillus aerogenes capsulatus	Perineal fistula. Meningitis, brain abscesses, death	Howard (1240i)	1899
Urethrotomy	1	Unidentified anaerobe (gram-positive rod ?)	Patient died	Jungano (1442)	1908
Removal of bladder calculi; prostatectomy	1	B. funduliformis	Poorly documented	Thompson and Beaver (1448)	1932
Nephrectomy	1	C. welchii	Wound infection with same organism	Mencher and Leiter (391)	1938
Cystoscopy	1	C. welchii	Prostatism, urinary retention	Baume [cited in Willis (1453)]	1960
Catheterization	2	Bacteroides		Vinke and Borghans (390b)	1963
Drainage of perineal abscess	1	Peptostreptococcus	?Bacteremia independent of surgery	Lufkin et al. (1462)	1966
Bilateral ureteroileocutaneous anastamosis	1	C. welchii	C. welchii also in bladder pus ; carcinoma of bladder	Rathbun (419)	1968
Transurethral prostatic resection	1	C. perfringens	Patient developed meningitis and died	Conomy and Dalton (362)	1969
Indwelling urethral catheter	3	1 B. fragilis, 2 Bacteroidaceae	All patients died	Gelb and Seligman (490)	1970
Bladder lavage	1	Ristella pseudoinsolita	"Staphylococcus" in blood also	Marcel et al. (1467)	1970
Not specified	?	Bacteroides		Marcoux et al. (420)	1970

Clinical condition	No.	Organisms	Associated condition	Reference	Year
Traumatic rupture of urethra	2	1 *Bacteroides variabilis*, 1 *Bacteroides* species		Felner and Dowell (168)	1971
Urethral dilatation	1	*Peptococcus*	Chills and fever	Pien et al. (390)	1972
In-dwelling urethral catheter	1	*Eubacterium alactolyticum*		Wilson et al. (105)	1972
Urethral dilatation	8	*Peptococcus* (3), *Peptostreptococcus* (2), *B. melaninogenicus* (1), *Bacteroides* species (1)		Sullivan et al. (1468)	1973
Manipulation other than urethral dilatation	3	*B. adolescentis* (1), *P. avidum* (1), *V. parvula* (1), Microaerophilic streptococci (1)			
Prostatectomy	1	*Bacteroides*		Okubadejo et al. (390a)	1973
2-"surgical manipulation for obstructive uropathy" 1-symphysiotomy	3	*Bacteroides fragilis* (2), Bacteroidaceae, *Clostridium* sp., *Eubacterium* sp. } 1, Anaerobic streptococcus		Chow and Guze (1573)	1974
Transurethral resection, bladder	1	*Bacteroides*	Carcinoma, bladder	Mackenzie and Litton (384a)	1974

The occasional presence of *Clostridium* as normal flora in the urethra ordinarily has no significance, but this carrier state may account for rare cases of gas gangrene following genitourinary tract surgery or manipulation [Bittner *et al.* (629)]. Note that most of the types of manipulation indicated in Table 11.2 involved the urethra, where anaerobes may be present in significant numbers as normal flora.

Pathogenesis of Anaerobic Urinary Tract Infection

With regard to pathogenesis, anaerobes may set up infection within the urinary tract as a result of pathologic processes permitting invasion by organisms normally present as indigenous flora. Anaerobes may also gain access to the urinary tract, other than the urethra, by the ascending route, by direct extension from adjacent organs, such as the uterus or bowel, or by way of the bloodstream. Bran *et al.* (44) showed that urethral trauma may introduce organisms from the urethra to the bladder. They studied three females without evidence of infection. Urine obtained by suprapubic bladder aspiration was cultured for anaerobes before and after "milking" of the urethra. One of the subjects had anaerobic diphtheroids present in both specimens, with a somewhat higher count in the second specimen. A second patient had 2×10^2 anaerobic diphtheroids following "milking," and a third less than 1 organism per milliliter, with *Fusobacterium gonidiaformans* plus *Enterococcus* and *Bacillus* species. The work of Alling *et al.* (634) indicating that patients with indwelling urethral catheters have a higher incidence of anaerobes recovered from urine than those who do not has already been cited. Sapico *et al.* (644) have shown that, on occasion, patients with indwelling Foley catheters will show anaerobes (*Clostridium*, *Bifidobacterium*, and *Veillonella*), along with aerobes and facultatives, in urine obtained by suprapubic bladder aspiration. The role of these anaerobes in initiating or perpetuating urinary tract infection is unclear. Once introduced into the bladder or other parts of the urinary tract certain anaerobes are capable of growing well in the urine itself [Finegold *et al.* (46)].

With regard to the possible role of anaerobes in pyelonephritis, it has been pointed out (645) that low medullary blood flow, plasma skimming, and countercurrent flow all promote lessened oxygen supply to the medullary tissues. Medullary tissues derive their metabolic energy from anaerobic glycolysis much more than does cortical tissue. Anaerobic glycolysis of the inner medulla is relatively unaffected by the hypertonic environment of the medulla. Other factors that may predispose to anaerobic infection of the urinary tract are noted by Leonhardt and Landes (646). The oxygen tension of the urine is sharply decreased in the dehydrated patient. The pO_2

of the renal parenchyma falls progressively from 100 mm Hg in the outer cortex of the kidney to 20 mm Hg in the deep medulla. The low parenchymal and urinary oxygen tensions are a reflection of the intrarenal arteriovenous shunt mechanism. In shock, there is a sharp fall in oxygen tension of the urine which is not reversed by norepinephrine. In unilateral chronic pyelonephritis, the oxygen tension of the urine from the renal pelvis is markedly decreased on the diseased side. Finally, acute ureteral obstruction leads to a sharply decreased oxygen tension of the trapped urine.

12

Female Genital Tract Infections

General

Anaerobic bacteria are involved in a wide variety of infections of the female genital tract, as indicated in Table 12.1. Conditions predisposing to or associated with such infections are listed in Table 12.2. The frequency with which anaerobic bacteria are involved in various obstetric and gynecologic infections varies greatly from one series to another, reflecting to some extent differences in patient material but probably primarily relating to differences in effectiveness of transport systems and anaerobic culture techniques.

Among 860 patients with bacteremia due to gram-negative organisms, DuPont and Spink (647) noted that 35 of the patients were from the obstetric–gynecology service and that 8 of these had infections with *Bacteroides*. There are several series reported of bacteremia due to anaerobic gram-negative bacilli, and it is of interest to note the frequency with which the female genital tract serves as a portal of entry for such bacteremia. Felner and Dowell (289) found that 41 of 250 cases of *Bacteroides* bacteremia originated from the genitourinary tract, half of these associated with post-partum or postabortal complications, with some additional cases related to cervical or uterine tumors. Bodner *et al.* (460) found seven cases of anaerobic gram-negative bacillary bacteremia originating from the female genital tract

TABLE 12.1

Types of Anaerobic Infection of the Female Genital Tract

Endometritis, pyometra
Myometritis
Parametritis
Pelvic cellulitis, abscess
Pelvic thrombophlebitis
Peritonitis, intraabdominal abscess
Bacteremia, with or without shock or intravascular hemolysis
Metastatic infection, particularly of lung, liver, brain, heart valves, kidneys
Wound infection or abscess
Vulvar abscess
Bartholinitis, Bartholin abscess
Skenitis, Skene's abscess
Paraclitoroidal abscess
Vaginitis, abscess of vaginal wall, paravaginal abscess
Salpingitis, tubal abscess
Ovarian or tubo-ovarian abscess
Abscesses of adjacent parts, groin, perirectal or isochiorectal areas, paraurethral, leg, abdominal
    wall
Chorioamnionitis
Fetal emphysema
Intrauterine or neonatal pneumonia, sepsis

---

among 39 cases in their study. In a study of 123 anaerobic gram-negative bacteremias, Marcoux *et al.* (420) found 16 originating in the female genital tract; over half of these patients had malignancy. In 35 anaerobic gram-negative bacillary bacteremias, Gelb and Seligman (490) found 18 originating in the female genital tract, all cases of incomplete septic abortion. In the

TABLE 12.2

Predisposing or Associated Conditions in Patients with Anaerobic Infections of the Female Genital Tract

---

Pregnancy
Puerperium, particularly with premature rupture of membranes, prolonged labor, or post-
    partum hemorrhage
Abortion, spontaneous or induced
Malignancy
Irradiation
Obstetrical or gynecologic surgery
Cervical cauterization
Endocervical or vaginal stenosis
Uterine fibroids
Secondary infection of old gonococcal salpingitis
Intrauterine contraceptive devices

---

series reported by Tynes and Frommeyer (489), 5 of 25 cases of anaerobic gram-negative rod bacteremia were related to the female genital tract, while in the series reported by Pearson and Anderson (648), 40 of 76 such bacteremias were of this origin. Thirty-four of these 40 were cases of incomplete septic abortion, whereas 6 were associated with puerperal infection. In this latter series, there were 13 additional cases of perinatal sepsis originating from the female genital tract. In a series of 173 cases of anaerobic gram-negative bacillary bacteremia encountered in a review of the literature from 1898 to 1956, Gunn (296) found 17 of female genital tract origin, whereas Love (649), reviewing the literature of such bacteremias from 1956 to 1971, found that 116 of 458 cases had the genital tract as the point of origin.

In four series of infection of all types due to anaerobic gram-negative bacilli, the female genital tract was involved in 3 of 112 patients [Saksena et al. (487)], in 15 of 47 cases [Beigelman and Rantz (282)], in 13 of 25 patients [Chapman (650)] and in 13 of 35 patients [McVay and Sprunt (621)]. In a series of 100 cultures that yielded anaerobic gram-negative bacilli, Smith et al. (651) found 48 isolates of these organisms from the female genital tract. In a similar series comprising 690 cultures, Lodenkämper and Stienen (287) found that at least 83 of these cultures were from the female genital tract.

Among 220 specimens of genital pus (defined as freshly opened pyosalpinx, Bartholin's abscess, and ischiorectal abscess), Stokes (108) found anaerobes of all types in 69 of the specimens; in 24 of these cases the anaerobes were recovered in pure culture. K. Ueno (personal communication) recovered anaerobes in pure culture from 15 specimens of pus among 86 from gynecologic patients and from 3 of 12 specimens of uterine discharge. In a series extending over a number of years, Oguri and Kozakai (359) noted that, among 1139 specimens of secretion from the female genital tract which were positive on culture, 473 (41.5%) yielded only anaerobic organisms. Carter (652) stated that non-spore-forming anaerobic bacilli are the most common isolates he has encountered from cases of bartholinitis, skenitis, pyometra, and postpartum and postabortal endometritis. He states that this is also true in the case of pelvic inflammatory disease if one eliminates from consideration cases due to the gonococcus and tubercle bacilli. Carter et al. (653) reported on 153 patients treated in their department of obstetrics and gynecology from whom various species of Bacteroides were isolated. Unfortunately, this study was poor from a bacteriological standpoint. Smears showing pleomorphic gram-negative bacilli thought to be typical for Bacteroides were accepted as evidence of infection with this organism even in the face of negative cultures. Patients whose cultures showed both non-pleomorphic Bacteroides and aerobic coliform bacteria were excluded because they found it difficult at that time to separate or identify such Bac-

teroides when they occurred in mixed cultures. Their primary media for isolation of anaerobes was liquid. Identification seemed to be based primarily or entirely on microscopic appearance of the organisms. Included in this series were a number of *Bacteroides* infections complicating malignancy, particularly of the cervix, uterus, and vulva, postoperative gynecologic infections, pelvic abscess, infections associated with pregnancy, and certain other infections including pyometra, Bartholin abscess, vulvar abscess, paraclitoroidal abscess, Skene's abscess, and paravaginal abscess. Clark and Wiersma (654) reported on 47 infections due to *Bacteroides* encountered on their obstetrical and gynecologic service over a period of 28 months.

Ledger *et al.* (655) commented on the significance of isolation of *Bacteroides* species in terms of the severity of infection in obstetric and gynecologic patients. *Bacteroides* were recovered from the infected adnexal tissue of 6 of 31 women with postoperative adnexal abscesses. In 3 of these 6 women, the abscess had ruptured before operation; one patient died and the other two experienced prolonged postoperative recoveries. In only 3 of the 25 women with adnexal abscesses caused by organisms other than *Bacteroides* was there prolonged convalescence. None of these 3 had rupture of the abscess prior to operation. Their experience with puerperal infections was similar. *Bacteroides* were recovered from 11 of 93 women with such infection. None of these infections was life-threatening, but convalescence was usually protracted. Seven of the 11 patients from whom *Bacteroides* were recovered required more than 48 hours to become afebrile after institution of antibiotics, as compared with 17 of the 82 patients with puerperal infection from whom *Bacteroides* was not isolated. These investigators had previously observed that one of the important factors related to prolonged morbidity in patients with postoperative adnexal abscesses was rupture of the abscess prior to surgery. The study just quoted, however, led these workers to suggest that the prolonged morbidity may actually be more closely related to the presence or absence of *Bacteroides* species. Gorbach and Bartlett (56b) also indicated that *Bacteroides* is the major isolate in serious infections of the female genital tract. *Bacteroides fragilis* and *B. melaninogenicus* are the most common species encountered. *Fusobacterium* is present in 20% of cases. Next in frequency are *Peptococcus* (especially *P. prevotii*) and *Peptostreptococcus* (particularly *P. anaerobius*). *Clostridium* is important in septic cases.

Other data on the incidence of anaerobic bacteria in various female genital tract infections are noted in Table 12.3. These studies of Mizuno and colleagues (656, 657) show that almost half of a large number of infections of a variety of types involving the female genital tract are caused by anaerobic bacteria. Anaerobic cocci and streptococci dominate, with *Bacteroides* also fairly commonly isolated. Non-spore-forming gram-positive

TABLE 12.3

Incidence of Anaerobes in Various Female Genital Tract Infections[a]

	Total positive cultures	Total positive for anaerobes	% of anaerobes	Total No. anaerobic strains	*Peptococcus*	*Peptostreptococcus*	*Veillonella*	*Clostridium*	*Bacteroides*	Other GNR[b]	NSF GPR[c]
Intrauterine infection	87	39	44.8	55	16	10	2	2	14	1	10
Postpartum fever ⎫ Pyometra ⎭	61	29	47.5	37							
Tubo-ovarian abscess ⎫ Pelvic abscess ⎭	36	17	47.2	26	8	6	1	1	9	0	1
Postoperative infections	81	39	48.1	61	15	10	2	2	25	0	7
Bartholin abscess ⎫ Vulvar abscess ⎭	29	14	48.3	20	5	2	0	0	8	1	4
Douglas' abscess	14	5									

[a] Data from Mizuno *et al.* (656) and Mizuno and Matsuda (657).
[b] GNR, gram-negative rods.
[c] NSF GPR, non-spore-forming gram-positive rods.

anaerobic bacilli were encountered more frequently than is true in many of the other series on these types of infections, but other organisms, including clostridia, were quite uncommon. Robinson (658) stated that streptococci, usually anaerobic or microaerophilic, were the most common organisms cultured from pelvic abscess (both cul-de-sac and tubo-ovarian). He also noted that these organisms were the rule in cases of puerperal or postabortal origin. Stone (24) obtained anaerobic streptococci from 25 of 40 uterine cultures, the great majority of which were from women with postabortal or puerperal infection; only 1 of the above 25 women had had a normal puerperium.

Postabortal Sepsis

From the data presented in Table 12.4, on the incidence of anaerobes in postabortal sepsis, it is clear that these organisms are of the greatest importance. The excellent study of Rotheram and Schick (401) involved detailed bacteriologic studies of both blood cultures and cervical cultures. From the 69 cervical cultures (6 of which were cultured only aerobically) a total of 129 anaerobic isolates was recovered. Anaerobes were found in abundance 67 times, whereas this was true for aerobes only 21 times. The analysis of these authors of their 22 cultures that showed little or no anaerobic growth is also of interest. In this group of 22 cultures, there were 4 with an abundant growth of microaerophilic organisms, 11 with scant growth of any kind of organism, and 1 which was overgrown with *Proteus*, the latter showing many gram-positive cocci on smear which were presumably anaerobic. This leaves only 6 cultures in which there was moderate or abundant growth of *E. coli*. Two of these latter patients had bacteremia, one with shock. The other patients did not have *E. coli* in the blood culture, but rather anaerobic streptococci and *Bacteroides fragilis*.

The presentation of patients with septic abortion is usually impressive with high fever, chills, leukocytosis, uterine tenderness, and foul-smelling cervical discharge that may be purulent and hemorrhagic. Nevertheless, most of these patients recover readily following evacuation of the uterus. The complications of septic abortion which may occur, however, are very severe. Bacteria migrate through the uterine wall by means of lymphatic channels and small venules to produce inflammation and abscesses in the broad ligaments and the retroperitoneal spaces of the pelvis. Such direct extension may result in tubal or tubo-ovarian abscesses (Fig. 12.1) which may rupture to produce local or remote intraabdominal or pelvic abscess or peritonitis. Septic thrombophlebitis of the pelvic and ovarian veins is common and should be presumed to be present in patients who are not

TABLE 12.4
Incidence of Anaerobes, Postabortal Sepsis

No. cases cultured	No. with anaerobes	No. sterile	No. with nonpathogens	Specific anaerobes found and comments	Reference
3	1	0	1	Anaerobic streptococci	Guilbeau et al. (1469)
2	2	0	0	Bacillus caducus and few aerobic pyogenic streptococci in both cases	Hallé (10)
4	4 (3?)[a]	0	0	Clostridium perfringens in all 4	Smith et al. (664)
11	4	0	0	2 anaerobic streptococci; 2 C. perfringens, the latter together with aerobic β-hemolytic streptococci; uterine lochia cultured	Harris and Brown (1470)
76	48	0	0	Cases with septic abortion and bacteremia; 38 anaerobic streptococci (31 alone), 14 Bacteroides (7 alone), 3 C. perfringens	Smith et al. (416)
30 fatal cases	11	0	0	Streptococcus putridus (7), C. perfringens (3), Clostridium sp. (1)	Schottmüller (258)
				8 additional cases yielded anaerobic streptococci (1 with C. perfringens); endometritis alone or with salpingitis, peritonitis, and/or bacteremia	
				Only 1 with aerobe (S. pyogenes) seen during same period	
430				7 with advanced clostridial infections	Eaton and Peterson (991)
1467				8 with advanced clostridial infections	Cavanaugh [discussion of Eaton and Peterson (991)]
				Author states that streptococci, usually anaerobic or microaerophilic, were the rule in pelvic abscesses of postabortal origin	Robinson (658)
600	304 (4)	?	?	All Streptococcus putridus	Schottmüller [cited in Schwarz and Dieckmann (1471)]

56 cases with blood cultures	25 (22) (47 isolates); only 9 grew aerobes alone (12 isolates)	22	1	Anaerobic streptococci (16), *B. fragilis* (9), *B. funduliformis* (4), *B. melaninogenicus* (3), anaerobic micrococci (2), microaerophilic streptococci (7), *Actinomyces* sp. (1)	Rotheram and Schick (401)
69 cases with cervical cultures (6 aerobes only and 63 anaerobes[b])	129 anaerobic isolates[c]	0?	52 strains	*B. fragilis* 35 (18 in abundance), anaerobic streptococci 33 (20), *C. perfringens* 2 (1), *B. melaninogenicus* 22 (11), *B. funduliformis* 13 (7), anaerobic micrococci 4 (3), anaerobic diphtheroid 2 (1), microaerophilic streptococci 17 (6), *Actinomyces* 1 (0)	
19	18 (2)	0	0	Anaerobes predominant in 13 of 16 mixed infections; anaerobic isolates—*B. perfringens* (12 cases), *B. thetoides* (6), anaerobic streptococci (6), *Staphylococcus parvulus* (6), *B. radiiformis* (4), *M. foetidus* (3), *B. caducus* (3), *B. ramosus* (2), undetermined coccobacillus (2), *B. nebulosus* (1), undetermined streptobacillus (1), undetermined rod (1)	Jeannin (11)
20				*C. welchii* from either blood or high vaginal swab in 10 cases, *Bacteroides* sp. in 4	Hawkins *et al.* (1471a)

[a] Numbers in parentheses are number in pure culture or number with only anaerobes.

[b] 30 of 63 (48%) gave abundant growth of anaerobes; 17 of 69 (25%) of anaerobic plates had abundant growth.

[c] Eliminated from consideration as microaerophiles following organisms which authors categorized as such: *H. vaginalis, Lactobacillus, pneumococcus,* and *N. gonorrheae.*

357

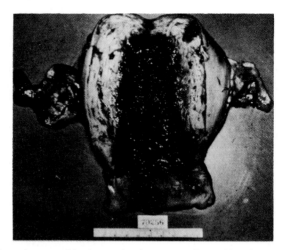

Fig. 12.1.   Opened uterus with adnexa, showing acute endometritis, edematous myometrium with myometritis, myometrial abscess (right upper corner of figure) and tubo-ovarian abscesses. From Schwarz (89a). Reproduced with permission.

responding to what should be appropriate antimicrobial therapy, particularly if there is no evidence of abscess formation or other obvious complication. This thrombophlebitis accounts for continued septicemia and may lead to metastatic infection of the lungs, brain, liver, kidney, and heart valves, although extensive spread in this manner is uncommon in the antimicrobial era. There are a number of reports of widespread dissemination of infection in connection with postabortal sepsis in the preantibiotic era. Two excellent reports are those of Schottmüller in 1910 (258) and Bingold in 1921 (659). Lehmann (660) reported three cases of anaerobic streptococcal postabortal sepsis which ended in endocarditis. Langeron *et al.* (661) reported a case of anaerobic staphylococcal endocarditis as a complication of postabortal sepsis. Other important complications of postabortal sepsis are hypotension or shock, oliguria, liver function abnormality, hemolytic anemia, and disseminated intravascular coagulation.

Although clostridia account for only a small percentage of postabortal infections, they are often prominent among the more dramatic and severe presentations. Clostridial uterine infection starts as a localized chorioamnionitis with invasion of a dead fetus and the placenta; this is known as fetal emphysema. This process is relatively benign, but the gaseous vaginal discharge, crepitus of the uterine wall, and striking X-ray findings may make it difficult to distinguish from more serious types of uterine involvement. The next stage is a low-grade endometritis. Here, one notes vaginal discharge and uterine tenderness, with or without gas formation. There is no toxemia. The most severe form of the infection extends beyond the endometrium to the uterine muscle and results in necrosis which may be accompanied by perforation, peritonitis, and sepsis (Figs. 12.2–12.4). The patient is very toxic, exhibits tachycardia out of proportion to the fever, and has uterine and abdominal pain and foul gaseous vaginal discharge that reveals numerous clostridia on gram stain (Fig. 12.5). These organisms on smear are frequently noted to have capsules, and white blood cells in the exudate may be disintegrated (662).

With severe clostridial sepsis, there is a most dramatic clinical picture consisting of hemolytic anemia, hemoglobinemia, hemoglobinuria, disseminated intravascular coagulation, bleeding tendency, bronze-colored skin, hyperbilirubinemia, hypotension or shock, and oliguria or anuria. The direct effect of the α-toxin of *C. perfringens* (a lecithinase) on the red blood cells (see Figs. 8.3 and 8.4) leads to progressive loss of red cell membrane phospholipid and thus to the formation of spherocytes and microspherocytes. Peripheral blood smears may reveal this type of red blood cell, with or without hemolysis (663). These spherocytes are extremely sensitive to osmotic lysis. The red blood cell swells and eventually disrupts, liberating the hemoglobin. Lysolecithin, which results from the action of α-toxin on lecithin, is also a hemolytic agent. The liberated hemoglobin soon saturates the haptoglobins and other transfer proteins, giving rise to methemoglobin and methemalbumin (664). Hemoglobin circulates in a free form in the plasma and extracellular fluid and part of it is deposited in cells giving the patients a peculiar bronze or magenta color (the "pink lady syndrome").

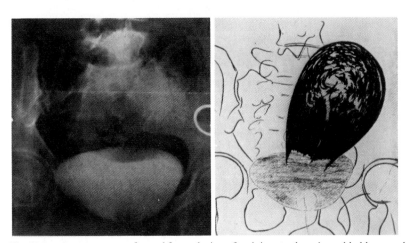

Fig. 12.2. A cystogram performed for exclusion of an injury to the urinary bladder revealed gas in an atypical arrangement. The gas formed onion peel-like layers. The arrangement of the layers conformed to the muscular layers of a moderately enlarged uterus. A semischematic sketch amplifies the radiographic findings. From Ref. (680) G. A. Doeher, K. G. Klinges, and B. J. Pisani, *Am. J. Obstet. Gynecol.* **79**, 542–544 (1960).

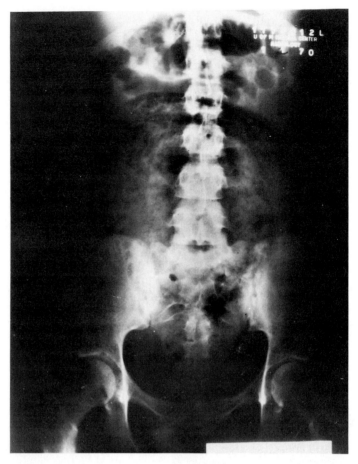

Fig. 12.3. X ray showing typical appearance of gas in myometrium in postabortal uterine infection. From Ref. (991), C. J. Eaton and E. P. Peterson, *Am. J. Obstet. Gynecol.* **109**, 1162–1166 (1971).

Part of the pigment is also excreted by the kidneys resulting in port wine-colored urine. Some of the hemoglobin is also transformed into bilirubin by the reticuloendothelial system. The lecithinase also causes hydrolysis of platelet phospholipid producing lysis with release of thromboplastin and resulting in thrombocytopenia. Intravascular coagulation will result from the release into the circulation of thromboplastins derived from both the platelets and the red blood cells. Rubenberg *et al.* (425) documented the presence of intravascular coagulation in a case of postabortal *Clostridium perfringens* septicemia. Acute renal failure results both from shock and from free hemoglobin. Hyperkalemia, related to liberation of large amounts of

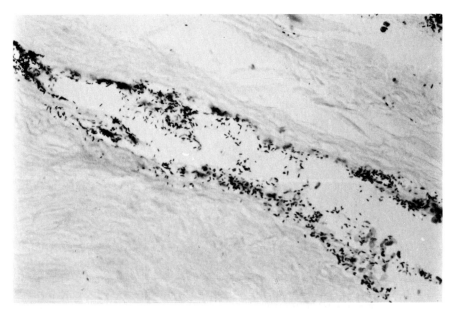

Fig. 12.4. Clostridial myometritis. Histological section showing aggregates of gram-positive bacilli (*Clostridium perfringens*) lining gas-filled space in necrotic myometrium. From Browne *et al.* (673). Reproduced with permission.

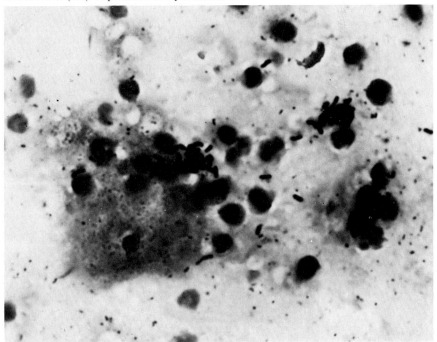

Fig. 12.5. Postabortal clostridial endomyometritis. Gram stain of cervical discharge showing large gram-positive bacilli (*Clostridium perfringens*). From Ref. (991), C. J. Eaton and E. P. Peterson, *Am. J. Obstet. Gynecol.* **109**, 1162–1166 (1971).

potassium from hemolyzed red blood cells, may be a problem, particularly if renal failure is also present. The bleeding tendency is related to depletion of fibrinogen stores during the course of intravascular coagulation, together with the thrombocytopenia and, at times, a circulating anticoagulant. There may also be a direct necrotizing effect of toxin on capillary walls.

Treatment of septic abortion depends on the nature and extent of the infection and the type of infecting organism present. Early evacuation of the uterus is of prime importance. With the more extensive clostridial infections, hysterectomy will be necessary. Early institution of therapy is crucial in the more serious infections such as sepsis. One cannot wait for definitive evidence of infection and recovery of specific etiologic agents. The clinician must have a high degree of suspicion and must depend on clinical judgment. Certainly the combination of fever, rapid hemolysis, and foul lochia, for example, would be an indication for immediate treatment of probable clostridial sepsis. For clostridial infections, penicillin is the drug of choice. It is also active against most of the other anaerobes that may be involved in post-abortal infection, the major exception being *Bacteroides fragilis*. When *B. fragilis* is present or suspected, chloramphenicol or clindamycin or another agent known to be active against the particular strain of *Bacteroides fragilis* involved will be necessary. Heparin may be very useful in the event of thrombophlebitis. Use of heparin for the management of disseminated intravascular coagulation is controversial. Exchange transfusion has been proposed for removal of spherocytic red blood cells, free hemoglobin, and fibrin split products from the circulation (425, 665). Supportive therapy, including careful management of fluid and electrolyte balance and treatment of shock and of renal failure are very important. Several other modalities of treatment are still under debate. Included here is the question of the need for administration of fibrinogen and of polyvalent gas gangrene antitoxin. Also in question is the value of hyperbaric oxygen in this type of anaerobic infection. However, the report of Perrin *et al.* (666) of successful treatment of a patient with postabortal gas gangrene of the uterus with clostridial septicemia is impressive. This patient was treated on 2 successive days, with dramatic improvement each time, and was ultimately cured after total abdominal hysterectomy.

We have isolated a gram-negative anaerobic bacillus, probably *Fusobacterium necrophorum*, from a blood culture of a young woman who had endometritis and pelvic abscesses together with bacteremia following a spontaneous abortion in the fifth month of pregnancy. From the uterine contents of a woman with endometritis secondary to septic abortion, we recovered anaerobic gram-positive cocci, *Bacteroides oralis*, and *Bacteroides melaninogenicus*, as well as *Proteus mirabilis* and *Escherichia coli*; this woman also had *E. coli* bacteremia.

Puerperal Sepsis

The incidence of anaerobic bacteria in puerperal sepsis other than amnionitis is noted in Table 12.5 from a number of series in the literature.
Again, the results are variable, but it is clear from the frequency with which
anaerobes were recovered in several large series that anaerobes play an
important role in this type of infection. Anaerobic cocci and streptococci
predominate, but *Bacteroides* and other gram-negative anaerobic bacilli,
clostridia, and other anaerobes may also be involved. Table 12.6 documents
the role of anaerobic bacteria in chorioamnionitis. Pearson and Anderson
(648) pointed out that bacteria normally present in the vagina cause amnionitis in approximately 10% of all deliveries. Very few of these cases are
clinically significant. However, such infections may be important in perinatal deaths, as Pearson and Anderson pointed out. Townsend *et al.* [cited
by MacVicar (1477a)] found that, in a series of cases of premature rupture
of the membranes, the infecting organism most commonly associated with
perinatal death was the anaerobic streptococcus; it was cultured from 6 of
12 babies.

It is generally conceded that bacteria involved in puerperal infection,
either amnionitis or endometritis, are originally from the vaginal flora.
However, Sen *et al.* (667) suggested that such organisms may be transmitted
from the throat of hospital personnel. Magara *et al.* (668) found that among
39 postpartum patients with fever, there were 29 who showed anaerobic
organisms in the uterus and/or vagina. In 12 patients, anaerobes were found
only in the uterus, in 14 in both the uterus and vagina, and in 3 in the vagina
only. Among the slightly larger group of postpartum women without fever,
only 27% had anaerobic organisms in these sites. Different results were
obtained in an extensive study of the bacterial flora of the vagina and uterus
reported by Hite *et al.* (669). Comparisons were made between uterine flora
in the absence of any evidence of endometritis and of this flora in patients
who had endometritis. In general, the flora was quite similar in the two
groups, the only significant difference being greater numbers of *Bacteroides
melaninogenicus* and other *Bacteroides* in the infected group. Gibbs *et al.*
(669a) did a similar comparison involving 27 endometritis patients and 47
controls and found no significant difference, either qualitatively or quanti-
tatively, between the two groups except that *Peptostreptococcus* was found
less frequently in the endometritis patients. Clearly, then, endometrial cultures will be very difficult to interpret and one should obtain anaerobic and
conventional blood cultures and culture any better sources of material that
may be available.

Although serious clostridial infections are much less common in connection with childbirth than following abortion, they may occur. Toombs and

TABLE 12.5

Incidence of Anaerobes, Puerperal Sepsis Other Than Amnionitis

No. cases cultured	No. with anaerobes	No. sterile	No. with non-pathogens	Specific anaerobes found and comments	Reference
108	77 (61)[a]	1		Acute endometritis	
22	15 (12)	0		Pelvic cellulitis	
14	12 (6)	0		Peritonitis	
10	6 (4)	0		Pelvic abscess	Brown (1472)
14	9 (9)	3		Pelvic thrombophlebitis	
25	15 (11)	0		Septicemia	
21 fatal cases	15 (12)	0	2	All anaerobes in this study were anaerobic streptococci	Schwarz and Dieckmann (1473)
166	66 (61)[b]	67	?	All "anaerobic streptococci"[a,b]	
76	17 (9+)	44	0	All blood cultures	Colebrook (1330)
				Pre-antibiotic era, anaerobic cocci and rods, 0% of puerperal fever cases	
				Sulfonamide era, anaerobic cocci and rods, 74% of puerperal fever cases	Takase [cited in Kozakai and Suzuki (265)]
				Antibiotic era, anaerobic cocci, 3% of puerperal fever cases	
1166 cultures	972[c]	68	465[c]	Uterine cultures	
1000 cases (1098 positive cultures)				Blood cultures	Douglas and Davis (402)
90+ blood cultures, 25 with 98 organisms			12	Blood cultures	
9 }	1	?	?	Episiotomy wound infection	
8 }				Abdominal wound infection (following Caesarean section)	Ledger et al. (1474)
40 (111 isolates)	6[d]	?	31	Postpartum endometritis	
3	2	0	1	Peritonitis, postpartum, all anaerobic streptococci	
5	4 (3)	0	1	Puerperal endometritis, all anaerobic streptococci	Guilbeau et al. (1469)
3	3 (1)	0	2	Parametritis, with or without endometritis and abscess, all anaerobic streptococci	

Reference	Comments			
Harris and Brown (1470)	Uterine lochia, culture, 1 had *C. welchii*; 1 *S. pseudonecrophorus*; 54 had anaerobic streptococci, 2 with *B. pseudonecrophorus*	102	56	0
Ledger *et al.* (655)	Amnionitis prior to delivery or endometritis following delivery, all *Bacteroides*	93	11	?
Schottmüller (258)	5 *Streptococcus putridus*, 1 *C. perfringens*; includes endometritis, parametritis, peritonitis, bacteremia, salpingitis	13	6 (4)	0
Robinson (658)	Author indicates that streptococci, usually anaerobic or microaerophilic, were the rule in pelvic abscesses of puerperal origin			1
Steinhorn (1475)	All anaerobic streptococci	37	17 (14)	8
Schottmüller [cited in Schwarz and Dieckmann (1471)]	All *Streptococcus putridus*	231 fatal cases	72 (65)	?
Jeannin (11)	10 mixed aerobes and anaerobes in equal numbers; 7 mixed but mostly anaerobes; 1 mixed with mostly aerobes. Anaerobes: *C. perfringens* (12 cases), *B. radiiformis* (9), anaerobic streptococci (9), *S. parvulus* (9), *M. foetidus* (6), *B. thetoides* (5), *B. ramosus* (4), anaerobic coccus (3), *B. caducus* (2), undetermined streptobacillus (2), undetermined coccobacillus (1), *B. nebulosus* (1), "pin bacillus" (1)	21	18 (1)	0
Hibbard *et al.* (1475a)	Postpartum infection involving either subgluteal tissues about hip joint, the retropsoal space, or both. Anaerobic streptococci in 2 cases, one with *Bacteroides* also present; two or more aerobes also present in each case	8	2	2
Wenger and Gitchell (1475b)	Postpartum subgluteal infection. One yielded *Bacteroides*, anaerobic streptococcus, and diphtheroids; the second grew *E. coli* and *Bacteroides*. Six other cases not discussed in detail but it was said that cultures usually grew anaerobic streptococcus plus *Bacteroides* plus *E. coli*	2	2	0

(continued)

TABLE 12.5 (*continued*)

No. cases cultured	No. with anaerobes	No. sterile	No. with non-pathogens	Specific anaerobes found and comments	Reference
244	113 strains			Postpartum endometritis. Anaerobes recovered included *Bacteroides* (41), *Peptostreptococcus* (37), *Peptococcus* (10), microaerophilic streptococcus (11), *Clostridium perfringens* (9), and *P. acnes* (5)	Sweet and Ledger (1475c)
?	4			Episiotomy wound infection. *Bacteroides fragilis* recovered; other organisms not mentioned	Leigh (1420f)
9	6 (in blood culture)	0	0	Series of 9 cases of postpartum endometritis (8 following Caesarean section) in whom there was bacteremia due to *Haemophilus vaginalis*. Six also yielded anaerobes on blood culture (2 others also had anaerobes on endometrial or endocervical culture). Blood cultures yielded *Bacteroides* sp. (1), *B. fragilis* (1), and anaerobic streptococcus (5)	Monif and Baer (387a)

[a] Figures in parentheses are anaerobes in pure culture.

[b] Schwarz and Dieckmann commented on a very small gram-negative anaerobic coccus or coccobacillus which produced black pigment on blood agar. It was found in 22 uterine cultures (of 36 classed as anaerobic streptococci) and in 9 blood cultures (of 39). Five cultures of this organism were later studied by Burdon (49) and confirmed as *B. melaninogenicus*.

[c] 135 anaerobic diphtheroids and 37 lactobacilli listed as anaerobic isolates were placed in nonpathogen category instead.

[d] Plus "some" viridans streptococci which the authors indicate might have been classified as anaerobic streptococci by others.

TABLE 12.6 Anaerobes in Chorioamnionitis

Anaerobes isolated	Aerobes isolated	Miscellaneous	Reference
5 cases putrid amnionitis		All mixed, anaerobes—aerobes, in one E. coli predominated[a]	Jeannin (11)
4 cases of amnionitis without membranes rupturing[b]			
Micrococcus foetidus	None	Amniotic fluid fetid, with much gas; metastases to lung and pleura; stillborn	Boez and Keller (1476)
Bacteroides	?		Clark and Wiersma (654)
B. fragilis (blood culture)	None	Fetid amniotic fluid and blood cultures; postpartum, baby survived	Delbove and Reynes (351)
Anaerobic streptococci, B. fragilis, (blood culture)		Fetid amniotic fluid; cervix gangrenous; postpartum, baby died	
C. perfringens		Foul, macerated, gas-filled fetus; membranes ruptured 5 days earlier	Nash et al. (1477)
B. fragilis (7 positive blood cultures)	None?	Localized chorioamnionitis—benign course	Ledger et al. (655)
Bacteroides (blood)	None	3 lb. 2 oz. hydrocephalic infant—died	Pearson and Anderson (648)
19 cases, 4 anaerobic cocci, 6 Bacteroides	6 aerobic streptococcus, 1 S. aureus, 2 S. albus	Two cases	Hirsch et al. (395)
21 cases, 15 yielded anaerobes (C. perfringens, M. foetidus, S. tenuis)	4 E. coli, 1 streptococcus, 1 staphylococcus		Krönig [cited in Jungano and Distaso (283)]
Clostridium welchii Bacteroides	None?	Chorioamnionitis associated with amniocentesis for hydramnios Chorioamnionitis. Patient severely ill	MacVicar (1477a)

[a] Anaerobes isolated: C. perfringens (4 cases). M. foetidus (3). Streptococcus tenuis (3). B. radiiformis (3). B. ramosus (2). B. caducus (2). and undetermined (3).

[b] E. coli (3 cases), S. pyogenes (1 case). Anaerobes in 3 cases were B. perfringens, B. nebulosus, Streptococcus tenuis, and 2 unidentified species.

367

Michelson (670) described a case of gas gangrene of the uterus and puerperal septicemia with *Clostridium perfringens* in a patient with myomata of the uterus. Ragan (671) described a case of gas gangrene complicating term pregnancy and reviewed the literature, noting a total of 70 cases reported up to 1959.

There are several reports of puerperal infection in association with Caesarean section. There are three such cases reported by Harris and Brown (672); these also involved abdominal wound infection and were caused by *Sphaerophorus pseudonecrophorus* either in pure culture or in association with anaerobic or microaerophilic streptococci and *Clostridium welchii*. Browne *et al.* (673) reported two cases of gas gangrene of the uterus following Caesarean section. There is a report of endometritis and peritonitis following Caesarean delivery [Ledger *et al.* (655)]; the organisms involved were *Bacteroides fragilis* and anaerobic streptococci. Tracy *et al.* (674) described a case of endometritis, peritonitis, wound infection, and bacteremia following Caesarean section due to anaerobic streptococci and *Bacteroides*.

Suppurative placentitis due to an "anaerobic enterococcus" is reported by Smith (675), and Bingold (676) reported peritonitis due to an anaerobic staphylococcus following a ruptured tubal pregnancy. McElin *et al.* (677) reported a case of postpartum ovarian vein thrombophlebitis due to microaerophilic streptococci. In order to effect a cure, it was necessary to resect a portion of the vena cava along with the ovarian vein.

The basic lesion of puerperal sepsis is endometritis and/or amnionitis. Infection may spread from the endometrium to produce parametritis and pelvic peritonitis or may cause pelvic thrombophlebitis and septicemia, with or without metastatic spread to the lungs and other organs. One important reason for the lesser virulence of aerobic pathogens that may be involved in puerperal sepsis is the fact that they produce thrombophlebitis less commonly than do anaerobic gram-negative bacilli and anaerobic streptococci. The clinical features of puerperal sepsis depend on the extent of the infection. With endometritis alone, there is usually only mild fever and some increase in lochia for the first several postpartum days. With spread of the infection, there is higher fever, chills, lower abdominal pain, and specific signs of involvement of the peritoneum and/or other structures. Putrid odor to the lochia is typical of anaerobic endometritis, and the unique morphology of the anaerobes involved may often guide the clinician to proper presumptive etiologic diagnosis.

The important conditions predisposing to puerperal sepsis are premature rupture of membranes, prolonged labor, and postpartum hemorrhage. Antimicrobial therapy is usually adequate, but heparin may be helpful in the case of thrombophlebitis. On occasion, it may be necessary to ligate the inferior vena cava when recurrent septic embolization occurs despite anticoagulation. We have seen one patient who manifested parametritis, peritonitis and tubo-ovarian abscess 4 weeks after a normal delivery, with no other apparent cause for the pelvic infection. *Bacteroides fragilis* was isolated in pure culture from the tubo-ovarian abscess, which was removed surgically. Postoperatively, the patient developed an abdominal wound infection due to this same organism.

A very serious and difficult to diagnose infection may occur in postpartum patients who have received either paracervical or transvaginal pudendal block anesthesia. Retropsoal or subgluteal infection may result (1475a,b) and the clinical picture may be very confusing, with predominant musculoskeletal symptoms. Death or serious residual such as paraplegia may result. Cases of this type are noted in Table 12.5.

Uterine Infections Due to Other Causes

Pyometra is a relatively frequent infection involving anaerobic bacteria. It is seen particularly in association with malignancy and especially when the lesion has been irradiated; however, it may also follow cervical cauterization or may be a consequence of endocervical stenosis. Table 12.7 shows several studies of the incidence of anaerobes in uterine infection other than that associated with abortion or childbirth. Again the results vary considerably from one series to another, but it is clear that anaerobes may play a very important role in this type of infection. Magara (668) found anaerobic cocci in 18% of cases of pyometra in the antibiotic era; 29% of cases failed to yield any growth. Sugiyama *et al.* (678) found that 39 of 46 cultures from patients with infection in association with carcinoma of the cervix contained anaerobic bacteria, 23 of these yielding anaerobes in the absence of any aerobes. These workers, and also Matsuda and Tanno (679), carried out an interesting study in which they were able to demonstrate that the foul odor associated with these anaerobic infections would disappear with therapy with metronidazole, a compound that is known to be very active against anaerobic bacteria. Typically the odor would disappear 3–5 days after treatment was started and would not reappear until 10–18 days after treatment had been discontinued.

Beerens and Tahon-Castel (115) reported an infection of a uterine hematoma due to *Staphylococcus anaerobius*. A similar background may have been present in the case reported by Doehner *et al.* (680) in which gas gangrene of the uterus followed a fall; in this case a coliform organism was isolated along with *Clostridium perfringens*. Kaufmann *et al.* (680a) reported a case of *Clostridium perfringens* septicemia complicating a degenerating leiomyoma of the uterus.

TABLE 12.7

Incidence of Anaerobes in Uterine Infection Other Than Postabortal or Puerperal (or Unspecified)

No. cases cultured	No. with anaerobes	No. sterile	No. with nonpathogens	Specific anaerobes found and comments	Reference
133	100 (49)[a]	21		Pyometra, 95 patients had malignancy Sulfonamide era, 8% of pyometra cases due to anaerobic cocci; 54% sterile Antibiotic era, 18% of pyometra cases due to anaerobic cocci; 29% sterile	Carter et al. (1478) Takase [cited in Kosakai and Suzuki (265)]
12	1	0	0	Pyometra Found B. melaninogenicus in great numbers, associated with other organisms, in 90% of uterine infections in his institution	Mizuno and Matsuda (657) Burdon (799)

[a] Numbers in parentheses are anaerobes in pure culture.

Pyometra is a benign condition. The symptoms include fever, local pain, bleeding, and discharge of foul-smelling pus that may contain gas bubbles. Treatment with drainage and antimicrobial compounds is usually effective.

A 44-year-old female seen in our hospital had vaginal bleeding of undetermined etiology. She underwent a dilatation and curettage that failed to reveal the underlying cause of the bleeding. Following this, she developed a purulent endometritis, complicated by pelvic abscess, peritonitis, and bacteremia. *Bacteroides fragilis* was isolated in pure culture from the pelvic abscess, the peritoneal fluid, and blood.

Although one would anticipate that insertion of intrauterine contraceptive devices (IUD's) would lead to introduction of certain elements of the normal vaginal flora and therefore set up infections, some of which would be anaerobic, the actual experience [Mishell and Moyer (681) and Sen *et al.* (682)] is that the incidence of infection following insertion of such devices is low. Cultures of the endometrial cavity obtained transfundally after hysterectomy (681, 683) are sterile in over 90% of patients if at least 24 hours has elapsed since the device was inserted. All such cultures were sterile 1 month following insertion. There is one report (684) of pelvic actinomycosis that started at the site of a chronic inflammatory process from erosion of the cervical mucosa by a metal endocervical device that had been inserted and not removed for 25 years. More recently there have been reports of 13 additional cases of actinomycosis associated with intrauterine contraceptive devices (684a, 684b, 684c). Some of these cases were localized to the endometrial cavity whereas others resulted in adnexal or pelvic infection. Kahn and Tyler (684d) reported a case of endometritis, salpingitis and pelvic peritonitis, complicated by *Fusobacterium necrophorum* sepsis, shock, and death 3 years following insertion of an IUD. Tubo-ovarian abscess and peritonitis involving *Bacteroides* following IUD insertion was described by Dawood and Birnbaum (684e).

Tubo-ovarian Infections

Table 12.8 is a summary of the literature with regard to the incidence of anaerobes in tubo-ovarian infections. The pattern is similar to that observed in other female genital tract infections, some series showing a very high incidence of anaerobes and others showing a considerably lower incidence with many instances of sterile specimens. In this connection, it is important to note that Altemeier (685) reviewed 31 series in the literature prior to his publication in 1940 and noted that among a total of 1179 cases, 53.6% were sterile on culture. Anaerobic cocci and streptococci predominate in tubo-ovarian abscesses, but anaerobic gram-negative bacilli are also encountered

TABLE 12.8

Incidence of Anaerobes, Tubo-ovarian Infections

No. cases cultured	No. with anaerobes	No. sterile	No. with nonpathogens	Specific anaerobes found and comments	Reference
36	6	15	0	4 *Bacteroides*, 2 anaerobic streptococci; ovarian abscess (no tubal involvement)	Willson and Black (1479)
25	22	2	2	Only 9 of 23 had "pathogenic" anaerobes, foul odor in 10 cases. 15 anaerobic streptococci, 7 microaerophilic streptococci, 4 anaerobic diphtheroids, 5 *B. melaninogenicus*, 2 yeastlike anaerobes, 1 unidentified gram-negative rod (anaerobic), 1 gram-positive anaerobic streptobacillus	Altemeier (685)
19	"Many"	3	?	14 were ovarian; this series deals with adnexal abscess only as a complication of pelvic surgery	Ledger *et al.* (1480)
121	6	53	8	2 each of anaerobic cocci, streptococci, and gram-positive rods	Mickal and Sellman (1481)
2 (Author)	1	1	0	Microaerophilic streptococci, anaerobic streptococci. Tubo-ovarian infection manifested during pregnancy	Friedman and Bobrow (1482)
4 (Lit.)	1	1	0?	Total of 17 cases of tubo-ovarian abscess with *B. funduliformis*; 4 in pure culture, rest mixed. Microaerophilic or anaerobic streptococci present at least three times. *Clostridium novyi* also present at least once)	Ernst (347) Hartl (249)
16	11 (3)[a] (all *Bacteroides*)	?	?	Associated with anaerobic streptococci 4 times, other streptococci (5), *E. coli* (1)	Pearson and Anderson (1465)
9	2	1	0	Cases occurring during pregnancy; literature review and one case of authors'	Dudley *et al.* (1483)
31	6 *Bacteroides*	2 (no data) ?	?	Postoperative adnexal abscess; 26 after gynecologic surgery	Ledger *et al.* (655)

372

Sample				Findings	Reference
49 patients 198 cervical cultures	42 of 135	?	30 of 135	Dominant anaerobe was anaerobic micrococcus; *Bacteroides* next, few clostridia 16 of 31 strains from control cervix cultures had anaerobes (12 patients)	Lip and Burgoyne (1484)
196 culdocentesis cultures	23 of 79	?	17 of 79	Only 1 isolate (aerobe) from culdocentesis in control group	Hirsch et al. (395)
80	37	35	6	14 anaerobic cocci, 23 anaerobic non-spore-forming rods	Mizuno and Matsuda (657)
24	4	4	0		
100	41 isolates	?	49 isolates	Microaerophilic cocci (6), *Bacteroides* (14), anaerobic cocci (19), clostridia (2)	Ringrose (1485)
(150 isolates) 18	3	8	0	2 microaerophilic or anaerobic streptococci, 1 *C. perfringens*	Nebel and Lucas (1486)
				Acute pelvic inflammatory disease.	Chow et al. (1486a, b)
Cervix-30	18	1	?	30 anaerobic strains and 17 isolates of *Neisseria gonorrhoeae* from cervical cultures. 16 anaerobic strains (13 cocci and 3 gram-positive bacilli) and 1 *N. gonorrhoeae* from cul-de-sac cultures	
Cul-de-sac-21	10	3	?		
Cul-de-sac cultures: salpingitis-54 controls-17	43 strains 5 strains	1 strain 3 strains		Well-documented and carefully studied cases of salpingitis. Patients yielded *Bacteroides fragilis* (14), other *Bacteroides* (5), *Fusobacterium* (3), *Peptostreptococcus* (7), *Peptococcus* (9), *Veillonella* (2), *Clostridium ramosum* (1), *Lactobacillus* (1), and *Propionibacterium acnes* (1). There were 51 aerobes, including 7 strains of gonococci	Eschenbach et al. (1486c)

[a] Numbers in parentheses are numbers in pure culture.

relatively frequently. Other anaerobes, including clostridia, are relatively uncommon.

Tubo-ovarian and other pelvic abscesses may be genital or nongenital in origin. Although infection may be hematogenous in origin, the majority arise by contiguity from septic abortion, uterine perforation, appendicitis, diverticulitis, rupture of any abdominal viscus, or following abdominal or pelvic surgery. Other background factors include trauma, abdominal or uterine malignancy, and degenerating fibroids. A gonococcal infection may predispose to subsequent (1486b) or concurrent infection with anaerobic bacteria; however, the excellent study of Eschenbach *et al.* (1486c) indicates that gonococcal and nongonococcal pelvic inflammatory disease are distinctive entities. The nongonococcal variety usually involves both anaerobes and aerobes, as noted in the data from their study presented in Table 12.8. The organisms involved are similar to those seen in other severe female pelvic infections, but *E. coli* is less common and clostridia and group A, B and D streptococci are apparently rare in acute pelvic inflammatory disease. The ovary itself is resistant to infection, but bacteria may gain entry to the substance of the ovary through a ruptured follicle or corpus luteum or by the hematogenous route. Tubo-ovarian infection secondary to intrauterine contraceptive devices has been described above.

Purulent vaginal discharge is relatively uncommon in tubo-ovarian abscess, and less than half of the patients have a history of prior pelvic infection. Culdocentesis is very important in establishing the diagnosis. With regard to therapy, aggressive surgical management is necessary. Immediate surgery is crucial in the event of ruptured tubo-ovarian abscess. At the time of surgery, it is important to explore the entire abdomen carefully, looking for concealed pockets of pus. In young patients, it is desirable to remove only the diseased organs. However, the nature of the disease is such that total abdominal hysterectomy and bilateral salpingo-oophorectomy is frequently necessary. Prolonged antimicrobial therapy is very important, as it is in virtually all anaerobic infections, because of a marked tendency for relapse. Patients should be treated for at least 2–3 weeks after all evidence of infection has disappeared.

Bartholin Gland Infection and Other Superficial Infections

Anaerobic bacteria are commonly involved in bartholinitis as indicated in Table 12.9. In this case, the anaerobic non-spore-forming gram-negative bacilli are more commonly encountered than are anaerobic or microaerophilic cocci. Skene's gland infection may also involve anaerobes.

TABLE 12.9
Incidence of Anaerobes in Bartholin Gland Infections

No. cases cultured	No. with anaerobes	No. sterile	No. with nonpathogens	Specific anaerobes found and comments	Reference
17	12 (2)[a]	1	1	Bartholin abscess, most fetid, quantitatively anaerobes predominated in 9 (others were gonococci)	Hallé (10)
173	26 *Bacteroides*	?	?	Associated organisms were microaerophilic or anaerobic streptococci (9), *Clostridium* (1), β-hemolytic streptococci (6), other streptococci (12), *E. coli* (1), gonococcus (1)	Pearson and Anderson (1465)
40	15	10	7	4 anaerobic cocci, 11 non-spore-forming anaerobic rods	Hirsch *et al.* (395)

[a] Numbers in parentheses are numbers in pure culture.

The difficulty in establishing the significance of anaerobic bacteria in superficial infections of the vulva and vagina where there is a large number of anaerobes normally as indigenous flora can well be appreciated. Certain features of these infections, such as the foul odor to discharges or lesions and the tissue necrosis, as well as increased numbers of anaerobes in the lesions, indicate that anaerobes are etiologically important. Such counterparts as ulcerative stomatitis and tonsillitis, which may be much more readily documented as anaerobic infections because of concurrent bacteremia with anaerobes, make it plausible that anaerobes play a significant role in the superficial infections of the vulva and vagina. Anaerobes in this type of infection have been reported by Arnold (686), Beigelman and Rantz (282), Chatillon (687), Greenbaum (688), Jump and Sperling (689), Pilot and Kanter (690) Robinson (691), and Ruiter and Wentholt (692).

More serious local infections such as noma have been reported (290). Other reports include that of an abscess of the external genitalia produced by a *Peptostreptococcus* [Hara *et al.* (176)], a vaginal abscess containing 300 ml of pus and yielding primarily *Bacteroides* on culture [Beigelman and Rantz (282)], an infected vaginal cyst which yielded *Bacteroides melaninogenicus, Fusobacterium*, anaerobic streptococci, and anaerobic diphtheroids [Weiss, (80)], and an infected cervical polyp [Pilot and Kanter, (690)]. Secondary infection of condylomata has also been reported (290).

Some workers have noted an increased number of anaerobic organisms in patients with nonspecific vaginitis [Matsuda and Tanno (679) and Müller *et al.* (693)] and have speculated as to a possible role of these organisms in this type of infection. Leigh (1420f) noted that *Bacteroides* was recovered in pure growth from a small number of patients with vaginitis, chiefly the senile and atrophic types, and that there was a response to therapy. The difficulties in establishing the significance of anaerobes in that setting will be readily appreciated. However, their role is susceptible to formal study. A study design that would incorporate therapy active only against anaerobic organisms, therapy active against aerobic organisms but not against anaerobes, the use of placebos, randomized assignment of patients to drug protocols, and careful bacteriologic studies should allow one to decide ultimately whether or not anaerobes may play a role in this type of infection. L. Ford (personal communication) has seen several cases of vaginitis with a foul discharge in young, sexually active females; there has been a good response to metronidazole, an agent active only against anaerobes. *Trichomonas* infection had been ruled out.

A number of workers have also commented on increased numbers of anaerobes recovered from patients with *Trichomonas* vaginitis [Müller *et al.* (693), Gordon *et al.* (694), Matsuda and Tanno (679), and Yamaji *et al.* (695)]. Whether the anaerobes play any role in *Trichomonas* infection or whether they are simply secondary invaders with no significance is difficult to decide. It is interesting to note however that two agents used for management of *Trichomonas vaginalis* infections, metronidazole and furazolidone, are quite active against anaerobic bacteria as well as against *Trichomonas*. Vaginitis emphysematosus is a relatively rare, benign, self-limited condition characterized by multiple gas-filled cystlike cavities of the vaginal and cervical mucosa. It is seen most often in pregnant patients. The gas recovered from the cysts has been noted to have a relatively high content of carbon dioxide (696, 697). Gardner and Fernet (697) have reviewed 145 cases from the literature and reported additional cases. In their own 10 cases, they note that 7 patients had associated *Trichomonas vaginalis* infections and 3 had associated *Haemophilus vaginalis* (*Corynebacterium vaginale*) infection. They also referred to a paper by Wilbanks and Carter in which trichomonads were noted in 3 of 4 patients with this disease and in which work was quoted showing that bacteria-free preparations of *Trichomonas vaginalis* can produce gas in subcutaneous tissues of guinea pigs. Gardner and Fernet noted rapid clearing of the emphysematous vaginitis in 3 of their patients who were treated for the *Trichomonas* infection, with metronidazole in two cases and furazolidone in the other. All of these things suggest the possibility of a role for anaerobes in the etiology of emphysematous vaginitis.

Postoperative Gynecologic Infections

Table 12.10 shows that anaerobic bacteria of all types are frequently involved in postoperative infections following various types of gynecologic surgery. As indicated earlier, gynecologic infections may result from other types of surgery as well. Arnar *et al.* (521) reported a case of an infected hysterotomy wound with muscle necrosis due to *Clostridium perfringens* and *Peptococcus*, along with aerobic streptococci (α-hemolytic) and *Pseudomonas* following Caesarean section. A number of other cases of infection following Caesarean section were referred to earlier.

An excellent discussion of postoperative pelvic infections is presented by Ledger (698). Oddly enough, young women seem to have the highest incidence of infection following surgery. To some extent this may reflect the amount of surgery done on people in this age group. However another important factor is the vaginal surgical approach. This is, of course, in turn related to the abundant bacterial flora of the vagina. A thorough surgical preparation of the vagina in the operating room, using an iodophor, is important in order to reduce the number of organisms. Duignan and Lowe (698a) have shown that *Bacteroides fragilis* is killed rapidly by dilute solutions of povidone–iodine and that a 2 minute preparation with this compound is

TABLE 12.10

Incidence of Anaerobes in Postoperative Gynecologic Infections

Type of surgery	No. infected cases	No. with anaerobes	No. sterile	No. with nonpathogens	Miscellaneous	Reference
Hysterectomy	64	63	0	9	2, anaerobes only; 61, mixed. Anaerobes isolated: anaerobic cocci (49), *Sphaerophorus* (5), *Bacteroides* (4), *C. perfringens* (1), *Clostridium* sp. (2)	Hall *et al.* (1487)
Miscellaneous gynecologic surgery	7	7	0	0	All had *Bacteroides*; 3 had anaerobic streptococci and 2, other streptococci also	Pearson and Anderson (1465)
See Table 12.8 8 vaginal hysterectomy	8	4 (3)[a]	2[b]	⎫		Ledger *et al.* (655)
2—one caesarean section, one resection of corpus luteum	2	2 (2)	0	0 ⎬	All infections adnexal	Ledger *et al.* (1488)

Procedure						Reference
19 major vaginal procedures, 7 abdominal procedures, 2 minor vaginal procedures, 1 radium insertion	29	2	6	3[b]		Lee and Turko (1489)
Vaginal hysterectomy	21 (4 had prophylaxis with cephaloridine)	8 of 45 isolates	?	11 of 45 isolates		Ledger et al. (1490)
Vaginal surgery	143	70	15	20	Anaerobic cocci (45), non-spore-forming anaerobic rods (25)	Hirsch et al. (395)
Abdominal surgery	177	74	54	35	Anaerobic cocci (46), non-spore-forming rods (28)	
6 cases postoperatively, surgery for cancer of cervix	All yielded anaerobes, no aerobes isolated				*B. fragilis* (3), *Bacteroides* sp. (1), *F. necrophorum* (2), *Peptococcus* (1)	Kamiya et al. (1491)

[a] Numbers in parentheses are numbers in pure culture (no aerobes).
[b] No infection.

very effective for preoperative disinfection of the vagina. Prior to opening the vagina to remove the cervix, the operative field should be well isolated by surgical draping. Following the vaginal closure, all towels, instruments, and gloves should be discarded and replaced with sterile equipment. If the interval between conization of the cervix and hysterectomy is too long, bacteria will proliferate in the devitalized conization site where poor tissue perfusion and clotted blood provide an excellent environment for bacterial growth (Fig. 12.6). When feasible, the surgery should be performed within 2 days of the time of conization or not until 6 weeks or more following it to allow adequate healing. The management of the ovarian pedicle may be an important factor increasing the morbidity after vaginal hysterectomy; the pedicle is sutured to the angle of the vaginal cuff rather than lying free within the pelvis as is true after abdominal hysterectomy. The close proximity of this ovarian pedicle to the infected postoperative vaginal cuff permits the spread of organisms to the adnexa and may contribute to later development of adnexal abscess (Fig. 12.7).

When Caesarean section is carried out in the presence of prolonged rupture of the membranes, there is a distinct hazard that amnionitis will be present and that there will be excessive contamination of the operative field with endogenous bacteria (Fig. 12.8). Aggressive obstetrical management should minimize delay from the onset of labor and thus make bacterial contamination of the operative field less likely. A low segment Caesarean section will usually minimize or prevent spillage of amniotic fluid or infected material from the uterus into the peritoneal cavity. When it appears that the operative field is likely to be contaminated, aerobic and anaerobic culture should be obtained at the time of operation so that in the event of

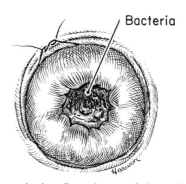

Fig. 12.6. Cervix after conization. Poor tissue perfusion and clotted blood provide an excellent environment for bacterial growth. This may result in excessive bacterial contamination of the operative field if the conization–hysterectomy interval is greater than 48 hours. From Ledger (698). Reproduced with permission.

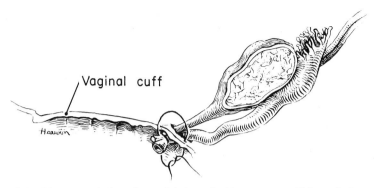

Fig. 12.7.   Management of ovarian pedicle in vaginal hysterectomy. This surgical maneuver places ovary close to infected vaginal cuff and may account in part for development of a postoperative adnexal abscess. From Ledger (698). Reproduced with permission.

subsequent infection intelligent therapy may be applied much sooner. In the event of overt infection, of course, therapy should be started without waiting for the culture report, based on the likely infecting organism and examination of the gram stain smear.

Postoperative infection is manifested either as a localized vaginal cuff abscess, diffuse pelvic cellulitis, or adnexal infection. Distinction between

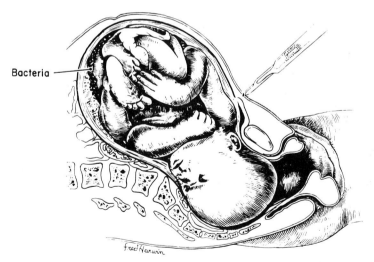

Fig. 12.8.   Cesarean section in the presence of amnionitis. There is excessive bacterial contamination of the surgical field and possibility of continued contamination of peritoneal cavity after operation. From Ledger (698). Reproduced with permission.

these different entities is not always clear-cut, and not infrequently post-operative infection will be manifested simply by fever. Nevertheless, management will be assisted by the precise assignment of a specific diagnostic category, whenever possible. Of course, more remote infections, such as urinary tract infection (particularly where an in-dwelling catheter has been used) or postoperative pneumonia, may be responsible for postoperative fever, in addition to the pelvic infections with which we are more immediately concerned.

Vaginal cuff infections manifest themselves in the early postoperative period, almost all of them being diagnosed within the first 10 postoperative days. The main presentation is fever with little in the way of systemic symptomatology. Drainage of the cuff collection results in rapid regression of the fever as a rule.

Pelvic cellulitis is usually recognized later in the postoperative period but before discharge from the hospital. Pelvic cellulitis may be distinguished from vaginal cuff infection either by absence of purulent material at the vaginal apex on pelvic examination or continuation of fever after adequate drainage of the vaginal cuff is established. A pelvic mass cannot be detected, and the patients respond clinically to appropriate systemic antimicrobial therapy.

Adnexal infections have already been discussed. Postoperative adnexal infections are usually discovered quite late in the postoperative period, almost always after the patient has been discharged from the hospital.

Treatment of urinary tract infections with agents such as sulfonamides or nitrofurantoin will minimize the possibility of masking other pelvic causes of fever or morbidity. In the absence of specific bacteriological data to guide therapy of pelvic postoperative infections, clindamycin or chloramphenicol would represent good starting choices, depending on the severity of the patient's illness. Surgical drainage is important for adnexal abscess, as has already been stressed.

Actinomycosis

Actinomycosis rarely affects the female genital tract. Involvement of the external female genitalia is extremely rare, but has been reported (210, 698b). Usually, primary involvement with actinomycosis is in the adnexa, and particularly the ovary. Most often such infection originates in the appendix or elsewhere in the intestinal tract and spreads by direct extension. Less commonly, the infection may spread by the lymphatics or occasionally by way of the bloodstream. Rarely, infection may be introduced from the outside in relation to criminal abortion or the use of an endocervical contraceptive device, as indicated earlier.

The involved ovary is always enlarged and, as a rule, firm. There may be formation of small abscesses. Adhesions are common, and fistulae leading to other organs or to the surface of the body may be seen as well. The body of the uterus is occasionally involved either as an isolated phenomenon or as part of involvement of the adnexa. Disease may spread to involve the entire pelvis (212, 699).

Miscellaneous

Septic thrombophlebitis should be suspected in women in the puerperal period or in the period following gynecologic surgery when there is unexplained fever, with or without vague abdominal symptoms, particularly where there are no physical findings and there has not been a response to antimicrobial agents. Such thrombophlebitis is commonly encountered in anaerobic infection. Blood cultures may be positive, particularly if appropriate technique is used to recover anaerobic organisms; however, blood cultures are not uncommonly negative. Pelvic thrombophlebitis may be difficult to diagnose and is basically a diagnosis of exclusion. Addition of heparin to the regimen will often bring about response and may actually serve as a diagnostic test (700–702).

Anaerobic bacteria are not uncommonly involved in septic shock in obstetrical and gynecologic patients. Table 12.11 lists a number of cases of septic shock in such patients; many others could be added. While endotoxin of certain gram-negative anaerobic bacilli may be involved in a number of these cases, it is clear that there are other mechanisms as well, as in aerobic bacteremia. A number of organisms not possessing endotoxin, notably anaerobic streptococci, may often be involved in the genesis of such shock.

Abscesses of adjacent parts, such as the groin, perirectal or ischiorectal areas, the upper leg, and the abdominal wall, may be related to anaerobic infections of female genital organs. Paraurethral abscesses have been discussed separately in Chapter 11. We have seen ischiorectal and upper thigh abscesses due to *Bacteroides fragilis* whose origin was in female genital organs.

Hare and Polunin (703) have raised the interesting possibility that anaerobic coccal postabortal or postpartum infections may be a major cause of infertility among certain tribes in North Borneo. In these populations, infertility following pelvic inflammation is common, and anaerobic cocci

TABLE 12.11 Anaerobes in Septic Shock in Obstetrics and Gynecology

Underlying condition	Anaerobe(s) isolated	Aerobes isolated	Miscellaneous	Reference
Abortion	B. fragilis	None	Blood pressure 60/50, marked hemolysis of serum, died. Foul odor—vagina; B. fragilis also cultured from endometrium	Dilworth and Ward (423)
Abortion	6, anaerobic streptococci; 1, anaerobic streptococci and Bacteroides; 1, C. perfringens	None		Smith et al. (416)
Abortion	B. necrophorum, anaerobic streptococci	None		Jones (1492)
Puerperal sepsis	3 Bacteroides	None		Pearson and Anderson (648)
Peritonitis following hysterectomy	B. fragilis	None		Hawbaker (343)
Abortion, 3 cases	1 C. perfringens (blood) 1 Bacteroides and anaerobic cocci (uterus). 1 anaerobic gram-negative cocci (blood)	γ-Hemolytic streptococcus, E. coli, S. albus Pseudomonas		Stevenson and Yang (1493)
Abortion	Bacteroides sp., anaerobic streptococci	None		Tynes and Frommeyer (489)
Endometritis	Bacteroides sp.	None		
Posthysterectomy	Anaerobic streptococci	None		Vermillion et al. (426)
Premature rupture of membranes, uterine infection and sepsis	Bacteroides sp.	Paracolon	Foul-smelling placenta	Adams and Pritchard (1494)
Abortion	B. fragilis	None	Shock for 3 days, acute tubular necrosis, profound thrombocytopenia	Rotheram and Schick (401)
Endometritis, salpingitis, pelvic peritonitis	Fusobacterium necrophorum (from blood)	None	Infection secondary to intrauterine contraceptive device	Kahn and Tyler (684d)

TABLE 12.12

Summary of Bacteriology of Anaerobic Obstetric–Gynecologic Infections from 15 Reports in the Literature[a]

Bacteria	No. of cases
Bacteroides and other gram-negative anaerobic bacilli only	52
Peptostreptococcus only	2
Peptococcus only	1
Two or more anaerobes	106
Mixed cultures, aerobes and anaerobes	175
Bacteroides and other gram-negative anaerobic rods	183 strains
Anaerobic cocci	120
Eubacterium	2
Actinomyces	1
Clostridium	9
Aerobes	202

[a] From K. Ueno [in Kozakai and Suzuki (265)].

of certain specific biochemical groups, as defined by Hare, were isolated from over one-third of the population that was infertile and only 10% of the fertile population.

Ueno [cited in Kozakai and Suzuki (265)] has summarized the bacteriology of anaerobic obstetrical and gynecologic infections from 15 reports in the literature. This data is presented in Table 12.12.

13

Infections of Skin, Soft Tissue, and Muscle

Infections Involving Skin or Skin Structures Primarily

There are many reports of *subcutaneous abscesses* in various sites in the body, usually representing metastatic spread of anaerobic infection, but occasionally appearing as a primary process. Inguinal abscesses are relatively common but any area of the body may be involved. In addition, various *pustular eruptions* may be noted during the course of anaerobic bacteremia; Beerens and Tahon-Castel (115) described 14 cases, 9 of which had various anaerobes in pure culture. Lesions of these types may also be seen as part of actinomycosis; fistula formation may also occur. Lesions of these types, and wound infections involving the skin and subcutaneous tissue as well, will not be discussed specifically since they are really secondary to another type of anaerobic infectious process. The role of anaerobic bacteria in acne is discussed in Chapter 21.

Several authors describe a chronic *pyoderma*, with or without draining sinuses, related to anaerobic infection. Bluefarb and Gecht (704) described a case involving the scalp, jaw, and neck. The discharge from these lesions had a fetid odor, and fusiform bacilli and spirilla were seen on direct smear. Rentsch (195) and Maibach (705) described scalp pustules due to *Corynebacterium acnes*, which are frequently misdiagnosed as staphylococcal folliculitis. Nobre and Caldeira (706) described cases of chronic nodular pustular lesions of the legs, some with draining sinuses, from which *C. acnes* was recovered in pure culture; only gram-positive rods were seen on direct smear of material from these lesions.

There are a number of reports of involvement of anaerobic bacteria in a conventional type of *cellulitis*. Wohlstein (707) described a case of a veterinarian who had manually removed a placenta from a cow, following which he developed a mild cellulitis and lymphangitis of the forearm. The gram stain of material from the lesion showed organisms typical of *S. necrophorus*, but the author was unable to recover the organism in culture. In 1932, Stolzova-Sutorisova and Kratochvil (708) described a patient who developed cellulitis and then lymphangitis following a pedicure. Subsequently, the patient developed several abscesses of the leg and thigh up to the groin from which *B. ramosus* plus a streptococcus were isolated; the patient also had bacteremia with *B. ramosus*. Herrell *et al.* (709) described three cases of severe cellulitis with anaerobic streptococci, Vincent's spirilli, and an aerobic α-hemolytic streptococcus. In 1951, Wright *et al.* (710) described a case of cellulitis of the buttock which yielded *Bacteroides* plus an anaerobic streptococcus on culture. Ruiter and Wentholt (711) described a case of low-grade cellulitis of the umbilicus in which the lesion was covered with a dirty gray membrane. Smear of material from this area revealed fusiform bacilli and spirochetes; culture yielded *F. necrophorum*. Brocard and Pham (712) discussed two cases of gangrenous erysipelas in which the lesions yielded both the expected streptococci and *Bacteroides terebrans*. Behrend and Krouse (713) described a very interesting case of fatal postoperative bacterial synergistic cellulitis (not gangrene) of the abdominal wall following herniorrhaphy. The lesion consisted of an intense boardlike edematous cellulitis without any evidence of gangrene. During the course of making several skin incisions, searching for the possibility of underlying pus, the surgeon was impressed with the tissue consistency, stating that it cut as though it were leather. The patient ultimately went into shock, manifested intravascular clotting, and then died. Cultures yielded an anaerobic nonhemolytic streptococcus and a gram-positive aerobe, *Bacillus subtilis*.

Primary cutaneous actinomycosis is rare, but several cases have been described (318, 714–716).

Buck and Kalkoff (717) described some interesting findings in patients with *perioral dermatitis*. In this condition, pseudopustulous papules of pinhead size are found about the mouth. Smears of these papules revealed fusobacteria-like organisms in 35 of 40 patients studied. The authors were unable to grow the organism, although they acknowledged that they were unable to use good anaerobic procedures. A good therapeutic response was obtained in this condition with demethylchlortetracycline. The authors did not find similar appearing organisms on smears from cases of acne or rosacea.

An unusual institutional outbreak of *subungual infection* in the newborn was described by Sinniah *et al.* (718). The illness passed through several stages, the first being a vesicular stage during which a little clear fluid appeared under the center of the nail. This later became purulent and then was absorbed, leaving a brown stain that gradually faded. The first stage lasted up to 1 day, the second 1 to 6 days, and the third, 2 to 6 weeks. There were a total of 42 newborn infants with this unusual illness, and the number of fingers affected ranged from 1 to 10; toenails were spared. The infants were otherwise healthy, and none of the mothers or staff had any clinical evidence of disease. From six infants with florid lesions, subungual pus was obtained by aseptic puncture of the nails. All six showed pus cells and some macrophages. Three of the specimens showed numerous tiny gram negative cocci about 0.4 μm in diameter; none were intracellular. Aerobic and anaerobic culture failed to recover the organisms, but the authors felt that the cocci morphologically resembled *Veillonella*. Early discharge of patients and/or barrier nursing led to an abrupt stop of the outbreak.

The report of Sandusky *et al.* (112) suggests that anaerobes may be involved in paronychia much more commonly than we have generally appreciated. They described seven cases from which were recovered six anaerobic streptococci and two microaerophilic streptococci, along with 14 aerobic bacteria. Lahelle (719) described one case in which fusiforms and spirochetes were seen on smear; the fusiforms were recovered in culture. Stokes (108) also described one case from which an anaerobe (type not specified) was recovered in pure culture.

Anaerobes may also be found with some frequency in *infected sebaceous or inclusion cysts*. Stokes (108) reported that 21 of 66 specimens that were positive for any organism yielded anaerobes on culture, 13 in pure culture (type not specified). Sandusky *et al.* (112) reported on seven infected sebaceous cysts, all of which yielded anaerobic streptococci, three in pure culture and four with one aerobe each in addition. Although the number of cases was not specified, Bornstein *et al.* (275) and Pien *et al.* (390) implied that they had several cases in which *Bacteroides* and anaerobic streptococci and anaerobic cocci were isolated. Beigelman and Rantz (282) described one case from which *Bacteroides* was isolated in pure culture. Saksena *et al.* (487) described recovery of Bacteroidaceae from infected sebaceous cysts (number not specified). Prévot (21) described one sebaceous cyst from which *Diplococcus plagarumbelli* was isolated and also an infected lipoma from which the same anaerobic coccus was recovered. Our group has also isolated anaerobes, chiefly anaerobic cocci, from infected sebaceous or inclusion cysts.

Judging from the foul odor of the pus obtained typically from patients with *hidradenitis suppurativa*, one would anticipate finding anaerobic bacteria as a cause of secondary infection very consistently. Somehow this has not been described very often. Smith and Ropes (226) recovered *Bacteroides funduliformis* mixed with either staphylococci or streptococci repeatedly from one patient. Beigelman and Rantz (282) described one case from which a heavy growth of *Bacteroides* was obtained, along with *Staphylococcus albus* and a few nonhemolytic streptococci. Parker and Jones (720) found mixed anaerobic cocci in one case. Hall *et al.* (721) noted that two patients with suppurating lesions had a good response to lincomycin therapy; no bacteriologic studies were done. Pien *et al.* (390) recovered anaerobic cocci from an unspecified number of patients with this condition. Our laboratory has recovered both anaerobic gram-negative rods and anaerobic cocci of various types from several patients with hidradenitis suppurativa.

Several reports indicate that anaerobes are commonly involved in *infection of pilonidal cysts*. Sandusky *et al.* (112) described 13 cases of pilonidal cyst abscess, all of which yielded anaerobic streptococci on culture (four in pure culture). Other organisms isolated from these 13 patients included 2 strains of *Clostridium perfringens*, 1 anaerobic coccus, 1 anaerobic gram-negative bacillus, and 19 aerobic or facultative bacteria. Beigelman and Rantz (282) described one case from which *Bacteroides* was isolated. Parker *et al.* (722) reported five cases of pilonidal cyst abscess in 1953; one yielded no growth on culture and two yielded only *S. albus*, suggesting that more fastidious organisms such as anaerobes might have been present. Bornstein *et al.* (275) recovered *Bacteroides* and anaerobic streptococci from an unspecified number of pilonidal cyst abscesses. Farquet *et al.* (723) reported one case from which *S. funduliformis* was recovered. Saksena *et al.* (487) recovered Bacteroidaceae from 12 pilonidal area infections. Rentsch (195) reported two sacral dermoid cysts infected with *Bacteroides fragilis*, two with *Bacteroides putidus*, one with *Sarcina ventriculi*, and an unspecified number with *Bacteroides funduliformis*. Schaffner (1240a) recovered *Eggerthella convexa* in pure culture from an infected pilonidal sinus. *Bacteroides* was recovered from 6 pilonidal sinuses by Mitchell (110a) and from 3 by Nobles (384a). Leigh (1420f) found *B. fragilis* in 4 cases of pilonidal sinus. *Peptococcus niger* was recovered from a pilonidal cyst (1435c). We have seen one patient with pilonidal cyst abscess from which anaerobes could be recovered; *Bacteroides fragilis* and anaerobic gram-positive cocci were grown in the absence of any aerobic or facultative organisms.

It is not generally appreciated that anaerobic bacteria are involved commonly in *infected diabetic foot ulcers*, with or without associated osteomyelitis. Meleney did stress the importance of anaerobes in diabetic gangrene in his discussion of Zierold's paper (724) in 1939. In a study from our group [Ziment *et al.* (725)], it was noted that seven of eight patients with anaerobic osteomyelitis of the foot, all of whom had acute or chronic soft tissue infection, were diabetic. We have also seen diabetic patients with anaerobic cellulitis and infected vascular gangrene due to anaerobic bacteria in which the process initially began with an infected ulcerated lesion of the foot. In addition, we have seen one patient who developed an infection in the stump of his right thigh following an amputation. A full gallon of pus was drained, from which we grew *Bacteroides fragilis* as well as a gram-positive non-spore-forming anaerobic rod, *Enterobacter aerogenes*, and α-hemolytic streptococci. Weiss (80) also reported an infected amputation stump in a diabetic from which he recovered *Bacteroides melaninogenicus*, anaerobic diphtheroids, *Proteus*, and hemolytic streptococci. Pien *et al.* (390) found anaerobic cocci in one patient with a diabetic foot ulcer with underlying osteomyelitis. Louie *et al.* (725a) studied the bacteriology of 8 diabetic foot ulcers (no coexisting osteomyelitis) and recovered anaerobes from all, a total of 23 strains. There were 20 aerobes recovered. One patient yielded a pure culture of *Clostridium perfringens*; all others had mixed infections. Principal anaerobic isolates were *Peptococcus* sp. (8), *Bacteroides* sp. (6), and *Clostridium* sp. (5). Haldane and van Rooyen (529) recovered *Bacteroides*, anaerobic streptococci, and *Staphylococcus aureus* from a gangrenous ulcer on the heel of a patient with arteriosclerosis obliterans.

Tropical ulcer is a chronic sloughing ulcer which may take on a more invasive character and which has hardly any tendency for spontaneous healing. It was described as early as 1792 by Hunter [quoted by Castellani and Chalmers (726)] as follows: "Sores . . . spread quickly, and form large ulcerated surfaces . . . . The granulations turn flaccid, and even mortify in part. The portion skinned over ulcerates afresh, and the sore becomes larger than ever. Ulcers of some standing . . . could not be healed in that country . . . . Opportunity was taken to send home men with ulcers." The disease is found in all tropical and subtropical regions of Africa, Asia, and America. It is particularly common in certain areas such as Aden, Cochin, Tonkin, and some islands of the Red Sea. The etiology is probably complex, but it is clear that fusiform bacilli and spirochetes, which are identical morphologically to *Treponema vincentii*, play a very important role. There are undoubtedly a number of predisposing causes, such as a hot damp climate, malnutrition (perhaps particularly involving vitamin A), local trauma, concurrent debilitating diseases, and poor hygiene. This is all summed up very graphically by Marsh and Wilson (727) by: "filth, food, friction, and fuso-spirillosis."

The disease is not highly contagious and indeed some self-inoculation experiments were unsuccessful. On the other hand, there is evidence of transmission to susceptible individuals under crowded and nonhygienic conditions. Strong (728) cited the case of a nurse whose arm was accidentally scratched by an infected knife with which a tropical ulcer had been excised 2 hours previously; despite washing and application of antiseptics, a tropical ulcer did eventually develop at the site.

Tropical ulcer generally affects the lower limbs, particularly the lower third of the leg, the ankle, and the dorsum of the foot. Occasionally, it appears on other uncovered parts of the body. In most cases there is a single ulcer, but two or more ulcers may be found in some patients. The disease begins with a small painful papule surrounded by a deeply infiltrated dusky red zone. This lesion becomes purulent and begins to slough to form an ulcer that gradually extends in depth and surface. Except in very old cases, the margins of the ulcer are not particularly raised nor thickened, nor are they undermined. The surrounding area is often edematous and may be somewhat painful on pressure, but it is notable that, in general, there is very little pain associated with the condition. The whole ulcer is generally covered with a thick dirty gray secretion with a highly offensive odor. On occasion, if left untreated, the ulcer may assume a real phagedenic character involving a large area and extending into the deeper structures to destroy muscle, tendon, and periosteum. Fusiform bacilli and spirochetes are found in abundance in the superficial layers of the ulcer, and there may be colonization with other bacteria, such as gram-positive cocci and gram-negative rods. Untreated lesions commonly reach a diameter of 5 to 10 cm and may persist for years without endangering the life of the patient. Favorable results have been achieved with topical therapy, including tyrothricin [Kolmer (729)] and combinations of streptomycin, bacitracin, and polymixin [Loughlin et al. (730)]. Agents effective against anaerobic bacteria of the types involved in this condition are very effective therapeutically. Thus, penicillin (731–733), tetracycline (734), chloramphenicol (735, 736), and metronidazole (737) have all proved to be effective.

Infections Involving Subcutaneous Tissue, with or without Skin Involvement

Anaerobic bacteria also commonly set up *infections in superficial ulcers* of various types. Pien et al. (390) found anaerobic cocci very commonly in ischemic foot or leg ulcers; 25% of the total isolates of anaerobic cocci in their series (85 isolates from 70 patients) were from this source. Manson (738) found that 6 of 27 infected ulcers yielded clostridia; two cases of gas gangrene followed amputation of legs with such lesions. Sandusky et al. (112) studied three varicose ulcers and six of other types, all of which were infected, and found that all of these had anaerobic streptococci; two also had microaerophilic streptococci, and there were a number of aerobic and facultative organisms isolated. Myers *et al.* (739) studied chronic cutaneous ulcers that were secondarily infected and found that 8 of 21 gave a very distinct red fluorescence under ultraviolet light. This characteristic finding is typical of infection with *Bacteroides melaninogenicus*. Cultures were made of 5 of these 8 lesions and all showed *Bacteroides melaninogenicus*, other *Bacteroides*, staphylococci, streptococci, and various aerobic gram-negative bacilli. Shevky *et al.* (281) also isolated *Bacteroides melaninogenicus* from one open leg ulcer. Beigelman and Rantz (282) recovered *Bacteroides* from one case of pyodermic ulcer. Mitchell (110a) recovered *Bacteroides* from a varicose ulcer. Prévot (21) recovered *Streptococcus anaerobius* from a varicose ulcer. Rathbun (419) recorded a case of *Clostridium perfringens* bacteremia secondary to a small ulcer near the anus.

Anaerobes are commonly involved in infection of *decubitus ulcers*. Such ulcers are commonly located in proximity to the anus, and anaerobic conditions are present because of the tissue necrosis and undermining. In addition, of course, patients with decubitus ulcers are typically debilitated. The significance of this type of infection has really been brought out in striking fashion in recent years by demonstration of the frequency with which such infected decubitus ulcers lead to anaerobic bacteremia. Felner and Dowell (168) reported seven cases of *Bacteroides fragilis* bacteremia secondary to infected decubitus ulcers. One of these patients also had a gangrenous foot, and three had positive cultures for *Bacteroides fragilis* from the local lesion. Marcoux *et al.* (420) reported six cases of *Bacteroides* sepsis secondary to infected decubitus ulcers. Gelb and Seligman (490) reported four fatal cases of Bacteroidaceae sepsis complicating infected decubitus ulcers; all of these patients were severely debilitated following cerebral vascular accidents. Rathbun (419) recorded four cases of infected decubitus ulcers leading to bacteremia; three of these were due to *Clostridium perfringens* (one also with *E. coli*) and one to *Clostridium paraputrificum*. Clostridial bacteremia in association with decubitus ulcers has also been noted by Gorbach and Thadepalli (486). Tynes and Frommeyer (489) recovered *Bacteroides* from two decubitus ulcers; both patients also had *Bacteroides* bacteremia. Kagnoff *et al.* (405) also reported *Bacteroides* sepsis following infected decubiti but did not specify the number of patients. Ellner and Wasilauskas (740) reported two fatal cases of bacteremia due to *Bacteroides* with the portal of entry being infected decubitus ulcers. Additional reports of bacteremia due to *Bacteroides fragilis* or other Bacteroidaceae which were secondary to decubitus ulcers are those of Schoutens *et al.* (1289a), 15 patients; Douglas and Kislak (740a), 4 patients; Chow and Guze (1573), 6 patients; Shimada *et al.* (1420c), 1 patient; and Nobles (384a), 1 patient.

There are also reports of infected decubitus ulcers without bacteremia, but the frequency with which this occurs is probably not generally appreciated. Manson (738) studied five cases of sacral decubitus ulcer and found that all harbored clostridia. Saksena *et al.* (487) reported a case of decubitus ulcer with Bacteroidaceae recovered from the lesion. Haldane and van Rooyen (529) recovered *Bacteroides*, *Proteus mirabilis*, and enterococci from a diabetic with a decubitus ulcer. Bazilevskaya and Polozova (741) cultured 100 infected decubitus ulcer patients and found that 28 yielded anaerobes, usually *Clostridium perfringens*. Schaffner (1240a) recovered 4 anaerobes, a *Vibrio* and 2 aerobes from an infected decubitus ulcer. Other reports are those of Chow *et al.* (1420e) with 5 cases (unspecified anaerobes); Mitchell (110a), 24 cases yielding *Bacteroides*; and Leigh (1420f), 3 cases with *B. fragilis*. We have recovered a wide variety of anaerobes from a number of patients with infected decubitus ulcers.

Noma, otherwise known as gangrenous stomatitis or cancrum oris (Fig. 13.1), is a term used to designate all forms of spontaneous gangrene involving mucous membranes or mucocutaneous orifices. It occurs most frequently about the mouth, but it may affect the nose, the auditory canal, the vulva, prepuce, or anus. In the early stages, constitutional symptoms are not severe, but with the development of noma the symptoms become

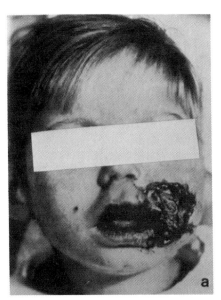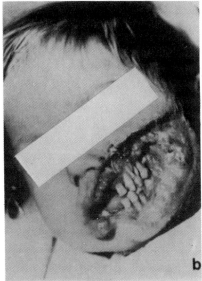

Fig. 13.1. (a) Noma: 4-year-old boy referred at the end of World War II with characteristic necrosis of left cheek tissue. (b) After detachment and removal of necrotic tissue. From Boering (208). Reproduced with permission.

severe suddenly. There is marked prostration, fever of 102°–103°F or higher, apathy, and weakness. The odor of the breath or a dull red spot on the cheek may be the first indication of the disease. Inspection of the mouth will reveal a dark, greenish-black area on the gum or the inside of the adjacent cheek. The surrounding tissues are reddened and edematous so that the affected part may be two to three times its normal size. There is an extremely offensive odor. The destructive process extends with great rapidity, attacking the periosteum and the underlying bone. The gums are destroyed, and the teeth fall out. Large bony sequestra may form and then come away. In untreated cases, the process may extend to the other cheek, into the nose, and may result in almost complete destruction of the face before death. The usual duration of the disease without therapy is 5–10 days. There is very little or no pain and children sometimes push their fingers through the necrotic area in the cheek. Hemorrhage is rare due to thrombosis of vessels in the area. If the patient succeeds in walling off the process spontaneously or if effective therapy is used, a line of demarcation appears, and the affected tissue is ultimately sloughed. Prior to the availability of antimicrobial therapy, the mortality was 70 to 100% [Blumer and MacFarlane (742)]. Emslie (743) pointed out that the effectiveness of therapy has actually presented a serious problem in certain parts of the world. Presently, many children who previously would not have survived are living and present facial defects that are virtually unreparable by plastic surgery.

Major factors predisposing to noma include systemic disease, malnutrition, and poor oral hygiene. Among the systemic diseases, measles stands out in terms of the frequency with which it is seen as a background problem in patients developing noma. Smallpox, malaria, and parasitic infestations may also predispose. Herpetic gingivostomatitis may predispose to acute ulcerative gingivitis and then to cancrum oris. Noma is quite rare in adults but a few cases have been noted in association with carcinoma (290) and in subjects in concentration camps (743).

Smears of lesions show a profusion of fusiform bacilli and spirochetes. In a study carried out by Emslie (743) in Nigeria, specimens were shipped to MacDonald at the Forsyth Dental Infirmary, Harvard School of Dental Medicine. Examinations of the smears before and after shipment revealed a significant loss of organisms during the shipping process; nevertheless, MacDonald was able to isolate *Bacteroides melaninogenicus* from some of the specimens. This, together with experimental studies from the Forsyth group, which suggest that *Bacteroides melaninogenicus* may be a key pathogen in so called mixed fusospirochetal or anaerobic infections, is of great interest. Intensive penicillin therapy is the treatment of choice, but other antimicrobial agents active against anaerobes may be suitable in penicillin-sensitive patients. Obviously, therapy should be started as early as possible in the course of the illness. Progressive oral gangrene has been described in patients with acatalasemia by Takahara (744). Emslie (743) also pointed out that undermining gangrene of the skin may be found occasionally in Nigeria. He described three patients with facial gangrene of this type which did not arise within the mouth. Prévot (21) reported isolation of *Streptococcus putridus* from a case of extensive cutaneous gangrene. Garrod (336) mentioned two cases of spreading subcutaneous gangrene of abdominal operative wounds. These wounds discharged a foul brown material from which he recovered *Bacteroides melaninogenicus*.

Anaerobic cellulitis is an acute anaerobic infection of soft tissue (Figs. 13.2 and 13.3). The term is not properly descriptive but persists because of common usage. It is synonymous with "gas abscess." It is sometimes known as clostridial cellulitis, but the clinical picture in the condition involving clostridia is not necessarily different from that involving non-spore-forming anaerobic bacteria. The process is a necrotizing one involving the epifascial, retroperitoneal, or other connective tissues of the extremities, perineum, abdominal wall, buttocks, hip, thorax, or neck which are easily contaminated by various discharges, including those from the intestinal, genitourinary, or respiratory tracts. The lesion may spread rapidly, but does not always do so. Predisposing factors include contamination of subcutaneous tissues from an operative or accidental wound or from preexisting localized infection. Many

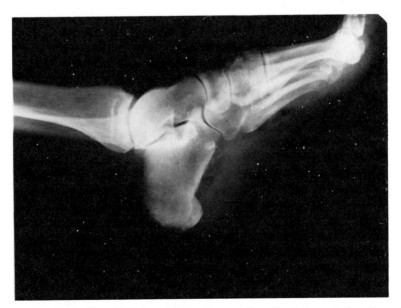

Fig. 13.2.   Anaerobic cellulitis in a diabetic. Note swelling of soft tissues and gas beneath the arch of the foot. The infection extended up to midthigh.

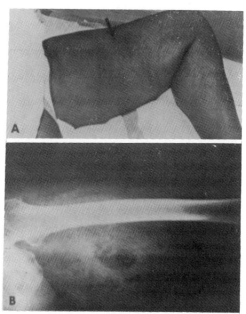

Fig. 13.3. (A) Nonclostridial crepitant cellulitis of left thigh recently traumatized by a fall in a 64-year-old acidotic diabetic with radiographic evidence (B) of gas in soft tissues. Cultures grew *Staphylococcus aureus*, aerobic and anaerobic nonhemolytic streptococci, and *Bacteroides* sp. From Pulaski (1589). Reproduced with permission.

of the patients with this process are diabetics. Certainly the high sugar content of the tissues and anaerobic conditions may contribute to extensive gas formation by either anaerobic or facultative organisms or both.

The onset of anaerobic cellulitis is ordinarily gradual, and systemic effects of the illness are usually not impressive, particularly in comparison with gas gangrene. Nevertheless, this is not a condition to be taken lightly. Spread of the infection may be rapid and extensive, and considerable morbidity or mortality may result if there is delay in initiating appropriate therapy. The deep fascia is not involved significantly, nor is there involvement of structures beneath it. Pathologically, the lesion is basically a wet inflammation of the subcutaneous tissues which progresses to necrosis with crepitation within 2–5 days of onset. When *Bacteroides melaninogenicus* is present, the subcutaneous tissues may present a brown to gray or black color [Altemeier and Culbertson (745)]. The skin may be grossly uninvolved unless the process started as a wound infection, but histologically thrombosis of nutrient vessels of the skin may be a prominent finding. Pain is often the first symptom, but it is relatively mild. Following this, there is swelling of the overlying skin and there may be erythema as well. Tenderness to touch may develop, and soon crepitation becomes noticeable. Some patients may ultimately manifest significant toxemia. Extension of the process may be relatively rapid, half of the torso wall being involved within 4 to 5 days in some cases [Altemeier and Culbertson (745)]. Gas may be detected by palpation or by X ray. It is important to keep in mind that the extent of the infection cannot be judged by the extent of crepitation. Examination of the wound when opened at the time of surgery will, in most instances, reveal a foul odor, gas, and variable quantities of pus, along with shreds of devitalized soft tissues. The wound may be lined with a shaggy gray-white pseudomembrane. By definition, muscle is not involved, although with long-standing disease, there may be some edema of the underlying muscle. In the case of muscle that has been devitalized by the initiating trauma, there may be gas present in that muscle tissue but not in adjacent healthy muscle.

The incidence of anaerobic cellulitis in war wounds may be as high as 5%, that is two to three times more common than clostridial myositis [Jergesen (746)]. Details of a number of reported cases, together with bacteriologic findings, are noted in Table 13.1. Table 13.2 presents the differential diagnosis of various gassy wounds [MacLennan (747)]. In addition to or instead of the clostridia (*Clostridium perfringens* included), a number of non-spore-forming anaerobic bacteria have been recovered from this type of process. The majority of these are anaerobic cocci or streptococci and anaerobic gram-negative bacilli. Coliform bacilli and aerobic streptococci or staphylococci, as well as other aerobic or facultative organisms, may be recovered along with the anaerobes.

In terms of diagnosis, the major condition to be differentiated from anaerobic cellulitis is gas gangrene. This may require surgery, which, in any case, is the definitive approach to therapy. The wound should be laid open widely, necrotic tissue debrided, all collections of pus and gas drained, and all involved fascial planes opened. The area should be left open initially. The condition of the fascia and muscle should be noted carefully; if these are healthy, no further surgical management is required. The differentiation of anaerobic cellulitis and gas gangrene is a crucial one, not only because of the urgency of providing definitive therapy in the case of gas gangrene but also to avoid unnecessarily severe treatment of the lesser infection—anaerobic cellulitis. Unnecessary amputation of a limb because of a mistaken diagnosis of gas gangrene is certainly a tragic error that has been made more than once. Antimicrobial therapy is also useful, the specific drugs to be chosen depending on the organisms present. Since *Bacteroides fragilis* may be encountered, it is advisable to use therapy that will cover this organism in patients who are quite ill. Topical therapy with agents such as zinc peroxide or hydrogen peroxide may also be useful.

TABLE 13.1
Anaerobic Cellulitis

Aerobes	Anaerobes	Diabetic?	Comments	Reference	Year
Hemolytic streptococcus, E. coli	2 cases with anaerobic streptococci	No		Marwedel and Wehrsig [cited by Spring and Kahn (1495)]	1915
E. coli	Anaerobic streptococcus	?		Meleney [cited by Altemeier and Culbertson (745)]	1936
E. coli	Anaerobic streptococcus	?		Terrell [cited by Altemeier and Culbertson (745)]	1940
0	Clostridium perfringens	No	Traumatic origin		
0	Clostridium perfringens	No	Traumatic origin		
0	Clostridium perfringens	No	Traumatic origin	Qvist (1496)	1941
S. aureus, α-hemolytic streptococcus	Anaerobic streptococcus	Yes	Foul odor, began at site of insulin injection (thigh)	Gillies (791)	1941
E. coli, nonhemolytic streptococcus	Anaerobic streptococcus, unidentified anaerobic cocci, Bacteroides melaninogenicus, Bacteroides fragilis, unidentified Bacteroides, unidentified anaerobic NSF GPR[a]	?	Origin, abdominoperineal wound	Altemeier and Culbertson (745)	1948
E. coli, nonhemolytic streptococcus	Anaerobic streptococcus, unidentified anaerobic cocci, B. melaninogenicus, B. thetoides, anaerobic diphtheroid	?	Involved perineum and scrotum		
0	Anaerobic streptococcus, B. melaninogenicus	?	Origin, peritonsillar abscess, involved deep fascial planes of neck and mediastinum		
Escherichia sendai, viridans streptococcus, S. albus, diphtheroid	Anaerobic staphylococcus, B. melaninogenicus	Yes	Foul pus, involved leg and foot		

Aerobic culture	Anaerobic culture	Foul odor	Clinical data	Reference	Year
Unidentified streptococci, Micrococcus	Unidentified anaerobic cocci, unidentified Bacteroides	?	Followed thoracotomy	Altemeier and Culbertson (745)	1948
Nonhemolytic streptococci, Micrococcus	B. melaninogenicus, unidentified NSF GPR[a]	No	Followed trauma, involved vulva, abdominal wall, flank		
E. coli, nonhemolytic streptococci	Anaerobic streptococcus, B. melaninogenicus	No	Followed appendectomy		
Alcaligenes faecalis, S. aureus, diphtheroid	Anaerobic staphylococcus, B. melaninogenicus	?	Followed trauma; buttock, hip, abdominal wall	Spring and Kahn (1495)	1951
E. coli	0	Yes	Malodorous fluid, therefore anaerobes present		
0	0	Yes	Putrescent odor, streptococci and gram-positive and gram-negative bacilli on smear; no growth		
0	B.funduliformis	No	Perirectal abscess extending to right buttock and thigh	McVay and Sprunt (621)	1952
E. coli (heavy growth)	Anaerobic streptococcus (moderate growth)	Yes	Died	Wills and Reece (1497)	1960
Paracolon	Gram-positive cocci on smear; not recovered			Stahlgren and Thabit (1460)	1961
0	C. perfringens	No		Eickhoff (514)	1962
0	Bacteroides (type B2)	No		Anderson (544)	1966
Klebsiella	Great many streptococci on smear; did not grow	Yes		Flynn et al. (1498)	1967
Proteus	Clostridium perfringens	Yes	Died	Meade and Mueller (1499)	1968
Few S. aureus, few enterococci	Bacteroides fragilis	Yes	Gas from foot to high thigh, followed trauma, foul odor	Ziment et al. (725)	1968
α-Hemolytic streptococci, β-hemolytic streptococci, Enterococcus, S. aureus, E. coli, Klebsiella–Enterobacter	Peptostreptococcus, 3 strains of Peptococcus (2 microaerophilic), Bacteroides fragilis, B. melaninogenicus	Yes	Foul odor		

(continued)

399

TABLE 13.1 (*continued*)

Aerobes	Anaerobes	Diabetic?	Comments	Reference	Year
α-Hemolytic streptococci, β-hemolytic streptococci, *Klebsiella-Enterobacter*	*Fusobacterium*	Yes	Foul odor	Ziment *et al.* (725)	1968
?	*Clostridium septicum*	No?	Patient had cyclic neutropenia, involvement of deltoids and thigh	Alpern and Dowell (103)	1969
?	*Clostridium septicum*	No?	Leukemia, abdominal wall and flank		
E. coli	*Peptostreptococcus*	No		Bartlett *et al.* (517)	1972
0	Anaerobic streptococcus	No	Fatal, extensive abdominal wall involvement	Anderson *et al.* (788)	1972
Coliform (1)	*Peptostreptococcus* (1), *Clostridium* (2)		4 cases	Nichols and Smith (1499a)	1975

^a NSF GPR, non-spore-forming, gram-positive rod.

TABLE 13.2

Differentiation of Gassy Infections of Skin and Soft Tissues[a]

Criterion	Anaerobic cellulitis	Infected vascular gangrene	Clostridial myonecrosis	Streptococcal myositis	Necrotizing fasciitis	Synergistic necrotizing cellulitis
Incubation	Almost always over 3 days	Over 5 days, usually longer	Usually under 3 days	3–4 days	1–4 days	Variable, 3–14 days
Onset	Gradual	Gradual	Acute	Subacute or insidious	Acute	Acute
Toxemia	Nil or slight	Nil or minimal	Very severe	Severe only after some time	Moderate to marked	Marked
Pain	Absent	Variable	Severe	Variable, as a rule fairly severe	Moderate to severe	Severe
Swelling	Nil or slight	Often marked	Marked	Marked	Marked	Moderate
Skin	Little change	Discolored, often black and dessicated	Tense, often very white	Tense, often with coppery tinge	Pale red cellulitis	Minimal change
Exudate	Nil or slight	Nil	Variable, may be profuse, serous, and blood stained	Very profuse, seropurulent	Serosanguinous	"Dishwater pus"
Gas	Abundant	Abundant	Rarely pronounced except terminally	Very slight	Usually not present	Not pronounced; present in 25% of cases
Smell	Foul	Foul	Variable, may be slight, often sweetish	Very slight, often sour	Foul	Foul
Muscle	No change	Dead	Marked change	At first little change but edema	Viable	Marked change

[a] Modified from MacLennan (747). Courtesy of Williams & Wilkins.

401

Infections similar to anaerobic cellulitis may be caused by aerobic or facultative gas-producing organisms, particularly in diabetics. For the most part, coliforms are involved in such infections, but extensive subcutaneous gas-forming infection in diabetics may be due to *Staphylococcus aureus* alone. Our overall experience is that this type of infection, particularly in diabetics, most often involves anaerobic bacteria, either in pure culture or mixed with facultative forms such as *E. coli*.

Infections that must be considered in the differential diagnosis will be discussed below. One must also consider nonbacterial causes of subcutaneous crepitation. Trauma to the chest or chest surgery may lead to extensive subcutaneous emphysema about the chest and even the neck and head. Extensive amounts of air may also be found in soft tissues after penetrating trauma, particularly in the case of a limb. Compressed air has accidentally been injected into wounds. Irrigation with hydrogen peroxide may generate gas chemically within a wound. Unduly vigorous irrigation of wounds may also result in air in tissues. Orbital emphysema may result from nose blowing [Kaplan and Winchell (748)]. Aelony (749), described a case of extensive crepitus over the lower abdomen, apparently due to rectal perforation secondary to a barium enema. Rubenstein *et al.* (750) reported three cases of subcutaneous gas following hand injuries in workers who were using finely powdered magnesium alloys. These authors were able to show experimentally in animals that subcutaneous injection of small amounts of powdered magnesium led to liberation of hydrogen. Filler *et al.* (751) gave a good discussion of noninfectious causes of gas in wounds.

Soft tissue abscesses may involve anaerobes. Schaffner (1240a) reported 3 such abscesses all of which grew *Eggerthella convexa*. One had only anaerobes and a second had 8 anaerobes plus enterococci.

Bacterial synergistic gangrene is a chronic gangrene of the skin which is usually a postoperative complication (Fig. 13.4A and B), particularly following abdominal or thoracic surgery or drainage of a peritoneal abscess or thoracic empyema. The disease is not always postoperative, however. Sometimes it develops slowly around a colostomy or ileostomy site or in a trivial accidental wound (Fig. 13.5A–G) or a skin lesion of long-standing due to some other cause (Fig. 13.6). The major symptom is the extraordinary pain and tenderness of the lesion. It usually appears at about the end of the first or second week after surgery, either as an infection of the whole wound or as a localized infection about retention sutures. At first the wound becomes red, swollen, and tender. Within the next few days, the wound margins or the area about the sutures develop an indurated appearance. The central area is purplish in color, and it is surrounded by a zone of erythema, varying from 1 to 10 cm. As the lesion progresses, the central

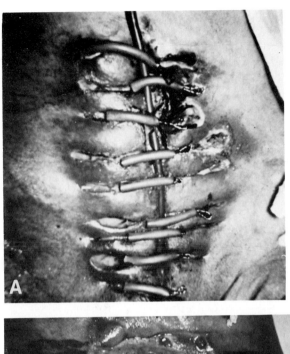

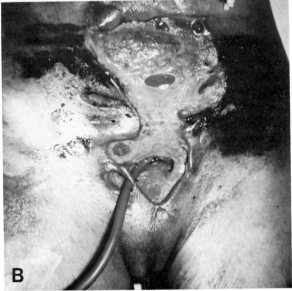

Fig. 13.4. Postoperative bacterial synergistic gangrene. Cultures yielded microaerophilic streptococci, enterococci, and *Proteus morganii*. (A) Abdominal wound after secondary closure. Wound edges are edematous and there is necrosis surrounded by inflammatory response about sutures. (B) Same wound 2 months later. There are four enterocutaneous fistulae and a vesicocutaneous fistula. From De Jongh *et al.* (757). Reproduced with permission.

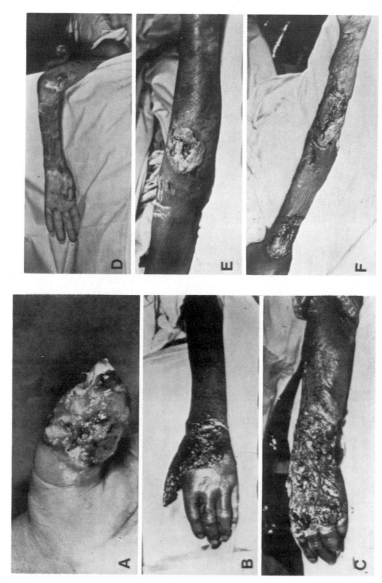

Fig. 13.5. Bacterial synergistic gangrene following injury. (A) Symbiotic infection. Compound fracture, left thumb, 5 days following injury showing infection advanced to proximal phalanx. (B) Spread of infection from thumb to wrist. (C) Infection 3 weeks following injury. (D) Skin grafts well healed on left forearm. (E) Development of infection at site of chlortetracycline injection right forearm, 4 weeks after initial injury. (F) Spread of infection to right arm and forearm. From Byrne (1590). Reproduced with permission.

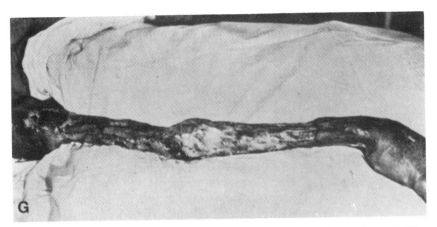

Fig. 13.5.   (G) Further spread of infection to entire right upper extremity. From Byrne (1590). Reproduced with permission.

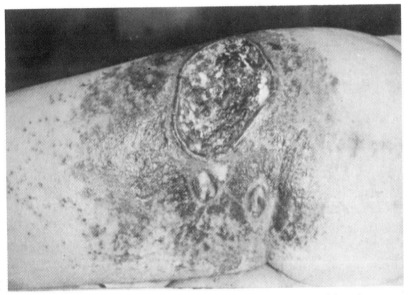

Fig. 13.6.   Bacterial synergistic gangrene. Central necrosis with undermining edges, a purple irregular border, and an outer zone of redness. Satellite lesions surround the ulcer. From Baxter (790). Reproduced with permission.

portion of the purplish area becomes frankly gangrenous, the color changing to a dirty gray-brown or yellow-green with a typical suede leather appearance. As the lesion spreads outwardly in all directions, the inner margin of the gangrenous zone becomes undermined and melts away. Eventually the center of the lesion becomes a granulating ulcer, and epithelium may eventually regenerate in this area. At this point, there is a central granulating area surrounded by gangrenous skin and then a raised purple zone with a sharp margin centrally blending into erythrema peripherally. Brewer and Meleney (752) described the process as a slowly advancing subcutaneous slough. The depth of the sloughing area varies considerably, sometimes being limited to the upper third of the subcutaneous fat and other times extending down to fascia. There is very little general reaction in terms of fever, and the patient remains in good general condition except for the pain.

Classically, the etiology is the combination of a microaerophilic non-hemolytic streptococcus that is found primarily in the spreading periphery of the lesion and *Staphylococcus aureus* that is found primarily in the zone of gangrene. The microaerophilic streptococcus keys out as *Streptococcus evolutus* in Prévot's classification scheme [Meleney (753)]. Touraine and Duperrat (754) summarized the literature on this condition up to the time of their report. Included in their paper was a summary of 81 bacteriologic observations. At times, the streptococcus is obligately anaerobic rather than microaerophilic, and a wide variety of other organisms may be seen instead of, or in addition to, the staphylococcus. *Proteus* is one of the more common organisms involved. Experimental animal studies have documented synergistic activity between microaerophilic streptococci and *Proteus* [Meleney *et al.* (755)] and between microaerophilic streptococci and *Aerobacter aerogenes* [Smith (756)]. De Jongh *et al.* (757) reported a case of postoperative synergistic gangrene due to a microaerophilic streptococcus (which subsequently grew aerobically after several passages) and *Proteus*. Other organisms were noted at various times. This particular case was fatal. Figure 13.4A and B shows the abdominal wound in their patient before and after dehiscense. The authors pointed out the difficulty of recovering the microaerophilic streptococcus from mixed cultures and stressed the importance of recognizing this organism as a primary agent in this type of infection. They stressed that confusion of this infection with an ordinary postoperative wound infection can lead to fatal delay in the administration of appropriate therapy.

Initially, wide excision along with antimicrobial therapy was considered necessary for cure. However, there are some reports of cures achieved with antimicrobial therapy alone (756, 758). The choice of antimicrobial therapy would depend on the specific organisms involved.

A recent paper by Güller (759) described a patient with progressive postoperative gangrene following a vein stripping. Initially, microbiological studies revealed branching fungi and nonpathogenic staphylococci. Subsequent cultures, perhaps because of antimicrobial therapy, proved sterile. The lesion continued to spread slowly despite therapy with penicillin, tetracycline, chloramphenicol, and sulfonamides at various times. Since the lesion failed to respond to antimicrobial therapy, despite the fact that cultures became negative, the author considered that the whole process might have been due to hypersensitivity. Accordingly, he treated the patient systemically and locally with corticosteroids to which the patient responded well. I have seen a somewhat similar case, a woman who developed a lesion typical of progressive bacterial synergistic gangrene following abdominal surgery. The patient had failed to respond to a variety of antimicrobial agents. At the time I saw her, cultures were sterile, both from the advancing edge and the gangrenous area. The patient was finally treated with corticosteroids at the suggestion of a dermatology consultant; she responded very well. Whether these latter two cases are simply hypersensitivity phenomena mimicking bacterial synergistic gangrene or whether the latter process may involve a hypersensitivity state remains to be determined.

Chronic undermining ulcer or "Meleney's ulcer" is a slowly progressive infection of the subcutaneous tissues associated with ulceration of the overlying skin. Gangrene of the skin is absent, but the rolled edges of the undermining skin may be cyanotic. The periphery of the lesion is erythematous and very tender. This ulcer may follow a wound or incision anywhere on the body, but most frequently is seen after lymph node surgery in the neck, axilla, or groin and after operations on the intestinal and genital tracts. As the lesion spreads, multiple ulcers and sinuses may develop at a distance from the original ulcer with undermining of the intervening skin. Epithelial strands and undermined bridges are characteristic. The advancing edge of the lesion is erythematous, painful, and tender. There is usually not much systemic reaction. There is little tendency for the lesions to heal spontaneously. They may go on for extended periods of time in the absence of effective therapy. The causative organism classically is a microaerophilic hemolytic streptococcus. However, Sandusky *et al.* (112) described three cases all of which yielded anaerobic streptococci; in addition, there were two microaerophilic streptococci isolated, one *C. perfringens*, one α-hemolytic aerobic streptococcus, one *S. aureus*, one *Proteus*, and two *Pseudomonas* strains. Altemeier and Culbertson (760) found anaerobic streptococci in two chronic burrowing ulcers. Prior to the availability of effective antimicrobial agents, meticulous debridement followed by local application of zinc peroxide was the treatment of choice. Debridement and drainage are still important but need not be radical. Penicillin ordinarily should be effective, the dosage depending on the susceptibility of the offending organism. Chloramphenicol has also been used successfully [Altemeier and Culbertson (760)], and other antimicrobial agents active *in vitro* against the organisms recovered should be satisfactory.

Infections Involving Fascia Primarily

Necrotizing fasciitis, otherwise known as hemolytic streptococcal gangrene or hospital gangrene, is a very serious infection that is important to recognize promptly. Interesting historical aspects are provided in the paper by Trendelenburg (761). He pointed out that the affliction was known in antiquity and was described by Hippocrates, Galen, and Avicenna. A famous surgeon, Ambroise Paré, reported it in great detail during the seige of Rouen. The disease was so widespread in both the beseiged and the beseigers that it was thought that the enemy bullets were poisoned. There were epidemics of it during the Napoleonic wars. It was prevalent during the Crimean War, particularly in Constantinople. It was noted during the United States Civil War in Washington and in Baltimore. Vincent (2) also presented interesting historical facets. Among 100 wounded patients in the Hotel-Dieu in Paris, 98 were attacked by this disease. The mortality was considerable. In English army hospitals in 1813, there were 512 deaths among 1614 cases of hospital gangrene. It was also seen in connection with the Spanish War and during the War of 1870 in Germany.

In general, necrotizing fasciitis is an acute process, but a chronic progressive form may be found as well (Fig. 13.7A and B). It was first described by Meleney (753) as a specific entity caused by hemolytic streptococci, hence the name hemolytic streptococcal gangrene. Subsequently, it was noted that staphylococci also commonly cause this infection, and some patients may have only aerobic or facultative gram-negative organisms. There have not been many reports of anaerobes involved in this type of infection, but it is likely that not all cases are reported, as with all other types of anaerobic infections. Table 13.3 summarizes information on the role of anaerobes in necrotizing fasciitis as reported in the literature. We have recently seen a case with extensive necrotizing fasciitis involving a good portion of the abdominal wall in which the dominant organism was a gram-negative anaerobic bacillus still not identified; *Klebsiella* was also isolated in smaller numbers.

The condition commonly originates in musculoskeletal wounds but may also appear in an operative wound or even after trivial injury. A pathognomonic feature is subcutaneous and fascial necrosis, with undermining of

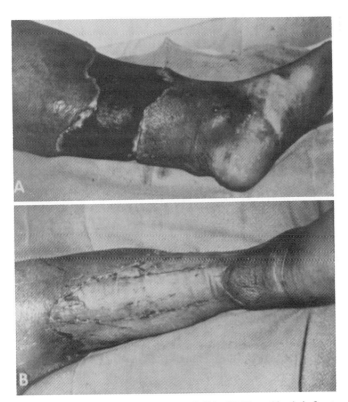

Fig. 13.7. (A) Acute necrotizing fasciitis and cellulitis. (B) Wound healed after two excisions of necrotic fascia, immobilization, antibiotic therapy, and skin graft. From Pulaski (1589). Reproduced with permission.

the skin. In the acute form, there is sudden onset of pain and swelling, with or without chills and fever. Within 24 hours, there may be considerable phlegmon, usually with erythema or cellulitis. Blue to brown ecchymotic skin discoloration is not uncommon, and cutaneous gangrene may be seen, particularly later in the illness. Prostration may be severe. Pain is gradually replaced by numbness or analgesia as a result of compression and destruction of cutaneous nerves. Fluid-filled vesicles appear in the area of cellulitis about the third to fifth day of the infection, following which the underlying skin becomes blue-black. This necrosis is due to thrombosis of nutrient vessels passing through the involved deeper tissues. Mild to massive edema is present in a majority of patients. This, along with calf tenderness, may simulate deep vein thrombosis [Rea and Wyrick (762)]. The most significant objective manifestation of this fasciitis is extensive undermining of the

TABLE 13.3

Necrotizing Fasciitis

Aerobes	Anaerobes	Diabetic?	Comments	Reference	Year
0	C. perfringens	No	Stinking odor, much gas, extended from scrotum to lower abdominal wall, died	Randall (764)	1920
S. faecalis	Very large numbers of Bacteroides varius	Yes	Amputation, stump became infected	Wills and Reece (1497)	1960
E. coli	Microaerophilic streptococci	Yes	Extended from knee joint to lower abdominal wall	Meade and Mueller (1499)	1968
E. coli	Bacteroides	?	Additional case with no growth	Rea and Wyrick (762)	1970
0	Bacteroides	No	Addict; fascial layers dissolved; infection burrowed between muscle bundles to bone	Rein and Cosman (768a)	1971
Enterococcus	Bacteroides	No	15 surgical procedures	Beazley et al. (394)	1972
Enterococcus, E. coli	Bacteroides incommunis	No	Extensive debridement anterior-abdominal wall		
E. coli, Enterobacter, Klebsiella, Citrobacter	Clostridium, B. melaninogenicum	No	Left lateral abdominal wall, associated perinephric abscess	Altemeier et al. (1466a)	1971
Proteus	Bacteroides, microaerophilic streptococcus	Yes	Arose in Bartholin gland or vulvar abscess. One additional case yielded Clostridium	Roberts and Hester (1499b)	1972
0	Bacteroides fragilis	?	2 cases, both with bacteremia	Nobles (56d)	1973
0	C. perfringens	No			
0	C. perfringens	No			
0	C. septicum	No	Necrotizing fasciitis was noted, along with abscess formation and local myositis, in 5 addicts with anaerobic subcutaneous abscesses	Gorbach and Thadepalli (486)	1975
0	C. septicum, B. fragilis	No			
E. aerogenes, E. cloacae	C. bifermentans, B. biacutus, Peptostreptococcus	No			
Klebsiella, E. coli	Bacteroides	No			
Klebsiella	Bacteroides	No	3 cases of Fournier's syndrome	Rudolph et al. (1499c)	1975
"Mixed coliforms"	Bacteroides	No			

skin. This can be demonstrated by passing a sterile instrument along the plane just superficial to the deep fascia. The instrument cannot be passed in ordinary cellulitis, but with necrotizing fasciitis there is extensive undermining of the skin. Because of the extensive edema, there may be a large deficit of extracellular fluid, and hypoproteinemia may result from exudation of serum from the large exposed surfaces resulting from surgical drainage [Grossman and Silen (763)]. Subcutaneous crepitation or other evidence of subcutaneous gas may be noted on occasion, but it is very uncommon. As indicated earlier, there is a chronic form of the disease that also involves subcutaneous undermining.

Since the mortality in necrotizing fasciitis approaches 50% or more without appropriate therapy, depending on the extent of involvement, and is probably 25% even with good therapy, it is important that the diagnosis be established as early as possible and that immediate surgical intervention be accomplished. Surgical therapy is the primary approach. There should be an extensive unroofing incision made through the area of involvement to the point where a hemostat can no longer separate skin and subcutaneous fascia from the deep fascia. Necrotic fat and fascia are all removed. Repeated debridement may be necessary, but obviously it is important to debride as much necrotic tissue as possible at the time of first surgery. The wound should be left open, but porcine grafts may be necessary in cases requiring extensive surgery since the lesion is much like a burn in terms of fluid and protein loss. Antimicrobial therapy is also important and will depend on the nature of the infecting organism.

A condition known as Fournier's gangrene, first accurately described by Fournier in 1883–1884 (764), usually starts in or around the scrotum and behaves like anaerobic cellulitis or necrotizing fasciitis (Figs. 13.8–13.10). This striking picture has also been called idiopathic gangrene of the scrotum, gangrenous erysipelas of the scrotum, streptococcus scrotal gangrene, and spontaneous fulminating gangrene of the scrotum. An excellent case report by Randall (764) emphasized certain highlights of the disease very effectively. This case report is cited herewith exactly as it appeared:

Albert M., age thirty-nine, a well developed and normal Negro, arose on the morning of January 2, 1920, in his usual good health. His occupation was that of chauffeur in a private family, driving only pleasure cars. About 3:00 PM on the above mentioned day, while driving a limousine in the city, he was seized with a chill which lasted half an hour, to be followed by a severe headache and fever. By 7:00 PM he was semi-delirious, and remembers but little of what transpired during the succeeding days. His physician was called and found his temperature to be 101°, and with a few rales in his chest naturally suspected grippe. An expectorant of ammonium carbonate and heroin was ordered, also a pill of camphor, quinine and Dover's powder. The following evening the evidence of pain first attracted his wife's attention to the genitalia, and his physician on the following, or third day of his illness, found the penis swollen to double its normal size, intensely inflamed, though neither indurated nor edematous. The swelling appeared to be proceeding from the distal end towards the pubis, stated his physician, and on reaching the hair margin it spread rapidly to and involved the entire scrotal wall. This scrotal involvement took place on January 6, 1920, the fourth day of the illness. An application of lead water and laudanum was applied to the tumefaction of the scrotum, while the condition on the penis, of little longer duration, had already developed a necrosing area near the prepuce, to which was applied aristol powder, and later zinc oxide ointment. The following day the scrotal swelling assumed the size of an infant's head, and was extremely painful; the patient suffered repeated chills, associated with profuse sweats and fever. Headache persisted and there were generalized pains over the chest and extremities. As the patient's condition seemed critical he was moved to a nearby hospital on January 7, and on the following day, the sixth of his illness, he was transferred to the Philadelphia General Hospital. The only other application to the local condition was olive oil and at no time was heat or an escharotic used. Urination remained undisturbed and normal. I first saw the patient on January 9, at which time the penile swelling had already become gangrenous and an enormous slough was separating, while at the bottom of the scrotum, on either side, were two areas, each about 5 by 10 cm in extent, where the skin was drying and puckering, blacker than the surrounding tissue, and extremely soft and boggy to the touch, the whole emitting a repulsive, fetid odor. The stench was horrible. Pain, on the other hand, was almost absent and on palpation there was a distinct emphysematous crepitation throughout the entire scrotal tumefaction.

I was anxious that the condition could be shown the student class on January 12, but by that day's arrival the necrotic mass had separated, the line of demarcation forming just below the scrotal attachment to the body, and three-fourths of the scrotal wall had sloughed off.

Although the disease may remain limited to the scrotum, it may also extend to the penis, the abdominal wall, and the perineum. Different patients may have manifestations consistent with anaerobic cellulitis or with necrotizing fasciitis, but the latter presentation is much more typical. Involvement of muscle may also be noted in more serious cases. Weiss (80) described a patient with an abscess involving the muscle and fascia of the leg along with gangrene of the scrotum. This diabetic yielded from culture of the abscess *Bacteroides melaninogenicus*, *Fusobacterium*, *Clostridium perfringens*, two aerobic streptococci, and *E. coli*. It has been associated with local trauma (mechanical, chemical, or thermal), underlying urinary tract disease (paraphimosis, penile erosion, inguinal adenitis, prostatoseminal vesiculitis, epididymitis, and periurethral extravasation of urine), distant acute inflammatory processes (perforative retrocecal appendicitis and acute hemorrhagic pancreatitis), operative procedures (circumcision, herniorrhaphy, hemorrhoidectomy, incision of perirectal abscess, injection of a hydrocele and abdominoperineal resection), and perianal abscess or fistula (765, 766).

The condition is not uncommon. Gibson, in 1930 (767), collected and analyzed 206 cases from the literature. The disease behaves as a phlegmon, frequently with gas in the tissues, which spreads along deep external fascial planes. Considerable necrosis of tissue results, and there is typically a foul odor. Gibson pointed out that few cases have been satisfactorily studied

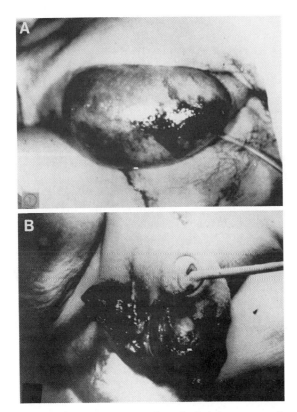

Fig. 13.8. (A) Appearance of scrotum on day of admission as seen from its posterior aspect and presenting between the patients thighs. The dark areas, purple in color, represent early changes of gangrene. The remaining scrotum is markedly erythematous. Microaerophilic staphylococci and fusiform bacilli cultured from scrotal fluid. (B) Appearance of genitalia 2 days following hospital admission. Extensive scrotal gangrene is evident, and the denuded testicles can be seen in the depth of the wound. From Dunaif (1591). *Plastic Reconstructive Surg.* **33**, 84–92 (1964). Copyright 1964 The Williams & Wilkins Co., Baltimore.

from the anaerobic standpoint and considered the possibility that all of these cases may be primarily anaerobic. The clinical picture is certainly consistent with and highly suggestive of this viewpoint. However, there is relatively little documentation of anaerobes in the process. In Gibson's review, there were two patients reported with fusiform bacilli (one with spirochetes as well), two with *Clostridium perfringens* and one with *Clostridium novyi*. One of the three cases reported by Himal and Duff in 1967 (766) as "endogenous gas gangrene" did grow out *Clostridium perfringens* along with *Streptococcus faecalis* and *Escherichia coli*. Culture results are not mentioned in the case of the other two patients, although one of these other two was given polyvalent gas gangrene antitoxin and all three cases

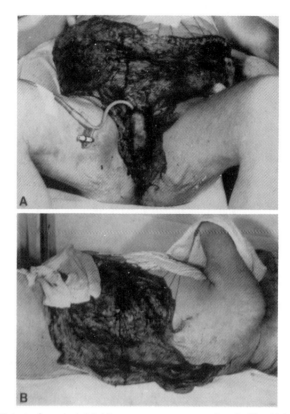

Fig. 13.9.   Extent of surgical debridement necessary on patient in Fig. 13.8 is shown. Un-involved skin and subcutaneous tissue reflected back and held in position by sutures. (A) The penis was sutured to the abdominal wall and the right testicle and cord were sutured to the inguinal region. The spermatic cord can be seen underneath the indwelling urethral catheter. (B) Lateral thoracic, abdominal, and lumbar areas were extensively involved. From Dunaif (1591). *Plastic Reconstructive Surg.* **33**, 84–92 (1964). Copyright 1964 The Williams & Wilkins Co., Baltimore.

were treated with hyperbaric oxygen (with results that cannot be regarded as impressive). Klotz (768) pointed out that the cases reported by Himal and Duff were really cases of Fournier's gangrene. Rein and Cosman (768a) described a case of necrotizing fasciitis due to *Bacteroides* in an addict (Fig. 13.11). Bras *et al.* (769) described a case of idiopathic gangrene of the scrotum in association with abdominal actinomycosis. The several cases that we have seen have all been in diabetics, who seem to be particularly prone to this problem, and have all yielded anaerobes, sometimes in asso-

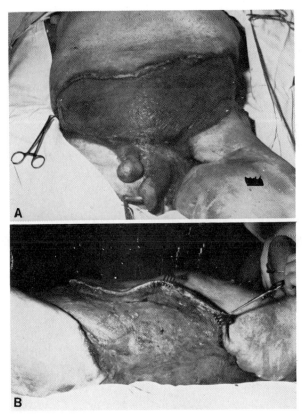

Fig. 13.10.   Appearance of patient's wounds 37 days after hospital admission (in Fig. 13.8) and just prior to skin grafting. Denuded testicle and spermatic cord as well as the remaining areas are covered with clean granulation tissue. From Dunaif (1591). *Plastic Reconstructive Surg.* **33**, 84–92 (1964). Copyright 1964 The Williams & Wilkins Co., Baltimore.

ciation with aerobes. Fairly complex mixtures may be recovered, as with many other types of anaerobic infections. For example, one of our patients yielded on culture, a heavy growth of *Bacteroides fragilis* ss. *thetaiotaomicron*, a moderate growth of two different types of *Bacteroides melaninogenicus*, a few colonies of an unidentified gram-negative anaerobic bacillus, a few colonies of *Fusobacterium nucleatum*, and a heavy growth of *E. coli*. Another yielded *Bacteroides* species, *E. coli*, *P. mirabilis*, and enterococci. In the preantimicrobial era, the mortality was 20–25%. Surgical debridement is still the main therapeutic approach. All necrotic tissue must be removed

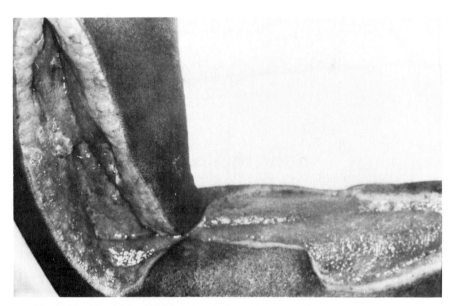

Fig. 13.11.   Necrotizing fasciitis in female drug addict who injected cocaine subcutaneously. Figure shows appearance 2 weeks after incision and drainage; the wound is granulating well. Culture grew *Bacteroides*. From Rein and Cosman (768a). *Plastic Reconstructive Surg.* **48**, 592–594 (1971). Copyright 1971 The Williams & Wilkins Co., Baltimore.

and all pus drained. Antimicrobial therapy suitable for the organisms isolated should also be used.

Certain *necrotizing dermogenital infections* do not always fit clearly into the various types of anaerobic infection which have been outlined. Yelderman and Weaver (770) described two patients with this type of infection. One, a 7-month-old boy, developed induration and erythema at the base of the penis following an episode of pharyngitis and otitis (Fig. 13.12). The area proceeded to become necrotic and the skin separated from the base of the penis. There was a large necrotic ulcer, almost encircling the base of the penis and extending down to Buck's fascia; this case resembles Fournier's gangrene or necrotizing fasciitis. Cultures yielded a microaerophilic streptococcus along with *Candida albicans*, *E. coli*, and *Pseudomonas aeruginosa*. The second patient had a history of ulcers of the glans penis for 3 months (Fig. 13.13). He was given corticosteroids because of erythema nodosum. Subsequently, he developed an erythematous painful papule on the glans penis which rapidly became punched out and reached a size of 4 × 8 mm. This healed in 1 month without therapy, but 2 weeks later two similar lesions developed on the glans penis and then coalesced. Aerobic culture revealed *Staphylococcus aureus*, and anaerobic culture of the ulcer margins yielded

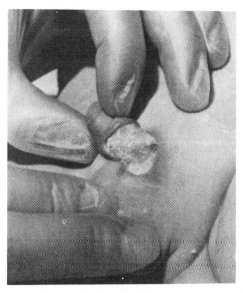

Fig. 13.12.   Penoscrotal ulcer due to microaerophilic streptococcus. Buck's fascia is exposed. From Yelderman and Weaver (770). *J. Urol.* **101**, 74–77 (1969). Copyright 1969, The Williams & Wilkins Co., Baltimore.

a microaerophilic streptococcus. The patient failed to respond to a 2-week course of penicillin followed by 2 weeks of therapy with chloramphenicol. Activated zinc peroxide was applied to the ulcer and gauze over the medication was kept moist. This resulted in arrest of the progress of the lesion and healing occurred over a 4-week period. This case may have represented a case of bacterial synergistic gangrene. These authors point out that Mair

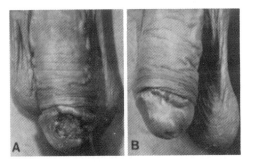

Fig. 13.13.   (A) Ulcer of glans penis due to microaerophilic streptococcus. (B) Complete healing of ulcer after treatment with zinc peroxide. From Yelderman and Weaver (770). *J. Urol.* **101**, 74–77 (1969). Copyright 1969, The Williams & Wilkins Co., Baltimore.

found that microaerophilic streptococci, either alone or with other organisms, was frequently associated with idiopathic gangrene of the scrotum or penis. They also referred to an interesting paper by Leibovitz concerning an outbreak of pyogenic penile ulcers associated with a microaerophilic streptococcus that was quite pleomorphic and resembled *Haemophilus ducreyi*. This organism was isolated from the vagina in a high percentage of prostitutes. Ulcers in the patients occurred 2 to 10 days after sexual contact, usually were single and ranged in size from 2 to 10 mm in diameter. Microaerophilic streptococci were found in 35% of cultures; 228 penile ulcers were described in that study.

Infections Involving Muscle Primarily

Historical aspects of gas gangrene (*clostridial myonecrosis*) are discussed by Millar (771) and MacLennan (747). Very early medical writings were not specific enough in their descriptions to permit us to judge whether or not they knew this condition. From the Middle Ages on there are descriptions of gangrenous infections with gas in tissues. Beginning with the eighteenth century, there have been a number of reports of illness clearly recognizable as gas gangrene. Proper understanding of the disease was held back early by the thinking that the disease was bacteria-specific. There was much progress and understanding of the disease during World War I. By 1919, it was recognized that malignant edema and gas gangrene were somewhat different clinical manifestations of the same general type of infection and that both might be caused by a wide variety of clostridia. One other important principle was established finally during World War II; that is that the term gas gangrene should be limited to those infections that invade muscle and produce massive necrosis of tissue. A much more appropriate name for the condition is clostridial myonecrosis. As MacLennan (747) stressed, clostridial wound infections are clinical conceptions rather than bacteriological entities, and therefore they must be defined in clinical terms.

The incidence of gas gangrene varies considerably under different circumstances. It was estimated that the disease killed at least 100,000 German soldiers in World War I. Nevertheless, it was rarely mentioned in most of the minor wars between 1919 and 1940. Usually the incidence of clostridial myonecrosis has been 1 to 8 cases per 1000 wounded. With a more or less static front, when surgical treatment of casualties can take place within a few hours of wounding, the disease becomes comparatively rare. The incidence of gas gangrene in peace time wounds is low for similar reasons. Dineen (772) states that the incidence in postoperative wounds in civilian trauma is now less than 0.1%.

Typically, and this refers primarily to infection with *Clostridium perfringens*, gas gangrene manifests itself with a sudden appearance of pain in the region of the wound. It may appear so suddenly as to suggest a vascular catastrophe. The pain steadily increases in severity but remains localized in the infected areas, spreading as the infection spreads. Soon afterward, local swelling and edema and a thin hemorrhagic exudate can be observed. There is a marked rise in the pulse rate, out of proportion to the slight elevation of temperature. The edematous area is very tender; the skin is tense, white, often marbled with blue, and somewhat colder than normal. There is some bronze discoloration that increases with time. The swelling, edema, and toxemia increase rapidly; the serous discharge becomes more profuse; the skin becomes more dusky or bronzed; and bullae filled with dark red or purplish fluid appear (Fig. 13.14). Gas becomes manifest frequently, although not so obviously as in anaerobic cellulitis. A peculiar

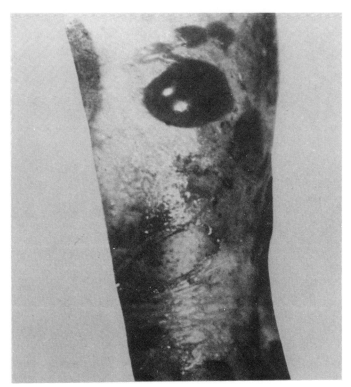

Fig. 13.14. Gas gangrene developing after compound fracture of tibia and fibula. Note necrotic skin and prominent bullae. From Altemeier and Fullen (1012). Reproduced with permission.

sweetish smell may be noted which, although suggestive, is not pathogno-monic of gas gangrene. It may also be absent in cases of definite clostridial myonecrosis.

Certain peculiar mental changes are noted. These may consist of intellec-tual clarity, full appreciation of the gravity of the disease, and a profound terror of impending death. Very characteristic is the toxic delirium that may precede any visible changes in the wound. The patient becomes in-coherent, disoriented, and obstreperous. Later there may be overwhelming prostration and toxemia. While some patients may be alert and apprehensive, others are apathetic and indifferent and are unaware of the seriousness of their condition.

Later in the course of the infection, there may be a drop in blood pressure. The patient may show a peculiar gray pallor, weakness, and profuse sweat-ing. In contrast to anaerobic cellulitis, the evidence of the disease at the skin surface and the clinically demonstrable gas are not as extensive as the involvement of the underlying muscle. Usually, changes in the muscle may be noted only at operation, and it is imperative that when gas gangrene is a clinical possibility, that there be very prompt surgical exploration. Early changes in the muscle consist primarily of edema and pallor, but later there is change in the color of the muscle, the blood supply is lost, contractility disappears, and gas may be demonstrable. In later stages of muscle involvement there is increased reddening, with mottled purple, and the consistency of the muscle may be pasty or mucoid. Still later, the muscle becomes diffusely gangrenous, dark greenish purple or black, friable, and even liquefied. Jaundice is rarely seen in gas gangrene of wounds, in con-trast to uterine infections, and when it does appear, it is associated with clostridial sepsis and intravascular hemolysis. Widespread fat embolism may be seen in clostridial infections [Govan (500)], possibly related to the lecithinase produced by *Clostridium perfringens*, but this is probably rarely of clinical importance.

Gas gangrene due to *Clostridium novyi* has a much longer incubation period than that due to *C. perfringens* (average of approximately 5 days compared to less than 24 hours). The earliest symptom, even before pain, is frequently a sense of heaviness in the affected part. Once pain appears, the course is very acute, with rapid appearance of profound toxemia and an extremely profuse golden yellow discharge from the wound. Gas is a minor feature, and odor is frequently absent. Edema is by far the most conspicuous feature of this type of infection. Toxemia is considerably out of proportion to the apparent local lesion. The patient is usually apathetic.

Gas gangrene due to *Clostridium septicum* tends to be intermediate in its clinical features between the disease due to *C. perfringens* and that due to *C. novyi*. However, its course is more acute than either of the others.

Clostridium histolyticum is uncommonly involved in gas gangrene and pure infections with it are extremely rare. Gas gangrene due to *Clostridium bifermentans* resembles that due to *C. novyi* with regard to the long incubation period, the firm gelatinous edema, and the absence of obvious gas. However, there is much less wound exudate, and it is invariably blood stained. There is commonly bronzing of the skin. The infected muscles are more hemorrhagic, and there is often a distinct odor of putrefaction. Infections due to *Clostridium fallax* are extremely rare.

Major predisposing causes of gas gangrene are extensive laceration or devitalization of muscles, particularly large muscle groups of extremities, impairment of the main blood supply to a limb or muscle group, contamination by various foreign bodies, and delay in prompt surgical management. Fractures, particularly compound fractures, predispose to gas gangrene because of damage to muscles and blood supply and perhaps also because of the presence of calcium salts, which favor growth of the anaerobic organisms. Clinical data suggest that diabetics are more susceptible to clostridial myonecrosis than others and that this may involve more than just the vascular problems that are seen commonly in diabetics. Experimental work by Adamkiewicz *et al.* (773) indicated that mice rendered diabetic by alloxan are much more susceptible to intramuscular injection of *Clostridium perfringens* than control mice.

The organisms most commonly involved in gas gangrene are *Clostridium perfringens* (80% of cases), *Clostridium novyi* (40%), *Clostridium septicum* (20%), and *Clostridium bifermentans*, with *Clostridium histolyticum* and *Clostridium fallax* as established but uncommon pathogens in this situation. The diagnosis of gas gangrene is primarily made clinically. Definitive identification of clostridia in the wound is not, in itself, of great value inasmuch as up to 88% of traumatic wounds may be colonized with clostridia without evidence of infection. Microscopic examination of exudate may reveal clostridia, usually as large gram-positive bacilli with square ends and seldom showing spores. Few intact polymorphonuclear leukocytes are seen in the exudate. Fluorescent antibody technique may facilitate rapid identification of these organisms. Special bacteriological procedures may permit tentative bacteriologic identification within hours by culture and biochemical techniques. X-ray films taken at intervals may help detect early or incipient gas gangrene. If gas increases in amount or spreads in a linear fashion along the muscle and fascial planes, this may suggest the diagnosis. Late in the course of illness, gas may be visualized in the muscles themselves.

In general, there are no satisfactory laboratory tests for diagnosis of gas gangrene, and one should not lose valuable time awaiting results of diagnostic tests. Rapid spread of the infection may require only 2–4 hours

[Altemeier (774)], and irreversible changes may develop in the tissues rapidly. Accordingly, immediate surgical exploration is indicated when there is clinical suspicion of the possibility of clostridial myonecrosis. The casts, splints, or large dressings that are necessarily used for treatment of major injuries may obscure wounds and make observation and interpretation of local signs difficult. Here, too, there must be an index of suspicion, particularly with patients who have had severe wounds associated with significant muscle damage. With the slightest suspicion of early symptoms of gas gangrene, dressings, casts, etc., should be removed or windowed for direct inspection of the wound.

Gas gangrene may also occur postoperatively, primarily in relation to surgery on the lower limb, bowel (774a), and gallbladder (774a). In this chapter, we will be concerned only with such infections of extremities. Parker, in 1969 (775), published results of a survey of various hospitals and public health laboratories in Great Britain over a period of 2 years. Bacteriologists in these facilities were asked to report cases of clostridial infection occurring following "clean" surgery. In all, 85 cases of clostridial infection were reported, and this included 56 cases of myonecrosis. Among the 56, there were 50 in which the surgery was performed on the patient's leg; 39 of these were amputations and 11 were other types of surgery including insertion of prostheses or nails, with or without plates, for arthritis of the hip or fractures of the femur. It was clear from this review that gas gangrene most commonly follows amputation of the lower limb in elderly patients, almost all of whom had arterial insufficiency.

Although *Clostridium perfringens* can be found readily in the hospital atmosphere, including operating rooms [Gye *et al.* (776)], studies indicate that these infections arise endogenously. The report of Ayliffe and Lowbury (777) involving typing of strains of *Clostridium perfringens* from four cases of postoperative gas gangrene, two of which occurred on successive days in one hospital, indicated that in all of these cases the infection was probably acquired from the patient's intestinal flora by way of fecal contamination of skin. The site of infection, as determined in Parker's survey (775), certainly supports this, as does the extensive sampling of skin flora in 76 patients carried out by Ayliffe and Lowbury which showed occasional heavy contamination of the thighs, groins, and buttocks with *Clostridium perfringens*. In general, this and other studies indicate that the skin of the extremity at some distance from the anus is much less likely to be contaminated with *Clostridium perfringens* and that patients who are incontinent of feces are more likely to have contamination of skin with *Clostridium perfringens*.

Postoperative gas gangrene may also follow vascular surgery on the lower limb [Gye *et al.* (776) and Parker (775)]. Three major predisposing factors in this setting are contamination of skin with *Clostridium perfringens* from the intestinal tract, severance of large muscle masses during surgery, and ischemia, which usually was the indication for the surgery. A number of patients with this type of postoperative problem are also diabetics.

The usefulness of iodophors in skin decontamination and the desirability of using penicillin prophylactically in this situation are discussed in Chapter 20. Problems in diagnosis of postoperative clostridial myonecrosis are emphasized by Gye *et al.* (776). The earliest signs of the infection in their patients were severe mental disturbance (usually confusion) and fever. The mental aberrations were originally considered to be related to the diabetic state. A high degree of suspicion is necessary for early diagnosis, which is so important for a successful outcome.

In addition to postoperative gas gangrene, there are on record quite a number of cases resulting from injections of one type or another. Koons and Boyden (778) reported a case of gas gangrene following intramuscular injection of epinephrine in oil. They cite a collection of 83 cases by Touraine, 33% of which were at the site of epinephrine injections and 22% of which were at the site of caffeine injections. Van Hook and Vandevelde (778a) recently reported another case following intramuscular injection of epinephrine in oil. A large group of other medications has also been incriminated [Berggren *et al.* (779)]: saline, camphor, quinine, morphine, procaine, ether in oil, sodium sulfathiazole, digitalis or similar preparations, liver extract, sodium amobarbital, scopolamine, insulin, ACTH, and others. Larcan *et al.* (1435b) reported four cases of gas gangrene related to injection or needling. Two followed intramuscular injection of antibiotics, one a vaso-constrictor administered subcutaneously, and one following puncture of a posttraumatic hematoma. Bishop and Marshall (780) showed in experimental animals that the oil played no role in *Clostridium perfringens* infections following injection of epinephrine in oil. Schreus (781) reported on 15 cases of gas gangrene in patients who inadvertently got injections of benzine rather than typhoid immunization which were intended. There are also reports of gas gangrene following ordinary venipuncture—both femoral venipuncture [Dikshit and Mehrotra (782)] and venipuncture in the antecubital fossa [Parker (775)]. It is well known that alcohol will not kill spores effectively, and *Clostridium perfringens* has been recovered from a container of alcohol sponges [Koons and Boyden (778)], and from alcohol used for sterilizing syringes [Magistris (783)]. Magistris also recovered *C. perfringens* from phenylmercuric borate solution used for skin disinfection and described three cases of gas gangrene following injection after the use of either alcohol or the mercuric compound as described for skin disinfection.

Gas gangrene may also arise spontaneously (nontraumatic or endogenous gas gangrene), chiefly in patients with malignancy or other debilitating disease and with ulcerating lesions in the gastrointestinal, biliary, or genito-urinary tracts. Marty and Filler (784) reported a case of gas gangrene of the forearm secondary to an ulcerating adenocarcinoma of the colon. Alpern and Dowell (103) reported 28 cases of *Clostridium septicum* infection, all probably arising from the patient's own intestinal tract. Most of the patients had malignancy. Four of these patients had gas gangrene of the buttock or extremities, and one had clostridial cellulitis of the deltoids and thigh; three of these had *C. septicum* bacteremia. Four had malignancy and one cyclic neutropenia; one also was a diabetic. Werner *et al.* (785) also reported two cases of spontaneous *C. septicum* gas gangrene (left arm and right foot in one case and thigh in the other). Both patients had non-malignant ulcerating lesions of the intestinal tract with numerous gram-positive rods noted in the histologic sections. One of the patients was a diabetic.

It should be appreciated that gas gangrene is not a rare phenomenon seen primarily in relation to wars. Brown and Kinman (785a) note that only 22 cases occurred during 8 years of combat in Viet Nam (among 139,000 combat casualties) and that in a recent 10 year period in Miami there were at least 27 cases. Trauma was involved in all of the Miami cases; 17 patients had an open fracture and 3 had damage to major arteries. A major differ-ence in management, accounting for the difference in incidence in the military and civilian populations, was the military policy of leaving wounds open and then employing secondary closure later. Very likely there was also more emphasis on thorough debridement.

Therapy of gas gangrene is considered in detail in Chapter 19. Gas gangrene of various visceral structures is considered in the chapters cor-responding to the organs involved.

Myositis or myonecrosis may also be caused by organisms other than clostridia. Chief among these is the anaerobic streptococcus. *Anaerobic streptococcal myonecrosis* may closely resemble subacute clostridial gas gangrene, and it is perhaps for this reason that it was not recognized until World War II. The incubation period is usually 3–4 days, and the pre-senting signs are swelling, edema, and a purulent or seropurulent wound exudate. Pain comes later in the course of illness (a distinct difference from gas gangrene) but may then be very severe. The edema (which is moist rather than gelatinous and not very hemorrhagic) progresses diffusely. Once pain becomes established as a symptom, the progress of the illness is rela-tively rapid, although not so rapid as with clostridial myonecrosis. Gas is present but not extensive. Involved muscles are at first pale and soft and later are bright red with typical regular purple barring. Later the muscles become dark purple, swollen, friable, and gangrenous. There is a peculiar sour odor to the wound and to the large quantities of seropurulent discharge.

The gas can be shown to be intramuscular as well as intermuscular in distribution. Fatal cases die after 1 week or longer with toxemia, disorientation or mild delirium, and shock as preterminal events. Although the condition is called anaerobic streptococcal myonecrosis, these organisms are always found in association with other organisms notable among which are *Streptococcus pyogenes* and *Staphylococcus aureus*. The character of the disease depends to some extent on the nature of the coinfecting organisms. When *S. pyogenes* is present, the disease is more acute with cutaneous erythema, bright red discoloration of the muscles, and frequently a terminal septicemia. Cases associated with staphylococci tend to be more insidious, with paler and more edematous muscles.

According to Jergesen (746) the muscle involvement may be focal rather than diffuse; however, this seems to be uncommon. A major difference from gas gangrene is that the involved muscle, although edematous and discolored, is still alive and reactive to stimuli. Microscopically a smear of the muscle tissue reveals vast numbers of streptococci, and perhaps other cocci, among masses of pus cells; no gram-positive bacilli are noted as would be seen in gas gangrene. Microscopically, the muscle lesion is characteristic and resembles that of clostridial myonecrosis very closely [MacLennan (786)]. There is an acute interstitial myositis with an extensive polymorphonuclear reaction. There are varying degrees of degeneration of muscle fibers.

Anaerobic streptococcal myonecrosis is apparently fairly uncommon. MacLennan (787) looked specifically for it over a 1-year period in the Middle East, at a time when the casualty rate was high, and encountered only 19 cases.

MacLennan (786) pointed out that anaerobic streptococci are notoriously resistant to sulfonamide drugs. Anderson *et al.* (788) described two cases of anaerobic streptococcal myositis in which the organisms were said to be resistant to penicillin and ampicillin; however, the sensitivities were done according to the Kirby–Bauer method which has not been standardized for anaerobic organisms. It is true, nevertheless, that certain anaerobic cocci may be relatively resistant to penicillin, the amount of drug required for inhibition being as high as 12 to 32 U/ml. Accordingly, penicillin would need to be given in very high dosage, at least until it was shown that the organisms were more sensitive.

Another condition involving muscles has been variously described as synergistic necrotizing cellulitis [Stone and Martin (789)] and gram-negative anaerobic *cutaneous gangrene* or *necrotizing cutaneous myositis* (Figs. 13.15 and 13.16) [Baxter (790)], among other things. A more appropriate name might be *synergistic nonclostridial anaerobic myonecrosis*. This is a highly virulent soft tissue infection involving skin, subcutaneous tissue, fascia,

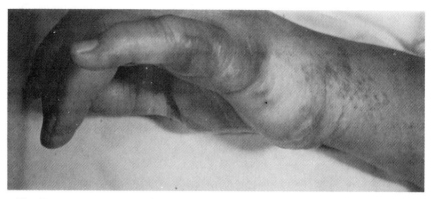

Fig. 13.15. Gram-negative (*Bacteroides*) necrotizing cutaneous myositis. Typical bluish-gray skin lesions overlying extensive tissue necrosis. From Baxter (790). Reproduced with permission.

and muscle. Unique to this infection are discrete large skin areas of blue-gray necrosis separated by normal skin, with much more extensive involvement of underlying tissues producing confluent necrotic liquefaction or dry gangrene in underlying muscle and fascia. Foul-smelling "dishwater" fluid may drain from skin ulcers. Although severe systemic toxicity occurs and may appear suddenly, there is usually extensive local necrosis before this is noted. There is exquisite local tenderness and severe pain. Gas formation, present in 25% of cases, is usually not pronounced. There may be

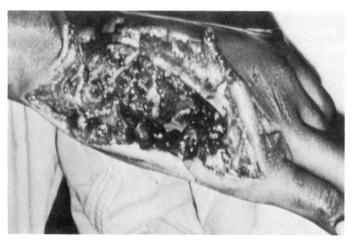

Fig. 13.16. Gram-negative necrotizing cutaneous myositis. Extensive muscle necrosis underlying small area of cutaneous gangrene. From Baxter (790). Reproduced with permission.

hemolysis. The incubation period is 3–14 days, with the onset acute. Three-fourths of patients with this infection have diabetes mellitus; other associated or predisposing factors include advanced age, renal disease, and either obesity or malnutrition. This infection occurs in the lower extremities and perineal areas primarily. Undoubtedly, some cases described as Fournier's gangrene represent this entity. It may follow inadequately treated perirectal or diverticular abscesses and other infections arising from perforations or entries into the gastrointestinal tract. Stone and Martin (789) recovered aerobic or facultative gram-negative bacilli in virtually all of their cases, associated with anaerobic streptococci and/or *Bacteroides*. They noted a high incidence (30%) of associated bacteremia involving aerobes and anaerobes. Baxter (790) considers the gram-negative anaerobic bacilli, especially *Bacteroides*, to be the key pathogens. Therapy requires radical debridement and appropriate antimicrobials. Best results are seen in patients in whom amputation of the infected area is possible. Mortality is as high as 75%.

In the fatal case of postoperative bacterial synergistic cellulitis of the abdominal wall, described by Behrend and Krouse (713) there was significant involvement of the rectus muscles of the abdominal wall. Whether the muscle involvement was due to the anaerobic streptococcus alone or represented synergism between this organism and the *Bacillus* which was also present, is uncertain. Prévot (21) described two cases of pectoral myositis, one due to *Staphylococcus aerogenes* and one due to *Streptococcus anaerobius*. Frumin and Fine (542) described a patient with a necrotic psoas muscle, the foul pus from which showed gram-positive cocci on smear. There was no mention of culture. It should also be appreciated that, on rare occasion, aerobic or facultative organisms may cause myonecrosis. Abscesses within muscle, or pyomyositis, is well known with staphylococci and certain gram-negative facultative bacilli; however, myonecrosis (usually not as acute or severe as that seen with anaerobic streptococci) may be seen rarely also. We have seen one such case in a diabetic with extensive muscle necrosis extending from the region of the ankle to a short distance below the knee with pain, edema, formation of bullae superficially, and gas in the tissues. Very careful study failed to reveal any organisms resembling clostridia, anaerobic streptococci, or any other anaerobes either on direct smear or on culture. Gillies (791) reported a similar although less extensive case. Myonecrosis or myositis due to gram-negative facultative bacilli is described by Bornstein et al. (275) and Myerowitz et al. (414). Gorbach et al. (113) found clostridia in three muscle abscesses in heroin addicts.

Infected vascular gangrene is another condition in which gas-producing anaerobes may be found. In this situation, the muscle (or frequently the entire limb) has already died as a result of circulatory insufficiency and the bacteria act primarily as saprophytes. There is little tendency for them to spread beyond the dead tissue into intact healthy muscle, and there is seldom a clinical picture of acute toxemia. There may be an extremely foul odor and considerable gas production, but this is the usual extent of the clinical picture. However, with neglect, cases may develop into true gas gangrene [MacLennan (747) and Bornstein *et al.* (275)]. Parker *et al.* (722) described a diabetic with infected vascular gangrene of the foot and leg. The cultures yielded β-hemolytic streptococci, *Proteus*, and a *Bacteroides*. Werner (523) described a diabetic with infected gangrene of the foot whose cultures yielded *Eggerthella* and *Sphaerophorus*. One of our patients, a diabetic with arteriosclerosis obliterans and diabetic neuropathy, had infected vascular gangrene with a gangrenous superficial ulcer and considerable gas and foul pus. Cultures yielded two strains of *Fusobacterium nucleatum*, *Bacteroides melaninogenicus*, two strains of anaerobic streptococci, *Proteus mirabilis*, and *Proteus morganii*. Rathbun (419) mentioned two patients with noninfected diabetic gangrene who nevertheless developed bacteremia due to *Clostridium perfringens*, the portal of entry apparently being the gangrenous extremity.

Infections Secondary to Trauma

Among the most impressive anaerobic infections secondary to trauma of relatively mild degree are those due to *bites*, particularly those involving human bites. Many of these are not actually bites in the strict sense, but rather injuries to the hand or fingers related to fighting and striking another person in the mouth or teeth. Good reviews and discussions of the general problem are provided in the papers of Barnes and Bibby (792) Boland (793), and Boyce (794, 795). When the clenched fist strikes the teeth, the tendon sheaths, tendons, and other tissues on the extensor surface of the fingers are stretched full length. The skin over the knuckles is penetrated and the tendons, and possibly the joint, are exposed. Then, as the fingers are straightened, the damaged parts relax and this carries implanted organisms from the mouth of the victim deep into the tissues of the person who has struck the blow. This leads to involvement of three spaces—the joint space, the dorsal subcutaneous space, and the dorsal subtendinous space between the tendon and the joint capsule. This environment, with the damaged tissue facilitates growth of anaerobic bacteria from the mouth of the victim. Human bites are distinctly more serious than animal bites. A number of cases on record document osteomyelitis, or joint space infection, or both, sometimes with considerable destruction of tissue. Fatal cases have been recorded. Bites of the ear, also relatively common, are benign in contrast to hand and finger wounds. The relative lack of subcutaneous tissue and the fibroelastic cartilage of the ear provide no locus for deep puncture wounds and thus anaerobic organisms cannot establish a good foothold [Brandt (796)]. Although most human bite infections involve bites per se or fist fights, significant anaerobic infections have also followed fingernail biting in children and occupational injuries to fingers of dentists and otolaryngologists. One of the author's sons fell while playing basketball as a youngster and imbedded one of his teeth in the soft tissue over his patella. An infection developed at the site from which *Fusobacterium nucleatum* and two strains of anaerobic cocci were recovered.

It is generally conceded that bite infections are primarily anaerobic, and a number of authors describe cases in which there was foul discharge or in which "fusiforms and spirilli" were seen on direct smear. However, there are relatively few cases of either human or animal bites in which anaerobic infection has been documented by appropriate culture. A summary of such reports is given in Table 13.4. Maier (797) noted that anaerobes were commonly cultured from human bite infections of the hand. He cited 17 cases of bite infections but gave no specific data regarding involvement of anaerobes. Lee and Buhr [cited by Meyers *et al.* (798)] isolated clostridia from some dog bite wounds in humans. Burdon (799) noted that *Bacteroides melaninogenicus* is part of the normal flora of the dog mouth. Anaerobic streptococci have also been found in the mouth of the dog. Pilot (800) described a fusospirochetal infection of the hand (documented by smear only) following a bite of a human by an orangutan. Williams (1556) described anaerobic bacteremia following a rat bite. Other animals have the potential to set up anaerobic infections following bites. Doering *et al.* (801) demonstrated that alligators have clostridia in their mouth flora. Ledbetter and Kutscher (802) isolated clostridia from 48% of venom samples and 86% of fangs of rattlesnakes.

Liebetruth (803) described recovery of *Bacteroides melaninogenicus* and *Fusobacterium* from an *infected hematoma*, site unspecified. Anaerobic infections of *burns* are probably not rare, but there are relatively few well documented cases in the literature. Sandusky *et al.* (112) described two burns infected with anaerobic streptococci. Parker *et al.* (722) recovered clostridia along with *Staphylococcus aureus* and β-hemolytic streptococci from a second-degree burn of a foot. Beerens and Tahon-Castel (115) recovered *Eggerthella convexa*, *Dialister*, anaerobic streptococci, and *Streptococcus zymogenes* from a postburn infection of the neck. Schaffner (1240a) recovered *Eggerthella convexa*, *Dialister pneumosintes* and *Streptococcus anaerobius* along with *Streptococcus zymogenes* and *Staphylococcus epidermidis* from an infected burn. Baird (324b) recovered *Bacteroides* from

TABLE 13.4 Documented Bite[a] Infections Involving Anaerobes

Aerobes	Anaerobes	Comments	Reference	Year
Human bites				
0?	Fusiforms		Peters (349)	1911
0	Fusiforms and spirochetes (both cultured)	Severe bite, fractured distal phalanx, severe infection—necrosis of phalanges and metacarpals, amputation of hand	Hennessy and Fletcher (1500)	1920
β-streptococci, diphtheroid	Fusiform		Pilot and Meyer (1501)	1925
0	Fusiforms, anaerobic cocci	Foul odor, spirochetes on smear	Bower and Lang (1502)	1930
Hemolytic streptococci, S. aureus	Anaerobic streptococci, anaerobic gram-positive rods, anaerobic gram-negative rods		Welch (1503)	1936
S. aureus	B. fusiformis	Foul odor, response to x-ray therapy	Smith and Manges (1504)	1937
0	Anaerobic streptococci, small anaerobic gram-negative coccus		Cohn (1505)	1940
S. aureus	Fusiforms, anaerobic streptococci	Foul odor, lost terminal phalanx	Butler (1506)	1944
Staphylococci	Fusiform (gram-positive!)	Metacarpal bone abscess	Andreasen (1507)	1947
0?	Fusiforms and spirochetes (smear)	Same organisms seen on smear from inoculated rabbit	Durante (1132)	1953
0?	Two positive blood cultures, Fusobacterium biacutum		Murphy et al. (383)	1963
0	Arachnia propionica		Brock et al. (304)	1973
Animal bites				
0	S. necrophorus	Veterinarian, abrasion from cow's tooth, developed ulcer extending almost entire length of arm	Van Wering (987)	1923
?	C. perfringens	Chimpanzee bite	Piette [cited by Pilot (800)]	1938
0	Fusiforms and spirochetes	Rat bite	Williams (1556)	1941
0	Anaerobic streptococci	Dog bite	Sandusky et al. (112)	1942
0	Sphaerophorus absceedens	Dog bite, calcaneal osteitis	Tardieux (1508)	1951
0	Bacteroides fragilis (identification questionable)	Monkey bite, draining sinuses, lymphangitis	Markham and Kershaw (1225)	1956

[a] Includes all injuries inflicted by teeth.

an infected burn of the neck. Leigh (1420f) found *Bacteroides fragilis* in a burn case. Felner and Dowell (168) reported bacteremia due to *Bacteroides fragilis* in a patient with an infection of toes affected by *frostbite*. Gorbach and Thadepalli (486) reported three cases of bacteremia due to *Clostridium* in patients with gas gangrene secondary to frostbite.

Soft tissue infections due to anaerobes have been described secondary to the use of narcotics by addicts. Oerther *et al.* (804) described an abscess of the neck due to *Bacteroides* and microaerophilic streptococci in a narcotic addict who was injecting paregoric intravenously into the jugular vein. Lerner and Oerther (805) described a similar infection at the same site due to *Bacteroides*. Subcutaneous injection of narcotics by "skin-poppers" may give rise to anaerobic infection as well. As noted earlier, Gorbach *et al.* (486) described isolation of clostridia from three addicts with muscle abscesses. A case of *Bacteroides* necrotizing fasciitis described by Rein and Cosman (768a) has already been mentioned. Brooks *et al.* (1531) recovered *Bacteroides* plus *Eikenella corrodens*, group C β-hemolytic streptococci and nonhemolytic streptococci from a thigh abscess in an addict. Geelhoed and Joseph (805a) commented that they have seen an anaerobic flora in the abscesses of some of their addicts related to drug injection. They noted that clostridia, especially *C. tetani*, are found more frequently in addicts from the eastern United States than from those from the West Coast. This is felt to be due to the use of quinine to cut heroin in the East. Quinine apparently lowers the redox potential. The recovery of clostridia from heroin and injection paraphernalia (432a) has already been noted. Lewis (805b) noted that addicts may present a clinical picture similar to that of tropical pyomyositis and that only 3 of 14 such patients he saw had a history of injection or trauma at the site of the lesions. Five of the 14 patients yielded anaerobes, three with clostridia (one of which grew *Bacteroides* from the blood) and two with both *Peptostreptococcus* and two strains of clostridia. Among 6 cases of drug related abscesses reported by Nichols and Smith (1499a), 4 yielded peptostreptococci and 1 *Bacteroides*. These authors noted that in the cases from which *Peptostreptococcus* was isolated there was usually a history of "lubricating" the needle with saliva prior to injection. The 5 subcutaneous abscesses with necrotizing fasciitis and local myositis in addicts reported by Gorbach and Thadepalli (486) have already been cited in Table 13.3. One of the cases of anaerobic streptococcal myositis reported by Anderson *et al.* (788) and cited earlier was in an addict who accidentally injected morphine into his femoral artery. We have seen large abscesses in the upper arm of addicts who injected narcotics locally. The foul pus from one such lesion yielded *Fusobacterium nucleatum*, *Bacteroides melaninogenicus* ss. *intermedius*, *Clostridium sordellii*, *Bacteroides oralis*, and *Eubacterium*. Another case yielded *Clostridium perfringens*,

Peptostreptococcus micros, Veillonella alcalescens, Bacteroides sp., *Lactobacillus fermentum, Streptococcus equisimilis, S. lactis, S. bovis, Klebsiella* sp., and *Enterobacter cloacae.* Addicts apparently commonly use water from toilet bowls in the course of preparing their narcotics, since they can do this with some privacy. This type of practice certainly might facilitate anaerobic infection.

Data on the involvement of anaerobic bacteria in both war time and civilian wounds vary considerably from study to study. This is particularly true with regard to the non-spore-forming anaerobes. Tissier, in 1916 (806), indicated that anaerobic bacteria really dominated war wounds and mentioned, in order, *Clostridium perfringens, Clostridium bifermentans, Clostridium putrificum, Clostridium septicum,* and *Coccobacillus preacutus* (the fusiform bacillus). Rustigian and Cipriani (807) recovered 214 species of bacteria from 36 war wounds from American troops in Italy. Included were *Clostridium sporogenes,* 22 strains, *C. putrificum* 8, *C. perfringens* 7, *C. tertium* 6, *C. novyi* 2, *C. septicum* 2, *C. sphenoides* 1, *C. bifermentans* 1, unidentified *Clostridium* 9, *Bacteroides melaninogenicus* 3, *Actinomyces* 4, anaerobic diphtheroids 2, anaerobic streptococci 9, anaerobic micrococcus 21, and microaerophilic micrococcus 14. Cruickshank (808) stated that local or generalized infection with anaerobic streptococci or anaerobic gram-negative bacilli was probably not uncommon in war wounds but did not give specific data. MacLennan (787), on the other hand, stated that both fusiform bacilli and anaerobic cocci were rare in war wounds in the Middle East. Lindberg and Newton (809) studied Korean War battle casualties. The most common clostridia isolated were *Clostridium perfringens* and *Clostridium sporogenes,* with *Clostridium novyi* and *Clostridium bifermentans* next in frequency. Pulaski *et al.* (810), studied fresh accidental civilian wounds; there were 102 clean wounds and 98 dirty wounds. Among the clean wounds, none yielded anaerobes in pure culture and 21.6% yielded anaerobes along with aerobes or facultatives. In the dirty wound group, in only 3 cases among the 98 were anaerobes isolated in pure culture; 42.8% of these patients had anaerobes isolated along with aerobic or facultative forms. Overall, clostridia were isolated from 23% of wounds, anaerobic streptococci from 6.5%, microaerophilic streptococci from 4.5%, and anaerobic gram-negative bacilli from 1.5%. Single anaerobic coccal infections following traumatic injuries are reported by Sandusky *et al.* (112) and Thomas and Hare (811).

14

Bone and Joint Infections

Most clinicians today, even orthopedic surgeons, infectious disease men, and rheumatologists, seem to be unaware of the role of anaerobic bacteria in bone and joint infections. It is true that anaerobic bacteria have not been documented with great frequency in purulent arthritis and that their role in osteomyelitis remains to be fully defined. Nevertheless, it is clear that anaerobic bacteria may cause significant infections of these types.

Osteomyelitis

In 1898, Rist (8) described eight cases of osteomyelitis of the skull, secondary to otitis media, in which anaerobic bacteria played a part. All but one of these cases occurred in young children. In five of them, anaerobes were the only bacteria recovered. The mastoid bone was involved in seven of the cases. One of these cases also had involvement of the sphenoid and the petrous bone. The eighth case had involvement of the petrous bone alone.

In 1917, Taylor and Davies (812) wrote a classic article which indicated that anaerobic bacteria could be recovered with some frequency in patients with osteomyelitis; they also developed experimental evidence to indicate that these organisms played an important role in the disease. These workers studied a total of 48 patients who developed fractures, primarily of long bones, secondary to war injuries. The particular cases studied constituted consecutive cases that had developed chronic osteomyelitis (varying in age from 45 to 300 days) and all of which showed sequestra on X-ray examination. All but two had draining sinuses. Despite the fact that they must have utilized techniques that were not very effective for growth of anaerobes compared to present techniques (they recovered only clostridia apparently, and some of them required 7 to 10 days incubation for growth), they recovered "gas-forming anaerobic bacilli" from 55% of the sequestra cultured at the time of surgery and "spore-forming anaerobes" from 42% of such sequestra. It is not clear how much overlap there was between the two groups of anaerobes described; no data is given on the total number of sequestra yielding anaerobic bacteria on culture. They compared cultures from the wound with cultures of the sequestra and found that, whereas with aerobic facultative bacteria the percentage recovery was only slightly higher in the case of the sequestra, there was a remarkable difference in the case of anaerobic bacteria. This was taken by the authors to mean that the bone itself appeared to be the most common site for persistence of anaerobic flora. However, it appears that their technique would have aerated the specimens used for wound culture, and it is possible or likely that their transport technique did not maintain anaerobiosis as far as the wound culture was concerned but anaerobiosis would be maintained, of course, in the case of the sequestra. In any case, "gas-forming anaerobes" were recovered nearly four times as often from the wound while "spore-forming anaerobes" were present six times more often in bone. The anaerobes most commonly recovered included *Bacillus aerogenes capsulatus* (*Clostridium perfringens*), *B. malignans oedematis* (*Clostridium septicum*), and Hibler Bacillus type IX.

Taylor and Davies (812) pointed out that bacteria of any type would be well protected within sequestra inasmuch as leukocytes are rarely seen within this dead bone, and organisms would be protected from body fluids and from various germicides that might be applied topically. They also pointed out that insoluble bone salts might combine with acids produced during metabolism of the bacteria present and thus neutralize the acids, which might otherwise result in self-sterilization. In *in vitro* experiments with dried pieces of beef bone, it was found that with the Hibler Bacillus and *Clostridium septicum* greater loss of weight of the bone resulted when the organisms and bone were incubated together than was true with the facultative bacteria (8 and 6% for the two anaerobes compared to 3.5 to 2% for streptococci, *Proteus*, and staphylococci). These workers also demonstrated that certain organic acids that are common end-products of metabolism of anaerobes (acetic, propionic, and butyric acids) exhibited significant erosive action on sequestra. Acetic acid was the most active of the three, and propionic next. A greater percentage of sequestra yielded anaerobes in older cases than in cases in early stages. Finally, Taylor and Davies showed that flares (fever and local inflammatory response) following sequestrectomy were much more likely to occur when anaerobic bacteria were present in the individual case.

A few reports in the antimicrobial era have also described several cases of osteomyelitis involving anaerobic bacteria. The paper by Heineman and Braude in 1963 (352) dealing primarily with brain abscess, described four cases of involvement of the mastoid bone with anaerobes secondary to otitis media, and two cases of frontal bone osteomyelitis due to anaerobes secondary to sinusitis. Beerens and Tahon-Castel, in 1965 (115) described five cases of osteomyelitis of the maxilla involving anaerobic bacteria. In three of these the infection followed trauma, and in a fourth it followed tooth extraction. The background in the fifth case is not clear. Paredes and Fernández (1404a) indicated that they recovered *Clostridium perfringens* from one or more cases of osteomyelitis, but no details were given. Hollin *et al.* (1268d) described a case of sinusitis with foul pus and with frontal bone osteomyelitis; culture yielded only *Staphylococcus aureus* and nonhemolytic streptococcus. Nobles (56d) mentioned a case of osteomyelitis (bone not specified) which served as the portal of entry for *Bacteroides* bacteremia. Levison *et al.* (1525g) recorded a case of osteomyelitis of the temporal bone following radical neck dissection and radiation therapy which yielded a mixture of anaerobes, Enterobacteriaceae, and *P. aeruginosa*.

In 1968, Ziment *et al.* (725) described 17 cases of osteomyelitis in which anaerobic bacteria were present. In most of those cases, anaerobic cultures were made because of features suggestive of infection with anaerobes. The patients were placed into three categories according to the site involved. Eight cases (four chronic and four acute) involved the bones of the feet; seven of these patients were diabetic. There were five cases (four chronic and one acute) of osteomyelitis involving bones of the skull; one of these patients was diabetic. Four cases (three chronic and one acute) involved long bones; one of these patients was a diabetic. In all but one case, anaerobes were associated with aerobic or facultative bacteria. However, when serial cultures were performed in three patients, persistence of anaerobes (and usually of the very same anaerobic types) was noted over a period of many months. One patient developed osteomyelitis following a ruptured appendix which persisted for 20 years. There was such extensive destruction of the femur that it was not feasible to remove all dead bone. The patient had chronic draining sinuses from which were cultured on numerous occasions *Staphylococcus aureus*, β-hemolytic streptococci, *Proteus mirabilis*, and five different anaerobic bacteria (gram-negative bacilli and cocci). Most of the cultures in this series were obtained directly from the lesion or a draining sinus. However, in one patient, culture of a specimen obtained at surgery revealed the same flora as that recovered earlier in the disease by the routine technique. The anaerobes recovered were as follows: *Bacteroides melaninogenicus* (11 cases), *B. fragilis* (10), *Peptostreptococcus* (6), *Peptococcus* (5), *Fusobacterium* (9), microaerophilic cocci (2), and *Bifidobacterium* (1). The one anaerobe obtained in pure culture was a *Bacteroides fragilis* obtained from a patient who had osteomyelitis of the head of the third metatarsal and a chronic draining ulcer of the sole of the foot. Eleven of the 14 patients had foul-smelling pus or discharge, and four had evidence of gas in soft tissues. Other clues to the presence of anaerobic bacteria in osteomyelitis in this series included sloughing of necrotic tissue (4 patients), black discharge (1 patient with *Bacteroides melaninogenicus*), failure to isolate potential pathogens on aerobic culture (5 patients), and suggestive findings on gram stain of discharges (8 cases of 12 in which this was done). In this series, 14 of the 17 infections resulted from invasion of the bone from a contiguous focus of infection; two cases were hematogenous in origin; and 1 case was a complication of treatment of a traumatic fracture. The strong association between presence of diabetes mellitus and mixed anaerobic osteomyelitis has not been stressed previously. Interestingly, one of our subsequent patients developed osteomyelitis and a soft tissue abscess in a finger following a cat scratch. The abscess yielded *Bacteroides melaninogenicus* and *Staphylococcus aureus* and debrided tendon and fascia yielded the same organisms plus *Escherichia coli*.

Bronner and Bronner (813) in a literature review, described eight cases of actinomycosis involving bone (four of the mandible, two of the maxilla, and one each of the cervical and thoracic vertebrae). Also in 1969, Nettles *et al.* (541) in a paper on musculoskeletal infections involving *Bacteroides*, described six cases of osteomyelitis from which *Bacteroides* were recovered. One of these also had anaerobic streptococci, and only one had a concurrent aerobe (*Staphylococcus aureus*) recovered. One of these occurred 13 years after a ruptured appendix. Pearson and Harvey, in 1971 (814), published an interesting article on *Bacteroides* infections in orthopedic conditions in which they described a total of 60 patients with various orthopedic conditions who developed *Bacteroides* infections. Among these were nine cases of osteomyelitis as well as two with pyarthrosis. One of the cases of osteomyelitis followed a human bite infection. Two alcoholics had osteomyelitis of a finger following fracture. The other six cases of osteomyelitis were of the foot, and four of these patients had diabetes mellitus. One of these patients had osteomyelitis involving several toes on both feet recurrently for over 4 years; *Bacteroides* and anaerobic streptococci were recovered periodically during this entire period of time. These authors noted that patients with *Bacteroides* infection (not necessarily restricted to bone and joints) had a distinctly higher incidence of diabetes, of alcoholism and of drug addiction than was true for patients in general on their Orthopedic Service.

Gupta (814a) recovered Bacteroidaceae from 12 cases of osteomyelitis or other bony lesions, twice in pure culture; no details were given. Mitchell *et al.* (814b) noted 1 isolate of *Bacteroides* from a wound infection following orthopedic surgery; there was no indication as to whether or not the bone became infected. DeHaan *et al.* (814c), in a summary of case reports submitted to The Upjohn Company in connection with clinical trials of clindamycin, recorded 10 cases of osteomyelitis involving anaerobes; an unspeciated *Bacteroides* was the only one recovered in pure culture, and only multiple anaerobes were obtained from 3 other cases. It is not known whether or not these 10 cases have subsequently been reported by the investigators concerned; accordingly, they are left out of Table 14.1. Chow *et al.* (1420e) mentioned 4 cases of osteomyelitis or septic arthritis from which anaerobes were recovered; no details were given. Sabbaj and Villanueva (557a) and Anguiano *et al.* (353a) indicated recovery of anaerobes from 1 or more cases of osteomyelitis, but no details were given.

The report of Lodenkämper and Rühlcke (814d) is impressive. These workers cultured material from 824 patients with osteomyelitis between 1956 and 1973. They recovered anaerobes alone from 43 of these patients, but 23 yielded only anaerobic *Corynebacterium*. *Bacteroides* was found in 12 of these, and anaerobic cocci in 10 (some had more than one anaerobe present). Anaerobes were recovered in association with aerobes from 55 patients; *Corynebacterium* was the only anaerobe in 28 of these. Specific details were given on a few of their cases and these are recorded in Table 14.1.

Our group has reviewed the possible role of anaerobes in osteomyelitis at Wadsworth Veterans Hospital since our previous report [Ziment *et al.* (725)]. In a 3 year period, 1973–1975 (R. P. Lewis, V. L. Sutter, and S. M. Finegold, unpublished data), 58 patients with osteomyelitis were admitted to our hospital for the first time or had osteomyelitis diagnosed for the first time. Among these 58, 22 had anaerobes recovered and 1 additional patient (not cultured) was felt to have anaerobes involved because of foul discharge. Three of these 22 cases with anaerobes recovered had these organisms as essentially the only isolates. One yielded only *Propionibacterium acnes* (a neurosurgical case with infection of the skull and a bone flap), a second (mandible) grew *Peptostreptococcus micros* and a *Lactobacillus*, and the third yielded 5 obligate anaerobes plus *Streptococcus dysgalactiae*. In the group of 22 cases yielding anaerobes on culture, there was a total of 92 aerobic or facultative isolates and 98 anaerobic isolates. Among the aerobes and facultatives, there were 14 isolates of *Staphylococcus aureus*, 32 Enterobacteriaceae, 9 *Pseudomonas aeruginosa*, and 22 streptococci. Anaerobes

TABLE 14.1 Overall Summary of Reported Cases of Osteomyelitis Involving

No. of bones involved	Bone involved, source	B. fragilis	B. melaninogenicus	Other Bacteroides (or unspecified)	F. necrophorum	F. nucleatum	Other Fusobacterium
46	Mastoid, ethmoid, sphenoid, temporal bone; all secondary to otitis	6 (352, 725, 1226)		8 (8, 282, 1233, 1245)	12 (8, 260–262, 266, 482, 621, 1247, 1258, 1301)	1 (350)	5 (8, 480, 725, 1224)
92	Skull—other sites and/or sources	4 (315, 352, 725, 1375)	8 (116, 315, 352, 725, 1525b)	7 (115, 135, 282, 1217, 1525a, 1532)	9 (146, 1173, 1521)		11 (115, 163, 248, 725, 734, 743, 1211, 1272)
44	Vertebrae	3 (1375)	5 (116)	1 (282)	1 (168)		3 (540, 1209)
11	Ribs, clavicle, scapula				1 (302)		
11	Long bones, hematogenous source of infection	3 (489, 725)	2 (725)		4 (164, 302, 1247, 1517)		3 (725, 1211, 1509)
43	Long bones, secondary to trauma	3 (226a, 1240e, 1266)	1 (725)	3 (541, 1522)			
2	Long bones, associated with vascular disease and/or neuropathy	1 (725)			1 (146)		
18	Long bones, etiology unclear	5 (266a, 1525e, f)	2 (116, 266a)	2 (541, 816)			
21	Hands and feet without vascular disease	3 (517, 1375)	2 (517)	3 (541, 814)			6 (1500, 1503, 1505, 1507, 1508)
19	Hands and feet with vascular disease and/or neuropathy	5 (725, 1525g)	7 (390, 725)	7 (541, 814)			5 (725)
2	Pelvis				1 (1517)		
26	Other	1 (1213)		4 (370, 416, 1525)	1 (490)		3 (21, 337, 1509)
Total 335		34	27	35	30	1	36

[a] Numbers are numbers of isolates. Numbers in parentheses are reference numbers (see Bibliography).

Anaerobic Bacteria[a]

Unidentified anaerobic gram-negative rod	Actinomyces and Arachnia	Other gram-positive non-spore-forming rods	C. perfringens	Other Clostridium	Anaerobic and microaerophilic cocci	Miscellaneous bacilli and spirochetes	Unidentified anaerobe
1 (8)	3 (278a, 352)	3 (8, 259, 1531)	1 (8)		14 (8, 266a, 268, 352, 480, 528, 1301)	3 (8, 268)	12 (8, 268)
	34 (143a, 212, 295, 315, 331, 337, 340, 517, 813, 1134a, 1138, 1142b, 1510, 1523, 1525b, d, j, k, l, 1529, 1530)	4 (115, 337, 558, 725)			31 (21, 115, 135, 143a, 146, 175, 176, 252, 291, 337, 352, 373, 709, 1217, 1240b, 1268e)	13 (115, 146, 248, 734, 743, 1272)	
	31 (212, 813, 1209, 1271b, 1511, 1526)	1 (1525i)			1 (1240o)	1 (21)	
	6 (212, 1267g, 1271b, 1524, 1526, 1528)				3 (390, 1331, 1334)		
	4 (212, 1267g, 1527)				3 (725, 1509)		
		2 (814d, 1525)	26 (812, 1525c)		7 (541, 725, 814d)	1 (1438)	
						1 (146)	
1 (714)	3 (714, 1515, 1518)				7 (266a, 1219, 1519, 1525e)		
2 (714, 1503)	5 (714, 716, 1513, 1516, 1520)	1 (1503)	1 (1525h)	1 (1466c)	3 (517, 1503)	4 (1500, 1503, 1505)	2 (1512, 1514)
					10 (390, 725, 814)		
					1 (659)		
		2 (167, 252)	1 (1525)		7 (112, 135, 337, 370, 416, 1509, 1525)		9 (287)
4	86	13	31		87	23	23

439

included 33 *Bacteroides*, 4 *Fusobacterium*, 29 *Peptococcus*, 8 *Peptostreptococcus*, and 6 *Actinomyces*. Our research laboratory cultured 18 of the patients yielding anaerobes. Among this group, there were 12 cultures of material from a draining sinus and 5 tissue biopsies, and these yielded an average of 4 anaerobes per sample. Eight cultures of bone or of abscess contents (obtained at surgery) yielded an average of 2.8 anaerobes per sample. In a 5 year period, we obtained bone for culture, at the time of surgery, on 15 patients (12 had anaerobic cultures). Eleven of these 12 grew anaerobes, a total of 29 strains (including 14 *Bacteroides*, 9 gram-positive anaerobic cocci, and 2 *Actinomyces*). Seventeen bone specimens from 15 patients yielded 28 nonanaerobes including 4 *S. aureus*, 11 gram-negative rods, and 9 streptococci.

Comparison of the 23 patients whose osteomyelitis involved anaerobes in the 1973–1975 study with the 35 patients involving only aerobes or facultatives yielded interesting information. The incidence of neuropathy was 43% in the anaerobic group to 28% in the other. Bacteremia was not detected in the anaerobic group, but occurred in 14% of the other. There were no impressive differences with regard to incidence of diabetes mellitus or vascular disease (surprisingly), renal failure, age of patient, or chronicity of infection. The mandible and ankle were each involved in 11.5% of the anaerobic cases and in none of the other group and toes were involved twice as frequently in the patients with anaerobes (31%) as in the others (16%). There was no involvement of the spine among anaerobic cases, but it was involved in 13.5% of the others. Osteomyelitis of the long bones of the lower extremity was less common in anaerobic cases (15%) than in the others (30%).

Table 14.1 is an overall summary of the reported cases of osteomyelitis involving anaerobic bacteria considered primarily in terms of the bones involved and the anaerobic bacteria recovered, with less emphasis on the source of the infection. In all, there were 335 bones involved, the skull alone accounting for 138 (Figs. 14.1–14.3). Involvement of the vertebrae and bones of the hands and feet was also common but it is notable that there were fewer cases involving long bones than in most series of osteomyelitis. There was a total of 430 anaerobes recovered. In the preantimicrobial era, most specimens did not yield aerobic or facultative bacteria concurrently with the anaerobes. Subsequently, however, most cultures yielding anaerobes also yielded these other organisms, except for the majority of cases of actinomycosis and many of the infections involving the bones of the skull. Over one-third of the anaerobic isolates were anaerobic gram-negative bacilli, chiefly *Bacteroides*. Among members of the *Fusobacterium* genus, *Fusobacterium necrophorum* was by far the most common isolate. The other two groups recovered most commonly were the anaerobic cocci and *Actinomyces*. Over half of the isolates of anaerobic cocci were recovered from osteomyelitis

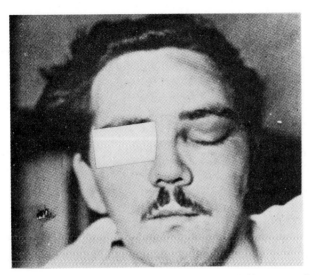

Fig. 14.1. Photograph of a patient with osteomyelitis of frontal bone secondary to frontal sinusitis. There is swelling, tenderness, and fluctuation over the left frontal area. From French and Chou (1592). "Cranial and Intracranial Suppuration" (E. S. Gurdjian, ed.), Thomas, Springfield, 1969.

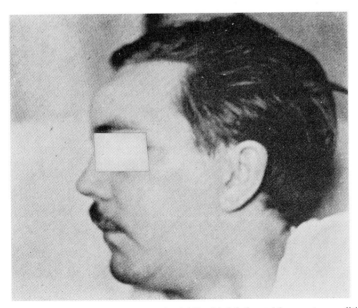

Fig. 14.2. Lateral view of patient (in Fig. 14.1) with left frontal bone osteomyelitis. From French and Chou (1592). "Cranial and Intracranial Suppuration" (E. S. Gurdjian, ed.), Thomas, Springfield, 1969.

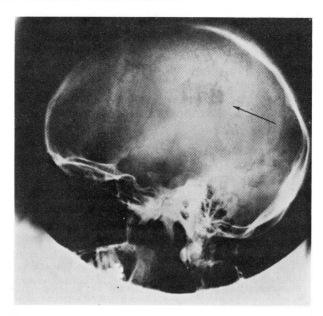

Fig. 14.3. Roentgenogram showing osteomyelitis of left parietal bone (arrow). The ragged, moth-eaten appearance is typical. From French and Chou (1592). "Cranial and Intracranial Suppuration" (E. S. Gurdjian, ed.), Thomas, Springfield, 1969.

involving bones of the skull. This was also a common site for involvement with *Actinomyces* and with *Fusobacterium necrophorum*. It is interesting to note, however, that 10 of the 34 isolates of *Bacteroides fragilis* also involved bones of the skull, despite the fact that this organism is not known to be part of the normal mouth or upper respiratory tract flora. In vertebral osteomyelitis involving anaerobes, *Actinomyces* was by far the most common isolate. Clostridia were particularly prevalent in osteomyelitis of long bones secondary to trauma. Anaerobic gram-negative bacilli predominated in osteomyelitis of small bones of the hands and feet.

It is clear that many patients with anaerobic organisms recovered from cases of osteomyelitis will have evidence of anaerobic infection elsewhere in the body and that this will be the source of the organisms involved in osteomyelitis—either contiguous infection extending to the bone or infection that reaches the bone by way of the bloodstream during the course of sustained or intermittent bacteremia. Thus, many or most patients with anaerobic osteomyelitis will give evidence of anaerobic infection elsewhere in the body. This can be of any type, but often will show characteristic features of anaerobic infection such as abscess formation, septic thrombophlebitis, production of foul odor and gas, and tissue necrosis (including necrosis of tendon adjacent to the area of osteomyelitis). Some of the patients with anaerobic osteomyelitis will also have arthritis involving anaerobic bacteria, usually in an adjacent joint. A certain number of patients will have positive blood cultures. Of course, the usual underlying diseases that may be seen with anaerobic infection of any type, such as diabetes mellitus and malignancy, may be seen in these patients as well. There may be draining sinuses.

While anaerobes are found in osteomyelitis with some frequency and undoubtedly would be found much more commonly if proper anaerobic cultures were utilized, the precise role of anaerobes in this condition remains to be established. Certainly those cases in which anaerobes are the only organisms recovered are prime examples of osteomyelitis due to these organisms. However, as indicated earlier, in the antimicrobial era most cases of osteomyelitis from which anaerobes are isolated also yield other types of bacteria. Whether the anaerobes are the prime pathogens, simply secondary colonizers of the wound, or might act synergistically with other organisms remains to be determined. This is a fertile field for additional experimental and clinical investigation.

Purulent Arthritis

A summary of several reviews of septic arthritis is given in Table 14.2. These 20 reviews comprise a total of 1103 patients from whom joint fluid was cultured. Although it is not always clear that cultures were performed for anaerobic bacteria or if they were, that they were performed in proper fashion with adequate anaerobic transport, very few anaerobic bacteria were recovered in these particular series. Of further interest is the fact that 200 cultures, or 18% of the total, yielded no growth whatever. While this might be partially related to prior use of antimicrobial therapy, it seems likely that at least a portion of these negative cultures involved fastidious organisms that were not recovered because of inadequate techniques utilized. Certainly the majority of these fastidious organisms would be anaerobes.

A review of the literature reveals a total of 180 joint infections involving anaerobic bacteria. These are summarized by the joints and organisms involved in Table 14.3. By far the most common isolate among the anaerobes was *Fusobacterium necrophorum*, accounting for over one-third of the total anaerobes recovered. This is in sharp contrast to osteomyelitis, which has been discussed earlier in this chapter. The majority of these cases were reported from the preantimicrobial era. As indicated in Chapters 6 and 9, this organism (in contrast to all other anaerobes) is seen much less commonly since the introduction of antimicrobial agents. It is not more sensitive to antimicrobial agents than a number of other anaerobes that are still seen with some frequency. The best explanation may be that this organism,

TABLE 14.2

Reviews of Septic Arthritis

Type of study	Number of cultures	Anaerobes isolated	Negative cultures	Reference	Year
36 cases of acute septic arthritis	35	None	3	Inge and Liebolt (1533)	1935
201 cases of septic arthritis	140	None	19	Heberling (1534)	1941
70 cases of septic arthritis	45	None	4	Altemeier and Largen (1535)	1952
51 cases of acute septic arthritis	51	None	None	Watkins et al. (1536)	1956
Suppurative arthritis complicating underlying arthritis	(a) 13 with rheumatoid arthritis (b) 10 without rheumatoid arthritis	None	1 with anaerobic streptococci and S. albus	2 (or e with gram-positive cocci on stain)	Kellgren et al. (1537)	1958
77 cases of septic arthritis	77	None	19	Chartier et al. (1538)	1959
24 cases of acute septic arthritis	24	None	3	Ward et al. (1539)	1960
19 cases of arthritis in in aged and chronically ill	19	None	5	Willkens et al. (1540)	1960
34 cases of septic arthritis	34	None	9	Ortiz and Miller (1541)	1961
66 cases of acute septic arthritis	66	None	10	Baitch (1542)	1962

444

52 cases of septic arthritis in childhood	52	None	17 (5 had antibiotics previously)	Borella et al. (1543)	1963
26 adults with hip arthritis	26	1 with anaerobic streptococci (ruptured appendix), 1 with anaerobic streptococci (appendectomy), 1 with anaerobic streptococci (steroid therapy?)	None	Kelly et al. (1519)	1965
42 cases of septic arthritis	42	None	None	Argen et al. (1544)	1966
50 cases of septic hip arthritis in adults	43	1 with C. welchii	4	Bulmer (1545)	1966
78 cases of septic arthritis in adults	78	1 with Bacteroides, 1 with anaerobic streptococci	None	Kelly et al. (1546)	1970
65 cases of hip and shoulder septic arthritis	27	None	5	Kolawole and Bohrer (1547)	1972
219 cases of septic arthritis in children[a]	219	1 with Clostridium novyi, 1 with Clostridium bifermentans, 1 with Bacteroides, funduliformis	74	Nelson (1548)	1972
27 cases of septic arthritis	10 without rheumatoid arthritis	None	1	Russell and Ansell (1549)	1972
	17 with rheumatoid arthritis	1 with B. fusiformis	1		
29 cases of septic arthritis	28	None	4	Lidgren and Lindberg (1550)	1973
47 cases of septic arthritis	47	None	20	Singh and Bhattacharyya (1551)	1973
Totals	1103		200 (18.1%)		

[a] Includes data of Nelson and Koontz (1562).

TABLE 14.3

Overall Summary of Reported Cases of Anaerobic Joint Infection by Joints and Organisms Involved

Joint involved[a]	Bacteroides fragilis	Bacteroides melaninogenicus	Bacteroides (other or unidentified)	Fusobacterium necrophorum	Fusobacterium nucleatum	Fusobacterium (other or unidentified)	Other unidentified anaerobic gram-negative bacilli	Clostridium perfringens	Clostridium (other or unidentified)	Actinomyces	Other non-spore-forming gram-positive bacilli	Anaerobic cocci	Miscellaneous anaerobic bacilli and spirochetes	Unidentified anaerobe
Temporomandibular [2]										1			1	
Cervical spine [4]	3			2		2								
Sternoclavicular [17]	3		1	14								1		
Shoulder [13]			1	7		4	3	1						
Acromioclavicular [2]			1	1										
Elbow [19]	5		1	12		2			1			1		
Wrist [1]				1										
Metacarpophalangeal [5]					1		1				1	5	1	
Hip [24]	3		2	7		2		1			1	7		
Knee [34]	2		6	18	1	1		3	1	1		5	1	1
Ankle [5]			1	4					1	1				
Metatarsophalangeal and phalangeal [2]						2								
Sacroiliac [8]			1	7										
Joint not specified [39]	4	1	4	3		5	15	1	4		4	9	3	1
Totals [180]	20	1	17	75	2	18	19	6	7	3	6	28	6	2
Total organisms = 210														

[a] Numbers in brackets are numbers of joints involved.

Fusobacterium necrophorum, was commonly found as a cause of tonsillitis or pharyngitis (Vincent's angina), and that routine antimicrobial therapy for this type of illness has eliminated most of the serious complications of this type of infection (which was often accompanied by septic thrombophlebitis and bacteremia). The genus *Bacteroides*, other members of the genus *Fusobacterium* aside from *F. necrophorum*, and unidentified anaerobic gram-negative bacilli also account for many of the cases of suppurative arthritis involving anaerobes. The other major group is the group of anaerobic cocci. Most prominent among the joints involved are the larger joints, particularly the hip and knee and, to a lesser extent, the elbow and shoulder.

Of particular interest is the predilection of the sternoclavicular and also the sacroiliac joints for purulent arthritis due to anaerobic bacteria. *Fusobacterium necrophorum* accounts for the majority of cases of arthritis of these two joints. Accordingly, septic arthritis of the sternoclavicular or sacroiliac joints in the course of the sepsis, particularly if the portal of entry is in the throat, should suggest the strong likelihood that *Fusobacterium necrophorum* is the offending organism.

Tables 14.4–14.6 summarize reported cases of anaerobic joint infection secondary, respectively, to hematogenous spread, to trauma or needling of the joint, and to contiguous infection. It will be noted that most of these cases of septic arthritis are secondary to hematogenous spread. It should also be noted that all of the cases of sternoclavicular arthritis and 7 of the 8 cases of sacroiliac septic arthritis are related to hematogenous spread. It should also be noted that 72 of the 75 isolates of *Fusobacterium necrophorum* came from joints involved with infection in the course of anaerobic sepsis. Many of the other gram-negative anaerobic bacilli were found with greatest frequency in cases of purulent arthritis secondary to sepsis. This was also true for approximately half of the anaerobic cocci recovered. Unfortunately, in many cases the joint involved was not specified, and/or the organism recovered was not well characterized. In contrast to osteomyelitis, it was unusual to recover more than a single anaerobe from the infected joint fluid. Furthermore the majority of the cases yielded anaerobes in pure culture. There were only 15 instances in which aerobic or facultative bacteria, other than skin contaminants, were found together with the anaerobes.

The majority of the cases of purulent arthritis involving anaerobes were reported in the preantimicrobial era, undoubtedly related to the frequency with which *Fusobacterium necrophorum* was seen at that time as compared to later. Nevertheless, about half of the total anaerobic arthritis cases found in the literature review were from the period after antimicrobial agents had been introduced. A number of the cases reported have followed surgery on various joints, including total joint replacement. Anaerobic diphtheroids (*Propionibacterium acnes*) and, to a lesser extent, anaerobic cocci have been

TABLE 14.4
Summary of Reported Cases of Anaerobic Joint Infection, Cases Secondary to

Joint involved[b]	Bacteroides fragilis	Bacteroides melaninogenicus	Bacteroides (other or unidentified)	Fusobacterium necrophorum	Fusobacterium nucleatum
Temporomandibular [1]					
Cervical spine [2]				2 (1159, 1160)	
Sternoclavicular [17]	3 (1146, 1150, 1375)			14 (261, 1148, 1159, 1160, 1162, 1297, 1301, 1554)	
Shoulder [11]	2 (1150, 1375)			7 (261, 262, 1152, 1173, 1554)	
Acromioclavicular [1]				1 (226)	
Elbow [17]	5 (168, 1144, 1150, 1375)			12 (226, 261, 489, 1152, 1154, 1173)	
Hip [12]	1 (415)		1 (816)	7 (261, 323, 621, 1152, 1159, 1247, 1555)	
Knee [21]	1 (168)		1 (816)	17 (7, 261, 262, 302, 489, 1154, 1157, 1159, 1160, 1162, 1173, 1175, 1554, 1555)	1 (1232)
Ankle [3]				3 (1160, 1162)	
Metatarsophalangeal and phalangeal [2]					
Sacroiliac [7]				7 (261, 323, 1157, 1159, 1162, 1297, 1554)	
Joint not specified [24]	2 (168)			2 (1159, 1162)	

[a] Numbers in parentheses are reference numbers (see Bibliography).
[b] Numbers in brackets are numbers of joints involved.

Hematogenous Spread[a]

Fusobacterium (other or unidentified)	Other unidentified anaerobic gram-negative bacilli	Clostridium perfringens	Clostridium (other or unidentified)	Actinomyces	Other non-spore-forming gram-positive bacilli	Anaerobic cocci	Miscellaneous anaerobic bacilli and spirochetes	Unidentified anaerobe
							1 (8)	
4 (1379, 1552)						1 (1301)		
	3 (1145, 1438)							
2 (168, 1558)						1 (659)		
1 (1374)		1 (1545)				1 (1555)		
1 (1265)						4 (1175, 1232, 1263, 1331)	1 (1232)	1 (1559)
2 (1379, 1552)								
1 (168)	14 (226, 259, 452)				1 (1203)	6 (452, 1263)	1 (8)	1 (452)

TABLE 14.5

Summary of Reported Cases of Anaerobic Joint Infection, Cases Secondary to Trauma or Need ing[a]

Joint involved[b]	Bacteroides fragilis	Bacteroides (other or unidentified)	Fusobacterium necrophorum	Fusobacterium nucleatum	Other unidentified anaerobic gram-negative bacilli	Clostridium perfringens	Clostridium (other or unidentified)	Actinomyces	Other non-spore-forming gram-positive bacilli	Anaerobic cocci	Miscellaneous anaerobic bacilli and spirochetes
Temporomandibular [1]								1 (1569)			
Shoulder [1]		1 (1560)									
Acromioclavicular [1]		1 (1560)									
Elbow [2]		1 (1574)					1 (1574)				
Wrist [1]			1 (1567)								
Metacarpophalangeal [5]				1 (1505)	1 (1503)				1 (1503)	5 (1503, 1505, 1537)	1 (1503)
Hip [3]	2 (1266, 1517)	1 (541)								1 (541)	
Knee [10]	1 (1568)	3 (814, 1560, 1564)	1 (1240e)			3 (1561, 1566)	1 (1571)			1 (1570)	
Ankle [1]		1 (1574)					1 (1574)				

[a] Numbers in parentheses are reference numbers (see Bibliography)
[b] Numbers in brackets are numbers of joints involved.

450

TABLE 14.6

Summary of Reported Cases of Anaerobic Joint Infection, Cases of Uncertain Source or Secondary to Contiguous Infection[a]

Joint involved[b]	Bacteroides fragilis	Bacteroides melaninogenicus	Bacteroides (other or unidentified)	Fusobacterium necrophorum	Fusobacterium (other or unidentified)	Other unidentified anaerobic gram-negative bacilli	Clostridium perfringens	Clostridium (other or unidentified)	Actinomyces	Other non-spore-forming gram-positive bacilli	Anaerobic cocci	Miscellaneous anaerobic bacilli and spirochetes
Cervical spine [2]					2 (1197)							
Shoulder [2]	1 (344a)						1 (1572)					
Hip [6][c]					1 (1455)					1 (1124)	5 (1124, 1519, 1553, 1563)	
Knee [3]			2 (544, 1557)						1 (1575)			
Ankle [1]									1 (1575)			
Sacroiliac [1][d]			1 (541)									
Joint not specified [21]	2 (168)	1 (116)	4 (361a, 1546, 1548, 1573)	1 (1562)	4 (21, 1197, 1549, 1556)	1 (1549)	1 (1565)	4 (1548, 1548a)		3 (167)	3 (361a)	2 (1546, 1556)

[a] Numbers in parentheses are reference numbers (see Bibliography).
[b] Numbers in brackets are numbers of joints involved.
[c] Three cases secondary to contiguous infection.
[d] Two cases secondary to contiguous infection.

451

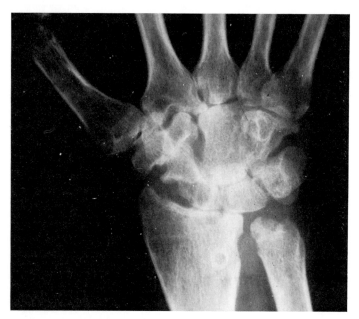

Fig. 14.4. X ray of wrist showing erosions of rheumatoid arthritis. From Karten (1567). Reproduced with permission.

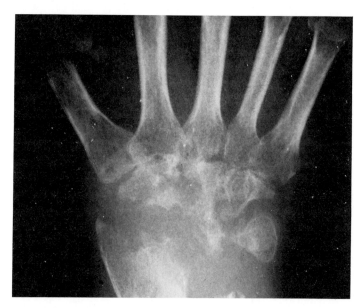

Fig. 14.5. Same patient as in Fig. 14.4. 20 months later following development of purulent arthritis due to *Bacteroides funduliformis* (*Fusobacterium necrophorum*). There is marked destruction of the carpal bones. Patient had received systemic and intraarticular corticosteroid therapy. From Karten (1567). Reproduced with permission.

452

implicated in recent reports on infection following total hip replacement [Kamme *et al.* (815)]. Since these organisms, especially the anaerobic diphtheroids, are prevalent as normal skin flora, one must be concerned as to whether an isolate may be only a skin contaminant. This may clearly present a problem in individual cases. L. J. Nastro and S. M. Finegold (unpublished data) have recovered a *Peptostreptococcus micros* from pus aspirated from a hip joint in a patient who had a total hip replacement one year earlier.

Multiple joint involvement in the same patient in the course of anaerobic sepsis is not at all rare. In this review, involvement of more than one joint of the same type (e.g., both knees) in the same patient was counted as a single joint, whereas involvement of two joints of different type in the same patient was counted as two. In the cases arising by way of the bloodstream, there was usually clear evidence of anaerobic infection elsewhere in the body. This was particularly evident in the lung where pneumonia or, more commonly, lung abscess and/or septic pulmonary infarction were noted. Multiple abscesses throughout the body and evidence of septic thrombophlebitis were seen often. Endocarditis was found on occasion. At times, anaerobic osteomyelitis was also present in the same patient. In recent years, the use of intraarticular corticosteroids has led to a number of cases of joint infection involving anaerobes (Figs. 14.4 and 14.5). Trauma of other types has been responsible in some cases.

The tendency of anaerobic bacteria to produce tissue destruction may be noted in this setting also. There may be destruction of joint cartilage and of the joint capsule and even periosteum. Joints may fuse as a result of the infection. Multiple draining sinuses may be noted rarely. On occasion, the joint fluid will have a foul odor in the case of anaerobic infection, and, on rare occasion, there may be gas under pressure in the joint. Other clues to suggest that a case of purulent arthritis may involve anaerobes would include failure to obtain organisms on routine culture, gram stain of the joint fluid showing organisms with the unique morphology of anaerobes, and evidence of anaerobic infection elsewhere in the body. Diabetes mellitus seems to be much less common as an underlying condition in the case of anaerobic arthritis as compared to osteomyelitis.

Therapy

Therapy of anaerobic osteomyelitis and purulent arthritis is not basically different from that required in other types of osteomyelitis and purulent arthritis. Included would be management of any underlying disease, appropriate drainage and debridement, temporary immobilization of the joint in the case of arthritis, and antimicrobial therapy pertinent to the bacteriology of the individual case (see Chapters 18 and 19). Intraarticular therapy should not be necessary and is generally undesirable. In the case of infection associated with an artificial prosthesis, removal of the prosthesis will ordinarily be necessary in order to cure the infection.

15

Pediatric Infections

Many pediatricians express the view that anaerobic infections must be quite rare in the pediatric age group. The data presented in this chapter suggests otherwise. There are several reports on anaerobic infections in the pediatric literature, and there are other reports whose titles indicate studies of anaerobic infections in a pediatric population. However, it is true that most of the reports concerning anaerobic infection in children are not identified as such and may therefore have escaped the attention of pediatricians.

There are two major types of infection that are unique to the pediatric age group—intrauterine infections and neonatal infections. Aside from these perinatal infections, there is generally nothing singular about anaerobic infections in children, either with regard to age distribution, clinical type of infection, or infecting organism. Conditions that interfere with host defense mechanisms may predispose to anaerobic infection in children as in adults; certain of these are unique to childhood. For example, anaerobic infections have been seen in children with agammaglobulinemia (816), sickle cell disease (817), leukemia (818), and periodic neutropenia (503). Felner and Dowell (168) noted three deaths due to *Bacteroides oralis* bacteremia in young children, all of whom were receiving adrenal corticosteroids or immunosuppressive therapy. *Clostridium tertium*, which has been considered to be a nonpathogen or a relatively unimportant pathogen, was recovered from two 7-year-old boys with bacteremia. These two children presented a picture clinically consistent with influenza during an influenza epidemic, although they did have high white blood cell counts, evidence of pneumonia, and positive blood cultures for *Clostridium tertium*. The studies that were performed do not permit a judgment as to the etiology of the pneumonic process. Children with Down's syndrome (trisomy 21) have a striking incidence of severe periodontal disease, often with bone destruction and with a greater prevalence of *Bacteroides melaninogenicus* in the gingival sulcus as compared to normal children (819).

An anaerobic bacterium (*Propionibacterium acnes*) plays an important role in acne, a disease that is very important in the pediatric population; this is discussed in Chapter 21.

Intrauterine Infections

Intrapartum infection is much more common than antenatal infection and is due to the ascent of vaginal bacteria into the amniotic cavity, usually after rupture of the membranes. Various possible routes of fetal infection are illustrated in Fig. 15.1. The inflammatory lesions observed in the membranes, umbilical cord, placenta, and fetus may be grouped under the name "amniotic infection syndrome." Transmission of infection to the fetus takes place primarily through its mouth and nose, and the most common and severe fetal infection is found in the lungs. Bacteremia in the fetus might in turn result from invasion of the bloodstream from the respiratory tract or the gastrointestinal tract or by direct invasion of the chorionic vessels on the surface of the placenta. Hematogenous dissemination of infection from the placenta to the fetus may take place in some cases of ascending infection with deciduitis. Infection of the amniotic fluid by way of the fallopian tube is a rarity. Hematogenous infections resulting from maternal sepsis will less commonly involve anaerobic bacteria. Ascending infections, either amniotic or placentofetal, usually originate from normal vaginal flora. Bacterial amniotic infection is the most common. Placentofetal infection is seldom demonstrated except in abortions and small premature infants.

Randall (361a) reported on cultures of amniotic fluids submitted to a clinical bacteriology laboratory; 25 of 158 isolates (16%) were anaerobes. Included were 15 *Lactobacillus*, 5 *Bacteroides*, 3 *Peptostreptococcus*, and 2 *Clostridium*. Appleman *et al.* (819a) noted that one of five amniotic fluids collected through an intrauterine fetal monitoring device yielded anaerobic bacteria. They found few septic complications despite this. *In vitro* studies by these workers demonstrated that amniotic fluid was bacteriostatic for two anaerobic cocci and a *Bacteroides*, but only for a period up to 20 hours.

Factors enhancing the likelihood of amniotic infection by organisms present in the vagina include premature rupture of the membranes with long

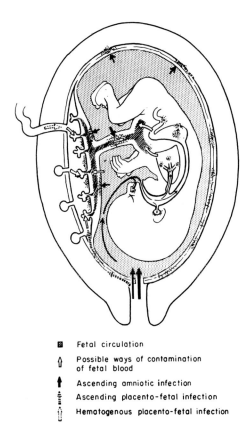

▨	Fetal circulation
⇧	Possible ways of contamination of fetal blood
↑	Ascending amniotic infection
⇡	Ascending placento-fetal infection
⇡	Hematogenous placento-fetal infection

Fig. 15.1.   Routes of fetal infection. From Blanc (820). Reproduced with permission.

delay before the onset of labor, prolonged labor even with late rupture of membranes, a prolonged period of cervical dilatation (even with membranes intact), and excessive manipulation. While organisms may be introduced from the outside by contaminated hands or instruments, the majority of infections involve organisms normally present in the vagina. Although examination of the cervix by way of the rectum does minimize introduction of organisms from the outside, even this type of examination promotes contact of the vaginal flora with the cervix and thus contamination of the membranes or fluid. The risk of exposure to bacteria increases sharply after 6 hours of labor with ruptured membranes. Prematurity also favors infection, as does the use of high forceps extraction in cases of uterine inertia. Swallowing and respiratory movements of the fetus undoubtedly expedite intake of bacteria that have reached the amniotic cavity. However, even without these movements, bacterial invasion probably takes place along the continuous film of fluid extending from the amniotic cavity to the tissues of the fetus. Excellent discussions of the pathogenesis of intrauterine infections will be found in the papers of Blanc (820) and Charles (821).

It is surprising, despite the general belief that the majority of these intrauterine infections are due to vaginal bacteria, specific discussion on the bacteriology of such infections focuses on *E. coli*, enterococci, staphylococci, and hemolytic streptococci with little or no mention of anaerobic bacteria. Since the pathway of ascending infection by vaginal bacteria is well established and since it is also now well established that anaerobic bacteria are prevalent in the vaginal flora, it is reasonable to expect that good anaerobic studies of intrauterine infections will reveal a much higher incidence of anaerobes than has been demonstrated up until now.

Pearson and Anderson (648) did find *Bacteroides* and other anaerobes in the lung of a dead aborted fetus with pneumonia; in this case there was maternal bacteremia with *Bacteroides* as well. They also noted *Bacteroides* bacteremia in five premature stillbirths and three term stillbirths. One of the latter was macerated with gas in the liver and blood vessels. Penner and McInnis (822) found anaerobic streptococci on both smear and culture of the lungs of a stillborn baby with diffuse pneumonia. Brim (823) recovered an anaerobic streptococcus from the liver of one macerated stillborn and a *Bacteroides* from another among eight cases studied in all. MacVicar (1477a) cited data from Webb, from Townsend *et al.*, and from McCallum and Govan on the role of anaerobic streptococci and *Clostridium welchii* in both maternal and perinatal infection related to premature rupture of the membranes or surgical induction of labor.

Evidence of inflammation of the placenta or cord would indicate an increased likelihood of infection in the infant. Navarro and Blanc (823a) confirmed this and described 16 cases of subacute necrotizing funisitis (inflammation of the umbilical cord). A bacteremia in one of these newborns involved *Peptostreptococcus* as well as *Klebsiella* and *Pseudomonas aeruginosa*. Neutrophils may be noted to accumulate beneath the amnion in chorioamnionitis. Stiehm *et al.* (824) pointed out that elevated levels of γ_M- or γ_A-globulin in the cord blood are presumptive evidence of an *in utero* infection. These authors noted this occurrence specifically in one case of *Bacteroides* sepsis in a newborn baby.

Neonatal Infections

A distinction between intrauterine infections and infections noted immediately after birth is obviously artificial. However, the neonatal period (the first month of life) is one in which the infant has less innate ability to overcome infection than is true later in life. The defect involves both the cellular and humoral immune systems. Thus any type of infection anywhere in the body may invade the bloodstream in such an infant. Pneumonia is a very important complication of amniotic infection in the fetus and in the newborn and may account for a significant percentage of fatal neonatal infections. About 50% of neonatal pneumonia is considered to be acquired *in utero*. Cases which appear within 24 hours of birth are almost always in this category, and this is probably true as well for most such infections up to 3 days of age. Again there is a curious lack of documentation of an important role for anaerobes in this type of infection despite the fact that the accepted means of acquisition of such infections indicates that anaerobes must be very commonly involved. Penner and McInnis (822) found anaerobic streptococci on culture of the lung of an infant that survived for 28 hours and was noted to have diffuse pneumonia at autopsy. However, the direct smear of the lung and the tissue gram stain were negative for organisms in this case. In this series, there were other cases in which gram-positive cocci were seen in tissue gram stains of the lungs but in which cultures failed to yield any organisms; some of these may have represented anaerobic infection.

There are a number of documented cases of anaerobic sepsis in the newborn period. Chow *et al.* (825) reported 23 cases of neonatal bacteremia and 57 additional cases of definite or presumptive perinatal bacteremia from a literature review. Their 23 cases were seen over a period of $3\frac{1}{2}$ years and represented 1.8 cases per 1000 live births and 26% of all cases of neonatal bacteremia. Their experience was that neonatal bacteremia had a favorable prognosis and tended to be self-limited regardless of antimicrobial therapy; only one of their 23 cases died. However, the mortality in the 43 cases of neonatal bacteremia that they reviewed from the literature was 26%. These authors noted that the clinical picture in anaerobic neonatal bacteremia was not different from that due to other organisms. Associated conditions included prolonged labor following rupture of membranes, maternal amnionitis, prematurity, fetal distress, foul odor at birth, and respiratory difficulty. Alpern and Dowell (409) studied 86 patients with bacteremia due to nonhistotoxic clostridia and noted that 18 of these were newborns. They seemed to be primarily or entirely of endogenous origin. Martin *et al.* (826), in a discussion of infections related to pediatric surgery, noted that the endogenous flora of the neonate is derived from the birth canal of the mother. They referred to a previous publication by Altemeier *et al.* in which *Bacteroides* sepsis was documented in the newborn period, although a number of other organisms were found more frequently. They also indicated that since the neonate has relatively few pathogenic organisms in his gastrointestinal tract, gastrointestinal surgery in this age group should be relatively free of hazard related to infection. Hall and O'Toole (827) supported this conclusion with a study of meconium passed by 50 newborn infants. The first specimens of meconium passed often failed to yield bacteria, although all did after 24 hours of life. The predominating organisms were micrococci and *E. coli* with rare clostridia and streptococci. *Clostridium perfringens* was found only once in 50 specimens. Nevertheless, in other studies (828–830), Hall and his colleagues isolated from the stool of newborn infants such organisms as *Clostridium paraputrificum*, *Clostridium difficile*, and *Clostridium tertium*.

Pearson and Anderson (648) isolated *Bacteroides* from blood cultures from five premature newborns with bacteremia. They also reported several cases of maternal *Bacteroides* bacteremia related to pregnancy; two of these cases were specifically noted to have amnionitis. Tynes and Frommeyer (489) reported *Bacteroides* septicemia in a 1½-day-old female. Robinow and Simonelli (385) reported a case of septicemia due to *Fusobacterium nucleatum* in a newborn. Septicemia due to group F streptococci was reported in the newborn by Nyhan and Fousek (830a). Harrod and Stevens (419c) described 3 cases of neonatal bacteremia due to *Bacteroides*. Two of these proved fatal and the third baby had hypotension. Two cases of double bacteremia in neonates due to anaerobic streptococci and *Haemophilus vaginalis* were described by Monif and Baer (387a).

Babes (200) reported a severe infection of the umbilical stump of a newborn baby. Hemorrhagic infected infarcts were found in the spleen and lungs and all involved areas gave a heavy growth of fusiform bacilli. Prévot (1199) also reported omphalitis in a newborn; the infecting organism was *Clostridium tertium*. Omphalitis in a newborn, due to *C. perfringens*, was reported by Isenberg (494). Bernstine *et al.* (1039) did a routine bacteriologic survey of ligated and nonligated umbilical cords. They found anaerobic bacteria (*Clostridium*, including some which were probably *C. perfringens*, *Bacteroides*, and anaerobic micrococci) on 24% of 63 ligated cords and 17% of 88 nonligated cords after 24 hours. When these groups were surveyed again after 72 hours, the percentage positive for anaerobes was 7 and 8%, respectively. There is no indication of overt infection of the cords in this series. The authors concluded from the overall bacteriological results (aerobic and anaerobic) that there are definite advantages in late severance and non-ligation of the umbilical cord.

Intestinal perforation is a rare complication of exchange transfusion in the newborn. Hardy *et al.* (552) referred to a case described by Hermann in which *Bacteroides* was recovered from the peritoneal cavity following such perforation. Isenberg (494) reported a case of *C. perfringens* sepsis in a newborn, probably related to exchange transfusion.

Cephalhematoma (a hematoma found between the bones of the skull and the periosteum) is usually a benign condition and bacterial infections of such hematomas are rare. Lee and Berg (1273) described an infected cephalhematoma from which *Bacteroides* was isolated in pure culture.

TABLE 15.1

Incidence of Patients in Pediatric Age Group (up to 15 Years) in Various Series of Anaerobic Infections

Total No. cases	No. in pediatric age group[a]	Types of infections in series	References
47	6 (12.8)	All types	Beigelman and Rantz (282)
17	10 (58.8)	Chronic otitis, mastoiditis, intracranial complications	Brisottq (268)
250	24 (9.6)	Bacteremia, with variable underlying disease	Felner and Dowell (168)
13	5 (38.5)	Putrid empyema with various underlying causes	Guillemot et al. (473)
11	1 (9.1)	Bacteremia, with variable underlying disease	Pham (261)
25	4 (16.0)	Bacteremia, with variable underlying disease	Tynes and Frommeyer (489)
45	1 (2.2)	Various surgical infections involving B. melaninogenicus	Weiss (80)
200 (with age indicated)	31 (15.5)	All types	Beerens and Tahon-Castel (115)

[a] Numbers in parentheses are percentages of cases in pediatric age group.

TABLE 15.2

Anaerobic Pediatric Infection (Literature Review)

Type of Infection	References
Dental–oral	115, 308, 735, 742, 743, 1137, 1199
Eye	168, 228, 242, 816
Head and neck, ear, nose, and throat	7–9, 115, 168, 251, 258–263, 268, 279, 302, 308, 309, 331, 350, 364, 473, 482, 816, 955, 1137, 1189, 1199, 1227, 1245, 1247, 1258, 1576–1581
Central nervous system	7, 8, 115, 168, 262, 268, 302, 350, 364, 473, 482, 489, 816, 1189, 1199, 1227, 1245, 1247, 1255, 1258, 1259, 1273
Cardiovascular	7, 8, 105, 168, 258–262, 296, 409, 420, 435, 489, 554, 647, 826, 1199, 1258, 1259, 1573, 1582
Chest	7–9, 115, 168, 258, 259, 262, 279, 296, 302, 349, 350, 473, 482, 489, 521, 1199, 1258, 1524, 1580, 1583, 1584
Abdominal and perineal	7, 80, 115, 168, 296, 387, 473, 489, 503, 523, 554, 581, 647, 723, 817, 1199, 1203, 1393, 1578, 1585, 1586
Female genital tract	1487
Skin, soft tissues, and muscles	8, 115, 168, 282, 521, 734, 735, 1199, 1247, 1587
Bone and joint	8, 168, 262, 282, 350, 816, 1124, 1247, 1562, 1564
Miscellaneous	345, 364
Intoxications	1199
No data on site	30, 390, 818
General	830b

Infections in Children beyond the Neonatal Period

It must be emphasized that children are susceptible to the same variety of anaerobic infections as adults are. Table 15.1 notes the incidence of patients in the pediatric age group in several reported series of anaerobic infections. While the percentage of anaerobic infections involving children varies greatly from series to series, it is clear that children are involved with some frequency. Table 15.2 lists a number of reports of anaerobic infections in the pediatric age group from the literature, organized primarily in terms of the region involved. This is not intended to be a complete list but will serve as an introduction to those who wish to pursue certain infections in a pediatric population in greater detail.

16

Miscellaneous Anaerobic Infections

Diseases of the Genitalia

It is beyond the scope of this book to discuss syphilis and the related diseases, yaws and pinta; however, it should be noted that *Treponema pallidum*, the causative organism of syphilis, is anaerobic. This is presumably true of the morphologically indistinguishable treponemes that cause yaws and pinta. Luetic chancres may be secondarily infected with anaerobic bacteria. Babes (200) and Brams *et al.* (47) mentioned secondary fuso-spirochetal infections, some of which may result in local gangrenous changes. We have seen one chancre secondarily infected with *Bacteroides melaninogenicus*.

The causative organism of granuloma inguinale is *Calymmatobacterium granulomatis*, a gram-negative encapsulated intracellular organism [Davis and Collins (831)] which requires a low oxidation–reduction potential [Goldberg (832)]. Goldberg (833) has isolated this organism from the feces of a patient with granuloma inguinale. In a later paper, Goldberg and Bernstein (834) presented two cases of granuloma inguinale in homosexuals, and from a review of other literature concluded that a significant percentage of anal lesions of granuloma inguinale are found in homosexuals who practice passive pederasty. They therefore postulate that this is not a venereal disease but rather a direct infection of the host by a fecal organism.

They point out further that most lesions are located so that fecal contamination could easily occur. Greenblatt and Wright (835) reported a secondary fusospirochetal infection of the lesions of granuloma inguinale in a patient. Costa [cited by Marples (836)], commented on secondary infection of the ulcers of granuloma inguinale with fusiforms, spirochetes, and gram-positive cocci.

There is a well-defined condition commonly known as venereal or genital fusospirochetosis which would better be called anaerobic balanoposthitis. This has also been called the "sixth venereal disease." A good review of the older literature on this condition is found in the book by Smith (290). This disease is not at all rare. Von Haam (837) found 37 cases of primary infection of this type among 622 patients studied in his venereal disease diagnostic laboratory. Twenty-three of these cases were observed in males; the majority were between 20 and 30 years of age, the period of greatest sexual activity. The principal complaints were pain and discharge. Itching was noted by some. The pain was a burning, stinging sensation that often radiated through the entire shaft of the penis. On occasion urination was very painful. The discharge was thick, gray in color, not profuse, and characterized by a foul, penetrating odor. Three advanced cases had significant bleeding, which in one case was so profuse as to require deep suturing. Systemic symptoms were usually mild or absent, but six patients did have fever up to 102°F, and three patients noted chills. Some patients had marked edema causing painful phimosis. The earliest lesion was a superficial destruction of the prepuce causing multiple small erosions. These were covered with a membranous exudate. Subsequently shallow ulcerations appeared, with considerable edema but little infiltration at the base of the ulcer. Thus, these ulcers resembled soft chancres. The lesions were noted primarily on the glans and the prepuce and were very commonly in the sulcus coronarius. When these infections are neglected, there may be extensive destruction, as was true in six cases in the series discussed by Von Haam (837). In two of his cases there was complete destruction of half of the penis after 6 to 9 months. Gangrenous necrosis of the penis was observed in one case; this was a fulminating process accompanied by shock and toxicity. A long, tight prepuce is a common predisposing factor, but on occasion lesions do develop on the exposed skin of the penis. There may be swelling of the inguinal lymph nodes and even palpable lymphangitis, but the lymph nodes are not painful nor tender. Poor hygiene is an important background factor. Salivary contact was noted in a small percentage of cases. In the report by Von Haam there were 14 infections of this type of the female genitalia. Again, the most important symptoms noted were pain and discharge. There may be involvement of the clitoris, vulva, vagina, or even uterus. In one of the women [Von Haam (837)] the infection spread rapidly in the fashion of noma to involve the vulva, vagina, cervix, and uterus and proved fatal. Smears or dark-field examination revealed spirochetes and fusiform bacilli. Cocci and vibrios may also be seen. More recent reports note the presence of a number of anaerobic bacteria of various types.

Corbus and Harris (838) reported three cases of erosive and gangrenous balanitis. Two patients responded, with excellent results, to hydrogen peroxide therapy locally. The third case was not treated early and ultimately suffered necrosis and slough of the entire glans, as well as $1\frac{1}{2}$ inches of the shaft of the penis. Brams et al. (47) described four cases of erosive balanitis with fusiforms, spirochetes, and cocci noted on smear. The fusiforms were recovered on culture, along with Staphylococcus albus [4 times], S. aureus [2], viridans streptococcus [1], and E. coli [1]. A fifth case started on the penile shaft following trauma sustained in the course of a fall. An ulcerated blackened area 1 × 2 cm was noted on the penile shaft. Following incision and drainage there was release of considerable foul, brown pus. The glans was eroded and ulcerated with necrosis of the entire prepuce. Although the infection spread to involve the entire shaft of the penis, it ultimately healed. Culture of the material from this case yielded fusiform bacilli and Staphylococcus albus. These workers then cultured smegma from the area behind the glans of 100 normal uncircumcised males. Fifty-one percent showed fusiforms and spirochetes on smear. This was most common in men with long, phimotic foreskins with poor hygiene. The fusiforms were cultured and aerobic culture also yielded S. albus in 86% of the cases, Staphylococcus aureus in 4%, and small numbers of streptococci in 17%. These authors drew a parallel to the situation in the mouth where these organism reside normally and only set up infection under certain conditions.

Lev and Tucker (839), found 14 cases of genital fusospirochetosis among 800 patients with penile lesions. Lahelle (840) cultured fusiforms and sometimes thread-like Bacteroides from several cases of gangrenous ulcer of the penis. Zinc oxide and potassium permanganate were used successfully locally to treat a case of fusospirochetal balanitis by Correa-Fuenzalida (617). Ruiter and Wentholt (841) reported a case of fusospirochetal infection of the glans following trauma from which they cultured fusiform bacilli, anaerobic streptococci, a diphtheroid, S. albus, and Mycoplasma. In this case, one-third of the glans became gangrenous. The same authors, in 1952 (842), reported three additional cases of fusospirochetal infection. One had paraphimosis and an 8 × 15 mm area of necrosis of the prepuce. From this patient they grew anaerobic streptococci, diphtheroids, S. aureus, Alcaligenes, as well as Mycoplasma. Another patient with phimosis and ulcerative balanitis showed on culture anaerobic streptococci, Corynebacterium acnes, and Mycoplasma. The third patient had relapsing erosive balanitis from which only Corynebacterium and staphylococci were recovered on culture. All of these patients showed fusiforms and spirochetes by smear. The mycoplasma that was recovered preferred anaerobic or micro-aerophilic conditions. Barile *et al.* (843) noted two cases of fusospirochetal infection in a study of 35 penile lesions among Armed Forces personnel in Japan. The incubation period was determined in these two cases to be 11 days. They examined the normal flora of the prepuce of 47 additional subjects and were unable to find fusiforms or spirochetes in these. *Bacteroides* was isolated from penile lesions in two patients by Smith *et al.* (651) and from five penile infections by Beazley *et al.* (394); no clinical details were given. Necrotizing dermogenital infections are also discussed in Chapter 13.

Secondary infection of chancroid with anaerobes has been noted by several workers [Brams *et al.* (47), Greenblatt and Wright (835), and Barile *et al.* (843)]. Brams *et al.* and Greenblatt and Wright mentioned fusiforms and spirochetes, and Barile *et al.* mentioned *Clostridium perfringens* (recovered from two of eight cases of chancroid). Most of the glans was destroyed in the subject mentioned by Greenblatt and Wright. Other types of genital lesions, such as papillomas [Brams *et al.* (47)] and condylomata acuminata [Hanson and Cannefax (844)], may be secondarily infected by anaerobes. Hanson and Cannefax recovered *Borrelia refringens* in pure culture from five of the ten patients whom they studied.

Infections of the Scrotum, Groin, and Buttock

Involvement of the scrotum in Fournier's gangrene is discussed in Chapter 13. Felner and Dowell (168) described a patient with scrotal abscesses due to *Bacteroides fragilis*. Mitchell (110a) recovered *Bacteroides* from a scrotal abscess.

Abscesses of the groin may reflect intraabdominal or retroperitoneal infection or infection of other nearby structures. Such infections are discussed elsewhere. Felner and Dowell (168) described two patients with bacteremia in association with groin abscesses. One of these patients had lymphosarcoma and was being treated with corticosteroids and cytotoxic drugs. The infecting organism was *Bacteroides fragilis*. The second patient had bacteremia with *Bacteroides* CDC F1.

Infections presenting at the buttock but reflecting infection of the perirectal or ischiorectal spaces are discussed in Chapter 10. Beerens and Castel (845) reported a voluminous abscess of the buttock related to injection of medication from which 11 anaerobic bacteria (six different anaerobic gram-negative bacilli, a vibrio, three different anaerobic cocci and *Corynebacterium liquefaciens*) and *Streptococcus faecalis* were recovered. Three additional cases of abscess of the buttocks are reported by Beerens and Tahon-Castel

(115). Two of these also followed intramuscular injection of medication at the site. One was an extensive abscess that yielded *Clostridium difficile* in pure culture, the second a gangrenous, gas-producing tumefaction that yielded *Streptococcus putridus* and *Staphylococcus anaerobius*. The third case was a voluminous cold abscess containing 250 ml in a 4-year-old child, with no apparent underlying cause. From the abscess contents *Leptotrichia innominata* and *Diplococcus plagarumbelli* were cultured. Schaffner (1240a) recovered *Eggerthella convexa* and *Fusiformis fusiformis*, with no aerobes, from an abscess of the buttock. Prévot (226a) obtained a pure culture of *Ristella pseudoinsolita* from a buttock abscess. Mitchell (110a) recovered *Bacteroides* from an abscess of the buttock. Gorbach and Thadepalli (344a) recovered *Bacteroides fragilis* and *Clostridium* sp., with no aerobes, from a gluteal abscess. We have seen two abscesses of the buttock without underlying pathology. From one of these were cultured *Bacteroides melaninogenicus* and *Staphylococcus epidermidis*, the latter undoubtedly being a skin contaminant. From the second three different anaerobic cocci—*Peptococcus prevotii*, *Peptococcus saccharolyticus* and an unidentified *Peptococcus*—as well as *Bacteroides fragilis* and *Staphylococcus epidermidis* were cultured.

Infections of the Breast

Puerperal breast abscesses are usually due to *Staphylococcus aureus*. Other breast abscesses, particularly those that are subareolar and recurrent, may often involve anaerobic bacteria. Pearson (846), in an institution which usually sees 20 breast abscesses per year, found nine breast abscesses associated with *Bacteroides* in a 4-year period. Relatively rapid development of large abscesses was noted in this series. Four of the nine patients had retracted nipples, and Pearson (846) cited the work of Caswell and Burnett in which it was postulated that retracted nipples are common precursors of this type of infection, with spread of the infection from the adjacent skin fold. Habif *et al.* (847) studied 152 women with acute, chronic, and recurring subareolar breast abscesses, with and without fistulae, caused by infection in or around a lactiferous duct lined by squamous epithelium and filled with keratin plugs. The common infecting organisms were *Staphylococcus aureus*, *Proteus*, *Bacteroides*, and nonhemolytic streptococci. Paredes and Fernández (1404a) noted recovery of *Clostridium perfringens* from mastitis, but no further details were given. A summary of various types of anaerobic infections of the breast from the literature is noted in Table 16.1. We have seen one breast abscess from which *Bacteroides melaninogenicus* was isolated in pure culture.

Actinomycosis of the breast may occur as a primary manifestation of the disease, as a result of extension from the underlying lung and thoracic wall,

TABLE 16.1

Anaerobic Infections of the Breast

Type of breast infection	Anaerobes isolated	Aerobes isolated	Comments	Reference	Year
Painful hard mass, pus from nipple	*Actinomyces*	?	Sinus tract to nipple	Sehrt [cited by Cope (210)]	1938
Suppuration, multiple sinuses	*Actinomyces*	?	Chronic, 10 years duration	De and Chatterjee [cited by Cope (210)]	1938
Recurrent mastitis	*Veillonella variabilis*			Prévot (1200)	1948
Abscess	*Bacteroides*, anaerobic streptococcus	0	Good response to drainage	Beigelman and Rantz (282)	1949
Abscess, sinus tract	Anaerobic streptococcus	0	Recurrent, over 3 year period	Altemeier and Culbertson (760)	1951
Abscess	*Corynebacterium*			Prévot (167)	1953
Suppurative mastitis	*Spherophorus absceedens*	?			
Suppurative mastitis	*Staphylococcus asaccharolyticus*	?		Prévot (21)	1955
Abscess	*Staphylococcus anaerobius*	0		Morin and Potvin (145)	1957
Abscess	*B. melaninogenicus*	*S. albus*		Pulverer (141)	1958
Abscess (2 patients)	Type not specified	0		Stokes (108)	1958
Postmastectomy wound infection and bacteremia	*C. perfringens*	?		Brummelkamp *et al.* (1002)	1963
3 breast abscess patients	Anaerobic streptococci	?		Bornstein *et al.* (275)	1964

Condition	Organism	Other organisms	Comments	Reference	Year
Abscess	*Bacteroides funduliformis*, *C. liquefaciens*, *Sarcina ventriculi*, *Staphylococcus anaerobius*	0		Beerens and Tahon-Castel (115)	1965
Abscess, areolar	*Bacteroides*	Nonhemolytic streptococci	Foul pus, 2 later recurrences—*Bacteroides* and *S. albus* and *Bacteroides* alone, inverted nipple	Pearson (846)	1967
Abscess, areolar	*Bacteroides*, anaerobic streptococci	*Staphylococcus albus*	Inverted nipple		
Nonareolar abscess	*Bacteroides*, anaerobic diphtheroid	*S. albus*, nonhemolytic streptococcus			
Cystic dilatation of glands, several small abscesses	*Bacteroides*	*S. albus*	Male, alcoholic		
Areolar abscess	*Bacteroides*, anaerobic diphtheroid	*S. albus*, nonhemolytic streptococcus	2 had inverted nipples		
Wound infection following radical mastectomy	Bacteroidaceae	?		Saksena *et al.* (487)	1968
Type unspecified (3 patients)	Bacteroidaceae	?			
Infected carcinoma	*Bacteroides fragilis*	Yes?		Werner and Pulverer (30)	1971

469

or as a metastatic development. Metastases to the breast are very rare and generally not clinically significant. Extension of thoracic disease to the breast is more important. Primary actinomycosis of the breast is also very rare; Cope (210) stated that only eight or nine documented cases had been recorded. The main characteristic of primary actinomycosis of the breast is a hard, slightly tender lump that may remain without change for many months or years. It ultimately softens and forms a fistula or an abscess. Induration may be so impressive as to stimulate carcinoma.

Other Miscellaneous Infections

Involvement of regional lymph nodes is relatively common in a variety of infections involving anaerobic bacteria. Lymph nodes may also be involved without any apparent primary focus. Suppurative involvement of lymph nodes by anaerobes has been reported on several occasions. Schottmüller, in 1910 (258) reported recovery of anaerobic streptococci from suppurative cervical lymph nodes that drained fetid pus. Beigelman and Rantz (282) recovered *Bacteroides* from an abscess in an inguinal lymph node. Prévot (167) reported four cases of suppurative adenitis, two with *Corynebacterium pyogenes* and one each with *Corynebacterium anaerobium* and *Corynebacterium liquefaciens*. Prévot (21) also noted recovery of *Streptococcus foetidus* and *Staphylococcus asaccharolyticus* from two different suppurating lymph nodes. *Staphylococcus asaccharolyticus* was also recovered from a case of submaxillary lymphadenopathy. Christ (300) recovered *C. parvum* from the pus from a lymph node at the angle of the jaw. Stokes (108) noted that 8 of 101 lymph nodes that were positive on culture yielded anaerobes, three times in pure culture (type not specified). Finally, Rentsch (195) recovered *F. fusiforme* from one abscessed lymph node and an anaerobic *Corynebacterium* from another. In 1964, Prévot (848) reported isolation of 10 strains of *Corynebacterium* from cases of lymphadenitis; included were 5 strains of *C. anaerobium*, 3 *C. parvum*, 1 *C. liquefaciens*, and 1 *C. pyogenes*.

Martin (532) recovered *Bacteroides necrophorum* in pure culture from pus from a subacromial bursa. This patient's illness had begun with an indolent ulcerative tonsillitis, following which empyema was noted.

Nettles *et al.* (541) grew *Bacteroides* and anaerobic streptococci from a necrotic Achilles tendon. The patient had had intermittent pain and drainage of the heel since a shrapnel injury almost 50 years previously.

Weese and Smith (113a) noted involvement of the adrenal gland in a patient with actinomycosis. The process in this patient originated in the tail of the pancreas.

Fever of Undetermined Origin

Anaerobic bacteria are likely candidates for the causation of fevers of undetermined origin in view of their propensity to produce septic thrombophlebitis and abscesses. They are also involved with a certain amount of frequency in bacterial endocarditis—not an uncommon cause of fever of undetermined origin. Furthermore, since anaerobes are relatively fastidious, they may not be recovered if proper precautions and techniques are not used. One should always obtain several anaerobic blood cultures in patients with fever of undetermined origin. There are several reports of anaerobic bacteremia in patients with fever of undetermined origin [Stokes (108), Marcoux *et al.* (420), Ellner and Wasilauskas, (740), and Felner and Dowell (168)].

17

Intoxications Due to Anaerobic Bacteria

Botulism

Botulism is a disease, usually acquired by food-poisoning, caused by the ingestion or release into the circulation of a protein neurotoxin elaborated by *Clostridium botulinum*. The organism is widely distributed in nature as a saprophyte and thus commonly contaminates vegetables, fruits, and marine products. Its spores, heat-resistant as a rule, survive food processing and then germinate. The resulting vegetative forms are the only forms that elaborate toxin.

The word botulism comes from the Latin "botulus" meaning sausage. This derivation loses some of its significance here now because plant products are more common vehicles than animal products in the United States. Interesting historical aspects of the disease were presented by Meyer (848a).

It is estimated that the minimum lethal dose of the toxin for a man is probably of the order of 10^{-8} gm. A fatal case is described which followed licking a fingertip covered with juice of spoiled preserves. As Dolman (849) stated. "Few diseases reveal more dramatically the ancient truth that in the midst of life we are in death" (Fig. 17.1). Botulism is apt to select the strong, healthy individual. It often follows a party or other festive occasion, or, ironically, a funeral wake.

Fig. 17.1. Funeral of family wiped out by botulism caused by home-canned string beans in Albany, Oregon, in 1924. Altogether there were 12 deaths. From Dolman (849), reproduced with permission.

Although the disease is relatively rare, in the period from 1899 through 1974 the Center for Disease Control recorded a total of 709 outbreaks involving 1816 patients with 985 deaths. After 1935, there was a gradual decline in the incidence of the disease, related to improved canning methods in industry and in the home. Between 1970 and 1973, there was a total of only 30 outbreaks with 91 cases and 21 deaths. However, in 1974 alone, there were 21 outbreaks with 32 cases and 7 deaths. This is the largest number of food-borne botulism outbreaks in any year since 1935, a period when home canning was at a peak because of the depression. The recent increase probably reflects an increase in home canning and emphasizes the need for educating home canners in proper methods of procedure.

Because each passing hour is critical, both for the survival of a patient with botulism and for prevention of other cases, it is crucial that prompt clinical, epidemiologic, and laboratory efforts are undertaken whenever botulism is considered a possible diagnosis.

There are seven antigenically distinct types of botulinal toxin designated by the letters A through G. Types A, B, E, and rarely F cause disease in man. Types C and D are usually associated with botulism in birds and nonhuman mammals, although rare cases of human disease due to these toxin types have been reported. Type G has not yet been associated with disease in either man or animals. The organism producing type A toxin and most type B organisms are putrefactive, so that food is obviously spoiled. This is not true in the case of type E and some type F organisms. Of the 688 food-borne botulism outbreaks in the United States since 1899, 23% were caused by type A toxin, 6% by type B, 3% by type E, and 0.1% by type F. In 67%, the type was not determined. In North America, type A is the most common, followed by types B, E, and F. About half of all outbreaks in Canada, Japan, and Scandinavia are caused by type E. Type B accounts for most of the outbreaks in Europe, where type A is rare. Table 17.1 shows the frequency of toxin types and geographic areas of occurrence of types associated with botulism in the United States. Type E botulism was not recognized as a major problem in the United States until 1963 when 22 cases were reported. Type E outbreaks have been reported from 10 states, with a predilection for Alaska and the Great Lakes area. To date, all outbreaks except one in Alaska with laboratory confirmation have been type E (849a). California, which has had more outbreaks of botulism than any other state, has had only one outbreak due to type E.

Botulinal toxin is heat-labile. Five minutes of boiling destroys the toxin, and little remains after 30 minutes at 80°C or 1 hour at 70°C. Toxin is produced by the organisms at all temperatures at which growth occurs (3.3° to 48°C). However, higher temperatures result in a slow drop in titer of the toxin because of its relative instability at temperatures greater than 30°C. Toxin is also formed at all pH values at which growth occurs (pH 4.8 to 8.5),

TABLE 17.1

Frequency of Toxin Types and Geographic Areas of Occurrence of Types Associated with Botulism in the United States[a,b]

Type	Outbreaks		Predominant in	Never reported in
	Number	Percent		
A	144	21.8	Western States (92%)	24 eastern states, Alaska and Hawaii
B	37	5.6	Eastern States (68%)	14 eastern states, Alaska, 5 western states and Hawaii
E	17	2.6	Alaska, Great Lakes region	40 eastern and western states; Hawaii
F	1	0.2	California	49 states
A–B	2	0.3		
Unknown	459	69.5		

[a] From Gangarosa (857) with the kind permission of the author, editor, and publisher.
[b] According to data from 659 outbreaks of botulism (1899–1970).

but the toxin is unstable and biological activity is reduced at pH values greater than 7. Organisms producing type E toxin are particularly treacherous because, in contrast to organisms producing other toxin types, they may produce toxin quickly in small fragments of fish exposed to air and at lower pH's and cooler temperatures than is true for other toxin-producing organisms.

Classically, *Clostridium botulinum* spores are heat-resistant (they resist boiling for several hours), and therefore germination of spores with subsequent toxin production by vegetative organisms follows prolonged storage at warm temperatures and under anaerobic conditions, especially at pH 6 or greater. As just mentioned, type E is an exception; also *C. botulinum* producing this toxin type forms spores that are heat-labile. The fact that type E spores are heat-labile makes it much more difficult to recover this organism from sources in nature or from vehicles of food poisoning in which other organisms may be present in large numbers as well. In type F *C. botulinum*, spores are heat-resistant, but the organism was not discovered sooner because it is sparsely distributed in nature. There are only two recorded outbreaks due to this organism: one from homemade liver paste in Denmark and the other due to home-prepared venison jerky in California.

Since 1899, 495 botulism outbreaks were traced to home-processed foods in the United States, and 62 to commercially processed foods. The source could not be determined in 131 food-borne outbreaks. The food products causing botulism outbreaks, by botulinum toxin type, are indicated in Table 17.2. Vegetables, fish, fruits, and condiments were the most important vehicles. In seven of 13 outbreaks due to vegetables in the United States in the period 1970–1973, peppers were involved. Beef, milk products, pork, poultry, and other vehicles caused relatively fewer outbreaks. While it is widely held that if botulism is caused by a marine product type E toxin is responsible, of the 29 outbreaks caused by fish products 19 were due to type E, 7 to type A, and 3 to type B. It is of interest to note that a survey of fresh fish caught in Lakes Michigan, Superior, Huron, and Erie showed that 1 to 9% of fish in each lake, and 59% of fish from certain areas of Lake Michigan, had toxin-producing *Clostridium botulinum* type E in their intestines. Among the 20 outbreaks of food-borne botulism in 1974, the offending vehicle was determined in 16 cases. Fifteen of these were associated with home-canned food, and one with commercially canned beef stew. Botulinal toxin has been detected in commercially canned tuna, but no illness has resulted to date (849b).

Commercial foods found to be contaminated with the organism or its toxin are noted in Table 17.3. Note that in the case of the chicken vegetable soup and the mushrooms, there was only one clinical case related to ingestion of such food, despite a number of batches found contaminated. This very likely relates to the probability that such foods would be heated before

TABLE 17.2
Food Products Causing Botulism Outbreaks 1899–1973[a,b]

Botulinum toxin type	Vegetables	Fish and fish products	Fruits	Condiments[c]	Beef[d]	Milk and milk products	Pork	Poultry	Other[e]	Total
A	96	7	22	16	3	2	2	1	5	154
B	24	3	5	4	1	2	1	2		42
E	1	19								20
F					1					1
A and B	2									2
Total	123	29	27	20	5	4	3	3	5	219

[a] From (855) with the kind permission of the authors and publisher.
[b] Includes only outbreaks in which the toxin type was determined.
[c] Includes outbreaks traced to tomato relish, chili peppers, chili sauce, and salad dressing.
[d] Includes one outbreak of type F in venison, and one outbreak of type A in mutton.
[e] Includes outbreaks traced to vichyssoise soup, spaghetti sauce, and to corn and chicken mash.

TABLE 17.3

Commercial Food Contaminated with Botulinal Toxin or *Clostridium botulinum*, 1970–1973[a]

Date	Food	Type	Cases
4/70	Mushrooms	B	0
8/70	Meatballs and spaghetti sauce[b]	A	4
6/71	Vichyssoise	A	2
8/71	Chicken vegetable soup	A[c], B[c]	0
2/73	Mushrooms	B	0
3/73	Mushrooms	B[c]	0
4/73	Mushrooms	B	0
5/73	Peppers	B	7
7/73	Mushrooms	B	1[d]
9/73	Mushrooms	B	0
12/73	Mushrooms[e]	B	0
12/73	Mushrooms[e]	B	0

[a] From Merson *et al.* (850). *J. Am. Med Assoc.* **229**, 1304 (1974). Copyright 1974, American Medical Association.

[b] Food epidemiologically incriminated; laboratory confirmation not obtained.

[c] Only *C. botulinum* organisms detected.

[d] Canadian case.

[e] Contamination detected late 1973; products recalled early January, 1974.

eating. Prior to 1960, there had been only one reported case of botulism from food commercially processed or canned in the United States since 1925.

Botulism may also result from production of toxin by *Clostridium botulinum* in an infected wound. This type of botulism may be much more common than had been previously appreciated. Between 1943 and 1973, 10 cases of wound botulism were reported to the Center for Disease Control in the United States. However, in 1974, an additional five cases were reported. In one outbreak in 1973 [Merson *et al.* (850)], type B toxin was detected in a patient's serum on the thirtieth day of illness, and both toxin and the organism itself were detected in feces on the twenty-second day. These are the longest time periods recorded in the United States for detection of toxin, and this supports the theory that toxin may be produced *in vivo* in the gastrointestinal tract.

Botulinal toxins are absorbed from the intestinal tract or the infected wound and transported via lymph and blood, and perhaps nerves, and ultimately attach to individual motor nerve terminals. Toxins differ in their affinity for nerve tissue, with type A having the greatest affinity, type E next, and type B the least. Accordingly, it is difficult to find type A toxin in the blood of patients a few days after exposure, but type E may be found even after 1 week and type B after more than 3 weeks. The toxin prevents acetylcholine release at the nerve endings. The cranial nerves are affected earliest and most severely. The central nervous system functions normally except for minor disturbances of spinal function. The severe deficit induced at the neuromuscular junction has been confirmed in man as well as in animals. Death is most often due to respiratory failure and at times to cardiac arrhythmias secondary to involvement of respiratory muscles and bulbar function. Patients may also die of aspiration and asphyxiation or pneumonia. The gross and histologic pathology at autopsy is nonspecific.

The nature of the epidemiology of the disease indicates that control is necessarily difficult. Nothing can be done to change the wide distribution of spores in soil and offshore waters. It would be very hard or impossible to change the dietary and culinary habits of people. Finally, even with commercial sources, there are significant difficulties—scientific, technologic, and even political and economic. The wide distribution patterns are facilitated by modern transportation. For example, the chicken vegetable soup found to be contaminated with botulinum toxin was distributed in 16 states. The vichyssoise was distributed nationally under 22 different brand names, including the canner's own name. That particular company also canned 89 other products, several of which revealed a high incidence of swollen cans.

In terms of the clinical picture, there are several cardinal features. Fever is absent early, although it of course may develop later with complicating pneumonia or other infection. Mental processes are clear. Some patients may be anxious, agitated, or somnolent, but most patients are responsive and oriented. The neurological manifestations consist of a descending symmetrical motor paralysis first affecting muscles supplied by the cranial nerves. There is no sensory disturbance or paresthesia, although vision may be impaired and hearing distorted from the involvement of cranial nerves. The pulse is normal or slow, although tachycardia may occur if hypotension develops.

Botulism should be strongly considered in patients who have acute cranial nerve impairment with symmetrical descending weakness or paralysis. Common symptoms include diplopia, dysarthria, and dysphagia. Pupils are often, but not invariably, dilated and fixed. Deep tendon reflexes may be depressed, but remain equal and symmetrical. There are no pyramidal tract signs. Mucous membranes of the mouth, tongue, and pharynx may be extremely dry and even painful so that a mistaken impression of pharyngitis may result. There may even be pharyngeal erythema. Dizziness or vertigo may occur. Urinary retention or incontinence is seen on occasion. Gastrointestinal symptoms are variable and are seen in only one-third of cases. In patients with type A or type B botulism, these manifestations are primarily abdominal pain, cramps, fullness, and diarrhea. Constipation or obstipation may be noted, rather than diarrhea or may supervene after an initial period of diarrhea. In contrast to patients with types A and B botulism, most patients with type E do show gastrointestinal symptoms as the first manifestations of illness. Included are nausea or vomiting, substernal burning or pain, abdominal distention, decreased bowel sounds, and dilated loops of small bowel visible on X-ray examination. Eisenberg and Bender (849a) described a diagnostic pentad in type E cases of nausea and vomiting, dysphagia, diplopia, dilated pupils, and dry throat; all of their cases had at least 3 of these 5 findings. Some patients have transitory diarrhea initially but later become constipated. Table 17.4 summarizes symptoms and signs noted in outbreaks of botulism reported to the Center for Disease Control, according to toxin type.

The usual incubation period for botulism is 18–36 hours. However, the onset of signs and symptoms may occur as soon as a few hours or as late as 8 days or longer after ingestion of contaminated food. In general, patients with short incubation periods, less than 24 hours, are more severely affected, have more protracted courses, and are more likely to die. The shortness of the incubation period and the severity of illness correlate with the amount of toxin ingested; however, patients have died of botulism after tasting only a small piece of beanpod or asparagus or, as indicated earlier, simply licking a finger.

Routine laboratory studies are of no aid in establishing the diagnosis of botulism. It is important to note that cerebrospinal fluid is entirely normal. The electrocardiogram may show nonspecific T wave changes.

Symptoms of wound botulism are similar to those of food botulism, but there may be important differences. Fever may be present in wound botulism, and gastrointestinal symptoms are not usually present in the wound form. Unilateral sensory changes may be found in wound botulism, in association with the trauma or infection. The incubation period in wound botulism is often 4–14 days. Finally, of course, there may be grossly purulent drainage in the wound itself, although on occasion the wound in cases of wound botulism has shown no evidence of infection.

The differential diagnosis of botulism includes such entities as myasthenia gravis, cerebrovascular accidents, Guillain–Barré syndrome, tick paralysis, chemical intoxication (e.g., carbon monoxide, belladonna, barium carbonate, methyl chloride, methyl alcohol, organic phosphorus compounds, and atropine), trichinosis, diphtheritic polyneuritis, psychiatric syndromes, and the Eaton–Lambert syndrome (usually associated with bronchogenic carcinoma). On occasion, the gastrointestinal symptoms and signs are so prominent that the illness may be misdiagnosed as appendicitis or bowel obstruction. Pharyngitis has already been commented upon. Common bacterial food poisonings are usually not a problem because of the absence of cranial nerve involvement. Signs of chemical food poisoning, which may cause neurologic manifestations, almost always appear within minutes or

TABLE 17.4

Outbreaks of Botulism Reported to Center for Disease Control in Which One or More Persons Were Affected by a Given Symptom or Sign, 1953–1973[a]

	Type A	Type B	Type E	Type F	Und.[b]	Total	Percent with symptom or sign
Outbreaks:	34	15	10	1	44	104	
Cases:	97	46	36	3	90	272	
Symptoms							
1. Blurred vision, diplopia, photophobia	31	13	9	1	40	94	90.4
2. Dysphagia	27	14	3		35	79	76.0
3. Generalized weakness	22	12	4		22	60	57.7
4. Nausea and/or vomiting	15	13	10	1	19	58	55.8
5. Dysphonia	25	8	5		19	57	54.8
6. Dizziness or vertigo	8	4	5		15	32	30.8
7. Abdominal pain, cramps, fullness	5	6	3		7	21	20.2
8. Diarrhea	5	6			5	16	15.4
9. Urinary retention or incontinence	2	2	1		2	7	6.7
10. Sore throat	4	2	1			7	6.7
11. Constipation	1	2			3	6	5.8
12. Paresthesias				1		1	1.0
Signs							
1. Respiratory impairment	32	7	7		30	76	73.1
2. Specific muscle weakness or paralysis	23	9	3		13	48	46.2
3. Eye muscle involvement, including ptosis	16	9	3	1	17	46	44.2
4. Dry throat, mouth, or tongue	7	6	2		7	22	21.2
5. Dilated, fixed pupils	3	4	2		8	16	15.4
6. Ataxia	3	1		1	4	9	8.7
7. Postural hypotension			1		2	3	2.9
8. Nystagmus	1		1		1	3	2.9
9. Somnolence			1			1	1.0

[a] From (855) with the kind permission of the authors and publisher. [b] Toxin type un-determined or unspecified.

at most hours after consumption of contaminated food. Atropine poisoning has a very rapid onset and is also distinctive by virtue of facial flushing and bizarre hallucinations. Shellfish and other fish poisonings have rapid onsets and often cause characteristic paresthesias, tremors, and other signs. Mushroom poisoning causes severe abdominal pain, violent vomiting, diarrhea, and coma.

Cerebrovascular accidents usually cause localized signs and may involve sensory loss. The absence of fever in botulism helps to exclude poliomyelitis, meningitis, and encephalitis. Myasthenia gravis can be differentiated by the presence of muscular fatigability (uncommon in botulism) and the response to the edrophonium (Tensilon) test (although rare exceptions in which botulism has apparently responded to this test have been described). The Guillain–Barré syndrome can mimic botulism closely, but usually it presents with ascending peripheral paralysis and later cranial nerve involvement. Muscle cramps, paresthesias, and elevated concentration of protein in the spinal fluid (in the absence of cells) help distinguish this disease, although protein elevation in the spinal fluid may not occur early. Electromyography may be extremely helpful in differentiating botulism from atypical cases of Guillain–Barré syndrome, the illness that is probably most frequently confused with botulism. In botulism, the amplitude of muscle action potential is usually decreased in response to a single supramaximal stimulus. However, when paired or repetitive supramaximal stimuli (at rates of 25 to 50/second) are applied, facilitation of the action potential occurs. This is seen most prominently in an affected limb and may not appear until a few days after the patient develops peripheral muscular weakness. This characteristic picture may also be seen in Eaton–Lambert syndrome. In tick paralysis, muscle weakness is usually ascending. Following surgery, certain antibiotics, particularly aminoglycosides, may induce symmetrical flaccid paralysis. Table 17.5 lists the final diagnoses in a number of outbreaks in which botulism was suspected.

Confirmation of diagnosis depends primarily on detection of toxin in the patient or in the implicated food. Efforts should also be made to recover the organism itself. Specimens to be examined for botulinal toxin include serum, gastric contents or vomitus, feces (at least 50 gm should be obtained when possible), and exudates from wounds and tissues. All specimens except those from wounds should be refrigerated (preferably not frozen) and examined as quickly as possible after collection. Whenever possible, food should be kept sealed in the original container. Sterile unbreakable containers should be used for other food samples. All specimens should be labeled carefully to allow prompt identification. Specimens to be shipped to distant laboratories should be placed in leak-proof containers, packed with ice in a second leakproof insulated container labeled "Medical Emergency" and

TABLE 17.5

**Final Diagnoses in Suspect Botulism Outbreaks Reported to
Center for Disease Control, 1964–1973[a]**

Final diagnosis	Total	Percent of total
Not botulism, but no final diagnosis made[b]	127	29.0
No illness, but concerned about possibility of botulism	99	22.6
Botulism	75	17.1
Food-borne	68	15.5
Wound	7	1.6
Guillain-Barré syndrome	46	10.5
Carbon monoxide poisoning	15	3.4
Food poisoning, unknown etiology	14	3.2
Staphylococcal food poisoning	13	3.0
Laboratory accident, no disease	5	1.1
Shigella or *Salmonella* gastroenteritis	4	0.9
Chemical food poisoning	4	0.9
Cerebrovascular accident	4	0.9
Neuropsychiatric disorder	4	0.9
Viral encephalitis	3	0.7
Phenothiazine idiosyncratic reaction	3	0.7
Subarachnoid hemorrhage	3	0.7
Myasthenia gravis	2	0.4
Acute alcoholic intoxication	2	0.4
Drug reaction	2	0.4
Miscellaneous[c]	14	3.2
	438	100.0

[a] From (855) with kind permission of the authors and publisher.

[b] Majority of cases in this group represent neuro-paralytic illness of unknown etiology and undiagnosed gastrointestinal illness.

[c] Includes 1 case each of *Clostridium perfringens* gangrene, *C. perfringens* sepsis, *C. perfringens* food poisoning, staphylococcal endocarditis, phenytoin sodium toxicity, methyl alcohol intoxication, Wernicke's encephalopathy, appendicitis, diabetes mellitus, brain tumor, seizure disorder, streptococcal food poisoning, parasympathetic blockage of unknown etiology, hyperventilation syndrome, and a drug reaction.

"Danger, Hazardous Material," and shipped by the most rapid means possible. The receiving laboratory should be notified in advance by phone or telegram as to how and when specimens were shipped, when they will arrive, and the billing or shipping number. Blood and other body materials should be obtained as soon as possible, but always before antitoxin is given. This is particularly important in the case of blood specimens. Thirty milliliters of blood should be obtained in large vacuum tubes and sent, without separation of serum, to the nearest laboratory capable of carrying out the mouse neutralization test and other tests for toxin. Laboratory confirmation of suspected botulism should be attempted even late in the clinical course.

Safety precautions are mandatory in handling materials which may contain botulinal toxin. Even minute quantities of toxin acquired by ingestion, inhalation, or by absorption through the eye or a break in the skin can cause profound intoxication and death. Therefore, all materials suspected of containing toxin must be handled with maximum precautions and only experienced personnel, preferably immunized with botulinal toxoid, should perform laboratory tests. If spills occur, toxin can be neutralized by the use of a strong alkaline solution, such as 0.1 N sodium hydroxide.

The most reliable technique for detection of toxin is the mouse neutralization test. Radioimmunoassay looks promising for toxin detection (850a). Electroimmunodiffusion has been claimed recently to be a rapid and effective means of detecting toxin [Miller and Anderson (851)]. Trypsin activation may be required for E and G toxin and for some F and B toxins. Detection of the organism itself may be done by culture, preferably using spore selection procedures, by fluorescent antibody technique, and, in a presumptive manner, by gas chromatography (851a).

With regard to therapy of botulism, all persons known or even suspected of having ingested toxin-contaminated food should be hospitalized for observation. If an individual has definitely consumed food which has been specifically incriminated in the disease in others (including a commercial production lot) and is seen within 7 days of exposure, he should receive antitoxin, even in the absence of evidence of the disease (851b). The most immediate danger to the patient is respiratory insufficiency. Patients with neurologic involvement should be managed in an intensive care unit with special surveillance of cardiac and respiratory functions. It is interesting to note that one patient required ventilatory assistance for 88 days and yet survived. Close observation is crucial because respiratory failure can develop precipitously. Tracheostomy should be performed early in patients with bulbar or respiratory involvement. Induced vomiting and gastric lavage should be carried out if exposure has occurred within several hours. An emetic such as ipecac (30–45 ml) should be given, except in the presence of signs of esophageal dysfunction. Purgation and high enemas are advisable, even after several days, to facilitate the elimination of unabsorbed toxin. Magnesium sulfate or another fast-acting cathartic should be given in large doses. Cathartics should not be used in patients with paralytic ileus.

Clear-cut evidence that antitoxin decreases mortality and shortens illness is available only for type E antitoxin. However, despite lack of proof, it may well help in other types of botulism as well and good practice demands that it be used. Only equine preparations of antitoxin are available. Accordingly, all of the usual reactions that may be encountered with horse serum may be seen in the management of botulism with antitoxin. The reaction rate is 21%, which is slightly higher than that observed for the equine preparations of tetanus antitoxin; this is possibly due to higher concentrations of antigenic material. The reaction rate is 32% with bivalent antitoxin and 17% with trivalent antitoxin (851c).

The trivalent ABE preparation of antitoxin is the therapy of choice in cases of botulism in which the toxin type has not yet been definitively identified. In this preparation the antitoxins are concentrated in a small volume and the three toxins most likely to be encountered will be neutralized. Monovalent E and bivalent AB preparations should be used only in cases in which the toxin has been definitely identified.

It is important that antitoxin be given as soon as possible after the clinical diagnosis is made. Therapy should never be delayed to obtain information on the specific type of toxin involved. The nature of the food exposure should not be utilized in determining which type of antitoxin to use. While it is true that most cases of type E botulism have been associated with fish products, and types A and B with foods of plant origin, there are enough exceptions that this type of information cannot be relied upon when the physician is faced with treating a given patient. While it is important to administer antitoxin as early in the course of the illness as possible, it should always be given to all patients with moderate or severe botulism regardless of the duration of their illness at the time of diagnosis. As indicated earlier, with type E and type B botulism, circulating toxin has been demonstrated in the blood for extended periods of time.

Polyvalent ABCDEF serum is available from the Center for Disease Control (CDC) in Atlanta, but is reserved exclusively for outbreaks due to types C, D, or F. Bivalent AB is available commercially (Lederle). All other antitoxin preparations are available through CDC. Table 17.6 lists information concerning the more common types of botulinal antitoxins that are

TABLE 17.6

Types of Botulinal Antitoxins Available[a]

Type	Manufacturer	Remarks	ml/vial	Potency (IU/vial)		Recommended dosage
ABE	Connaught	Distributed by Center for Disease Control	8 ml	A B E	7,500 5,500 8,500	Contents of one vial i.v. and one i.m., repeat once in 2–4 hours if symptoms persist
AB	Lederle	Distributed by Lederle	30 ml	A B	10,000 10,000	
E	Connaught	Distributed by Center for Disease Control	1 ml	E	5,000	

[a] From (855) with kind permission of the authors and publisher.

available and gives information concerning specific dosage recommendations. Because of the risk of anaphylaxis and serum sickness, botulinum antitoxin should be given only to patients who have clinical signs of botulism or to asymptomatic persons who have been exposed to a known contaminated vehicle. Consumption of food from a swollen can or from a recalled lot of a commercial product is not, in itself, sufficient to justify antitoxin prophylaxis. These patients may be purged and have emesis induced, and should be kept under close surveillance. Testing for sensitivity to horse serum should always precede the use of antitoxin, and, because of the threat of anaphylaxis, epinephrine should be immediately available at the time of administration of antitoxin.

The use of guanidine hydrochloride therapy in botulism is controversial. The drug has been shown to have a favorable effect on the neuromuscular block of botulism in animals, and there are a few reports of apparent benefit in man (851d,e). However, there are also negative reports in human cases. It must be remembered that guanidine is a powerful macromolecular denaturant and may have a deleterious effect upon mucosal surfaces after nasogastric administration. Reported side effects include diarrhea, salivation, vomiting, hypoglycemia, tremors, muscle twitching, and paresthesias.

Penicillin has been recommended by some because of the theoretical possibility that toxin may be released *in vivo* following germination of spores. As indicated previously, there is some evidence that this may take place in the gastrointestinal tract. Certainly penicillin should be utilized, in addition to thorough debridement and irrigation of the wound, in cases of wound botulism. *In vitro* data (852) indicate that tetracycline and chloramphenicol should be active against *Clostridium botulinum* in patients with penicillin allergy.

Again it should be emphasized that others who have been exposed to a toxin source but who are not yet ill should be closely observed, and if the exposure to the toxin source is definite, they should be treated as outlined above. Emergency assistance is available on a round-the-clock basis from the Center for Disease Control in Atlanta for problems in the therapy of botulism or anything else concerned with possible cases of this disease.

Improvements in diagnosis and the availability of modern respiratory support and of antitoxin have probably been the major factors responsible for the reduction of mortality rates from 60 to 25% in the 20-year period from 1945–1964. In the period 1960 to 1973, the mortality has been 23%. The mortality is lower with type B disease (9%) than with type A (22%) or E (20%). Cases with unknown toxin type have had a mortality of 47%. Several other factors affecting prognosis are important. The mortality rate is lower in individuals under 20 years of age than in those who are older (10% for those under 20, compared with 25–40% for adult groups). A very important factor is the dose of toxin ingested, as reflected by the length of the incubation period. Those with delayed onset of symptoms usually have a better prognosis. Obviously, a most important factor is the speed with which the diagnosis is established, or strongly suspected, and treatment given. If the index case in an outbreak can be detected early, other cases that might have been exposed from the same vehicle will have a much more favorable prognosis. The severity of disease will vary significantly among various people exposed to a given source contaminated with toxin simply by virtue of irregular distribution of toxin throughout the vehicle. It has been stated that those who drink alcoholic beverages along with the vehicle containing botulinal toxin suffer less serious consequences than those who do not. There is no basis whatsoever for this statement, except insofar as one who drinks heavily may eat less food.

Type A botulinal toxin probably does not cross the placenta because of its large molecular weight. There is one report of a woman with type A botulism who gave birth 4 days after onset of symptoms and 2 days following antitoxin. The infant was alive and did not have or develop botulism.

With regard to prevention, local, state, and federal health authorities should be notified immediately of all suspected outbreaks so that appropriate investigations can be undertaken. The possibility that a commercial product may be responsible should always be considered. Control measures developed and initiated by the smoked fish industry after three outbreaks in the mid-1960's serve as a model for responsible action. New federal regulations for canning of low-acid foods are crucial in preventing processing errors by the canning industry. There is a continuing need to instruct home canners as to how to sterilize containers and food before preserving, and how to cook these foods adequately before serving. In canning, a pressure cooker must be used to obtain temperatures well above boiling which are necessary to destroy spores of *Clostridium botulinum* types A and B; a temperature of 116°C is recommended for certain foods (852a). According to Cherington (851a), spores of *C. botulinum* are destroyed at 120°C after 30 minutes, and pressure cookers set at 15 lb will achieve this temperature. Cherington stressed that, in canning at higher altitudes, one should add $\frac{1}{2}$ lb to the pressure gauge for each additional 1000 ft above sea level.

Three botulism outbreaks have now been traced to a high-acid home-canned food (pH ≤ 4.5) since October, 1973 (852b). It had been thought that *Clostridium botulinum* could not grow and produce toxin in such foods. Two hypotheses have been suggested to explain the phenomenon. One is that other organisms present in food inadequately heated during canning may allow germination of *C. botulinum* spores and toxin production. The other is that the distribution of acid is unequal and that multiplication and toxin production may take place in less acid portions of the food. The time and temperature prescribed in booklets accompanying pressure cookers are designed to safeguard the product from botulism. Home-canned foods should be boiled for 10 minutes before serving. Ingelfinger (852c) called attention to scare tactics used by a vendor of bacon containing no nitrites or nitrates (and thus presumably not carcinogenic). Nitrites and nitrates are used in bacon, canned ham, frankfurters, and other products to prevent botulism. They are effective, and there is no evidence that they pose any health hazard. All laboratory technicians working with botulinal toxin should be immunized. A pentavalent toxoid vaccine is available from the Center for Disease Control for active immunization of such individuals. However, antibody production is not always evoked by botulinus toxoid and may not even appear after recovery from clinical botulism (853).

Tetanus

Tetanus was described by Hippocrates approximately 30 centuries ago. Over the years, subsequently, it has been documented as a scourge of child-bearing women, newborn babies, and wounded soldiers. In the eighteenth century, one of every six infants born in the Rotunda Hospital in Dublin died of tetanus neonatorum.

The impact of the disease can be appreciated from the vivid description by Aretaeus about the second or third century A.D. (858).

> Tetanus, in all its varieties, is a spasm of an exceedingly painful nature, very swift to prove fatal, but neither easy to be removed . . . . There is a pain and tension of the tendons and spine, and of the muscles connected with the jaws and cheek; for they fasten the lower jaw to the upper, so that it could not easily be separated even with levers or a wedge. But if one, by forcibly separating the teeth, pour in some liquid the patients do not drink it but squirt it out, or retain it in the mouth, or it regurgitates by the nostrils; for the isthmus faucium is strongly compressed, and the tonsils being hard and tense, do not coalesce so as to propel that which is swallowed. The face is ruddy and of mixed colours, the eyes almost immovable, or are rolled about with difficulty; strong feeling of suffocation; respiration bad, distension of the arms and legs; subsultus of the muscles; the countenance variously distorted; the cheeks and lips tremulous; the jaw quivering, and the teeth rattling, and in certain rare cases even the ears are thus affected. I myself have beheld this and wondered! The urine is retained, so as to induce strong dysuria, or passes spontaneously from contraction of the bladder. These symptoms occur in each variety of the spasms.
>
> But there are peculiarities in each; in Tetanus there is tension in a straight line of the whole body, which is unbent and inflexible; the legs and arms are straight.
>
> Opisthotonos bends the patient backward, like a bow, so that the reflected head is lodged between the shoulder-blades; the throat protrudes; the jaw sometimes gapes, but in some rare cases it is fixed in the upper one; respiration stertorous; the belly and chest prominent, and in these there is usually incontinence of urine; the abdomen stretched, and resonant if tapped; the arms strongly bent back in a state of extension;

the legs and thighs are bent together, for the legs are bent in the opposite direction to the hams.

But if they are bent forwards, they are protuberant at the back, the loins being extruded in a line with the back, the whole of the spine being straight; the vertex prone, the head inclining towards the chest; the lower jaw fixed upon the breast bone; the hands clasped together, the lower extremities extended; pains intense; the voice altogether dolorous; they groan, making deep moaning. Should the mischief then seize the chest and the respiratory organs, it readily frees the patient from life; a blessing this, to himself, as being a deliverance from pains, distortion, and deformity; and a contingency less than usual to be lamented by the spectators, were he a son or a father. But should the powers of life still stand out, the respiration, although bad, being still prolonged, the patient is not only bent up into an arch but rolled together like a ball, so that the head rests upon the knees, while the legs and back are bent forwards, so as to convey the impression of the articulation of the knee being dislocated backwards.

An inhuman calamity! an unseemly sight! a spectacle painful even to the beholder! an incurable malady! owing to the distortion, not to be recognized by the dearest friends; and hence the prayer of the spectators, which formerly would have been reckoned not pious, now becomes good, that the patient may depart from life, as being a deliverance from the pains and unseemly evils attendant on it. But neither can the physician, though present and looking on, furnish any assistance, as regards life, relief from pain or from deformity. For if he should wish to straighten the limbs, he can only do so by cutting and breaking those of a living man. With them, then, who are overpowered by this disease, he can merely sympathise. This is the great misfortune of the physician.

In the early nineteenth century, Sir Charles Bell, a Scottish surgeon, made a drawing of a British soldier who had been wounded during the Peninsular War in Spain. This is shown in Figure 17.2.

In 1889, Kitasato isolated the bacterium *Clostridium tetani*, responsible for the disease (Fig. 17.3). The next year, Faber confirmed a theory of Nicolaier by reproducing the disease in experimental animals injected with filtrates of cultures of the organism, thus demonstrating that a cell-free toxin was responsible for the manifestations. The discovery of tetanus antitoxin was made by Behring and Kitasato in 1890. In 1920, Glenny and Ramon independently discovered that a harmless toxoid could be made by combining the toxin with formaldehyde; this toxoid could still stimulate antibody formation.

The toxin in tetanus, tetanospasmin, is extremely potent, being roughly equivalent in potency to botulinus toxin. As little as 130 μg of purified tetanospasmin may be lethal for man. On the basis of theoretical comparisons utilizing injected toxin, tetanus and botulinus toxins would each be roughly one million times as powerful as rattlesnake poison. Again, assuming that the toxin were injected and that man and the mouse would be equally susceptible, an amount of tetanus or botulinus toxin weighing no more than the ink in the period at the end of this sentence would be enough to kill 30 grown men. An ounce could kill 30,000,000 tons of living matter and $\frac{1}{2}$ lb would be more than enough to destroy the entire human population of the world.

Fig. 17.2. Bell's portrait of a soldier with tetanus.

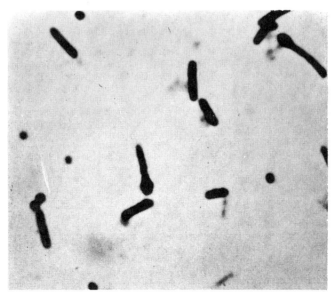

Fig. 17.3. *Clostridium tetani* from distal phalanx of right index finger of patient with tetanus; typical drumstick appearance. From Braude (1021), reproduced with permission.

   Since tetanus affects individuals and does not cause outbreaks, it is less noticed than certain other infectious diseases, such as smallpox, malaria, and cholera. Nevertheless, despite the availability of simple benign protective measures, tetanus ranks high among the infectious diseases as a cause of death throughout the world (Fig. 17.4). Available statistics are poor because tetanus is not a reportable disease in many countries. The World Health Organization lists it among the ten leading causes of death in a number of tropical countries. The correlation is not with climate, however, but rather with social and economic conditions. In Europe, tetanus kills more people each year than smallpox, rabies, malaria, and diphtheria combined. Table 17.7 shows the estimated mortality rates due to tetanus throughout the world in 1961–1965, as indicated by B. Bytchenko (personal communication) of the World Health Organization. He estimates that about half a million deaths occur from tetanus each year and that approximately 80 to 90% of these are newborn children. Even this still does not present a true picture because of very incomplete registration of deaths in developing countries. Where the national income per capita is about $100 or less per year, it is estimated that about 10% of newborns die from tetanus. In such countries, mortality rates related to posttraumatic tetanus are close to 10 to 15 per 100,000 inhabitants. One state in Brazil with a population of 15 million loses more than 1000 babies per year because of umbilical tetanus. Veronesi (859) noted in 1956 that in the state of São Paulo, Brazil tetanus

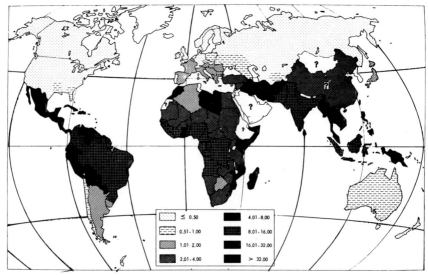

Fig. 17.4.   Tetanus approximate expected mortality rates per 100,000 inhabitants in the world during the period 1951–1960. Courtesy of Dr. B. Bytchenko, World Health Organization.

TABLE 17.7

Estimated Mortality Rates Due to Tetanus in the World (1961–1965)[a]

Continent	No. of countries reporting tetanus	No. of reported deaths from tetanus (per year)	Mean population of countries reporting tetanus (in thousands)	Mean mortality rate for this population	Mean population of the continent (in thousands)	Expected No. of deaths per year per continent
Africa	32	56,732	222,399	25.5	306,000	78,030
North America	2	284	214,288	0.13	214,288	284
Central and South America	28	73,554	238,248	30.9	243,512	75,245
Asia	22	133,586	893,176	14.9	1,746,400	261,261
Europe	22	2,651	422,782	0.63	444,800	2,802
Australia and New Zealand	2	28	14,051	0.2	14,051	28
Oceania	7	5,002	2,915	171.5[b]	3,449	5,908
Total	115	271,837	2,007,859	14.2	2,972,500	423,558

[a] Courtesy of Dr. B. Bytchenko, WHO, Geneva.   [b] Papua and New Guinea.

is responsible for 52.6% of the deaths from specific infantile diseases (as distinct from those of multiple etiology such as gastroenteritis or pneumonia).

In the United States, less than 200 cases of tetanus have been reported annually since 1968 (859b), but this figure is probably not accurate. From 1950 to 1966 in the United States, there has been a reduction of about one-half in incidence of tetanus. In 1973, 88 cases were reported. However, this is not an impressive drop as compared to the precipitous drops in diphtheria, whooping cough, poliomyelitis, and measles. As a mortality problem, tetanus now exceeds measles. On a national basis, the incidence is approximately one case per million per year. The annual incidence rate for neonatal tetanus in the United States declined significantly from 0.73 cases per 100,000 live births in 1965–1967 to 0.19 in 1968–1969 and to 0.14 in 1970–1971 (859a). About two-thirds of cases in the United States occur between May and November. This is probably a function of greater outdoor activity and exposure to soil in the spring and summer. The areas of highest incidence in the United States are the South Atlantic, East South Central, and West South Central regions (Fig. 17.5). Approximately three males are affected for every two females, undoubtedly reflecting differences in exposure. A fivefold lower frequency of tetanus in Caucasians is most likely attributable to lower immunization levels in non-Caucasians. In New York City, in a

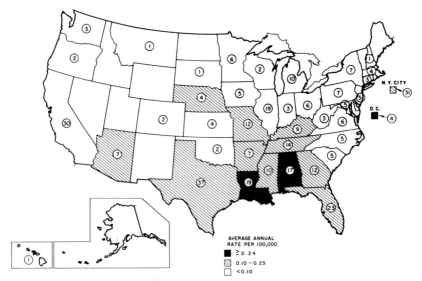

Fig. 17.5. Geographic distribution of tetanus cases and incidence of rates by state, 1968–1969 (excluding neonatal tetanus). Courtesy Center for Disease Control, United States Public Health Service, Atlanta, Georgia.

recent 11-year period, 72% of cases were in heroin addicts. In the entire United States in 1970–1971, 12.7% of all tetanus cases were in drug addicts (859a).

The spores of *Clostridium tetani* may survive for many years away from sunlight. These spores are ubiquitous—in soil and street dust throughout the world. *Clostridium tetani* spores resist phenol, boiling, and even autoclaving at 120°C for 15–20 minutes (859c). Different studies have indicated an incidence in the intestinal tract of man of 10 to 25%; they are also found in feces of horses, dogs, cats, and various other animals. The toxin is produced by vegetative cells of the organism after the spores have germinated.

Clostridium tetani enters the body through wounds of various types, the sort that are commonly acquired by everyone; all types of wounds may provide a suitable site for growth of the organism; included are lacerations, slivers, compound fractures, gunshot wounds, burns, frostbite, penetrating wounds, human and animal bites, decubitus and varicose ulcers, superficial scratches, the puncture of a drug addict's needle, etc. In 1970 and 1971, 4% of the reported cases of tetanus in the United States were associated with decubitus or diabetic cutaneous ulcers (859d). No primary wound is evident in 10 to 20% of cases. At times, tetanus is related to various infectious processes or to surgery. Thus, infected gums and teeth, gangrenous extremities, or surgery performed for these conditions may lead to tetanus. Tetanus may follow middle ear infection or mastoiditis, childbirth, abortion, and surgery of various types—particularly surgery involving the gastrointestinal tract. It may also be associated with infected tumors. Table 17.8 lists the various injuries associated with tetanus in the United States in 1968

TABLE 17.8

Classification of Injuries Associated with Tetanus, United States, 1970 and 1971[a,b]

Type of wound	Cases with a given injury	% of total cases	Fatal cases	Case–fatality ratio
Puncture	74	46.0	43	58.1
Laceration	56	34.8	27	48.2
Abrasion	6	3.7	3	50.0
No injury	19		11	57.9
Crush	4	2.5	1	25.0
Surgical	4	2.5	2	50.0
Dental	2	1.2	1	50.0
Injection	15	9.3	10	66.7
	161	100.0	87	54.0

[a] Modified from Center for Disease Control (876a).

[b] Excludes 10 neonates, 65 persons of unknown age, and other types of injury or unknown outcome.

and 1969. The route of infection in a large series of cases from Bombay, India, is shown in Table 17.9. Tetanus is an uncommon but calamitous postoperative complication. Among 91 tetanus patients treated at the Mayo Clinic prior to 1957, only three were postoperative cases. It has also been demonstrated that tetanus may result from passage of the organisms or toxin through gangrenous but unperforated bowel [Clay and Bolton (860)]. The disease in addicts is related to subcutaneous injections of narcotic, primarily by women, and to the injection along with the narcotic of quinine, which produces anaerobic conditions at the site of injection. About two-thirds of all injuries leading to tetanus occur in the home, and about 20% take place on farms and in gardens. It is important to note that spores may persist in normal tissues for months to years and then germinate at a later time when another injury may provide anaerobic conditions.

The organism grows at low oxidation–reduction potentials in the wound and then elaborates toxin. The local infection per se is of negligible proportions as an infection—there is no invasion. It seems likely at present that two routes accommodate the spread of toxin from the site of infection to the central nervous system: (a) adsorption at myoneural junctions followed by migration through perineural tissue spaces of nerve trunks to the central nervous system and (b) passage from tissue spaces to lymphatics and blood and thus to the central nervous system. The toxin then becomes bound to gangliosides within the central nervous system. Early involvement of the lower cranial nerves has been suggested to result from their anatomical relationship to the area postrema on the floor of the fourth ventricle where the blood–brain barrier does not exist. In fatal cases, there may be widespread central nervous system involvement with brainstem lesions parti-

TABLE 17.9

Route of Infection of *Clostridium tetani* in a Large Series of Tetanus Cases at the King Edward Memorial Hospital, Bombay, India[a]

Route	Total	Percent
Injury	892	44.5
Otorrhea	406	20.2
Umbilicus	292	14.5
Unknown	283	14.2
Puerperal	67	3.3
Injection	32	1.6
Vaccination	20	1.0
Operation	15	0.7

[a] Modified from Furste and Wheeler (867) with kind permission of the authors and publisher.

cularly. The action of tetanospasmin is very much like that of strychnine, in suppressing the central inhibitory balancing influences on motor neuron activity leading to intensified reflex response to afferent stimuli and thus to spasticity and convulsions. The toxin also acts on the sympathetic nervous system and on neurocirculatory and neuroendocrine systems.

The incubation period for tetanus varies from 1 to at least 54 days, but is usually 6 to 15 days, with a median of 7 or 8 days. In one series [LaForce *et al.* (861)], 88% of all cases began within 14 days of the day of injury.

The full-blown picture of tetanus is extremely dramatic and presents no difficulty in diagnosis. However, it is important to diagnose patients as early as possible. Early evidences of the disease are indicated in Table 17.10. Tension or cramps in the muscles about a wound, increased reflexes in the wounded extremity, and mild pains in the facial muscles and/or stiffness of the muscles of the jaw are the most important early signs of the disease.

Localized, relatively benign forms of the disease may occur rarely. Thus, the disease may be limited to a wounded extremity, particularly in a partially immunized individual. The manifestations in so-called cephalic tetanus are limited to the head. Cephalic tetanus is characterized by a short incubation period, facial paralysis, and dysphagia (there may be dysfunction of cranial nerves III, IV, VII, IX, X, and XII) associated with infection on the face or head. Rarely, one may find disease limited to the trunk (thoracoabdominal tetanus). However, as already noted, it is important to recognize that evidences of muscle spasm in the region of a wound may simply be the earliest manifestations of generalized tetanus. Generalized tetanus usually presents in a descending manner beginning with trismus and then spasticity of the neck, trunk, and limbs. Spasticity occurring in the reverse order is rare.

Virtually all neonatal tetanus is generalized. The first sign is difficulty in sucking beginning 3 to 10 days after birth and progressing to total inability to suck. In adult patients, the most typical complaint in generalized tetanus is lockjaw or trismus—the inability to open the mouth because of spasm of

TABLE 17.10

Early Signs and Symptoms of Tetanus

Tension or cramps and twitching in muscles around wound
Increased reflexes in wounded extremity
Stiffness of jaw muscles; mild pains in facial muscles
Slight difficulty in swallowing
Stiffness of neck
Constipation
Headache, backache, general irritability and restlessness
Sweating, tachycardia
Anxious facies

the masseter muscles. The characteristic grotesque, grinning facial expression known as risus sardonicus is a consequence of spasm of the facial muscles. Spasms of the muscles of the trunk and extremities may be widespread and may result in opisthotonos and boardlike rigidity of the abdomen and other portions of the body. In patients with moderate to severe degrees of generalized tetanus, there are acute, paroxysmal, uncoordinated widespread spasms of muscles. These tonic convulsions occur intermittently and unpredictably, lasting for a few seconds to several minutes. As these continue, they become severe and painful and exhaust the patient. Such paroxysms may occur spontaneously, but are often precipitated by various stimuli such as drafts of cold air, minor noises, turning the light on in the room, attempting to drink, and attempting to move or turn the patient. They may also be precipitated by such conditions as a distended bowel or bladder, or mucus plugs in the bronchi. Spasms of the pharyngeal and laryngeal musculature may lead to difficulty in swallowing, cyanosis, and even sudden death from respiratory arrest. There is no fever in the absence of complications. There are profuse sweats. At least initially, the patient is conscious and mentally clear, but apprehensive and in enormous pain. Delirium may occur in later stages of the disease and is generally a good prognostic sign. The patient babbles a great deal and occasionally may become violent. This condition usually only lasts for 1 or 2 days.

Patients with severe tetanus may develop a characteristic syndrome consisting of, in addition to the above-mentioned features, labile hypertension and tachycardia, irregularities of cardiac rhythm, peripheral vascular constriction, fever, increased carbon dioxide output, increased urinary catecholamine excretion, and, at times, the late development of hypotension. There may also be sudden changes in central venous pressure. These all apparently point to autonomic dysfunction, but a role for medullary damage and myocarditis due to the disease has been claimed by some investigators.

Complications in patients with tetanus include those related to the pulmonary system [atelectasis, aspiration pneumonia, pulmonary emboli, and ventilation-perfusion problems (861a)], sepsis, acute gastric ulcer, fecal impaction, urinary retention, urinary tract infection, decubiti, compression fractures, deformities or subluxation of vertebrae (particularly in thoracic vertebrae and in children), and spontaneous rupture of muscles and intramuscular hematoma.

The diagnosis of tetanus is a clinical one. In typical, fully developed cases, the diagnosis is easy. However, it is important to increase our index of suspicion in order to be able to detect cases early. Patients will commonly give no definite history of previous tetanus toxoid immunization, and most will have evidence of a wound, usually sustained within the previous 2 weeks. Laboratory studies are not particularly helpful. The cerebrospinal fluid is normal in patients with tetanus, although spinal fluid pressure may be elevated due to muscular contractions. There is usually a moderate leukocytosis in the peripheral blood. Neither electroencephalography nor electromyography is helpful. Wound cultures are positive for *Clostridium tetani* in only about one-third of patients with this disease. Furthermore, it is important to remember that isolation of this organism from contaminated wounds does not mean that the patient has or will contract tetanus. Heating a specimen to get rid of other organisms that are not spore formers in mixed cultures may facilitate recovery of *C. tetani*. The fluorescent antibody technique may be useful in identifying the organism as well. One should be very skeptical of diagnosing tetanus in individuals with reliable histories of having received two or more injections of tetanus toxoid in the past. In this situation, serum should be obtained for assay of antitoxin level. The presence of 0.01 IU of antitoxin per milliliter of serum is generally considered protective.

Many different conditions may be considered in the differential diagnosis of tetanus. Trismus due to tetanus may be erroneously attributed to alveolar abscess, peritonsillar abscess, temporomandibular joint disease, or functional causes. Phenothiazine reactions may cause trismus, but the associated tremors, athetoid movements, and torticollis should alert one to this possibility. Phenothiazine reactions may even cause spasm of the back muscles sufficient to produce opisthotonos. Administration of 100 mg of diphenhydramine hydrochloride (Benadryl) intravenously over 15 seconds will cause subsidence of the tetanus-like reaction to phenothiazine drugs within 1 or 2 minutes in the great majority of cases. Encephalitis is occasionally associated with trismus and muscle spasm, but the sensorium of such patients is clouded. Meningitis usually presents no problems in differential diagnosis. In rabies, the muscular spasms occur early in the course of the disease and involve the muscles of respiration and swallowing. Strychnine poisoning may simulate tetanus very closely; however, there may be a history of accidental or intentional ingestion of the drug; there is no wound, trismus tends to appear late, and symptoms and signs develop much more rapidly than in tetanus. There is muscular relaxation between crises of convulsions. In tetany, spasms are characteristically located in the muscles of the feet and hands, leading to typical posturing. The absence of trismus and a low serum calcium further establish the diagnosis of tetany. Abdominal rigidity in tetanus may suggest an acute intraabdominal process. This may present a special problem if the tetanus has occurred following gastrointestinal surgery. Hysterical reactions must be considered in the differential diagnosis.

The three major objectives of therapy in tetanus are (a) to provide supportive care until the toxin fixed to nervous tissue has been metabolized,

(b) to neutralize circulating toxin, and (c) to remove the source of toxin. Patients should be in a respiratory intensive care unit, and when indicated, they should be treated by means of tracheostomy, curarization, and artificial respiration. Tracheostomy is indicated in all heroin addicts because there is a high incidence of cardiorespiratory dysrhythmia and hyperthermia (859c). Agents such as meprobamate and diazepam are very safe and serve as tranquilizers and sedatives, as well as to abolish muscle spasm. The latter effect is not necessary, of course, when a curare-like agent is given. The effect of both of these agents may be prolonged by addition of small doses of phenobarbital. Milder cases can usually be managed with supportive care and a combination of muscle relaxants and sedatives. The patient's room must be kept quiet and dark, with the number of examinations and manipulations minimized. Intravenous fluids are given; electrolyte and blood gas studies are important for guidance of therapy. In the case of urinary retention, intermittent catheterization is preferable to an in-dwelling catheter. Management of fecal impactions may be required. Dedicated nursing care is absolutely essential. The use of β-adrenergic blocking agents, such as propanolol (and an α-blocker, bethanidine, in the case of hypertension), has been reported to be very helpful in patients with evidence of sympathetic nervous system hyperactivity (862, 862a, 863). The suggestion has been made that there may be a place in therapy for cholinesterase-restoring agents (pralidoxime and cyanocobalamin) (863a).

With regard to neutralization of tetanus toxin, the use of tetanus immune globulin (TIG) of human origin is indicated. A total dose of 3000 U injected as three equal portions into three sites intramuscularly is recommended, although some workers recommend as much as 6000 U or more. There is no need for repeated doses. It is desirable to give TIG in the proximal portion of an extremity where the inciting wound is located when this is feasible. This preparation must not be given intravenously. Analysis of accumulated reports for a 7-year period (1965–1971) by Buchanan et al. (864) and Blake et al. (864a) showed a significantly lower case fatality ratio for those patients treated with tetanus antitoxin as compared to others not receiving antitoxin. TIG does not penetrate the blood–brain barrier and has no effect on toxin already fixed to nervous tissue. The intrathecal administration of antitoxin was abandoned some time ago because of severe reactions. However, it appears now that the concomitant use of corticosteroids and/or the use of an antitoxin preparation that has been dialyzed to remove the formalin preservative may possibly safely permit such therapy. Additional studies are required (865, 866, 866a, 866b). Early reports that the use of hyperbaric oxygen might be useful in treating tetanus have not been substantiated.

The final aspect of therapy is to eradicate the infected site. Proper drainage and debridement of wounds must be carried out. Antibiotics are typically recommended, but there is no significant difference in case fatality rates

TABLE 17.11

Incidence of Tetanus in the United States Army[a]

	Admissions for wounds and injuries	Cases of tetanus	Cases/100,000 wounds and injured
Civil War	280,040	505	18.03
World War I	523,158	70	13.4
1920–1941[b]	580,283	14	2.4
World War II	2,734,819	12	0.44
Korean War		6–8	
1956–Nov. 1971 (including Vietnam conflict)		0	

[a] From Furste and Wheeler (867) with the kind permission of the authors and publisher.
[b] Inclusive.

between patients receiving antibiotics and those not (864). Nonetheless, antibiotics may be indicated because of the wound infection itself. In terms of the *Clostridium tetani*, penicillin would be the drug of choice. Tetracyclines are no longer reliable in the absence of susceptibility data because many clostridia are resistant. Chloramphenicol is another alternative.

Prognosis is still very poor in tetanus. Over the 17-year interval from 1950 to 1967, the case fatality rate has remained unchanged at about 65% (861). Mortality is maximal at the extreme of age, with patients between the ages of 10 and 20 years having less than one-tenth the incidence and mortality of those less than 1 year or over 60 years of age. When neonatal cases are excluded, the case–fatality ratio of patients under 50 years of age is 55%, as compared with 76% for those over 50 years. The mortality in neonatal cases is 77%. Incidentally, the median age of non-neonatal patients in the United States is 53 years. In general, there is a tendency for patients with relatively short incubation periods to have a poor prognosis. Paroxysmal spasms are associated with a significantly higher mortality. The more rapid the evolution of symptoms and signs, the worse the prognosis. High fever or a rising temperature is an ominous sign. The quality of the supportive care is certainly extremely important in prognosis.

During early World War I, nearly 8 British soldiers of every 1000 wounded died of tetanus. The dramatic impact of prophylaxis and, in particular that associated with the use of tetanus toxoid, is indicated in Table 17.11. Note that in World War II, there were only 0.44 cases of tetanus per 100,000 wounds and injuries. Among the small numbers of soldiers who did develop tetanus, there were several who had not received any toxoid at any time and others who had failed to get boosters after being wounded. A general guide to prophylaxis against tetanus in wound management as proposed by Furste and Wheeler (867) is given in Table 17.12. The term "tetanus-prone" should be reserved for compound fractures, gunshot wounds, burns, crush injuries, and Wheeler (867) is given in Table 17.12. The term "tetanus-prone" should be reserved for compound fractures, gunshot wounds, burns, crush injuries, wounds with retained foreign bodies, deep puncture wounds, wounds contaminated with soil or feces, wounds untended for more than 24 hours, wounds infected with other bacteria, wounds with devitalized or avascular tissue, and induced abortions. Immediate thorough surgical treatment of wounds is an extremely important aspect of tetanus prevention. Good wound care implies thorough cleansing of the wound and removal of all foreign bodies and devitalized tissue. It should be appreciated that although TIG is certainly effective, nearly 5% of cases of tetanus seen in the United States occur in persons given TIG at the time of injury (866).

In the United States, at present, elderly individuals represent a great reservoir of unimmunized persons. Most young and middle-aged males have been immunized while in the military service. Programs for immunization of children have also been established since World War II so that children and young adults of both sexes are now fairly well protected, particularly those living in cities. Middle-aged females are probably the next greatest reservoir of unimmunized persons.

TABLE 17.12

A Guide to Prophylaxis against Tetanus in Wound Management[a]

General principles
I. The attending physician must determine for each patient with a wound what is required for adequate prophylaxis against tetanus
II. Regardless of the active immunization status of the patient, meticulous surgical care, including removal of all devitalized tissue and foreign bodies, should be provided immediately for all wounds. Such care is essential as part of the prophylaxis against tetanus
III. Each patient with a wound should receive adsorbed tetanus toxoid[b] intramuscularly at the time of injury, either as an initial immunizing dose, or as a booster for previous immunization, unless he has received a booster or has completed his initial immunization series within the past 5 years. As the antigen concentration varies in different products, specific information on the volume of a single dose is provided on the label of the package
IV. Whether or not to provide passive immunization with homologous tetanus immune globulin (human) must be decided individually for each patient. The characteristics of the wound, conditions under which it was incurred, its treatment, its age, and the previous active immunization status of the patient must be considered
V. To every wounded patient give a written record of the immunization provided, instructing him to carry the record at all times, and if indicated, to complete active immunization. For precise tetanus prophylaxis, an accurate and immediately available history regarding previous active immunization against tetanus is required
VI. Basic immunization with adsorbed toxoid requires 3 injections. A booster of adsorbed toxoid is indicated 10 years[c] after the third injection or 10 years[c] after an intervening wound booster. All individuals, including pregnant women, should have basic immunization and indicated booster injections

TABLE 17.12 (*Continued*)

Specific measures for patients with wounds
 I. Previously immunized individuals
  A. When the patient has been actively immunized within the past 10 years[c]
   1. To the great majority, give 0.5 ml of adsorbed tetanus toxoid[b] as a booster *unless it is certain that the patient has received a booster within the previous 5 years*
   2. To those with severe, neglected, or old (more than 24 hours) tetanus-prone wounds, give 0.5 ml of adsorbed toxoid[b] *unless it is certain that the patient has received a booster within the previous year*
  B. When the patient has been actively immunized more than 10 years[c] previously
   1. To the great majority, give 0.5 ml of adsorbed tetanus[b] toxoid
   2. To those with severe, neglected, or old (more than 24 hours) tetanus-prone wounds
    a. Give 0.5 ml of adsorbed toxoid[b,d]
    b. Give 250 U[e] of tetanus immune globulin (human)[d]
    c. Consider giving penicillin
 II. Individuals *not* previously immunized
  A. With clean minor wounds in which tetanus is most unlikely, give 0.5 ml of adsorbed tetanus toxoid[b] (initial immunizing dose)
  B. With all other wounds
   1. Give 0.5 ml of adsorbed tetanus toxoid[b] (initial immunizing dose)[d]
   2. Give 250 U[e] of tetanus immune globulin (human)[d]
   3. Consider giving penicillin

[a] Modified from Furste and Wheeler (867) with kind permission of the authors and publisher.
[b] The Public Health Service Advisory Committee on Immunization Practices in 1969 recommended DTP (diphtheria and tetanus toxoids combined with pertussis vaccine) for basic immunization in infants and children from 2 months through the sixth year of age and Td (combined tetanus and diphtheria toxoids: adult type) for basic immunization of those over 6 years of age. For the latter group, Td toxoid was recommended for routine or wound boosters, but if there is any reason to suspect hypersensitivity to the diphtheria component, tetanus toxoid (T) should be substituted for Td.
[c] Some authorities advise 6 rather than 10 years, particularly for patients with severe, neglected, or old (more than 24 hours) tetanus-prone wounds such as may be sustained by military personnel in combat.
[d] Use different syringes, needles, and sites of injection.
[e] In severe, neglected, or old (more than 24 hours) tetanus-prone wounds, 500 U of tetanus immune globulin (human) are advisable.

When wound contamination and tissue destruction have been very great, for example, in cases of extensive third degree burns, both active and passive immunization at the time of injury may be beneficial, even if the patients have received active immunization previously. In such cases, tetanus may develop more rapidly than in the 4 to 7 days that are required to obtain maximal response to a booster dose of tetanus toxoid. Furste (866a) suggested that TIG should be considered in certain situations in which an actively immunized individual might fail to have an anamnestic response: patients with agammaglobulinemia or carcinoma of the breast, patients on immunosuppressive drugs, and people recently exposed to high doses of radiation. Only adsorbed or precipitated tetanus toxoid preparations should be used, since these are more immunogenic than fluid toxoid and produce an equally rapid recall.

Physicians should offer active immunization with tetanus toxoid to all patients whenever feasible. It should be part of the routine periodic health examination and checkup. Another logical opportunity for immunizing is when a patient seeks advice before overseas or domestic travel. Still another good occasion for initiating active immunization is at the time of treatment for an injury, whether or not antitoxin is required at that time. The benefit of active immunization is greatly decreased unless the patient is fully informed as to the nature of the immunization he has been given; he should also receive a written record of his immunization. If this is not done, a physician treating an injury at a later date may feel compelled to administer antitoxin if he has no evidence of the immunity status of the patient.

Neonatal tetanus can be prevented entirely by providing at least two doses of tetanus toxoid to pregnant women, training midwives in aseptic techniques, and administering TIG to infants born under unsterile conditions outside of the hospital to unimmunized mothers. Unimmunized or incompletely immunized mothers should receive TIG intramuscularly immediately before or at the time of delivery. This is especially important if they are not delivered in the hospital. Immunization with tetanus toxoid can be started or continued at this same time by using separate syringes and different extremities for injection.

We must keep in mind that surviving an attack of tetanus does not result in immunity. Second attacks months or years after recovery have occurred in patients not given toxoid after the first attack. Apparently the tetanospasmin is so toxic that an amount sufficient to cause clinical tetanus is not adequate to produce immunity.

Reactions to tetanus toxoid are uncommon and are seen most often in those who have received too many doses of toxoid in the past. Intramuscular injection is said to result in significantly fewer local reactions than is true for the subcutaneous route [White et al. (866b)]. The reaction rate is higher in women than in men. Almost all reactors have a satisfactory serum antitoxin concentration or develop it within 1 to 6 months. Local reactions of edema, erythema, and pain may be accompanied by fever and start a few hours after the injection of toxoid. They resemble an Arthus phenomenon. These reactions may occur in as many as 30% of persons who have received toxoid previously, but for the most part the discomfort is trivial. Persons known to have had moderately severe reactions to previous injections of toxoid (urticaria, arthralgia) can be given reduced dosages (one-fifth to one-tenth of the usual dose) if an additional injection is really required. In such circumstances, patients should be observed closely with adequate equipment and drugs on hand to treat severe reactions. Delayed hypersensitivity to tetanus toxoid has been observed 6 to 7 days after administration, with pain, discomfort, and local irritation at the injection site. Individuals with known severe hypersensitivity to toxoid (nephrosis, anaphylaxis) may be treated with TIG, if indeed they need any further prophylaxis. Studies of antitoxin titers in the patient's serum should be done. There are several excellent general references concerning tetanus (861, 864, 866–868).

Necrotizing Enteritis and Colitis Due to Clostridium

Several types of enteritis or enterotoxemia in animals are caused by *Clostridium perfringens* of different toxin types. Thus, *C. perfringens* type A may, rarely, cause enterotoxemia in unweaned lambs and newborn alpacas. Type B is the usual cause of lamb dysentery (necrotic enteritis) and may cause enterotoxemia of foals, sheep, goats, and calves. Type C *C. perfringens* produces enterotoxemia of sheep, calves, lambs, and piglets and necrotic enteritis of fowl. Enterotoxemia of sheep, goats, and cattle is related to overeating and involves *C. perfringens* type D. Furthermore, *Clostridium septicum* causes "braxy," an enterotoxemia of sheep.

With this impressive array of animal intestinal diseases involving *Clostridium perfringens* of diverse toxin types and *Clostridium septicum*, it would be reasonable to anticipate that similar intestinal problems might occur in man. There is indeed considerable evidence for this, but the nonspecificity of the clinical and pathologic picture, lack of definitive studies in many cases, and the fact that other types of pathologic processes, such as obstruction and vascular impairment, may lead to somewhat similar clinical and pathological conditions makes this whole area very difficult to interpret.

The best studied of these conditions in man is the necrotic enteritis or pig-bel found in New Guinea in association with traditional pig-feasting activities where large quantities of pork are consumed. The disease is characterized by abdominal cramps, vomiting, shock, diarrhea (sometimes bloody), and acute inflammation of the small intestine, with areas of necrosis and gangrene particularly in the jejunum. The mortality rate is high. The evidence that *Clostridium perfringens* type C (specifically the β-toxin) is the cause of the disease is good. It is as follows: type C *Clostridium perfringens* is frequently isolated from resected bowel and from stool of patients; there is a rise and subsequently fall in levels of antitoxin to β-toxin in the serum of patients; β-antitoxin levels are significantly higher in patients than in healthy individuals; mortality rate is reduced significantly by treatment with β-antitoxin; and fatal necrotic enteritis is reproduced in guinea pigs with type C strains from cases of pig-bel. The source of the type C organisms is unknown, as they have not been recovered from the intestines of pigs in the area where the disease occurs or from the processed pork. Overeating is thought to play a role in this disease—perhaps by means of producing temporary stasis of intestinal contents related to overdistension. Type C *Clostridium perfringens* has not been found in individuals without pig-bel in New Guinea, although detectable levels of β-antitoxin were found in the serum of over 70% of the

TABLE 17.13

Necrotizing Bowel Disease Involving *Clostridium*

Country	Patient's age	Underlying disease	Other predisposing factors	Gastrointestinal manifestations
Germany	?	?	?	Enteritis necroticans— primarily jejunal involvement?
Germany	71	None?	Ate home-tinned rabbit	Necrotic jejunitis plus several small ulcers in ileum and rectum
U.S.A.	35	Chronic diarrhea, psychogenic vs. non-tropical sprue	Weight 75 lbs. Hypo-motility, jejunum	Marked inflammation jejunum, ileum, proximal ascending colon. Jejunal ulcers, edema and gas bubbles in mucosa and submucosa and mesenteric nodes. Bleeding from jejunal ulcers
England	41	"Mild post-dysenteric colitis"	None	Profuse diarrhea (14–16 stools/day) with mucus and pus
England	53	Strangulation of ileum (adhesion)	None	Those of strangulation obstruction

Type of *Clostridium*	Source of *Clostridium*	Outcome	Comments	Reference and year
C. perfringens, type F, heat-resistant	? Bowel lumen, stool	8 of 12 died?	No specific clinical details given; primarily a bacteriologic paper	Zeissler and Rassfeld-Sternberg (876b), 1949
C. perfringens, type F, heat-resistant	Tinned rabbit meat, patient's intestinal contents	Died	Abdominal pain, diarrhea, vomiting, shock. Same organism recovered from stools of two persons who shared meal with patient and had mild illness	Hain (876c), 1949
C. perfringens, not typed	Blood, jejunal ulcer, also bowel wall (stain)	Died	Sudden acute picture superimposed on chronic (7 years duration) diarrhea. Prior to this, barium study showed hypomotility and stagnation in jejunum	Patterson and Rosenbaum (876d), 1952
C. perfringens, type D	Feces— during illness and again 4 weeks later	Recovered	Direct smear of feces revealed many clostridia. 4 weeks after illness, serum contained 0.2 U ε antitoxin/ml (none detected in 9 healthy persons)	Kohn and Warrack (872), 1955
C. perfringens, type D (and type A) (few colonies of D, heavy growth of A)	Ileal contents	Died (of hemorrhage)	Authors postulate toxemia of obstruction may be caused by clostridial toxins. At least 5000 mouse MLD of epsilon toxin demonstrated in ileal contents (20 MLD/gm). (Study of ileal contents from 33 randomly selected autopsies failed to reveal epsilon toxin)	Gleeson-White and Bullen (871), 1955

(*continued*)

TABLE 17.13 (*continued*)

Country	Patient's age	Underlying disease	Other predisposing factors	Gastrointestinal manifestations
England	56–78 (six cases)		Four had varying degrees of constipation	Necrotizing colitis, patchy in four, extensive in two. Sigmoid always spared. Mottled gray-green necrosis—full thickness gangrene of bowel wall (patchy); adjacent bowel congested and edematous. Necrosis mostly limited to mucosa
U.S.A.	21	Acute lymphocytic leukemia, no intestinal involvement	Corticosteroids, 6-MP	Hemorrhagic necrosis of ileum and cecum. Confluent mucosal ulceration. Numerous clostridia seen on stain
Canada	70	Incarcerated inguinal hernia	None	Necrotizing colitis—from neck of hernial sac to ileocecal valve
New Guinea	4 to 45 (17 cases)	14 ate pork (3-no data), often cooked 1–2 days or more earlier. Sometimes noted as "stale"	Fair–poor nutritional state in 10	Jejunum and upper ileum involved with edema, hemorrhage, mucosal and submucosal ulceration. Pseudomembrane formed at times. Inflammation and necrosis of vessels, and thrombi seen in some. Gas in subserosa and mesenteric glands in 4 cases. Perforation in some cases
New Guinea, Papua	No specific case details given			"Necrotic enteritis"

Type of Clostridium	Source of Clostridium	Outcome	Comments	Reference and year
Unknown, presumably C. perfringens (large gram-positive bacilli in colonies in mucosa and submucosa), no cultures	Bowel wall (stain only)	3 died	Vomiting. No diarrhea. Four shocky. Much putrid peritoneal fluid. Vascular disease said to be excluded, but possibly present in two cases. No ulceration. Intense polymorpho-nuclear infiltration	Killingback and Williams (876e), 1961
C. perfringens	Bowel (wall?)	Died	Disseminated C. perfringens infection with gas cysts in many viscera	Jarkowski and Wolf (604), 1962
C. perfringens, not typed	Lumen of obstructed bowel; invaded all layers of colon also	Died	Gangrene of entire colon developed in 6 hours	Ghent et al. (876f), 1963
C. perfringens, not typed	Samples of pork consumed by 4 patients; feces of 3 patients. Gram-positive bacilli in abundance in small bowel contents and on mucosal surfaces	6 died. Two recovered but had malab-sorption syndrome	Most had surgical resection of affected bowel. Onset 5 to 24 hours (usually 12 or less) after eating pork. Pig feasts not hygienic. Pork flesh readily con-taminated with bowel contents. Cooking not thorough. Intestines also cooked and eaten. Leftover pork often not recooked, not refriger-ated. Huge volume of food eaten at feasts	Murrell and Roth (876g), 1963
C. perfringens, type C	Bowel lumen	—	Similar to cases described by Murrell and Roth	Egerton and Walker (876h), 1964

(continued)

TABLE 17.13 (*continued*)

Country	Patient's age	Underlying disease	Other predisposing factors	Gastrointestinal manifestations
England	66	None	Prednisone for 5 months (rheumatoid arthritis)	Gangrenous colitis—from upper sigmoid to ileocecal valve. Appendix not involved. Serosa involved in places, along with mucosa and submucosa. No ulceration or membrane formation
England	45–85 (13 cases)	Heart disease with recent heart failure in most of 13 cases	Six had received antimicrobial drugs	Hemorrhagic necrosis, often patchy, involving colon primarily but stomach and small bowel at times. Hemorrhage into mucosa, superficial necrosis. Ulceration and inflammatory exudate at times
Canada	Newborn	None	—	Necrotizing colitis, transverse colon. Purulent inflammatory infiltrate localized primarily in muscularis and serosal layers. Many gas bubbles in mucosa and submucosa. Microulcerations of mucosa
U.S.A.	24	Crohn's disease	None	Necrotizing jejunitis and ileitis with multiple ileal perforations

healthy population. Murrell *et al.* (869) pointed out that clinically the diagnosis of pig-bel is difficult to differentiate from other causes of food poisoning or dysentery, on the one hand, and from other causes of strangulation and peritonitis, on the other.

Necrotic enteritis has also been described in Germany, where it was called Darmbrand. Again, excessive eating of rich food seemed to be an important factor. Consumption of contaminated canned meats was also notable. Type C *Clostridium perfringens* was again implicated, although the toxin type was originally described as type F (subsequent consideration indicates that it is a variant of type C). Injection of cultures of type C isolated from human

Type of Clostridium	Source of Clostridium	Outcome	Comments	Reference and year
C. perfringens, type A, heat-resistant	Lumen of bowel, bowel wall	Died	Predominance of clostridia in bowel lumen by smear. Shock, no diarrhea, and no remarkable quantities of fluid in bowel lumen. Specifically ruled out vascular cause for pathology	Tate et al. (876i), 1965
All 13 had C. perfringens; 3 type A, others not tested. Not heat-resistant	Involved bowel lumen	All died	Some had no gastro-intestinal symptoms, others diarrhea (4), vomiting (6), abdominal distention (4)	McKinnell and Kearney (874), 1967
Clostridium, probably C. perfringens	Peritoneal exudate, stool, and intra-abdominal abscess	Hemi-colectomy, survived	Masses of gram-positive bacilli on stain of bowel wall. S. aureus also found at all sites yielding Clostridium	Pereira et al. (876j), 1968
C. perfringens	Purulent peritoneal exudate (also yielded E. coli)	Died	Two similar cases did not yield clostridia. Acute process distinct from Crohn's disease. No mention of organisms seen in bowel wall.	Mogadam and Priest (876k), 1969

cases into the small intestine of guinea pigs produced necrotic enteritis experimentally. Hain (870) studied stools from 108 healthy individuals in Hamburg and recovered type F (type C) C. perfringens from one-sixth of these. However, these strains were much less pathogenic for animals than those isolated from cases of enteritis necroticans. The spores of the strains of C. perfringens from Germany were heat-resistant, but this has not been described with all types of C. perfringens involved in necrotic enteritis. Type C C. perfringens has also been isolated from similar disease in Uganda and in England. In animals, type C strains show evidence of bacterial adhesion to intestinal epithelial cells. Paredes and Fernández (1404a) mentioned recovery of *Clostridium perfringens* from enteritis necroticans, but gave no details.

Clostridium perfringens type D has been isolated from man on two occasions, but the relationship to the enteric disease is questionable in at least one of these (see Table 17.13). Type D *C. perfringens* was isolated from the contents of the ileum of over half of 33 autopsy cases without evidence of enteric disease sampled at random (871), but the organism is also found in the intestine of healthy sheep and goats, where it is established as a definite cause of enterotoxemia (872).

A summary of a number of cases of necrotizing disease of the small or large bowel or both involving *Clostridium* is presented in Table 17.13. Aside from the cases already discussed which seemed to involve food poisoning and/or overeating of rich food, a number of the cases presented in Table 17.13 must be regarded as questionable. At times the papers do not give adequate clinical or pathologic data, and the identification of the organisms is often poor. In the case of those organisms that have not been toxin-typed, one cannot even be sure that they are actually *Clostridium perfringens*. Even legitimate isolates of *Clostridium perfringens* may really have nothing to do with the disease. It is true that, in general, overt causes of gangrene of the bowel, such as strangulation, volvulus, intussusception, and mesenteric arterial or venous occlusion, have been excluded. Morson (873) described acute or transient ischemic disease of the colon, not always grossly demonstrable and sometimes involving a reflex shutdown of vessels, and McKinnell and Kearney (874) described impaired intestinal blood flow related to cardiovascular problems. Both of these papers indicate that *Clostridium perfringens* may then invade secondarily and either contribute to the pathology or not. Foci of leukemic infiltration or solid tumor in the bowel may also permit invasion by clostridia and production of disease locally or systemically. Other disease of the bowel, and particularly diseases that lead to stasis, obstruction, or decreased splanchic blood flow (874a, 874b), may predispose to overgrowth by anaerobes (and clostridia in particular) in the involved segment; these clostridia may then invade. Accordingly, it seems reasonable to say at this point that clostridial disease of the bowel may be either primary or secondary, and that there also may be postmortem or other invasion of clostridia into the bowel wall without necessarily any associated disease. When clostridia are involved in disease, it may be related to production of toxin, direct infection, or both.

It is also reasonable to speculate about the possibility of *C. perfringens* or other clostridia being involved in certain cases of pseudomembranous enterocolitis, colitis arising postoperatively, or in relation to antimicrobial therapy, particularly clindamycin, to which various clostridia may be resistant. Similarly one may wonder about the role of *C. perfringens* or other clostridia in neonatal necrotizing enterocolitis or in toxic megacolon, particularly in view of the fact that pneumatosis cystoides intestinalis may be seen with some frequency in neonatal necrotizing enterocolitis (874c).

Clostridium Perfringens Food Poisoning

Certain strains of type A *Clostridium perfringens* are an important cause of a mild type of food poisoning. In England and Wales, there were 47 outbreaks with approximately 1500 cases in 1969 and 32 outbreaks with almost 1300 cases in 1970 [Hobbs, (875)]. In the United States, in the years 1971–1974, *Clostridium perfringens* was always the second or third most common cause of food poisoning outbreaks and second or third in terms of the total number of cases affected (875a, 875b). Since confirmation of *Clostridium perfringens* food-poisoning outbreaks requires special techniques for isolation and serotyping of the organisms, it is likely that a number of outbreaks are overlooked (875c). The median number of individuals involved per outbreak is usually larger than for other common causes of food poisoning, such as *Staphylococcus*, *Salmonella*, and *Shigella*. At one banquet in New York City in 1968, of a total of 1800 individuals at the banquet, over 900 became ill with *Clostridium perfringens* food poisoning. Outbreaks are most often associated with meats of various types, poultry, or gravy, but fish may be involved at times and, less commonly, dairy products, fruit, vegetables, and Chinese food. Most occurrences are noted in homes or restaurants, but institutional outbreaks, including outbreaks in hospitals, not uncommonly may also involve *Clostridium perfringens*. The organism is widespread—it may be isolated from the feces of man and of most domestic and farm animals. Samples of raw meat are often contaminated with the organism. It has also been found in flies, soil, and dirt from kitchens.

The incubation period is 8 to 12 hours after ingestion of the contaminated food, occasionally as little as 6 hours or as long as 24 hours. The main symptoms are crampy abdominal pain and diarrhea. Stools are liquid but do not contain blood or mucus. A small percentage of patients may have nausea, but vomiting is rare. The illness is usually not severe and lasts less than 24 hours.

It is now clear that outbreaks may be caused by heat-resistant strains of *C. perfringens* which may be either α-hemolytic or nonhemolytic and also by so-called "heat-sensitive" strains, which are mostly β-hemolytic on horse blood agar. The spores of "heat-sensitive" strains are less resistant to heat than the others, but they can nevertheless survive cooking. Type A strains may be further subdivided on the basis of serotyping. The spores of the organism survive in food after cooking, can be activated to germinate by the heat treatment ("heat shock"), and then grow at various rates in slowly cooling food. Large numbers of organisms are required to initiate the disease. The characteristics of the organism are important in relation to the food poisoning; temperature limits for its growth are about 25° and 50°C, and the optimum temperature for growth in cooked meat is between 43° and 47°C. The generation time can be as short as 10 to 12 minutes. The spores of heat-sensitive strains may often germinate even without heat activation. With the resistant strains, activation is usually detected at 75°–80°C. The heat resistance, heat activation, and rapid generation time are the factors that predispose the organism to survival, followed by rapid multiplication in slowly cooling masses of meat and poultry. When meat is cooked in bulk, the heat gain is slow, and subsequent cooling is slow. The heat drives off dissolved oxygen maintaining an anaerobic environment. Heating induces germination of spores and then when the temperature drops to below 50°C, vegetative cells multiply rapidly. Only enterotoxin-producing strains of C. perfringens may give rise to food poisoning. The toxin that is responsible is produced in the large intestine of the victim during sporulation. The potential for sporulation in the food itself is poor. After ingestion, spores are readily formed in the intestine and then toxin is produced. The enterotoxin may be confirmed by the use of ligated intestinal loops in rabbits, by intradermal injection into guinea pigs and rabbits (producing erythema), or by electroimmunodiffusion (876). The mode of action of the toxin appears to be similar to that of *Vibrio cholerae*.

To confirm *Clostridium perfringens* as the cause of an outbreak of food poisoning, several types of tests are useful. Included are demonstration of the same serotype of isolates from the stools of ill individuals and from the implicated food, presence of large numbers (greater than 10^6/gm) of C. perfringens in the implicated food, demonstration of larger numbers of C. perfringens in the stools of affected individuals as compared with control subjects, and demonstration of enterotoxin production by C. perfringens from food or affected individuals.

Addendum

Recently five cases of botulism have been identified in infants in an 18-month period [Botulism in infants—California. *Morbidity and Mortality Weekly Rept.* Center for Disease Control, **25**, 269 (1976)] suggesting that the disease in infants may be more common than previously recognized and raising the possibility that this disease may account for some cases of unexplained sudden infant death syndrome. In these infants, toxin was identified in the stool but not in the serum. It is postulated that the infants may have ingested C. *botulinum* spores on raw foods or even well-cooked food (since the spores are heat-resistant) and that the spores may have germinated in the infants' intestinal tracts resulting in the production of toxin at that site. Botulism should be considered in infants with unexplained weakness, ophthalmoplegia, dysphagia, or respiratory arrest.

18

Antimicrobial Agent Susceptibility of Anaerobic Bacteria

Techniques for Susceptibility Testing
of Anaerobic Bacteria

A number of specialized types of antimicrobial susceptibility tests have been devised specifically for anaerobic bacteria. These seem unnecessary, however, since it is perfectly feasible to use the tests commonly used with aerobic or facultative bacteria under the conditions of anaerobiosis. The tube dilution test is convenient for use with single strains or small numbers of strains, and provides an easy way to determine bactericidal endpoints. One may also use automatic or semiautomatic dilution apparatus to do large numbers of tests of this type with anaerobes in small wells (876l). The agar dilution technique is convenient for study of large numbers of strains of anaerobes using a replica inoculator. For the most part, agar tests are more easily read. Simplified and economical broth and agar dilution techniques for susceptibility testing of anaerobes have been described (876m, 876n). The paper disc diffusion technique (97, 877–882, 882a) and the broth–disc technique (883) may also be used provided that one adheres closely to the techniques and standards described or that one develops his own standards. The use of standards designed for aerobic and facultative bacteria [such as those of Bauer *et al.* (884)] with anaerobes is entirely inappropriate and may lead to erroneous results.

Specific details on setting up various types of susceptibility tests for anaerobes are given in the second edition of the "Wadsworth Anaerobic Bacteriology Manual" (97).

Factors Affecting Anaerobic Susceptibility Test Results

There are many problems associated with susceptibility testing of anaerobic bacteria. In addition to the factors that may influence results of tests with aerobic or facultative bacteria, such as age and density of the inoculum and incubation period, certain conditions inherent in or commonly used for growth of anaerobic bacteria may affect results of susceptibility tests. The components of the media used for anaerobic susceptibility testing may significantly influence the results. Anaerobiosis itself, the technique used to obtain anaerobiosis, the presence and amount of carbon dioxide in the atmosphere, humidity, and the pH of the media (both the initial pH and the ultimate pH in the anaerobic atmosphere) all undoubtedly influence the results (885, 886). Furthermore, the degree of influence appears to vary with different drugs and different organisms. Exact definition and significance of such effects remain to be determined.

In the absence of definitive information on most of these points, and because of the necessity to use many of these conditions in order to obtain satisfactory growth of many anaerobes, a reasonable approach is to use standardized techniques so that results of several laboratories may be compared until enough information has accumulated to permit intelligent interpretation of *in vitro* susceptibility test results. Clinical correlation will ultimately be required. Efforts are underway to develop reference standard dilution techniques for anaerobic susceptibility testing.

Indications for Susceptibility Testing with Anaerobic Bacteria

Many anaerobes have predictable patterns of susceptibility to antimicrobial drugs, so that if the organism is well identified one may usually predict its susceptibility pattern fairly accurately. Susceptibility testing will be required in serious infections, in infections failing to respond to what was thought to be appropriate therapy, and in patients who relapse after an initial response to therapy. In most cases of anaerobic infection, antimicrobial therapy must be started before availability of definitive bacteriologic data and susceptibility testing.

Data on Susceptibility of Anaerobes
to Antimicrobial Agents

There will be no attempt made here to review the extensive literature on *in vitro* susceptibility patterns of various anaerobic bacteria. Only data developed in the Wadsworth Anaerobic Bacteriology Laboratory will be presented. There are many excellent papers in the literature to which the reader may wish to refer. There are good reviews by Del Bene and Farrar (887) and Phillips (887a). References dealing with susceptibility patterns of a variety of anaerobic bacteria to a number of different antimicrobial drugs include the following (115, 195, 287, 321, 336, 359, 631, 888–894, 894a, 1404a). Other references dealing primarily with susceptibility of various anaerobes to one drug or a group of related drugs include the following (895–914, 914a–914g). There are also a number of references dealing primarily with the susceptibility of one species or genus of anaerobes to one or more antimicrobial agents; included are the following (390, 405, 587, 809, 915–932, 932a–932d, 1289a). Three of these papers do warrant special comment inasmuch as they point out resistance in anaerobes to two of the most effective drugs for the management of anaerobic infection—chloramphenicol and clindamycin. In the paper by Kagnoff *et al.* (405), significant resistance to chloramphenicol was reported. This has not been reported from other centers in the United States to date. Paredes and Fernández (1404a), in Chile, noted that 18 of 60 strains of non-spore-forming anaerobes were resistant to chloramphenicol. However, they used a disc diffusion technique, and it is not clear that this was standardized against a dilution procedure. If this resistance is not spurious, it may relate to availability of the drug without prescription in that country. Wilkins and Thiel (915) found that 16 of 180 strains of 18 different clinically important species of *Clostridium* had a minimal inhibitory concentration of greater than 12.5 μg/ml of clindamycin; an additional 7 strains required 12.5 μg/ml for inhibition. Species having two or more strains with this degree of resistance to clindamycin included: *Clostridium tertium, C. subterminale, C. difficile, C. ramosum, C. innocuum,* and *C. malenominatum.* Our group has previously published a review of our *in vitro* susceptibility studies with anaerobic gram-negative bacilli (933).

Table 18.1 presents the criteria currently utilized in our laboratory for determination of susceptibility to antimicrobial agents among the anaerobic bacteria. Drugs which may be clinically useful in management of anaerobic infections are listed, along with the blood levels that are readily achieved with conventional dosage (susceptible), levels achieved with higher dosage (intermediate resistance), and levels that are not readily achieved without hazard of toxicity (resistant). Tables 18.2–18.16 give specific data for a large number of anaerobic isolates of various types within the range of concentrations likely to be useful clinically with the various compounds. All strains tested were human isolates and most were from clinical infections. All were studied by comparable technique in the Wadsworth Anaerobic Bacteriology Laboratory (agar dilution technique, as described in the "Wadsworth Anaerobic Bacteriology Manual"), and in many or most cases the very same strains were tested against all or most of the drugs.

It is clear from the data in Tables 18.2–18.5 and Figs. 18.1 and 18.2 that various penicillins have differences in activity against anaerobic bacteria. Drugs such as penicillin V and phenethicillin, both of which are thought to be quite comparable to penicillin G, may show less activity than penicillin G against various anaerobic bacteria (Fig. 18.1) (933, 934) but penicillin V is usually comparable, considering levels achievable with oral therapy. Susceptibility patterns vary from species to species and from strain to strain within the same species. Methicillin is rather consistently less active against various anaerobic bacteria than is penicillin G (934). Nafcillin is less active than penicillin G versus *Bacteroides fragilis* and anaerobic cocci and is distinctly less active than penicillin G against most strains of *B. melaninogenicus* and other *Bacteroides* and anaerobic gram-positive bacilli (933, and

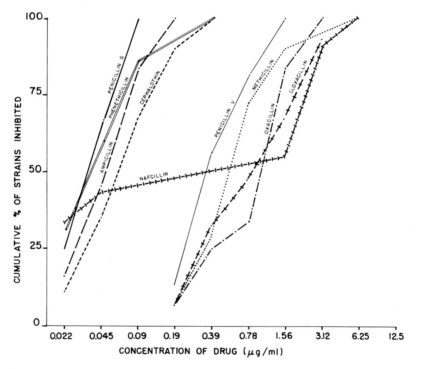

Fig. 18.1.   Activity of various penicillins and cephalosporins against *Bacteroides melaninogenicus.*

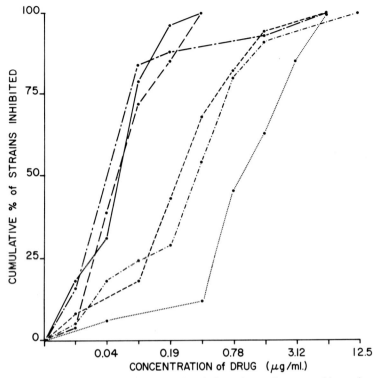

Fig. 18.2.   Activity of penicillins and cephalosporins against anaerobic cocci.

D. F. Busch, V. L. Sutter, and S. M. Finegold, unpublished data). Isoxazolyl penicillins such as cloxacillin and dicloxacillin are less active than penicillin G against many or most anaerobic bacteria. Carbenicillin is quite active against most anaerobes, including *B. fragilis*, considering the high serum levels achievable (935). Ticarcillin is quite comparable to carbenicillin in its level of activity against anaerobic bacteria (V. L. Sutter, Y. Y. Kwok, and S. M. Finegold, unpublished data). An experimental penicillin, BLP 1654, showed good activity against anaerobic bacteria *in vitro* (V. L. Sutter, Y. Y. Kwok, and S. M. Finegold, unpublished data); 90% of strains of *Bacteroides fragilis* were inhibited by 32 μg/ml or less, and most other anaerobes were inhibited by 16 μg/ml or less.

As with the penicillins, various cephalosporins differ in their activity against anaerobic bacteria. This is noted in Tables 18.6–18.8 for cephalothin, cefazolin and cefoxitin. Cephaloridine is roughly equivalent to penicillin G in activity against *Bacteroides fragilis*. Cephapirin, like cephalothin, has good activity against most anaerobic bacteria other than *Bacteroides fragilis*.

Cephalexin and cephradine are somewhat less active than the two cephalo-sporins just mentioned but should be superior to penicillin G or V for oral therapy because of higher achievable serum levels. Cefamandole is a little more active than cephalothin against gram-negative anaerobes and a little less active against gram-positive anaerobes. Cefoxitin is the most active of the cephalosporins (935).

Many anaerobes are now resistant to tetracycline. However, doxycycline (936) and minocycline are more active than the other tetracyclines (Tables 18.9–18.11).

Thiamphenicol, an analogue of chloramphenicol, is quite comparable to chloramphenicol which shows excellent activity against anaerobic bacteria (Table 18.12). However, three strains susceptible to achievable levels of chloramphenicol were resistant to thiamphenicol. One strain of *Bacteroides corrodens* had an MIC of 32 μg/ml, an unidentifiable *Bacteroides* species had an MIC of 128 μg/ml, and a strain of *Clostridium difficile* had an MIC of 64 μg/ml (G. K. Harding, V. L. Sutter, Y. Y. Kwok, and S. M. Finegold, unpublished data).

Erythromycin (Table 18.13) shows inconsistent activity, and lincomycin (Table 18.14) is inferior to clindamycin (Table 18.15), a drug with excellent activity. With clindamycin (Table 18.15), we noted two strains in the "other *Bacteroides*" group with minimal inhibitory concentrations of 16 μg/ml. One of these was a *Bacteroides corrodens*, and the other an unidentifiable species of *Bacteroides*. There were five strains of *Peptococcus* with MIC's equal to or greater than 64 μg/ml. Two of these were *Peptococcus variabilis*; both had MIC's greater than 128 μg/ml. There were two strains of *Pepto-coccus prevotii*, one of which had an MIC of 128 μg/ml, and the other 64 μg/ml. The final strain was *Peptococcus magnus*; its MIC was 128 μg/ml.

Metronidazole (Table 18.16) also has excellent activity against anaerobes except for certain gram-positive organisms (*Actinomyces, Propionibacterium,* and *Eubacterium*) and certain other microaerophilic organisms.

Rifampin is very active against most anaerobic bacteria (937, 938). An-aerobes that are usually exquisitely sensitive to rifampin include anaerobic and microaerophilic cocci and streptococci, *Bacteroides melaninogenicus, Bifidobacterium, Propionibacterium,* and some clostridia. Other anaerobes potentially within the therapeutic range of rifampin include *Bacteroides fragilis* (highest MIC among 32 strains was 3.2 μg/ml), *Bacteroides oralis, Fusobacterium* strains, *Veillonella,* and some clostridia. Anaerobes that are highly resistant to rifampin include *Fusobacterium varium* and *F. mortiferum,* certain strains of *Eubacterium,* and certain clostridia, including *Clostridium ramosum* (118). It should be pointed out that there is no clinical experience with the use of rifampin in treatment of anaerobic infections, and it is not approved by the United States Food and Drug Administration for this purpose. Furthermore, there are resistant mutants within populations of anaerobes, just as have been demonstrated commonly with rifampin for other organisms; it is likely that if this agent were used as a single drug, resistance would be a serious problem.

We have tested only a moderate number of anaerobes against nitrofurantoin, but this compound appears to be quite active against anaerobes, including *Bacteroides fragilis*. We have not studied cotrimoxazole ourselves; reports in the literature give variable results (see Chapter 19).

Vancomycin is completely inactive against *Bacteroides fragilis* at achievable blood levels (the lowest MIC is 16 μg/ml and only 55% of strains are inhibited by as much as 64 μg/ml). At a cutoff point of 8 μg/ml, 40% of strains of *Bacteroides melaninogenicus* and of *Fusobacterium* would be inhibited. Vancomycin is quite active against most gram-positive anaerobes, including *Peptococcus*, *Peptostreptococcus*, *Arachnia*, *Propionibacterium*, *Actinomyces*, and most clostridia. *Eubacterium* may also be susceptible, but not consistently (V. L. Sutter, Y. Y. Kwok, and S. M. Finegold, unpublished data).

Polymyxin B and colistin (polymyxin E) are inactive against gram-positive anaerobic bacteria and against *Bacteroides fragilis*. These compounds show activity against a number of strains of *B. melaninogenicus*, against some strains of *B. oralis*, and against *Fusobacterium nucleatum* (933, V. L. Sutter, Y. Y. Kwok, and S. M. Finegold, unpublished data). Bacitracin shows variable activity against gram-negative anaerobic bacilli (933). Most anaerobes require at least 10 μg/ml of fucidin for inhibition (S. M. Finegold and L. G. Miller, unpublished data). Rosamicin shows very good activity against most anaerobes except for a percentage of strains of *Peptococcus* (939). Spiramycin was quite active against Bacteroidaceae *in vitro* (disc technique, apparently not standardized) according to Gupta (814a).

Anaerobic bacteria are rather consistently resistant to various aminoglycoside agents, such as kanamycin, gentamicin (940), and amikacin, and to nalidixic acid (941). Twenty-nine *Fusobacterium* strains all had MIC's of >100 μg/ml with amikacin. Most, but not all, strains of anaerobic cocci were resistant; 21/31 had an MIC of 50 μg/ml or greater but 5 strains had an MIC of 12.5 μg/ml or less (V. L. Sutter, Y. Y. Kwok and S. M. Finegold, unpublished data). Most strains of *Bacteroides fragilis* were inhibited by 16 μg/ml of spectinomycin [Phillips and Warren (939a)].

Information should also be given on selected anaerobic bacteria not mentioned per se in the tabular data presented in this chapter. *Bacteroides oralis* is very sensitive to penicillin G (all 21 strains inhibited by 3.1 μg/ml or less). It is also very sensitive to erythromycin (maximum MIC 1.6 μg/ml), to lincomycin (maximum MIC 6.2 μg/ml) and to clindamycin, metronidazole, and rifampin. Ampicillin, carbenicillin, cephapirin, and cephalexin are all quite active against *Bacteroides oralis*. Some strains of *B. oralis* are resistant to tetracycline; minocycline shows greater activity than tetracycline (933). *Bacteroides corrodens* has unique susceptibility patterns. Drugs that are quite active include penicillin G, tetracycline, chloramphenicol, clindamycin, and metronidazole. Lincomycin is moderately active; some strains are resistant to erythromycin. Vancomycin is inactive. In contrast to what is found with most other anaerobic bacteria, aminoglycosides show definite activity against *Bacteroides corrodens*. Inhibition was achieved at 25 μg/ml with neomycin and kanamycin; lower concentrations were not tested (933). *Fusobacterium necrophorum* is very sensitive to most antimicrobial agents, including penicillin G, tetracycline, chloramphenicol, lincomycin, clindamycin, and metronidazole. Erythromycin shows only moderate activity and vancomycin is inactive (933).

Bactericidal Activity

Bactericidal activity is important in endocarditis and perhaps in certain other serious infections in which host defense mechanisms may be impaired. In general, penicillins and cephalosporins are the only compounds among those used clinically for anaerobic infections which consistently show bactericidal activity. Metronidazole, still experimental for anaerobic infections in the United States, is very active against anaerobes (942) and also shows excellent bactericidal activity—even against *Bacteroides fragilis*, which is resistant to most penicillins and cephalosporins (938, 942a). Under certain conditions, there is a lag before metronidazole begins its killing phase (942b). The bactericidal activity of metronidazole depends on the concentration of the drug, the growth pattern of the test organism, and experimental conditions affecting oxygen exposure of the organism (942c).

Combinations of Antimicrobial Agents

The effect of nine pairs of antimicrobial agents against *Bacteroides fragilis* was studied (943). Neither synergism nor antagonism was noted with metronidazole plus erythromycin, carbenicillin, or cefoxitin; with clindamycin plus choramphenicol; with penicillin plus tetracycline; or with amikacin plus carbenicillin, cefoxitin, or clindamycin. Synergism was noted with clindamycin plus metronidazole. Videau (906) has shown synergism between metronidazole and spiramycin.

Resistance to Antimicrobial Agents

There are several anaerobic bacteria that show significant resistance to antimicrobial agents. The most notable of these is *Bacteroides fragilis*, not only because it is resistant to several antimicrobial agents but also because it is the most commonly encountered of all anaerobes in clinical infections. *Fusobacterium varium* may also be quite resistant, but is not commonly encountered clinically. Certain newly described anaerobic species or groups, such as *Bacteroides splanchnicus*, *Bacteroides* group I (*Bacteroides disiens*) and *Bacteroides* group PS (*Bacteroides bivius*) often show resistance to penicillin G and presumably to similar compounds. Aside from *Bacteroides fragilis*, the other anaerobic bacterium that is relatively commonly encountered and which shows significant resistance is *Clostridium ramosum* (118) (see Table 18.17). Mitre and Rotheram (943a) have described bacteremia due to an unidentified anaerobic gram-negative bacillus that was uniquely resistant both *in vitro* and clinically. Its MIC with penicillin G, tetracycline, clindamycin, and chloramphenicol was > 20 $\mu g/ml$; the MIC with metronidazole was 0.625 $\mu g/ml$, and the patient was successfully treated with this compound.

Resistance of *Clostridium perfringens* to antibiotics has been shown [Sebald *et al.* (943b)] to be plasmid-mediated. Two different extrachromosomal DNA plasmids were demonstrated, one responsible for resistance to tetracycline and chloramphenicol and the other for resistance to erythromycin and clindamycin. Sebald and Bréfort (943c) subsequently demonstrated that the tetracycline–chloramphenicol plasmid was transferable to sensitive strains of *C. perfringens*. M. Sebald (personal communication) has also found that tetracycline resistance in 2 strains of *C. perfringens* isolated in our laboratory was transferable to sensitive strains. Thus this may be a widespread phenomenon and suggests the importance of restricting the use of this drug. Del Bene *et al.* (943d) failed in attempts at transferring antibiotic resistance from *Bacteroides* to *Escherichia coli* and vice versa. S. J. Burt and D. R. Woods (personal communication), however, were able to effect R factor transfer from *E. coli* to *Bacteroides* sp., *B. fragilis*, and *Fusobacterium* sp. following heat treatment (50°C) which presumably inhibited the restriction systems of the anaerobes.

Development of resistance has not been a major problem among anaerobes to date. However, there are two definite examples of such resistance which must be noted. One is the significant resistance of *Bacteroides fragilis* and many other anaerobes to tetracycline. Secondly, we are now seeing some strains of *Bacteroides melaninogenicus* that are moderately or highly resistant to penicillin G and similar compounds. Recently documented resistance to clindamycin in certain *Peptococcus* species should also be noted.

With regard to mechanisms of resistance of anaerobes to antimicrobial agents, penicillinase production by *Bacteroides fragilis* and other gram-negative anaerobic bacilli has been described (944, 945, 945a). Not all investigators have been able to demonstrate this, and penicillinase has not always been found in penicillin-resistant strains. However, Braude (945) states that there is a close correlation between penicillinase activity and relative penicillin resistance. Del Bene and Farrar (887) were unable to demonstrate penicillinase activity, but did demonstrate small amounts of cephalosporinase activity in some strains of *Bacteroides fragilis*. In a later paper, however, Weinrich and Del Bene (945b) did find β-lactamases with predominantly penicillinase activity in 2 strains of anaerobes, *Bacteroides clostridiiformis* and *Clostridium ramosum;* both of these were inducible with penicillin, and the former with cefoxitin as well. Ten *Bacteroides fragilis* strains produced β-lactamases predominantly active against cephalosporins. These were inhibited by cloxacillin and cefoxitin. With all 12 strains, hydrolysis of cefoxitin and carbenicillin was absent or minimal.

Hackman and Wilkins (945c) noted that a mixed experimental infection in mice with *Fusobacterium necrophorum* and *Bacteroides fragilis* was resistant to penicillin therapy, even though *F. necrophorum* infection alone responded to this drug and *B. fragilis* did not produce infection alone. They concluded that the protection afforded by the presence of *B. fragilis* appeared to be due mainly to production of penicillinase by this organism.

Kanazawa *et al.* (946) found that only 2 of 10 strains of *Bacteroides* inactivated penicillin G and cephaloridine. However, they noted that many anaerobes of various genera and species were capable of inactivating chloramphenicol and that all 90 strains of various anaerobes that they studied were capable of inactivating a nitrofuran derivative. The significance of this in terms of clinical use of antimicrobial agents, if any, is uncertain, inasmuch as there was no correlation between this inactivation and the susceptibility of the anaerobic bacteria to the compound inactivated.

TABLE 18.1

Classification of Susceptibility to Antimicrobials Based on MIC[a]

Antimicrobial agent	MIC (μg/ml)		
	Susceptible	Intermediate	Resistant
Penicillin G[b]	2	4 to 32–64	64–128 or more
Ampicillin	3	6–25	50 or more
Amoxicillin (oral)	4	8	16 or more
Carbenicillin	64	128–256	256 or more
Cephalothin	8	16–32	32–64 or more
Cefazolin	12–25	25–50	100 or more
Cefoxitin	8	16–32	64–128 or more
Chloramphenicol	16	32	64 or more
Lincomycin	3	6–25	25–50 or more
Clindamycin	2–4	4 to 8–16	16–32 or more
Tetracycline	1–2	4–8	16 or more
Doxycycline	1–2	4[c]	8 or more
Minocycline	1–2	4–8	16 or more
Vancomycin	4	8	16 or more
Erythromycin	1–2	2–4	8 or more
Metronidazole	8	16 to 32–64	64–128 or more

[a] This arbitrary classification is based on the assumption that strains are susceptible when the MIC is equivalent to or less than blood levels readily achieved with conventional dosage, that strains are of intermediate susceptibility when blood levels equivalent to the MIC can be achieved only with higher doses of drug, and resistant when the MIC is such that blood levels equivalent to it could be achieved only with doses carrying a definite risk of toxicity.

[b] Penicillin G is expressed in U/ml.

[c] With present dosage recommendations.

TABLE 18.2

Activity of Penicillin G versus Anaerobic Bacteria[a]

Organism	No. strains	Cumulative percent susceptible to indicated concentration (U/ml)							
		<1.0	2.0	4.0	8.0	16.0	32.0	64.0	128.0
Bacteroides fragilis	76	2.6	5.3	6.6	13.2	42.1	72.4	90.8	90.8
Bacteroides melaninogenicus	66	83.3	86.4	89.4	93.9	97.0	100.0		
Other Bacteroides	70	52.1	56.3	63.4	71.8	76.1	83.1	94.4	97.2
Fusobacterium mortiferum	2	50.0	50.0	50.0	50.0	50.0	50.0	50.0	100.0
Fusobacterium nucleatum	18	94.4	94.4	100.0					
Other Fusobacterium	16	75.0	87.5	87.5	87.5	87.5	87.5	87.5	93.8
Peptococcus	60	100.0							
Peptostreptococcus	29	89.7	89.7	89.7	96.6	96.6			
Gram-negative cocci	26	84.6	88.5	92.3	100.0		100.0		
Eubacterium	8	50.0	75.0	87.5	100.0				
Arachnia	2	100.0							
Propionibacterium	4	75.0	100.0						
Actinomyces	16	100.0							
Clostridium perfringens	51	100.0							
Other Clostridium	76	80.3	88.2	94.7	97.4	97.4	97.4	97.4	97.4

[a] From V. L. Sutter, Y. Y. Kwok, and S. M. Finegold, unpublished data, and F. L. Sapico, *et al.* (878).

TABLE 18.3

Activity of Ampicillin against Anaerobic Bacteria[a]

Organism	No. strains	Cumulative percent susceptible to indicated concentration (μg/ml)				
		≤0.8	3.1	6.2	12.5	25
Bacteroides fragilis	30	0	10.0	20.0	36.7	56.7
Bacteroides melaninogenicus	15	66.7	100.0			
Other *Bacteroides*	19	68.4	78.9	89.5	100.0	
Fusobacterium mortiferum	1	100.0				
Fusobacterium nucleatum	19	100.0				
Fusobacterium varium	2	0	0	100.0		
Other *Fusobacterium*	4	25.0	25.0	50.0	50.0	50.0
Peptococcus	43	100.0				
Peptostreptococcus	13	84.6	84.6	92.3	100.0	
Gram-negative cocci	12	83.3	91.7	100.0		
Eubacterium	1	100.0				
Propionibacterium	3	100.0				
Actinomyces	3	100.0				
Clostridium	8	85.7	100.0			

[a] V. L. Sutter, and S. M. Finegold, unpublished data.

TABLE 18.4

Activity of Amoxicillin against Anaerobic Bacteria[a]

Organism	No. strains	Cumulative percent susceptible to indicated concentration (μg/ml)		
		≤1.0	4.0	8.0
Bacteroides fragilis	43	9.3	25.6	30.2
Bacteroides melaninogenicus	60	86.7	95.0	100.0
Other *Bacteroides*	20	65.0	80.0	80.0
Fusobacterium mortiferum	2	50.0	50.0	50.0
Fusobacterium nucleatum	8	87.5	100.0	
Other *Fusobacterium*	10	70.0	70.0	70.0
Peptococcus	21	95.2	100.0	
Peptostreptococcus	19	89.5	89.5	89.5
Gram-negative cocci	7	85.7	85.7	100.0
Eubacterium	7	100.0		
Arachnia	2	100.0		
Propionibacterium	4	100.0		
Actinomyces	16	100.0		
Clostridium perfringens	8	87.5	87.5	87.5
Other *Clostridium*	26	92.3	96.2	100.0

[a] V. L. Sutter, Y. Y. Kwok, and S. M. Finegold, unpublished data.

TABLE 18.5

Activity of Carbenicillin against Anaerobic Bacteria[a]

Organism	No. strains	Cumulative percent susceptible to indicated concentration (μg/ml)						
		≤1.0	8.0	32.0	64.0	128.0	256.0	≥512.0
Bacteroides fragilis	76	5.3	25.0	47.4	80.3	94.7	94.7	94.7
Bacteroides melaninogenicus	67	89.6	100.0					
Other *Bacteroides*	70	53.5	81.7	93.0	97.2	97.2	97.2	97.2
Fusobacterium mortiferum	2	0	100.0					
Fusobacterium nucleatum	18	83.3	100.0					
Other *Fusobacterium*	16	50.0	87.5	93.8	93.8	93.8	93.8	93.8
Peptococcus	61	86.9	100.0					
Peptostreptococcus	29	69.0	89.7	93.1	96.6	100.0		
Gram-negative cocci	26	42.3	88.5	96.2	96.2	100.0		
Eubacterium	8	25.0	50.0	100.0				
Arachnia	2	0	100.0					
Propionibacterium	4	75.0	75.0	100.0				
Actinomyces	16	100.0						
Clostridium perfringens	9	100.0						
Other *Clostridium*	74	24.3	73.0	97.3	100.0			

[a] V. L. Sutter, Y. Y. Kwok, and S. M. Finegold, unpublished data.

TABLE 18.6

Activity of Cephalothin against Anaerobic Bacteria[a]

Organism	No. strains	Cumulative percent susceptible to indicated concentration (μg/ml)				
		≤1.0	4	8	16	32
Bacteroides fragilis	124	0	5.0	6.0	8.0	11.0
Bacteroides melaninogenicus	9	78.0	89.0	89.0	89.0	89.0
Other *Bacteroides*	50	40.0	44.0	46.0	52.0	66.0
Fusobacterium mortiferum	11	27.0	73.0	73.0	90.0	100.0
Fusobacterium nucleatum	10	100.0				
Fusobacterium varium	8	0	88.0	100.0		
Fusobacterium necrophorum	4	100.0				
Peptococcus	44	95.5	97.7	100.0		
Peptostreptococcus	14	78.6	92.9	100.0		
Gram-negative cocci	19	84.2	100.0			
Eubacterium	8	37.0	50.0	63.0	75.0	100.0
Propionibacterium	1	100.0				
Clostridium	28	36.0	61.0	86.0	89.0	93.0

[a] From Sutter and Finegold (935), and V. L. Sutter, Y. Y. Kwok, and S. M. Finegold (unpublished data).

TABLE 18.7

Activity of Cefazolin against Anaerobic Bacteria[a]

Organism	No. strains	Cumulative percent susceptible to indicated concentration (μg/ml)				
		≤0.8	3.1	12.5	25	50	100
Bacteroides fragilis	32	0	3.0	13.0	13.0	34.0	53.0
Fusobacterium mortiferum	15	53.0	100.0				
Fusobacterium varium	10	0	10.0	100.0			
Other *Fusobacterium*	6	83.0	100.0				
Peptococcus and *Peptostreptococcus*	28	96.0	100.0				
Gram-negative cocci	3	100.0					
Eubacterium	19	47.0	58.0	74.0	95.0	100.0	
Clostridium	31	45.0	61.0	81.0	100.0		

[a] From Sutter and Finegold (935).

TABLE 18.8

Activity of Cefoxitin against Anaerobic Bacteria[a]

Organism	No. strains	Cumulative percent susceptible to indicated concentration (μg/ml)						
		≤1.0	4.0	8.0	16.0	32.0	64.0	128.0
Bacteroides fragilis	164	2.4	29.4	66.5	80.5	96.3	99.4	99.4
Bacteroides melaninogenicus	64	93.8	98.4	100.0				
Other Bacteroides	21	47.6	71.4	81.0	81.0	85.7	85.7	90.5
Fusobacterium mortiferum	11	35.0	100.0					
Fusobacterium nucleatum	8	75.0	100.0					
Other Fusobacterium	10	40.0	50.0	70.0	80.0	80.0	80.0	80.0
Peptococcus	19	89.5	94.7	94.7	100.0			
Peptostreptococcus	19	63.2	89.5	89.5	100.0			
Gram-negative cocci	7	100.0						
Eubacterium	7	42.9	71.4	85.7	85.7	100.0		
Arachnia	2	100.0						
Propionibacterium	4	75.0	75.0	75.0	100.0			
Actinomyces	16	93.8	100.0					
Clostridium perfringens	8	50.0	75.0	75.0	100.0			
Other Clostridium	51	13.7	45.1	51.0	54.9	58.8	74.5	94.1

[a] V. L. Sutter, Y. Y. Kwok, and S. M. Finegold, unpublished data, and Sutter and Finegold (935).

TABLE 18.9

Activity of Tetracycline against Anaerobic Bacteria[a]

Organism	No. strains	Cumulative percent susceptible to indicated concentration (μg/ml)			
		≤ 1.0	2.0	4.0	8.0
Bacteroides fragilis	76	39.5	39.5	42.1	46.1
Bacteroides melaninogenicus	67	76.1	79.1	86.6	94.0
Other *Bacteroides*	71	35.2	43.7	50.7	60.6
Fusobacterium nucleatum	18	100.0			
Other *Fusobacterium*	16	93.8	93.8	93.8	93.8
Peptococcus	61	31.1	37.7	37.7	39.3
Peptostreptococcus	29	41.4	48.3	51.7	72.4
Gram-negative cocci	26	69.2	73.1	73.1	73.1
Eubacterium	8	50.0	50.0	50.0	62.5
Arachnia	2	100.0			
Propionibacterium	4	75.0	75.0	75.0	75.0
Actinomyces	16	68.8	93.8	93.8	93.8
Clostridium perfringens	51	53.0	62.7	82.4	96.1
Other *Clostridium*	71	53.5	59.2	63.4	69.0

[a] From V. L. Sutter, Y. Y. Kwok, and S. M. Finegold, unpublished data, and Sapico *et al.* (878).

TABLE 18.10

Activity of Doxycycline against Anaerobic Bacteria[a]

Organism	No. strains	Cumulative percent susceptible to indicated concentration (μg/ml)			
		≤ 1.0	2.0	4.0	8.0
Bacteroides fragilis	76	42.1	50.0	75.0	88.2
Bacteroides melaninogenicus	67	77.6	89.6	95.5	97.0
Other *Bacteroides*	71	43.7	53.5	69.0	78.9
Fusobacterium nucleatum	18	100.0			
Other *Fusobacterium*	16	87.5	87.5	87.5	100.0
Peptococcus	62	37.1	51.6	71.0	93.5
Peptostreptococcus	29	44.8	65.5	79.3	96.6
Gram-negative cocci	26	69.2	73.1	80.8	96.2
Eubacterium	8	50.0	75.0	75.0	75.0
Arachnia	2	100.0			
Propionibacterium	4	75.0	75.0	75.0	75.0
Actinomyces	16	68.8	93.8	100.0	
Clostridium perfringens	51	96.1	96.1	98.0	100.0
Other *Clostridium*	71	64.8	67.6	80.3	91.5

[a] From V. L. Sutter, Y. Y. Kwok, and S. M. Finegold, unpublished data, and Sapico *et al.* (878).

TABLE 18.11

Activity of Minocycline against Anaerobic Bacteria[a]

Organism	No. strains	Cumulative percent susceptible to indicated concentration (µg/ml)			
		≤1.0	2.0	4.0	8.0
Bacteroides fragilis	76	56.6	69.7	81.6	97.4
Bacteroides melaninogenicus	65	81.5	89.2	96.9	98.5
Other *Bacteroides*	69	56.5	71.0	79.7	89.9
Fusobacterium nucleatum	18	100.0			
Other *Fusobacterium*	16	93.8	93.8	100.0	
Peptococcus	60	60.0	75.0	93.3	96.7
Peptostreptococcus	28	75.0	82.1	92.9	100.0
Gram-negative cocci	26	69.2	69.2	88.5	96.2
Eubacterium	7	57.1	71.4	71.4	85.7
Arachnia	2	100.0			
Propionibacterium	4	75.0	75.0	75.0	75.0
Actinomyces	16	87.5	93.8	100.0	
Clostridium perfringens	51	96.1	96.1	96.1	96.1
Other *Clostridium*	26	61.5	73.1	76.9	96.2

[a] From V. L. Sutter, Y. Y. Kwok, and S. M. Finegold, unpublished data, and Sapico *et al.* (878).

TABLE 18.12

Activity of Chloramphenicol against Anaerobic Bacteria[a]

Organism	No. strains	Cumulative percent susceptible to indicated concentration (µg/ml)				
		≤1.0	4.0	8.0	16.0	32.0
Bacteroides fragilis	76	1.3	30.3	85.5	98.7	100.0
Bacteroides melaninogenicus	68	55.9	100.0			
Other *Bacteroides*	71	38.0	87.3	97.2	100.0	
Fusobacterium nucleatum	18	94.4	100.0			
Other *Fusobacterium*	16	87.5	100.0			
Peptococcus	61	41.0	98.4	100.0		
Peptostreptococcus	29	65.5	96.6	100.0		
Gram-negative cocci	25	68.0	100.0			
Eubacterium	8	0	100.0			
Arachnia	2	50.0	100.0			
Propionibacterium	4	75.0	75.0	100.0		
Actinomyces	16	81.3	93.8	100.0		
Clostridium perfringens	51	0	98.0	100.0		
Other *Clostridium*	76	6.6	40.8	92.1	100.0	

[a] From V. L. Sutter, Y. Y. Kwok, and S. M. Finegold, unpublished data, and Sapico *et al.* (878).

TABLE 18.13

Activity of Erythromycin against Anaerobic Bacteria[a]

Organism	No. strains	Cumulative percent susceptible to indicated concentration (μg/ml)			
		$\leq$0.5	1.0	2.0	4.0
Bacteroides fragilis	76	10.5	19.7	38.2	60.5
Bacteroides melaninogenicus	68	94.1	100.0		
Other *Bacteroides*	71	43.7	74.6	85.9	95.8
Fusobacterium nucleatum	18	5.6	11.1	11.1	27.8
Other *Fusobacterium*	16	6.3	12.5	25.0	31.3
Peptococcus	61	11.5	39.3	70.5	80.3
Peptostreptococcus	28	82.1	89.3	96.4	96.4
Gram-negative cocci	26	3.8	34.6	61.5	80.8
Eubacterium	8	62.5	87.5	100.0	
Arachnia	2	100.0			
Propionibacterium	4	100.0			
Actinomyces	16	100.0			
Clostridium perfringens	52	0	5.8	23.0	82.7
Other *Clostridium*	75	64.0	84.0	88.0	88.0

[a] From V. L. Sutter, Y. Y. Kwok, and S. M. Finegold, unpublished data, and Sapico *et al.* (878).

TABLE 18.14

Activity of Lincomycin against Anaerobic Bacteria[a]

Organism	No. strains	Cumulative percent susceptible to indicated concentration (μg/ml)				
		$\leq$0.8	3.1	6.2	12.5	25.0
Bacteroides fragilis	122	3.3	13.2	57.3	89.4	98.4
Bacteroides melaninogenicus	15	100.0				
Fusobacterium mortiferum	10	100.0				
Fusobacterium nucleatum	18	100.0				
Fusobacterium varium	8	25.0	25.0	37.5	75.0	100.0
Other *Fusobacterium*	6	100.0				
Gram-positive cocci	21	47.6	100.0			
Gram-negative cocci	2	100.0				
Eubacterium	45	44.4	80.0	88.0	95.5	95.5
Actinomyces	13	84.6	100.0			
Clostridium perfringens	43	46.5	58.1	90.7	97.7	100.0

[a] Finegold *et al.* (976), Bartlett *et al.* (517), V. L. Sutter, Y. Y. Kwok, and S. M. Finegold, unpublished data, and Sapico *et al.* (878).

TABLE 18.15

Activity of Clindamycin against Anaerobic Bacteria[a]

Organism	No. strains	Cumulative percent susceptible to indicated concentration (μg/ml)			
		≤ 1.0	4.0	8.0	16.0
Bacteroides fragilis	123	79.7	96.0	100.0	
Bacteroides melaninogenicus	64	100.0			
Other *Bacteroides*	71	94.4	95.8	97.2	100.0
Fusobacterium mortiferum	10	100.0			
Fusobacterium nucleatum	8	100.0			
Fusobacterium varium	8	25.0	50.0	75.0	87.5
Other *Fusobacterium*	10	100.0			
Peptococcus	61	80.3	83.6	83.6	83.6
Peptostreptococcus	29	100.0			
Gram-negative cocci	25	100.0			
Eubacterium	45	80.0	93.3	95.5	95.5
Arachnia	2	100.0			
Propionibacterium	4	100.0			
Actinomyces	16	100.0			
Clostridium perfringens	43	58.1	100.0		
Other *Clostridium*	34	52.9	79.4	88.2	91.2

[a] V. L. Sutter, Y. Y. Kwok, and S. M. Finegold, unpublished data, Bartlett *et al.* (517), and Sapico *et al.* (878).

TABLE 18.16

Activity of Metronidazole against Anaerobic Bacteria[a]

Organism	No. strains	Cumulative percent susceptible to indicated concentration (μg/ml)				
		≤ 1.0	4.0	8.0	16.0	32.0
Bacteroides fragilis	76	27.6	85.5	98.7	100.0	
Bacteroides melaninogenicus	69	92.8	100.0			
Other *Bacteroides*	70	68.6	94.3	97.1	97.1	98.6
Fusobacterium mortiferum	17	100.0				
Fusobacterium varium	10	100.0				
Fusobacterium nucleatum	18	94.4	100.0			
Other *Fusobacterium*	15	100.0				
Peptococcus	60	86.7	96.7	96.7	96.7	96.7
Peptostreptococcus	29	75.9	86.2	93.1	96.6	96.6
Gram-negative cocci	25	92.0	100.0			
Eubacterium	50	58.0	64.0	70.0	70.0	74.0
Arachnia	2	0	0	0	0	0
Propionibacterium	4	25.0	25.0	25.0	25.0	25.0
Actinomyces	16	0	12.5	12.5	18.8	50.0
Clostridium perfringens	11	63.6	100.0			
Other *Clostridium*	157	65.0	82.8	92.4	98.1	100.0

[a] V. L. Sutter, Y. Y. Kwok, and S. M. Finegold, unpublished data.

TABLE 18.17

Susceptibility of *Clostridium ramosum* to Antimicrobial Agents[a]

Antimicrobial Agent	No. Strains	Cumulative percent susceptible to indicated concentration (μg/ml)						
		≤1.6	3.1	6.2	12.5	25	50	100
Penicillin G[b]	49	82.0	94.0	100.0				
Carbenicillin	49	20.0	40.0	62.0	83.0	96.0	100.0	
Chloramphenicol	49	6.0	60.0	100.0				
Lincomycin	49	0	0	2.0	34.0	65.0	81.0	85.0
Clindamycin	49	12.0	30.0	84.0	88.0	88.0	88.0	88.0
Erythromycin	48	69.0	84.0	84.0	84.0	84.0	84.0	84.0
Vancomycin	49	0	49.0	98.0	100.0			
Metronidazole	48	20.0	50.0	76.0	94.0	100.0		
Rifampin	49	0	0	0	0	0	0	0
Gentamicin	49	0	0	0	0	0	0	72.0
Tetracycline	47	44.0	49.0	53.0	53.0	55.0	76.0	93.0
Doxycycline	47	58.0	58.0	64.0	92.0	97.0	100.0	

[a] Tally *et al.* (118), Sutter *et al.* (936).
[b] Penicillin G expressed in U/ml.

533

19

Therapy and Prognosis in Anaerobic Infections

General

In our therapeutic approach to anaerobic infections, we must not be overly preoccupied with "the bug and the drug." While antimicrobial therapy is unquestionably of great importance in the proper management of this type of infection (56d, 1573), it is but one aspect of the multifaceted approach necessary for the successful management of these problems. The principles of treatment are (a) making the environment such that anaerobic bacteria find it difficult to proliferate, (b) checking the spread of anaerobic bacteria into healthy tissues, and (c) neutralizing the toxins of anaerobes. To control the local environment, useful measures include removal of dead tissue, drainage of collections of pus, elimination of obstructions, decompression of tissues where indicated, release of trapped gas, improvement of circulation to the part, and improved oxygenation of tissues. Antimicrobial agents are a very important factor in limiting the spread of anaerobes into healthy tissues. Toxin neutralization has been achieved primarily with specific antitoxins, but new approaches are being studied.

With anaerobic infections, there are special bacteriological problems not encountered in other types of infections which may make the therapeutic approach more difficult. Generally, bacteriological results will not be available as quickly as in aerobic infections, particularly if the infection is mixed (as it will be in about two-thirds of cases). Some laboratories may fail to recover certain or all of the anaerobes present in a specimen. This is particularly true if the specimen is not very promptly put under anaerobic conditions for transport to the laboratory. If care is not taken to avoid "contamination" of the specimen with normal flora, anaerobes may be recovered which have nothing to do with the patient's illness. Since not all laboratories are equipped to identify anaerobes accurately, presumptive results may be very misleading. Finally, standardized disc susceptibility tests are not yet available for all anaerobic bacteria, although a great deal of progress has been made in this area (see Chapter 18; and 947). Some laboratories have used a disc test without measurement of zones of inhibition or have measured zones applying the Kirby–Bauer–Sherris standards that were developed for aerobic and facultative organisms; results obtained with these techniques are completely undependable. Conventional tube or plate dilution tests with incubation in an anaerobic jar are reliable.

Therapy of specific disease entities will not be discussed in this chapter per se (see appropriate chapters). Special characteristics of specific diseases, of course, will significantly affect the approach to therapy. In gas gangrene, it is absolutely essential that every particle of dead muscle be removed. In bacterial endocarditis, a bactericidal drug must be used. In urinary tract infections, the higher levels of most drugs which may be achieved in the urine (and therefore in the renal medulla as well) may permit therapy with drugs that are active against anaerobes but do not ordinarily provide satisfactory serum and tissue levels for management of systemic infections. The spontaneous bronchial drainage of lung abscesses permits effective therapy by medical means alone. Furthermore, the lung is very vascular, and it possesses mechanical properties that permit collapse during healing. The brain, on the other hand, is encased in the skull which is relatively rigid so that cavities tend to stay open once formed and high intracerebral pressures develop which may interfere with blood flow at the periphery of an already devitalized area (352). Furthermore, the blood–brain barrier makes certain drugs unsuitable and others less effective for treatment of brain abscess or meningitis. Ledger (947a) emphasizes that curettage is the primary therapy for septic abortion.

Surgical Therapy

Surgical therapy is of the greatest importance in anaerobic infections and, in lesser infections, may be all that is required. Types of surgical therapy required include drainage of abscesses; excision of necrotic tissue; relief of obstructions; decompression of compartments where excessive pressure has built up as a result of edema, gas formation, and other factors; removal of associated malignancies or foreign bodies; removal of thrombi and/or ligation of veins; and obtaining material by biopsy or aspiration for diagnostic purposes. In general, abscess formation, which is very common in anaerobic infections, is a specific indication for surgical therapy. A distinct exception is lung abscess, for reasons already mentioned. Surgery for lung abscess is ordinarily not necessary and may present a particular hazard of spillage of abscess contents into other pulmonary segments. Liver abscess has generally been considered a specific indication for surgical therapy. Gilbert (565) has suggested that antimicrobial therapy may suffice; he briefly presented a case of multiple liver abscesses due to *Fusobacterium* species which responded to penicillin therapy, with the patient remaining well after a 4-month follow-up. However, the relatively high mortality seen in some series of liver abscess patients [Barbour and Juniper (948)], particularly when surgery was not performed [Altemeier *et al.* (949) and Sabbaj *et al.* (387)], suggest the need for caution. An intermediate approach that is certainly deserving of further study is suggested by the report of McFadzean, *et al.* (564) who reported 14 consecutive cases of solitary pyogenic liver abscess that were treated successfully by a combination of closed aspiration and antibiotic therapy. Prolonged high dose therapy, preferably with bactericidal agents, together with closed aspiration of abscess contents deserves further trial. Patients treated in this manner should be watched very closely and observed for an extended period of time (preferably at least 2 years) after therapy is completed.

Location of abscesses in critical areas, such as the speech area of the brain, may warrant a trial of simple aspiration and intensive antimicrobial therapy in order to avoid the hazard of radical surgical excision [Spatz (950)]. More conservative drainage procedures may also be appropriate for certain patients with serious bleeding tendencies or other factors that would make them a formidable surgical risk. On the other hand, a patient who has a lesion that constitutes a definite indication for surgical therapy should not ordinarily be given antimicrobial therapy for a period prior to surgery in hope of improving his general condition so that he will become a better surgical risk. Such patients will usually not respond to antimicrobial therapy, and their condition may deteriorate even further without definitive surgical management.

The good results that may be obtained by a good surgical approach despite extensive tissue damage are exemplified by the report of McNally *et al.* (515), in which good results were obtained in three patients despite involvement with gas gangrene of large areas of the anterior abdominal wall and even part of the chest wall in one of the patients. Only one of these patients had supplementary hyperbaric oxygen therapy. Morgan *et al.* (951)

reported a similar result as did Phillips *et al.* (951a). When extensive debridement is necessary, it should be done in the operating room under general anesthesia; repeated debridement may be required [Duff *et al.* (952)].

Adherence to important surgical principles such as achievement of good hemostasis, avoidance of dead space, gentle handling of tissues during surgery, and avoidance of contamination of clean body cavities (for example, using an extraperitioneal approach for drainage of adnexal abscess when possible) is very important.

Antimicrobial Therapy

In vitro susceptibility tests, properly performed, serve as a good guide to drug therapy for anaerobic infections. Table 19.1 lists the most useful drugs and their activity against various anaerobic bacteria. This and the detailed data in Chapter 18 should serve as reliable guides when results of susceptibility tests are not available. At the present time, I would regard chloramphenicol as the drug of choice for seriously ill patients with suspected or established anaerobic infection in whom there is not yet adequate bacteriological data.

As in therapy of other types of infections, various factors will influence the choice of drugs—the toxicity of chloramphenicol and clindamycin, whether or not the patient is allergic to penicillin and the severity of such allergy, the lack of penetration of the central nervous system by clindamycin, the excellent bactericidal activity of metronidazole against *Bacteroides fragilis* and other anaerobes, the renal and hepatic functional status of the patient, etc. Table 19.2 summarizes a number of important factors with regard to the drugs most active against anaerobic bacteria. Obviously, given the choice, the physician will wish to choose the agent with the least toxicity and impact on the normal flora of the body and with the best pharmacologic characteristics for the particular patient and the specific type of infection to be treated.

The nature of the various organisms in a mixed infection will also influence the choice of drugs. Drugs active against anaerobic bacteria may be quite inactive against accompanying aerobic or facultative organisms in mixed infections. In mixed infections involving several organisms, two, three, or more drugs may be required to provide effective coverage for each of the organisms in the mixture. Is it really necessary to provide coverage for each single organism present, particularly if some are present only in small numbers? In a case of necrotizing pneumonia, the transtracheal aspirate yielded *Fusobacterium*, *Bacteroides fragilis*, *B. oralis*, *Peptococcus*, and *Propionibacterium* as well as *Klebsiella pneumoniae*. Three of the anaerobes

TABLE 19.1

Susceptibility of Anaerobes to Antimicrobial Agents[a]

Agent	Microaerophilic and anaerobic cocci	Bacteroides fragilis	Bacteroides melaninogenicus	Fusobacterium varium	Other Fusobacterium species	Eubacterium and Actinomyces	Clostridium perfringens	Other clostridia
Penicillin G	+++ to ++++	+	+++ [b]	+++ [b]	++++	++++	++++ [b]	+++
Lincomycin	+++	+ to ++	+++	+ to ++	+++	++ to +++	++ to +++	+
Clindamycin	++ to +++	+++	+++	+ to ++	+++	++ to +++	+++ [c]	++
Metronidazole	++	+++	+++	+++	+++	+ to ++	+++	++ to +++
Chloramphenicol	+++	+++	+++	+++	+++	+++	+++	++ to +++
Tetracycline	++	+ to ++	++ to +++	++	+++	++ to +++	++ to +++	++
Erythromycin	++ to +++	+ to ++	+++	+	+	+++	++ to +++	++ to +++
Vancomycin	+++	+	+	+	+	++ to +++	+++	++ to +++

[a] ++++, Drug of choice; +++, good activity; ++, moderate activity; +, poor or inconsistent activity. There is no difference in activity between drugs rated +++ and those rated ++++. The ++++ indicates a drug with good activity, good pharmacologic characteristics, and low toxicity.
[b] A few strains are resistant.
[c] Rare strains resistant.

538

TABLE 19.2

Comparison of Antimicrobial Agents Active against Anaerobic Bacteria

Agent	Approved by FDA for anaerobic infections	Toxicity (including hypersensitivity reactions)	Central nervous system penetration	Effect on normal flora	Oral dosage form available	Parenteral dosage form available	Bactericidal activity
Penicillin	Yes	Low–moderate	Good	Minimal	Yes[a]	Yes	Very good
Lincomycin	No	Moderate–high	Poor	Major	Yes[a]	Yes	Little
Clindamycin	Yes	Moderate–high	Poor	Major	Yes	Yes	Moderate
Metronidazole	No	Low[c]	Good	Minimal–moderate	Yes	No[b]	Excellent
Chloramphenicol	Yes	High	Excellent	Minimal	Yes	Yes	None
Tetracycline	Yes	Low	Good	Minimal	Yes	Yes	None
Erythromycin	No	Low	Moderately good	Minimal–moderate	Yes	Yes	None
Vancomycin	No	Moderate–high	Poor	?Minimal	No	Yes	Fair to good[d]

[a] Parenteral form generally preferable for anaerobic infections (to facilitate higher dosage and higher blood levels).

[b] Experimental parenteral form available.

[c] Carcinogenicity in animals on prolonged therapy; mutagenicity in bacteria.

[d] Not studied with anaerobes.

were present in distinctly larger numbers than the *Klebsiella*, but the latter organism was present in a count of 10^6/ml. The patient was treated with lincomycin (which is inactive against *Klebsiella*) and responded very well, indicating that in this case the anaerobes were significant pathogens [Bartlett and Finegold (161)]. We have had similar results in a number of other cases of mixed pulmonary infection. Thadepalli and Gorbach (485) reported on 12 patients with intraabdominal or pelvic abscesses containing mixtures of aerobes and anaerobes who were treated with clindamycin alone. Although aerobes usually persisted in the drainage, all patients had a good clinical response without any septic complications. Leigh (110b) indicated that his experience in a small group of patients was that it was frequently necessary to treat only the *Bacteroides* in mixed infections. Gorbach and Bartlett (56b) surveyed the literature and found 31 cases of mixed aerobic–anaerobic infection which responded to therapy with clindamycin alone despite the presence of various aerobes resistant to this agent. This indicates that the anaerobes may have been the key pathogens in those infections. While it must be acknowledged that some patients who are treated with antimicrobial agents active only against the aerobic or facultative organisms present in such a mixture might respond well clinically, the data of Thadepalli *et al.* (495, 953) indicated clearly that in patients with penetrating abdominal trauma and mixed infections the patients receiving clindamycin and kanamycin did distinctly better than those receiving cephalothin and kanamycin (a regimen not very effective against anaerobes from the bowel). Much more information is needed on this point; the work cited will undoubtedly encourage others to continue this line of investigation in patients who are not too ill. Ledger *et al.* (953a) noted a failure of response to a penicillin–kanamycin regimen in women with severe *Bacteroides fragilis* obstetrical and gynecologic infections; a regimen of clindamycin–kanamycin was effective. At present, it is my feeling that therapy should be directed against all significant components of a mixed flora in seriously ill patients.

Weinstein *et al.* (954, 954a) and Bartlett *et al.* (68b) reported on experimental intraabdominal infections in rats which simulated intestinal perforation. In this model, with rat fecal flora (as modified by a meat diet), animals often have *E. coli* and *Proteus* bacteremia early, with significant mortality. Surviving animals develop intraabdominal abscesses in which anaerobes predominate. Early therapy with gentamicin significantly reduced the acute mortality, and with clindamycin the incidence of intraabdominal abscess was greatly reduced. Combination therapy with these drugs led to significant reduction in both complications. These important studies should, and undoubtedly will, be extended. At present, the significance of these data with regard to management of disease in humans must be considered uncertain.

While Rotheram and Schick (401) found that anaerobic bacteria were the dominant infecting organisms in septic abortion, *Escherichia coli* may be solely responsible for this infection on occasion and is the most frequent cause of endotoxic shock in such patients. Holm (955) mentioned several cases of actinomycosis in which organisms other than *Actinomyces* persisted after penicillin therapy and accounted for a poor clinical result. These organisms included *Actinobacillus actinomycetemcomitans*, the "corroding bacillus," and other anaerobic gram-negative bacilli. Bartlett *et al.* (517) described a patient with actinomycosis of the jaw in whom lincomycin therapy eliminated the *Actinomyces* and resulted in a partial clinical response. However, *Actinobacillus actinomycetemcomitans* emerged (this organism is resistant to lincomycin), and the patient was not cured until tetracycline therapy was instituted.

Arsphenamines were widely used systemically and topically in treatment of so-called fusospirochetal infections years ago. Clinical opinion was divided as to efficacy. Rosebury *et al.* (956) performed a controlled study in guinea pigs with experimental infection and concluded that neither neo-arsphenamine nor sulpharsphenamine was useful by systemic routes. Topical application was effective for accessible infections.

Chrysotherapy was used many years ago with some beneficial results. Christiaens *et al.* (540) reported a case with widespread infection due to an organism identified as *Fusiformis fusiformis*. This organism was not sensitive to sulfonamides, penicillin, or streptomycin, the only antimicrobials available at that time. Ultimately, in desperation, they resorted to therapy with gold thiosulfate with results they described as allowing them to speak of resurrection without exaggeration. Favorable results in anaerobic pulmonary infection and *in vitro* susceptibility data by Benard are also cited in this report. Stilbamidine was used successfully in one case of actinomycosis after inadequate response to procaine penicillin and sulfonamides [Miller, *et al.* (957)].

When sulfonamides became available, these were used with variable success. Among the clostridia, *Clostridium perfringens* was the most susceptible to treatment, with *Clostridium septicum* less susceptible; and *Clostridium novyi* relatively resistant. Graybill and Silverman (333) described a patient with pulmonary actinomycosis with striking subcutaneous metastic abscesses who showed a very good response to sulfonamide therapy given under the mistaken initial impression that the patient had nocardiosis. The patient was subsequently cured with penicillin. The authors point out that *Actinomyces israelii* is usually sensitive to sulfonamides. Sulfonamide treatment of infection with anaerobic gram-negative bacilli gave variable results. In 1939 Hemmens and Dack (958) showed that experimental infections of rabbits with *Bacteroides necrophorum* could be treated effectively with sulfanilamide if the treatment was started on or before the third day of infection. The first instance of recovery from *Bacteroides* septicemia at the Mayo Clinic was reported by Brown *et al.* (959) in 1941; this was a case of bacteremia and widespread infection with *Bacteroides funduliformis* which responded to sulfapyridine. Topical sulfathiazole was used to effect cure in Vincent's angina in a report by Manson and Craig (960) in 1945. In 1951, Willich (961) reported effective results with sulfonamides in two cases of infection due to Buday bacillus and noted that four strains of this organism were susceptible to sulfonamides *in vitro*. Gunn (296), in 1956, reviewed the literature dealing with approximately 150 patients with "*Bacteroides*" septicemia. Before the use of sulfonamides the mortality rate was 81%; treatment with sulfonamides reduced the mortality rate to 64%. Tynes and Frommeyer (489) cured a patient with "*Bacteroides*" septicemia with sulfadiazine after he had failed to respond to tetracycline. Another one of their patients with *Bacteroides fragilis* bacteremia and osteomyelitis showed improvement with sulfonamide therapy. In a large series of patients with bacteremia due to gram-negative anaerobic bacilli, Felner and Dowell (168) noted that patients treated with tetracyclines, sulfonamides, incision, and drainage, or a combination of these showed the highest survival rate. This type of broad generalization seems unwarranted. In general, sulfonamides have not been considered to have a place in the present day management of anaerobic infections. However, we recently encountered a patient with a large lung abscess whose transtracheal aspirate yielded a large number of different anaerobic organisms. The patient's fever and purulent sputum persisted in spite of what should have been appropriate antibiotic therapy until sulfadiazine was added to the regimen. Sulfadiazine was added because of the possibility that the patient might have nocardiosis, but *Nocardia* was never recovered.

Recently there is some interest in the possibility that combinations of trimethoprim and sulfonamides will be useful in the management of anaerobic infections. Okubadejo *et al.* (913) in a brief report, indicated that all 60 strains of *Bacteroides* (nearly all *B. fragilis*) tested were inhibited by the combination of trimethoprim and sulfamethoxazole. Phillips and Warren (962) noted synergism of the two compounds versus *B. fragilis* in plate dilution but not disc diffusion tests, and suggested that at least some infections with this organism might respond to therapy with cotrimoxazole. Rosenblatt and Stewart (963), however, found that sulfamethoxazole and trimethroprim were not active against most anaerobes, either individually or in combination; they studied 98 strains encompassing a number of different genera and species. Studies by Darrell *et al.* (914) indicated that most strains of *Clostridium perfringens* and of *Bacteroides* were resistant to trimethoprim *in vitro*. Bushby (963a) found relatively low activity of trimethoprim and the combination against 3 strains of *Bacteroides* and 1 *Clostridium* and cited the work of Naff in which it was found that cotrimoxazole eliminated Enterobacteriaceae from feces but did not affect *Bacteroides* or anaerobic lactobacilli. Knothe (963b) did find elimination of the latter two groups from the fecal flora of some individuals, but this effect was not consistent. Strauss and Pochi (964) found that the combination of trimethoprim and sulphafurazole produced a 30% decrease in the titratable acidity of sebum in acne patients, whereas neither compound alone produced any change. Okubadejo (964a) mentioned that 5 patients with *Bacteroides fragilis* infections were treated with cotrimoxazole and responded well. A case of cystitis due to *Clostridium perfringens* was treated successfully with the combination of trimethoprim and sulfamethoxazole; this may reflect the high levels of drugs achievable in the urine. Darrell *et al.* (914) noted that a patient with *Bacteroides* septicemia and liver abscess did not improve on trimethoprim plus sulfonamide and that 2 patients with *E. coli* bacteremia complicating septic abortion developed *Bacteroides* bacteremia as the original organism was eliminated by this therapy.

Penicillin G is well established as being effective clinically in infections with anaerobes which are susceptible *in vitro*. Many or most strains of *Bacteroides fragilis* are resistant *in vitro*. Ten percent of *B. fragilis* strains are even resistant to 256 U/ml or more. Accordingly, even high dose therapy is not feasible for a number of strains. Penicillin neurotoxicity would be a threat if one attempted to treat the more resistant strains (964b). *Bacteroides fragilis* produces a penicillinase, and, as noted earlier, Braude (945) stated that there is a close correlation between the penicillinase activity and the relative resistance to penicillin. Clinically, penicillin G has not generally been useful in infection with *Bacteroides fragilis*, even when very high doses were used [Bodner *et al.* (460)]. However, when *B. fragilis* is part of a mixed flora in anaerobic pulmonary infection, the patient may respond to penicillin [(Bartlett (477)]. I would still favor use of an agent active against *B. fragilis* in seriously ill patients with anaerobic pleuropulmonary infection of uncertain type. On occasion, penicillin G has been useful in endocarditis due to *Bacteroides fragilis* (433). Recently, we have noted some strains of *B. melaninogenicus* to be resistant to penicillin G. Certain clostridia other than *C. perfringens* may be highly resistant to this drug also. Occasional strains of *C. perfringens*, of other species of *Bacteroides*, and of *Fusobacterium varium* and *F. mortiferum* are also resistant to penicillin G. With these exceptions, penicillin G is very active against anaerobes. Some anaerobic cocci, however, require as much as 32 U/ml for inhibition, so large doses may be required. Ordinarily, doses of 2 to 10 million U or less per day are adequate for penicillin-susceptible anaerobic infections. Serious infections should be treated initially with 25 to 30 million U daily.

Other penicillins are not always as active as penicillin G, as indicated in Chapter 18. Generally, ampicillin, amoxicillin, and penicillin V are roughly comparable to penicillin G in activity, whereas other penicillins tend to be less active, and sometimes they are distinctly less active. Holloway (964c) reported good results with amoxicillin therapy of 3 anaerobic infections. Two involved *Actinomyces*, with *Bacteroides melaninogenicus* coexisting in one case. The third infection was due to an anaerobic streptococcus. Tyldesley (317) and Graybill and Silverman (333) reported successful use of penicillin V in patients with actinomycosis. White and Varga (965) reported a treatment failure in a patient with anaerobic streptococcal bacterial endocarditis treated with dimethoxyphenyl penicillin (the organism was relatively resistant *in vitro*). Isoxazolyl penicillins show relatively poor activity against anaerobes, and nafcillin's activity is variable and generally poor. Carbenicillin, because of the very high levels that may be achieved safely, is active against most anaerobes, including 95% of the *Bacteroides fragilis* strains. Clinical experience with this drug [Meny *et al.* (966) and Fiedelman and Webb (967)] has been favorable. Labowitz and Holloway (968) reported the successful treatment of a patient with *Bacteroides fragilis* bacteremia with carbenicillin. Marks and Eickhoff (969) reported a fatal superinfection with *Bacteroides* in a patient receiving carbenicillin for management of another type of infection.

Cephalosporins are also quite active against many anaerobes, as noted in Chapter 18, but most have relatively poor activity against *Bacteroides fragilis*. Wilson *et al.* (105), in a study of bacteremia due to anaerobic organisms, noted that over half of their patients were receiving antimicrobial agents at the time the first positive anaerobic culture was obtained. Cephalothin was the most frequently used antimicrobial. On the other hand, Walters *et al.* (970) reported a good response to cephalothin in a patient critically ill with generalized peritonitis due to a *Bacteroides* species not further identified. Bran *et al.* (44) reported a poor result in the treatment of an anaerobic streptococcal pneumonia with cephapirin, an antibiotic very closely related to cephalothin. Cefoxitin, a β-lactamase resistant cephamycin, has good activity against anaerobes *in vitro*, as noted in Chapter 18. Hackman and Wilkins (970a) noted that cefoxitin was more effective than cephalothin in an experimental mixed infection involving *Bacteroides fragilis* and *Fusobacterium necrophorum*. Very early clinical experience with cefoxitin in therapy of anaerobic infections is encouraging (970b).

Tetracycline used to be the antibiotic of choice for most anaerobic infections; however, the development of significant resistance on the part of many strains of a variety of different anaerobes in recent years has relegated tetracycline to a minor role in the treatment of anaerobic infections. Some workers have questioned the clinical significance of *in vitro* resistance to tetracycline, but our experience is that these strains are resistant in the clinical setting as well, and others have noted this too (970c). The problem of episomally transferable tetracycline resistance in *C. perfringens*, detected in both Paris and Los Angeles, was noted in Chapter 18. Doxycycline and minocycline are more active than tetracycline *in vitro* against anaerobes, but a number of strains of various types are still resistant. These latter drugs are effective clinically against susceptible strains (S. M. Finegold, D. F. Busch, S. E. Wilson, and G. Becker, unpublished data).

Chloramphenicol has the greatest *in vitro* activity against anaerobes of all drugs, and it is well established as effective in clinical practice over the years. There is one disturbing report, that of Kagnoff *et al.* (405) indicating significant resistance of *Bacteroides* to chloramphenicol. The *in vitro* studies were done by a three disc sensitivity technique which was not standardized against quantitative techniques. By this technique, only 21% of 48 isolates studied were totally sensitive and 31% partially sensitive to chloramphenicol. Unfortunately, no tube dilution or plate dilution results were given. While the significance of the disc sensitivity tests, in this case, must be considered questionable, it is important to note that 11 patients had a positive blood culture for *Bacteroides* species while receiving chloramphenicol in daily doses ranging from 1 to 3 gm by the parenteral route. Nevertheless, even this type of data is not necessarily significant. Seven of the 11 isolates from these patients were studied *in vitro*, and 3 were found to be susceptible to chloramphenicol. Bacteremia may persist in the face of appropriate medical therapy when the underlying process is such that surgical therapy is required for its eradication. Others have also noted chloramphenicol failures (56b), so one must be aware of this possibility. However, failures may occur with any drug; in our long experience with chloramphenicol it has been highly effective. Only rare individual strains of anaerobes of various types have been reported to be resistant to chloramphenicol by other workers except for the report from Chile noted in Chapter 18. We have encountered only rare strains of any type of anaerobe resistant to chloramphenicol in our 20 years of working with these organisms. I regard chloramphenicol as the drug of choice in serious anaerobic infections, particularly in central nervous system infections, which are not yet defined bacteriologically. Although a number of strains of anaerobes have been found to inactivate chloramphenicol [Kanazawa *et al.* (946)], these strains nevertheless remain susceptible to the drug.

Erythromycin shows its greatest activity against *Bacteroides melaninogenicus*, microaerophilic and anaerobic streptococci, gram-positive non-spore-forming anaerobic bacilli, and certain clostridia. It shows relatively good activity against *Clostridium perfringens* and poor or inconsistent activity against gram-negative anaerobic bacilli. There is, however, some disagreement in the literature on this point which seems to relate to the technique used for obtaining anaerobiosis during the performance of susceptibility studies. Ingham *et al.* (885) pointed out that if anaerobic susceptibility tests are done without carbon dioxide in the environment, anaerobes are much more susceptible to erythromycin (and lincomycin) than if the tests are done with $10\% \, CO_2$ in the atmosphere. This is not just a pH effect, but may relate to a smaller inoculum resulting from poor growth of some anaerobes in the absence of carbon dioxide. It will be important to perform clinical studies and to correlate *in vitro* susceptibility by the various techniques with achievable blood and body fluid levels and clinical response. Shoemaker and Yow (971) reported on treatment of 2 patients with *Bacteroides funduliformis* infections. One patient with empyema recovered completely after drainage and erythromycin therapy. The second patient had multiple lung abscesses and bilateral empyema secondary to *Bacteroides* septicemia; this patient responded well to erythromycin in terms of fever, but since the organism persisted in the empyema fluid chlortetracycline was substituted after 2 weeks. This patient also recovered. Herrell *et al.* (331) found that *Actinomyces* was quite susceptible to erythromycin *in vitro* and treated 4 patients with actinomycosis with this drug. Two cases of cervicofacial actinomycosis and one of extensive perirectal abscess responded well to therapy. The fourth case, also cervicofacial actinomycosis, responded well but recurred some time later and was treated with penicillin on the second occasion. A case reported by Ingham *et al.* (885) is difficult to interpret. This was a brain abscess due to *Bacteroides fragilis* with a scant growth also of *Proteus* and coliforms. There was a very good initial response to drainage, mastoidectomy, and chloramphenicol therapy for 10 days. Following a switch to sulfamethazine, the patient relapsed and reaccumulated pus, which yielded *Bacteroides fragilis* on culture. Reinstitution of chloramphenicol therapy and frequent surgical drainage failed to eradicate the infection. Further drainage was carried out and erythromycin was added in addition to the chloramphenicol. Within 10 days, the patient became afebrile and there was no further accumulation of pus. A case of mixed endocarditis involving *Bacteroides fragilis*, among other organisms [Child *et al.* (972)] failed to respond to three courses of erythromycin. However, it should be pointed out that erythromycin is not ordinarily bactericidal for *Bacteroides fragilis*.

Bacitracin is very active against *Bacteroides melaninogenicus* but is generally inactive against *Bacteroides fragilis* and *Fusobacterium nucleatum* [Finegold and Sutter (933)]. It is active against most gram-positive cocci and many gram-positive bacilli, although there are few studies specifically with anaerobes. Levin and Longacre (973) reported on two human bite infections with foul-smelling discharge treated with intramuscular bacitracin; both patients did well. No bacteriological studies were reported. Meleney *et al.* (755) reported on the use of systemic bacitracin in the treatment of five cases of progressive bacterial synergistic gangrene, four of which had failed to respond to penicillin. Three of the cases revealed the typical organisms on culture—microaerophilic streptococci and *Staphylococcus aureus*. The other two cases yielded only the *S. aureus*. All five cases responded promptly to the bacitracin therapy. Meleney *et al.* (758), in 1952, had good to excellent results in 22 of 25 patients with a variety of infections involving anaerobic or microaerophilic cocci or *Clostridium perfringens*. Bacitracin was used systemically and locally.

Vancomycin is quite active against most gram-positive anaerobes of all types. By dilution technique, it is not very active against gram-negative anaerobic bacilli, but it appears more active by disc technique. There are no good published data concerning possible clinical efficacy of this drug used systemically. One would anticipate that it would certainly be effective in many gram-positive anaerobic infections. There is one report [Mitchell and Baker (192)] concerning topical use of vancomycin for necrotizing gingivitis. This was a double-blind study comparing vancomycin and a placebo ointment. Vancomycin proved to be quite effective in this study; the amount of vancomycin in the ointment is not specified. I am aware of one case of endocarditis due to *Peptostreptococcus* which was cured with vancomycin.

Polymyxin B is very active against *Fusobacterium nucleatum in vitro* and also against a number of strains of *Bacteroides* other than *Bacteroides fragilis*. Polymyxin E (colistimethate) was used effectively in therapy of a patient who had septicemia with an unspecified species of *Bacteroides* [Holloway and Scott (974)].

Nitrofurantoin is very active against *Bacteroides fragilis*, *Bacteroides melaninogenicus*, and *Fusobacterium nucleatum* [Finegold and Sutter (933)]. Christenson *et al.* (975), in a summary of reports submitted by investigators to Eaton Laboratories, mentioned a case of clostridial gas gangrene of the bladder which showed improvement after therapy with intravenous nitrofurantoin; the organism was sensitive *in vitro*. Tynes and Frommeyer (489) indicated that 9 of 10 strains of *Bacteroides* isolated from patients with septicemia were susceptible to nitrofurantoin *in vitro*. However, one patient who had been on this drug (it was not specified whether this was by the oral or intravenous route) developed a positive blood culture with a *Bacteroides* while on the drug and subsequently died. Oral nitrofurantoin should ordinarily be suitable for urinary tract infection due to anaerobes.

In 1966, Finegold *et al.* (976) first demonstrated that lincomycin was very active against a variety of anaerobic bacteria. Clostridia, *Bacteroides fragilis*, and *Fusobacterium varium* are all relatively resistant to lincomycin, whereas the drug is quite active against other types of anaerobic bacteria.

The comments concerning differences in results relating to differences in type of anaerobic atmosphere for testing noted with erythromycin earlier apply as well to lincomycin. Clinical experience in a variety of anaerobic infections with susceptible organisms was favorable [Tracy *et al.* (674) and Bartlett *et al.* (517)]. Mohr *et al.* (977) reported the successful treatment of four cases of actinomycosis with lincomycin. Hall *et al.* (721) reported good results with lincomycin in the treatment of acne as well as in 2 of 3 cases of hidradenitis suppurativa, a condition which in our experience frequently involves infection with anaerobes.

The 7-chloro derivative of lincomycin, clindamycin, represents a significant advance. This compound is distinctly more active than lincomycin against anaerobic bacteria, including *Bacteroides fragilis*. Two strains of *B. fragilis* resistant to clindamycin were recently described [Salaki *et al*, (977a)]. A number of strains of *Fusobacterium varium* are resistant to clindamycin, and 15–20% of strains of certain species of clostridia other than *Clostridium perfringens* are resistant to this agent. Rare strains of *Clostridium perfringens* and of gram-positive non-spore-forming anaerobes are also resistant to clindamycin. Recently, we (V. L. Sutter and S. M. Finegold, unpublished data) have noted clindamycin resistance in the genus *Peptococcus* as well. But for these exceptions, clindamycin is very active against all other anaerobic bacteria. Tracy *et al.* (674) treated one patient with bacteremia with clindamycin in a dosage of 150 mg four times per day, with a transient clinical improvement. A complicating infection with *Pseudomonas*, however, made it difficult to judge the results, and the patient ultimately died. Bartlett *et al.* (517) treated fourteen patients with clindamycin and all responded well. Subsequently, Haldane and van Rooyen (529) reported on the use of parenteral clindamycin in 18 *Bacteroides* infections, all but one of which showed a rapid favorable response. Four of these patients were treated concurrently with chloramphenicol and one with carbenicillin. Clindamycin also proved very effective alone and in combination with kanamycin in intraabdominal infections following trauma [Thadepalli *et al.* (495, 953)]. It was also very effective in 22 severely ill patients with anaerobic infection or mixed aerobic and anaerobic infection in the abdominal or pelvic cavities [Thadepalli and Gorbach (485). Rose and Rytel (978) reported a good response to clindamycin in a patient with thoracic actinomycosis. Thadepalli *et al.* (953) reported that one of their patients treated with clindamycin had repeatedly positive blood cultures for both *Clostridium perfringens* and *Clostridium tertium* while receiving the drug. He was subsequently cured with chloramphenicol. Patients treated with clindamycin orally lose *Bacteroides fragilis*, all other gram-negative anaerobic bacilli, all anaerobic cocci, lactobacilli, most of the

Bifidobacterium, and sometimes the *Clostridium perfringens* from the normal fecal flora. The feces becomes heavily populated with a variety of other clostridia and *Eubacterium*. There is significant ingrowth of *Klebsiella–Enterobacter–Serratia* [Sutter and Finegold (979)]. For serious anaerobic infections, parenteral clindamycin should be used in a dosage of at least 300 mg every 6 hours. It may be given either intravenously or intramuscularly. The maximum approved dosage is 2700 mg/day. It is generally considered that clindamycin does not cross the blood–brain barrier and should not be used for central nervous system infections. One case report of *Bacteroides* brain abscess (1240f) apparently treated successfully with clindamycin in addition to surgical drainage and other antimicrobials is difficult to assess, as the authors acknowledge.

Metronidazole (Flagyl), first introduced for infection caused by *Trichomonas vaginalis* in 1959, has subsequently been shown to have significant activity against anaerobic bacteria. Tally *et al.* (942) found that all but 1 of 54 strains of anaerobic bacteria (a microaerophilic coccus) were sensitive to 25 μg/ml or less of metronidazole, an achievable level. Anaerobes of all types, with the exception of certain cocci (particularly those that are microaerophilic), *Actinomyces*, *Propionibacterium*, and some strains of *Eubacterium*, are very susceptible to this agent. Metronidazole is unique in having activity only against anaerobic microorganisms of all types—bacteria, protozoa, and even a threadworm. It is the only compound that we have found with consistent bactericidal activity against *Bacteroides fragilis* [Nastro and Finegold (938)] as well as other anaerobes. Metronidazole has been found to be active in experimental anaerobic infections in animals [Freeman, *et al.* (980) and Ueno *et al.* (981)]. In 1962, Shinn (982) reported the unexpected cure of acute gingivitis in a patient who received metronidazole for trichomonal vaginitis, which was also cured. He subsequently treated six other gingivitis patients with good results. A number of other reports attest to the value of metronidazole in Vincent's gingivitis. (185–189) Tropical ulcer, an anaerobic skin infection, has also been shown to respond well to metronidazole [Lindner and Adeniyi-Jones (737)]. Sugiyama *et al.* (678) noted decreased discharge and a decrease in the foul odor of discharge of patients with cancer of the cervix and secondary anaerobic infection when treated with metronidazole. Other nitroimidazole compounds, such as nitrimidazine, have also been shown to be effective in acute ulcerative gingivitis [Lozdan *et al.* (190)]. The first use of metronidazole in the treatment of systemic anaerobic infections in man was reported from the Wadsworth Veterans Hospital by Tally *et al.* (942). A patient with *Fusobacterium necrophorum* sepsis of uncertain etiology and several patients with anaerobic pulmonary infection showed good responses to metronidazole.

There was one instance of pneumococcal superinfection which responded subsequently to the addition of penicillin. One patient developed granulo-cytopenia that necessitated discontinuation of the drug after which the patient's marrow promptly returned to normal. A subsequent report by Tally *et al.* (93) detailed the results of therapy in 10 patients. Results were generally good. One patient relapsed after discontinuing the medication on his own accord and then failed to respond to a second course, presumably related to involvement of *Haemophilus influenzae* in his pneumonic process. Ledger *et al.* (91) have reported favorable results with metronidazole in 80% of 25 women with postpartum pelvic infection; the treatment failures were apparently related to concurrent or subsequent involvement of faculta-tive bacteria (which are uniformly resistant to metronidazole). Ingham *et al.* (984) reported three instances of serious *B. fragilis* infection which responded to metronidazole. One was a recurrent cerebellar abscess with two species of *Bacteroides* and an anaerobic streptococcus; this case re-sponded to drainage, instillation of lincomycin locally and oral metroni-dazole. The other two cases were *B. fragilis* septicemia secondary to urinary tract infection; both responded well, but one developed *Streptococcus faecalis* bacteremia while on therapy. Two additional cases of brain abscess treated with metronidazole were reported subsequently by Ingham *et al.* (1240c); these patients responded well, but the abscesses were drained and other chemotherapy was used. Metronidazole was used with good results in 2 cases of *Bacteroides fragilis* bacteremia reported by Baron *et al.* (596a). The successful use of metronidazole in a case of bacteremia due to an unusually resistant gram-negative anaerobic bacillus [Mitre and Rothcram (943a)] has been cited in Chapter 18.

Metronidazole apparently crosses the blood–brain barrier, but few specific data are available. A parenteral form is available for investigation. The United States Food and Drug Administration has not yet approved this drug for anaerobic infections. Dosage of metronidazole used in anaerobic infec-tions has varied between 250 and 750 mg three times daily.

Rifampin is also very active against many anaerobic bacteria, notable exceptions being *Fusobacterium mortiferum* and *F. varium, Clostridium ramosum*, and certain strains of *Eubacterium*. Certain clostridia other than *C. ramosum* may also be resistant. As in the case of facultative bacteria, populations of anaerobes not uncommonly contain mutants resistant to rifampin; it is therefore not likely that this compound would be suitable as a single agent for treatment of anaerobic infection. There is no clinical experience with this agent in anaerobic infections.

Aminoglycoside antibiotics (such as streptomycin, kanamycin, neomycin, gentamicin, and amikacin) are generally quite inactive against anaerobic bacteria (888, 933, 940). Indeed, anaerobic infections are not uncommon as superinfections in patients who have been treated with aminoglycosides by any route.

The activity of several other miscellaneous antimicrobial compounds against anaerobes has been noted in Chapter 18. There are no clinical data with regard to these other agents, and in general they do not look promising.

The penicillins, cephalosporins, and vancomycin all show good bactericidal activity but are inactive against *Bacteroides fragilis*. Clindamycin is bactericidal against a portion of *Bacteroides fragilis* strains; only metronidazole shows consistently good killing activity against this species.

Certain anaerobic organisms are of particular interest because of their relative resistance to antimicrobial therapy. *Bacteroides fragilis* is the most important of these inasmuch as it is not only the most resistant to antimicrobials of all anaerobes but is also much more commonly recovered from infections than any other anaerobe. Only clindamycin, chloramphenicol, and metronidazole are consistently active against it. The clinical dictum that anaerobic infections above the diaphragm may usually be treated effectively with penicillin is not really helpful in the individual case if the patient is quite ill. *Bacteroides fragilis* is not uncommonly involved in pleuropulmonary infections, central nervous system infections (particularly brain abscess), and chronic otitis media and sinusitis.

Clostridium ramosum is the second most resistant anaerobe, next to *B. fragilis*. Although all strains are susceptible to penicillin G, it may require as much as 6.2 U/ml to inhibit the more resistant strains. About 15% of strains of *C. ramosum* are highly resistant to clindamycin. Lincomycin is considerably less active than clindamycin. Many strains of this species are resistant to tetracycline and erythromycin. All strains are susceptible to achievable levels of chloramphenicol, vancomycin, and metronidazole, although as much as 25 μg/ml of metronidazole may be required for inhibition. This organism undoubtedly has frequently been overlooked in infections or misclassified, particularly as a non-spore-former.

The final organism requiring special mention is *Fusobacterium varium*, which is also relatively resistant compared to other anaerobes. Occasional strains are resistant to penicillin. Over one-third of 17 strains tested in our laboratory required 12.5 μg/ml or more of clindamycin for inhibition (over 10% required 50 μg/ml or more). Lincomycin is even slightly less active. Chloramphenicol and metronidazole are very active against *F. varium*, although relatively few strains (8 and 10, respectively) have been tested. This organism is encountered much less frequently than the two previously mentioned but may cause serious infection in essentially any part of the body.

In general, antimicrobial therapy of anaerobic infections must be prolonged to avoid relapse. This is related to the marked tendency for anaerobes to cause tissue necrosis, abscess formation, and septic thrombophlebitis.

Antitoxin Therapy

There is considerable disagreement as to whether or not it is desirable to give gas gangrene antitoxin. There is no unanimity of opinion as to its effectiveness, and those who oppose its use cite this as well as side effects that may attend its administration. This author is convinced, after reviewing the subject, that the efficacy of gas gangrene antitoxin was clearly demonstrated both during World War I and World War II. The question now would seem to be whether more modern methods of therapy such as more intensive penicillin therapy, better surgical and general care, and hyperbaric oxygen therapy might have improved the situation to the point where one could no longer demonstrate beneficial effects of antitoxin. Altemeier's studies (774) indicate that there is still a definite place for antitoxin therapy in gas gangrene.

Weinberg and Séguin [as quoted by Reed and Orr (985)], demonstrated beneficial effects of gas gangrene antitoxin during World War I. Only 11 of 30 patients with gas gangrene given polyvalent serum died, whereas among 66 other gas gangrene patients not given serum, 35 died. MacLennan (747), in his classical review of clostridial infections in 1962, stated unequivocally that the data from World War II concerning British wounded (as developed by MacFarlane and by MacLennan) provide definite statistical evidence of the benefit of antitoxin; the decrease in mortality approximated 40%. This was true even in patients in whom an immediate amputation was carried out and also in patients who received penicillin. MacLennan further stated that he was not aware of any thorough clinical investigation of gas gangrene which concluded that the use of antitoxin was unnecessary or undesirable. He stated that the adverse reports dealt either with small groups of cases or with cases that could not be documented as true gas gangrene, either on clinical or bacteriological grounds. MacLennan advocated local instillation of antitoxin where it is necessary to leave remnants of infected muscle. Altemeier (774) reported that among 42 patients with gas gangrene, 13 received gas gangrene antitoxin with a mortality of 30%, whereas a similar group of patients receiving no serotherapy had a 50% mortality.

The main purpose of using antitoxin is to prevent death from toxemia until surgery and antimicrobial therapy can control the infection. This presumes that antitoxin would neutralize toxin elaborated by multiplying organisms and provide time for the normal defense mechanisms and other therapy to deal with the organism. However, it must be kept in mind that the circulation in infected areas is very poor and that clostridial toxins are rapidly fixed. It is rare to find toxin in the blood in gas gangrene, and, indeed, it has been suggested that the toxemia is related to a secondary substance produced by a reaction between toxin and muscle tissue. If the latter is the case, then antiserum might act to prevent the evolution of this factor by specific inhibition of the clostridial toxins involved. While the mode of action of antitoxin is not clear, the fact remains that its use has been associated with a reduction in the mortality from gas gangrene, and it should therefore be used as early as possible in the course of the illness. The exact dosage will depend somewhat on the preparation available. In general, the polyvalent gas gangrene antitoxin contains antitoxin to *Clostridium perfringens, C. novyi,* and *C. septicum,* and sometimes to *C. histolyticum* and *C. bifermentans* also. Dosage is individualized according to the severity of the disease, response to treatment, and tolerance. In severe cases, eight to ten vials ("therapeutic doses") should be given initially, followed by two to four or more vials every 4 to 6 hours for 24 to 48 hours until toxemia is controlled. Initial therapy should be intravenous in a 1 : 10 dilution in normal saline or 5% dextrose, given as a slow drip. Subsequently it may be given intramuscularly. There is no tetanus antitoxin in the preparation so that either tetanus antitoxin or a toxoid booster must be given independently where this is indicated. The gas gangrene antiserum is a horse serum preparation and precautions must be taken against allergic reactions. Sensitivity testing must be carried out even though this may not be reliable in severely ill patients. The use of antitoxin may be associated with either local or systemic anaphylactic reactions and such delayed reactions as serum sickness or serum neuritis. Epinephrine and corticosteroids are useful in combating hypersensitivity reactions. Corticosteroids are beneficial in the serum neuritis as well [Crocker and Cunningham (986)]. The amount of fluid necessary as a vehicle for i.v. administered antitoxin may present problems in anuric patients.

There is less controversy about the need for gas gangrene antitoxin in the case of postabortal or other sepsis with *Clostridium perfringens* with intravascular hemolysis. The shock and hemolysis in this clinical entity is caused directly by the circulating α-toxin of *Clostridium perfringens.* Antitoxin should be given as early as possible in the course of the illness.

Antitoxin therapy, of course, is indicated in tetanus and botulism. This is covered in Chapter 17.

Miscellaneous Therapy

General supportive measures required in anaerobic infections will vary with the nature and severity of the infection. In serious infections, such as spreading gas gangrene, the need for good supportive measures is great. These may include blood transfusions, maintenance of fluid and electrolyte balance, adequate immobilization of the infected injured part, treatment of shock, relief of pain, management of renal failure, etc. Thrombophlebitis is frequently a prominent manifestation of anaerobic infection. Use of anticoagulants is indicated here. Because certain anaerobes may produce a heparinase, very large doses of heparin may be required [Tracy *et al.* (674)]. Persistent fever, despite adequate antimicrobial therapy, may respond to anticoagulant administration, particularly in pelvic thrombophlebitis [Schulman (702)]. As indicated under surgical therapy above, excision of clots and/or ligation of veins may be indicated at times. Heparin therapy may also be of value in the case of disseminated intravascular coagulation, but this is controversial.

Exchange transfusion has been recommended in postabortal or other septicemia due to *Clostridium perfringens* when there is significant intravascular hemolysis [Strum *et al.* (665)]. The exchange transfusion may markedly reduce the plasma hemoglobin concentration, effectively remove other toxic red blood cell breakdown products, and reduce the concentration of α-toxin of *Clostridium perfringens* in the plasma. Exchange transfusion may also be very useful in the case of disseminated intravascular coagulation [Rubenberg *et al.* (425)] by removing spherocytic red blood cells and fibrin split products from the circulation, in addition to free hemoglobin. Since renal failure may also be an important problem in *Clostridium perfringens* sepsis and other serious anaerobic infections, Smith *et al.* (664) have recommended early placement of an arteriovenous shunt, which may be used both for exchange transfusion and for hemodialysis.

Oxygen therapy is often useful. Systemic administration of oxygen by nasal mask or nasal prongs may be helpful. Hyperbaric oxygen will be discussed in a separate section below. Topically applied oxygen may be remarkably beneficial. This may be carried out by direct irrigation of the affected area with oxygen or, more commonly, by use of hydrogen peroxide or zinc peroxide locally. Oxygen is released by tissue catalases when peroxides are used locally. Most people have used 3% hydrogen peroxide but concentrated solutions of 30% have been used in the past [Van Wering (987)]. Meleney (988), championed the use of zinc peroxide. This type of simple therapy, often overlooked in the antimicrobial era, will at times make a distinct difference in a stubborn anaerobic infection. Hydrogen peroxide may also be used systemically, although there is hazard. Finney *et al.* (989) found that intraarterial infusion of hydrogen peroxide was beneficial in treating experimental clostridial myositis in rabbits. Regional intraarterial hydrogen peroxide infusion has been used in man in connection with radiation in the treatment of malignancies [Mallams *et al.* (990)]. Eaton and Peterson (991) have suggested the use of intraarterial infusion of 0.2% hydrogen peroxide in normal saline, by means of a transfemoral catheter placed just above the bifurcation of the aorta, in cases of serious clostridial sepsis complicating abortion when a hyperbaric oxygen chamber is not available. Arterial spasm and hemolysis are potential hazards of intraarterial hydrogen peroxide therapy.

The future may see new therapeutic approaches in anaerobic infections. The studies of McCabe *et al.* (992) suggest that it may become possible to immunize against endotoxin of gram-negative organisms. If this is so, it may also be possible to use preformed antiserum in passive therapy of endotoxin shock in the course of selected cases of gram-negative anaerobic bacteremia. Lynch and Moskowitz (993) and Senff and Moskowitz (994) have used various chelates in the chemotherapy of experimental gas gangrene toxemia. Ethylenediamine tetraacetate (EDTA) is found to inhibit the activity of the α-toxin of *Clostridium perfringens*, apparently by binding certain metal ions that normally enhance the activity of the toxin. This compound has been used in humans as a chelating agent. Another compound, diethylenetriamine pentaacetic acid (DTPA), is 10 to 20 times more efficient than EDTA.

Hyperbaric Oxygen Therapy

Since other forms of oxygen therapy have proved of value in anaerobic infections, it was natural to consider that additional benefit might be obtained by the use of oxygen under increased pressure. This would permit a significant increase in the amount of oxygen physically dissolved in the plasma and tissue fluids.

Several workers, but not all, have shown inhibitory and even bactericidal effects on *Clostridium perfringens* exposed to oxygen under increased pressure. Nuckolls and Osterhout (995) showed that log phase cultures of *C. perfringens* were more rapidly inactivated by oxygen than stationary phase cultures and that this effect of oxygen was reduced by the presence of blood. Kaye (996) showed that the bactericidal effect of oxygen on *C. perfringens* was decreased by the presence of blood or muscle and was less at 25°C than at 37°C. *Clostridium tetani* was similarly affected by high pressure oxygen. Fredette (997) also showed an effect of oxygen on *C. perfringens*, but showed that *C. septicum* was virtually resistant to this effect and that tetanus toxin was not affected by hyperbaric oxygenation. He also showed that an anaerobic streptococcus, *Streptococcus parvulus*, was not affected by hyperbaric oxygen. In a later study, Fredette (998) showed some inhibition of *Streptococcus parvulus* by oxygen under pressure and that another anaerobic streptococcus, *Streptococcus micros*, was more sensitive to oxygen. Hill and Osterhout (999) found hyperbaric oxygen to be bactericidal for *Clostridium perfringens*, *Clostridium novyi*, *Clostridium histolyticum*, and *Clostridium tetani*. *Clostridium bifermentans* and *Clostridium septicum* were more resistant. Spore suspensions of *C. perfringens*, *C. histolyticum*, and *C. bifermentans* were resistant to oxygen at 3 atm absolute pressure for a period of 18 hours. The α-toxin of *Clostridium perfringens* is not inactivated by hyperbaric oxygen, but Van Unnik [as quoted by Hill and Osterhout (999)] did show that production of this toxin is inhibited during exposure of the organisms to hyperbaric oxygen.

In animal experiments, Nuckolls and Osterhout (995) found a decrease in the growth of *Bacteroides* in agar disks implanted subcutaneously in mice when the animals were exposed to hyperbaric oxygen. Hunt *et al.* (1000) reported on the use of hyperbaric oxygen in experimental infections in rabbits. A lethal infection was obtained by using a mixture of *C. perfringens* and *C. novyi.* Hyperbaric oxygen at 2 atm failed to modify the mortality rate but studies with animals treated at 3 atm of oxygen, started immediately after initiation of the infection, revealed some protection. Kaye (996) found that mice injected intramuscularly (in normal muscle) with *Clostridium perfringens* could be protected by hyperbaric oxygen; this was true even if earth was injected along with the organism. However, if calcium chloride was injected in place of earth along with *C. perfringens* into either normal or crushed muscle, there was no longer a protective effect of oxygen. Hill and Osterhout (999) found that *Clostridium perfringens* inside small agar discs implanted in mice were killed when the mice were exposed to hyperbaric oxygen at 3 atm. Inclusion of blood or muscle in the discs partially protected the organism. The mortality of mice given intramuscular injections of *C. perfringens* with epinephrine was markedly reduced by intensive exposure to hyperbaric oxygen at 2 or 3 atm. When calcium chloride was injected along with the organisms, there was less protection from hyperbaric oxygen. Hitchcock *et al.* (774a) cited earlier studies by their group (Demello *et al.*) which involved intramuscular injection of mice with spores of *C. perfringens* plus epinephrine and delay of treatment until 12 hours following onset of infection. A combination of surgical debridement, antibiotics, and hyperbaric oxygen produced a survival rate of 95%, whereas the three modalities of therapy used individually or in pairs gave survival rates of 0 to 70%. Hill (1000a) obtained definite benefit, and at times cure, from the use of hyperbaric oxygen in mice with experimental liver abscesses due to gram-negative anaerobic bacilli. Bornside *et al.* (1000b) found that hyperbaric oxygen raised intracolonic oxygen tension significantly in the rat and that prolonged exposure (4 hours) prevented subsequent recovery of both anaerobes and coliforms from the rat colonic flora.

The foregoing studies suggest that hyperbaric oxygen might be useful in the treatment of gas gangrene, and possibly other anaerobic infections, on the basis of some interference with the organism itself and with the production of one major toxin by the organism. Raising the oxygen tension in the area of an infection and adjacent tissues may be even more important in combating or limiting the spread of this type of infection.

In 1962, Brummelkamp (1001) reported on the treatment of 21 cases of gas gangrene with oxygen under a pressure of 3 atm. This series was expanded in future reports [Brummelkamp *et al.* (1002) and Brummelkamp (1003)]. The 1966 report is all inclusive and covers 55 patients. Various types of clostridial infection, including postabortal sepsis, were treated. It is stated that most patients had clostridial myositis, although the author acknowledged that, since surgery was not performed before submitting the patients to hyperbaric oxygen therapy, they could not tell initially which patients had myositis and which had only clostridial cellulitis. Altogether, 10 of the 55 patients died. In two cases, death was due to gas gangrene, and in two others it was felt that death was probably indirectly attributable to gas gangrene. There were three cases of oxygen toxicity, all in patients with high fever. While most patients with extensive or deep-seated gas gangrene died of one cause or another, there were a number of cures in patients with serious disease, such as endometritis plus sepsis and gas gangrene extending to the groin, buttock, shoulder, and abdominal and thoracic wall. Slack and co-workers (1004, 1005) reported on the treatment of a number of cases of clostridial infection, including gas gangrene. They utilized a one-man transparent oxygen chamber at pressures of 2 or 2.5 atm. Results were generally good. In the 1969 report, 9 of 32 patients with various clostridial infections died, although none died from the clostridial infection per se.

Duff *et al.* (1006) reported on the use of hyperbaric oxygen for a variety of indications in 83 patients. Six of these patients had gas gangrene, and four survived. In a later report from this group [Duff *et al.* (952)] they discussed a series of 29 patients with gas gangrene. They concluded in this later study that surgical debridement was the mainstay of treatment and seemed to influence the course of infection more than other forms of therapy, including hyperbaric oxygen and antibiotics. Van Zyl *et al.* (1007) reported on 29 patients treated with hyperbaric oxygen in a one-man chamber. No clostridia could be cultured from 13 of these patients, but they were nevertheless selected for treatment in the chamber because the authors felt that the clinical picture was so suggestive of gas gangrene. Among the 16 cases considered to be proved clostridial gas gangrene, five died, three of causes other than gas gangrene. Van Zyl (1008) referred to a survey of the use of hyperbaric oxygen in the treatment of clostridial gas gangrene made by questionnaires sent to various workers active in the field. A total of 170 cases was collected, 43 of whom died. It was felt that 23 of these died as a direct result of the gas gangrene. Other reported series of clostridial infections, including gas gangrene, treated by hyperbaric oxygen include those of Van Elk *et al.* (1009) and Arnar *et al.* (521). Davis *et al.* (1010) reported on 23 cases of diffuse, spreading clostridial myonecrosis treated with hyperbaric oxygen; five of these, moribund on arrival at the treatment facility, died. Davis *et al.* (1010) summarized a total of 267 cases of gas gangrene treated with hyperbaric oxygen from the literature. The overall mortality was

23%; the mortality in "salvageable" cases was 9%. Jackson and Waddell (1010a) felt that hyperbaric oxygen was responsible for the lower mortality in 15 patients with clostridial myonecrosis receiving this therapy plus surgery and antibiotics, as compared with 9 patients treated in an earlier period with just the latter two modes of therapy. Hitchcock *et al.* (774a) also felt hyperbaric oxygen was very effective in 89 patients they treated (as compared with 44 patients not receiving such therapy). Many of their patients had "minimal involvement of muscle" and probably represented necrotizing fasciitis and/or what has variously been described as cutaneous gangrene, necrotizing cutaneous myositis, or synergistic necrotizing cellulitis. Hart *et al.* (1011) reported on 54 cases of histologically and bacteriologically proved clostridial myonecrosis treated with hyperbaric oxygen in a single person chamber. The mortality in 36 patients with no underlying disease was 5.5%; it was 27% in 15 patients who suffered the disease as a postoperative complication. Six patients with spontaneous occurrence of the disease, five of whom were diabetics, had an 85% mortality. Mortality in patients over 30 years of age was 50%.

Unfortunately, there are no controlled studies evaluating the effect of hyperbaric oxygen in gas gangrene, and it might prove impossible to do properly controlled studies. Therefore, it is very difficult to evaluate the reports in the literature. One of the primary reasons for this is the fact that satisfactory clinical and bacteriological evidence of gas gangrene is often not provided. It is well known that clostridia may be present in a wound without contributing to infection and that clostridial infection may occur without producing muscle necrosis which is the *sine qua non* of gas gangrene. Also, on occasion, organisms other than clostridia may produce muscle necrosis in the course of infection. Altemeier and Fullen (1012) stated that the difficulties in diagnosis of gas gangrene are such that without operative intervention and bacteriological studies, such cases must be considered unproved. Furthermore, almost invariably (and appropriately), patients who have been treated with hyperbaric oxygen have also received many other modalities of therapy, so that it becomes difficult to judge the relative importance of each of the components of the therapeutic regimen. Furthermore, it is difficult to predict the likelihood of response in a given case of gas gangrene. In addition to all this, a number of workers have achieved good results in gas gangrene without the use of hyperbaric oxygen. The most outstanding series in this connection is that of Altemeier and Fullen (1012) in which 54 cases of gas gangrene were treated with a mortality of 14.8%. These excellent results were achieved by early adequate debridement of wounds together with penicillin or other antimicrobial therapy, polyvalent gas gangrene antitoxin, and general supportive therapy. The study of Duff *et al.* (952) in which these workers concluded that surgical debridement was the mainstay of treatment and influenced the course of gas gangrene more than other forms of therapy, including hyperbaric oxygen, has been referred to; these workers had had extensive experience with hyperbaric oxygen themselves. In postabortal *Clostridium perfringens* sepsis, Smith *et al.* (664) concluded from a small series of patients that hyperbaric oxygen therapy was not satisfactory in their hands. Finally, it is of interest to note the excellent results achieved by McNally *et al.* (515) and others in the management of gas gangrene of the anterior abdominal wall by surgical debridement without hyperbaric oxygen therapy.

The hazards of hyperbaric oxygen therapy should not be underestimated. During compression, ear equilibration has been the most disturbing problem. When necessary, this can be handled by needle myringotomy; some centers use myringotomy routinely. Particularly at higher pressures, there may be problems with acute oxygen intoxication with such impressive cerebral complications as confusion and convulsions. Pulmonary side effects related to oxygen toxicity may also be significant. With decompression, there is the danger of "bends." While the Naval Diving Decompression tables may be used as a guide for decompression, it must be remembered that these tables were derived from experience with normal healthy men. Presently, many hyperbaric units have limited facilities for patient care. Prolonged treatment schedules tend to isolate the patient from conventional hospital diagnostic and therapeutic equipment. The workload for the attending staff is enormous if a major recompression is required. Personnel in the chamber may be confined for extended periods of time. Also, certain of the hazards to the patient such as the "bends" are assumed by personnel in the chamber as well. Some of these hazards may be avoided by the use of one-man chambers in which only the patient is housed and by the use of lower pressures. However, some workers still feel that 3 atm of pressure are necessary for good results. Central nervous system oxygen toxicity may appear some time after termination of the exposure [Fuson *et al.* (1013)]. Management of toxic reactions may be complicated by the fact that central nervous system signs of oxygen toxicity are nonspecific and can be misinterpreted readily, particularly in patients who have serious underlying diseases. There are similarities between the neurologic manifestations of acute oxygen toxicity and decompression illness. The therapy is distinctly different for each condition.

The report of Perrin *et al.* (666) concerning the use of hyperbaric oxygen in postabortal clostridial sepsis, is noteworthy. However, the use of hyperbaric oxygen in other types of anaerobic infections, has not always been as impressive. In the study of Slack *et al.* (1005) three of seven patients with crepitant infections mimicking clostridial gas gangrene died in spite of hyperbaric oxygen therapy. Rō *et al.* (1014) reported a case of gas-forming cellulitis of an extremity in a diabetic due to *Bacteroides melaninogenicus* and enterococci; although the patient seemed to show some response to hyperbaric oxygen therapy, he ultimately died. Parker and Jones, (720) discussed two cases of nonclostridial anaerobic infection treated with hyperbaric oxygen therapy. One of these was a patient with hidradenitis of the vulva with anaerobic cocci recovered on multiple cultures; this patient did improve some with hyperbaric oxygen therapy, but the results were not particularly dramatic. The second patient was a 17-year-old girl who had a ruptured pelvic abscess and generalized peritonitis and shock due to *Sphaerophorus*. This patient had both central nervous system and pulmonary toxicity due to the hyperbaric oxygen treatment and died with extensive pulmonary changes. *Sphaerophorus* was recovered from the peritoneal cavity at autopsy. This patient was also reported by Fuson *et al.* (1013). Wallyn *et al.* (1015) used hyperbaric oxygen therapy in treatment of a patient with a very large liver abscess from which was grown *Bacteroides*, *E. coli*, and a possible *Clostridium*. Although there was a distinct reponse to this therapy, the patient subsequently developed sepsis with staphylococci and died. Irvin and Smith (1016) obtained excellent results with hyperbaric oxygen therapy in a case of progressive bacteral synergistic gangrene of an abdominal wound from which a microaerophilic streptococcus and *Streptococcus faecalis* were recovered on culture. Schreiner *et al.* (1011a) reported the successful use of hyperbaric oxygen in two patients with *Bacteroides fragilis* infections, one a bacteremia and the other a wound infection also involving anaerobic streptococci; both patients had failed to respond to antibiotic therapy (of inappropriate type in at least one of the cases).

Although Pascale *et al.* (1017) reported favorable results with the use of hyperbaric oxygen in nine patients with tetanus, subsequent studies [Milledge (1018) and others reviewed by Meljne (1019)] indicated that hyperbaric oxygen is really not effective in this disease.

Fischer (1020) obtained some beneficial results with the use of topical hyperbaric oxygen in patients with decubitus ulcers that are frequently secondarily infected with anaerobic bacteria (no specific cultural data was given in the cases cited).

In summary, there are no definitive studies to document the effectiveness of hyperbaric oxygen therapy in anaerobic infections. There are many difficulties associated with the use of this therapy, and the hazards are not to be overlooked. Hyperbaric oxygen is not a substitute for any of the conventional modalities of therapy. Gas gangrene, particularly of the acute spreading variety, and postabortal clostridial sepsis may be indications for hyperbaric oxygen therapy. However, surgical debridement and decompression remain by far the most important elements of therapy of gas gangrene and should not be delayed pending hyperbaric oxygen treatment. Intensive antimicrobial therapy and polyvalent gas gangrene antitoxin should also be used. While hyperbaric oxygen therapy must be considered as experimental still, its use is justified when proper facilities are available along with trained personnel and when it is used only as a supplement to conventional therapy for these infections. It may be most important in helping to demarcate and limit the spread of gas gangrene and may permit delay in excision of tissues of uncertain viability (1021).

Prognosis

The prognosis of anaerobic infections, as with infections in general, varies with the nature and site of the infection, the age and general condition of the infected patient, the nature and resistance of the organisms causing the infection, and the speed with which the infection is diagnosed and treated. Some average mortality figures for selected types of serious anaerobic infections during the antimicrobial era follow.

The mortality with gas gangrene ordinarily varies between 15 and 35% [Duff et al. (952) and Altemeier and Fullen (1012)]. The mortality in gas gangrene is worse with involvement of large muscle groups, such as those of the buttock, thigh, and shoulder, or involvement of other areas that are hard to debride such as the viscera or the pelvis. The mortality in postabortal clostridial sepsis is still 50 to 85% [Smith et al. (664), Strum et al. (665), and Rubenberg et al. (425)]. In tetanus, the mortality is still 55 to 65%; in botulism, the mortality varies widely but averages about 25%.

Brain abscess has a mortality greater than 50% [Braude (93a)]. Anaerobic pulmonary infections have an overall mortality of 15 to 20%, with a distinctly higher mortality in the case of necrotizing pneumonia—greater than 30% [Bartlett and Finegold (161)]. Barbour and Juniper (948) summarized several series of liver abscess from the literature (including two series from the preantibiotic era) and concluded that the average mortality was 55%. The mortality in their own series, which was in the antibiotic era, was 79%. Altemeier et al. (949) noted a mortality of 28% in cases discovered in their institution from 1965 to 1969. However, four of these patients were discovered only at autopsy. There was only one fatality among 14 cases diagnosed during life. Similar findings were noted by Sabbaj et al. (387); 6 of 25 patients in this series died, but only 1 of 20 diagnosed during life had a fatal outcome.

In a report published in 1956, Gunn (296) reviewed approximately 150 patients with Bacteroides sepsis from the literature. In the presulfonamide era, the mortality was 81%. With sulfonamide therapy, the mortality was reduced to 64%, and with the use of multiple antibiotics in later years, the mortality was 33%. The latter figure is verified by a number of reports in the recent literature. Ten of 35 patients with *Bacteroides* sepsis studied by DuPont and Spink (647) died. In the series reported by Bodner *et al.* (460) 15 of 39 patients (38%) died. Kagnoff *et al.* (405) reported a series of *Bacteroides* sepsis from a cancer hospital. Of the patients with *Bacteroides* sepsis 91% had a neoplasm; 25 of 55 patients died, 17 of sepsis. Tynes and Frommeyer (489) noted a 20% mortality in 25 patients with *Bacteroides* sepsis. In a study encompassing 123 patients, Marcoux *et al.* (420) had a mortality rate of 28.5%. In a larger series collected from various hospitals that submitted organisms to the Center for Disease Control for identification, Felner and Dowell (168) noted a 32% mortality in 250 patients. Ellner and Wasilauskas (740) reported a group of patients with *Bacteroides* sepsis and analyzed results in terms of age. Fourteen young females all survived, whereas among 18 patients over 45 years of age, seven (39%) died. The impact of appropriate therapy is emphasized in a report by Chow and Guze (1022). In this report, 16% of 37 patients treated with chloramphenicol for *Bacteroides* sepsis died, whereas 62% of 29 patients receiving inappropriate therapy succumbed. Wilson *et al.* (105) reported 67 patients with sepsis due to anaerobic bacteria; 78% of the organisms recovered were from the family Bacteroidaceae. The overall mortality in the series, and among those with *Bacteroides* sepsis as well, was 31%. The prognosis was better in those patients who had incision and drainage of the lesions which were the portals of entry for the bacteremia. Patients with shock had a much poorer prognosis, 15 of 19 dying.

In patients with endocarditis due to gram-negative anaerobic bacilli the overall mortality was 35%, whereas mortality in cases due to *Bacteroides fragilis* was 45% [Nastro and Finegold (433)]. In cases due to gram-negative anaerobic bacilli sensitive to penicillin *in vitro* the mortality was only 15% [Felner and Dowell (289)].

Addendum

The sole manufacturer of polyvalent gas gangrene antitoxin in the United States had decided to stop manufacturing this product. There is no other source for the product in this country and the Center for Disease Control has no plans to make it available.

20

Prophylaxis

General

In view of the significant mortality and morbidity attending anaerobic infection, it is clearly desirable to prevent such infections whenever possible. The major principles in prophylaxis are (a) to avoid conditions that reduce the redox potential of the tissues and (b) to prevent the introduction of anaerobes from the normal flora of the body or from the outside into wounds, closed cavities, or other sites where they may set up infection. In tetanus, prophylaxis is also aimed at protection against the toxin produced by the organism. Proper treatment of already established anaerobic infections may in itself prevent further problems with anaerobic infection. Similarly, adequate medical treatment of anaerobic infection will prevent metastatic disease due to these organisms. Sandler (1023) described two cases of fatal brain abscess in patients who previously had anaerobic pulmonary infection and who were discharged as cured.

Bathing causes a significant reduction in skin organisms of the transient type (such as *Clostridium perfringens*). A shower seems to bring about an even greater reduction in skin bacterial count than tub bathing [Drewett *et al.* (1024)]. Various germicides may be useful in certain situations. Unpublished data from our laboratory reveals that benzalkonium chloride is effective against non-spore-forming anaerobic bacteria. Formaldehyde rapidly sterilizes anaerobic cultures, including spore formers, but phenol acts very slowly [Scott (1025)]. Iodophors [such as povidone iodine (Betadine)] are sporicidal and may be used effectively in removing *Clostridium perfringens* from skin, provided there is sufficient contact time. Drewett *et al.* (1024) showed that the use of iodophor compresses for 15 minutes resulted in inability to isolate *Clostridium perfringens* from 60 patients from whom the organism had previously been recovered from the skin over the hips. Mironova (1026) found that 6% hydrogen peroxide solution was an effective disinfectant with good activity against clostridia.

In traumatic wounds, the most effective prophylaxis is thorough debridement and cleansing of the wound, elimination of foreign bodies and dead space, and reestablishment of good circulation. This type of wound management must be carried out as early as possible following injury. Prophylactic antimicrobial agents given systemically have not ordinarily proved helpful except where there must be significant delay before definitive debridement and cleansing of the wound can take place, as in battlefield wounds. Experience during the war has shown that the use of antibiotics prophylactically offered a valuable 6 hours of grace, permitting transport of the wounded to areas where effective debridement could be carried out. Prophylaxis of gas gangrene and tetanus are discussed later in this chapter. At the time of debridement, the material from the wound should be gram stained and cultured, both aerobically and anaerobically. In this way, if infection does occur, information concerning likely infecting organisms will be at hand early in the course of the infectious process. Topical disinfectants may be useful in treating traumatic wounds but should not be used as a substitute for good surgical management and vigorous cleansing. In the case of penetrating wounds of the abdomen, particularly with wounds of the colon, two recent studies have shown beneficial effects from the use of antibiotics [Fullen *et al.* (1027) and Thadepalli *et al.* (953)]. This, in effect, may be considered therapy rather than prophylaxis. In the paper by Fullen and co-workers, the results were much better in those patients who received their initial antimicrobial therapy preoperatively as compared to those receiving it during surgery or postoperatively. As would be expected in this situation, regimens including drugs active against anaerobes proved most effective.

It is remarkable to note the extent to which the normal flora of the body may be reduced by a combination of antimicrobial therapy, a diet low in bacterial count, and isolation of the patient in a special protected environment [Levine *et al.* (1028)]. In this study, patients with acute leukemia who were to be treated intensively with antitumor chemotherapy had their normal flora reduced by the means just described. One hundred and thirty-nine stool cultures were performed on 26 patients handled this way, with 88% showing no anaerobic growth. Actually, the antimicrobial regimen alone (oral genta-micin, vancomycin, and nystatin) effected a comparable reduction in anaer-obic flora, but the effect on the aerobic and fungus flora was not as good. With the total regimen (antimicrobial agents, isolation, low bacteria diet, as well as topical and orificial treatments), it was possible to eliminate an-aerobic organisms from 62% of the cultures of the inner nose, 78% of those of the inguinal area, 44% in the anal area, and 68% of vaginal cultures. However, only 3% of the cultures of the mouth and oropharynx failed to yield anaerobes. There were significantly fewer infections among patients receiving the full regimen of prophylactic antibiotics and protected environ-ment as well as topical applications. While anaerobic infections were not mentioned specifically in this particular paper, it is clear that anaerobic infections are important in this population.

Prevention of Specific Anaerobic Infections

The paper of Sandler (1023) referred to above emphasizes the need for intensive and prolonged therapy of anaerobic infections in order to minimize the possibility of relapse or metastatic infection. This is particularly true if there is to be surgical drainage, since this may also increase the risk of spread. There is no evidence that routine prophylactic use of antimicrobial agents after clean neurosurgical procedures of any type has resulted in decreased infection rates [Wright (1029)]. Even in types of neurosurgical procedures with a high risk of infection (reexploration of a supratentorial craniotomy, lengthy supratentorial operative procedures, or cases performed in the hypothermic state), there has been no evidence to indicate that prophylactic antibiotic therapy reduces the risk of infection [Wright (373)]. When the paranasal sinuses are entered during craniotomy, these openings must be closed meticulously to prevent infection by this route later. Wright (1029) recommends reflecting the mucosal lining downward into the exposed sinus, then packing the cavity with gelatin foam and reflecting a flap of adjacent pericranium across the opening and suturing this to the dura.

Vigorous treatment of chronic otitis media, sinusitis, and mastoiditis may prevent subsequent serious intracranial anaerobic infection. The presence of cholesteatoma is a grave sign, since this provides both anaerobic conditions and necrotic tissue.

It is well established that anaerobic bacteremia very commonly follows dental manipulations of one type or another [Rogosa et al. (374) and Khairat (1030)]. Bacteremia of all types following dental surgery or manipulation is, for the most part, not clinically significant as such. Its major importance lies in the fact that people with certain types of underlying cardiac disease or valvular prostheses may develop endocarditis as a result of such bacter-emia. In general, the incidence of bacteremia is higher in the face of poor oral hygiene or periodontal disease and with more extensive surgical proce-dures, although there does not seem to be general agreement on these points. Various antimicrobial agents, including penicillin, ampicillin [Tolman *et al.* (381)], erythromycin [Khairat (1030)], tetracycline, and lincomycin [Francis and DeVries (149)], have been used to reduce the incidence of bacteremia. Of these penicillin, ampicillin, and lincomycin have been the most effective. Clindamycin would also be an appropriate choice. It is my feeling that such prophylactic therapy should start only 1 to 2 hours before dental manipula-tion, that these agents should be given in full dosage (procaine penicillin G, 600,000 U 1 hour before dental manipulation and 12 and 24 hours later, lincomycin 2 gm per day, or clindamycin 1.2 gm per day) and that this prophylaxis be kept up for a total of 24 to 48 hours. The combination of penicillin plus an aminoglycoside or the use of vancomycin provided the most effective regimens for prevention of experimental endocarditis (1030a). Lincomycin was also shown in one study to decrease fever, malaise, edema, necrosis, and pain following periodontal surgery, as compared with a placebo [Ariaudo (1031)].

Precautions to minimize the possibility of aspiration will be helpful in preventing anaerobic pulmonary infection. Thus, patients should be kept in the head-down position during and immediately following tonsillectomy. Care should be observed in feeding feeble or confused patients and patients with swallowing difficulties. The head of the bed should be elevated for a time after feeding by gastric tube. When surgery is required in the face of, or for, bronchiectasis or suppurative pulmonary infection, prophylactic penicillin lowers the morbidity significantly [White *et al.* (1032)].

With regard to postoperative peritonitis or other intraabdominal infection, good surgical technique is highly important in minimizing the risk of such infection. Injury or devitalization of intraperitoneal tissues must be reduced to a minimum by gentle handling of the intestines and peritoneum and by avoidance of strangulation and of large bites of tissue by ligature. The use of sharp dissection, when possible, and preliminary decompression before intestinal resection in the presence of obstruction are important considera-tions. Careful aseptic technique during surgery includes the use of secondary isolation procedures during colonic resections, the use of closed methods of resection and anastomosis when possible, preservation of adequate blood supply, and the avoidance of tension at anastomotic sites.

Presently there is much disagreement concerning the role of prophylactic antimicrobial therapy in this setting. Everyone agrees that prior to bowel surgery patients should undergo mechanical bowel preparation with low residue diet and then, subsequently, a liquid diet, cathartics, and enemas.

Nichols *et al.* (1033) noted that, although this does reduce the total fecal mass, the numbers of anaerobic and microaerophilic intestinal bacteria are not significantly altered in the remaining intestinal contents. Many people recommend the use of so-called intestinal antisepsis prior to bowel surgery. The agent most commonly used for this purpose in recent years has been oral neomycin. From the standpoint of the anaerobic bacteria in the bowel, this is clearly inappropriate, since the anaerobes are very resistant to neomycin (which is actually used in selective media to isolate anaerobes from mixtures in the laboratory), and anaerobes may persist in high count following such therapy [Finegold *et al.* (56), Baird (324b), Azar and Drapanas (511), Washington *et al.* (56j), and Nichols *et al.* (1035a)]. Anaerobic infections, including bacteremia, have also been noted in patients receiving oral neomycin or oral kanamycin before colon surgery (419d, 492). Furthermore, there is the hazard of overgrowth of resistant organisms such as *Staphylococcus aureus* [Altemeier *et al.* (1034) and Gaylor *et al.* (1035)]. Other potential pathogens such as *Pseudomonas* and *Klebsiella* appeared in significant numbers either during the period of administration of oral neomycin (or kanamycin) or immediately following discontinuation of this agent, before the normal flora was restored [Gaylor *et al.* (1035)]. With regard to the persistence of anaerobes following oral neomycin preoperative bowel preparation, one could add an additional agent, such as erythromycin or lincomycin, to suppress significantly or eliminate these forms. Two groups [Washington *et al.* (56j) and Nichols *et al.* (1035a)] have studied this matter and obtained many fewer anaerobes from the bowel lumen of subjects receiving either tetracycline or erythromycin in addition to neomycin than from subjects receiving only neomycin. As noted in Chapter 10, Washington *et al.* found that the incidence of anaerobes in postoperative wound infections paralleled their presence in the lumen. In the study of Nichols *et al.*, none of 69 patients treated preoperatively with oral neomycin and erythromycin developed wound infection. However, it must be kept in mind that prophylactic antimicrobial therapy, in general, has not proved effective over the long run, except in special situations where a very sensitive organism such as the group A β-hemolytic streptococcus can be treated with a benign agent that is consistently highly effective against it (such as penicillin G), particularly when this agent can be given in a dosage that produces minimal or no change in the normal endogenous flora of the body. Our experience with combinations of antimicrobial agents in an attempt to eliminate the bowel flora totally (for short periods of time) [Finegold *et al.* (56)] indicated that one type or another of resistant organism is likely to colonize the bowel in such a situation.

This author does not subscribe to the use of intraperitoneal kanamycin or other antimicrobial agents for the purpose of preventing or treating peritonitis. In general, when indicated, systemic antimicrobial agents should be more effective and less likely to produce side effects. What about the question of prophylactic systemic antibiotics then? We have already cited studies showing striking beneficial effects from the use of "prophylactic" antibiotics in the case of penetrating wounds of the abdomen, particularly the bowel. While, as indicated, this may really represent therapy rather than prophylaxis, the analogy to bowel surgery is evident. Certainly if there is spillage of bowel contents into the peritoneal cavity during surgery, or if there is a penetrating carcinoma of the bowel, prophylactic antimicrobial therapy is indicated. The study of Fullen *et al.* (1027) indicates that such therapy would best be given shortly prior to surgery when the difficulties can be anticipated. In terms of the anaerobic flora, clindamycin would be a good choice for prophylaxis supplemented perhaps with penicillin G or ampicillin to cover resistant anaerobes such as certain clostridia. Coverage against facultative bacteria should also be considered. Whether or not a short course of systemic therapy of this type, starting 2 hours prior to surgery and continuing for 2 or 3 days postoperatively, would be beneficial in the absence of any complications of bowel surgery is open to question. A prospective double-blind randomized study of this possibility is certainly in order. Stone *et al.* (384d) failed to obtain better results in patients undergoing emergency celiotomy with the prophylactic use of clindamycin than with cephalothin (3% abdominal infection rate and 14–15% wound infection rate with each). However, in an earlier period, cephalothin was given routinely (not randomized with clindamycin); the abdominal infection rate was 8% and the wound infection rate 26%. Assuming the patient populations given each drug in the randomized trial were truly comparable, the results can be interpreted in several ways: (a) reduction of anaerobes and of aerobes is of equal importance in preventing postoperative infection (and therefore use of agents to cover both types of organisms might give still better results); (b) neither regimen was effective; (c) the surgical technique was so good that results could not be improved with the use of antimicrobials; or (d) the study does not have statistical validity (particularly in view of the different results obtained with cephalothin in the two time periods).

Gilmore *et al.* (1035b) and Gilmore (1035c) found that routine use of a povidone iodine dry powder aerosol spray prior to appendectomy reduced the incidence of wound infection postoperatively (450 cases). Leigh *et al.* (585a) noted that the rate of wound infection in patients with perforated appendix was 50% if no chemotherapy or ineffective drugs were used and only 15% when appropriate therapy was used.

In gallbladder surgery, one should always keep in mind the possibility of serious clostridial complications. The hazard of this may be minimized by avoiding trauma to the liver bed and by the prophylactic use of penicillin, which carries little risk of superinfection with resistant organisms.

The important conditions predisposing to puerperal sepsis are premature rupture of the membranes, prolonged labor, and postpartum hemorrhage. If labor continues long enough with ruptured membranes, clinical signs of infection will always occur, even in the absence of pelvic examinations or other possible outside sources of infection. On the other hand, if no labor develops following premature rupture of the membranes and the patient is kept at rest in bed, infection is unlikely to occur. Prolonged labor should be anticipated when possible and analgesics employed judiciously. With rare exceptions, Caesarean section should be performed electively or after not more than 8 to 12 hours of labor. During labor, particularly with ruptured membranes, pelvic and rectal examinations should be kept to a minimum. During delivery, trauma should be minimized and lacerations repaired with accepted surgical principles. Retained portions of the placenta should be removed immediately following the third stage of labor. Excessive blood loss or anemia from other cause should be corrected immediately postpartum.

Prophylaxis against postoperative pelvic infection, aside from the question of antimicrobials, is discussed in detail in Chapter 12. Ledger (1036) presented an excellent discussion on the use of prophylactic antimicrobial therapy in obstetrics and gynecology. He pointed out that the new strategy of using the drugs prior to surgery, as well as postoperatively, has led to good results in selected cases. Because of the risk of developing an antimicrobial-resistant flora in a closed hospital environment, he favored using prophylaxis only in situations that represent a high risk for development of serious infections. He included in this category premenopausal women undergoing vaginal hysterectomy, Caesarean sections in women with rupture of membranes and in labor, and patients undergoing radical pelvic surgery for gynecological malignancy. Several reports indicate that one may decrease the number of serious postoperative pelvic abscesses following vaginal hysterectomy by using prophylactic ampicillin, cephaloridine, tetracycline, clindamycin, or metronidazole. Prophylactic metronidazole given to 56 gynecologic patients reduced the vaginal carriage rate of nonclostridial anaerobes from 41 to 9% [Willis et al. (1037)]. Anaerobic infection did not develop in any of 38 treated patients undergoing hysterectomy but was seen in 6 of 37 control patients in this study. There is no convincing evidence that antimicrobial prophylaxis has been useful in emergency Caesarean section. This failure is probably related to the amount of bacterial contamination of the operative field, the difficulty in delivering adequate levels of drug to the uterine wound, and the problem of trauma to the uterine muscle during surgery.

Neary *et al.* (1037a) noted that bacteria, particularly *Bacteroides*, were found in the vagina more often in the first half of the menstrual cycle than in the last and suggested, therefore, that abdominal hysterectomy would be safer when performed during the second half of the cycle.

Bacteremia with anaerobes is relatively common following urethral dilatation [Sullivan *et al.* (388)]. In the case of patients with valvular heart disease undergoing urethral dilatation, ampicillin would be a good choice for prevention of endocarditis, since it would cover all of the anaerobes likely to be involved as well as the enterococcus and most strains of *Escherichia coli.*

The studies of Drewett *et al.* (1024) are pertinent to the prevention of infection following orthopedic surgery. The counts of *Clostridium perfringens* on the skin over the hip area and elsewhere on the limbs are reduced by bathing, and particularly by showering. Application of a compress of an iodophor for 15 minutes resulted in the elimination of this organism from patients in their study.

Debridement is the most important approach to the prevention of infection of the skin and soft tissue wounds; all devitalized tissue and foreign bodies must be removed within 6 to 8 hours after injury. Adequate hemostasis with maintenance of an adequate blood supply, suturing of tissues without tension and with a minimum amount of buried suture, and apposition of live tissue to live tissue are all important for promotion of healing without infection. Infection following potentially contaminated operations, such as open intestinal resection and anastomosis, may usually be prevented or ameliorated by delayed primary closure of the skin and subcutaneous tissue. This probably relates to the fact that anaerobic bacteria are an important part of the bacterial flora that may produce infection in such cases and delayed closure may modify anaerobic conditions that would otherwise exist in the contaminated closed wound [Bernard and Cole (518)]. Bite infections are handled much like other traumatic wounds. The wound is cleaned, dressed, and splinted as soon as possible after the bite. The anaerobic state is eliminated by adequate debridement. Injured tendons are not repaired primarily, but if the joint capsule has been entered it should be drained. The injured part (usually a hand) is elevated and splinted in the position of maximum function, and wet dressings of hydrogen peroxide are applied [Shamblin (1038)].

Care of the newborn's umbilical cord has been discussed in Chapter 15. There is a definite advantage to late severance and nonligation of the umbilical cord. The amount of handling of the cord should be minimized in order to keep down the bacterial flora [Bernstine *et al.* (1039)].

Indwelling intravenous catheters should not be placed in the inguinal area, since this predisposes to bacteremia with anaerobes from the fecal flora.

Prevention of Gas Gangrene

Although clostridia are widely distributed in nature, in soil and sand, and clostridial spores are regularly found on wool clothing, there is very strong evidence to indicate that the primary source of clostridia in gas gangrene and other infections following trauma or surgery is the human gastrointestinal tract. The evidence for this is reviewed in detail in the excellent monograph of MacLennan (747). Wounds incurred in the deserts of North Africa, which were essentially free of clostridia, were followed by gas gangrene with approximately the same frequency as in other areas where the soil is regularly contaminated with these organisms. Henry [quoted by MacLennan (747)] indicated that the best prophylaxis against gas gangrene during wartime was a soap bath and clean clothing the night before an attack.

Wounds involving areas of deep muscle are the most vulnerable. Therefore, wounds involving the buttock, thigh, leg, and shoulder are particularly prone to gas gangrene, as are compound fractures, severe crushing injuries, and injuries secondary to high-velocity missiles. The lower limb and buttock are involved in three-fourths of cases of gas gangrene, and most of the other cases are in the upper extremity, two-thirds of these involving the shoulder. There is a characteristically extensive laceration or devitalization of the muscles. Other predisposing factors are impairment of the main blood supply to a limb or large muscle group, contamination by dirt, clothing, or other foreign body, and delay in proper surgical treatment.

By far the most important aspect of prophylaxis is adequate surgical management as soon as possible following injury. As MacLennan (747) pointed out, probably the most important reason why gas gangrene was so much less common following industrial or auto accidents than in war wounds was that surgical treatment could be given more quickly in the first two types of injury. This situation has now changed. The speed with which military casualties were evacuated from the battlefront and treated definitively in recent conflicts was such that the incidence of gas gangrene was quite low. Brown and Kinman (785a) noted that only 22 cases of gas gangrene occurred in United States forces during 8 years of combat in Viet Nam, whereas there were 27 cases in a recent 10 year period in Miami. The primary reason for the relatively high rate in the civilians involved in various types of accidents in Miami was that all of them had primary closure of their wounds. In Viet Nam, delayed closure was practiced routinely. The most important aspect of the surgical management is thorough debridement, removing every last fragment of devitalized tissue. Meticulous hemostasis is also very important. Deep irregular wounds must be kept widely open. Tight packing of the wound should be avoided, and if the wound is prone to gas gangrene, there should not be primary closure. There must be adequate drainage of all aspects of the wound. On occasion, injudicious use of tourniquets or application of tight casts (or casts that become tight due to swelling of the underlying limb) may contribute to the problem. If a wound is one that is prone to gas gangrene and a cast must be applied, it should be bivalved from the outset. Wounds involving extensive injury to the buttocks are obviously the hardest to clean and to keep clean. Many years ago colostomy was used prophylactically in these patients, but it later fell into disfavor. This may still be worth considering in selected cases.

Altemeier concluded long ago that *prophylactic* administration of gas gangrene antitoxin was of little or no practical value and has not used it since 1943 (774). The subcommittee on Trauma of the National Research Council and the United States Department of Defense have also recommended that gas gangrene antitoxin not be given prophylactically because of its ineffectiveness.

On the other hand, antimicrobial prophylaxis is very definitively indicated. Penicillin is the drug of choice and should be given as early as possible following injury, since even a 2 hour delay may result in markedly decreased efficacy. It must be strongly emphasized that antimicrobial prophylaxis is strictly adjunctive and far from adequate by itself. The surgical management of wounds is the major approach of importance. Penicillin should be given parenterally in high dosage. In severe wounds prone to gas gangrene, at least 10,000,000 U daily should be used. Chloramphenicol is a good alternative drug [MacLennan (747)]. Early trials with tetracycline suggested that this compound was less effective, and, considering the recent development of resistance of clostridia and other anaerobes to this drug, it would certainly not be appropriate. In view of the relatively common resistance of many clostridia to clindamycin and the occasional resistance of *Clostridium perfringens* itself to this agent, clindamycin would probably not be a suitable choice for prophylaxis. There is no experience with vancomycin or metronidazole in this regard, but in theory these agents would seem to promise good results.

Even in animal studies, hyperbaric oxygen was not effective in preventing gas gangrene {Irvin *et al.* [quoted by Altemeier and Fullen (1012)]}.

Gas gangrene acquired in the hospital is usually postoperative, particularly following amputation of a lower limb. It may also be seen after other types of surgery of the lower limb, particularly hip surgery. It has been very clearly established by the excellent studies of Ayliffe and Lowbury (777) that the source of the organisms is the patient's fecal flora. Bathing, particularly showering, reduces the number of organisms significantly and, as indicated earlier, application of a wet compress with an iodophor for 15 minutes is very effective against spores of *Clostridium perfringens* [Drewett *et al.* (1024)]. Lower limb surgery, particularly in an elderly patient and especially if there is impaired circulation and/or diabetes, is a definite indication for antimicrobial prophylaxis in addition. Penicillin is the drug of choice. Ordinarily, a dosage of 500,000 U intramuscularly of aqueous penicillin G every 6 hours would be appropriate.

Epinephrine or epinephrine-containing compounds should never be injected into the buttock or into areas with already compromised circulation, since this will, on occasion, predispose the patient to gas gangrene at the site. There are also relatively rare reports of gas gangrene due to inadequate sterilization of instruments—either in alcohol or another nonsporicidal substance, or as a result of poor autoclaving technique (e.g., overloading of autoclaves, running a shorter than normal cycle). Even boiling of instruments is not reliable for killing of spores.

Prevention of Tetanus

Prophylaxis of tetanus is also considered in Chapter 17. Better sanitation, theoretically an important means of preventing tetanus, is not likely to come soon, depending as it does on improved economic conditions. It is simply not realistic to anticipate good sanitary conditions throughout the world in the foreseeable future.

The surgical prophylaxis of tetanus involves the same principles that have have described for the prevention of gas gangrene. If devitalized tissue is removed and anaerobic conditions are prevented, any *Clostridium tetani* organisms that are present in the wound will not produce toxin, and therefore tetanus will not ensue. The availability of good immunizing agents against tetanus should not minimize the importance of proper wound management. It must be appreciated that tetanus not uncommonly follows small and easily overlooked wounds in the unimmunized. Antibiotics play a minor role in the prevention of tetanus.

People who have been actively immunized with toxoid maintain good recall to a booster dose of toxoid for at least 15 years following primary immunization. Fluid toxoid, rather than alum-precipitated material, should be used for booster doses, since it results in a more rapid rise in antibody titer. If the patient has not been immunized previously, tetanus immune globulin (human) is the best material for passive protection. Two hundred and fifty units intramuscularly is adequate for routine prophylaxis, but larger doses are desirable in severe injuries. Active immunization with toxoid (given in the other arm) should be started at the same time. In this way, long-term immunity can be set up. It should also be appreciated that the protection afforded by the antitoxin is relatively limited in duration, and the hazard of the development of tetanus may actually remain beyond this time in serious injuries. It has been suggested that when wound contamination and tissue destruction have been great, as, for example, in cases of extensive third-degree burns, that both active and passive immunization may be indicated at the time of injury, even if the patients have received active immunization previously. This is because in such cases tetanus may develop more rapidly than in the 4 to 7 days that may be required to obtain maximal response to a booster dose of tetanus toxoid. Intensive efforts should be made to immunize high risk groups routinely. This would include pregnant women, addicts, and people of poor economic circumstances.

Prospects for the Future

The studies of McCabe et al. (992) indicate that antibody to the core portion of the lipopolysaccharides of various gram-negative bacteria protects against heterologous gram-negative bacilli. Should this work be substantiated, one might look forward to both active and passive immunization against such endotoxin so that bacteremic shock from certain gram-negative anaerobic (and aerobic and facultative as well) bacilli might be combated effectively.

In his review article in 1962, MacLennan (747) summarized efforts toward preparing toxoids of the organisms involved in gas gangrene so that prophylactic immunization might be achieved. He indicated that at that time satisfactory preparations had readily been obtained against *Clostridium novyi*, *C. septicum* and *C. histolyticum*, but that it had not been possible to produce a toxoid of *C. perfringens* which was consistently antigenic and protective. However, Altemeier and Fullen (1012) indicated that they have developed and tested toxoids for both *C. perfringens* and *C. novyi* which have proved very effective in preventing experimental gas gangrene in animals and which have provided adequate levels of circulating antitoxin in volunteer medical students.

The experimental studies of Senff and Moskowitz (994) suggest that agents such as diethylenetriamine pentaacetate and EDTA might be useful in protecting against the toxemia of *Clostridium perfringens* α-toxin.

Role of Anaerobes in
Miscellaneous Pathologic Conditions

Dental Caries and Periodontal Disease

Many bacteria may be involved in these processes, which must still be regarded as rather poorly understood. In the case of *caries*, organisms frequently characterized as "microaerophilic" (but truly facultative organisms) (such as *Streptococcus mutans* and other streptococci) are clearly important in coronal or multisurface caries in view of their ability to produce and survive in large amounts of acid and extracellular polysaccharide and in view of their relative tolerance of large amounts of dietary sucrose [Newman and Poole (1040)]. The unique water-insoluble glucans produced from sucrose are considered to be responsible for the ability of *S. mutans* to develop adherent plaques on tooth surfaces. Staat and Schachtele (1040a) recently isolated several dextranase-producing organisms from dental plaque (one identified as *Actinomyces israelii*) and suggested that such organisms could be natural antagonists to *S. mutans* since they would suppress the ability of *S. mutans* to adhere to teeth (and thus its cariogenic potential). Lactobacilli and gram-positive anaerobic bacilli are important in pit and fissure lesions and root caries (Fig. 21.1a and b) [Genco *et al.* (1041) and Jordan and Hammond (1042)].

In *periodontal disease* (Fig. 21.1a and b), oral bacteria of various types (including anaerobes) clearly play a role. There are several types of periodontal disease, with different manifestations and with different bacteria

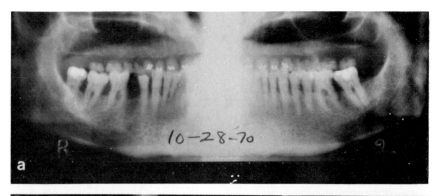

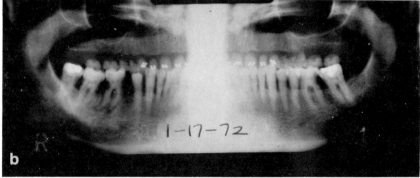

Fig. 21.1.   (a) Early root caries lower left bicuspid. (b) Over 1 year later same individual now shows extension of root caries on left, root caries lower right bicuspid, and periodontal disease with loss of alveolar bone about roots of tooth.

implicated in the processes as noted by Gibbons and van Houte (1042a). Periodontitis is the most prevalent form of periodontal disease and is characterized by chronicity and slow tissue destruction over a period of years. Periodontitis entails gingivitis, pathologically deep gingival sulci (pockets), and loss of alveolar bone. Gingivitis per se denotes gingival inflammation without bone loss, but acute necrotizing ulcerative gingivitis (ANUG) leads to loss of supporting bone, if untreated.

Periodontosis is characterized by minimal gingivitis, minimal plaque formation, and rapid loss of supporting bone (3 to 4 times faster than in periodontitis). Periodontosis is generally localized to the first permanent molars and the incisors and is seen mostly in individuals 25 years of age or younger [Gibbons and van Houte (1042a) and Newman and Socransky (1042b)]. Recent work [Newman and Socransky (1042b) and Newman *et al.* (1042c)] implicates in periodontosis 5 distinct groups of saccharolytic gram-negative anaerobic bacilli, many of which cannot be classified at present. These organisms are not found in clinically normal sites in the same individuals, are pathogenic in animal models, and initiate osteoclastic destruction when introduced into germfree rats. Tetracycline has shown effectiveness in periodontosis (1042a).

In the more prevalent forms of periodontal disease, gingivitis and chronic periodontitis, filamentous organisms (notably *Actinomyces viscosus* and *A. naeslundii*) are numerous in gingival plaque, itself an important feature of the process. Morphologically similar organisms have been noted invading and eroding human cementum. This type of organism exhibits pathogenic potential in animals. For all of these reasons, these organisms are probably of etiologic significance in gingivitis and chronic periodontitis. A similar case can be made for *Streptococcus mutans*, *S. mitis*, and *S. sanguis* [Gibbons and van Houte (1042a)]. There is some evidence that in more rapidly destructive periodontitis cases various asaccharolytic gram-negative organisms, chiefly anaerobic rods, contribute to the process (1042a, 1042c). As noted in Chapter 4, unusual spirochetal forms appear to be specifically associated with ANUG. The ability of organisms to accumulate in large masses and to adhere to teeth in the form of plaque appears to be a prerequisite for initiation of periodontitis [Socransky (1043)]. Mechanical cleansing of plaque from tooth surfaces and topical use of agents with antimicrobial properties [Löe (1044); Davies (1045)] definitely ameliorate the problem, while neglect of oral hygiene certainly aggravates it. In the crevice between tooth and gum where plaque organisms are in contact with various proteinaceous materials, bacteria that tolerate an alkaline environment and which secrete proteolytic substances are likely to survive. Changes in plaque formation and distribution related to modern diet have modified bacterial metabolism in plaque and led to disease. The mechanism by which bacteria cause destruction of the periodontium is uncertain. Possible mechanisms include production of lytic enzymes (1042d) and "cytoxic" metabolites, or initiation of acute and chronic inflammation by way of hypersensitivity [Socransky (1043)], or endotoxin activation of host effector systems [Mergenhagen and Snyderman (1046)].

Skin Diseases

Acne is a condition of uncertain origin in which several factors may operate. It is clear that the presence of *Propionibacterium* (*Corynebacterium*) *acnes* does contribute to the process and that suppression of this organism with antimicrobial agents leads to significant, although temporary, clinical improvement. *Propionibacterium acnes* has been shown to be capable of cleaving triglycerides, liberating free fatty acids [Reisner *et al.* (1047)]. It is the production of free fatty acids in sebum that leads to the characteristic inflammatory response in acne. Tetracycline has been demonstrated to inhibit the lipase of *P. acnes* [Weaber *et al.* (1048)], as well as the organism itself; use of this drug in humans has been demonstrated to decrease the concentration of free fatty acids in human sebum [Freinkel *et al.* (1049)]. Accordingly, low dose long-term therapy with an agent such as a tetracycline is commonly employed, with good effect. Erythromycin may also be effective. An *ad hoc* committee of the American Academy of Dermatology's National Program for Dermatology has reviewed studies of antimicrobial therapy of acne and concluded that tetracycline and erythromycin are safe and effective drugs (1049a, 1049b). They also noted that lincomycin, sulfadimethoxine, and trimethoprim–sulfamethoxazole were approximately as effective and that penicillin was not as effective. They indicated that lincomycin and clindamycin would probably best be avoided for acne pending more information on the question of their toxicity in this setting. Krueger and Christian (1049c), however, noted that rare cases of severe forms of acne may respond well to clindamycin after failure with other agents.

Again, in the case of *hidradenitis suppurativa*, bacteria play a secondary but very important role. It is our experience that anaerobes of various types, primarily anaerobic cocci and gram-negative bacilli, are the major offenders in the infectious complications of this process. Abscess formation is common and may be widespread, as the basic disease often is. The foul odors produced by the anaerobes in these infections may be of serious concern to these patients.

Bowel Bacterial Overgrowth Syndromes

Haenel and Bendig (1049d) have provided an excellent review of the makeup of the gastrointestinal flora in health and disease. There is also an excellent monograph on this subject by Drasar and Hill (1049e).

In children with intestinal obstruction distal to the duodenal–jejunal junction, a profuse flora of fecal type is found above the obstruction. An abnormally profuse but nonspecific bacterial flora is found in the upper small bowel of infants with protracted diarrhea and carbohydrate intolerance [Anderson *et al.* (1050)]. Infants and young children with malnutrition and diarrhea showed gross microbial contamination of the upper gastrointestinal tract including, in several cases, large numbers of anaerobes (53a).

Obvious sources of contamination of the small bowel with large numbers of bacteria occur primarily in adults. Among the causes are gastrocolic or enterocolic fistula, stagnation of bowel contents secondary to surgical procedures creating blind loops, disease leading to impaired motility or stricture, diverticula, pernicious anemia where lack of gastric acid permits colonization of the small bowel, and an infected biliary tract draining into the duodenum.

Small intestinal overgrowth of bacteria may be clinically benign or may cause one or more manifestations of malabsorption—diarrhea, steatorrhea, vitamin B_{12} deficiency with its clinical manifestations, protein malnutrition, and impaired absorption of sugars [Neale et al. (1051)]. Although many studies of blind loop contents did not employ optimum bacteriological techniques, there are several reports in which anaerobes were recovered from an area of small bowel stasis. McKenna et al. (1052) found heavy growth of Clostridium and moderate numbers of Bacteroides, as well as facultative bacteria, in a patient with multiple jejunal diverticula. Lyall and Parsons (1053) found heavy growth of Bacteroides and lesser numbers of coliforms in a patient with jejunal enteroenterostomy and malabsorption. Kinsella et al. (1054) isolated an "anaerobic gram-positive bacillus" along with aerobic gram-negative bacilli from the afferent loop of one patient. Drasar et al. (1055) grew Bacteroides from the jejunum in ten patients with a variety of gastrointestinal disorders, including the blind loop syndrome, but gave no further details. Tabaqchali et al. (1056) cultured Bacteroides from a small percentage of 50 patients with partial gastrectomy or blind loop syndrome. Polter et al. (1057) reported a case of bacterial overgrowth in the small intestine in a patient who had both a surgically created afferent loop and multiple small bowel diverticula (Fig. 21.2). Culture of the afferent loop yielded almost 10^{10}/ml anaerobes (Bacteroides fragilis, Bacteroides species, and a Fusobacterium) and about 10^9/ml facultative bacteria (E. coli, Klebsiella–Enterobacter, Streptococcus lactis, and Lactobacillus). This patient had elevated fecal fat and vitamin B_{12} malabsorption. Oral neomycin effected a modest reduction in facultative count, with no effect on anaerobes or on the clinical picture or laboratory evidence of malabsorption. Subsequent treatment with lincomycin eliminated all anaerobes without affecting the facultatives; the patient improved clinically and the laboratory parameters returned to normal coincident with this treatment. Goldstein et al. (1058) found large numbers of various anaerobes and smaller numbers of facultatives in two patients with small bowel diverticulosis. Pearson et al. (1059) described a patient with intestinal pseudoobstruction and malabsorption whose duodenal juice contained 10^5–10^6 coliforms/ml plus numerous gram-positive bacilli, the latter failing to grow (? anaerobic). The patient did not respond to oral neomycin but did improve on tetracycline therapy. Farrar et al. (1060) reported on several cases of malabsorption secondary to stagnation-inducing lesions of the small intestine. Quantitative cultures of small bowel content, study of bile acids in the small bowel, and response to therapy indicated a significant role for anaerobes, especially Bacteroides, in the pathogenesis of the malabsorption. Challacombe et al. (1060a) described an 8-year-old girl with two stagnant loops in

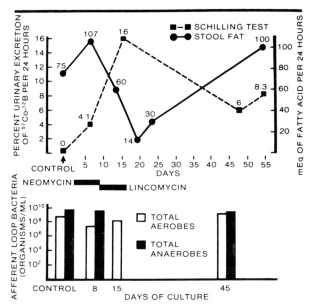

Fig. 21.2. Patient with afferent "blind loop" following gastric surgery and multiple jejunal diverticula. Prior to therapy fecal fat was high and vitamin B_{12} excretion on Schilling test was low; high counts of both anaerobes and facultatives (with anaerobes predominating) were noted in afferent loop contents. Oral neomycin therapy lowered facultative count somewhat but did not correct absorptive abnormalities. Oral lincomycin treatment eliminated anaerobes from loop with virtually no effect on aerobes; malabsorption of fat and vitamin B_{12} was corrected. From Polter *et al. Gastroenterology* **54**, 1148 (1968). Copyright 1968. The William & Wilkins Co., Baltimore.

the terminal ileum. Anaerobes outnumbered facultatives by between one to two logs in both loops; *Bacteroides* predominated, but anaerobic cocci and lactobacilli were also present. The proximal ileum had the same bacteriologic picture except for absence of lactobacilli. In rats with surgically created self-filling upper small bowel blind loops, increased counts of *Bacteroides* and *E. coli* were found [Kent *et al.* (1060b)]. Therapy which reduced *Bacteroides* counts, but not those of *E. coli*, led to weight gain and reduced fecal fat excretion.

In addition to the cultural data cited above, the following points support the suggestion that anaerobes may play an important role in the blind loop syndrome in man: (a) material from the area of stasis has been noted to be foul-smelling [Kinsella *et al.* (1054), Kinsella and Hennessey (1061), and Strauss *et al.* (1062)]; (b) many anaerobes, but few facultatives, are able to deconjugate bile acids [Shimada *et al.* (37) Gustaffson *et al.* (1063), Hill and Drasar (38), Aries and Hill (1063a), Lewis and Gorbach (1064), and Midtvedt (1063b)], (c) oral neomycin is generally relatively ineffective in eliminating intestinal anaerobes and is generally not useful in the blind loop syndrome [Doig and Girdwood (1065), Halsted *et al.* (1066), and Goldstein *et al.* (1067)], whereas tetracyclines are active against anaerobes and are usually effective clinically in this syndrome [Doig and Girdwood (1065), Wirts and Goldstein (1068), Paulk and Farrar (1069)]. The difference in the activity of neomycin and tetracycline against anaerobic bacteria may explain the difference in their clinic effectiveness in the blind loop syndrome. The suppression of anaerobic growth would explain the paradoxical persistence of large numbers of coliforms in the area of small bowel stasis despite clinical remission induced by tetracycline [Wirts and Goldstein (1068) and Paulk and Farrar (1069)]. Bacterial deconjugation of bile salts, rendering them ineffective in micelle formation, appears to account for the steatorrhea in most cases. Vitamin B_{12} malabsorption is probably related to binding of the vitamin, which may be caused by *Bacteroides fragilis*, as well as by *E. coli* (Schjönsby and Hofstad (1070)]. Subsequent studies by Schjönsby *et al.* (1070a) showed that there was minimal uptake of labeled vitamin B_{12} bound to human gastric juice by *Bifidobacterium* and *Clostridium* and none by enterococci, whereas there was 3.0% uptake by coliforms and 7.3% by *Bacteroides*. These workers treated a patient with jejunal diverticulosis and B_{12} malabsorption with neomycin and noted a drop in anaerobic count of almost 3 logs, no change in aerobic count, and improved B_{12} absorption. Schjönsby *et al.* (1070a) also did quantitative bacteriology on the upper jejunal and upper ileal fluid of 7 patients with jejunal diverticulosis; all had anaerobes (all but one in counts of $> 10^6/\text{ml}$). Anaerobes outnumbered facultatives in 3 patients.

Bacterial overgrowth may be responsible for the malabsorption which may be seen in intestinal scleroderma and diabetic neuropathy; this may respond to therapy with tetracycline. A patient reported by Hines and Davis (1071) had Ehlers–Danlos syndrome with malabsorption, megaduodenum, and bacterial overgrowth. The flora included *Bacteroides*. Jejunal aspirates from tropical sprue patients, however, do not show anaerobes; the colonization is with Enterobacteriaceae (1071a, 1071b).

The cause of diarrhea and malabsorption following peptic ulcer surgery is unknown but may be related to bacterial overgrowth in the jejunum (Greenlee *et al.* (1072)]. On occasion, antibiotics have proved beneficial in this situation, but studies to date are inadequate.

Other Gastrointestinal Conditions

Whipple's disease is more likely a specific infection than a syndrome resulting from derangement of normal bowel flora. Although tetracycline therapy is effective, there is much disagreement as to the specific identity of the bacillary bodies seen in the gut on electron microscopy. Prévot and co-workers (1073, 1074) recovered anaerobic corynebacteria in this condition, but many other workers reported a variety of other bacteria on culture.

While some anaerobic bacteria, such as some *Bacteroides, Clostridium, Bifidobacterium*, and anaerobic streptococci, produce ammonium and may have urease activity [Phear and Ruebner (1075), Huet and Aladame (1076), Sabbaj *et al.* (1077)], the well-established efficacy of oral neomycin and kanamycin in patients with *hepatic coma* or encephalopathy indicates that facultative urea splitters, such as *Klebsiella* and *Proteus*, are probably much more important sources of the offending ammonia and other metabolic products which may be important. The possibility that these drugs might affect the metabolism of these anaerobes without effectively suppressing their growth has not been explored, however. Recent evidence [Varel *et al.* (1077a)] indicates that *Peptostreptococcus productus* is the most numerous of the ureolytic species found in human feces; the average count in 5 individuals was 10^9/gm of feces. Ammonium and more complex organic nitrogen sources strongly repress production of urease by this organism.

Hermans' syndrome (lymphonodular hyperplasia of the small bowel with immunoglobulin A deficiency) may have malabsorption related to either bacterial overgrowth in the small bowel, or giardiasis, or both. We have had occasion to study one patient with this syndrome and found that the upper small bowel contents had a high count of organisms, including anaerobes, commonly found in the oral flora, as well as *Giardia lamblia*.

The possibility should be considered that changes in normal flora, rather than ingrowth of specific pathogens such as *Staphylococcus aureus*, may account for some instances of *antibiotic-induced diarrhea*. This might be true also for *neonatal necrotizing enterocolitis* and *toxic megacolon*. Among the anaerobes, clostridia would be prime suspects because of their known involvement in various types of enteric pathology in man and animal. The relative resistance of some clostridia to clindamycin is also of interest in this regard. Increased numbers of such normal flora components as *Clostridium perfringens* (which may be implicated in food poisoning) and *C. bifermentans* might account for antibiotic-induced diarrhea—a sort of "endogenous food poisoning." Marr *et al.* (1077b) reported that patients with clindamycin-associated diarrhea and pseudomembranous colitis showed a striking qualitative and quantitative decrease in anaerobic bacteria in feces. In our study of the effect of clindamycin on the normal human fecal flora [Sutter and Finegold (979)], however, we noted elimination of all anaerobes except *Eubacterium* and *Clostridium*; *Bifidobacterium* also persisted in reduced count in one individual. None of our five patients had gastrointestinal side effects. We had found similar results earlier in studies with lincomycin [Finegold *et al.* (976)]. It is interesting to note that volatile amines are thought to be possible bowel irritants. Other possible mechanisms of diarrhea that may be related to quantitative changes in normal flora include increased amounts of lactic acid and volatile fatty acids, increased amounts of deoxycholic and chenodeoxycholic acids (which inhibit resorption of water) or of lithocholic acid (which, if not sulfated, might induce an inflammatory reaction), and production of such hydroxy fatty acids as 10-hydroxystearic acid (castor oil is thought to work through the action of such an acid, i.e., ricinoleic acid). Production of hydroxy fatty acids from dietary fatty acids is carried out by a number of different intestinal bacteria, including various anaerobes (1077c, 1077d).

The possibility of hypoprothrombinemic hemorrhage secondary to antibiotic suppression of vitamin K production by intestinal bacteria has long been a matter of concern. It rarely has been a problem, partly because vitamin K is readily available in the normal adult diet and the liver stores are very large.

Crohn's disease (regional enteritis) has been noted recently [Ursing and Kamme (1078)] to respond to metronidazole, an agent active only against anaerobic microorganisms. If this is confirmed, it remains to be determined whether the beneficial effect was related to improvement in the basic process itself or just to reduction of bacterial overgrowth in the lumen or tissues of the small bowel; increased luminal counts have been noted by several workers and this may contribute to the symptomatology in these patients. Overgrowth of organisms in the bowel may, under appropriate circumstances, lead to inflammatory disease of the bowel per se in ileal bypass for obesity. Antibodies to *Bacteroides fragilis* ss. *fragilis* and another unidentified anaerobe have been found in intestinal tissue homogenates from patients with Crohn's disease, and serum antibodies to certain anaerobes have also been demonstrated [Persson and Danielsson (1079)]. Wensinck (1080) studied the fecal flora of subjects with Crohn's disease (clinical features of the disease were not given in this preliminary report) and healthy subjects and noted "statistically significant" differences—higher total counts, counts of gram-negative rods, and "coccoid gram-positive rods" (including *Eubacterium contortum*, *Eubacterium* species, and perhaps some anaerobic cocci) in the patients with Crohn's disease. He also noted agglutinating serum antibodies against *Eubacterium contortum* to be highly specific for Crohn's disease. Similarly, antibodies reacting with anaerobes but not aerobes have been found in rectal and colonic mucosa in patients with *chronic ulcerative colitis* [Monteiro *et al.* (1081)]. Brown and Lee (1081a) noted that serum antibodies within the various immunoglobulin classes against *Bacteroides fragilis*, *Escherichia coli*, and enterococcus were of the same order of magnitude in healthy subjects. IgG antibodies to *B. fragilis* and *E. coli* were elevated in subjects with selective IgA deficiency. In patients with chronic inflammatory bowel disease (regional enteritis and ulcerative colitis), anti-bodies of all three classes to *B. fragilis* were elevated, as much as 14 times compared to normals. They considered that the elevated antibody levels were probably a consequence of the diseases, but felt that it was possible they played a pathogenetic role. Hammarström *et al.* (1081b) found that germfree rats, monocontaminated with *Clostridium difficile* or another, unidentified, *Clostridium* produced autoantibodies to colon antigen. The rats also de-veloped antibodies to *C. difficile* but there was no cross-reaction between *C. difficile* antigen and colon antigen. Goldman and Peppercorn (1081c) noted that the effectiveness of sulfasalazine in management of chronic ulcerative colitis may relate to the activity of this drug against anaerobes, particularly clostridia.

Ileal bypass for obesity is intended to produce certain controlled absorp-tive defects and is expected to lead to steatorrhea. However, in addition, a number of patients have developed unexpected complications leading to a syndrome that has been termed *bypass enteropathy* [Drenick *et al.* (1082)]; it resembles the intestinal and systemic manifestations of chronic inflamma-tory bowel diseases. A nonspecific chronic inflammatory process originating in the excluded loops of small bowel is associated in various patients with fever, diarrhea, accumulation of large volumes of fluid and gas in distended loops of bowel, electrolyte imbalance, *pneumatosis cystoides intestinalis*, rupture of the gas-filled blebs leading to gas under the diaphragm, ileus, and pseudoobstruction of the colon simulating the clinical picture of toxic megacolon, and protein-losing enteropathy. Several patients have come to surgery because an acute surgical abdomen was simulated. Associated with all of the above, and presumably primarily responsible for it, is stag-nation of intestinal contents and bacterial overgrowth in the excluded loop. *Clostridium perfringens* had been found to produce pneumatosis cystoides intestinalis in animals [Yale *et al.* (1083)] and was found in an ileal bypass patient's bypassed loop in a count of 10^5/ml, along with a number of other gas-producing organisms, most of them anaerobic [Drenick *et al.* (1082)]. Coliforms were found in a count of 5×10^8/ml, along with some group D streptococci. Four subspecies of *Bacteroides fragilis* were found in counts of 10^9/ml along with eleven other anaerobic species, many in high counts. This patient had pneumatosis cystoides intestinalis, severe diarrhea, elec-trolyte disturbances, fever, chronic nonspecific inflammation of the bypassed bowel segment, and erythema nodosum. Yale (1083a) has recently noted that *Clostridium tertium* can also produce pneumatosis cystoides intestinalis in germfree rats. Extraintestinal manifestations of bypass enteropathy include, in addition to erythema nodosum, various hepatic disorders (fatty infiltration, fibrosis and cirrhosis, and granulomatous hepatitis), hyper-oxaluria, renal oxalate stones, arthritis, and perhaps aortic valvulitis. Various possibilities exist as to mechanisms of these several disorders. Included are toxins produced by the bacteria, circulating immune complexes relating to the bacteria and cross-reacting with various tissues, disturbed bile acid metabolism, and excessive absorption of oxalates and/or their precursors (e.g., glycine liberated by bacterial deconjugation of bile acids). Studies of material from the bypassed segment in two other patients yielded only *E. coli* in one case and *E. coli*, non-group A β-hemolytic streptococci, *Bacteroides fragilis* ss *fragilis*, and *Peptostreptococcus anaerobius*. The similarity of the above clinical picture to that seen in inflammatory bowel disease and possible relationships to Whipple's disease, Reiter's syndrome, and the arthritis and other extraintestinal manifestations of bacillary dysentery, ulcerative colitis, etc., is of great interest. Wands *et al.* (1083b) noted that the serum of 3 patients with acute arthritis following intestinal bypass for obesity contained circulating cryoprotein complexes which comprised IgG, IgM, IgA; complement components C_3, C_4, C_5; and IgG antibody against *E. coli* and *B. fragilis*. The latter antibody was concentrated several times in the complexes as compared to serum levels. The cryoprotein complexes activate both the classical and alternate complement pathways. Wands *et al.* cited work by Mezey *et al.* in which the blind loop resulting from the intestinal bypass procedure was found to be heavily colonized with *E. coli* and *Bacteroides* species. We (E. J. Drenick and S. M. Finegold, unpublished data) have noted very good responses in patients with enteric manifestations of bypass enteropathy and in one patient with arthritis related to intestinal bypass for obesity treated with metronidazole, an agent active only against anaerobes. Others have also noted response to drugs active against anaerobes [Faloon (1083c)]. O'Leary *et al.* (1083d) created jejunoileal bypasses in dogs. The excluded loops in these animals grew either *Bacteroides fragilis*, *Bacteroides* species, or both. Treatment with doxycycline prevented appearance of the anaerobes and prevented hepatic fatty metamorphosis and death.

Anderson (1083e) found that *R factor transfer* between donor and recipient strains of *E. coli* in broth culture was completely inhibited in the presence of dense suspensions of *Bacteroides fragilis*. This may explain the almost total inhibition of conjugation in the human colon in the absence of antibiotics.

Low-Beer and Pomare (1083f) found that cholesterol content of biliary lipids increased significantly in volunteers ingesting deoxycholic acid. A relative excess of cholesterol over bile salts and lecithin in bile predisposes to cholesterol *gall stones*. Since deoxycholic acid is formed in the colon (by action of anaerobic bacteria on cholic acid), Low-Beer and Pomare suggested that high rates of absorption of deoxycholate from the colon would predispose to cholesterol gall stones.

Biswas (1083g) claimed a beneficial effect of metronidazole in patients with *hemorrhoids*—subsidence of pain, bleeding, and inflammation. If this can be confirmed it would suggest that anaerobes in the fecal flora are implicated in these complications.

Eyssen and Parmentier (1083h) have indicated that biohydrogenation of unsaturated fatty acids and 5β-H reduction of Δ^5-3β-hydroxy steroids are carried out in the intestinal tract by strict anaerobes. A *Eubacterium* which they isolated is active in some of these transformations. This type of activity may have implications with regard to *atherosclerosis*.

Rheumatoid Arthritis

Our studies [Sapico *et al.* (1084)] failed to confirm those of Mansson and Colldahl (1085) which indicated a relationship between high counts of atypical *Clostridium perfringens* in the stool and rheumatoid arthritis.

Bartholomew and Nelson (1086) found *Propionibacterium acnes* and L forms thereof in 13% of blood cultures, 6% of synovial fluid cultures, and 43% of synovial tissue cultures of patients with rheumatoid arthritis, and an antigen cross-reacting with *P. acnes* (and with no other organisms tested) in 88% of rheumatoid synovial fluid leukocytes. They indicated that the antigen might be involved in formation of soluble immune complexes. Earlier, Duthie and co-workers (1087) had found diphtheroids in synovial tissue of rheumatoid arthritis patients. As Clasener and Biersteker (1088) noted, interest in these organisms had been heightened by the findings that a certain anerobic diphtheroid (*Corynebacterium parvum*) had such immunologic effects as inhibition of graft versus host disease and prevention of the induction of immunological tolerance. However, these latter workers, using germfree inoculation technique, found that the frequency of isolation of diphtheroids was reduced significantly and concluded that these organisms should be considered as contaminants.

Cancer

Possible interrelationships between diet, bowel flora, and bowel cancer are of great interest. Japan is one of a number of countries with a low incidence of large bowel cancer. However, Japanese migrating to the United States and adopting the Western diet develop this cancer with increased frequency, approaching that of native Americans [Berg *et al.* (1089)].

Hill *et al.* (1090) have postulated that intestinal bacteria may produce carcinogens from bile acids or other substrates and that differing incidences of colon cancer may depend, at least partly, on differences in bowel flora related to diet. Deoxycholic acid, itself a product of intestinal bacterial action on a primary bile acid, is weakly carcinogenic and might be important considering the large amount of deoxycholic acid the bowel is exposed to over a period of many years. The diets consumed in areas with a high incidence of bowel cancer are high in fat and animal protein content. High-fat diets result in a high fecal concentration of bile acids, thus providing more substrate for conversion to carcinogens. Studies comparing feces of subjects from Scotland, England, and the United States (high bowel cancer incidence) with feces from subjects of India, Japan, and Uganda (low incidence of bowel cancer) revealed that in the subjects from countries of high cancer incidence there was a higher concentration of fecal acidic steroids, that a greater percentage of these steroids was present in degraded form, and that intestinal bacteria recovered (especially certain anaerobes) exhibited a much greater capacity for degrading steroids [Drasar and Hill (1091)]. Furthermore, there is evidence that colonic neoplasia begins on the surface of the bowel in response to an exogenous stimulus [Cole (1092)].

Clostridium paraputrificum, C. tertium, and other clostridia may be of particular significance, since these organisms have been shown to carry out certain modifications of bile acids which might represent transformation of these compounds toward carcinogenic agents [Aries *et al.* (1093), Goddard and Hill (1094), and Goddard *et al.* (1094a)]. To date, however, our own studies have failed to confirm the presence of higher counts of *C. paraputrificum* and related organisms in the feces of subjects with a high risk of developing colon cancer [Finegold *et al.* (35, 483)].

Breast cancer incidence is closely correlated with that of bowel cancer. Intestinal bacteria produce estrogens from steroid substances in the colon [Drasar and Hill (1091)]. Thus diet, bowel flora, and breast cancer may also be interrelated.

The theory [Burkitt (1095)] that fiber in the diet is important in terms of intestinal transit time and time for interaction of bacteria and substrate (bile acids, for example) for carcinogen production is not inconsistent with the above speculations. Certainly relative stasis of intestinal contents may lead to overgrowth of anaerobes.

Drasar and Hill (1091), pointed out that intestinal bacteria might be involved in carcinogenesis without the mediation of carcinogen production—through an effect on the immune system or by other mechanisms.

It is interesting to note that preoperative bowel preparation with oral neomycin before operation for cancerous lesions may predispose to subsequent recurrence of the cancer at the suture line [Herter and Slanetz (1096)].

As indicated earlier, neomycin is relatively inactive against anaerobes. Whether the influence of such a short-term exposure might be related to the drug's antibacterial spectrum or not is unknown. A possible role for indole-acetic acid, produced by intestinal anaerobes, is noted below.

Corynebacterium (*Propionibacterium*) *parvum* has been shown to stimulate the reticuloendothelial system, in terms of phagocytic function [Prévot *et al.* (1097)]. Cummins and Johnson (1097a) have determined that *C. parvum* should be regarded as a synonym of *Propionibacterium acnes*. Other anaerobic corynebacteria, especially *C. anaerobium*, *C. avidum*, and *C. granulosum* also show such activity (1098, 1098a). Prévot (848) stated that primary focal (tonsillar, periodontal, etc.) infection with anaerobic *Corynebacterium* strains may spread throughout the recticuloendothelial system and that in over 10% of cases, such infections precede the onset of a malignant reticulosis. Paradoxically, perhaps, tumor growth may be inhibited by injection of *C. parvum* [Halpern *et al.* (1099) and Woodruff and Boak (1100)].

Several workers [Möse (1101), Kayser (1102), Kretschmer (1103)] have reported oncolytic effects produced by injection, by various routes, of spores of nonpathogenic clostridia into both animals and man. Reactions are minor. Large tumors have responded best. It is clear that this therapy, by itself, will not be definitive, however.

Miscellaneous

Finally, as indicated earlier, the normal intestinal flora (and presumably the anaerobes in particular, in view of their numerical superiority) may exert important effects in terms of *metabolizing drugs* that were orally ingested (or intestinally excreted) so as to make them available for reabsorption, to make them biologically active or inactive, etc. This is a very important field that is just now coming under active investigation. Peppercorn and Goldman (1104) showed that several anaerobes and other intestinal bacteria were able to metabolize sulfasalazine (salicylazosulfapyridine). Certain anaerobes, as well as other intestinal bacteria, produce various glycosidases [Hawksworth *et al.* (1105)]. Soleim and Scheline (1105a) noted that intestinal anaerobes metabolized xenobiotics. Bokkenheuser *et al.* (1105b) found that fecal anaerobes metabolized deoxycorticosterone. Other examples were cited in Chapter 1.

Hare and Polunin (703) have speculated that previous pelvic infection with anaerobic cocci may be an important cause of *infertility* in some populations. In British North Borneo, they found much higher carrier rates of these organisms among infertile women than among fertile.

Chung *et al.* (1105c) found that 2 strains of *Bacteroides fragilis* ss. *thetaio-taomicron* and one of *Citrobacter* sp., among 23 intestinal strains (chiefly anaerobes) studied, were able to produce indoleacetic acid from tryptophan. Indoleacetic acid may be important in the etiology of cancer. Mice given this compound developed malignant tumors of the lymphoreticular system or leukemia and rats receiving it developed bladder cancer. The compound is found in increased quantities in certain hereditary disorders, including phenylketonuria. Indoleacetic acid can also induce pulmonary edema and emphysema in cattle.

Inglis *et al.* (1105d, 1105e) and Bird *et al.* (1105f) have described a new type of polyagglutination in which the T and Tk receptors are both exposed on the erythrocyte membrane. This was associated *in vivo* with *Bacteroides fragilis* infection and possibly contributes to a poorer prognosis. There is no evidence of an associated hemolytic process but if patients require transfusion, difficulties may be encountered with cross-matching. The phenomenon might prove useful as an aid in the diagnosis of *B. fragilis* infection.

A recent study (1105g) indicated that idiopathic *trigeminal neuralgia* is often associated with bony cavities at sites of previous dental extraction. Such cavities, sometimes containing necrotic bone, yielded 3 to 14 bacteria on culture, with anaerobes the most common isolates (1105h). The vast majority of anaerobes recovered were gram-positive anaerobic bacilli and cocci.

Bibliography

1. H. C. Plaut, V. Studien zur bacteriellen Diagnostik der Diphtherie und der Anginen. *Dtsch. Med. Wochenschr.* **20**, 920 (1894).
2. H. Vincent, Sur l'etiologie et sur les lesions anatomo-pathologiques de la pourriture d'hôpital. *Ann. Inst. Pasteur, Paris* **10**, 488 (1896).
3. E. Levy, Ueber einen Fall von Gasabscess. VI. *Dtsch. Z. Chir.* **32**, 248 (1891).
4. M, A. Veillon, Sur un microcoque anaérobie trouvé dans des suppurations fetides. *C. R. Seances Soc. Biol. Ses Fil.* **45**, 807 (1893).
5. W. Lubinski, Ueber die Anaerobiose bei der Eiterung. Zweiter Teil der Arbeit "Ueber Anaerobe Eiterungsmikroben." *Zentralbl. Bakteriol., Parasitenkd.* **16**, 769 (1894).
6. A. Veillon and A. Zuber, Sur quelques microbes strictement anaérobies et leur rôle dans la pathologie humaine. *C. R. Seances Soc. Biol. Ses Fil.* **49**, 253 (1897).
7. A. Veillon and A. Zuber, Recherches sur quelques microbes strictement anaérobies et leur rôle en pathologie. *Arch. Med. Exp. Anat. Pathol.* **10**, 517 (1898).
8. E. Rist, Etudes bactériologiques sur les infections d'origine otique. Thesis, Faculty of Medicine, University of Paris, Paris (1898).
9. L. D. Guillemot, Recherches sur la gangrène pulmonaire. Thesis, Faculty of Medicine, University of Paris, Paris (1898–1899).
10. J. Hallé, Recherches sur la bactériologie du canal génital de la femme (etat normal et pathologique). Thesis, Faculty of Medicine, University of Paris, Paris (1898).
11. C. Jeannin, Etiologie et pathogénie des infections puerperales putrides (recherches cliniques et bactériologiques). Thesis, Faculty of Medicine, University of Paris, Paris (1902).
12. J. Cottet, Recherches bactériologiques sur les suppurations périurethrales. Thesis, Faculty of Medicine, University of Paris, Paris (1899).
13. L. DS. Smith, Anaerobes and oxygen. *In* "The Anaerobic Bacteria" (V. Fredette, ed.), p. 13. Inst. Microbiol. Hyg., Montreal University, Laval-des-Rapides, Quebec, 1968.
14. J. M. McCord, B. B. Keele, Jr., and I. Fridovich, An enzyme-based theory of obligate anaerobiosis: The physiological function of superoxide dismutase. *Proc. Natl. Acad. Sci. U.S.A.* **68**, 1024 (1971).
14a. L. DS. Smith, "The Pathogenic Anaerobic Bacteria," 2nd ed. Thomas, Springfield, Illinois, 1975.
14b. J. Hewitt and J. G. Morris, Superoxide dismutase in some obligately anaerobic bacteria. *FEBS Lett.* **50**, 315 (1975).
14c. F. P. Tally, N. V. Jacobus, N. Sullivan, S. L. Gorbach, and B. R. Goldin, Superoxide dismutase activity in anaerobic bacteria. *Clin. Res.* **23**, 108A (1975).
14d. F. P. Tally, N. V. Jacobus, B. R. Goldin, and S. L. Gorbach, Superoxide dismutase in anaerobic bacteria. *Clin. Res.* **23**, 418A (1975).

591

14e. W. C. Walden and D. J. Hentges, Differential effects of oxygen and oxidation-reduction potential on the multiplication of three species of anaerobic intestinal bacteria. *Appl. Microbiol.* **30**, 781 (1975).

15. MacLeod, Variations in the periods of exposure to air and oxygen necessary to kill anaerobic bacteria. *Acta Pathol. Microbiol. Scand., Suppl.* **3**, 255 (1930).

16. T. Rosebury, "Glove-Box Procedures for Cultivation of Spirochetes and Other Fastidious Anaerobes. With Preliminary Data on Isolation, Cultivation and Maintenance of Oral Spirochetes and on Limiting Oxygen Concentrations for Surface Growth of These and Other Anaerobic Bacteria," Final Rep., Studies under National Science Foundation Grants GB465, GB654, and GB3823; US Public Health Service, Department of Health, Education and Welfare, Washington, DC., Grant CC00139, 1966.

17. W. J. Loesche, Oxygen sensitivity of various anaerobic bacteria. *Appl. Microbiol.* **18**, 723 (1969).

18. V. Fredette, C. Plante, and A. Roy, Numerical data concerning the sensitivity of anaerobic bacteria to oxygen. *J. Bacteriol.* **93**, 2012 (1967).

19. F. P. Tally, P. R. Stewart, V. L. Sutter, and J. E. Rosenblatt, Oxygen tolerance of fresh clinical anaerobic bacteria. *J. Clin. Microbiol.* **1**, 161, 1975.

20. L. V. Holdeman and W. E. C. Moore, "Anaerobe Laboratory Manual," 3rd ed. V.P.I. Anaerobe Lab., Virginia Polytechnic Institute and State University, Blacksburg, Virginia, 1975.

21. A. R. Prévot, "Biologies des maladies dues aux anaérobies." Editions Médicales Flammarion, Paris, 1955.

22. R. Austrian and P. Collins, Importance of carbon dioxide in the isolation of pneumococci. *J. Bacteriol.* **92**, 1281 (1966).

23. G.-B. Roemer, II. Das serologische und biologische Verhalten der fur den Menschen pathogenen Streptokokken. *Ergeb. Hyg., Bakteriol., Immunitaetsforsch. Exp. Ther,* **26**, 139 (1949).

24. M. L. Stone, Studies on the anaerobic streptococcus. I. Certain biochemical and immunological properties of anaerobic streptococci. *J. Bacteriol.* **39**, 559 (1940).

25. R. F. Anders, D. M. Hogg, and G. R. Jago, Formation of hydrogen peroxide by group N streptococci and its effect on their growth and metabolism. *Appl. Microbiol.* **19**, 608 (1970).

26. R. S. Breed, E. G. D. Murray, and N. R. Smith, eds., "Bergey's Manual of Determinative Bacteriology," 7th ed., Williams & Wilkins, Baltimore, Maryland, 1957.

27. R. J. Duma, A. N. Weinberg, T. F. Medrek, and L. J. Kunz, Streptococcal infections: A bacteriologic and clinical study of streptococcal bacteremia. *Medicine (Baltimore)* **48**, 87 (1969).

27a. J. Carlsson, A numerical taxonomic study of human oral streptococci. *Odontol. Revy.* **19**, 137 (1968).

28. J. B. Evans and M. A. Kerbaugh, Recognition of *Aerococcus viridans* by the clinical microbiologist. *Health Lab. Sci.* **7**, 76 (1970).

29. H. Werner, Das serologische Verhalten von Stammen der Species *Bacteroides convexus, B. thetaiotaomicron, B. vulgatus,* und *B. distasonis. Zentralbl. Bakteriol., Parasitenkd., Infektionskr. Hyg. I. Orig.* **210**, 192 (1969).

30. H. Werner and G. Pulverer, Haufigkeit und medizinische Bedeutung der eitererregenden *Bacteroides*—und *Sphaerophorus*—Arten. *Dtsche. Med. Wochenschr.* **96**, 1325 (1971).

31. H. Werner and G. Rintelen, Untersuchungen uber die Konstanz der Kohlenhydratspaltung bei intestinale: *Bacteroides (Eggerthella)*-Arten. *Zentralbl. Bakteriol., Parasitenkd., Infektionskr. Hyg. I. Orig.* **208**, S521 (1968).

32. R. J. Gibbons, Aspects of the pathogenicity and ecology of the indigenous oral flora

of man. *In* "Anaerobic Bacteria: Role in Disease" (A. Balows *et al.*, eds.), Chapter XXI. Thomas, Springfield, Illinois, 1974.

32a. R. J. Gibbons and J. van Houte, Bacterial adherence in oral microbial ecology. *Ann. Rev. Microbiol.* **29**, 19 (1975).

32b. K. Okuda and I. Takazoe, Haemagglutinating activity of *Bacteroides melaninogenicus*. *Arch. Oral Biol.* **19**, 415 (1974).

32c. W. C. Noble, Skin as a microhabitat. *Postgrad. Med. J.* **51**, 151 (1975).

33. R. J. Gibbons, S. S. Socransky, W. C. De Araujo, and J. van Houte, Studies of the predominant cultivable microbiota of dental plaque. *Arch. Oral Biol.* **9**, 365 (1964).

34. S. M. Finegold, V. L. Sutter, J. D. Boyle, and K. Shimada, The normal flora of ileostomy and transverse colostomy effluents. *J. Infect. Dis.* **122**, 376 (1970).

35. S. M. Finegold, H. R. Attebery, and V. L. Sutter, Effect of diet on human fecal flora: Comparison of Japanese and American diets. *Am. J. Clin. Nutr.* **27**, 1456 (1974).

35a. H. Werner, Differentiation and medical importance of saccharolytic intestinal *Bacteroides*. *Arzneim.-Forsch.* **24**, 340 (1974).

36. R. J. Gibbons and L. P. Engle, Vitamin K compounds in bacteria that are obligate anaerobes. *Science* **146**, 1307 (abstr.) (1964).

37. K. Shimada, K. S. Bricknell, and S. M. Finegold, Deconjugation of bile acids by intestinal bacteria: Review of literature and additional studies. *J. Infect. Dis.* **119**, 273 (1969).

38. M. J. Hill and B. S. Drasar, Degradation of bile salts by human intestinal bacteria. *Gut* **9**, 22 (1968).

39. M. Bohnhoff, C. P. Miller, and W. R. Martin, Resistance of the mouse's intestinal tract to experimental *Salmonella* infection. I. Factors which interefere with the initiation of infection by oral inoculation. *J. Exp. Med.* **120**, 805 (1964).

39a. D. J. Hentges, Inhibition of *Shigella flexneri* by the normal intestinal flora. *J. Bacteriol.* **93**, 1369 (1967).

39b. M. E. Levison, Effect of colon flora and short-chain fatty acids on growth *in vitro* of *Pseudomonas aeruginosa* and Enterobacteriaceae. *Infect. Immun.* **8**, 30 (1973).

40. P. B. Mead and D. B. Louria, Antibiotics in pelvic infections. *Clin. Obstet. Gynecol.* **12**, 219 (1969).

41. S. Suzuki and K. Ueno, Methods of isolation and identification of anaerobes from clinical material. *In* "First Symposium on Anaerobic Bacteria and Their Infectious Diseases" (S. Ishiyama *et al.*, eds.), p. 81. Eisai Co., Tokyo, Japan, 1971.

42. J. de Louvois, V. C. Stanley, R. Hurley, J. B. Jones, and J. E. B. Foulkes, Microbial ecology of the female lower genital tract during pregnancy. *Postgrad. Med. J.* **51**, 156 (1975).

43. S. L. Gorbach, K. B. Menda, H. Thadepalli, and L. Keith, Anaerobic microflora of the cervix in healthy women. *Am. J. Obstet. Gynecol.* **117**, 1053 (1973).

43a. M. J. Ohm and R. P. Galask, Bacterial flora of the cervix from 100 prehysterectomy patients. *Am. J. Obstet. Gynecol.* **12**, 683 (1975).

43b. C. V. Sanders, A. Mickal, A. C. Lewis and J. Torres, Anaerobic flora of the endocervix in women with normal versus abnormal Papanicolaou (Pap) smears. *Clin. Res.* **23**, 30A (1975).

43c. F. J. Bartizal, J. C. Pacheco, G. D. Malkasian, Jr., and J. A. Washington II, Microbial flora found in the products of conception in spontaneous abortions. *Obstet. Gynecol.* **43**, 109 (1974).

44. J. L. Bran, M. E. Levison, and D. Kaye, Entrance of bacteria into the female urinary bladder. *N. Engl. J. Med.* **286**, 626 (1972).

45. S. S. Ambrose, W. W. Taylor, and E. J. Josefiak, Flora of the male lower genitourinary tract. *J. Urol.* **85**, 365 (1961).

46. S. M. Finegold, L. G. Miller, S. L. Merrill, and D. J. Posnick, Significance of anaerobic and capnophilic bacteria isolated from the urinary tract. *In* "Progress in Pyelonephritis" (E. H. Kass, ed.), p. 159. Davis, Philadelphia, Pennsylvania, 1965.

47. J. Brams, I. Pilot, and D. J. Davis, Studies of fusiform bacilli and spirochetes. II. Their occurrence in normal preputial secretions and in erosive and gangrenous balanitis. *J. Infect. Dis.* **32**, 159 (1923).

48. D. J. Davis and I. Pilot, Studies of *Bacillus fusiformis* and Vincent's spirochete. I. Habitat and distribution of these organisms in relation to putrid and gangrenous processes. *J. Am. Med. Assoc.* **79**, 944 (1922).

49. K. L. Burdon, *Bacterium melaninogenicum* from normal and pathologic tissues. *J. Infect. Dis.* **42**, 161 (1928).

49a. M. E. Levison, L. Corman and D. Kaye, Bacterial interference against *Neisseria gonorrhoeae*. *Clin. Res.* **22**, 448A (1974).

50. V. Hurst, Fusiforms in the infant mouth. *J. Dent. Res.* **36**, 513 (1957).

50a. D. A. M. Geddes and G. N. Jenkins, Intrinsic and extrinsic factors influencing the flora of the mouth. *In* "The Normal Microbial Flora of Man" (F. A. Skinner and J. G. Carr, eds.), p. 85. Academic Press, London, 1974.

51. B. S. Drasar, M. Shiner, and G. M. McLeod, Studies on the intestinal flora. I. The bacterial flora of the gastrointestinal tract in healthy and achlorhydric persons. *Gastroenterology* **56**, 71 (1969).

52. J. S. Clarke, Surgical considerations. In Bacteriology of the gut and its clinical implications. *West. J. Med.* **121**, 390 (1974).

52a. R. L. Nichols, B. Miller, and J. W. Smith, Septic complications following gastric surgery: Relationship to the endogenous gastric microflora. *Surg. Clin. N. Amer.* **55**, 1367 (1975).

52b. S. Ishiyama, T. Sakabe, I. Nakayama, H. Iwamoto, S. Iwai, F. Oshima, M. Takatori, T. Kawabe, I. Sakota, T. Kimizaka, and S. Tanaka. Bacteriological studies of the stomach following gastric surgery. *In* "Third Symposium on Anaerobic Bacteria and Their Infectious Diseases," (S. Ishiyama *et al.*, eds.), p. 75. Eisai, Tokyo, Japan, 1973.

53. P. Bhat, S. Shantakumari, D. Rajan, V. I. Mathan, C. R. Kapadia, C. Swarnabai, and S. J. Baker, Bacterial flora of the gastrointestinal tract in southern Indian control subjects and patients with tropical sprue. *Gastroenterology* **62**, 11 (1972).

53a. M. Gracey, Suharjono, Sunoto, and D. E. Stone, Microbial contamination of the gut: Another feature of malnutrition. *Am. J. Clin. Nutr.* **26**, 1170 (1973).

54. D. M. Parkin, R. R. O'Moore, D. J. Cussons, R. R. G. Warwick, P. Rooney, I. W. Percy-Robb, and D. J. C. Shearman, Evaluation of the "breath test" in the detection of bacterial colonisation of the upper gastrointestinal tract. *Lancet* **2**, 777 (1972).

54a. W. R. Brown, D. C. Savage, R. S. Dubois, M. H. Alp, A. Mallory, and F. Kern, Jr., Intestinal microflora of immunoglobulin-deficient and normal human subjects. *Gastroenterology* **62**, 1143 (1972).

55. S. M. Finegold, Interaction of antimicrobial therapy and intestinal flora. *Am. J. Clin. Nutr.* **23**, 1466 (1970).

56. S. M. Finegold, D. J. Posnick, L. G. Miller, and W. L. Hewitt, The effect of various antibacterial compounds on the normal human fecal flora. *Ernährungsforschung* **10**, 316 (1965).

56a. R. C. Hays and G. L. Mandell, pO_2, pH, and redox potential of experimental abscesses. *Proc. Soc. Exp. Biol. Med.* **147**, 29 (1974).

56b. S. L. Gorbach and J. G. Bartlett, Anaerobic infections. *N. Engl. J. Med.* **290**, 1177, 1237, 1289 (1974).

56c. E. B. Kenney and M. M. Ash, Jr., Oxidation reduction potential of developing plaque, periodontal pockets and gingival sulci. *J. Periodontol.* **40**, 14 (1969).

56d. E. R. Nobles, Jr., *Bacteroides* infections. *Ann. Surg.* **177**, 601 (1973).

56e. J. W. Wynne and D. Armstrong, Clostridial septicemia. *Cancer* **29**, 215 (1972).

56f. D. Armstrong, L. S. Young, R. D. Meyer, and A. H. Blevins, Infectious complications of neoplastic disease. *Med. Clin. N. Amer.* **55**, 729 (1971).

56g. R. D. Meyer, Anaerobic gram-negative rod bacteremia. *In*: UCLA Conference on Gram-negative Bacteremia. (L. S. Young, ed.) *Ann. Intern. Med.* in press.

56h. G. T. Keusch, Opportunistic infections in colon carcinoma. *Am. J. Clin. Nutr.* **27**, 1481 (1974).

56i. J. S. Remington, J. D. Gaines, R. B. Griepp, and N. E. Shumway, Further experience with infection after cardiac transplantation. *Transplant. Proc.* **4**, 699 (1972).

56j. J. A. Washington II, W. H. Dearing, E. S. Judd and L. R. Elveback, Effect of pre-operative antibiotic regimen on development of infection after intestinal surgery. Prospective, randomized, double-blind study. *Ann. Surg.* **180**, 567 (1974).

56k. C. A. Rotilie, R. J. Fass and R. L. Perkins, Gentamicin potentiation of *Bacteroides fragilis* infection in rabbits. *Clin. Res.* **22**, 452A (1974).

56l. G. L. Mandell: Bactericidal activity of aerobic and anaerobic polymorphonuclear neutrophils. *Clin. Res.* **22**, 35A (1974).

56m. G. T. Keusch and S. D. Douglas, Intraleukocytic survival of anaerobic bacteria. *Clin. Res.* **22**, 445A (1974).

57. P. R. Courant, I. Paunio, and R. J. Gibbons, Infectivity and hyaluronidase activity of gingival crevice debris. *Arch. Oral Biol.* **10**, 119 (1965).

58. H. Shpuntoff and T. Rosebury, Infectivity of fusospirochetal exudates for guinea pigs, hamsters, mice and chick embryos by several routes of inoculation. *J. Dent. Res.* **28**, 7 (1949).

59. T. Rosebury, A. R. Clark, S. G. Engel, and F. Tergis, Studies of fusospirochetal infection. I. Pathogenicity for guinea pigs of individual and combined cultures of spirochetes and other anaerobic bacteria derived from the human mouth. *J. Infect. Dis.* **87**, 217 (1950).

60. T. Rosebury, A. R. Clark, F. Tergis, and S. G. Engel, Studies of fusospirochetal infection. II. Analysis and attempted quantitative recombination of the flora of fusospirochetal infection after repeated guinea pig passage. *J. Infect. Dis.* **87**, 226 (1950).

61. T. Rosebury, A. R. Clark, J. B. MacDonald, and D. C. O'Connell, Studies of fusospirochetal infection. III. Further studies of a guinea pig passage strain of fusospirochetal infection, including the infectivity of sterile exudate filtrates, of mixed cultures through ten transfers, and of recombined pure cultures. *J. Infect. Dis.* **87**, 234 (1950).

62. J. B. MacDonald, S. S. Socransky, and R. J. Gibbons, Aspects of the pathogenesis of mixed anaerobic infections of mucous membranes. *J. Dent. Res.* **42**, 529 (1963).

63. S. S. Socransky and R. J. Gibbons, Required role of *Bacteroides melaninogenicus* in mixed anaerobic infections. *J. Infect. Dis.* **115**, 247 (1965).

64. K. E. Hite, M. Locke, and H. C. Hesseltine, Synergism in experimental infections with nonsporulating anaerobic bacteria. *J. Infect. Dis.* **84**, 1 (1949).

65. F. L. Meleney, Bacterial synergism in disease processes with a confirmation of the synergistic bacterial etiology of a certain type of progressive gangrene of the abdominal wall. *Ann. Surg.* **94**, 961 (1931).

66. J. B. MacDonald, J. B. Sutton, M. L. Knoll, E. M. Madlener, and R. M. Grainger, The pathogenic components of an experimental fusospirochetal infection. *J. Infect. Dis.* **98**, 15 (1956).

67. W. A. Altemeier, The pathogenicity of the bacteria of appendicitis peritonitis. *Surgery* **11**, 374 (1942).

67a. I. Takazoe and T. Nakamura, Experimental mixed infection by human gingival crevice material. *Bull. Tokyo Dent. Coll.* **12**, 85 (1971).

68. A. B. Onderdonk, W. M. Weinstein, N. M. Sullivan, J. G. Bartlett, and S. L. Gorbach, Experimental intra-abdominal abscesses in rats: Quantitative bacteriology of infected animals. *Infect. Immun.* **10**, 1256 (1974).

68a. A. B. Onderdonk, J. G. Bartlett, T. Louie, N. Sullivan-Seigler, and S. L. Gorbach, Microbial synergy in experimental intraabdominal abscess. *Infect. Immun.* in press.

68b. J. G. Bartlett, A. B. Onderdonk, T. Louie, and S. L. Gorbach, Experimental intra-abdominal sepsis. *Proc. Intern. Cong. Chemotherapy, IX* in press.

69. F. L. Meleney, "Treatise on Surgical Infections." Oxford Univ. Press, London and New York, 1948.

70. D. A. Casciato, J. E. Rosenblatt, L. S. Goldberg, and R. Bluestone, In vitro interaction of *Bacteroides fragilis* with polymorphonuclear leukocytes and serum factors. *Infect. Immun.* **11**, 337 (1975).

71. I. Takazoe and T. Nakamura, Experimental mixed infection due to various oral anaerobes. *In* "First Symposium on Anaerobic Bacteria and Their Infectious Diseases" (S. Ishiyama *et al.*, eds.), p. 70. Eisai, Tokyo, Japan, 1971.

71a. K. Okuda and I. Takazoe, Antiphagocytic effects of the capsular structure of a pathogenic strain of *Bacteroides melaninogenicus*. *Bull. Tokyo Dent. Coll.* **14**, 99 (1973).

71b. D. L. Kasper, The polysaccharide capsule of *Bacteroides fragilis* subspecies *fragilis*: Immunochemical and morphologic definition. *J. Infect. Dis.* **133**, 79 (1976).

71c. C. B. Cox, C. Hardegree, and R. Fornwald, Effect of tetanolysin on platelets and lysosomes, *Infect. Immun.* **9**, 696 (1974).

71d. C.-E. Nord, T. Wadström, K. Dornbusch, and B. Wretlind, Extracellular proteins in five clostridial species from human infections. *Med. Microbiol. Immunol.* **161**, 145 (1975).

72. S. E. Mergenhagen, E. G. Hampp, and H. W. Scherp, Preparation and biological activities of endotoxins from oral bacteria. *J. Infect. Dis.* **108**, 304 (1961).

72a. D. L. Kasper and M. W. Seiler, Immunochemical characterization of the outer membrane complex of *Bacteroides fragilis* subspecies *fragilis*. *J. Infect. Dis.* **132**, 440 (1975).

72b. T. Kristoffersen and T. Hofstad, Immunochemistry of fusobacteria. *Res. Immunochem. Immunobiol.* **3**, 253 (1973).

73. T. Hofstad, Chemical characteristics of *Bacteroides melaninogenicus* endotoxin. *Arch. Oral Biol.* **13**, 1149 (1968).

74. T. Hofstad and T. Kristoffersen, Chemical characteristics of endotoxin from *Bacteroides fragilis* NCTC 9343. *J. Gen. Microbiol.* **61**, 15 (1970).

75. T. Hofstad and T. Kristoffersen, Lipopolysaccharide from *Bacteroides melaninogenicus* isolated from the supernatant fluid after ultracentrifugation of the water phase following phenol-water extraction. *Acta Pathol. Microbiol. Scand., Sect. B* **79**, 12 (1971).

76. T. Hofstad, Preparation and chemical characteristics of endotoxic lipopolysaccharide from three strains of *Sphaerophorus necrophorus*. *Acta Pathol. Microbiol. Scand., Sect. B* **79**, 385 (1971).

77. T. Hofstad, T. Kristoffersen, and K. A. Selvig, Electron microscopy of endotoxic lipopolysaccharide from *Bacteroides*, *Fusobacterium*, and *Sphaerophorus*. *Acta Pathol. Microbiol. Scand., Sect. B* **80**, 413 (1972).

78. H. Schwabacher, D. R. Lucas, and C. Rimington, *Bacterium melaninogenicum*—a misnomer. *J. Gen. Microbiol.* **1**, 109 (1947).

79. S. J. Sawyer, J. B. MacDonald, and R. J. Gibbons, Biochemical characteristics of *Bacteroides melaninogenicus*. *Arch. Oral Biol.* **7**, 685 (1962).

79a. G. Dahlen and A. Linde, Screening plate method for detection of bacterial β-glucuronidase. *Appl. Microbiol.* **26**, 863 (1973).

79b. R. J. Sharbaugh and W. M. Rambo, Serum prognostic indicators in experimental *Bacteroides* peritonitis. *Arch. Surg.* **110**, 1146 (1975).

80. C. Weiss, The pathogenicity of *Bacteroides melaninogenicus* and its importance in surgical infections. *Surgery* **13**, 683 (1943).

81. G. Pulverer and S. Heinrich, Infektionsversuche an Laboratoriumstieren und *in vitro*-Untersuchungen zur Fermentausstattung des *Bacteroides melaninogenicus*. *Z. Hyg.* **146**, 341 (1960).

82. K. L. Burdon, Isolation and cultivation of *Bacterium melaninogenicum*. *Proc. Soc. Exp. Biol. Med.* **29**, 1144 (1932).

83. R. J. Gibbons and J. B. MacDonald, Degradation of collagenous substrates by *Bacteroides melaninogenicus*. *J. Bacteriol.* **81**, 614 (1961).

84. F. A. Waldvogel and M. N. Swartz, Collagenolytic activity of bacteria. *J. Bacteriol.* **98**, 662 (1969).

85. E. Hausmann, P. R. Courant, and D. S. Arnold, Conditions for the demonstration of collagenolytic activity in *Bacteroides melaninogenicus*. *Arch. Oral Biol.* **12**, 317 (1967).

86. R. C. Kestenbaum, J. Massing, and S. Weiss, The role of collagenase in mixed infections containing *Bacteroides melaninogenicus*. *Int. Assoc. Dent. Res., 42nd Gen. Meet., 1964* Abstract No. 9 (1964).

87. E. J. Kaufman, P. A. Mashimo, E. Haussman, C. T. Hanks, and S. A. Ellison, Fusobacterial infection: Enhancement by cell free extracts of *Bacteroides melaninogenicus* possessing collagenolytic activity. *Arch. Oral Biol.* **17**, 577 (1972).

88. B. M. Gesner and C. R. Jenkin, Production of heparinase by *Bacteroides*. *J. Bacteriol.* **81**, 595 (1961).

89. H. S. Bjornson and E. O. Hill, Bacteroidaceae in thromboembolic disease: Effects of cell wall components on blood coagulation *in vivo* and *in vitro*. *Infect. Immun.* **8**, 911 (1973).

89a. R. H. Schwarz, "Septic Abortion." Lippincott, Philadelphia, Pennsylvania, 1968.

89b. W. A. Altemeier, W. R. Culbertson, and J. P. Fidler, Giant horseshoe intraabdominal abscess. *Ann. Surg.* **181**, 716 (1975).

89c. T. W. Huber, E. G. Macias, P. Holmes, and H. D. Bredthauer, A clinical isolate of *Salmonella typhi* requiring anaerobic conditions for primary isolation. *Amer. J. Clin. Pathol.* **63**, 117 (1975).

89d. J. Armata, R. Cyklis, and A. Scieslicki, Thrombocythaemia in actinomycosis. *Lancet* **2**, 261 (1973).

89e. P. D. Hoeprich, Etiologic diagnosis of lower respiratory tract infections. *Calif. Med.* **112**, 1 (1970).

90. J. G. Bartlett, J. E. Rosenblatt, and S. M. Finegold, Percutaneous transtracheal aspiration in the diagnosis of anaerobic pulmonary infection. *Ann. Intern. Med.* **79**, 535 (1973).

90a. J. A. Guilbeau, Jr. and I. G. Schaub, Uterine culture technique. A simple method for avoiding contamination by cervical and vaginal flora. *Am. J. Obstet. Gynecol.* **58**, 407 (1949).

91. W. J. Ledger, C. Gee, P. A. Pollin, W. L. Lewis, V. L. Sutter, and S. M. Finegold, A new approach to patients with suspected anaerobic post partum pelvic infections. Transabdominal uterine aspiration for culture and metronidazole for treatment. *Am. J. Obstet. Gynecol.* **126**(1), 1 (1976).

92. H. R. Attebery and S. M. Finegold, Combined screw-cap and rubber-stopper closure for Hungate tubes (pre-reduced anaerobically sterilized roll tubes and liquid media). *Appl. Microbiol.* **18**, 558 (1969).

93. H. R. Attebery and S. M. Finegold, A miniature anaerobic jar for tissue transport or for cultivation of anaerobes. *Am. J. Clin. Pathol.* **53**, 383 (1970).

93a. A. I. Braude, Anaerobic infection: Diagnosis and therapy. *Hosp. Pract.* **3**, 42 (1968).
93b. H. O. Hallander, A. Flodström and K. Holmberg, Influence of the collection and transport of specimens on the recovery of bacteria from peritonsillar abscesses. *J. Clin. Microbiol.* **2**, 504 (1975).
93c. K. Holmberg, "Studies on the Actinomycetaceae by means of numerical taxonomy, immunofluorescence, and crossed immuno-electrophoresis." Dissertation thesis. National Bacteriological Laboratory and Karolinska Institute, Stockholm, Göteborgs offsettrycker, Stockholm, 1975.
93d. M. H. Griffin, Fluorescent antibody techniques in the identification of the gram-negative nonsporeforming anaerobes. *Health Lab. Sci.* **7**, 78 (1970).
93e. G. L. Lombard and V. R. Dowell, Jr., Preparation of fluorescent antibody reagents for identification of *Bacteroides. Abstr. Annual Meet. Amer. Soc. Microbiol. 1972,* Abstract M93, p. 95.
93f. R. K. Porschen and E. H. Spaulding, Fluorescent antibody study of the gram-positive anaerobic cocci. *Appl. Microbiol.* **28**, 851 (1974).
93g. M. Sterne and I. Batty, "Pathogenic Clostridia." Butterworths, London, 1975.
93h. S. L. Gorbach, J. W. Mayhew, J. G. Bartlett, H. Thadepalli, and A. B. Onderdonk, Rapid diagnosis of *Bacteroides fragilis* infections by direct gas liquid chromatography of clinical specimens. *Clin. Res.* **22**, 442A (1974).
94.   G. E. Killgore, S. E. Starr, V. E. Del Bene, D. N. Whaley, and V. R. Dowell, Jr., Comparison of three anaerobic systems for the isolation of anaerobic bacteria from clinical specimens. *Am. J. Clin. Pathol.* **59**, 552 (1973).
95.   J. E. Rosenblatt, A. Fallon, and S. M. Finegold, Comparison of methods for isolation of anaerobic bacteria from clinical specimens. *Appl. Microbiol.* **25**, 77 (1973).
95a. J. H. Brewer and D. L. Allgeier, Safe self-contained carbon dioxide-hydrogen anaerobic system. *Appl. Microbiol.* **14**, 985 (1966).
96.   W. J. Martin, Practical method for isolation of anaerobic bacteria in the clinical laboratory. *Appl. Microbiol.* **22**, 1168 (1971).
97.   V. L. Sutter, V. L. Vargo, and S. M. Finegold, "Wadsworth Anaerobic Bacteriology Manual," 2nd ed. Ext. Div., Univ. of California, Los Angeles, 1975.
98.   V. R. Dowell, Jr. and T. M. Hawkins, "Laboratory Methods in Anaerobic Bacteriology," CDC Lab. Manual. US Dept. of Health, Education, and Welfare, Pub. Health Serv., Atlanta, Georgia, 1974. (DHEW Publ. No. (CDC) 74–8272. US Govt. Printing Office, Washington, D.C., 1974.)
99.   E. H. Lennette, E. H. Spaulding, and J. P. Truant, "Manual of Clinical Microbiology," 2nd ed. Am. Soc. Microbiol., Washington, D.C., 1974.
99a. J. E. Rosenblatt and P. R. Stewart, Anaerobic bag culture method. *J. Clin. Microbiol.* **1**, 527 (1975).
99b. D. C. Shanson and M. Barnicoat, An experimental comparison of Thiol broth with Brewer's thioglycolate for anaerobic blood cultures. *J. Clin. Pathol.* **28**, 407 (1975).
100. E. D. Hoare, The suitability of "Liquoid" for use in blood culture media, with particular reference to anaerobic streptococci. *J. Pathol. Bacteriol.* **48**, 573 (1939).
101. M. H. Graves, J. A. Morello, and F. E. Kocka, Sodium polyanethol sulfonate sensitivity of anaerobic cocci. *Appl. Microbiol.* **27**, 1131 (1974).
102. F. E. Kocka, E. J. Arthur, and R. L. Searcy, Comparative effects of two sulfated polyanions used in blood culture on anaerobic cocci. *Am. J. Clin. Pathol.* **61**, 25 (1974).
103. R. J. Alpern and V. R. Dowell, Jr., *Clostridium septicum* infections and malignancy. *J. Am. Med. Assoc.* **209**, 385 (1969).
103a. D. M. Poretz, L. Wood and C. Park, Adenocarcinoma of the colon presenting as *Clostridium septicum* cellulitis of the left thigh. *South. Med. J.* **67**, 862 (1974).

103b. K. Holmberg, C.-E. Nord, and T. Wadström: Serologic studies of *Actinomyces israelii* by crossed immunoelectrophoresis: Standard antigen-antibody system for *A. israelii*. *Infect. Immun.* **12**, 387 (1975).

103c. J. D. Quick, H. S. Goldberg, and A. C. Sonnenwirth, Human antibody to Bacteroidaceae. *Am. J. Clin. Nutr.* **25**, 1351 (1972).

103d. T. Hofstad, Antibodies reacting with lipopolysaccharides from *Bacteroides melaninogenicus*, *Bacteroides fragilis*, and *Fusobacterium nucleatum* in serum from normal human subjects. *J. Infect. Dis.* **129**, 349 (1974).

103e. J. P. Rissing, J. G. Crowder, J. W. Smith, and A. White, Detection of *Bacteroides fragilis* infection by precipitin antibody. *J. Infect. Dis.* **130**, 70 (1974).

103f. D. W. Lambe, Jr., D. Danielsson, D. H. Vroon, and R. K. Carver, Immune response in eight patients infected with *Bacteroides fragilis*. *J. Infect. Dis.* **131**, 499 (1975).

104. M. C. Robson and J. P. Heggers, Surgical infection. I. Single bacterial species or polymicrobic in origin? *Surgery* **65**, 608 (1969).

105. W. R. Wilson, W. J. Martin, C. J. Wilkowske, and J. A. Washington, II, Anaerobic bacteremia. *Mayo Clin. Proc.* **47**, 639 (1972).

106. G. M. Dack, Non-sporeforming anaerobic bacteria of medical importance. *Bacteriol. Rev.* **4**, 227 (1940).

107. H. Lödenkamper and G. Stienen, The treatment of anaerobic infections. *Ger. Med. Mon.* **1**, 233 (1956).

108. E. J. Stokes, Anaerobes in routine diagnostic cultures. *Lancet* **1**, 668 (1958).

109. W. J. Martin, Isolation and identification of anaerobic bacteria in the clinical laboratory. A 2-year experience. *Mayo Clin. Proc.* **49**, 300 (1974).

110. K. Hoffmann and F. W. Gierhake, Postoperative infection of wounds by anaerobes. *Ger. Med. Mon.* **14**, 31 (1969).

110a. A. A. B. Mitchell, Incidence and isolation of *Bacteroides* species from clinical material and their sensitivity to antibiotics. *J. Clin. Pathol.* **26**, 738 (1973).

110b. D. A. Leigh, *Bacteroides* infections. *Lancet* **1**, 1081 (1973).

110c. L. P. Clark, H. A. Marshall, and N. B. Ackerman, The role of *Bacteroides* as an infectious organism. *Surg., Gynecol. Obstet.* **138**, 562 (1974).

111. L. H. Mattman, G. Senos, and E. D. Barrett, The anaerobic micrococci: Incidence, habitat, growth requirements. *Am. J. Med. Technol.* **35**, 167 (1958).

112. W. R. Sandusky, E. J. Pulaski, B. A. Johnson, and F. L. Meleney, The anaerobic nonhemolytic streptococci in surgical infections on a general surgical service. *Surg., Gynecol. Obstet.* **75**, 145 (1942).

113. S. L. Gorbach, H. Thadepalli, and J. Norsen, Isolation of 25 clostridia species from clinical sources. *Abstr., 12th Interscience Conf. Antimicrob. Agents & Chemotherapy, 1972*, p. 186 (1972).

113a. W. C. Weese and I. M. Smith, A study of 57 cases of actinomycosis over a 36-year period. *Arch. Intern. Med.* **135**, 1562 (1975).

113b. G. Pulverer, Problems of human actinomycosis. *Postepy. Hig. Med. Dosw.* **28**, 253 (1974).

114. K. Nakamura, K. Saito, K. Inoue, A. Hayashi, K. Sawatari, and C. Mochida, Present status of isolation of anaerobic bacteria from clinical material. *Jpn. J. Clin. Pathol.* **19**, Suppl., 127 (1971).

115. H. Beerens and M. Tahon-Castel, "Infections humaines a bactéries anaérobies nontoxigénes." Présses Acad. Eur., Bruxelles, 1965.

115a. W. J. Ledger and K. A. Hackett, Significance of clostridia in the female reproductive tract. *Obstet. Gynecol.* **41**, 525 (1973).

115b. B. Duflo, A. Schaeffer, A.-M. Durivage, B. Christoforov, J. Guerre and H. Pequignot,

Infections nonpuerpérales, sans hémolyse ni anurie à *Welchia perfringens*. A propos de 7 observations. *Sem. Hôp.* **49**, 991 (1973).

116. S. Heinrich and G. Pulverer, Uber den Nachweis des *Bacteroides melaninogenicus* in Krankheitsprozessen bei Mensch und Tier. *Z. Hyg.* **146**, 331 (1960).

117. H. Werner, G. Rintelen, and H. Kunstek-Santos, A new butyric-acid-producing *Bacteroides* species: *B. splanchnicus* n. sp. *Zentralbl. Bakteriol. Parasitenkd., Infektionskr. Hyg., Abt. 1: Orig., Reihe A* **231**, 133 (1975).

118. F. P. Tally, A. Y. Armfield, V. R. Dowell, Jr., Y.-Y. Kwok, V. L. Sutter, and S. M. Finegold, Susceptibility of *Clostridium ramosum* to antimicrobial agents. *Antimicrob. Agents & Chemother.* **5**, 589 (1974).

119. A. Y. Armfield, J. M. Felner, F. S. Thompson, L. M. McCroskey, and A. Balows, Comparison of clinical and bacteriologic data from *Eubacterium filamentosum* and *Clostridium innocuum. Bacteriol. Proc.* (Abst. M266) p. 109 (1971).

120. L. K. Georg. G. W. Robertstad, S. A. Brinkman, and M. D. Hicklin, A new pathogenic anaerobic *Actinomyces* species. *J. Infect. Dis.* **115**, 88 (1965).

121. A. J. R. Möller, Microbiological examination of root canals and periapical tissues of human teeth. Methodological studies. *Odontol. Tidskr.* **74**, 1 (1966).

122. T. H. Melville and R. H. Birch, Root canal and periapical floras of infected teeth. *Oral Surg., Oral Med. Oral Pathol.* **23**, 93 (1967).

123. J. J. Crawford and R. J. Shankle, Application of newer methods to study the importance of root canal and oral microbiota in endodontics. *Oral Surg., Oral Med. Oral Pathol.* **14**, 1109, 1961.

124. E. Baumgartner, Kulturmethoden. *Vierteljahrsschr. Zahnheilkd.* **24**, 362 (1908).

125. F. P. Hadley and U. G. Rickert, Studies on the etiology of root canal infections. II. The nature of the causative agents and means for their detection. *Dent. Cosmos* **69**, 665 (1927).

126. E. G. Hampp, Isolation and identification of spirochetes obtained from unexposed canals of pulp-involved teeth. *Oral Surg., Oral Med. Oral Pathol.* **10**, 1100 (1957).

127. L. R. Brown, Jr. and C. E. Rudolph, Jr., Isolation and identification of microorganisms from unexposed canals of pulp-involved teeth. *Oral Surg., Oral Med. Oral Pathol.* **10**, 1094 (1957).

128. M. A. Mazzarella, W. J. Hedman, Jr., and L. R. Brown, Jr., "Classification of Microorganisms from the Pulp Canal of Nonvital Teeth," Res. Rep., Proj. NM008 015. 10.01. US Naval Dental School, Bethesda, Maryland, 1955.

129. B. Engstrom and G. Frostell, Bacteriological studies of the non-vital pulp in cases with intact pulp cavities. *Acta Odontol. Scand.* **19**, 23 (1960).

130. B. Engstrom, Bacteriological cultures in root canal therapy. *Trans. R. Sch. Dent.* **11**, 5 (1964).

131. K. C. Winkler and J. van Amerongen, Bacteriologic results from 4,000 root canal cultures. *Oral Surg., Oral Med. Oral Pathol.* **12**, 857 (1959).

132. R. J. Matusow, Microbiology of the pulp and periapical tissues: Culture control. *Dent. Clin. North Am.* p. 549 (1967).

133. J. Fourestier, J. Gacon, H. Le Diascorn, F. Pierre, A. Le Stir, A. Ransan, H. Curcier, and J. Comby, Contribution à l'étude des foyers infectieux buccodentaires chez le sujet jeune. *Rev. Stomatol. Chir. Maxillo-Fac. Paris* **71**, 411 (1970).

134. A. R. Prévot, Diversite clinique et pluralité étiologique des maladies causées par les anaérobies du genre *Ramibacterium. Rev. Med. Univ. Montreal* **1**, 162 (1949).

135. H. Dietrich, W. Klein, and H. Nickel, Uber Auftreten und Resistenzbestimmung Anaerober Sporenloser Infektionserreger. *Zentralbl. Bakteriol., Parasitenkd., Infektionskr. Hyg., Abt. 1: Orig.* **197**, 515 (1965).

136. D. J. Davis and F. B. Moorehead, Studies of fusiform bacilli and spirochaetes. VIII. Their occurrence in alveolar abscesses. *J. Dent. Res.* **5**, 1 (1923).

137. J. Seitz, *Bacillus hastilis. Z. Hyg. Infektionskr.* **30**, 47 (1899).

138. T. L. Gilmer and A. M. Moody, A study of the bacteriology of alveolar abscess and infected root canals. *J. Am. Med. Assoc.* **63**, 2023 (1914).

139. A. Bulleid, Bacteriological studies of apical infection. *Br. Dent. J.* **52**, 105 (1931).

140. K. L. Pesch and C. Ruland, Anaerobe Streptokokken in Zahnwurzelgranulomen. Variabilitat der Anaerobiose. *Zentralbl. Bakteriol., Parasitenkd. Infektionskr. Hyg., Abt. 1: Orig.* **134**, 1 (1935).

141. G. Pulverer, Zur Morphologie, Biochemie und Serologie des *Bacteroides melaninogenicus. Z. Hyg.* **145**, 293 (1958).

142. A. Grumbach and H. W. Hotz, *Bacillus anaerobius diphtheroides* als Erreger einer Septikopyamie. *Schweiz. Z. Pathol. Bakteriol.* **2**, 230 (1939).

143. R. M. Browne and B. C. O'Riordan, A colony of *Actinomyces*-like organisms in a periapical granuloma. *In* "Actinomycosis" (M. Bronner and M. Bronner, eds.), p. 154. Williams & Wilkins, Baltimore, Maryland, 1969.

143a. D. Schlegel and E. Hieber, Praktische Erfahrungen mit 7-Chlor-7-desoxy-Lincomycin in der Kieferchirurgie. *Zahnaerztl. Praxis.* **20**, 37 (1969).

144. A. Bulleid, Bacteriological studies of apical infection. *Br. Dent. J.* **52**, 197 (1931).

145. J. E. Morin and A. Potvin, Remarques sur trente-cinq infections a petits anaérobies. *Laval Med.* **22**, 625 (1957).

146. M. H. Vincent, Sur les propriétés pyogènes du *Bacille fusiforme. C. R. Seances Soc. Biol. Ses Fil* **58**, 772 (1905).

147. C. C. Alling, Postextraction osteomyelitic syndrome. *Dent. Clin. North Am.* p. 621 (1959).

148. J. Schroff and H. A. Bartels, Painful sockets after extractions. A preliminary report on the investigation of their etiology, prevention, and treatment. *J. Dent. Res.* **9**, 81 (1929).

149. L. E. Francis and J. A. de Vries, Therapeutics and the management of common infections. *Dent. Clin. North Am.* p. 243 (1968).

150. G. W. Stevenson and H. H. Gossman, Dental and intracranial actinomycosis. *Br. J. Surg.* **55**, 830 (1968).

151. P. Kapsimalis and G. E. Garrington, Actinomycosis of the periapical tissues. *Oral Surg., Oral Med. Oral Pathol.* **26**, 374 (1968).

152. R. M. Coleman, L. K. Georg, and A. R. Rozzell, *Actinomyces naeslundii* as an agent of human actinomycosis. *Appl. Microbiol.* **18**, 420 (1969).

153. V. Bezjak and O. P. Arya, Actinomycosis of the face associated with *Actinomyces naeslundii. Mykosen* **14**, 9 (1971).

154. A. R. Prévot, P. Tardieux, L. Joubert, and F. de Cadore, Recherches sur *Fusiformis nucléatus* (Knorr) et son pouvoir pathogène pour l'homme et les animaux. *Ann. Inst. Pasteur, Paris* **91**, 787 (1956).

155. D. Veszprémi, Kultur- und Tierversuche mit dem *Bacillus fusiformis* und dem *Spirillum. Zentralbl. Bakteriol., Parasitenkd., Infektionskr. Hyg., Abt. 1: Orig.* **38**, 136 (1905).

156. L. E. Thompson, A fatal case of brain abscess from Vincent's angina following extraction of a tooth under procaine hydrochloride. *J. Am. Med. Assoc.* **93**, 1063 (1929).

157 H. Ikemoto, N. Hazato, and F. Matsumura, Actinomycosis of the central nervous system. *Naika.* **21**, 754 (1968).

158. D. A. Mason, Steroid therapy and dental infection. Case report. *Br. Dent. J.* **128**, 271 (1970).

159. E. Schulz, Ueber Bacillus Buday-Sepsis, insbesondere Endokarditis dentaler Genese. *Zentralbl. Bakteriol., Parasitenkd., Infektionskr. Hyg., Abt. 1:* **134**, 466 (1935).

160. D. K. Crystal, S. W. Day, C. L. Wagner, and J. M. Kranz, Emergency treatment in Ludwig's angina. Surg., Gynecol. Obstet. **129**, 755 (1969).

161. J. G. Bartlett and S. M. Finegold, Anaerobic pleuropulmonary infections. Medicine (Baltimore) **51**, 413 (1972).

162. E. Witebsky and G. E. Miller, A common infection commonly overlooked. GP **18**, 111 (1958).

163. W. P. Larson and M. Barron, Report of a case in which the fusiform bacillus was isolated from the blood stream. J. Infect. Dis. **13**, 429 (1913).

164. D. Gröschel, Akute Anaerobier—Osteomyelitis nach Extraktion von Milchzahnen. ZWR—Zahnaerztl. Welt, Zahnaerztl. Rundsch., Zahnaerztl. Reform, Stoma **81**, 488 (1972).

165. A. Laporte, H. Brocard, and M. Bouvier, Bactériémie d'origine dentaire a Fusobacterium biacutum. Bull. Mem. Soc. Med. Hôp. Paris **57**, 667 (1941).

166. L. Boez., A. Kehlstadt, and J. Schreiber, Les bactériémies anaérobies à Bacillus ramosus (sept observations). Ann. Med. (Paris) **23**, 340 (1928).

167. A. R. Prévot, Morphologie, physiologie, pouvoir pathogène et systématique des Actinomycétales anaérobies, in "Union Internationale des Sciences Biologiques," Ser. B, No. 4. Actinomycétales. Morfologia, Biologia, et Sistematica. Secrétariat Général de l'U.I.S.B., Paris, 1953.

168. J. M. Felner and V. R. Dowell, Jr., "Bacteroides" bacteremia. Am. J. Med. **50**, 787 (1971).

169. M. Vic-Dupont, J. Laufer, and F. Cartier, Onze maladies d'Osler d'origine dentaire. Bull. Mem. Soc. Med. Hôp. Paris [3] **114**, 869 (1963).

170. S. Mutermilch and P. Séguin, Un cas d'infection fuso-spirillaire mortelle chez un sujet atteint d'épithelioma du plancher de la bouche et traité par le radium. C. R. Seances Soc. Biol. Ses. Fil. **88**, 28 (1923).

171. A. R. Prévot, Recherches sur les associations microbiennes de quelques infections buccales. Rev. Stomatol. Chir. Maxillo-Fac. **49**, 1 (1948).

172. J. Caroli, A. Paraf, J. Etévé, and A. Prévot, Hépatite ictérigéne necrosante pseudo-augiocholitique à Fusiformis nucleatus. Rev. Med.-Chir. Mal. Foie **31**, 29 (1956).

173. A. R. Prévot, M. Digeon, M. Peyre, J. Pantaleon, and J. Senez, Étude de quelques. bactéries anaérobies nouvelles ou mal connues. Ann. Inst. Pasteur, Paris **73**, 409 (1947).

174. R. Vinzent and V. Reynes, Étude d'un nouvel anaérobie: Ramibacterium dentium. Ann. Inst. Pasteur, Paris **73**, 594 (1947).

175. E. J. Degnan, E. C. Hinds, and A. H. Sills, Role of anaerobic streptococci in oral surgery. J. Oral Surg., Anesth. Hosp. Dent. Serv. **18**, 464 (1960).

176. K. Hara, A. Saito, K. Inoue, A. Hayashi, T. Mochida, and K. Sawaturi, 31 cases of anaerobie infectious diseases. In "First Symposium on Anaerobic Bacteria and Their Infectious Diseases" (S. Ishiyama et al., eds.), p. 26. Eisai, Tokyo, Japan, 1971.

177. G. A. Caron and I. Sarkany, Cervicofacial actinomycosis. In "Actinomycosis" (M. Bronner and M. Bronner, eds.), p. 163. Williams & Wilkins, Baltimore, Maryland, 1969.

178. S. Heinrich and G. Pulverer, Zur Ätiologie und Mikrobiologie der Aktinomykose. III. Die pathogene Bedeutung des Actinobacillus actinomycetem-comitans unter den "Begleitbakterien" des Actinomyces israelii. Zentrlbl. Bakteriol., Parasitenkd., Infektionskr. Hyg., Abt. 1: Orig. **176**, 91 (1959).

178a. J. Linhard, Sur une nouvelle Actinomycétale anaérobie des cellulites jugales: Actinobacterium cellulitis n. sp. Ann. Inst. Pasteur, Paris **76**, 478 (1949).

179. J. E. Hamner, III and M. E. Schaefer, Anterior maxillary actinomycosis. In "Actinomycosis" (M. Bronner and M. Bronner, eds.), p. 196. Williams & Wilkins, Baltimore, Maryland, 1969.

180. W. G. Sprague and W. G. Shafer, Presence of actinomyces in dentigerous cyst. *In* "Actinomycosis" (M. Bronner and M. Bronner, eds.), p. 208, Williams & Wilkins, Baltimore, Maryland, 1969.

181. K. H. Thoma and H. M. Goldman, "Oral Pathology," 5th ed. Mosby, St. Louis, Missouri, 1960.

182. F. M. Wentz and R. J. Pollock, Jr., Periodontics. *Dent. Clin. North Am.* **13**, 495 (1969).

183. G. P. Barnes, W. F. Bowles, III, and H. G. Carter, Acute necrotizing ulcerative gingivitis: A survey of 218 cases. *J. Periodontol.* **44**, 35 (1973).

184. D. B. Giddon, P. Goldhaber, and J. M. Dunning, Prevalence of reported cases of acute necrotizing ulcerative gingivitis in a university population. *J. Periodontol.* **34**, 366 (1963).

184a. M. A. Listgarten and D. W. Lewis, The distribution of spirochetes in the lesion of acute necrotizing ulcerative gingivitis: An electron microscopic and statistical survey. *J. Periodontol.* **38**, 379 (1967).

185. A. H. Davies, J. A. McFadzean, and S. Squires, Treatment of Vincent's stomatitis with metronidazole. *Br. Med. J.* **1**, 1149 (1964).

186. C. L. S. Shinn, S. Squires, and A. McFadzean, The treatment of Vincent's disease with metronidazole. *Dent. Pract.* **15**, 275 (1965).

187. A. B. Wade, G. C. Blake, and K. B. Mirza, Effectiveness of metronidazole in treating the acute phase of ulcerative gingivitis. *Dent. Pract.* **16**, 440 (1966).

188. R. Duckworth, J. P. Waterhouse, D. E. R. Britton, A. Shciham, R. Winter, and G. C. Blake, Acute ulcerative gingivitis. *Br. Dent. J.* **2**, 599 (1966).

189. J. P. Fletcher and C. G. Plant, An assessment of metronidazole in the treatment of acute ulcerative pseudomembranous gingivitis (Vincent's disease). *Oral Surg., Oral Med. Oral Pathol.* **22**, 729 (1966).

190. J. Lozdan, A. Sheiham, B. A. Pearlman, B Keiser, C. C. Rachanis, and R. Meyer, The use of nitrimidazine in the treatment of acute ulcerative gingivitis. A double-blind controlled trial. *Br. Dent. J.* **130**, 294 (1971).

191. S. M. Kozol and H. V. Shuster, A description of the antibiotic bacitracin: Its topical use in the treatment of Vincent's infection. *Oral Surg., Oral Med. Oral Pathol.* **5**, 717 (1952).

192. D. F. Mitchell and B. R. Baker, Topical antibiotic control of necrotizing gingivitis. *J. Periodontol.* **39**, 21 (1968).

193. A. B. Wade, G. C. Blake, J. D. Manson, J. K. Berdon, F. Mathieson, and D. M. Bate, Treatment of the acute phase of ulcerative gingivitis (Vincent's type). *Br. Dent. J.* **115**, 372 (1963).

194. A. B. Wade and K. B. Mirza, The relative effectiveness of sodium peroxyborate and hydrogen peroxide in treating acute ulcerative gingivitis. *Dent. Pract.* **14**, 185 (1964).

195. R. Rentsch, Zur Diagnostik nicht sporenbildender Anaerobier. Inaugural-Dissertation, University of Zurich (1963).

196. C. Weiss and D. G. Mercado, Demonstration of type specific proteins in extracts of fusobacteria. *J. Exp. Med.* **67**, 49 (1938).

197. A. R. Prévot. Actinomycétale anaérobie stricte nouvelle: *Spherophorus ridiculosus. Ann. Inst. Pasteur, Paris* **75**, 387 (1948).

198. R. A. Kyle and J. W. Linman, Gingivitis and chronic idiopathic neutropenia: Report of two cases. *Mayo Clin. Proc.* **45**, 494 (1970).

199. K. Buday, Zur Pathogenese der gangranosen Mund—und Rachenentzundungen. *Beitr. Pathol. Anat. Allg. Pathol.* **38**, 255 (1905).

200. V. Babes, Spindelformige Bazillen. *In* "Handbuch der pathologischen Mikroorganismen". (W. Kolle and A. von Wassermann, eds.), Suppl. Vol. 1, p. 271. Fischer, Jena, 1906–1907.

201. F. Campbell and F. W. Shaw, A case of gangrenous stomatitis probably caused by the *Bacillus necrophorus. J. Kans. Med. Soc.* **11**, 77 (1911).

202. P. Carnot and P. Blamoutier, Stomato-rhino-conjonctivite à fusospirilles. *Arch. Int. Laryngol.* **31**, 1 (1924).

203. O. Lahelle, Four cases of fusospirochetal infections investigated with respect to production of antibodies. *Acta Pathol. Microbiol. Scand.* **22**, 34 (1945).

204. H. H. Lichtenberg, M. Werner, and E. V. Lueck, The pathogenicity of the fusiform bacillus and spirillum of Plaut-Vincent. *J. Am. Med. Assoc.* **100**, 707 (1933).

205. A. P. Stephens, Acute fusospirochaetal stomatitis associated with a partial upper denture. *Dent. Pract.* **22**, 305 (1972).

206. D. B. Jelliffe, Infective gangrene of the mouth. *Pediatrics* **9**, 544 (1952).

207. D. B. Jelliffe, Antibiotic treatment of infective gangrene of the mouth. *J. Trop. Med. Hyg.* **56**, 53 (1953).

208. G. Boering, "Diseases of the Oral Cavity and Salivary Glands." Wright, Bristol, 1971.

209. L. H. Roth, Aureomycin in overcoming oral odors. *W. Virg. Dent. J.* **25–26**, 98, (1951).

210. Z. Cope, "Actinomycosis." Oxford Univ. Press, London and New York, 1938.

211. L. F. Barker and S. R. Miller, Perforating ulcer of the hard palate resembling tertiary syphilis but due to a fusospirillary invasion (so-called Vincent's angina). *J. Am. Med. Assoc.* **71**, 793 (1918).

211a. C. D. Fritsche, Die Klinik und Therapie der Bacteroidaceae. *Zbl. Bakt. Hyg. I. Abt. Orig. A.* **228**, 80 (1974).

212. J. C. Harvey, J. R. Cantrell, and A. M. Fisher, Actinomycosis: Its recognition and treatment. *Ann. Intern. Med.* **46**, 868 (1957).

213. M. Cecchi, Il problema della simbiosi fungino-microbica e quello delle forme atipiche dell'actinomices a proposito di due osservazioni di actinomicosi linguale. *Arch. Vecchi Anat. Patol.* **52**, 345 (1968).

214. R. Sodagar and E. Kohout, Actinomycosis of tongue as pseudotumor. *Laryngoscope* **82**, 2149 (1972).

215. J. Quérangal, Sur trois cas d'ulcère necrosant de la langue a fuso-spirilles. *Bull. Soc. Pathol. Exot.* **26**, 720 (1933).

215a. R. A. Storring and M. E. Gerken, Mucosal ulcers and anaerobic organisms. *Lancet* **1**, 1138 (1975).

216. H. Matsuura, Anaerobes in the conjunctiva and outer eye diseases. *Rinsho Ganka* **25**, 123 (1971).

217. K. Scholtz, Gangran der Bindehaut. *Klin. Monatsbl. Augenheilkd.* **9**, 62 (1910).

218. J. H. Dunnington and D. Khorazo, Conjunctivitis due to fusospirochetal infection. *Arch. Ophthalmol.* **16**, 252 (1936).

219. H. Rocha, Conjunctivite fuso-espirillar. *Bras.-Med.* **51**, 983 (1937).

220. F. B. Bowman: A case of Vincent's infection involving mouth, eyes and penis. *Lancet* **2**, 536 (1917).

221. F. Herrenschwand, Spirochaeten und *Bacillus fusiformis* bei akuter Konjunktivitis. *Z. Augenheilkd.* **62**, 370 (1927).

222. W. G. Goudie and J. R. Sutherland, Case of corneal ulceration associated with the presence of spirilla and fusiform bacilli. *Proc. R. Soc. Med.* **6**, 142 (1913).

223. C. Dejean and J. Temple, Conjunctivite avec fuso-spirilles de Vincent. *Ann. Ocul.* **164**, 198 (1927).

224. E. H. Gutierrez, Bacterial infections of the eye. *In* "Microbiology of the Eye" (D. Locatcher-Khorazo and B. C. Seegal, eds.), p. 63. Mosby, St. Louis, Missouri, 1972.

225. P. Henkind and H. Fedukowicz, *Clostridium welchii* conjunctivitis. *Arch. Ophthalmol.* **70**, 791 (1963).

226.   W. E. Smith and M. W. Ropes, Bacteroides infections. An analysis based on a review of the literature and a study of twenty cases. *N. Engl. J. Med.* **232**, 31 (1945).

226a. A. R. Prévot, Une nouvelle entite bacterio-clinique: l'infection à *Ristella pseudoinsolita. Acad. Nat. Med. (Paris).* Suppl. **149**, 689 (1965).

226b. R. E. Perkins, R. B. Kundsin, M. V. Pratt, I. Abrahamsen, and H. M. Leibowitz, Bacteriology of normal and infected conjunctiva. *J. Clin. Microbiol.* **1**, 147 (1975).

227.   M. Mikuni, M. Ohishi, M. Imai, T. Takahashi, and T. Takizawa, Anaerobes isolated from infectious diseases of the eye. *In* "Second Symposium on Anaerobic Bacteria and their Infectious Diseases" (S. Ishiyama *et al.*, eds.), p. 133, Eisai, Tokyo, 1972.

228.   L. Pine, H. Hardin, L. Turner, and S. S. Roberts, Actinomycotic lacrimal canaliculitis: A report of two cases with a review of the characteristics which identify the causal organism, *Actinomyces israelii. Am. J. Ophthalmol.* **49**, 1278 (1960).

229.   L. Pine and H. Hardin, *Actinomyces israelii,* a cause of lacrimal canaliculitis in man. *J. Bacteriol.* **78**, 164 (1959).

230.   B. B. Buchanan and L. Pine, Characterization of a propionic acid producing actinomycete, *Actinomyces propionicus,* sp. nov. *J. Gen. Microbiol.* **28**, 305 (1962).

231.   L. Pine and L. K. Georg, Reclassification of *Actinomyces propionicus. Int. J. Syst. Bacteriol.* **19**, 267 (1969).

232.   M. A. Gerencser and J. M. Slack, Isolation and characterization of *Actinomyces propionicus. J. Bacteriol.* **94**, 109 (1967).

233.   L. K. Georg, Diagnostic procedures for the isolation and identification of the etiologic agents of actinomycosis. *Proc. Int. Symp. Mycoses* Sci. Publ. No. 205, W.H.O., Wash., D.C. p. 71 (1970).

234.   M. Bonnet, M. B. Gilot, and S. Duborgel, Pseudo-mycose des voies lacrymales. *Bull. Soc. Ophthalmol. Fr.* **70**, 524 (1970).

235.   S. R. Gifford, Fusiform bacilli on the conjunctiva and in the Meibomian glands. *Arch. Ophthalmol.* **49**, 477 (1920).

235a. M. Oishi, M. Imai, Takahasi, and G. Takizawa, Two cases of anaerobic eye infections. *In* "Third Symposium on Anaerobic Bacteria and Their Infectious Diseases," (S. Ishiyama *et al.*, eds.), p. 99. Eisai, Tokyo, Japan, 1973.

236.   Wakisaka, Bakteriologische Untersuchung der Tränensackentzündung. *Klin. Monatsbl. Augenheilkd.* **8**, 797 (1909).

237.   W. Löhlein, Spirochaten und *Bacillus fusiformis* bei Dacryocystitis. *Arch. Augenheilkd.* **89**, 201 (1921).

238.   M. Martres, E. R. Brygoo, and H. Thouvenot, Étude d'une espèce nouvelle du genre *Zuberella: Z. constellata* n. sp. *Ann. Inst. Pasteur, Paris* **83**, 139 (1952).

239.   C. Brons, Die anaeroben Bazillen in der Augenbakteriologie. *Zentralbl. Bakteriol., Parasitenkd., Infektionskr. Hyg., Abt. 1:* Referate. **42**, 625 (1909).

240.   H. H. Slansky, M. C. Gnadinger, M. Itoi, and C. H. Dohlman, Collagenase in corneal ulcerations. *Arch. Ophthalmol.* **82**, 108 (1969).

241.   W. D. Gingrich and M. E. Pinkerton, Anaerobic *Actinomycosis bovis* corneal ulcer. *Arch. Ophthalmol.* **67**, 549 (1962).

242.   A. Bertozzi, Un caso di ottalmia metastatica da bacillo fusiforme di Vincent durante il decorso di un infezione morbillosa. *Ann. Ottalmol. Clin. Ocul.* **36**, 138 (1907).

243.   C. Weiss, Bacteriologic observations on infections of the eye. *Arch. Ophthalmol.* **30**, 110 (1943).

243a. J. W. Wahl: Vibrio endophthalmitis. *Arch. Ophthalmol.* **91**, 423 (1974).

244.   G. Goldsand and A. I. Braude, "Anaerobic Infections." *Dis.-Mon.* Nov. (1966).

245.   B. Golden, R. C. Watzke, S. S. Lindell, and A. P. McKee, Treponemal-like organisms in the aqueous of nonsyphilitic patients. An immunologic study. *Arch. Ophthalmol.* **80**, 727 (1968).

246.   R. B. Leavelle, Gas gangrene panophthalmitis. Review of the literature; report of new cases. *AMA Arch. Ophthalmol.* [ns] **53**, 634 (1955).

247.   J. M. Levitt and J. Stam, *Clostridium perfringens* panophthalmitis. *Arch. Ophthalmol.* **84**, 227 (1970).

247a.   J. F. Frantz, M. A. Lemp, R. L. Font, R. Stone, and E. Eisner: Acute endogenous panophthalmitis caused by *Clostridium perfringens*. *Am. J. Ophthalmol.* **78**, 295 (1974).

248.   D. P. Seecof, Vincent's organisms. In chronic sinusitis, osteomyelitis of frontal bone, orbital cellulitis, meningitis and pulmonary gangrene. Report of a case. *Arch. Otolaryngol.* **10**, 384 (1929).

248a.   D. Sevel, B. Tobias, S. L. Sellars, and A. Forder, Gas in the orbit associated with orbital cellulitis and paranasal sinusitis. *Br. J. Ophthalmol.* **57**, 133 (1973).

249.   H. Hartl, Klinische beobachtungen bei Funduliformis-infektionen in der Gynakologie. *Schweiz. Med. Wochenschr.* **80**, 1136 (1950).

249a.   S. Klein, A. Klein und D. Krause: Befunde am Augenhintergrund bei Gasbrand. *Ophthalmologica* **170**, 334 (1975).

250.   H. Lodenkämper and G. Stienen, Uber das Auftreten und die Resistenzbestimmung anaerober sporenloser Infektionserreger. *Z. Hyg.* **142**, 371 (1956).

251.   L. H. Barenberg and J. M. Lewis, Vincent's (fusospirillary) infection of the ear. *J. Am. Med. Assoc.* **94**, 1065 (1930).

252.   A. Saito, K. Inoue, A. Hayashi, K. Hara, K. Sawatari, and S. Mochida, 31 cases with anaerobic bacteria as the only isolates. *Jpn. J. Clin. Pathol.* **19**, Suppl., 133 (1971).

253.   V. M. Howie, J. H. Ploussard, and R. L. Lester, Jr., Otitis media: A clinical and bacteriological correlation. *Pediatrics* **45**, 29 (1970).

254.   E. A. Mortimer, Jr. and R. L. Watterson, Jr., A bacteriologic investigation of otitis media in infancy. *Pediatrics* **17**, 359 (1956).

255.   G. B. Stickler and J. B. McBean, The treatment of acute otitis media in children. II. A second clinical trial. *J. Am. Med. Assoc.* **187**, 85 (1964).

256.   E. Rist, Neue Methoden und neue Ergebnisse im Gebiete der bakteriologischen untersuchung gangranoser und fötider Eiterungen. *Zentralbl. Bakteriol., Parasitenkd. Infektionskr., Abt. 1.* **30**, 287 (1901).

257.   E. Wirth, Seltene gramnegative, anaerobe Bazillen als ungewohnlich bosartige Erreger von akuten Mittelohrentzundungen. *Zentralbl. Bakteriol., Parasitenkd. Infektionskr., Abt. 1: Orig.* **105**, 201 (1927–1928).

258.   H. Schottmüller, Zur Bedeutung einiger Anaeroben in der Pathologie, insbesondere bei puerperalen Erkrankungen. *Mitt. Grenzgeb. Med. Chir.* **21**, 450 (1910).

259.   R. Massini, Uber anaerobe Bakterien. *Z. Gesamte Exp. Med.* **2**, 81 (1913).

260.   L. Jame and C. Jaulmes, Un cas de septicemie à "*Bacillus funduliformis.*" *Bull. Soc. Med. Hôp. Paris* [3] **49**, 970 (1933).

261.   H. C. Pham, Les septicémies dués au *Bacillus funduliformis*. M. D. Thesis, Faculty of Medicine, University of Paris, Paris (1935).

262.   A. Grumbach, A. Lemierre, and J. Reilly, Zur Bakteriologie der Anaeroben-Sepsis die Identitat von *Fusobacterium nucleatum*, *Bacillus funduliformis*, und *Bacterium pyogenes anaerobium* Buday. *Schweiz. Med. Wochenschr.* **17**, 834 (1936).

263.   A. R. Prévot and J. Senez, Recherches biochimiques sur *Micrococcus grigoroffi*. *C. R. Seances Soc. Biol. Ses Fil.* **138**, 35 (1944).

264.   A. M. Fisher and V. A. McKusick, Bacteroides infections: Some clinical and therapeutic features. *Trans. Am. Clin. Climatol. Assoc.* **4**, 1 (1952).

265.   N. Kozakai and S. Suzuki, "Anaerobes in Clinical Medicine." Igaku Shoin Ltd., Tokyo, 1968.

266.   R.-G. Lin and A. E. Arcala, *Fusobacterium* septicemia with otitis media and mastoiditis. *Postgrad. Med.* **57**, 159 (1975).

266a. R. J. Fass, J. F. Scholand, G. R. Hodges, and S. Saslaw, Clindamycin in the treatment of serious anaerobic infections. *Ann. Intern. Med.* **78**, 853 (1973).

267. I. Pilot and S. J. Pearlman, Studies in fusiform bacilli and spirochetes. V. Occurrence in otitis media chronica. *J. Infect. Dis.* **33**, 139 (1923).

268. P. Brisottq, Dei germi anaerobi in otologia e di alcune ricerche batteriologiche sull'ozena. *Riforma Med.* **39**, 769 (1923).

269. G. Busacca, Alcune ricerche sul reperto batteriologico della otite cronica suppurata, con speciale riguardo alla presenza di spirochete e di *Bacilli fusiformi*. Tentativi di cura col neosalvarsan. *Arch. Ital. Otol., Rinol. Laringol.* **34**, 414 (1923).

270. T. Palva and O. Hallstrom, Bacteriology of chronic otitis media. Results of analyses from the ear canal and from the operative cavity. *Arch. Otolaryngol.* **82**, 359 (1965).

271. T. Palva, J. Karja, A. Palva, and V. Raunio, Bacteria in the chronic ear, pre- and postoperative evaluation. *Pract. Oto-Rhino-Laryngol.* **31**, 30 (1969).

272. S. Baba, K. Mamiya, and A. Suzuki, Recent experience with anaerobic infections in otorhinolaryngology. *In* "Second Symposium on Anaerobic Bacteria and Their Infectious Diseases" (S. Ishiyama *et al.*, eds.), p. 136. Eisai, Tokyo, Japan, 1972.

273. C. Krumwiede, Jr. and J. S. Pratt, Fusiform bacilli: Cultural characteristics. *J. Infect. Dis.* **13**, 438 (1913).

274. A. Hansen, "Nogle Undersogelsen over Gramnegative Anaerobe IkkeSporeDannende Bacterier Isolerede fra Peritonsilloere Abscesser bos Mennesker." Munksgaard, Copenhagen, 1950.

275. D. L. Bornstein, A. N. Weinberg, M. N. Swartz, and L. J. Kunz, Anaerobic infections—review of current experience. *Medicine (Baltimore)* **43**, 207 (1964).

276. F. O. Black and C. C. Atkins, Tetanus from tympanomastoiditis. *Arch. Otolaryngol.* **96**, 76 (1972).

277. L. Boez, R. Keller, and A. Kehlstadt, Bacteriémies anáerobies à *B. fragilis* (trois observations) *Bull. Mem. Soc. Med. Hôp. Paris* [3] **51**, 1184 (1927).

278. M. De Vos, M. Van Der Straeten, M. Blaauw, and R. Hombrouck, *Bacteroides fragilis* septicemia originating in the middle ear and the lungs. *Infection* **3**, 19 (1975).

278a. J. H. Leek, Actinomycosis of the tympanomastoid. *Laryngoscope* **84**, 290 (1974).

279. E. Rist and L. Ribadeau-Dumas, Abces du foie et angiocholite au cours de septicémies expérimentales à microbes anaérobies. *C. R. Seances Soc. Biol. Ses Fil.* **63**, 538 (1907).

280. A. Plaut, Ueber Fusospirilläre Assoziation im Mittelohr. *Zentralbl. Bakteriol., Parasitenkd. Infektionskr. Abt. 1: Orig.* **83**, 537 (1919).

281. M. Shevky, C. Kohl, and M. S. Marshall, *Bacterium melaninogenicum*. *J. Lab. Clin. Med.* **19**, 689 (1934).

282. P. M. Beigelman and L. A. Rantz, Clinical significance of *Bacteroides*. *Arch. Intern. Med.* **84**, 605 (1949).

283. M. Jungano and A. Distaso, "Les Anaerobies." Masson, Paris, 1910.

284. K. Urbal and P. Berdal, The microbial flora in 81 cases of maxillary sinusitis. *Acta Oto-Laryngol.* **37**, 20 (1949).

285. V. Fredette, A. Auger, and A. Forget, Anaerobic flora of chronic nasal sinusitis in adults. *Can. Med. Assoc. J.* **84**, 164 (1961).

286. S. Baba, Anaerobic bacteria isolated from chronic nasal sinusitis. *In* "Clinical Significance of Non-Sporeforming Bacteria in Clinical Bacteriology" (S. Suzuki *et al.*, eds.), Symp. No. 40, Jpn. Soc. Microbiol., p. 44. Nagoya City, Japan, 1966.

287. H. Lodenkämper and G. Stienen, Importance and therapy of anaerobic infections. *Antibiot. Med.* **1**, 653 (1955).

288. K. Gullers, C. Lundberg, and A.-S. Malmborg, Penicillin in paranasal sinus secretions. *Chemotherapy (Basel)* **14**, 303 (1969).

289. J. M. Felner and V. R. Dowell, Jr., Anaerobic bacterial endocarditis. *N. Engl. J. Med.* **283**, 1188 (1970).

290. D. T. Smith, "Oral Spirochetes and Related Organisms in Fuso-Spirochetal Disease." Williams & Wilkins, Baltimore, Maryland, 1932.

291. H. L. Williams and D. R. Nichols, Spreading osteomyelitis of the frontal bone treated with penicillin. *Proc. Staff Meet. Mayo Clin.* **18**, 467 (1943).

292. K. von Gusnar and H. Globig, Uber eine besondere Form der Sepsis. *Dtsch. Z. Chir.* **221**, 263 (1929).

293. H. Nathan, Uber den Ausbreitungsweg septischer, metastasierender Infektionen. *Virchows Arch. Pathol. Anat. Physiol.* **281**, 430 (1931).

294. A. R. Hollender, Delayed healing of septal resections due to Vincent's infections. Report of three cases. *Arch. Otolaryngol.* **9**, 422 (1929).

295. G. G. Thomas, R. J. Toohill, and R. H. Lehman, Nasal actinomycosis following heterograft. A case report. *Arch. Otolaryngol.* **100**, 377 (1974).

296. A. A. Gunn, *Bacteroides* septicaemia, *J. R. Coll. Surg. Edinburgh* **2**, 41 (1956).

297. M. H. Vincent, Sur une forme particulière d'angine dipthéroide (angine à bacilles fusiformes). *Bull. Mem. Soc. Med. Hôp.Paris* [3] **15**, 244 (1898).

298. Aretaeus, On ulcerations about the tonsils. From "The Extant Works of Aretaeus, the Cappadocian" (F. Adams, ed. and transl.). Sydenham Society, London, 1856 (as reproduced in R. H. Major, "Classic Descriptions of Disease." Thomas, Springfield, Illinois, 1932).

299. L. Haymann, Erhebungen uber die tonsillogene Sepsis (ein Beitrag zur Pathogenese, Pathologie und Klinik). *Z. Hals-, Nasen- Ohrenheilkd.* **35**, 288 (1934).

300. P. Christ, Uber Erkrankungen durch anaerobe, nicht sporenbildende Bakterien. *Dtsch. Arch. Klin. Med.* **203**, 186 (1956).

301. C. E. Ware, A case of Vincent's angina. *J. Am. Med. Assoc.* **86**, 450 (1927).

301a. P. M. Sprinkle, R. W. Veltri, and L. M. Kantor, Abscesses of the head and neck. *Laryngoscope* **84**, 1142 (1974).

302. J. M. Alston, Necrobacillosis in Great Britain. *Br. Med. J.* **2**, 1524 (1955).

303. O. P. N. Grüner, Actinomyces in tonsillar tissue. A histological study of a tonsillectomy material. *Acta Pathol. Microbiol. Scand.* **76**, 239 (1969).

304. D. W. Brock, L. K. Georg, J. M. Brown, and M. D. Hicklin, Actinomycosis caused by *Arachnia propionica:* Report of 11 cases. *Am. J. Clin. Pathol.* **59**, 66 (1973).

305. T. Thjötta and J. Jonsen, Studies on *Bacteroides*. III. Investigation and discussion of methods. *Acta Pathol. Microbiol. Scand.* **26**, 538 (1949).

306. A. R. Prévot, G. J. Beal, and P. Tardieux, Rôle des anaérobies appartenant aux genres *Actinobacterium, Corynebacterium* et *Ramibacterium* dans l'étiologie des suppurations des régions jugales et cervico-faciales. *Ann. Inst. Pasteur, Paris* **79**, 763 (1950).

307. H. Beeuwkes, J. B. M. Vismans, and A. H. Smeets, *Bacteroides—*infecties. *Ned. Tijdschr. Geneesk.* **95**, 1143 (1951).

308. P. Ingelrans, Infection grave de la face a "*Spherophorus varius*" d'origine dentaire avec empyème du sinus maxillaire. *Mem. Acad. Chir.* **73**, 433 (1947).

309. H. Schaller, Zur Lokalisation der Aktinomykose im Zerviko-Fazialgebiet. *Pract. Oto-Rhino-Laryngol.* **30**, 282 (1968).

310. H. A. Gins, Die Bedeutung der anaeroben Mundflora fur die Zahnkrankheiten. *Dtsch. Zahnarztl. Wochenschr.* **39**, 884 (1936).

311. M. Modjallal, Les bactériémies à *Bacillus ramosus*. Thesis, Faculty of Medicine, University of Paris, Paris (1937).

312. A. R. Prévot and H. Thouvenot, A propos du pouvoir pathogène de *Staphylococcus asaccharolyticus* Distaso. *Ann. Inst. Pasteur, Paris* **86**, 667 (1954).

313. A. R. Prévot and J. Taffanel, Recherches sur une nouvelle espèce anaérobie *Ramibacterium alactolyticum* (nov. spec.). *Ann. Inst. Pasteur, Paris* **68**, 259 (1942).

314. A. R. Prévot, F. Magnin, J. Levrel, D. Duby, and F. de Cadore, Recherches sur le pouvoir pathogène de *Ramibacterium pleuriticum*. *Ann. Inst. Pasteur, Paris* **95**, 241 (1958).

315. D. L. Leake, Bacteroides osteomyelitis of the mandible. A report of two cases. *Oral Surg., Oral Med. Oral Pathol.* **34**, 585 (1972).

315a. L. J. Monaldo, J. Bellome, D. J. Zegarelli, and V. E. Ragaini, *Bacteroides* infection of the mandible with secondary spread to the neck. *J. Oral Surg.* **32**, 370 (1974).

316. J. Graham, K. Malmberg, R. Patey, and A. Rubenstein, Facial actinomycosis misdiagnosed as tetanus. *Ill. Med. J.* **115**, 271 (1959).

317. W. R. Tyldesley, Chronic actinomycosis treated by oral penicillin. *Br. Dent. J.* **126**, 359 (1969).

318. H. Schubert and H. D. Tauchnitz, Ein Beitrag zur menschlichen Aktinomykose der Haut. *Mykosen* **12**, 427 (1969).

319. R. M. Coleman and L. K. Georg, Comparative pathogenicity of *Actinomyces naeslundii* and *Actinomyces israelii*. *Appl. Microbiol.* **18**, 427 (1969).

320. F. E. Koch and F. Rinsche, Erstzüchtungen von *Bact. pneumosintes* aus Eiterungen. *Zentralbl. Bakteriol., Parasitenkd. Infektionskr. Hyg. Abt. 1* **134**, 367 (1935).

321. G. T. Keusch and C. J. O'Connell, The susceptibility of *Bacteroides* to the penicillins and cephalothin. *Am. J. Med. Sci.* **251**, 428 (1966).

322. J. Bφe, "Fusobacterium. Studies on its Bacteriology, Serology, and Pathogenicity." Skrifter utg. Det. Norske Videnskaps- Akad. Kommisjon Hos Jacob Dybwad, Oslo, 1941.

323. J. S. Cunningham, Human infection with *Actinomyces necrophorus*. Bacteriologic and pathologic report of two cases terminating fatally. *Arch. Pathol.* **9**, 843 (1930).

324. W. Wey, Zervikale Gasphlegmone mit letalem Ausgang (*Bacteroides melaninogenicus*). *Monatsschr. Ohrenheilkd. Laryngo-Rhinol.* **104**, 73 (1970).

324a. J. D. Richardson, G. L. Fox, F. L. Grover, and A. B. Cruz, Jr., Necrotizing fasciitis complicating dental extraction. *Arch. Surg.* **110**, 129 (1975).

324b. R. M. Baird, Postoperative infections from *Bacteroides. Amer. Surg.* **39**, 459 (1973).

325. G. W. Levitt, Cervical fascia and deep neck infections. *Laryngoscope* **80**, 409 (1970).

326. D. T. Smith, Etiology of primary bronchiectasis. *Arch. Surg.* (*Chicago*) **21**, 1173 (1930).

326a. R. B. Marks, R. Akin, P. P. Walters and D. J. Ellis, Ludwig's angina: Report of case. *J. Oral Surg.* **32**, 462 (1974).

327. L. V. McVay, F. Guthrie, and D. H. Sprunt, Aureomycin in the treatment of actinomycosis. *N. Engl. J. Med.* **245**, 91 (1951).

328. S. L. Lane, A. H. Kutscher, and R. Chaves, Oxytetracycline in the treatment of orocervical facial actinomycosis. *J. Am. Med. Assoc.* **151**, 986 (1953).

329. P. Bramley and H. S. Orton, Cervicofacial actinomycosis. *In* "Actinomycosis" (M. Bronner and M. Bronner, eds.), p. 143. Williams & Wilkins, Baltimore, Maryland, 1969.

330. P. E. B. Holmes, Cervicofacial actinomycosis in relation to dental treatment. *In* "Actinomycosis" (M. Bronner and M. Bronner, eds.), p. 148. Williams & Wilkins, Baltimore, Maryland, 1969.

331. W. E. Herrell, A. Balows, and J. S. Dailey, Erythromycin in the treatment of actinomycosis. *Antibiot. Med.* **1**, 507 (1955).

332. H. Beerens, Étude de 25 souches d'une bactérie anaérobie non sporulée: *Ramibacterium pleuriticum. Ann. Inst. Pasteur, Lille* **6**, 116 (1953–1954).

333. J. R. Graybill and B. D. Silverman, Sulfur granules. Second thoughts. *Arch. Intern. Med.* **123**, 430 (1969).

334. K. Macoul and P. T. Souliotis, Actinomycosis: Solitary nodule. *Postgrad. Med.* **42**, 288 (1967).

335. T. Sugano, M. Fujii, and T. Goto, Actinomycotic infection of the neck simulating cervical malignancy—isolation of *Actinomyces* and its taxonomic position. *Otolaryngology (Tokyo)* **42**, 83 (1970).

336. L. P. Garrod, Sensitivity of four species of *Bacteroides* to antibiotics. *Br. Med. J.* **2**, 1529 (1955).

337. A. R. Prévot, F. de Cadore, and H. Thouvenot, Recherches sur le pouvoir pathogène et la pigmentation de *Micrococcus niger*. *Ann. Inst. Pasteur, Paris* **97**, 860 (1959).

337a. S. C. Deresinski and D. A. Stevens, Anterior cervical infections: Complications of transtracheal aspirations. *Amer. Rev. Resp. Dis.* **110**, 354 (1974).

337b. B. Lourie, B. McKinnon, and L. Kibler, Transtracheal aspiration and anaerobic abscess. *Ann. Intern. Med.* **80**, 417 (1974).

337c. T. T. Yoshikawa, A. W. Chow, J. Z. Montgomerie, and L. B. Guze, Paratracheal abscess: An unusual complication of transtracheal aspiration. *Chest* **65**, 105 (1974).

338. W. E. Heck and R. C. McNaught, Periauricular *Bacteroides* infection, probably arising in the parotid. *J. Am. Med. Assoc.* **149**, 662 (1952).

339. E. Bock, Ueber isolierte Entzundung der Glandula sublingualis durch Plaut-Vincentsche Infektion. *Muench. Med. Wochenschr.* **85**, 786 (1938).

340. L. Sazama, Actinomycosis of the parotid gland. Report of five cases. *Oral Surg., Oral Med. Oral Pathol.* **19**, 197 (1965).

341. A. Lemierre, On certain septicemias due to anaerobic organisms. *Lancet* **1**, 701 (1936).

342. C. P. W. Warren and B. J. Mason, *Clostridium septicum* infection of the thyroid gland. *Postgrad. Med.* **46**, 586 (1970).

343. E. L. Hawbaker, Thyroid abscess. *Am. Surg.* **37**, 290 (1971).

344. W. D. Leers, J. Dussault, J. E. Mullens, R. Volpe, and K. Arthurs, Suppurative thyroiditis: An unusual case caused by *Actinomyces naeslundi*. *Can. Med. Assoc. J.* **101**, 714 (1969).

344a. S. L. Gorbach and H. Thadepalli, Clindamycin in pure and mixed anaerobic infections. *Arch. Intern. Med.* **134**, 87 (1974).

344b. E. Blanc and M. Jenny, Thyroidite à actinomycose. *Schweiz. Med. Wochenschr.* **104**, 1094 (1974).

345. R. K. Sharma and R. H. Rapkin, Acute suppurative thyroiditis caused by *Bacteroides melaninogenicus*. *J. Am. Med. Assoc.* **229**, 1470 (1974).

346. M. C. Myerson, Anaerobic retropharyngeal abscess. *Ann. Otol., Rhinol., & Laryngol*, **41**, 805 (1932).

347. O. Ernst, Zur Bedeutung des *Bac. funduliformis* als Infektions-erreger. *Z. Hyg.* **132**, 352 (1951).

348. I. P. Janecka and R. M. Rankow, Fatal mediastinitis following retropharyngeal abscess. *Arch. Otolaryngol.* **93**, 630 (1971).

348a. C. S. Bryan, B. G. King, Jr., and R. E. Bryan, Retropharyngeal infection in adults. *Arch. Intern. Med.* **134**, 127 (1974).

349. W. H. Peters, Hand infection apparently due to *Bacillus fusiformis*. *J. Infect. Dis.* **8**, 455 (1911).

350. A. Grumbach and C. Verdan, *Fusobacterium nucleatum* als Erreger von septischer Angina und Mastoiditis purulenta acuta. *Arch. Hyg. Bakteriol.* **115**, 115 (1936).

351. P. Delbove and V. Reynes, A propos de trois cas de septicèmie postpartum à *B. fragilis*. *Bull. Soc. Pathol. Exot.* **34**, 58 (1941).

352. H. S. Heineman and A. I. Braude, Anaerobic infection of the brain. *Am. J. Med.* **35**, 682 (1963).

353. J. G. Bartlett, V. L. Sutter, and S. M. Finegold, Anaerobic pleuropulmonary disease:

Clinical observations and bacteriology in 100 cases. *In* "Anaerobic Bacteria: Role in Disease" (A. Balows *et al.*, eds.), Chapter XXV. Thomas, Springfield, Illinois, 1974.

353a.   B. Anguiano, E. Flora, J. Sabbaj, C. Camargo, L. Gonzalez, and J. V. Ordonez, Espectro clinico de las infecciones por anaerobios. *XVI Congr. Med. Centroamer. Hoja Res. Trabajos Libres.* (1975) Abst.

354.   D. J. Prolo and J. W. Hanbery, Secondary actinomycotic brain abscess. *Arch. Surg.* (*Chicago*) **96**, 58 (1968).

355.   H. S. Heineman, A. I. Braude, and J. L. Osterholm, Intracranial suppurative disease. Early presumptive diagnosis and successful treatment without surgery. *J. Am. Med. Assoc.* **218**, 1542 (1971).

356.   M. N. Swartz and P. R. Dodge, Bacterial meningitis—A review of selected aspects. I. General clinical features, special problems and unusual meningeal reactions mimicking bacterial meningitis. *N. Engl. J. Med.* **272**, 725 (1965).

357.   M. N. Swartz, Anaerobic bacteria in central nervous system infections. *J. Fl. Med. Assoc.* **57**, 19 (1970).

358.   A. Lutz, O. Grootten, and M. A. Berger, Considerations à propos des germes isolés dans 309 cas de meningites suppurées. *Strasbourg Med.* **8**, 610 (1962).

359.   T. Oguri and N. Kozakai, Anaerobes isolated from clinical specimens for ten year period and their antibacterial susceptibility. *In* "First Symposium on Anaerobic Bacteria and Their Infectious Diseases" (S. Ishiyama *et al.*, eds.), p. 1. Eisai, Tokyo, Japan, 1971.

360.   N. Kozakai, Anaerobic infections from the standpoint of the clinical laboratory. *In* "Clinical Significance of Non-Spore-Forming Bacteria in Clinical Bacteriology" (S. Suzuki *et al.*, eds.), Symp. No. 40, p. 28. Jpn. Soc. Microbiol., Nagoya City, Japan, 1966.

361.   O. Nakamura, A. Saito, K. Inoue, A. Hayashi, K. Sawatari, and S. Mochida, Anaerobes isolated from clinical specimens and their clinical significance. *In* "First Symposium on Anaerobic Bacteria and Their Infectious Diseases" (S. Ishiyama *et al.*, eds.), p. 7. Eisai, Tokyo, Japan, 1971.

361a.   E. L. Randall, Organisms isolated from body fluids. *In* "Pathogenic Microorganisms from Atypical Clinical Sources," (A. von Graevenitz and T. Sall, eds.). Dekker, New York, 1975.

361b.   M. E. Sharpe, L. R. Hill, and S. P. Lapage, Pathogenic lactobacilli. *J. Med. Microbiol.* **6**, 281 (1973).

362.   J. P. Conomy and J. W. Dalton, *Clostridium perfringens* meningitis. *Arch. Neurol.* (*Chicago*) **21**, 44 (1969).

363.   C. Worster-Drought, A case of meningitis associated with the presence of *Bacillus fusiformis*, and in which, at autopsy, no primary focus beyond pyorrhoea alveolaris was apparent. *Br. Dent. J.* **39**, 377 (1918).

364.   V. Frühwald, Der *Bacillus fusiformis* als Erreger von Meningitis und Hirnabsess nach Fremdkorperverletzung des Pharynx. *Monatsschr. Ohrenheilkd. Laryngo-Rhinol.* **47**, 1021 (1913).

365.   J. G. Alexander, A case of fatal *Clostridium welchii* toxaemia due to ward infection. *J. Clin. Pathol.* **22**, 508 (1969).

366.   E. Hitchcock and A. Andreadis, Subdural empyema: A review of 29 cases. *J. Neurol., Neurosurg. Psychiatry.* **27**, 422 (1964).

367.   E. S. Gurdjian and L. M. Thomas, Surgical treatment of cranial and intracranial suppuration. *In* "Cranial and Intracranial Suppuration" (E. S. Gurdjian, ed.), Chapter I, p. 3. Thomas, Springfield, Illinois, 1969.

368.   T. T. Yoshikawa, A. W. Chow, and L. B. Guze, Role of anaerobic bacteria in subdural empyema. *Am. J. Med.* **58**, 99 (1975).

369. F. Schiller, H. Cairns, and D. S. Russell, The treatment of purulent pachymeningitis and subdural suppuration with special reference to penicillin. *J. Neurol., Neurosurg. Psychiatry* **11**, 143 (1948).

370. R. Wiesmann, Über subdurale Empyeme. *Acta Neurochir.* **20**, 153 (1969).

371. R. L. Parker and G. H. Collins, Intramedullary abscess of the brain stem and spinal cord. *South. Med. J.* **63**, 495 (1970).

372. H. Cairns, C. A. Calvert, P. Daniel, and G. B. Northcroft, Complications of head wounds, with especial reference to infection. *Br. J. Surg.* **1**, 198 (1947).

373. R. L. Wright, "Postoperative Craniotomy Infections." Thomas, Springfield, Illinois, 1966.

374. M. Rogosa, E. G. Hampp, T. A. Nevin, H. N. Wagner, E. J. Driscoll, and P. N. Baer, Blood sampling and cultural studies in the detection of postoperative bacteremias. *J. Am. Dent. Assoc.* **60**, 171 (1960).

375. A. Marseille, Bacteriaemie na Kiesextractie. *Geneeskd. Tijdschr. Ned.-Indie* **77**, 2491 (1937).

376. O. Khairat, The non-aerobes of post-extraction bacteremia. *J. Dent. Res.* **45**, 1191 (1966).

377. M. G. McEntegart and J. S. Porterfield, Bacteraemia following dental extractions. *Lancet* **2**, 596 (1949).

378. E. H. Müller, Bakteriamie nach Zahnextraktionen. *Schweiz. Monatsschr. Zahnheilkd.* **72**, 283 (1962).

379. H. D. Conner, S. Haberman, C. K. Collings, and T. E. Winford, Bacteremias following periodontal scaling in patients with healthy appearing gingiva. *J. Periodontol.* **38**, 466 (1967).

380. A. Schirger, W. J. Martin, R. Q. Royer, and G. M. Needham, Bacterial invasion of blood after oral surgical procedures. *J. Lab. Clin. Med.* **55**, 376 (1960).

381. D. E. Tolman, A. Schirger, W. J. Martin, and J. A. Washington, II, Ampicillin administered prophylactically in oral surgery. *Northwest Dent.* **51**, 9 (1972).

381a. F. A. Berry, Jr., S. Yarbrough, N. Yarbrough, C. M. Russell, M. A. Carpenter, and J. O. Hendley, Transient bacteremia during dental manipulation in children. *Pediatrics* **51**, 476 (1973).

381b. A. A. DeLeo, F. D. Schoenknecht, M. W. Anderson, and J. C. Peterson, The incidence of bacteremia following oral prophylaxis on pediatric patients. *Oral Surg.* **37**, 36 (1974).

381c. J. J. Crawford, J. R. Sconyers, J. D. Moriarty, R. C. King, and J. F. West, Bacteremia after tooth extractions studied with the aid of prereduced anaerobically sterilized culture media. *Appl. Microbiol.* **27**, 927 (1974).

381d. S. A. Berger, S. Weitzman, S. C. Edberg, and J. I. Casey, Bacteremia after the use of an oral irrigation device. A controlled study in subjects with normal-appearing gingiva: Comparison with use of toothbrush. *Ann. Intern. Med.* **80**, 510 (1974).

382. J. E. Puklin, G. A. Balis, and D. W. Bentley, Culture of an Osler's node. A diagnostic tool. *Arch. Intern. Med.* **127**, 296 (1971).

383. R. Murphy, S. Katz, and D. Massaro, *Fusobacterium* septicemia following a human bite. *Arch. Intern. Med.* **111**, 51 (1963).

383a. S. O. Burman, Bronchoscopy and bacteriemia. *J. Thorac. Cardiov. Surg.* **40**, 635 (1960).

383b. J. L. LeFrock, C. A. Ellis, J. B. Turchik, J. K. Zawacki, and L. Weinstein, Transient bacteremia associated with percutaneous liver biopsy. *J. Infect. Dis. Suppl.* **131**, S104 (1975).

383c. J. Le Frock, C. A. Ellis, A. S. Klainer, and L. Weinstein, Transient bacteremia associated with barium enema. *Arch. Intern. Med.* **135**, 835 (1975).

384. J. L. LeFrock, C. A. Ellis, J. B. Turchik, and L. Weinstein, Transient bacteremia associated with sigmoidoscopy. *N. Engl. J. Med.* **289**, 467 (1973).

384a. I. Mackenzie and A. Litton, *Bacteroides* bacteriaemia in surgical patients. *Br. J. Surg.* **61**, 288 (1974).

384b. R. M. Swenson, B. Lorber, T. C. Michaelson, and E. H. Spaulding, The bacteriology of intra-abdominal infections. *Arch. Surg.* **109**, 398 (1974).

384c. H. H. Stone, L. D. Kolb, and C. E. Geheber, Incidence and significance of intra-peritoneal anaerobic bacteria. *Ann. Surg.* **181**, 705 (1975).

385. M. Robinow and F. A. Simonelli, *Fusobacterium* bacteremia in the newborn. *Am. J. Dis. Child.* **110**, 92 (1965).

386. A. Ansari, Spontaneous acute peritonitis with bacteremia in patients with decompensated Laennec's cirrhosis. *Am. J. Gastroenterol.* **55**, 265 (1971).

387. J. Sabbaj, V. L. Sutter, and S. M. Finegold, Anaerobic pyogenic liver abscess. *Ann. Intern. Med.* **77**, 629 (1972).

387a. G. R. G. Monif and H. Baer, *Haemophilus* (*Corynebacterium*) *vaginalis* septicemia. *Am. J. Obstet. Gynecol.* **120**, 1041 (1974),

388. N. M. Sullivan, V. L. Sutter, W. T. Carter, H. R. Attebery, and S. M. Finegold, Bacteremia after genitourinary tract manipulation: Bacteriological aspects and evaluation of various blood culture systems. *Appl. Microbiol.* **23**, 1101 (1972).

389. C. I. Biorn, W. H. Browning, and L. Thompson, Transient bacteremia immediately following transurethral prostatic resection. *J. Urol.* **63**, 155 (1950).

390. F. D. Pien, R. L. Thompson, and W. J. Martin, Clinical and bacteriologic studies of anaerobic gram-positive cocci. *Mayo Clin. Proc.* **47**, 251 (1972).

390a. O. A. Okubadejo, P. J. Green, and D. J. H. Payne, *Bacteroides* infection among hospital patients. *Br. Med. J.* **1**, 212 (1973).

390b. B. Vinke and J. G. A. Borghans, *Bacteroides* as a cause of suppuration and septicaemia. *Trop. Geogr. Med.* **15**, 76 (1963).

391. W. H. Mencher and H. E. Leiter, Anaerobic infections following operations on the urinary tract. *Surg., Gynecol. Obstet.* **66**, 677 (1938).

392. R. S. Fishbach and S. M. Finegold, Anaerobic prostatic abscess with bacteremia. *Am. J. Clin. Pathol.* **59**, 408 (1973).

393. W. A. Altemeier, J. J. McDonough, and W. D. Fullen, Third day surgical fever. *Arch. Surg.* (*Chicago*) **103**, 158 (1971).

394. R. M. Beazley, S. H. Polakavetz, and R. M. Miller, *Bacteroides* infections on a university surgical service. *Surg., Gynecol. Obstet.* **135**, 742 (1972).

395. H. A. Hirsch, G. Branscheidt, and C. Datwyler, Epidemiologie und Antibiotikaempfindlichkeit der Erreger von gynakologisch-geburtshilflichen Infektionen. *Geburtshilfe Frauenheilkd.* **31**, 923 (1971).

396. H. Smits and L. R. Freedman, Prolonged venous catheterization as a cause of sepsis. *N. Engl. J. Med.* **276**, 1229 (1967).

396a. K. Crossley and J. M. Matsen, Intravenous catheter-associated bacteremia: Role of the diagnostic microbiology laboratory. *Appl. Microbiol.* **26**, 1006 (1973).

396b. Center for Disease Control, Morbidity and Mortality Weekly Report: Epidemiologic notes and reports. Recall of contaminated intravenous cannulae—United States (1974). Vol. 23, p. 57.

397. P. E. Hermans and J. A. Washington, II, Polymicrobial bacteremia. *Ann. Intern. Med.* **73**, 387 (1970).

398. A. von Graevenitz and W. Sabella, Unmasking additional bacilli in gram-negative rod bacteremia. *J. Med.* **2**, 185 (1971).

398a. M. C. McHenry, T. L. Gavan, W. A. Hawk, C. Ma, and J. N. Berrettoni, Gram-negative bacteremia: Variable clinical course and useful prognostic factors. *Cleveland Clin. Quart.* **42**, 15 (1975).

398b. J. P. Sanford, Generalised gram-negative bacillary infections. *J. Roy. Coll. Phys. (London)* **6**, 189 (1972).

399. W. R. Cole, M. H. Witte, and C. L. Witte, Lymph culture: A new tool for the investigation of human infections. *Ann. Surg.* **170**, 705 (1969).

400. W. J. Martin and M. C. McHenry, Bacteremia due to gram-negative bacilli. *J.—Lancet* **84**, 385 (1964).

401. E. B. Rotheram, Jr. and S. F. Schick, Nonclostridial anaerobic bacteria in septic abortion. *Am. J. Med.* **46**, 80 (1969).

402. R. G. Douglas and I. F. Davis, Puerperal infection. Etiologic, prophylactic and therapeutic considerations. *Am. J. Obstet. Gynecol.* **51**, 352 (1946).

403. J. A. Washington, II and W. J. Martin, Comparison of three blood culture media for recovery of anaerobic bacteria. *Appl. Microbiol.* **25**, 70 (1973).

404. A. Cabrera, Y. Tsukada, and J. W. Pickren, Clostridial gas gangrene and septicemia in malignant disease. *Cancer* **18**, 800 (1965).

405. M. F. Kagnoff, D. Armstrong, and A. Blevins, *Bacteroides* bacteremia. *Cancer* **29**, 245 (1972).

406. G. P. Bodey, B. A. Nies, and E. J. Freireich, Multiple organism septicemia in acute leukemia. *Arch. Intern. Med.* **116**, 266 (1965).

406a. M. C. McHenry, R. B. Turnbull, Jr., F. L. Weakley, and W. A. Hawk, Septicemia in surgical patients with intestinal diseases. *Dis. Colon Rectum* **14**, 195 (1971).

406b. A. A. Medeiros, *Bacteroides* bacillemia. *Arch. Surg.* **105**, 819 (1972).

406c. A. von Graevenitz, Gram-negative rods as agents of nosocomial disease: Some recent developments, with comments on mixed cultures. *In* "Gram-Negative Bacterial Infections and Mode of Endotoxin Actions," (B. Urbaschek, R. Urbascheck, and E. Neter, eds.), p. 30. Springer-Verlag, New York, (1975).

407. S. S. Donaldson, M. R. Moore, S. A. Rosenberg, and K. L. Vosti, Characterization of postsplenectomy bacteremia among patients with and without lymphoma. *N. Engl. J. Med.* **287**, 69 (1972).

408. P. Schain, A. De Stefano, and J. P. Kazlowski, *Actinomyces bovis* in tissues and body fluids. *J. Lab. Clin. Med.* **34**, 677 (1949).

409. R. J. Alpern and V. R. Dowell, Jr., Nonhistotoxic clostridial bacteremia. *Am. J. Clin. Pathol.* **55**, 717 (1971).

410. C. F. Dixon and J. L. Deuterman, Postoperative *Bacteroides* Infections. Report of six cases. *J. Am. Med. Assoc.* **108**, 181 (1937).

411. J. G. Sinkovics and J. P. Smith, Septicemia with *Bacteroides* in patients with malignant disease. *Cancer* **25**, 663 (1970).

412. J. S. Goodman, *Bacteroides* sepsis: Diagnosis and therapy. *Hosp. Pract.* **6**, 121 (1971).

413. D. R. Boggs, E. Frei, and L. B. Thomas, Clostridial gas gangrene and septicemia in four patients with leukemia. *N. Engl. J. Med.* **259**, 1255 (1958).

413a. R. A. Clift, C. D. Buckner, A. Fefer, K. G. Lerner, P. E. Neiman, R. Storb, M. Murphy, and E. D. Thomas, Infectious complications of marrow transplantation. *Transplant. Proc.* **6**, 389 (1974).

414. R. L. Myerowitz, A. A. Medeiros, and T. F. O'Brien, Bacterial infection in renal homotransplant recipients. A study of fifty-three bacteremic episodes. *Am. J. Med.* **53**, 308 (1972).

415. V. A. Fulginiti, R. Scribner, C. G. Groth, C. W. Putnam, L. Brettschneider, S. Gilbert, K. A. Porter, and T. E. Starzl, Infections in recipients of liver homografts. *N. Engl. J. Med.* **279**, 619 (1968).

415a. M. C. Bach, A. P. Monaco, and M. Finland, Pulmonary nocardiosis. Therapy with minocycline and with erythromycin plus ampicillin. *J. Am. Med. Assoc.* **224**, 1378 (1973).

415b. J. A. Washington, II, Relative frequency of anaerobes. *Ann. Intern. Med.* **83**, 908 (1975).

416. J. W. Smith, P. M. Southern, Jr., and J. D. Lehmann, Bacteremia in septic abortion: Complications and treatment. *Obstet. Gynecol.* **35**, 704 (1970).

417. J. A. Washington, II, Comparison of two commercially available media for detection of bacteremia. *Appl. Microbiol.* **22**, 604 (1971).

417a. P. M. Southern, Jr., Bacteremia due to *Succinivibrio dextrinosolvens*. *Am. J. Clin. Pathol.* **64**, 540 (1975).

418. V. Reynes, Densité de l'infection sanguine dans les bacteriémies et septicémies. *C. R. Seances Soc. Biol. Ses Fil.* **141**, 261 (1947).

419. H. K. Rathbun, Clostridial bacteremia without hemolysis. *Arch. Intern. Med.* **122**, 496 (1968).

419a. J. H. Christy, Treatment of gram-negative shock. *Am. J. Med.* **50**, 77 (1971).

419b. R. H. Dietzman, R. A. Ersek, J. M. Bloch, and R. C. Lillehei, High-output, low-resistance gram-negative septic shock in man. *Angiology* **20**, 691 (1969).

419c. J. R. Harrod and D. A. Stevens, Anaerobic infections in the newborn infant. *J. Pediat.* **85**, 399 (1974).

419d. M. R. B. Keighley, D. W. Burdon, W. T. Cooke, and J. Alexander-Williams, Influence of prophylaxis against *Bacteroides* on incidence of sepsis after large-bowel surgery. *Proc. Roy. Soc. Med.* **68**, 27 (1975).

419e. E. J. Winslow, H. S. Loeb, S. H. Rahimtoola, S. Kamath, and R. M. Gunnar, Hemodynamic studies and results of therapy in 50 patients with bacteremic shock. *Am. J. Med.* **54**, 421 (1973).

420. J. A. Marcoux, R. J. Zabransky, J. A. Washington, II, W. E. Wellman, and W. J. Martin, *Bacteroides* bacteremia. *Minn. Med.* **53**, 1169 (1970).

420a. T. Hofstad, The distribution of heptose and 2-keto-3-deoxyoctonate in Bacteroidaceae. *J. Gen. Microbiol.* **85**, 314 (1974).

421. R. F. Wilson, A. D. Chiscano, E. Quadros, and M. Tarver, Some observations on 132 patients with septic shock. *Anesth. Analg. (Cleveland)* **46**, 751 (1967).

422. A. Litton, Haemovascular changes in septic shock. *Postgrad. Med. J.* **45**, 551 (1969).

423. E. E. Dilworth and J. V. Ward, Bacteremic shock in pyelonephritis and criminal abortion. *Obstet. Gynecol.* **17**, 160 (1961).

424. A. E. Nathan, Alvorlige Infektioner med Anaerobe, gramnegative ikke sporedannende Stave. *Ugeskr. Laeg.* **130**, 1350 (1968).

425. M. L. Rubenberg, L. R. I. Baker, J. A. McBride, L. H. Sevitt, and M. C. Brain, Intravascular coagulation in a case of *Clostridium perfringens* septicaemia: Treatment by exchange transfusion and heparin. *Br. Med. J.* **4**, 271 (1967).

426. S. E. Vermillion, J. A. Gregg, A. H. Baggenstoss, and L. G. Bartholomew, Jaundice associated with bacteremia. *Arch. Intern. Med.* **124**, 611 (1969).

427. M. Rapin, A. Hirsch, J. R. Legall, A. Barois, and M. Goulon, Les ictères au cours des septicémies à-propos de 17 cas observés chez l'adulte. *Rev. Fr. Etud. Clin. Biol.* **14**, 472 (1969).

428. W. R. McCabe, Serum complement levels in bacteremia due to gram-negative organisms. *N. Engl. J. Med.* **288**, 21 (1973).

429. S. Attar, A. R. Mansberger, Jr., B. Irani, W. Kirby, Jr., C. Masaitis, and R. A. Cowley, Coagulation changes in clinical shock. II. Effect of septic shock on clotting times and fibrinogen in humans. *Ann. Surg.* **164**, 41 (1966).

430. A. C. Sonnenwirth, E. T. Yin, E. M. Sarmiento, and S. Wessler, Bacteroidaceae endotoxin detection by limulus assay. *Am. J. Clin. Nutr.* **25**, 1452 (1972).

431. W. A. Altemeier, J. C. Todd, and W. W. Inge, Gram-negative septicemia: A growing threat. *Ann. Surg.* **166**, 530 (1967).

432. W. A. Altemeier, E. O. Hill, and W. D. Fullen, Acute and recurrent thromboembolic disease: A new concept of etiology. *Ann. Surg.* **170**, 547 (1969).

432a. C. U. Tuazon, R. Hill, and J. N. Sheagren, Microbiologic study of street heroin and injection paraphernalia. *J. Infect. Dis.* **129**, 327 (1974).

433. L. J. Nastro and S. M. Finegold, Endocarditis due to anaerobic gram-negative bacilli. *Am. J. Med.* **54**, 482 (1973).

434. P. I. Lerner and L. Weinstein, Infective endocarditis in the antibiotic era. *N. Engl. J. Med.* **274**, 199 (1966).

435. A. E. Dormer, Bacterial endocarditis. Survey of patients treated between 1945 and 1956. *Br. Med. J.* **1**, 63 (1958).

436. L. Campeau, P. Leclair, P. David, and M. Lefebvre, Treatment of bacterial endocarditis: A review of 35 cases. *Can. Med. Assoc. J.* **83**, 933 (1960).

437. J. E. Cates and R. V. Christie, Subacute bacterial endocarditis. A review of 442 patients treated in 14 centres appointed by the Penicillin Trials Committee of the Medical Research Council. *Q. J. Med.* **20**, 93 (1951).

438. A. S. Werner, C. G. Cobbs, D. Kaye, and E. W. Hook, Studies on the bacteremia of bacterial endocarditis. *J. Am. Med. Assoc.* **202**, 199 (1967).

439. J. Wedgwood, Early diagnosis of subacute bacterial endocarditis. *Lancet* **2**, 1058 (1955).

440. R. Tompsett, Diagnosis and treatment of bacterial endocarditis. *Dis.-Mon.*, Sept. (1964).

440a. E. J. Harder, C. J. Wilkowske, J. A. Washington II, and J. E. Geraci, *Streptococcus mutans* endocarditis. *Ann. Intern. Med.* **80**, 364 (1974).

440b. W. R. Lockwood, L. A. Lawson, D. L. Smith, K. M. McNeill, and F. S. Morrison, *Streptococcus mutans* endocarditis. Report of a case. *Ann. Intern. Med.* **80**, 369 (1974).

440c. C. N. Baker and C. Thornsberry, Antimicrobial susceptibility of *Streptococcus mutans* isolated from patients with endocarditis. *Antimicrob. Ag. Chemother.* **5**, 268 (1974).

441. A. F. Masri and M. H. Grieco, Bacteroides endocarditis. Report of a case. *Am. J. Med. Sci.* **263**, 357 (1972).

442. K. Bingold, Uber eine durch anaerobe Streptokokken verursachte Endokarditisform. *Dtsch. Med. Wochenschr.* **58**, 443 (1932).

442a. A. R. Prévot, Les corynebacterioses anaerobies. *Laval Med.* **39**, 308 (1968).

443. P. Cayeux, J. F. Acar, and Y. A. Chabbert, Bacterial persistence in streptococcal endocarditis due to thiol-requiring mutants. *J. Infect. Dis.* **124**, 247 (1971).

443a. R. B. Carey, K. C. Gross, and R. B. Roberts, Vitamin B_6- dependent *Streptococcus mitior (mitis)* isolated from patients with systemic infections. *J. Infect. Dis.* **131**, 722 (1975).

443b. R. H. George, The isolation of symbiotic streptococci. *Med. Microbiol.* **7**, 77 (1974).

444. W. P. Dutton and A. P. Inclan, Cardiac actinomycosis. *Dis Chest* **54**, 463 (1968).

445. J. D. Boyle, M. L. Pearce, and L. B. Guze, Purulent pericarditis: Review of literature and report of eleven cases. *Medicine (Baltimore)* **40**, 119 (1961).

446. E. L. Coodley, Actinomycosis: Clinical diagnosis and management. *Postgrad. Med.* **46**, 73 (1969).

447. G. Verme and L. Contu, L'actinomicosi del cuore. Osservazioni su un caso clinico e rassegna della letteratura. *Arch. Vecchi Anat. Patol. Med. Clin.* **37**, 13 (1962).

448. K. Mohan, S. I. Dass, and E. E. Kemble, Actinomycosis of pericardium. *J. Am. Med. Assoc.* **229**, 321 (1974).

448a. J. S. Datta and M. J. Raff, Actinomycotic pleuropericarditis. *Am. Rev. Resp. Dis.* **110**, 338 (1974).

449. E. Vaucher and P. Worlinger, Bactériémies et septicémies dués à des anaérobies. *J. Med. Fr.* **14**, 141 (1925).

450. D. Leys, A peculiar case of pneumopyopericardium. *Lancet* **2**, 1004 (1926).

451. F. Avierinos and J. Turries, Pyo-pneumo-péricarde. D'apparence primitive avec vomique. La forme péricardique des infections gangréneuses. *Arch. Mal. Coeur Vaiss. Sang* **19**, 670 (1926).

452.   L. Weyrich, Beitrag zur Sepsis. *Mitt. Grenzgeb. Med. Chir.* **44**, 459 (1936).

452a. K. Gould, J. A. Barnett, and J. P. Sanford, Purulent pericarditis in the antibiotic era. *Arch. Int. Med.* **134**, 923 (1974).

453.   D. E. Ross, Suppurative pericarditis. *Am. J. Surg.* **43**, 134 (1939).

453a. R. H. Rubin and R. C. Moellering, Jr., Clinical, microbiologic and therapeutic aspects of purulent pericarditis. *Am. J. Med.* **59**, 68 (1975).

453b. F. Guneratne, Gas gangrene (abscess) of heart. *N.Y. State J. Med.* **75**, 1766 (1975).

454.   R. Tennant and H. W. Parks, Myocardial abscesses. A study of pathogenesis with report of a case. *AMA Arch. Pathol.* **68**, 456 (1959).

455.   D. E. Pittman, L. P. Merkow, and L. B. Brent, Myocardial abscess causing occlusion of the coronary ostium. *Arch. Intern. Med.* **126**, 294 (1970).

455a. J. F. Lewis, Myocardial infarction during pregnancy: With associated myocardial *Bacteroides* abscess. *South. Med. J.* **66**, 379 (1973).

456.   Case Records of the Massachusetts General Hospital, Case 27–1970. *N. Engl. J. Med.* **282**, 1477 (1970).

457.   J. R. Schenken and W. C. Heibner, Acute isolated myocarditis. Report of a case due to micro-aerophylic *Streptococcus hemolyticus*. *Am. Heart J.* **29**, 754 (1945).

458.   W. C. Roberts and C. W. Berard, Gas gangrene of the heart in clostridial septicemia. *Am. Heart J.* **74**, 482 (1967).

459.   R. L. Warkel and W. F. Doyle, Clostridia septicemia in malignant disease. *Mil. Med.* **135**, 889 (1970).

460.   S. J. Bodner, G. Koenig, and J. S. Goodman, Bacteremic *Bacteroides* infections. *Ann. Intern. Med.* **73**, 537 (1970).

461.   D. E. Szilagyi, R. F. Smith, J. P. Elliott, and M. P. Vrandecic, Infection in arterial reconstruction with synthetic grafts. *Ann. Surg.* **176**, 321 (1972).

461a. H. J. Knoepfli and B. Friedli, Systemic-to-pulmonary artery fistula following actino-mycosis. *Chest* **67**, 494 (1975).

461b. C. B. Anderson, H. R. Butcher, Jr., and W. F. Ballinger, Mycotic aneurysms. *Arch. Surg.* **109**, 712 (1974).

462.   L. Waitzkin, Latent *Corynebacterium acnes* infection of bone marrow. *N. Engl. J. Med.* **281**, 1404 (1969).

463.   H. Brocard, M. Bouvier, and G. Péan, Infection post-angineuse à *"Bacillus funduli-formis"* avec pleurésie purulentr et phlébite d'un membre inférieur. *Bull. Mem. Soc. Med. Hôp. Paris* **70**, 223 (1954).

464.   R. J. Fass, R. L. Perkins, and S. Saslaw, Cephalexin—a new oral cephalosporin: Clinical evaluation in sixty-three patients. *Am. J. Med. Sci.* **259**, 187 (1970).

465.   M. N. Eade and B. N. Brooke, Portal bacteraemia in cases of ulcerative colitis sub-mitted to colectomy. *Lancet* **1**, 1008 (1969).

466.   J. G. Bartlett and S. M. Finegold, Clinical features and diagnosis of anaerobic pleuro-pulmonary infections. *Antimicrob. Ag. Chemother.* p. 78 (1971).

467.   J. G. Bartlett and S. M. Finegold, Improved technique to culture coughed sputum. *Bacteriol. Proc.-1971* p. 91, Abstract M160 (1971).

468.   J. G. Bartlett, S. L. Gorbach, and S. M. Finegold, The bacteriology of aspiration pneumonia. *Am. J. Med.* **56**, 202 (1974).

469.   J. G. Bartlett, S. L. Gorbach, H. Thadepalli, and S. M. Finegold, The bacteriology of empyema. *Lancet* **1**, 338 (1974).

470.   J. G. Bartlett, S. L. Gorbach, F. P. Tally, and S. M. Finegold, Bacteriology and treatment of primary lung abscess. *Am. Rev. Respir. Dis.* **109**, 510 (1974).

471.   J. G. Bartlett and S. M. Finegold, Anaerobic infections of the lung and pleural space. *Am. Rev. Respir. Dis.* **110**, 56 (1974).

472.   S. M. Finegold, B. Smolens, A. A. Cohen, W. L. Hewitt, A. B. Miller, and A. Davis,

Necrotizing pneumonitis and empyema due to microaerophilic streptococci. *N. Engl. J. Med.* **273**, 462 (1965).

473. L. Guillemot, J. Hallé, and E. Rist, Recherches bactériologiques et expérimentales sur les pleurésies putrides. *Arch. Med. Exp.* **16**, 571 (1904).

473a. H. M. Goldstein, R. A. Castellino, L. Wexler, and E. B. Stinson: Roentgenologic aspects of cardiac transplantation. *Am. J. Roentgenol.* **111**, 476 (1971).

473b. P. D Bandt, N. Blank and R. A. Castellino: Needle diagnosis of pneumonitis. *J. Am. Med. Assoc.* **220**, 1578 (1972).

473c. N. Blank, R. A. Castellino, and V. Shah. Radiographic aspects of pulmonary infection in patients with altered immunity. *Radiol. Clin. N. Amer.* **11**, 175 (1973).

473d. K. Ries, M. Levison and D. Kaye, Experience with aerobic and anaerobic cultures in transtracheal aspirations (TTA). *Clin. Res.* **21**, 732 (1973).

473e. J. P. O'Keefe, J. G. Bartlett, F. P. Tally, and S. L. Gorbach., An heuristic approach to hospital-acquired pneumonia. *Clin. Res.* **23**, 589A (1975).

474. S. Rona, Zur Ätiologie und Pathogenese der Plaut-Vincentschen Angina, der Stoma-kace, der Stomatitis gangraenosa idiopathica, beziehungsweise der Noma, der Stomatitis mercurialis gangraenosa und der Lungengangran. *Arch. Dermatol. Syph.* **74**, 171 (1905).

475. D. T. Smith, Fuso-spirochaetal diseases of the lungs. *Tubercle* **9**, 420, (1928).

476. B. Lorber and R. M. Swenson, Bacteriology of aspiration pneumonia. A prospective study of community- and hospital-acquired cases. *Ann. Intern. Med.* **81**, 329 (1974).

476a. K. M. Sullivan, R. D. O'Toole, R. H. Fisher, and K. N. Sullivan, Anaerobic empyema thoracis. *Arch. Intern. Med.* **131**, 521 (1973).

476b. B. Lorber: "Bad breath": Presenting manifestation of anaerobic pulmonary infection. *Am. Rev. Respir. Dis.* **112**, 875 (1975).

477. S. M. Finegold, J. G. Bartlett, A. W. Chow, D. J. Flora, S. L. Gorbach, E. J. Harder, and F. P. Tally, Management of anaerobic infections, UCLA Conference. *Ann. Intern. Med.* **83**, 375 (1975).

477a. G. A. Wong, P. D. Hoeprich, A. L. Barry, T. H. Peirce and D. C. Rausch, Lower respiratory tract flora in chronic bronchitis: Fiberoptic bronchoscopy): Transtracheal aspiration. *Clin. Res.* **22**, 127A (1974).

478. P. H. Greey, The bacteriology of bronchiectasis. An analysis based on nine cases in which lobectomy was done. *J. Infect. Dis.* **50**, 203 (1932).

479. H. T. Ballantine, Jr. and J. C. White, Brain Abscess. Influence of the antibiotics on therapy and mortality. *N. Engl. J. Med.* **248**, 14 (1953).

480. G. F. Dick, Fusiform bacilli associated with various pathological processes. *J. Infect. Dis.* **12**, 191 (1913).

481. A. J. Moon, K. G. Williams, and W. I. Hopkinson, Infective gangrene surrounding an empyema wound treated with hyperbaric oxygen. *Br. J. Dis. Chest* **58**, 198 (1964).

481a. A. V. Thomas, T. H. Sodeman, and R. R. Bentz, *Bifidobacterium* (*Actinomyces*) *eriksonii* infection. *Am. Rev. Respir. Dis* **110**, 663 (1974).

482. A. W. Franklin, An anaerobic gram-negative bacillus as a cause of pyaemia. *Lancet* **2**, 645 (1933).

482a. B. Varkey, F. B. Landis, T. T. Tang, and H. D. Rose, Thoracic actinomycosis. *Arch. Intern. Med.* **134**, 689 (1974).

482b. J. F. Morris and P. Kilbourn, Systemic actinomycosis caused by *Actinomyces odontoly-ticus*. *Ann. Intern. Med.* **81**, 700 (1974).

483. S. M. Finegold, D. J. Flora, H. R. Attebery, and V. L. Sutter, Fecal bacteriology of colonic polyp patients and control patients. *Cancer Res.* **35**, 3407 (1975).

484. S. L. Gorbach, H. Thadepalli, and J. Norsen, Anaerobic microorganisms in intra-abdominal infections. *In* "Anaerobic Bacteria: Role in Disease" (A. Balows *et al.*, eds.), Chapter XXXII. Thomas, Springfield, Illinois, 1974.

484a. B. Lorber and R. M. Swenson, The bacteriology of intra-abdominal infections. *Surg. Clin. N. Amer.* **55**, 1349 (1975).

485. H. Thadepalli and S. L. Gorbach, Clindamycin in pure and mixed anaerobic infections. *Arch. Intern. Med.* **134**, 87 (1974).

486. S. L. Gorbach and H. Thadepalli, Isolation of *Clostridium* in human infections: Evaluation of 114 cases. *J. Infect. Dis.* **131**, Suppl., S81 (1975).

487. D. S. Saksena, M. A. Block, M. C. McHenry, and J. P. Truant, Bacteroidaceae: Anaerobic organisms encountered in surgical infections. *Surgery* **63**, 261 (1968).

488. J. A. Spittel, W. J. Martin, and D. R. Nichols, Bacteremia owing to gram-negative bacilli: Experiences in the treatment of 137 patients in a 15-year period. *Ann. Intern. Med.* **44**, 302 (1956).

489. B. S. Tynes and W. B. Frommeyer, Jr., *Bacteroides* septicemia. Cultural, clinical, and therapeutic features in a series of 25 patients. *Ann. Intern. Med.* **56**, 12 (1962).

490. A. F. Gelb and S. J. Seligman, Bacteroidaceae bacteremia. Effect of age and focus of infection upon clinical course. *J. Am. Med. Assoc.* **212**, 1038 (1970).

491. A. G. Kuklinca and T. L. Gavan, Anaerobic bacteria in postmortem blood cultures. Correlation with lesions of the gastrointestinal tract. *Cleveland Clin. Q.* **38**, 5 (1971).

492. S. M. Finegold, V. H. Marsh, and J. G. Bartlett, Anaerobic infections in the compromised host. *Proc. Int. Conf. Nosocomial Infections., 1970,* p. 123 (1971).

493. L. E. Jones, W. A. Wirth, and C. C. Farrow, Clostridial gas gangrene and septicemia complicating leukemia. *South. Med. J.* **53**, 863 (1960).

494. A. Isenberg, *Clostridium welchii* infection. A clinical evaluation. *Arch. Surg.* (*Chicago*) **92**, 727 (1966).

495. H. Thadepalli, S. L. Gorbach, P. Broido, and J. Norsen, A prospective study of infections in penetrating abdominal trauma. *Am. J. Clin. Nutr.* **25**, 1405 (1972).

496. R. L. Nichols, H. Thadepalli, and S. L. Gorbach, Infections following abdominal trauma. *Ill. Med. J.* 42:50, 1972.

497. Case Records of the Massachusetts General Hospital, Case 41–1968. *N. Engl. J. Med.* **279**, 819 (1968).

498. S. Ishiyama, K. Sakabe, K. Kawakami, K. Nakayama, H. Iwamoto, S. Iwai, K. Oshima, M. Katori, T. Kawabe, K. Suzuki, and F. Murakami, Bleeding stomach ulcer with intestinal complications. One case and related material. *In* "Second Symposium on Anaerobic Bacteria and Their Infectious Diseases" (S. Ishiyama *et al.*, eds.), p. 101. Eisai, Tokyo, Japan, 1972.

499. J. G. Duncan and E. Samuel, Extra-abdominal abscesses of intestinal origin. *Br. J. Radiol.* **33**, 627 (1960).

499a. R. Mzabi, H. S. Himal, and L. D. MacLean, Gas gangrene of the extremity: The presenting clinical picture in perforating carcinoma of the caecum. *Br. J. Surg.* **62**, 373 (1975).

500. A. D. Govan, An account of the pathology of some cases of *Cl. welchii* infection. *J. Pathol. Bacteriol.* **58**, 423 (1946).

501. J. B. Fethers, Gas gangrene as a complication of haematemesis. *Br. J. Surg.* **47**, 187 (1959–1960).

502. T. L. Canipe and A. S. Hudspeth, Gas gangrene septicemia. Report of an unusual case. *Arch. Surg.* (*Chicago*) **89**, 544 (1964).

503. V. J. Felitti, Primary invasion by *Clostridium sphenoides* in a patient with periodic neutropenia. *Calif. Med.* **113**, 76 (1970).

504. B. Heyworth, S. Basu, and J. Clegg, Crohn's disease and pregnancy. *Br. Med. J.* **1**, 363 (1970).

505. J. F. Wiot and B. Felson, Gas in the portal venous system. *Am. J. Roentgenol., Radium Ther. Nucl. Med.* [ns] **86**, 920 (1961).

506. A. F. Barrett, Gas in the portal vein. Diagnostic value in intestinal gangrene. *Clin. Radiol.* **13**, 92 (1962).

507. H. L. Fred, C. G. Mayhall, and T. S. Harle, Hepatic portal venous gas. A review and report on six new cases. *Am. J. Med.* **44**, 557 (1968).

508. D. J. Palmisano and D. J. Russin, Primary actinomycosis of the transverse mesocolon. *Mil. Med.* **134**, 281 (1969).

509. P. Holm, Studies on the aetiology of human actinomycosis. I. The "other microbes" of actinomycosis and their importance. *Acta Pathol. Microbiol. Scand.* **27**, 736 (1950).

510. E. H. Spaulding, V. Vargo, T. C. Michaelson, R. Vitagliano, R. M. Swenson, and E. Forsch, Value of a prereduced anaerobic method in the bacteriology of abdominal wounds. *Bacterial. Proc.* p. 109, Abstract M269. (1971).

511. H. Azar and T. Drapanas, Relationship of antibiotics to wound infection and enterocolitis in colon surgery. *Am. J. Surg.* **115**, 209 (1968).

512. W. A. Gillespie and J. Guy, Bacteroides in intra-abdominal sepsis. Their sensitivity to antibiotics. *Lancet* **1**, 1039 (1956).

513. L. J. Pyrtek and S. H. Bartus, *Clostridium welchii* infection complicating biliary-tract surgery. *N. Engl. J. Med.* **266**, 689 (1962).

514. T. C. Eickhoff, An outbreak of surgical wound infections due to *Clostridium perfringens*. *Surg., Gynecol. Obstet.* **114**, 102 (1962).

515. J. B. McNally, W. R. Price, and M. Wood, Gas gangrene of the anterior abdominal wall. *Am. J. Surg.* **116**, 779 (1968).

516. E. Witebsky and G. E. Miller, The surgical wound. *G.P.* **20**, 111 (1959).

517. J. G. Bartlett, V. L. Sutter, and S. M. Finegold, Treatment of anaerobic infections with lincomycin and clindamycin. *N. Engl. J. Med.* **287**, 1006 (1972).

518. H. R. Bernard and W. R. Cole, Wound infections following potentially contaminated operations. *J. Am. Med. Assoc.* **184**, 290 (1963).

519. D. Fromm and W. Silen, Postoperative clostridial sepsis of the abdominal wall. *Am. J. Surg.* **118**, 517 (1969).

520. J. Bittner, N. Munteanu-Ivanus, D. Radulesco, V. Tataru, V. Voinesco, and E. Milea, Gangrène de la paroi abdominale à "*Clostridium bifermentans*." *Sem. Hôp.* **47**, 1900 (1971).

521. O. Arnar, J. E. Bitter, J. J. Haglin, and C. R. Hitchcock, Hyperbaric oxygen therapy in *Clostridium perfringens* and *Peptococcus* infections. *Proc. Third Int. Conf. Hyperbaric Med. N.A.S.—N.R.C., Publ.* 1401, p. 508 (1966).

522. K. Furuta and T. Tsuchiya, Abdominal wall abscesses due to *Bacteroides fragilis*. *J. Jpn. Assoc. Infect. Dis.* **47**, 44 (1973).

523. H. Werner, Eggerthella-Arten aus pathologischem Material und aus Stuhlproben. *Arch. Hyg. Bakteriol.* **151**, 492 (1967).

524. H. S. Weens, Gas formation in abdominal abscesses: A roentgen study. *Radiology* **47**, 107 (1946).

525. H. Werner and C. Reichertz, Buttersaurebildende *Bacteroides*-Kulturen. *Zentralbl. Bakteriol. Parasitenkd., Infektionskr. Hyg., Abt. 1: Orig., Reihe A* **217**, 206 (1971).

526. K. K. Kazarian and I. M. Ariel, Etiology and pathogenesis of intra-abdominal abscesses. *In* "Diagnosis and Treatment of Abdominal Abscesses" (I. M. Ariel and K. K. Kazarian, eds.), p. 51. Williams & Wilkins, Baltimore, Maryland, 1971.

527. W. A. Altemeier, W. R. Culbertson, W. D. Fullen, and C. D. Shook, Intra-abdominal abscesses. *Am. J. Surg.* **125**, 70 (1973).

528. D. J. Magilligan, Jr., Suprahepatic abscess. *Arch. Surg.* (*Chicago*) **96**, 14 (1968).

529. E. V. Haldane and C. E. van Rooyen, Treatment of severe *Bacteroides* infections with parenteral clindamycin. *Can. Med. Assoc. J.* **107**, 1177 (1972).

530. M. C. McHenry, W. E. Wellman, and W. J. Martin, Bacteremia due to *Bacteroides*. *Arch. Intern. Med.* **107**, 572 (1961).

531. M. V. Shoemaker, Gram negative anaerobic bacilli: Isolation, incidence, and cultural characteristics. *Am. J. Med. Technol.* **26**, 279 (1960).

532. W. B. Martin, Human infection with *B. necrophorum. Am. J. Clin. Pathol.* **10**, 567 (1940).

533. W. M. R. Henderson, Three cases of suppuration due to *Fusiformis* infection. *Br. Med. J.* **1**, 975 (1953).

534. J. R. McDonald, J. C. Henthorne, and L. Thompson, Role of anaerobic streptococci in human infections. *Arch. Pathol.* **23**, 230 (1937).

535. J. B. Jacobs, W. G. Hammond, and J. L. Doppman, Arteriographic localization of suprahepatic abscesses. *Radiology* **93**, 1299 (1969).

536. H. E. Müller, Retroperitonealer *Clostridium-perfringens*-Abszess. *Med. Klin. (Munich)* **66**, 885 (1971).

537. M. Ortmayer, Bilateral non-tuberculous iliopsoas abscess. *Surg., Gynecol. Obstet.* **66**, 778 (1938).

538. R. Debré, H. Bonnet, and J. Haguenau, Infection à allurés septicémiques dué à un microbe anaérobie. *Bull. Soc. Med. Hôp. Paris* [3] **47**, 1578 (1923).

539. A. L. Wyman, Endogenous gas gangrene complicating carcinoma of colon. *Br. Med. J.* **1**, 266 (1949).

540. L. Christiaens, M. Linquette, and J. Diculouard, Multiples suppurations chroniques à "*Fusiformis fusiformis*" resultats remarquables de la chrysothérapie. *Bull. Mem. Soc. Med. Hôp. Paris* [3] **66**, 224 (1950).

541. J. L. Nettles, P. J. Kelly, W. J. Martin, and J. A. Washington II, Musculoskeletal infections due to *Bacteroides. J. Bone. Jt. Surg., Am. Vol.* **51**, 230 (1969).

542. M. J. Frumin and E. Fine, Post mortem or post hoc. *J. Am. Med. Assoc.* **208**, 519 (1969).

543. A. Ghon and B. Roman, Berichte über Krankheitsfälle und Behandlungsverfahren. Zu den Infektionen mit fusiformen Bakterien. *Med. Klin. (Munich)* **12**, 177 (1916).

544. K. Anderson, "The Clinical Practice of Bacteriology," p. 53. Blackwell, Oxford, 1966.

545. F. L. Meleney, J. Olpp, H. D. Harvey, and H. Zaytseff-Jern, Peritonitis. II. Synergism of bacteria commonly found in peritoneal exudates. *Arch. Surg. (Chicago)* **25**, 709 (1932).

546. W. A. Altemeier, The bacterial flora of acute perforated appendicitis with peritonitis. *Ann. Surg.* **107**, 517 (1938).

547. P. L. Friedrich, Zur bacteriellen Aetiologie und zur Behandlung der diffusen Peritonitis. *Dtsch. Ges. Chir.* **31**, 608 (1902).

548. F. L. Meleney, H. D. Harvey, and H. Zaytseff-Jern, Peritonitis. I. The correlation of the bacteriology of the peritoneal exudate and the clinical course of the disease in one hundred and six cases of peritonitis. *Arch. Surg. (Chicago)* **22**, 1 (1931).

549. W. E. Schatten, Intraperitoneal antibiotic administration in the treatment of acute bacterial peritonitis. *Surg., Gynecol. Obstet.* **102**, 339 (1956).

550. L. T. Wright, W. I. Metzger, E. B. Shapero, S. J. Carter, H. Schreiber, and J. W. Parker, Treatment of acute peritonitis with aureomycin. *Am. J. Surg.* **78**, 15 (1949).

551. S. Shibata, S. Tsuruga, T. Ohnishi, and N. Shinagawa, On Anaerobic Infections in Surgery. *In* "First Symposium on Anaerobic Bacteria and Their Infectious Diseases" (S. Ishiyama *et al.*, eds.), p. 37. Eisai, Tokyo, Japan, 1971.

552. J. D. Hardy, T. R. Savage, and C. Shirodaria, Intestinal perforation following exchange transfusion. *Am. J. Dis. Child.* **124**, 136 (1972).

553. M. L. Snyder and I. C. Hall, *Bacillus capitovalis*, a new species of obligate anaerobe encountered in post mortem materials, in a wound infection and in the feces of infants. *Zentralbl. Bakteriol. Parasitenkd. Infektionskr. Hyg., Abt. I: Orig.* **135**, 290 (1935).

554. L. DS. Smith and E. O. King, Occurrence of *Clostridium difficile* in infections of man. *J. Bacteriol.* **84**, 65 (1962).

555. M. J. Silverstein, C. R. Silverstein, and S. Shulman, Spontaneous gas peritonitis: A case report. *Mil. Med.* **138**, 160 (1973).

555a. J. P. Correia and H. O. Conn, Spontaneous bacterial peritonitis in cirrhosis: endemic or epidemic? *Med. Clin. N. Amer.* **59**, 963 (1975).

555b. S. I. Kahn, S. Garella, and J. A. Chazan, Nonsurgical treatment of intestinal perforation due to peritoneal dialysis. *Surg., Gynecol. Obstet.* **136**, 40 (1973).

556. W. A. Altemeier, Recent trends in the management of hepatic abscess. *Del. Med. J.* **43**, 327 (1971).

557. M. Sparberg, A. Gottschalk, and J. B. Kirsner, Liver abscess complicating regional enteritis. *Gastroenterology* **49**, 548 (1965).

557a. J. Sabbaj and M. Villanueva, Infeccion por anaerobios. *Rev. Col. Med. (Guatemala)* **25**, 17 (1974).

558. K. Kaplan and L. Weinstein, Diphtheroid infections of man. *Ann. Intern. Med.* **70**, 919 (1969).

559. S. P. Kahn, S. Lindenauer, R. S. Wojtalik, and D. Hildreth, Clostridia hepatic abscess. *Arch. Surg. (Chicago)* **104**, 209 (1972).

560. H. Legrand and E. Axisa, Ueber Anaerobien im Eiter dysenterischer Leber-und Gehirn-abscesse in Aegypten. *Dtsch. Med. Wochenschr.* **31**, 1959 (1905).

561. F. Deve and M. Guerbet, Suppuration gazeuse spontanée d'un kyste hydatique du foie. Presence exclusive de germes anaérobies. *C. R. Seances Soc. Biol. Ses Fil.* **63**, 305 (1907).

562. E. Chabrol, Sterboul, and Fallot, L'ictère anaérobie du kyste hydatique. *Bull. Acad. Natl. Med., Paris* **132**, 562 (1948).

563. H. C. Patterson, Open aspiration for solitary liver abscess. *Am. J. Surg.* **119**, 326 (1970).

564. A. J. S. McFadzean, K. P. S. Chang, and C. Wong, Solitary pyogenic abscess of the liver treated by closed aspiration and antibiotics. A report of 14 consecutive cases with recovery. *Br. J. Surg.* **41**, 141 (1953).

565. V. E. Gilbert, Anaerobic liver abscess: Medical treatment. *Ann. Intern. Med.* **78**, 303 (1973).

566. F. G. Marson, M. J. Meynell, and F. Welsh, Pylephlebitis and septicaemia treated with aureomycin. *Br. Med. J.* **1**, 764 (1953).

567. E. Andrews and L. D. Henry, Bacteriology of normal and diseased gallbladders. *Arch. Intern. Med.* **56**, 1171 (1935).

568. Y. A. Edlund, B. O. Mollstedt, and O. Ouchterlony, Bacteriological investigation of the biliary system and liver in biliary tract disease correlated to clinical data and microstructure of the gallbladder and liver. *Acta Chir. Scand.* **116**, 461 (1958/1959).

569. R. J. Flemma, L. M. Flint, S. Osterhout, and W. W. Shingleton, Bacteriologic studies of biliary tract infection. *Ann. Surg.* **66**, 563 (1967).

569a. J. Engstrom, K. Hellstrom, L. Hogman, and B. Lonnqvist, Microorganisms of the liver, biliary tract and duodenal aspirates in biliary diseases. *Scand. J. Gastroenterol.* **6**, 177 (1971).

570. L. E. Schottenfeld, Anaerobic infection of the biliary tract. *Surgery* **27**, 701 (1950).

571. H. Werner and H. P. R. Seeliger, Kulturelle Untersuchungen über den Keimgehalt der Appendix unter besonderer Berücksichtigung der Anaerobier. *Zentralbl. Bakteriol., Parasitenkd., Infektionskr. Hyg. Abt. 1: Orig.* **188**, 345 (1963).

572. O. Lanz and E. Tavel, Bacteriologie de l'appendicite. *Rev. Chir.* **30**, 43 (1904).

573. M. Heyde, Bakteriologische und experimentelle Untersuchungen zur Aetiologie der Wurmfortsatzentzundung (mit besonderer Berücksichtigung der anaeroben Bakterien). *Beitr. Klin. Chir.* **76**, 1 (1911).

574. H. Brütt, Die Bedeutung der anaeroben Streptokokken fur die destruktive Appendicitis. *Bruns' Beitr. Klin. Chir.* **129**, 175 (1923).

575. M. Weinberg, C. Renard, and J. Davesne, Présence des anaérobies de la gangrène gazeuse dans la flore microbienne de l'appendicité. *C. R. Seances Soc. Biol. Ses Fil.* **94**, 813 (1926).

576. M. Weinberg, A. R. Prévot, J. Davesne, and C. Renard, Flore microbienne des appendicités aigues. *C. R. Seances Soc. Biol. Ses Fil.* **98**, 749 (1928).

577. M. Weinberg, A. R. Prévot, J. Davesne, and C. Renard, Recherches sur la bactériologie et la sérothérapie des appendicités aigues. *Ann. Inst. Pasteur, Paris, Suppl.* **42**, 1167 (1928).

578. H. Schmitz, Zur Bakteriologie der Appendicitis. *Zentralbl. Bakteriol., Parasitenkd. Infektionskr., Abt. 1:Orig.* **117**, 378 (1930).

579. W. Löhr and L. Rabfeld, "Die Bakteriologie der Wurmfortsatzentzundung und der Appendikularen Peritonitis." Thieme, Stuttgart, 1931.

580. A. F. Maccabe and J. Orr, A study of 200 cases of appendicitis with special reference to their bacteriology. *Edinburgh Med. J.* **59**, 100 (1952).

581. L. O. Holgersen and E. G. Stanley-Brown, Acute appendicitis with perforation. *Am. J. Dis. Child.* **122**, 288 (1971).

582. T. Ohnishi, M. Kanamori, H. Miyaji, H. Otani, T. Kato, N. Shinagawa, N. Tsuruga, and S. Shibata, Anaerobic isolations from surgical infections and their clinical and bacteriological characteristics. *In* "Second Symposium on Anaerobic Bacteria and Their Infectious Diseases" (S. Ishiyama *et al.*, eds.), p. 89. Eisai, Tokyo, Japan, 1972.

583. P. Cazzamali and R. Miglierina, La batteriologia delle peritoniti acute. *Arch. Ital. Chir.* **34**, 573 (1933).

584. J. E. Jennings, The relation of the Welch bacillus to appendicitis and its complications. *Ann. Surg.* **93**, 828 (1931).

585. Perrone, Contribution à l'étude de la bacteriologie de l'appendicité. *Ann. Inst. Pasteur, Paris* **19**, 367 (1905).

585a. D. A. Leigh, K. Simmons, and E. Norman, Bacterial flora of the appendix fossa in appendicitis and postoperative wound infection. *J. Clin. Pathol.* **27**, 997 (1974).

585b. H. Werner, H. Kunštek-Santos, C. Schockemöhle und M. Gündürewa, *Bacteroides* und Appendizitis. *Pathol. Microbiol.*, **42**, 110 (1975).

586. M. Weinberg and A. R. Prévot, Recherches sur la flore microbienne de l'appendicité. *Fusobacterium biacutum* (n. sp.) *C. R. Seances Soc. Biol. Ses. Fil.* **95**, 519 (1926).

587. H. Werner, Das kulturell-biochemische Verhalten und die Antibiotikaemfindlichkeit des *Bacteroides putredinis* (Weinberg *et al.*, 1937) Kelly 1957. *Zentralbl. Bakteriol., Parasitenkd., Infektionskr. Hyg., Abt. 1: Orig.* **215**, 327 (1970).

588. E. C. Nash, Colitis—its etiology, diagnosis and treatment. *Med. Ann. D.C.* **38**, 197 (1969).

589. A. L. Warshaw, Pancreatic abscesses. *N. Engl. J. Med.* **287**, 1234 (1972).

590. R. A. Steedman, R. Doering, and R. Carter, Surgical aspects of pancreatic abscess. *Surg., Gynecol. Obstet.* **125**, 757 (1967).

591. C. Norris, Suppurative pylephlebitis associated with anaerobic microorganisms. *J. Med. Res.* **6**, 97 (1901).

591a. R. Kodesch and H. L. DuPont, Infectious complications of acute pancreatitis. *Surg., Gynecol. Obstet.* **136**, 763 (1973).

591b. C. E. Jones, H. C. Polk, Jr., and R. L. Fulton, Pancreatic abscess. *Am. J. Surg.* **129**, 44 (1975).

592. V. E. Siler and J. H. Wulsin, Acute pancreatitis. *J. Am. Med. Assoc.* **2**, 78 (1950).

593. W. A. Altemeier and J. W. Alexander, Pancreatic abscess—A study of 32 cases. *Arch. Surg. (Chicago)* **87**, 96 (1963).

594. M. Serrano-Rios, V. Navarro, J. Fontan, H. Oliva, and J. Ramirez, Isolated hepato-pancreatic actinomycosis. Report of a case simulating an acute abdomen of fatal course. *Digestion* **2**, 262 (1969).

595. T. T. Irvin, A. J. Donaldson, and G. Smith, *Clostridium welchii* infection in gastric surgery. *Surg., Gynecol. Obstet.* **124**, 77 (1967).

596. W. E. Herrell, F. R. Heilman, W. E. Wellman, and L. G. Bartholomew, Terramycin: Some pharmacologic and clinical observations. *Proc. Staff Meet. Mayo Clin.* **25**, 183 (1950).

596a. D. Baron, H. Drugeon, F. Nicolas and A.-L. Courtieu, Intérêt du métronidazole dans les septicémies à *Bacteroides fragilis*. Deux observations. *Nouv. Presse Med.* **4**, 667 (1975).

597. H. D. Rose and R. J. Bukosky, *Clostridium perfringens* septicemia following perforation of a duodenal ulcer. *J. Am. Med. Assoc.* **198**, 1368 (1966).

598. H. S. Weens, Emphysematous gastritis. *Am. J. Roentgenol. Radium Ther.* [ns] **55**, 588 (1946).

599. V. de Lavergne and P. Florentin, Fuso-spirochetose à localisation rectale. *C. R. Seances Soc. Biol. Ses Fil.* **92**, 1097 (1925).

600. A. G. Shera, Specific granular lesions associated with intestinal spirochaetosis. *Br. J. Surg.* **50**, 68 (1962).

601. J. W. S. Macfie and H. F. Carter, The occurrence of *Spirochaeta eurygyrata* in Europeans in England with a note on a second species of *Spirochaeta* from the human intestine. *Ann. Trop. Med. Parasitol.* **11**, 75 (1917).

601a. W. D. Leach, A. Lee, and R. P. Stubbs, Localization of bacteria in the gastrointestinal tract: a possible explanation of intestinal spirochaetosis. *Infect. Immun.* **7**, 961 (1973).

602. W. A. Harland and F. D. Lee, Intestinal spirochaetosis. *Br. Med. J.* **3**, 718 (1967).

603. F. D. Lee, A. Kraszewski, J. Gordon, J. G. R. Howie, D. McSeveney, and W. A. Harland, Intestinal spirochaetosis. *Gut* **12**, 126 (1971).

603a. T. Olsson, Angiography in actinomycosis of the abdomen. *Am. J. Roentgenol., Radium Ther. Nucl. Med.* **122**, 278 (1974).

604. T. L. Jarkowski and P. L. Wolf, Unusual gas bacillus infections including necrotic enteritis. *J. Am. Med. Assoc.* **181**, 845 (1962).

605. G. D. Amromin and R. D. Solomon, Necrotizing enteropathy. A complication of treated leukemia or lymphoma patients. *J. Am. Med. Assoc.* **182**, 23 (1962).

606. S. D. Henriksen, Studies in gram-negative anaerobes. II. Gram-negative anaerobic rods with spreading colonies. *Acta Pathol. Microbiol. Scand.* **25**, 368 (1948).

607. J. D. McCarthy and J. G. Picazo, Diverticulitis—A case reflecting upon pathogenesis. *Dis. Colon Rectum* **12**, 451 (1969).

608. G. A. Fry, W. J. Martin, W. H. Dearing, and C. E. Culp, Primary actinomycosis of the rectum with multiple perianal and perineal fistulae. *Mayo Clin. Proc.* **40**, 296 (1965).

608a. N. S. Brewer, R. J. Spencer, and D. R. Nichols, Primary anorectal actinomycosis. *J. Am. Med Assoc.* **228**, 1397 (1974).

609. J. W. S. Macfie, Urethral spirochaetosis. *Parasitology* **9**, 274 (1917).

610. W. E. Coutts, Contribucion al estudio de la presencia y significado de espiroquetos y espirilos en la orina asepticamente extraida. *Rev. Soc. Chil. Urol.* **2**, 224 (1938).

611. R. Atlas, Discussion on non-bacterial infection of the urinary tract. *Br. J. Vener. Dis.* **24**, 120 (1948).

612. W. E. Coutts, Non-bacterial infection of the urinary tract. *Br. J. Vener. Dis.* **24**, 109 (1948).

613. W. E. Coutts and R. Vargas-Zalazar, Abacterial pyuria with special reference to infection by spirochaetes. *Br. Med. J.* **2**, 982 (1946).

614. W. E. Coutts, O. Barthet, and J. M. Herrera, The pathological significance of certain naso-buco-pharyngeal organisms in the genital tract of males and females. *Urol. Cutaneous Rev.* **41**, 434 (1937).

615. W. E. Coutts, R. Vargas-Zalazar, and E. Silva, Fungal, viral, spirochetal, proto- or metazoan infection or infestation of the urinary tract, particularly in relation to abacterial pyuria. *Urol. Cutaneous Rev.* **55**, 148 (1951).

616. B. Bustamante, V. Zalazar, and O. Castro, V. Espirilosis urinaria. *Rev. Chil. Urol.* **13**, 90 (1950).

617. O. Correa-Fuenzalidá, D. Infecciones génito-urinarias por espiroquetoideos. *In* "Infecciones Génito-Urinarias no Bactericas," p. 58. Stanley, Santiago, Chile, 1950.

618. J. L. Stoddard, The ocurrence of spirochaetes in the urine. *Br. Med. J.* **2**, 416, 1917.

619. Y. Kon and T. Watabiki, The presence of spirochetes in the kidney. *J. Am. Med. Assoc.* **70**, 1522 (1918).

620. A. Castellani, Notes on non-gonorrhoeal urethrites. *J. Trop. Med. Hyg.* **28**, 250 (1925).

621. L. V. McVay, Jr. and D. H. Sprunt, *Bacteroides* infections. *Ann. Intern. Med.* **36**, 56 (1952).

622. W. I. Metzger, L. T. Wright, F. R. DeLuca, C. E. Ford, and F. Katske, Aureomycin in urinary tract infections. *J. Urol.* **67**, 374 (1952).

623. A. Lutz, M. A. Witz, C. Grad, and A. Schaeffer, Étude sur les infections urinaires I. Aspects microbiologiques. *Strasbourg Med.* **10**, 203 (1959).

624. J. T. Headington and B. Beyerlein, Anaerobic bacteria in routine urine culture. *J. Clin. Pathol.* **19**, 573 (1966).

625. I. J. Slotnick and W. F. Mackey, Observations on anaerobic bacteria in the female urinary tract. *Am. J. Obstet. Gynecol.* **99**, 413 (1967).

626. R. Rosner, The effect of increased incubation time on the number of positive urine cultures obtained from patients with a clinical diagnosis of chronic pyelonephritis. *J. Urol.* **99**, 688 (1968).

627. H. Lodenkämper, Voraussetzung und Bedeutung der bakteriologischen Untersuchungen bei der Pyelonephritis. *Therapiewoche* **24**, 1101 (1969).

628. A. G. Kuklinca and T. L. Gavan, The culture of sterile urine for detection of anaerobic bacteria—not necessary for standard evaluation. *Cleveland Clin. Q.* **36**, 133 (1969).

629. J. Bittner, C. Racovita, J. Ardeleanu, S. Masek, E. Tomas, A. Radulesco, and E. Sorea, Incidence des anaérobies du genre *Clostridium* dans les urocultures. *Arch. Roum. Pathol. Exp. Microbiol.* **28**, 519 (1969).

630. F. H. Caselitz, V. Freitag, T. Kreitz, and E. Rheinwald, Anaerobe Bakterien im Urin bei Harnwegsinfektionen. *Muench. Med. Wochenschr.* **112**, 1940 (1970).

631. I. Nakamura, Y. Takahira, and K. Inoue, Anaerobic bacteria in clinical material and their antibiotic susceptibility. *Saishin Igaku* **25**, 1170 (1970).

632. K. Furuta, E. Nakano, S. Kaniwa, T. Tsuchiya, H. Kaya, K. Okuyama, A. Otsuka, and A. Horikoshi, Significance of anaerobic bacteria isolated from the urinary tract. *Jpn. J. Clin. Pathol.* **19**, Suppl., 129 (1971).

633. J. Kumazawa, H. Kiyohara, and T. Momose, Anaerobic bacteria in urinary tract infections. *Jpn. J. Clin. Pathol.* **19**, Suppl., 131 (1971).

633a. Y. Shimizu, K. Ninomiya, and K. Watanabe, Anaerobic organisms isolated from urine specimens (bacteriological study). *In* "Third Symposium on Anaerobic Bacteria and Their Infectious Diseases," (S. Ishiyama *et al.*, eds.), p. 86. Eisai, Tokyo, Japan, 1973.

633b. J. Kumazawa, H. Kiyohara, K. Narahashi, M. Hikaka, and S. Momose, Significance of anaerobic bacteria isolated from the urinary tract. I. Clinical studies. *J. Urol.* **112**, 257 (1974).

634. B. Alling, A. Brandberg, S. Seeberg, and A. Svanborg, Aerobic and anaerobic microbial flora in the urinary tract of geriatric patients during long-term care. *J. Infect. Dis.* **127**, 34 (1973).

635. T. A. Stamey, "Urinary Infections." Williams & Wilkins, Baltimore, Maryland, 1972.

636. K. O. Ghormley, E. N. Cook, and G. M. Needham, Bacterial flora in chronic prostatitis. *Am. J. Clin. Pathol.* **24**, 186 (1954).

636a. T. Justesen, M. L. Nielsen, and T. Hattel, Anaerobic infections in chronic prostatitis and chronic urethritis. *Med. Microbiol. Immunol.* **158**, 237 (1973).

636b. P.-A. Mårdh and S. Colleen, Search for uro-genital tract infections in patients with symptoms of prostatitis. *Scand. J. Urol. Nephrol.* **9**, 8 (1975).

637. M. L. Nielsen and H. Laursen, Clostridial infection in the urinary tract. *Scand. J. Urol. Nephrol.* **6**, 120 (1972).

638. J. W. Segura, P. P. Kelalis, W. J. Martin, and L. H. Smith, Anaerobic bacteria in the urinary tract. *Mayo Clin. Proc.* **47**, 30 (1972).

639. S. L. Daines and N. B. Hodgson, Spontaneous rupture of a necrotic bladder. *J. Urol.* **102**, 431 (1969).

640. O. Salvatierra, Jr., W. B. Bucklew, and J. W. Morrow, Perinephric abscess: A report of 71 cases. *J. Urol.* **98**, 296 (1967).

640a. G. Furness, M. H. Kamat, Z. Kaminski, and J. J. Seebode, An investigation of the relationship of nonspecific urethritis corynebacteria to the other microorganisms found in the urogenital tract by means of a modified chocolate agar medium. *Invest. Urol.* **10**, 387 (1973).

640b. S. Hafiz, R. S. Morton, M. G. McEntegart, and S. A. Waitkins, *Clostridium difficile* in the urogenital tract of males and females. *Lancet* **1**, 420 (1975).

640c. S. R. Shapiro and L. C. Breschi, Acute epididymitis in Vietnam: Review of 52 cases. *Mil. Med.* **138**, 643, (1973).

641. G. Goldsand, Actinomycosis. *In* "Infectious Diseases" (P. D. Hoeprich, ed.), p. 395. Harper & Row, New York, 1972.

642. M. Anhalt and R. Scott, Jr., Primary unilateral renal actinomycosis: Case report. *J. Urol.* **103**, 126 (1970).

643. J. C. Henthorne, L. Thompson, and D. C. Beaver, Gram-negative bacilli of the genus *Bacteroides. J. Bacteriol.* **31**, 255 (1936).

644. F. L. Sapico, P. A. Wideman, and S. M. Finegold, Aerobic and anaerobic bladder urine flora of patients with indwelling urethral catheters. *Urology* **7**, 382 (1976).

645. Editorial, Oxygen tension of the urine and renal function. *N. Engl. J. Med.* **269**, 159 (1963).

646. K. O. Leonhardt and R. R. Landes, Oxygen tension of the urine and renal structures. *N. Engl. J. Med.* **269**, 115 (1963).

647. H. L. DuPont and W. W. Spink, Infections due to gram-negative organisms: An analysis of 860 patients with bacteremia at the University of Minnesota Medical Center, 1958–1966. *Medicine (Baltimore)* **48**, 307 (1969).

648. H. E. Pearson and G. V. Anderson, Perinatal deaths associated with *Bacteroides* infections. *Obstet. Gynecol.* **30**, 486 (1967).

649. W. C. Love, Clinical aspects of *Bacteroides* infections. *In* "Host Resistance to Commensal Bacteria" (T. Macphee, ed.), p. 298. Churchill Livingstone, Edinburgh, 1972.

650. O. D. Chapman, *Bacteroides* infections. *Proc. N.Y. State Assoc. Pub. Health Lab.* **30**, 15 (1950).

651. C. D. Smith, C. Deane, J. Montgomery, R. Williams, and C. C. Sampson, *Bacteroides. J. Natl. Med. Assoc.* **60**, 215 (1968).

652. B. Carter, Anaerobic infections in obstetrics and gynaecology. *Proc. Roy. Soc. Med.* **56**, 1095 (1963).

653. B. Carter, C. P. Jones, R. L. Alter, R. N. Creadick, and W. L. Thomas, *Bacteroides* infections in obstetrics and gynecology. *Obstet. Gynecol.* **1**, 491 (1953).

654. C. E. Clark and A. F. Wiersma, *Bacteroides* infections of the female genital tract. *Am. J. Obstet. Gynecol.* **63**, 371 (1952).

655. W. J. Ledger, R. L. Sweet, and J. T. Headington, *Bacteroides* species as a cause of severe infections in obstetric and gynecologic patients. *Surg., Gynecol. Obstet.* **133**, 837 (1971).

656. J. Mizuno, S. Matsuda, and M. Enno, The clinical significance of anaerobic infections in gynecology. *In* "First Symposium on Anaerobic Bacteria and Their Infectious Diseases" (S. Ishiyama *et al.*, eds.), p. 53. Eisai, Tokyo, Japan, 1971.

657. S. Mizuno and S. Matsuda, Chemotherapy of pelvic inflammatory diseases. *Antimicrob. Ag. Chemother* p. 679 (1966).

658. S. C. Robinson, Pelvic abscess. *Am. J. Obstet. Gynecol.* **81**, 250 (1961).

659. K. Bingold, Die Bedeutung anaerober Bakterien als Infektionserreger septischer interner Erkrankungen. *Virchow's Arch. Pathol. Anat. Physiol.* **234**, 332 (1921).

660. W. Lehmann, Die Bedeutung anaerober Streptokokken fur die Aetiologie der akuten septischen Endokarditis. *Muench. Med. Wochenschr.* **1**, 233 (1926).

661. L. Langeron, P. Michaux, and J. Liefooghe, Septicémie "post abortum" à staphylocoque anaérobie. Endocardite maligné végétante. *Bull. Mem. Soc. Med Hôp. Paris* **64**, 61 (1948).

662. H. M. Butler, The examination of cervical smears as a means of rapid diagnosis in severe *Clostridium welchii* infections following abortion. *J. Pathol. Bacteriol.* **54**, 39 (1942).

663. G. G. Hadley and R. D. Ekroth, Spherocytosis as a manifestation of postabortal *Clostridium welchii* infections. *Am. J. Obstet. Gynecol.* **67**, 691 (1954).

664. L. P. Smith, A. P. McLean, and G. B. Maughan, *Clostridium welchii* septicotoxemia. A review and report of 3 cases. *Am. J. Obstet. Gynecol.* **110**, 135 (1971).

665. W. B. Strum, J. R. Cade, D. L. Shires, and A. de Quesada, Postabortal septicemia due to *Clostridium welchii.* Treatment with exchange transfusion. *Arch. Intern. Med.* **122**, 73 (1968).

666. L. E. Perrin, D. R. Ostergard, and D. R. Mishell, Jr., The use of hyperbaric oxygen in the treatment of clostridial septicemia complicating septic abortion. *Am. J. Obstet. Gynecol.* **106**, 666 (1970).

667. R. Sen, S. N. Saxena, and L. R. Dasgupta, Anaerobic streptococci in pathological processes (with special reference to puerperal sepsis). *Indian J. Med. Res.* **55**, 799 (1967).

668. M. Magara, Z. Takase, I. Konno, Y. Koshino, S. Saeki, M. Yamada, and K. Ebina, Clinical significance of non-sporeforming anaerobes in gynecologic patients. *In* "Clinical Significance of Non-Sporeforming Bacteria in Clinical Bacteriology" (S. Suzuki *et al.*, eds.), Symps. No. 40 p. 41. Jpn. Soc. Microbiol., Nagoya City, Japan, 1966.

669. K. E. Hite, H. C. Hesseltine, and L. Goldstein, A study of the bacterial flora of the normal and pathologic vagina and uterus. *Am. J. Obstet. Gynecol.* **53**, 233 (1947).

669a. R. S. Gibbs, T. N. O'Dell, R. R. MacGregor, R. H. Schwarz, and H. Morton, Puerperal endometritis: A prospective microbiologic study. *Am. J. Obstet. Gynecol.* **121**, 919 (1975).

670. P. W. Toombs and I. D. Michelson, *Clostridium welchii* septicemia complicating prolonged labor due to obstructing myoma of uterus, with report of case. *Am. J. Obstet. Gynecol.* **15**, 379 (1928).

671. W. D. Ragan, Gas gangrene complicating term pregnancy. *Obstet. Gynecol.* **15**, 332 (1960).

672. J. W. Harris and J. H. Brown, Description of a new organism that may be a factor in the causation of puerperal infection. *Bull. Johns Hopkins Hosp.* **40**, 203 (1927).

673. J. T. Browne, A. H. Vanderhor, T. S. McConnell, and J. W. Wiggins, *Clostridium perfringens* myometritis complicating cesarean section. Report of 2 cases. *Obstet. Gynecol.* **28**, 64 (1966).

674. O. Tracy, A. M. Gordon, F. Moran, W. C. Love, and P. McKenzie, Lincomycins in the treatment of *Bacteroides* infections. *Br. Med. J.* **1**, 280 (1972).

675. J. F. Smith, Acute suppurative placentitis with fetal distress. Report of a case. *Obstet. Gynecol.* **30**, 834 (1967).

676. K. Bingold, Über septischen Ikterus. *Z. Klin. Med.* **92**, 140 (1921).

677. T. W. McElin, R. E. LaPata, G. O. Westenfelder, and R. P. Hohf, Postpartum ovarian vein thrombophlebitis and microaerophilic streptococcal sepsis. *Obstet. Gynecol.* **35**, 632 (1970).

678. Y. Sugiyama, K. Yamaji, and K. Seiga, Anaerobic bacterial isolations in the materials of cancer of the cervix. *In* "Second Symposium on Anaerobic Bacteria and Their Infectious Diseases". (S. Ishiyama *et al.*, eds.), p. 125. Eisai, Tokyo, Japan, 1972.

679. S. Matsuda and M. Tanno, The clinical and bacteriological significance of anaerobes in leucorrhea (whites). *In* "Second Symposium on Anaerobic Bacteria and Their Infectious Diseases" (S. Ishiyama *et al.*, eds.), p. 109. Eisai, Tokyo, Japan, 1972.

680. G. A. Doehner, K. G. Klinges, and B. J. Pisanti, The x-ray diagnosis of gas gangrene of the uterus. *Am. J. Obstet. Gynecol.* **79**, 542 (1960).

680a. B. M. Kaufmann, J. M. Cooper, and P. Cookson, *Clostridium perfringens* septicemia complicating degenerating uterine leiomyomas. *Am. J. Obstet. Gynecol.* **118**, 877 (1974).

681. D. R. Mishell, Jr. and D. L. Moyer, Association of pelvic inflammatory disease with the intrauterine device. *Clin. Obstet. Gynecol.* **12**, 179 (1969).

682. R. Sen, K. Ray, C. Bose, G. Kripalani, and N. Purkayastha, An assessment of the safety of inserting Lippes "loop" as intra-uterine contraceptive device in respect to microbial infection. *Indian J. Med. Res.* **56**, 668 (1968).

683. D. R. Mishell, Jr., J. H. Bell, R. G. Good, and D. L. Moyer, The intrauterine device: A bacteriologic study of the endometrial cavity. *Am. J. Obstet. Gynecol.* **96**, 119 (1966).

684. R. W. Brenner and S. W. Gehring, II, Pelvic actinomycosis in the presence of an endocervical contraceptive device. Report of a case. *Obstet. Gynecol.* **29**, 71 (1967).

684a. F. E. Dische, L. J. M. Burt, N. J. H. Davidson, and S. Puntambekar, Tubo-ovarian actinomycosis associated with intrauterine contraceptive devices. *J. Obstet. Gynecol.* **81**, 724 (1974).

684b. P. K. O'Brien, Abdominal and endometrial actinomycosis associated with an intrauterine device. *Can. Med. Assoc. J.* **112**, 596 (1975).

684c. M. A. Schiffer, A. Elguezabal, M. Sultana, and A. C. Allen, Actinomycosis infections associated with intrauterine contraceptive devices. *Obstet. Gynecol.* **45**, 67 (1975).

684d. H. S. Kahn and C. W. Tyler, Jr., Mortality associated with use of IUDs. *J. Am. Med. Assoc.* **234**, 57 (1975).

684e. M. Y. Dawood and S. J. Birnbaum, Unilateral tubo-ovarian abscess and intrauterine contraceptive device. *Obstet. Gynecol.* **46**, 429 (1975).

685. W. A. Altemeier, The anaerobic streptococci in tubo-ovarian abscess. *Am. J. Obstet. Gynecol.* **39**, 1038 (1940).

686. C. H. Arnold, Plaut-Vincent's infection of the vagina. Report of case. *J. Am. Med. Assoc.* **94**, 1461 (1930).

687. F. Chatillon, Ulcère vulvaire à association fuso-spirillaire de Vincent. *Rev. Fr. Gynecol. Obstet.* **25**, 473 (1930).

688. S. S. Greenbaum, Fusospirillary dermatitis. Its occurrence in acute pellagra. *Arch. Dermatol. Syphilol.* **15**, 678 (1927).

689. H. D. Jump and S. J. Sperling, Fusospirochetal (Vincent's) infection of pleura and vagina. *J. Am. Med. Assoc.* **98**, 219 (1932).

690. I. Pilot and A. E. Kanter, Studies of fusiform bacilli and spirochetes. III. Occurrence in normal women about the clitoris and significance in certain genital infections. *J. Infect. Dis.* **32**, 204 (1923).

691. H. R. Robinson, Infection of the vagina by Vincent's spirillae, complicating pregnancy and the puerperium. *Tex. State J. Med.* **22**, 687 (1927).

692. M. Ruiter and H. M. M. Wentholt, Isolation of a pleuropneumonia-like organism (G-strain) in a case of fusospirillary vulvovaginitis. *Acta Derm.-Venereol.* **33**, 123 (1953).

693. W. A. Müller, J. Holtorff, and R. Blaschke-Hellmessen, Untersuchungen über das Vorkommen und die Haufigkeit verschiedenartiger Mikroorganismen (anaerobe Bakterien, Trichomonaden, Mycoplasmen und Sprosspilse) in der menschlichen Vagina bei Scheidenentzundungen. *Arch. Hyg. Bakteriol.* **151**, 609 (1967).

694. A. M. Gordon, H. E. Hughes, and G. T. D. Barr, Bacterial flora in abnormalities of the female genital tract. *J. Clin. Pathol.* **19**, 429, (1966).

695. K. Yamaji, Y. Sugiyama, and K. Seiga, The clinical significance of anaerobes in *Trichomonas vaginalis* infections. *In* "First Symposium on Anaerobic Bacteria and Their Infectious Diseases" (S. Ishiyama *et al.*, eds.), p. 61. Eisai, Tokyo, Japan, 1971.

696. D. B. Hoffman and P. Grundfest, Vaginitis emphysematosa. *Am. J. Obstet. Gynecol.* **78**, 428 (1959).

697. H. L. Gardner and P. Fernet, Etiology of vaginitis emphysematosa. *Am. J. Obstet. Gynecol.* **88**, 680 (1964).

698. W. J. Ledger, Postoperative pelvic infections. *Clin. Obstet. Gynecol.* **12**, 265 (1969).

698a. N. M. Duignan and P. A. Lowe, Pre-operative disinfection of the vagina. *J. Antimicrobial Chemother.* **1**, 117 (1975).

698b. H. Wagman, Genital actinomycosis. *Proc. Roy. Soc. Med.* **68**, 228 (1975)

699. P. Bagović, A. Bunarević, J. Zadjelović, and I. Puharić, Actinomycosis of the female genital organs. *Lijec. Vjesn.* **89**, 143 (1967).

700. W. J. Ledger and E. P. Peterson, The use of heparin in the management of pelvic thrombophlebitis. *Surg., Gynecol. Obstet.* **131**, 1115 (1970).

701. L. J. Dunn and L. W. Van Voorhis, Enigmatic fever and pelvic thrombophlebitis. Response to anticoagulants. *N. Engl. J. Med.* **276**, 265 (1967).

702. H. Schulman, Use of anticoagulants in suspected pelvic infection. *Clin. Obstet. Gynecol.* **12**, 240 (1969).

703. R. Hare and I. Polunin, Anaerobic cocci in the vagina of native women in British North Borneo. *J. Obstet. Gynecol. Br. Emp.* **67**, 985 (1960).

704. S. M. Bluefarb and M. Gecht, Pyoderma associated with keloids. Infestation with fusospirillary organisms. *Q. Bull. Northwest. Univ. Med. Sch.* **35**, 290 (1961).

705. H. I. Maibach, Scalp pustules due to *Corynebacterium acnes*. *Arch. Dermatol.* **96**, 453 (1967).

706. G. Nobre and J. B. Caldeira, Chronic granulomatous infection of the legs associated with *Corynebacterium acnes*. *Br. J. Dermatol.* **81**, 548 (1969).

707. E. Wohlstein, Ein Fall von Nekrophorus-Infektion der Haut beim Menschen. *Dermatol. Z.* **56**, 415 (1929).

708. M. Stolzova-Sutorisova and J. Kratochvil, O sepsi zpusobene striktnim anaerobem *Bacilem ramosem*. *Cas. Lek. Cesk.* **48**, 1528 (1932).

709. W. E. Herrell, D. R. Nichols, and D. H. Heilman, Penicillin. Its usefulness, limitations, diffusion and detection, with analysis of 150 cases in which it was employed. *J. Am. Med. Assoc.* **125**, 1003 (1944).

710. L. T. Wright, J. W. Parker, F. R. Allen, and M. S. Beinfield, Terramycin in soft tissue infections. *Antibiot. Chemother.* (*Washington, D. C.*) **1**, 165 (1951).

711. M. Ruiter and H. M. M. Wentholt, Isolation of a pleuropneumonia-like organism from a skin lesion associated with a fusospirochetal flora. *J. Invest. Dermatol.* **24**, 31 (1955).

712. H. Brocard and H. C. Pham, Sur un bacille anaérobie isolé dans deux cas d'erysipèlas gangrénéux: *Bacillus terebrans. C. R. Seances Soc. Biol. Ses Fil.* **117**, 997 (1934).

713. M. Behrend and T. B. Krouse, Postoperative bacterial synergistic cellulitis of abdominal wall. *J. Am. Med. Assoc.* **149**, 1122 (1952).

714. R. M. Montgomery and W. A. Welton, Primary actinomycosis of the upper extremity. *AMA Arch. Dermatol.* **79**, 578 (1959).

715. R. R. Briney, Primary cutaneous actinomycosis. *J. Am. Med. Assoc.* **194**, 209 (1965).

716. E. S. Mahgoub and A. A. A. Yacoub, Primary actinomycosis of the foot and leg. Report of a case. *J. Trop. Med. Hyg.* **71**, 256 (1968).

717. A. Buck and K. W. Kalkoff, Zum Nachweis von Fusobakterien aus Effloreszenzen der perioralen Dermatitis. *Hautarzt* **22**, 433 (1971).

718. D. Sinniah, B. R. Sandiford, and A. E. Dugdale, Subungual infection in the newborn. *Clin. Pediatr. (Philadelphia)* **11**, 690 (1972).

719. O. Lahelle, *Necrobacterium.* A study of its bacteriology, serology and pathogenicity, and its relation to *Fusobacterium. Acta Pathol. Microbiol. Scand., Suppl.* **67**, 1 (1947).

720. R. T. Parker and C. P. Jones, Anaerobic pelvic infections and developments in hyperbaric oxygen therapy. *Am. J. Obstet. Gynecol.* **96**, 645 (1966).

721. J. H. Hall, J. P. Tindall, J. L. Callaway, and J. G. Smith, Jr., The use of lincomycin in dermatology. *South. Med. J.* **61**, 1287 (1968).

722. J. W. Parker, Jr., J. C. Lord, Jr., J. C. DiLorenzo, L. T. Wright, and B. A. Shidlovsky, Further observations on soft tissue infections treated with terramycin. *Antibiot. Chemother. (Washington, D.C.)* **3**, 122 (1953).

723. J. J. Farquet, A. Roupas, J. de Freudenreich, and O. Koralnik, Étude clinique et bactériologique des infections à *Bacteroides. Schweiz. Med. Wochenschr.* **101**, 809 (1971).

724. F. L. Meleney, Discussion of paper by A. A. Zierold. Gangrene of the extremity in the diabetic. *Ann. Surg.* **110**, 723 (1939).

725. I. Ziment, L. G. Miller, and S. M. Finegold, Nonsporulating anaerobic bacteria in osteomyelitis. *Antimicrob. Ag. Chemother.* p. 77 (1968).

725a. T. J. Louie, F. P. Tally, J. G. Bartlett, and S. L. Gorbach, The microbiology of diabetic foot ulcers. *Clin. Res.* **23**, 107A (1975).

726. A. Castellani and A. J. Chalmers, "Manual of Tropical Medicine," 3rd ed., Part III. Baillière, London, 1919.

727. F. Marsh and H. A. Wilson, Tropical ulcer. *Trans. Roy. Soc. Trop. Med. Hyg.* **38**, 259 (1945).

728. R. P. Strong, Spirochaetal infections of man. *United Fruit Co., Med. Dep., 14th Annu. Rep.* p. 218 (1925).

729. J. A. Kolmer, "Penicillin Therapy." Appleton, New York, 1945.

730. E. H. Loughlin, C. W. Price, and A. A. Joseph, Streptomycin—bacitracin—polymyxin combinations. The application of multiple antibiotic combinations in prophylaxis of skin infections with common bacteria and the treatment of chronic (tropical) ulcers. "Antibiotics Annual 1953–1954," p. 291. Med. Encycl., New York, 1953.

731. C. F. Gutch, Local penicillin therapy for tropical ulcer. *US Nav. Med. Bull.* **47**, 801 (1947).

732. W. G. Hamm and G. Ouary, Penicillin therapy in phagedenic ulcer (tropical sloughing phagedena). *US Nav. Med. Bull.* **43**, 981 (1944).

733. J. M. Pinkerton, Tropical ulcer, as seen in south Iran, and its treatment with penicillin. *J. Trop. Med. Hyg.* **50**, 243 (1947).

734. O. Ampofo and G. M. Findlay, Oral aureomycin in the treatment of tropical ulcers and cancrum oris. *Trans. Roy. Soc. Trop. Med. Hyg.* **44**, 307 (1950).

735. O. Ampofo and G. M. Findlay, Chloramphenicol in the treatment of yaws and tropical ulcer. *Trans. Roy. Soc. Trop. Med. Hyg.* **44**, 315 (1950).

736. E. H. Payne, A. Bellerive, and J. Lafont, Chloromycetin as a treatment for yaws and tropical ulcer. *Antibiot. Chemother.* (*Washington, D.C.*) **1**, 88 (1951).

737. R. R. Lindner and C. Adeniyi-Jones, The effect of metronidazole on tropical ulcers. *Trans. Roy. Soc. Trop. Med. Hyg.* **62**, 712 (1968).

738. M. H. Manson, Pathogenic gas-producing anaerobic bacilli in chronic ulcers. *Arch. Surg.* (*Chicago*) **24**, 752 (1932).

739. M. B. Myers, G. Cherry, B. B. Bornside, and G. H. Bornside, Ultraviolet red fluorescence of *Bacteroides melaninogenicus*. *Appl. Microbiol.* **17**, 760 (1969).

740. P. D. Ellner and B. L. Wasilauskas, *Bacteroides* septicemia in older patients. *J. Am. Geriatr. Soc.* **19**, 296 (1971).

740a. R. L. Douglas and J. W. Kislak, Treatment of *Bacteroides fragilis* bacteremia with clindamycin. *J. Infect. Dis.* **128**, 569 (1973).

741. Z. Bazilevskaya and I. Polozova, Bacteriologic control in decubitus ulcer. *Khirurgiia* (*Moscow*). **47**, 120 (1971).

742. G. Blumer and A. MacFarlane, An epidemic of noma: Report of sixteen cases. *Am. J. Med. Sci.* **122**, 527 (1901).

743. R. D. Emslie, Cancrum oris. *Dent. Pract.* **13**, 481 (1963).

744. S. Takahara, Progressive oral gangrene probably due to lack of catalase in the blood (actalasaemia). *Lancet* **2**, 1101 (1952).

745. W. A. Altemeier and W. R. Culbertson, Acute non-clostridial crepitant cellulitis. *Surg., Gynecol. Obstet.* **87**, 206 (1948).

746. F. H. Jergesen, Anaerobic infections. *Med. Bull. N. Afr. Theat. Op.* **1**, 2 (1944).

747. J. D. MacLennan, The histotoxic clostridial infections of man. *Bacteriol. Rev.* **26**, 232 (1962).

748. K. Kaplan and G. D. Winchell, Orbital emphysema from nose blowing. *N. Engl. J. Med.* **278**, 1234 (1968).

749. Y. Aelony, Post-traumatic crepitation. *N. Engl. J. Med.* **279**, 660 (1968).

750. A. D. Rubenstein, I. R. Tabershaw, and J. Daniels, Pseudo-gas gangrene of the hand. *J. Am. Med. Assoc.* **129**, 659 (1945).

751. R. M. Filler, N. T. Griscom, and A. Pappas, Post-traumatic crepitation falsely suggesting gas gangrene. *N. Engl. J. Med.* **278**, 758 (1968).

752. G. E. Brewer and F. L. Meleney, Progressive gangrenous infection of the skin and subcutaneous tissues, following operation for acute perforative appendicitis. *Ann. Surg.* **84**, 438 (1926).

753. F. L. Meleney, A differential diagnosis between certain types of infectious gangrene of the skin. *Surg., Gynecol. Obstet.* **56**, 847 (1933).

754. A. Touraine and R. Duperrat, La gangrène post-opératoire progressive de la peau. *Ann. Dermatol.* **10**, 257 (1919).

755. F. L. Meleney, P. Shambaugh, and R. S. Millen, Systemic bacitracin in the treatment of progressive bacterial synergistic gangrene. *Ann. Surg.* **131**, 129 (1950).

756. J. G. Smith, Progressive bacterial synergistic gangrene: Report of a case treated with chloramphenicol. *Ann. Intern. Med.* **44**, 1007 (1956).

757. D. S. De Jongh, J. P. Smith, and G. W. Thoma, Postoperative synergistic gangrene. *J. Am. Med. Assoc.* **200**, 557 (1967).

758. F. L. Meleney, B. A. Johnson, and P. Teng, Further experiences with local and systemic bacitracin in the treatment of various surgical and neurosurgical infections and certain related medical infections. *Surg., Gynecol. Obstet.* **94**, 401 (1952).

759. R. Güller, Progressive postoperative Gangraen und Dermatitis ulcerosa. *Hautarzt* **19**, 408 (1968).

760. W. A. Altemeier and W. R. Culbertson, Chloramphenicol (chloromycetin) and aureomycin in surgical infections. *J. Am. Med. Assoc.* **145**, 449 (1951).

761. H. Trendelenburg, Ueber Nosokomialgangran. *Berl. Klin. Wochenschr.* **51**, 1967 (1914).

762. W. J. Rea and W. J. Wyrick, Jr., Necrotizing fasciitis. *Ann. Surg.* **172**, 957 (1970).

763. M. Grossman and W. Silen, Serious post-traumatic infections with special reference to gas gangrene, tetanus and necrotizing fasciitis. *Postgrad. Med.* **32**, 110 (1962).

764. A. Randall, Idiopathic gangrene of the scrotum. *J. Urol.* **4**, 219 (1920).

765. S. L. Moschella, The clinical significance of necrosis of the skin. *Med. Clin. N. Am.* **53**, 259 (1969).

766. H. S. Himal and J. H. Duff, Endogenous gas gangrene: A report of three cases. *Can. Med. Assoc. J.* **97**, 1541 (1967).

767. T. E. Gibson, Idiopathic gangrene of the scrotum with a report of a case and review of the literature. *J. Urol.* **23**, 125 (1930).

768. P. G. Klotz, Correspondence—Endogenous gas gangrene. *Can. Med. Assoc. J.* **98**, 264 (1968).

768a. J. M. Rein and B. Cosman, Bacteroides necrotizing fasciitis of the upper extremity. *Plast. Reconstr. Surg.* **48**, 592 (1971).

769. G. Bras, K. P. Clearkin, H. Annamunthodo, and F. H. Caselitz, Abdominal actinomycosis associated with idiopathic gangrene of scrotum. *West Indian Med. J.* **3**, 137 (1954).

770. J. J. Yelderman and R. G. Weaver, The urologist and necrotizing dermogenital infections. *J. Urol.* **101**, 74 (1969).

771. W. M. Millar, Gas gangrene in civil life. *Surg., Gynecol. Obstet.* **54**, 232 (1932).

772. P. Dineen, Gas Gangrene *In* "Infectious Diseases" (P. D. Hoeprich, ed.), p. 1223, Harper, New York, 1972.

773. V. W. Adamkiewicz, B. Kopacka, and V. Fredette, Rôle du diabete alloxanique dans la gangrène gazeuse expérimentale de la souris blanche. *Rev. Can. Biol.* **26**, 153 (1967).

774. W. A. Altemeier, Diagnosis, classification, and general management of gas-producing infections, particularly those produced by *Clostridium perfringens. Proc. Third Int. Conf. Hyperbaric Med., N.A.S.—N.R.C., Publ.* 1401, p. 481 (1966).

774a. C. R. Hitchcock, F. J. Demello, and J. J. Haglin, Gangrene infection. New approaches to an old disease. *Surg. Clin. N. Amer.* **55**, 1403 (1975).

775. M. T. Parker, Postoperative clostridial infections in Britain. *Br. Med. J.* **3**, 671 (1969).

776. R. Gye, P. M. Rountree, and J. Loewenthal, Infection of surgical wounds with *Clostridium welchii. Med. J. Aust.* **48**, 761 (1961).

777. G. A. J. Ayliffe and E. J. L. Lowbury, Sources of gas gangrene in hospital. *Br. Med. J.* **2**, 333 (1969).

778. T. A. Koons and G. M. Boyden, Gas gangrene from parenteral injection. *J. Am. Med. Assoc.* **175**, 46 (1961).

778a. R. Van Hook and A. G. Vandevelde, Gas gangrene after intramuscular injection of epinephrine: Report of a fatal case. *Ann. Intern. Med.* **83**, 669 (1975).

779. R. B. Berggren, T. D. Batterton, G. McArdle, and W. H. Erb, Clostridial myositis after parenteral injections. *J. Am. Med. Assoc.* **188**, 1044 (1964).

780. R. F. Bishop and V. Marshall, The enhancement of *Clostridium welchii* infection by adrenaline-in-oil. *Med. J. Aust.* **47**, 656 (1960).

781. Schreus, Felderfahrungen über die Chemotherapie der Anaerobenwundinfektion insbesondere mit Globucid nebst Bemerkungen über Chemoprophylaxe. *Dtsch. Med. Wochenschr.* **69**, 73 (1943).

782. S. K. Dikshit and S. N. Mehrotra, Gas gangrene of the lower limb following femoral venepuncture. *J. Trop. Med. Hyg.* **71**, 162 (1968).

783. F. Magistris, Iatrogene Gasbrandinfektionen. *J. Int. Coll. Surg.* **37**, 489 (1962).

784. A. T. Marty and R. M. Filler, Recovery from non-traumatic localised gas gangrene and clostridial septicaemia. *Lancet* **2**, 79 (1969).

785. H. Werner, U. Gott, and G. Rintelen, Zur Kasuistik der enterogenen, nichttraumatischen Gasodeminfektionen durch *Clostridium septicum*. *Z. Med. Mikrobiol. Immunol.* **156**, 265 (1971).

785a. P. W. Brown and P. B. Kinman, Gas gangrene in a metropolitan community. *J. Bone Joint Surg.* **56-A**, 1445 (1974).

786. J. D. MacLennan, Streptococcal infection of muscle. *Lancet* **1**, 582 (1943).

787. J. D. MacLennan, Anaerobic infections of war wounds in the Middle East. *Lancet* **2**, 63, 94, and 123 (1943).

788. C. B. Anderson, J. J. Marr, and B. M. Jaffe, Anaerobic streptococcal infections simulating gas gangrene. *Arch. Surg. (Chicago)* **104**, 186 (1972).

789. H. H. Stone and J. D. Martin, Jr., Synergistic necrotizing cellulitis. *Ann. Surg.* **175**, 702 (1972).

790. C. R. Baxter, Surgical management of soft tissue infections. *Surg. Clin. N. Amer.* **52**, 1483 (1972).

791. C. L. Gillies, Interstitial emphysema in diabetes mellitus due to colon bacillus infection. *J. Am. Med. Assoc.* **117**, 2240 (1941).

792. M. N. Barnes and B. G. Bibby, A summary of reports and bacteriologic study of infections caused by human tooth wounds. *J. Am. Dent. Assoc.* **26**, 1163 (1939).

793. F. K. Boland, Morsus humanus. Sixty cases of human bites in Negroes. *J. Am. Med. Assoc.* **116**, 127 (1941).

794. F. F. Boyce, Human bites. An analysis of 90 (chiefly delayed and late) cases from Charity Hospital of Louisiana at New Orleans. *South. Med. J.* **35**, 631 (1942).

795. F. F. Boyce, Human bites. A study of a second series of 93 (chiefly delayed and late) cases from Charity Hospital of Louisiana at New Orleans. *South. Surg.* **14**, 690 (1948).

796. F. A. Brandt, Human bites of the ear. *Plast. Reconstr. Surg.* **43**, 130 (1969).

797. R. L. Maier, Human bite infections of the hand. *Ann. Surg.* **106**, 423 (1937).

798. B. R. Meyers, S. Z. Hirschman, and W. Sloan, Generalized Shwartzman reaction in man after a dog bite. *Ann. Intern. Med.* **73**, 433 (1970).

799. K. L. Burdon, The occurrence of a black pigment producing anaerobic microorganism on the normal mucous membranes and skin in pathological conditions. *J. Missouri State Med. Assoc.* **24**, 381 (1927).

800. I. Pilot, Fusospirochetal infection from the bite of an orangutan. *Arch. Pathol.* **25**, 601 (1938).

801. E. J. Doering, III, C. T. Fitts, W. M. Rambo, and G. B. Bradham, Alligator bite. *J. Am. Med. Assoc.* **218**, 255 (1971).

802. E. O. Ledbetter and A. E. Kutscher, The aerobic and anaerobic flora of rattlesnake fangs and venom. *Arch. Environ. Health* **19**, 770 (1969).

803. E. Liebetruth, Untersuchungen über das *Bacterium melaninogenicum*. *Z. Hyg. Infektionskr.* **116**, 611 (1934).

804. F. J. Oerther, J. L. Goodman, and A. M. Lerner, Infections in paregoric addicts. *J. Am. Med. Assoc.* **190**, 683 (1964).

805. A. M. Lerner and F. J. Oerther, Characteristics and sequelae of paregoric abuse. *Ann. Intern. Med.* **63**, 1019 (1966).

805a. G. W. Geelhoed and W. J. Joseph, Surgical sequelae of drug abuse. *Surg., Gynecol. Obstet.* **139**, 749 (1974).

805b. R. J. Lewis, Bone, muscle and fascial infections in addicts. *In* "Medical Aspects of Drug Abuse," (R. W. Richter, ed.). Harper, New York, 1975.

806. H. Tissier, Recherches sur la flore bacteriënne des plaies de guerre. *Ann. Inst. Pasteur, Paris* **30**, 681 (1916).

807. R. Rustigian and A. Cipriani, The bacteriology of open wounds. *J. Am. Med. Assoc.* **133**, 224 (1947).

808. R. Cruickshank, Bacteriology of infected wounds. *Lancet* **1**, 704 (1940).

809. R. B. Lindberg and A. Newton, Sensitivity to antibiotics of clostridia from Korean battle casualties. "Antibiotics Annual 1954–1955," p. 1059. Med. Encycl., New York, 1955.

810. E. J. Pulaski, F. L. Meleney, and W. L. C. Spaeth, Bacterial flora of acute traumatic wounds. *Surg., Gynecol. Obstet.* **72**, 982 (1941).

811. C. G. A. Thomas and R. Hare, The classification of anaerobic cocci and their isolation in normal human beings and pathological processes. *J. Clin. Pathol.* **7**, 300 (1954).

812. K. Taylor and M. Davies, Persistence of bacteria within sequestra. *Ann. Surg.* **66**, 522 (1917).

813. M. Bronner and M. Bronner, eds., "Actinomycosis." Williams & Wilkins, Baltimore, Maryland, 1969.

814. H. E. Pearson and J. P. Harvey, Jr., *Bacteroides* infections in orthopedic conditions. *Surg., Gynecol. Obstet.* **132**, 876 (1971).

814a. U. Gupta, A study of Bacteroidaceae from clinical material. *Indian J. Med. Res.* **61**, 1002 (1973).

814b. C. Mitchell, M. G. Ehrlich, and R. S. Siffert, Comparative bacteriology of early and late orthopedic infections. *Clin. Orthoped. Related Res.* **96**, 277 (1973).

814c. R. M. DeHaan, D. Schellenberg, and R. T. Pfeifer, Bacterial etiology of some common anaerobic infections. *Infect. Dis. Rev.* **3**, 59 (1974).

814d. H. Lodenkämper and U. Rühlcke, Anaerobier. *In* "Die Behandlung der Sekundarchronischen Osteomyelitis. Bucherei des Orthopaden." (Plaue, ed.), Vol. 13, p. 37. Enke, Stuttgart, 1974.

815. C. Kamme, L. Lidgren, L. Lindberg, and P. A. Mårdh, Anaerobic bacteria in late infections after total hip arthoplasty. *Scand. J. Infect. Dis.* **6**, 161 (1974).

816. D. Y. Sanders and J. Stevenson, *Bacteroides* infections in children. *J. Pediatr.* **72**, 673 (1968).

817. S. T. Shulman and M. O. Beem, An unique presentation of sickle cell disease: Pyogenic hepatic abscess. *Pediatrics* **47**, 1019 (1971).

818. W. T. Hughes, Fatal infections in childhood leukemia. *Am. J. Dis. Child.* **122**, 283 (1971).

819. L. H. Meskin, E. M. Farsht, and D. L. Anderson, Prevalence of *Bacteroides melaninogenicus* in the gingival crevice area of institutionalized trisomy 21 and cerebral palsy patients and normal children. *J. Periodontol.* **39**, 326 (1968).

819a. M. D. Appleman, H. Thadepalli, J. E. Maidman, J. J. Arsi, and E. C. Davidson, Jr., Effect of amniotic fluid on anaerobic bacteria. *Clin. Res.* **23**, 300A (1975).

820. W. A. Blanc, Pathways of fetal and early neonatal infection. *J. Pediatr.* **59**, 473 (1961).

821. D. Charles, Infection and the obstetric patient. *In* "Modern Treatment. Female Genital Infections" (D. Charles, ed.), Vol. 7, p. 789. Harper, New York, 1970.

822. D. W. Penner and A. C. McInnis, Intrauterine and neonatal pneumonia. *Am. J. Obstet. Gynecol.* **69**, 147 (1955).

823. A. Brim, A bacteriologic study of 100 stillborn and dead newborn infants. *J. Pediatr.* **15**, 680 (1939).

823a. C. Navarro and W. A. Blanc, Subacute necrotizing funisitis. *J. Pediatr.* **85**, 689 (1974).

824. E. R. Stiehm, A. J. Ammann, and J. D. Cherry, Elevated cord macroglobulins in the diagnosis of intrauterine infections. *N. Engl. J. Med.* **275**, 971 (1966).

825. A. W. Chow, R. D. Leake, T. Yamauchi, B. F. Anthony, and L. B. Guze, The significance of anaerobes in neonatal bacteremia: Analysis of 23 cases and review of the literature. *Pediatrics* **54**, 736 (1974).

826. L. W. Martin, W. A. Altemeier, and P. M. Reyes, Jr., Infections in pediatric surgery. *Pediatr. Clin. N. Am.* **16**, 735 (1969).

827. I. C. Hall and E. O'Toole, Bacterial flora of first specimens of meconium passed by fifty new-born infants. *Am. J. Dis. Child.* **47**, 1279 (1934).

828. I. C. Hall and M. L. Snyder, Isolation of an obligately anaerobic bacillus from the feces of newborn infants and from other human sources, and its probable identity with the "Kopfchenbackterien" of Escherich, Rodella's "Bacillus III," and *Bacillus paraputrificus* (Bienstock). *J. Bacteriol.* **28**, 181 (1934).

829. I. C. Hall and E. O'Toole, Intestinal flora in new-born infants with a description of a new pathogenic anaerobe, *Bacillus difficilis*. *Am. J. Dis. Child.* **49**, 390 (1935).

830. I. C. Hall and K. Matsumura, Recovery of *Bacillus tertius* from stools of infants. *J. Infect. Dis.* **35**, 502 (1924).

830a. W. L. Nyhan and M. D. Fousek, Septicemia of the newborn. *Pediatrics* **22**, 268 (1958).

830b. L. M. Dunkle, T. J. Brotherton, and R. D. Feigin, Anaerobic infections in children: A prospective study. *Pediatrics* **57**, 311 (1976).

831. C. M. Davis and C. Collins, Granuloma inguinale: An ultrastructural study of *Calymmatobacterium granulomatis*. *J. Invest. Dermatol.* **53**, 315 (1969).

832. J. Goldberg, Studies on granuloma inguinale. IV. Growth requirements of *Donovania granulomatis* and its relationship to the natural habitat of the organism. *Br. J. Vener. Dis.* **35**, 266 (1959).

833. J. Goldberg, Studies on granuloma inguinale. V. Isolation of a bacterium resembling *Donovania granulomatis* from the faeces of a patient with granuloma inguinale. *Br. J. Vener. Dis.* **38**, 99 (1962).

834. J. Goldberg and R. Bernstein, Studies on granuloma inguinale. VI. Two cases of perianal granuloma inguinale in male homosexuals. *Br. J. Vener. Dis.* **40**, 137 (1964).

835. R. B. Greenblatt and J. C. Wright, The significance of fusospirillosis in genital lesions. *Am. J. Syph., Gonorrhea, Vener. Dis.* **20**, 654 (1936).

836. M. J. Marples, "The Ecology of the Human Skin." Thomas, Springfield, Illinois, 1965.

837. E. von Haam, Venereal fuso-spirochetosis. *Am. J. Trop. Med.* **18**, 595 (1938).

838. B. C. Corbus and F. G. Harris, Erosive and gangrenous balanitis. The fourth venereal disease. *J. Am. Med. Assoc.* **52**, 1474 (1909).

839. M. Lev and E. B. Tucker, A note on genital fusospirochetosis. *Am. J. Clin. Pathol.* **16**, 401 (1946).

840. O. Lahelle, Penicillin-sensitivity of *Fusobacterium* and *Bacteroides funduliformis*. *Acta Pathol. Microbiol. Scand.* **24**, 567 (1947).

841. M. Ruiter and H. M. M. Wentholt, A pleuopneumonia-like organism in primary fusospirochetal gangrene of the penis. *J. Invest. Dermatol.* **15**, 301 (1950).

842. M. Ruiter and H. M. M. Wentholt, The occurrence of a pleuopneumonia-like organism in fuso-spirillary infections of the human genital mucosa. *J. Invest. Dermatol.* **18**, 313 (1952).

843. M. F. Barile, J. M. Blumberg, C. W. Kraul, and R. Yaguchi, Penile Lesions among US Armed Forces personnel in Japan. *Arch. Dermatol.* **86**. 273 (1962).

844. A. W. Hanson and G. R. Cannefax, Isolation of *Borrelia refringens* in pure culture from patients with condylomata acuminata. *J. Bacteriol.* **88**, 111 (1964).

845. H. Beerens and M. M. Castel, Procède simplifié de culture en surface des bactéries anaérobies. Comparison avec la technique utilisant la culture en profondeur. *Ann. Inst. Pasteur, Lille* **10**, 183 (1958–1959).

846. H. E. Pearson, Bacteroides in areolar breast abscesses. *Surg., Gynecol. Obstet.* **125**, 1 (1967).

847. D. V. Habif, K. H. Perzin, R. Lipton, and R. Lattes, Subareolar abscess associated with squamous metaplasia of lactiferous ducts. *Am. J. Surg.* **119**, 523 (1970).

848. A. R. Prévot, Système reticulo-endothelial et corynébactérioses anaérobies. *RES, J. Reticuloendothel. Soc.* **1**, 115 (1964).

848a. K. F. Meyer, The rise and fall of botulism. *Calif. Med.* **118**, 63 (1973).

849. C. E. Dolman, Botulism as a world health problem. *In* "Botulism—Proceedings of a Symposium" PHS Publ. No. 999-FP-1, (K. H. Lewis and K. Cassel, Jr., eds.), p. 5. US Dept. of Health, Education & Welfare, Public Health Service, Cincinnati, Ohio, 1964.

849a. M. S. Eisenberg and T. R. Bender, Botulism in Alaska, 1947 through 1974. *J. Am. Med. Assoc.* **235**, 35 (1976).

849b. Botulinal toxin in an opened can of commercial tuna fish. *Morbidity Mortality Weekly Rept. Center for Disease Control*, **23**, 176 (1974).

850. M. H. Merson, J. M. Hughes, V. R. Dowell, Jr., A. Taylor, W. H. Barker, and E. J. Gangarosa, Current trends in botulism in the United States. *J. Am. Med. Assoc.* **229**, 1305 (1974).

850a. D. A. Boroff and G. Shu-Chen, Radioimmunoassay for type A toxin of *Clostridium botulinum. Appl. Microbiol.* **25**, 545 (1973).

851. C. A. Miller and A. W. Anderson, Rapid detection and quantitative estimation of type A botulinum toxin by electroimmunodiffusion. *Infect. Immun.* **4**, 126 (1971).

851a. J. W. Mayhew and S. L. Gorbach, Rapid gas chromatographic technique for presumptive detection of *Clostridium botulinum* in contaminated food. *Applied Microbiol.* **29**, 297 (1975).

851b. M. H. Merson, E. J. Gangarosa, and V. R. Dowell, Jr., More on botulism. *Calif. Med.* **119**, 65 (1973).

851c. Probable botulism—Maryland. *Morbidity Mortality Weekly Rept. Center for Disease Control*, **22**, 255 (1973).

851d. M. Cherington, Botulism. Ten-year experience. *Arch. Neurol.* **30**, 432 (1974).

851e. Y. Sudre, B. Becq-Giraudon, and P. Boutaud, Le botulisme. Conduite thérapeutique à propos de 36 cas. *Sem. Hôp. Paris.* **51**, 807 (1975).

852. A. A. Andersen, H. D. Michener, and H. S. Olcott, Effect of some antibiotics on *Clostridium botulinum. Antibiot. Chemother.* **3**, 521 (1953).

852a. Botulism and improper home canning—California. *Morbidity Mortality Weekly Rept., Center for Disease Control*, **24**, 236 (1975).

852b. Wound Botulism—Idaho, Utah, California, *Morbidity Mortality Weekly Rept., Center for Disease Control*, **23**, 246 (1974).

852c. F. J. Ingelfinger, Cancer! Alarm! Cancer! *N. Engl. J. Med.* **293**, 1319 (1975).

853. M. G. Koenig, D. J. Drutz, A. I. Mushlin, W. Schaffner, and D. E. Rogers, Type B botulism in man. *Am. J. Med.* **42**, 208 (1967).

854. K. H. Lewis and K. Cassel, Jr., eds., "Botulism—Proceedings of a Symposium." PHS Publ. No. 999-FP-1. US Dept. of Health, Education & Welfare, Public Health Service, Cincinnati, Ohio, 1964.

855. "Botulism in the United States, 1899–1973" (Handbook for Epidemiologists, Clinicians and Laboratory Workers.) DHEW Publ. No. (CDC) 74–8279. Center for Disease Control, US Dept. of Health, Education and Welfare, Atlanta, 1974.

856. J. A. Donadio, E. J. Gangarosa, and G. A. Faich, Diagnosis and treatment of botulism. *J. Infect. Dis.* **124**, 108 (1971).

857. E. J. Gangarosa, Botulism. *In* "Infectious Diseases" (P. D. Hoeprich, ed.), p. 1031. Harper, New York, 1972.

858. Aretaeus, On tetanus. *From* "The Extant Works of Aretaeus, the Cappadocian" (F. Adams, ed. and transl.), Sydenham Society, London, 1856 (as reproduced in R. H. Major, "Classic Descriptions of Disease." Thomas, Springfield, Illinois, 1932).

859. R. Veronesi, Clinical observations on 712 cases of tetanus subject to four different methods of treatment: 18.2% mortality rate under a new method of treatment. *Am. J. Med. Sci.* **232**, 629 (1956).

859a. P. A. Blake and R. A. Feldman: Tetanus in the United States, 1970–1971. CDC news. *J. Infect. Dis.* **131**, 745 (1975).

859b. W. Furste, Tetanus statistics. *J. Am. Med. Assoc.* **228**, 28 (1974).

859c. L. Weinstein, Tetanus. *N. Engl. J. Med.* **289**, 1293 (1973).

859d. D. W. Fraser, Preventing tetanus in patients with wounds. *Ann. Intern. Med.* **84**, 95 (1976).

860. R. C. Clay and J. W. Bolton, Tetanus arising from gangrenous unperforated small intestine. *J. Am. Med. Assoc.* **187**, 856 (1964).

861. F. M. LaForce, L. S. Young, and J. V. Bennett, Tetanus in the United States (1965–1966). Epidemiologic and clinical features. *N. Engl. J. Med.* **280**, 569 (1969).

861a. D. Femi-Pearse, Blood gas tensions, acid-base status, and spirometry in tetanus. *Am. Rev. Resp. Dis.* **110**, 390 (1974).

862. M. Lazar, Zur Pathogenese und Therapie des Tetanus. *Schweiz. Med. Wochenschr.* **100**, 1486 (1970).

862a. C. Prys-Roberts, Treatment of cardiovascular disturbances in severe tetanus. *Proc. Roy. Soc. Med.* **62**, 662 (1969).

863. K. Tsueda, P. B. Oliver, and R. W. Richter, Cardiovascular manifestations of tetanus. *Anesthesiology* **40**, 588 (1974).

863a. G. Leonardi, K. G. Nair, F. D. Dastur, J. S. Kamat, and B. T. Desai, Evaluation of cholinesterase-restoring agents in the treatment of tetanus in man. *J. Infect. Dis.* **128**, 652 (1973).

864. T. M. Buchanan, G. F. Brooks, S. Martin, and J. V. Bennett, Tetanus in the United States, 1968 and 1969. *J. Infect. Dis.* **122**, 564 (1970).

864a. P. A. Blake, R. A. Feldman, T. M. Buchanan, G. F. Brooks, and J. V. Bennett, Serologic therapy of tetanus in the United States, 1965–1971. *J. Am. Med. Assoc.* **235**, 42 (1976).

865. I. Ildirim, A. R. Meira, and M. L. Furcolow, Tetanus. *N. Engl. J. Med.* **280**, 1243 (1969).

865a. W. Furste, Third international conference on tetanus: A report. *J. Trauma* **11**, 721 (1971).

865b. W. Furste, Treatment of acute tetanus. *J. Am. Med. Assoc.* **234**, 761 (1975).

866. J. V. Bennett, Tetanus. *In* "Infectious Diseases" (P. D. Hoeprich, ed.), p. 1021. Harper, New York, 1972.

866a. W. Furste, Four keys to 100 percent success in tetanus prophylaxis. *Am. J. Surg.* **128**, 616 (1974).

866b. W. G. White, G. M. Barnes, E. Barker, D. Gall, P. Knight, A. H. Griffith, R. M. Morris-Owen, and J. W. G. Smith, Reactions to tetanus toxoid. *J. Hyg.* **71**, 283 (1973).

867. W. Furste and W. L. Wheeler, "Tetanus: A Team Disease." Current Problems in Surgery. Yearbook Publ., Chicago, Illinois, 1972.

868. N. A. Christensen, Treatment of the patient with severe tetanus, Surg. Clin. N. Amer. **49**, 1183 (1969).

869. T. G. C. Murrell, L. Roth, J. Egerton, J. Samels, and P. D. Walker, Pig-Bel: Enteritis necroticans. A study in diagnosis and management. *Lancet* **1**, 217 (1966).

870. E. Hain, On the occurrence of *Cl. welchii* type F in normal stools. *Br. Med. J.* **1**, 271 (1949).

871. M. H. Gleeson-White and J. J. Bullen, *Clostridium welchii* epsilon toxin in the intestinal contents of man. *Lancet* **1**, 384 (1955).

872. J. Kohn and G. H. Warrack, Recovery of *Clostridium welchii* type D from man. *Lancet* **1**, 385 (1955).

873. B. C. Morson, Ischaemic colitis. *Postgrad. Med. J.* 44: 665, 1968.

874. J. S. McKinnell and M. S. Kearney, Haemorrhagic necrosis of the intestine. *Br. Med. J.* **2**, 460 (1967).

874a. S. E. Cohen, M. I. Feldman, and E. F. Wolfman, Nonocclusive hemorrhagic necrosis of the intestine. *West. J. Med.* **121**, 449 (1974).

874b. T. E. Tully and S. B. Feinberg, Those other types of enterocolitis. *Am. J. Roentgenol. Radium Ther. Nucl. Med.* **121**, 291 (1974).

874c. T. V. Santulli, J. N. Schullinger, W. C. Heird, T. D. Gongaware, J. Wigger, B. Barlow, W. A. Blanc, and W. E. Berdon, Acute necrotizing enterocolitis in infancy: A review of 64 cases. *Pediatrics* **55**, 376 (1975).

875. B. C. Hobbs, *Clostridium welchii* and *Bacillus cereus* infection and intoxication. *Postgrad. Med. J.* **50**, 597 (1974).

875a. J. M. Hughes, M. H. Merson, and R. A. Pollard, Jr., Food-borne disease outbreaks in the United States, 1973. *J. Infect. Dis.* **132**, 224 (1975).

875b. Center for Disease Control, "Foodborne and waterborne disease outbreaks—Annual Summary 1974." US Public Health Service, Atlanta, Georgia, 1976.

875c. *Clostridium perfringens* food poisoning—Tennessee. *Morbidity Mortality Weekly Rept., Center for Disease Control,* **23**, 19 (1974).

876. C. L. Duncan and E. B. Somers, Quantitation of *Clostridium perfringens* type A enterotoxin by electroimmunodiffusion. *Appl. Microbiol.* **24**, 801 (1972).

876a. Center for Disease Control, "Tetanus Surveillance," Rep. No. 4. CDC, US Public Health Service, Atlanta, Georgia, 1974.

876b. J. Zeissler and L. Rassfeld-Sternberg, Enteritis necroticans due to *Clostridium welchii* type F. *Br. Med. J.* **1**, 267 (1949).

876c. E. Hain, Origin of *Cl. welchii* type F infection. *Br. Med. J.* **1**, 271 (1949).

876d. M. Patterson and H. D. Rosenbaum, Enteritis necroticans. *Gastroenterology* **21**, 110 (1952).

876e. M. J. Killingback and K. L. Williams, Necrotizing colitis. *Br. J. Surg.* **49**, 175 (1961).

876f. W. R. Ghent, L. R. Maclean, and R. Ceballos, Necrotizing colitis. *Can. Med. Assoc. J.* **89**, 1327 (1963).

876g. T. G. C. Murrell and L. Roth, Necrotizing jejunitis: A newly discovered disease in the highlands of New Guinea. *Med. J. Aust.* **50**, 61 (1963).

876h. J. R. Egerton and P. D. Walker, The isolation of *Clostridium perfringens* type C from necrotic enteritis of man in Papua-New Guinea. *J. Pathol. Bact.* **88**, 275 (1964).

876i. G. T. Tate, H. Thompson, and A. T. Willis, *Clostridium welchii* colitis. *Br. J. Surg.* **52**, 194 (1965).

876j. V. L. Pereira, R. Turcot, and C. Brunet, Colite necrotique du nouveau-né avec survie. *Can. Med. Assoc. J.* **98**, 600 (1968).

876k. M. Mogadam and R. J. Priest, Necrotizing enteritis in Crohn's disease of the small bowel. *Gastroenterology* **56**, 337 (1969).

876l. C. A. Rotilie, R. J. Fass, R. B. Prior, and R. L. Perkins, Microdilution technique for antimicrobial susceptibility testing of anaerobic bacteria. *Antimicrob. Ag. Chemother.* **7**, 311 (1975).

876m. R. J. Fass, R. B. Prior, and C. A. Rotilie, Simplified method for antimicrobial susceptibility testing of anaerobic bacteria. *Antimicrob. Ag. Chemother.* **8**, 444 (1975).

876n. K. J. Hauser, J. A. Johnston, and R. J. Zabransky, Economic agar dilution technique for susceptibility testing of anaerobes. *Antimicrob. Ag. Chemother.* **7**, 712 (1975).

877. V. L. Sutter, Y.-Y. Kwok, and S. M. Finegold, Standardized antimicrobial disc suscepti-

bility testing of anaerobic bacteria. I. Susceptibility of *Bacteroides fragilis* to tetracycline. *Appl. Microbiol.* **23**, 268 (1972).

878. F. L. Sapico, Y.-Y. Kwok, V. L. Sutter, and S. M. Finegold, Standardized antimicrobial disc susceptibility testing of anaerobic bacteria: *In vitro* susceptibility of *Clostridium perfringens* to nine antibiotics. *Antimicrob. Ag. Chemother.* **2**, 320 (1972).

879. V. L. Sutter, Y.-Y. Kwok, and S. M. Finegold, Susceptibility of *Bacteroides fragilis* to six antibiotics determined by standardized antimicrobial disc susceptibility testing. *Antimicrob. Ag. Chemother.* **3**, 188 (1973).

880. V. L. Sutter and S. M. Finegold, Antibiotic susceptibility testing of anaerobes. *In* "Current Techniques for Antibiotic Susceptibility Testing" (A. Balows, ed.), Chapter XI. Thomas, Springfield, Illinois, 1974.

881. V. L. Sutter, Y.-Y. Kwok, and S. M. Finegold, *In vitro* susceptibility testing of anaerobes: Standardization of a single disc test. *In* "Anaerobic Bacteria: Role in Disease" (A. Balows *et al.*, ed.), Chap. 36, p. 457. Thomas, Springfield, Illinois, 1974.

882. Y.-Y. Kwok, F. P. Tally, V. L. Sutter, and S. M. Finegold, Disk susceptibility testing of slow growing anaerobic bacteria. *Antimicrob. Ag. Chemother.* **7**, 1 (1975).

882a. A. L. Barry and G. D. Fay Evaluation of four disk diffusion methods for antimicrobic susceptibility tests with anaerobic gram-negative bacilli. *Am. J. Clin. Pathol.* **61**, 592 (1974).

883. T. D. Wilkins and T. Thiel, Modified broth-disk method for testing the antibiotic susceptibility of anaerobic bacteria. *Antimicrob. Ag. Chemother.* **3**, 350 (1973).

884. A. W. Bauer, W. M. M. Kirby, J. C. Sherris, and M. Turck, Antibiotic susceptibility testing by a single disc method. *Am. J. Clin. Pathol.* **45**, 493 (1966).

885. H. R. Ingham, J. B. Selkon, A. A. Codd, and J. H. Hale. The effect of carbon dioxide on the sensitivity of *Bacteroides fragilis* to certain antibiotics *in vitro*. *J. Clin. Pathol.* **23**, 254 (1970).

886. J. E. Rosenblatt and F. Schoenknecht, Effect of several components of anaerobic incubation on antibiotic susceptibility test results. *Antimicrob. Ag. Chemother.* **1**, 433 (1972).

887, V. E. Del Bene and W. E. Farrar, Jr., Antimicrobial therapy of infections due to anaerobic bacteria. *Semin. Drug Treat.* **2**, 295 (1972).

887a. I. Phillips, Antibiotic sensitivity of non-sporing anaerobes. *In* "Infection With Non-Sporing Anaerobic Bacteria," (I. Phillips and M. Sussman eds.), Chap. 4, p. 37, Livingstone, Edinburgh, 1974.

888. W. J. Martin, M. Gardner and J. A. Washington, II, *In vitro* antimicrobial susceptibility of anaerobic bacteria isolated from clinical specimens. *Antimicrob. Ag. Chemother.* **1**, 148 (1972).

889. H. Beerens, M. M. Castel, and A. Modjadedy, Recherches sur la détermination de la sensibilité des bactéries anaérobies non sporulées a onze antibiotiques par la méthode des disques. Resultats obtenus avec 180 souches. *Ann. Inst. Pasteur, Lille* **13**, 105 (1962).

890. H. Lodenkämper and G. Stienen, Zur Therapie anaerober Infektionen. *Dtsch. Med. Wochenschr.* **81**, 1226 (1956)

891. K. Mitsu, T. Oguri, and N. Kozakai, Anaerobes isolated from clinical specimens in 1971 and their antibiotic susceptibility. *In* "Second Symposium on Anaerobic Bacteria and Their Infectious Diseases" (S. Ishiyama *et al.*, eds.), p. 23, Eisai, Tokyo, Japan, 1972.

892. E. de Lavergne, J. C. Burdin, J. Schmitt, and M. T. Le Moyne, Étude du comportement de 17 espèces anaérobies strictes vis-a-vis de 11 antibiotiques. *Ann. Inst. Pasteur, Paris* **91**, 631 (1956).

893. G. F. Thornton and J. A. Cramer, Antibiotic susceptibility of *Bacteroides* species. *Antimicrob. Ag. Chemother.* p. 509 (1971).

894.   G. Willich, Die bakteriostatische wirksamkeit von Sulfonamiden und Antibiotics gegen die pathogenen Anaerobier. *Z. Hyg. Infektionskr.* **134**, 573 (1952).

894a.   J. L. Staneck and J. A. Washington II, Antimicrobial susceptibilities of anaerobic bacteria: Recent clinical isolates. *Antimicrob. Ag. Chemother.* **6**, 311 (1974).

895.   U. Berger, Die Empfindlichkeit einiger sporenfreier Anaerobier gegen Spiramycin *in vitro. Z. Hyg.* **145**, 160 (1958).

896.   U. Berger, Die Wirkung des Vancomycin gegenuber oralen Spirochaten und Fuso-bakterien. *Arch. Hyg. Bakteriol.* **143**, 316 (1959).

897.   M. Füzi and Z. Csukás, Das antibakterielle Wirkungsspektrum des Metronidazols. *Zentralbl. Bakteriol, Parasitenkd. Infektionskr. Hyg. Abt. 1: Orig.* **213**, 258 (1970).

898.   M. C. Dodd, The chemotherapeutic properties of 5-nitro-2-furaldehyde semicarbazone (Furacin). *J. Pharmacol. Exp. Ther.* **86**, 311 (1946).

899.   J. A. Yurchenco, M. C. Yurchenco, and C. R. Piepoli, Antimicrobial properties of furoxone (N-5-nitro-2-furfurylidene-3-amino-2-oxazolidone). *Antibiot. Chemother. (Washington, D.C.)* **3**, 1035 (1953).

900.   K. Suk and M. Pohunek, Behandlung der Scheidenentzundungen mit Metronidazol. *Zentralb. Gynaekol.* **40**, 1380 (1971).

901.   K. Ueno, K. Ninomiya, F. Ohtani, S. Kosaka, and S. Suzuki, Antianaerobic bacterial activity of thiamphenicol (neomyson). *In* "First Symposium on Anaerobic Bacteria and Their Infectious Diseases" (S. Ishiyama *et al.*, eds.), p. 103. Eisai, Tokyo, Japan, 1971.

902.   H. N. Prince, Specific inhibition of obligate anaerobes. *Nature (London)* **186**, 817 (1960).

903.   H. N. Prince, E. Grunberg, E. Titsworth, and W. F. DeLorenzo, Effects of 1-(2-nitro-1-imidazolyl)-3-methoxy-2-propanol and 2-methyl-5-nitro-imidazole-1-ethanol against anaerobic and aerobic bacteria and protozoa. *Appl. Microbiol.* **18**, 728 (1969).

904.   F. Benazet, L. Lacroix, C. Godard, L. Guillaume, and J.-P. Leroy, Laboratory studies of the chemotherapeutic activity and toxicity of some nitroheterocycles. *Scand. J. Infect. Dis.* **2**, 139 (1970).

905.   A. L. Courtieu, J. J. Monnier, P. de Lajudie, and F. N. Guillermet, Spectre antibacterien de la colistine vis-à-vis de 1,200 souches. *Ann. Inst. Pasteur, Paris* **100**, 14 (1961).

906.   D. Videau, Association metronidazole-spiramycine sur les germes anaérobies. *Pathol. Biol.* **19**, 661 (1971).

907.   Evaluations on new drugs. Trimethoprim-sulphamethoxazole. *Drugs* **1**, 7 (1971).

908.   K. Ueno, K. Ninomiya, H. Kamiya, K. Watanabe, and S. Suzuki, Activity of sulfo-benzylpenicillin on strictly anaerobic bacteria. *Chemotherapy* **19**, 875 (1971).

909.   K. Ueno, K. Ninomiya, F. Otani, T. Kozaka, and S. Suzuki, Antibacterial activity of thiamphenicol against anaerobes. *Chemotherapy* **19**, 115, (1971).

910.   K. Ueno, K. Ninomiya, and S. Suzuki, Antibacterial activity of metronidazole against anaerobic bacteria. *Chemotherapy* **19**, 111 (1971).

911.   K. Ueno, K. Ninomiya, H. Shimizu, and S. Suzuki, Antibacterial action of rifampicin against anaerobes. *Shinryo* **23**, 994 (1970).

912.   T. Marui, Susceptibility of anaerobes to cephaloridine. *Iyaku-no-mon* **7**, 7 (1967).

913.   O. A. Okubadejo, P. J. Green, and D. J. H. Payne, *Bacteroides* in the blood. *Lancet* **1**, 147 (1973).

914.   J. H. Darrell, L. P. Garrod, and P. M. Waterworth, Trimethoprim: Laboratory and clinical studies. *J. Clin. Pathol.* **21**, 202 (1968).

914a.   F. P. Tally, N. V. Jacobus, J. G. Bartlett, and S. L. Gorbach, *In vitro* activity of penicil-lins against anaerobes. *Antimicrob. Ag. Chemother.* **7**, 413 (1975).

914b.   F. P. Tally, N. V. Jacobus, J. G. Bartlett, and S. L. Gorbach, Susceptibility of anaerobes to cefoxitin and other cephalosporins. *Antimicrob. Ag. Chemother.* **7**, 128 (1975).

914c.   A. W. Chow, V. Patten, and L. B. Guze, Comparative susceptibility of anaerobic bacteria to minocycline, doxycycline and tetracycline. *Antimicrob. Ag. Chemother.* **7**, 46 (1975).

914d. K. Ninomiya, K. Watanabe, K. Ueno, S. Suzuki, I. Mou, Y. Ban, Y. Shimizu, K. Isogai, and T. Nishiura, Susceptibility of anaerobic bacteria to lincomycin and clindamycin. *Jap. J. Antibiot.* **26**, 157 (1973).

914e. H. Werner and G. Böhm, Susceptibility of non-sporing anaerobes of the genera *Bacteroides, Fusobacterium, Sphaerophorus, Peptococcus* and *Peptostreptococcus* to clindamycin. *Zbl. Bakt. Hyg. I. Abt. Orig. A.* **229**, 401 (1974).

914f. K. Dornbusch and C.-E. Nord, *In vitro* effect of metronidazole and tinidazole on anaerobic bacteria. *Med. Microbiol. Immunol.* **160**, 265 (1974).

914g. A. W. Chow, V. Patten, and L. B. Guze, Susceptibility of anaerobic bacteria to metronidazole: relative resistance of non-spore-forming gram-positive bacilli. *J. Infect. Dis.* **131**, 182 (1975).

915. T. D. Wilkins and T. Thiel, Resistance of some species of *Clostridium* to clindamycin. *Antimicrob. Ag. Chemother.* **3**, 136 (1973).

916. G. C. Blake, Sensitivities of colonies and suspensions of *Actinomyces israelii* to penicillins, tetracyclines and erythromycin. *Br. Med. J.* **1**, 145 (1964).

917. U. Berger, Zur Empfindlichkeit der Fusobakterien gegen einige neuere Antibiotica und Chemotherapeutica *in vitro. Arch. Hyg. Bakteriol.* **140**, 288 (1956).

918. J. W. Kislak, The susceptibility of *Bacteroides fragilis* to 24 antibiotics. *J. Infect. Dis.* **125**, 295 (1972).

919. S. J. Bodner, M. G. Koenig, L. L. Treanor, and J. S. Goodman, Antibiotic susceptibility testing of *Bacteroides. Antimicrob. Ag. Chemother.* **2**, 57 (1972).

920. J. F. Collins and H. M. Hood, Sensitivity of oral microorganisms to vancomycin—an *in vitro* study. *J. Oral Ther. Pharmacol.* **4**, 214 (1967).

921. G. Pulverer and S. Heinrich, Die *in vitro*-Empfindlichkeit des *Bacteroides melaninogenicus* genen die gebrauchlichen Antibiotica und Sulfonamide. *Z. Hyg.* **146**, 1 (1959).

922. P. I. Lerner, Susceptibility of *Actinomyces* species to lincomycin and its 7-halogenated analogues. *Antimicrob. Ag. Chemother.* **8**, 461 (1968).

923. R. J. Fitzgerald, M. L. Parramore, and M. E. MacKintosh, Antibiotic sensitivity of strains of *Veillonella. Antibiot. Chemother. (Washington, D.C.)* **9**, 145 (1959).

924. A. Newton, J. Strawitz, R. B. Lindberg, J. M. Howard, and C. P. Artz, Sensitivities of ten species of clostridia to penicillin, aureomycin, terramycin and chloramphenicol. *Surgery* **37**, 392 (1955).

925. G. B. Roemer, Untersuchungen über die antibakterielle Wirkung neuerer Antibiotika auf Welch-Fraenkelsche Gasbrandbazillen (*Cl. perfringens*). *Dtsch. Med. Wochenschr.* **93**, 2205 (1968).

926. M. Füzi and Z. Csukás, *Veillonella*-törzsek Metronidazol-érzékenysége. *Fogorv. Sz.* **62**, 324 (1969).

927. J. L. Hoogendijk, Resistance of some strains of *Bacteroides* to ampicillin, methicillin and cloxacillin. *Antonie van Leeuwenhoek* **31**, 383 (1965).

928. R. R. Omata, *In vitro* sensitivity of fusobacteria to various antibiotics. *J. Dent. Res.* **30**, 799 (1951).

929. L. Reinhold, Resistenzprufungen an Stammen des Genus *Bacteroides* und *Sphaerophorus* im Agardiffusionstest. *Z. Gesamte Hyg. Ihre Grenzgeb.* **11**, 711 (1965).

930. H. Werner and G. Boll, Die Antibiotikaempfindlichkeit von *Bacteroides* (*Eggerthella*)-Stämmen. *Zentralbl. Bakteriol. Parasitenkd., Infektionskr. Hyg., Abt. 1: Orig.* **208**, 437 (1968).

931. H. R. Ingham, J. B. Selkon, A. A. Codd, and J. H. Hale, A study *in vitro* of the sensitivity to antibiotics of *Bacteroides fragilis. J. Clin. Pathol.* **21**, 432 (1968).

932. P. I. Lerner, Susceptibility of *Actinomyces* to cephalosporins and lincomycin. *Antimicrob. Ag. Chemother.* p. 730 (1968).

932a. K. Dornbusch, C.-E. Nord, and T. Wadström, Biochemical characterization and *in vitro* determination of antibiotic susceptibility of clinical isolates of *Bacteroides fragilis*. *Scand. J. Infect. Dis.* **6**, 253 (1974).

932b. J. V. A. Robinson and A. L. James, *In vitro* susceptibility of *Bacteroides corrodens* and *Eikenella corrodens* to ten chemotherapeutic agents. *Antimicrob. Ag. Chemother.* **6**, 543 (1974).

932c. D. E. Mahony, Antibiotic sensitivity of *Clostridium perfringens* and L-forms of *C. perfringens* induced by bacteriocin. *Can. J. Microbiol.* **19**, 735 (1943).

932d. P. I. Lerner, Susceptibility of pathogenic actinomycetes to antimicrobial compounds. *Antimicrob. Ag. Chemother.* **5**, 302 (1974).

933. S. M. Finegold and V. L. Sutter, Antimicrobial susceptibility of anaerobic gram-negative bacilli. *In* "Host Resistance to Commensal Bacteria" (T. MacPhee, ed.), p. 275. Churchill Livingstone, Edinburgh, 1972.

934. S. M. Finegold, O. T. Monzon, E. E. Sweeney, H. G. Dangerfield, B. T. Blackman, and W. L. Hewitt, Laboratory experiences with dimethoxyphenyl penicillin. "Dimethoxyphenyl Penicillin Symposium," p. 26. Upstate Medical Center, Syracuse, New York, 1961.

935. V. L. Sutter and S. M. Finegold, Susceptibility of anaerobic bacteria to carbenicillin, cefoxitin and related drugs. *J. Infect. Dis.* **131**, 417 (1975).

936. V. L. Sutter, F. P. Tally, Y.-Y. Kwok, and S. M. Finegold, Activity of doxycycline and tetracycline versus anaerobic bacteria. *Clin. Med.* **80**, 31 (1973).

937. S. M. Finegold, V. L. Sutter, and P. T. Sugihara, Susceptibility of anaerobic bacteria to rifampicin. *Bacteriol. Proc.—1969*, Abstract M36. p. 73 (1969).

938. L. J. Nastro and S. M. Finegold, Bactericidal activity of five antimicrobial agents against *Bacteroides fragilis*. *J. Infect. Dis.* **126**, 104 (1972).

939. V. L. Sutter and S. M. Finegold, Rosamicin: *In vitro* activity against anaerobes and comparison with erythromycin. *Antimicrob. Ag. Chemother.* **9**, 350 (1976).

939a. I. Phillips and C. Warren: Susceptibility of *Bacteroides fragilis* to spectinomycin. *J. Antimicrob. Chemother.* **1**, 91 (1975).

940. S. M. Finegold and V. L. Sutter, Susceptibility of gram-negative anaerobic bacilli to gentamicin and other aminoglycosides. *J. Infect. Dis.* **124**, Suppl. S56, (1971).

941. S. M. Finegold, L. G. Miller, D. Posnick, D. K. Patterson, and A. Davis, Nalidixic acid: Clinical and laboratory studies. *Antimicrob. Ag. Chemother.* p. 189 (1967).

942. F. P. Tally, V. L. Sutter, and S. M. Finegold, Metronidazole versus anaerobes. *In vitro* data and initial clinical observations. *Calif. Med.* **117**, 22 (1972).

942a. J. P. F. Whelan and J. H. Hale, Bactericidal activity of metronidazole against *Bacteroides fragilis*. *J. Clin. Pathol.* **26**, 393 (1973).

942b. E. D. Ralph and W. M. M. Kirby, Unique bactericidal action of metronidazole against *Bacteroides fragilis* and *Clostridium perfringens*. *Antimicrob. Ag. Chemother.* **8**, 409 (1975).

942c. P. Corrodi, D. F. Busch, P. A. Wideman, D. M. Citronbaum, V. L. Sutter, and S. M. Finegold, Factors affecting the *in vitro* bactericidal activity of metronidazole against *Bacteroides fragilis*. *Abstr. Ann. Meet. Am. Soc. Microbiol., 1976*, p. 42 (Abstr. C99), (1976).

943. D. F. Busch, V. L. Sutter, and S. M. Finegold, Activity of combinations of antimicrobial agents against *Bacteroides fragilis*. *J. Infect. Dis.* **133**, 321 (1976).

943a. R. J. Mitre and E. B. Rotheram, Anaerobic septicemia from thrombophlebitis of the internal jugular vein. Successful treatment with metronidazole. *J. Am. Med. Assoc.* **230**, 1168 (1974).

943b. M. Sebald, D. Bouanchaud, and G. Bieth, Nature plasmidique de la resistance à plusieurs antibiotiques chez *C. perfringens* type A, souche 659. *C.R. Acad. Sci. Paris* **280**, 2401 (1975).

943c. M. Sebald and M. G. Brefort: Transfert du plasmide (TET-CHL) chez *Clostridium perfringens*. *C. R. Acad. Sci. Paris* **281**, 317 (1975).

943d. V. E. Del Bene, M. Rogers, and W. E. Farrar, Jr., Failure to demonstrate transfer of antibiotic resistance between *Bacteroides* and *Escherichia coli*. *Clin. Res.* **22**, 31A (1974).

944. G. Pinkus, G. Veto, and A. I. Braude, *Bacteroides* penicillinase. *J. Bacteriol.* **96**, 1437 (1968).

945. A. I. Braude, Bacteroides in the blood. *Lancet* **1**, 377 (1973).

945a. J. D. Anderson and R. B. Sykes, Characterisation of a β-lactamase obtained from a strain of *Bacteroides fragilis* resistant to β-lactam antibiotics. *J. Med. Microbiol.* **6**, 201 (1973).

945b. A. E. Weinrich and V. E. Del Bene, Characterization of β-lactamase activity in anaerobic bacteria. *Clin. Res.* **24**, 27A (1976).

945c. A. S. Hackman and T. D. Wilkins, *In vivo* protection of *Fusobacterium necrophorum* from penicillin by *Bacteroides fragilis*. *Antimicrob. Ag. Chemother.* **7**, 698 (1975).

946. Y. Kanazawa, T. Kuramata, and S. Miyamura, Inactivation of chemotherapeutic agents by anaerobes. *Prog. Antimicrob. Anticancer Chemother., Proc. 6th Int. Cong. Chemother., 1969* p. 387 (1970).

947. T. D. Wilkins, L. V. Holdeman, I. J. Abramson, and W. E. C. Moore, Standardized single-disc method for antibiotic susceptibility testing of anaerobic bacteria. *Antimicrob. Ag. Chemother.* **1**, 451 (1972).

947a. W. J. Ledger, Anaerobic infections. *Am. J. Obstet. Gynecol.* **123**, 111 (1975).

948. G. L. Barbour and K. Juniper, Jr., A clinical comparison of amebic and pyogenic abscess of the liver in sixty-six patients. *Am. J. Med.* **53**, 323 (1972).

949. W. A. Altemeier, C. G. Schowengerdt, and D. H. Whiteley, Abscesses of the liver: Surgical considerations. *Arch. Surg.* (*Chicago*) **101**, 258 (1970).

950. E. L. Spatz, Central nervous system infections of surgical importance. *Am. J. Surg.* **107**, 678 (1964).

951. A. Morgan, W. Morain, and A. Eraklis, Gas gangrene of the abdominal wall: Management after extensive debridement. *Ann. Surg.* **173**, 617 (1971).

951a. J. Phillips, D. M. Heimbach, and R. C. Jones, Clostridial myonecrosis of the abdominal wall. *Am. J. Surg.* **128**, 436 (1974).

952. J. H. Duff, A. P. H. McLean, and L. D. MacLean, Treatment of severe anaerobic infections. *Arch. Surg.* (*Chicago*) **101**, 314 (1970).

953. H. Thadepalli, S. L. Gorbach, P. W. Broido, J. Norsen, and L. Nyhus, Abdominal trauma, anaerobes, and antibiotics. *Surg., Gynecol. Obstet.* **137**, 270 (1973).

953a. W. J. Ledger, T. J. Kriewall, R. L. Sweet, and F. R. Fekety, Jr., The use of parenteral clindamycin in the treatment of obstetric-gynecologic patients with severe infections. *Obstet. Gynecol.* **43**, 490 (1974).

954. W. M. Weinstein, A. B. Onderdonk, J. G. Bartlett, T. J. Louie, and S. L. Gorbach, Antimicrobial therapy of experimental intraabdominal sepsis. *J. Infect. Dis.* **132**, 282 (1975).

954a. W. M. Weinstein, A. B. Onderdonk, J. G. Bartlett, and S. L. Gorbach, Experimental intra-abdominal abscesses in rats: Development of an experimental model. *Infect. Immunity* **10**, 1250 (1974).

955. P. Holm, Studies on the aetiology of human actinomycosis. II. Do the other microbes of actinomycosis possess virulence? *Acta Pathol. Microbiol. Scand.* **28**, 391 (1951).

956. T. Rosebury, G. Foley, and F. L. Rights, Effects of neoarsphenamine and sulpharsphenamine on experimental fuso-spirochetal infection in guinea pigs. *J. Infect. Dis.* **65**, 291 (1939).

957. J. M. Miller, P. H. Long, and E. B. Schoenbach, Successful treatment of actinomycosis with "Stilbamidine." *J. Am. Med. Assoc.* **150**, 35 (1952).

958. E. S. Hemmens and G. M. Dack, The effect of sulfanilamide on experimental infections with *Bacterium necrophorum* in rabbits. *J. Infect. Dis.* **64**, 43 (1939).

959. A. E. Brown, H. L. Williams, and W. E. Herrell, *Bacteroides* septicemia. Report of a case with recovery. *J. Am. Med. Assoc.* **116**, 402 (1941).

960. W. W. Manson and I. T. Craig, Treatment of Vincent's angina with sulfathiazole. *J. Am. Med. Assoc.* **127**, 277 (1945).

961. G. Willich, Zur Bedeutung des *Bacterium pyogenes anaerobium* (Buday). *Zentralbl. Bakteriol., Parasitenkd., Infektionskr. Hyg. Abt. 1: Orig.* **157**, 99 (1951).

962. I. Phillips and C. Warren, Susceptibility of *Bacteroides fragilis* to trimethoprim and sulphamethoxazole. *Lancet* **1**, 827 (1974).

963. J. E. Rosenblatt and P. R. Stewart, Lack of activity of sulfamethoxazole and trimethoprim against anaerobic bacteria. *Antimicrob. Ag. Chemother.* **6**, 93 (1974).

963a. S. R. M. Bushby, Trimethoprim-sulfamethoxazole: *In vitro* microbiological aspects. *J. Infect. Dis.* **128** Suppl. S442 (1973).

963b. H. Knothe, The effect of a combined preparation of trimethoprim and sulfamethoxazole following short-term and long-term administration on the flora of the human gut. *Chemotherapy* **18**, 285 (1973).

964. J. S. Strauss and P. E. Pochi, The effect of sulfisoxazole-trimethoprim combination on titratable acidity of human sebum. *Br. J. Dermatol.* **82**, 493 (1970).

964a. O. A. Okubadejo, Susceptibility of *Bacteroides fragilis* to co-trimoxazole. *Lancet* **1**, 1061 (1974).

964b. B. Fossieck and R. H. Parker, Neurotoxicity during intravenous infusion of penicillin. A review. *J. Clin. Pharmacol.* **14**, 504 (1974).

964c. W. J. Holloway, Clinical experience with amoxicillin—A preliminary report. *Infect. Dis. Rev.* **2**, 245 (1973).

965. A. White and D. T. Varga, Antistaphylococcal activity of dimethoxyphenyl penicillin. *In* "Dimethoxyphenyl Penicillin Symposium," p. 146. Upstate Medical Center, Syracuse, New York, 1961.

966. R. Meny, C. D. Webb, and W. Fiedelman, Carbenicillin therapy of anaerobic infections. *Curr. Ther. Res.* **17**, 478 (1975).

967. W. Fiedelman and C. D. Webb, Clinical evaluation of carbenicillin in the treatment of infection due to anaerobic bacteria. *Curr. Ther. Res.* **18**, 441 (1975).

968. R. Labowitz and W. J. Holloway, Carbenicillin in the treatment of severe infections. *Curr. Ther. Res.* **11**, 143 (1969).

969. M. I. Marks and T. C. Eickhoff, Carbenicillin: A clinical and laboratory evaluation. *Ann. Intern. Med.* **73**, 179 (1970).

970. E. W. Walters, M. J. Romansky, and A. C. Johnson, Cephalothin—Laboratory and clinical studies in 109 patients. *Antimicrob Ag. Chemother.* p. 247 (1964).

970a. A. S. Hackman and T. W. Wilkins, Comparison of cefoxitin and cephalothin therapy of a mixed *Bacteroides fragilis* and *Fusobacterium necrophorum* infection in mice. *Antimicrob. Ag. Chemother.* **8**, 224 (1975).

970b. P. N. R. Heseltine, D. F. Busch, R. D. Meyer, and S. M. Finegold, Clinical experience with cefoxitin. *Clin. Res.* **24**, 113A (1976).

970c. D. J. H. Payne and O. A. Okubadejo, *Bacteroides* infections. *Lancet* **2**, 845 (1973).

971. E. H. Shoemaker and E. M. Yow, Clinical evaluation of erythromycin. *Arch. Intern. Med.* 93, 397 (1954).

972. J. A. Child, J. H. Darrell, N. R. Davies, and L. Davis-Dawson, Mixed infective endocarditis in a heroin addict. *J. Med. Microbiol.* **2**, 293 (1969).

973. I. A. Levin and A. B. Longacre, Antibacterial therapy in infections resulting from human bites. Review of 27 cases, with report of four cases in which bacitracin was administered. *J. Am. Med. Assoc.* **147**, 815 (1951).

974. W. J. Holloway and E. G. Scott, Colistimethate sodium: A clinical study. *J. Urol.* **89**, 264 (1963).

975. P. J. Christenson, P. F. MacLeod, W. R. Bett, and L. Arce, Intravenous use of nitrofurantoin. A review. *Curr. Ther. Res.* **2**, 458 (1960).

976. S. M. Finegold, N. E. Harada, and L. G. Miller, Lincomycin: Activity against anaerobes and effect on normal human fecal flora. *Antimicrob. Ag. Chemoth.* p. 659 (1966).

977. J. A. Mohr, E. R. Rhoades, and H. G. Muchmore, Actinomycosis treated with lincomycin. *J. Am. Med. Assoc.* **212**, 2260 (1970).

977a. J. S. Salaki, R. Black, F. P. Tally, and J. W. Kislak, *Bacteroides fragilis* resistant to the administration of clindamycin. *Am. J. Med.* **60**, 426 (1976).

978. H. D. Rose and M. W. Rytel, Actinomycosis treated with clindamycin. *J. Am. Med. Assoc.* **221**, 1052 (1972).

979. V. L. Sutter and S. M. Finegold, The effect of antimicrobial agents on human fecal flora: Studies with cephalexin, cyclacillin and clindamycin. *In* "The Normal Microbial Flora of Man" (F. A. Skinner and J. G. Carr, eds.), p. 229. Academic Press, New York, 1974.

980. W. A. Freeman, J. A. McFadzean, and J. P. F. Whelan, Activity of metronidazole against experimental tetanus and gas gangrene. *J. Appl. Bacteriol.* **31**, 443 (1968).

981. K. Ueno, K. Ninomiya, S. Koosaka, H. Kamiya, and S. Suzuki, Antibacterial activity of metronidazole (Flagyl) against anaerobes *in vivo*. *Clin. Rep.* **5**, 737 (1971).

982. D. L. S. Shinn, Metronidazole in acute ulcerative gingivitis. *Lancet* **1**, 1191 (1962).

983. F. P. Tally, V. L. Sutter, and S. M. Finegold, Treatment of anaerobic infections with metronidazole. *Antimicrob. Ag. Chemother.* **7**, 672 (1975).

984. H. R. Ingham, G. E. Rich, J. B. Selkon, J. H. Hale, C. M. Roxby, M. J. Betty, R. W. G. Johnson, and P. R. Uldall, Treatment with metronidazole of three patients with serious infections due to *Bacteroides fragilis*. *J. Antimicrob. Chemother.* **1**, 235 (1975).

985. G. B. Reed and J. H. Orr, Gas gangrene. *Am. J. Med. Sci.* **206**, 379 (1943).

986. L. G. Crocker and M. Cunningham, Gas gangrene infection, antitoxin, and serum neuritis. *Arch. Intern. Med.* **115**, 173 (1965).

987. F. Van Wering, Over een Geval van Besmetting met den Necrose-Bacil van Jensen. *Ned. Tijdschr. Geneeskd.* Part I, Sect. 2, No. 26, p. 2892 (1923).

988. F. L. Meleney, Present role of zinc peroxide in treatment of surgical infections. *J. Am. Med. Assoc.* **149**, 1450 (1952).

989. J. W. Finney, S. Haberman, G. J. Race, G. A. Balla, and J. T. Mallams, Local and regional applications of hydrogen peroxide in the control of clostridial myositis in rabbits. *J. Bacteriol.* **93**, 1430 (1967).

990. J. T. Mallams, G. A. Balla, and J. W. Finney, Regional intra-arterial hydrogen peroxide infusion and irradiation in the treatment of head and neck malignancies: A progress report. *Trans. Am. Acad. Ophthalmol. Otolaryngol.* **67**, 546 (1963).

991. C. J. Eaton and E. P. Peterson, Diagnosis and acute management of patients with advanced clostridial sepsis complicating abortion. *Am. J. Obstet. Gynecol.* **109**, 1162 (1971).

992. W. R. McCabe, B. E. Kreger, and M. Johns, Type-specific and cross-reactive antibodies in gram-negative bacteremia. *N. Engl. J. Med.* **287**, 261 (1972).

993. K. L. Lynch and M. Moskowitz, Effects of chelates in chemotherapy of experimental gas-gangrene toxemia. *J. Bacteriol.* **96**, 1925 (1968).

994. L. M. M. Senff and M. Moskowitz, Relation of *in vitro* inhibition by chelates of *Clostridium perfringens* α-toxin to their ability to protect against experimental toxemia. *J. Bacteriol.* **98**, 29 (1969).

995. J. G. Nuckolls and S. Osterhout, The effect of hyperbaric oxygen on anaerobic bacteria. *Clin. Res.* **12**, 244 (1964).

996. D. Kaye, Effect of hyperbaric oxygen on clostridia *in vitro* and *in vivo*. *Proc. Soc. Exp. Biol. Med.* **124**, 360 (1967).

997. V. Fredette, Effect of hyperbaric oxygen on anaerobic bacteria and toxins. *Ann. N.Y. Acad. Sci.* **117**, 700 (1965).

998. V. Fredette, Effect of hyperbaric oxygen upon anaerobic streptococci. *Can. J. Microbiol.* **13**, 423 (1967).

999. G. B. Hill and S. Osterhout, Experimental effects of hyperbaric oxygen on selected clostridial species. I. *In vitro* studies. *J. Infect. Dis.* **125**, 17 (1972).

1000. T. K. Hunt, I. M. Ledingham, and J. G. P. Hutchison, Effect of hyperbaric oxygen on experimental infections in rabbits. *Proc. Intern. Conf. Hyperbaric Med., 3rd NAS—NRC* Publ. No. 1401, p. 572 (1966).

1000a. G. B. Hill, Hyperbaric oxygen exposures for intrahepatic abscesses produced in mice by non-spore-forming anaerobic bacteria. *Antimicrob. Ag. Chemother.* **9**, 312 (1976).

1000b. G. H. Bornside, G. W. Cherry, and M. Myers, Intracolonic oxygen tension and *in vivo* bactericidal effect of hyperbaric oxygen on rat colonic flora. *Aerospace Med.* **44**, 1282 (1973).

1001. W. H. Brummelkamp, Anaerobic infections. *Bull. Soc. Int. Chir.* **21**, 481 (1962).

1002. W. H. Brummelkamp, I. Boerema, and L. Hoogendyk, Treatment of clostridial infections with hyperbaric oxygen drenching. A report on 26 cases. *Lancet* **1**, 235 (1963).

1003. W. H. Brummelkamp, Treatment of anaerobic infections with hyperbaric oxygen. *Proc. Intern. Conf. Hyperbaric Med., 3rd NAS–NRC* Publ. No. 1401, p. 492 (1966).

1004. W. K. Slack, D. A. Thomas, G. C. Hanson, H. E. R. Chew, R. H. Maudsley, and M. R. Colwill, Hyperbaric oxygen in infection. *Proc. Intern. Conf. Hyperbaric Med., 3rd NAS—NRC*. Publ. No. 1401, p. 521 (1966).

1005. W. K. Slack, G. C. Hanson, and H. E. R. Chew, Hyperbaric oxygen in the treatment of gas gangrene and clostridial infection. *Br. J. Surg.* **56**, 505 (1969).

1006. J. H. Duff, H. R. Shibata, L. Vanschaik, R. Usher, R. A. Wigmore, and L. D. MacLean, Hyperbaric oxygen: A review of treatment in eighty-three patients. *Can. Med. Assoc. J.* **97**, 510 (1967).

1007. J. J. W. Van Zyl, P. R. Maartens, and F. D. Du Toit, Gas gangrene treated in a one-man hyperbaric chamber. *Proc. Intern. Conf. Hyperbaric Med., 3rd NAS—NRC*. Publ. No. 1401, p. 515 (1966).

1008. J. J. W. Van Zyl, Discussion. *Proc. Intern. Conf. Hyperbaric Med., 3rd NAS—NRC*. Publ. No. 1401, p. 552 (1966).

1009. J. Van Elk, O. Trippel, A. Ruggie, and C. Staley, High pressure oxygen therapy for patients with gas gangrene. *Proc. Intern. Conf. Hyperbaric Med., 3rd NAS—NRC*. Publ. No. 1401, p. 526 (1966).

1010. J. C. Davis, J. M. Dunn, C. O. Hagood, and B. E. Bassett, Hyperbaric medicine in the US Air Force. *J. Am. Med. Assoc.* **224**, 205 (1973).

1010a. R. W. Jackson and J. P. Waddell, Hyperbaric oxygen in the management of clostridial myonecrosis (gas gangrene). *Clin. Orthoped. Related Res.* **96**, 271 (1973).

1011. G. B. Hart, R. R. O'Reilly, R. H. Cave, N. D. Broussard, and D. B. Goodman, Clostridial myonecrosis: The constant menace. *Mil. Med.* **140**, 461 (1975).

1011a. A. Schreiner, S. Tönjum, and A. Digranes, Hyperbaric oxygen therapy in *Bacteroides* infections. *Acta Chir. Scand.* **140**, 73 (1974).

1012. W. A. Altemeier and W. D. Fullen, Prevention and treatment of gas gangrene. *J. Am. Med. Assoc.* **217**, 806 (1971).

1013. R. L. Fuson, H. A. Saltzman, W. W. Smith, R. E. Whalen, S. Osterhout, and R. T. Parker, Clinical hyperbaric oxygenation with severe oxygen toxicity. *N. Engl. J. Med.* **273**, 415 (1965).

1014. J. Rö, G. Odland, J. A. Maeland, and O. I. Haavelsrud, Hyperbar Oxygenbehandling ved Anaerob Infeksjon. *Nord. Med.* **72**, 1077 (1964).

1015. R. J. Wallyn, S. H. Gumbiner, S. Goldfein, and L. R. Pascale, The treatment of anaerobic infections with hyperbaric oxygen. *Surg. Clin. N. Amer.* **44**, 107 (1964).

1016. T. T. Irvin and G. Smith, Treatment of bacterial infections with hyperbaric oxygen. *Surgery* **63**, 363 (1968).

1017. L. R. Pascale, R. J. Wallyn, S. Goldfein, and S. H. Gumbiner, Treatment of tetanus by hyperbaric oxygenation. *J. Am. Med. Assoc.* **189**, 408 (1964).

1018. J. S. Milledge, Hyperbaric oxygen therapy in tetanus. *J. Am. Med. Assoc.* **203**, 875 (1968).

1019. N. G. Meljne, "Hyperbaric Oxygen and Its Clinical Value," Chapt. 15. Thomas, Springfield, Illinois, 1970.

1020. B. H. Fischer, Topical hyperbaric oxygen treatment of pressure sores and skin ulcers. *Lancet* **2**, 405 (1969).

1021. L. Weinstein and M. A. Barza, Gas gangrene. *N. Engl. J. Med.* **289**, 1129 (1973).

1022. A. W. Chow and L. B. Guze, More on antibiotic susceptibility of *Bacteroides*. (Letter to the editor). *Ann. Intern. Med.* **75**, 810 (1971).

1023. B. P. Sandler, The prevention of cerebral abscess secondary to pulmonary suppuration. *Dis. Chest* **18**, 32 (1965).

1024. S. E. Drewett, D. J. H. Payne, W. Tuke, and P. E. Verdon, Skin distribution of *Clostridium welchii*: Use of iodophor as sporicidal agent. *Lancet* **1**, 1172 (1972).

1025. J. P. Scott, The action of phenol and formol on aerobic and anaerobic organisms. *J. Infect. Dis.* **43**, 90 (1928).

1026. T. A. Mironova, Experimental study of disinfecting agents used on objects infected with the causative agents of anaerobic infections. *Zh. Mikrobiol., Epidemiol. Immunobiol.* **10**, 138 (1968).

1027. W. D. Fullen, J. Hunt, and W. A. Altemeier, Prophylactic antibiotics in penetrating wounds of the abdomen. *J. Trauma* **12**, 282 (1972).

1028. A. S. Levine, S. E. Siegel, A. D. Schreiber, J. Hauser, H. Preisler, I. M. Goldstein, F. Seidler, R. Simon, S. Perry, J. E. Bennett, and E. S. Henderson, Protected environments and prophylactic antibiotics. *N. Engl. J. Med.* **288**, 477 (1973).

1029. R. L. Wright, Complications of intracranial surgery. *In* "Cranial and Intracranial Suppuration" (E. S. Gurdjian, ed.), Chapter VI. Thomas, Springfield, Illinois, 1969.

1030. O. Khairat, An effective antibiotic cover for the prevention of endocarditis following dental and other post-operative bacteraemias. *J. Clin. Pathol.* **19**, 561 (1966).

1030a. D. T. Durack, Current practice in prevention of bacterial endocarditis. *Brit. Heart J.* **37**, 478 (1975).

1031. A. A. Ariaudo, The efficacy of antibiotics in periodontal surgery: A controlled study with lincomycin and placebo in 68 patients. *J. Periodontol.-Periodontics* **40**, 30 (1969).

1032. W. L. White, W. E. Burnett, C. P. Bailey, G. P. Rosemond, C. W. Norris, G. O. Favorite, E. H. Spaulding, A. Bondi, Jr., and R. H. Fowler, Use of penicillin in prevention of postoperative empyema following lung resection. *J. Am. Med. Assoc.* **126**, 1016 (1944).

1033. R. L. Nichols, S. L. Gorbach, and R. E. Condon, Alteration of intestinal microflora following preoperative mechanical preparation of the colon. *Dis. Colon Rectum* **14**, 123 (1971).

1034. W. A. Altemeier, W. R. Culbertson, and R. P. Hummel, Surgical considerations of endogenous infections—sources, types, and methods of control. *Surg. Clin. N. Amer.* **48**, 227 (1968).

1035. D. W. Gaylor, J. S. Clarke, Z. Kudinoff, and S. M. Finegold, Preoperative bowel "sterilization"—a double-blind study comparing kanamycin, neomycin and placebo. *Antimicrob. Ag. Ann.* p. 392 (1961).

1035a. R. L. Nichols, P. Broido, R. E. Condon, S. L. Gorbach, and L. M. Nyhus: Effect of preoperative neomycin–erythromycin intestinal preparation on the incidence of infectious complications following colon surgery. *Ann. Surg.* **178**, 453 (1973).

1035b. O. J. A. Gilmore, T. D. M. Martin, and B. N. Fletcher, Prevention of wound infection after appendicectomy. *Lancet* **1**, 220 (1973).

1035c. O. J. A. Gilmore, Prevention of wound infection in acute appendicitis. *Lancet* **2**, 448 (1973).

1036. W. J. Ledger, Infections in obstetrics and gynecology. New developments in treatment. *Surg. Clin. N. Amer.* **52**, 1447 (1972).

1037. A. T. Willis *et al.*, Preliminary communication. Report by a study group. Metronidazole in the prevention and treatment of *Bacteroides* infections in gynaecological patients. *Lancet* **2**, 1540 (1974).

1037a. M. P. Neary, J. Allen, O. A. Okubadejo, and D. J. H. Payne, Preoperative vaginal bacteria and postoperative infections in gynaecological patients. *Lancet* **2**, 1291 (1973).

1038. W. R. Shamblin, The diagnosis and treatment of acute infections of the hand. *South. Med. J.* **62**, 209 (1969).

1039. J. B. Bernstine, A. Ludmir, and M. A. Fritz, Bacteriologic studies in ligated and non-ligated umbilical cords. *Am. J. Obstet. Gynecol.* **78**, 69 (1959).

1040. H. N. Newman and D. F. G. Poole, Structural and ecological aspects of dental plaque. *In* "The Normal Microbial Flora of Man" (F. A. Skinner and J. G. Carr, eds.), p. 111. Academic Press, New York, 1974.

1040a. R. H. Staat and C. F. Schachtele, Characterization of a dextranase produced by an oral strain of *Actinomyces israelii. Infect. Immunity* **12**, 556 (1975).

1041. R. J. Genco, R. T. Evans, and S. A. Ellison, Dental research in microbiology with emphasis on periodontal disease. *J. Am. Dent. Assoc.* **78**, 1016 (1969).

1042. H. V. Jordan and B. F. Hammond, Filamentous bacteria isolated from human root surface caries. *Arch. Oral Biol.* **17**, 1333 (1972).

1042a. R. J. Gibbons and J. van Houte, Bacteriology of periodontal disease. *In* "The Textbook of Oral Biology," (J. H. Shaw, C. Cappucino, S. Meller, and E. Sweeney, eds.). Saunders, Philadelphia, (in press).

1042b. M. G. Newman and S. S. Socransky, Predominant cultivable microbiota in periodontosis. *J. Periodontal Res.* in press.

1042c. M. G. Newman, S. S. Socransky, E. D. Savitt, D. A. Propas, and A. Crawford, Studies of the microbiology of periodontosis. *J. Periodontol.* **47**, 373 (1976).

1042d. W. J. Loesche, K. U. Paunio, M. P. Woolfolk, and R. N. Hockett, Collagenolytic activity of dental plaque associated with periodontal pathology. *Infect. Immunity* **9**, 329 (1974).

1043. S. S. Socransky, Relationship of bacteria to the etiology of periodontal disease. *J. Dent. Res.* **49**, 203 (1970).

1044. H. Löe, Present day status and direction for future research on the etiology and prevention of periodontal disease. *J. Periodontol. Periodontics* **40**, 678 (1969).

1045.   R. M. Davies, Control of oral flora by hibitane and other antibacterial agents. *In* "The Normal Microbial Flora of Man" (F. A. Skinner and J. G. Carr, eds.), p. 101. Academic Press, New York, 1974.

1046.   S. E. Mergenhagen and R. Snyderman, Periodontal disease: A model for the study of inflammation. *J. Infect. Dis.* **123**, 676 (1971).

1047.   R. M. Reisner, D. Z. Silver, M. Puhvel, and T. H. Sternberg, Lipolytic activity of *Corynebacterium acnes*. *J. Invest. Dermatol.* **51**, 190 (1968).

1048.   K. Weaber, R. Freedman, and W. W. Eudy, Tetracycline inhibition of a lipase from *Corynebacterium acnes*. *Appl. Microbiol.* **21**, 639 (1971).

1049.   R. K. Freinkel, J. S. Strauss, S. Y. Yip, and P. E. Pochi, Effect of tetracycline on the composition of sebum in acne vulgaris. *N. Engl. J. Med.* **273**, 850 (1965).

1049a.   D. K. Chalker and J. G. Smith, Jr., Systemic antibiotics and acne. *J. Am. Med. Assoc.* **234**, 1058 (1975).

1049b.   P. E. Pochi, Antibiotics in acne. *N. Engl. J. Med.* **294**, 43 (1976).

1049c.   G. G. Krueger and G. L. Christian, Clindamycin for acne. *J. Am. Med. Assoc.* **235**, 250 (1976).

1049d.   H. Haenel and J. Bendig, Intestinal flora in health and disease. *Progr. Food Nutr. Sci.* **1**, 21 (1975).

1049e.   B. S. Drasar and M. J. Hill, "Human Intestinal Flora." Academic Press, London, 1974.

1050.   C. M. Anderson, D. N. Challacombe, and J. M. Richardson, The bacterial flora of the upper gastrointestinal tract in children both in health and disease. *In* "The Normal Microbial Flora of Man" (F. A. Skinner and J. G. Carr, eds.), p. 197. Academic Press, New York, 1974.

1051.   G. Neale, D. Gompertz, H. Schonsby, S. Tabaqchali, and C. C. Booth, The metabolic and nutritional consequences of bacterial overgrowth in the small intestine. *Am. J. Clin. Nutr.* **25**, 1409 (1972).

1052.   R. D. McKenna, I. T. Beck, and H. Epstein, Clinical significance of small bowel function. *Can. Med. Assoc. J.* **83**, 896 (1960).

1053.   I. G. Lyall and P. J. Parsons, Some aspects of the blind-loop syndrome. *Med. J. Aust.* **2**, 904 (1961).

1054.   V. J. Kinsella, W. B. Hennessy, and E. P. George, Studies on post-gastrectomy malabsorption: The importance of bacterial contamination of the upper small intestine. *Med. J. Aust.* **2**, 257 (1961).

1055.   B. S. Drasar, M. J. Hill, and M. Shiner, The deconjugation of bile salts by human intestinal bacteria. *Lancet* **1**, 1237 (1966).

1056.   S. Tabaqchali, O. A. Okubadejo, G. Neale, and C. C. Booth, Influence of abnormal bacterial flora on small intestinal function. *Proc. Roy Soc. Med.* **59**, 1244 (1966).

1057.   D. E. Polter, J. D. Boyle, L. G. Miller, and S. M. Finegold, Anaerobic bacteria as cause of the blind loop syndrome. *Gastroenterology* **54**, 1148 (1968).

1058.   F. Goldstein, W. Wirts, G. Salen, and R. J. Mandle, Diverticulosis of the small intestine. Clinical, bacteriologic and metabolic observations in a group of seven patients. *Am. J. Dig. Dis.* **14**, 170 (1969).

1059.   A. J. Pearson, A. Brzechwa-Ajdukiewicz, and C. F. McCarthy, Intestinal pseudo-obstruction with bacterial overgrowth in the small intestine. *Am. J. Dig. Dis.* **14**, 200 (1969).

1060.   W. E. Farrar, Jr., N. M. O'Dell, J. L. Achord, and H. A. Greer, Intestinal microflora and absorption in patients with stagnation-inducing lesions of the small intestine. *Am. J. Dig. Dis.* **17**, 1065 (1972).

1060a.   D. N. Challacombe, J. M. Richardson, S. Edkins, and I. F. Hay, Ileal blind loop in childhood, *Am. J. Dis. Child.* **128**, 719 (1974).

1060b. T. H. Kent, R. W. Summers, L. DenBesten, J. C. Swaner, and M. Hrouda, Effect of antibiotics on bacterial flora of rats with intestinal blind loops. *Proc. Soc. Exp. Biol. Med.* **132**, 63 (1969).

1061. V. J. Kinsella and W. B. Hennessey, Gastrectomy and the blind loop syndrome. *Lancet* **2**, 1205 (1960).

1062. E. W. Strauss, R. M. Donaldson, Jr., and F. H. Gardner, A relationship between intestinal bacteria and the absorption of vitamin B_{12} in rats with diverticula of the small bowel. *Lancet* **2**, 736 (1961).

1063. B. E. Gustafsson, T. Midtvedt, and A. Norman, Isolated fecal microorganisms capable of 7α-dehydroxylating bile acids. *J. Exp. Med.* **123**, 413 (1966).

1063a. V. Aries and M. J. Hill, Degradation of steroids by intestinal bacteria. I. Deconjugation of bile salts. *Biochim. Biophys. Acta* **202**, 526 (1970).

1063b. T. Midtvedt, Microbial bile acid transformation. *Am. J. Clin. Nutr.* **27**, 1341 (1974).

1064. R. Lewis and S. Gorbach, Modification of bile acids by intestinal bacteria. *Arch. Intern. Med.* **130**, 545 (1972).

1065. A. Doig and R. H. Girdwood, The absorption of folic acid and labelled cyanocobalamin in intestinal malabsorption. *Q. J. Med.* **29**, 333 (1960).

1066. J. A. Halsted, P. M. Lewis, and M. Gasster, Absorption of radioactive vitamin B_{12} in the syndrome of megaloblastic anemia associated with intestinal stricture or anastomosis. *Am. J. Med.* **20**, 42 (1956).

1067. F. Goldstein, C. W. Wirts, and S. Kramer, The relationship of afferent limb stasis and bacterial flora to the production of post-gastrectomy steatorrhea. *Gastroenterology* **40**, 47 (1961).

1068. C. W. Wirts and F. Goldstein, Studies of the mechanism of post-gastrectomy steatorrhea. *Ann. Intern. Med.* **58**, 25 (1963).

1069. E. A. Paulk and W. E. Farrar, Jr., Diverticulosis of the small intestine and megaloblastic anemia. *Am. J. Med.* **37**, 473 (1964).

1070. H. Schjönsby and T. Hofstad, Effect of bacteria on intestinal uptake of vitamin B_{12}. II. The consequences of *in vitro* preincubation of B_{12} with pure bacterial populations. *Scand. J. Gastroenterol.* **7**, 353 (1972).

1070a. H. Schjönsby, B. S. Drasar, S. Tabaqchali, and C. C. Booth, Uptake of vitamin B_{12} by intestinal bacteria in the stagnant loop syndrome. *Scand. J. Gastroenterol.* **8**, 41 (1973).

1071. C. Hines, Jr. and W. D. Davis, Jr., Ehlers-Danlos syndrome with megaduodenum and malabsorption syndrome secondary to bacterial overgrowth. *Am. J. Med.* **54**, 539 (1973).

1071a. F. A. Klipstein, L. V. Holdeman, J. J. Corcino, and W.E.C. Moore, Enterotoxigenic intestinal bacteria in tropical sprue. *Ann. Intern. Med.* **79**, 632 (1973).

1071b. A. M. Tomkins, B. S. Drasar, and W. P. T. James, Bacterial colonisation of jejunal mucosa in acute tropical sprue. *Lancet* **1**, 59 (1975).

1072. H. B. Greenlee, R. Vivit, J. Paez, and A. Dietz, Bacterial flora of the jejunum following peptic ulcer surgery. *Arch. Surg. (Chicago)* **102**, 260 (1971).

1073. A. R. Prévot and C. Morel, Nouveau cas de maladie de Whipple à *Corynebacterium anaerobium* guérie par antibiothérapie. *Bull. Acad. Natl. Med., Paris* **148**, 540 (1964).

1074. J. Caroli, A. R. Prévot, C. Julien, L. Gueritat, and H. Stralin, L'étiologie bactérienne de la maladie de Whipple. III. A-propos d'une nouvelle observation. Isolement de *Corynebacterium anaerobium*. *Arch. Mal. Appar. Dig. Mal. Nutr.* **52**, 177 (1963).

1075. E. A. Phear and B. Ruebner, The *in vitro* production of ammonium and amines by intestinal bacteria in relation to nitrogen toxicity as a factor in hepatic coma. *Br. J. Exp. Pathol.* **37**, 253 (1956).

1076. M. Huet and N. Aladame, Recherches sur l'uréase des bactéries anaérobies. *Ann. Inst. Pasteur, Paris* **82**, 766 (1952).

1077. J. Sabbaj, V. L. Sutter, and S. M. Finegold, Urease and deaminase activities of fecal bacteria in hepatic coma. *Antimicrob. Ag. Chemother.* p. 181 (1971).

1077a. V. H. Varel, M. P. Bryant, L. V. Holdeman, and W. E. C. Moore, Isolation of ureolytic *Peptostreptococcus productus* from feces using defined medium; failure of common urease tests. *Appl. Microbiol.* **28**, 594 (1974).

1077b. J. J. Marr, M. D. Sans, and F. J. Tedesco, Bacterial studies of clindamycin-associated colitis. *Gastroenterology* **69**, 352 (1975).

1077c. J. R. Pearson, Alteration of dietary fatty acid by human intestinal bacteria. *Proc. Nutr. Soc.* **32**, 8A (1973).

1077d. J. R. Pearson, H. S. Wiggins, and B. S. Drasar, Conversion of long-chain unsaturated fatty acids to hydroxy acids by human intestinal bacteria. *J. Med. Microbiol.* **7**, 265 (1974).

1078. B. Ursing and C. Kamme, Metronidazole for Crohn's disease. *Lancet* **1**, 775 (1975).

1079. S. Persson and D. Danielsson, Immunologiska Studier vid Crohns Sjukdom. *Nord. Med.* **23**, 1590 (1971).

1080. F. Wensinck, The faecal flora of patients with Crohn's disease. *Antonie van Leeuwenhoek* **41**, 214 (1975).

1081. E. Monteiro, J. Fossey, M. Shiner, B. S. Drasar, and A. C. Allison, Antibacterial antibodies in rectal and colonic mucosa in ulcerative colitis. *Lancet* **1**, 249 (1971).

1081a. W. R. Brown and E. Lee, Radioimmunological measurements of bacterial antibodies. *Gastroenterology* **66**, 1145 (1974).

1081b. S. Hammarström, P. Perlmann, B. E. Gustaffson, and R. Lagercrantz, Autoantibodies to colon in germfree rats monocontaminated with *Clostridium difficile. J. Exper. Med.* **129**, 747 (1969).

1081c. P. Goldman and M. A. Peppercorn, Drug therapy, sulfasalazine. *New Engl. J. Med.* **293**, 20 (1975).

1082. E. J. Drenick, M. E. Ament, S. M. Finegold, P. Corrodi, and E. Passaro, Jr., Bypass enteropathy: Intestinal and systemic manifestations following small bowel bypass. *J. Am. Med. Assoc.* **236**, 269 (1976).

1083. C. E. Yale, E. Balish, and J. P. Wu, The bacterial etiology of pneumatosis cystoides intestinalis. *Arch. Surg.* (*Chicago*) **109**, 89 (1974).

1083a. C. E. Yale, Etiology of pneumatosis cystoides intestinalis. *Surg. Clin. N. Amer.* **55**, 1297 (1975).

1083b. J. R. Wands, J. T. LaMont, E. Mann, and K. J. Isselbacher, Arthritis associated with intestinal-bypass procedure for morbid obesity. *N. Engl. J. Med.* **294**, 121 (1976).

1083c. W. W. Faloon, An evaluation of risks—bypass versus obesity. *N. Engl. J. Med.* **294**, 159 (1976).

1083d. J. P. O'Leary, J. W. Maher, J. I. Hollenbeck, and E. R. Woodward, Pathogenesis of hepatic failure following jejunoileal bypass. *Gastroenterology* **66**, 859 (1974).

1083e. J. D. Anderson, Factors affecting transfer of antibiotic resistance between gram-negative bacteria in the human intestine. *J. Clin. Pathol.* **26**, 1985 (1973).

1083f. T. S. Low-Beer and E. W. Pomare, Can colonic bacterial metabolites predispose to cholesterol gallstones? *Br. Med. J.* **1**, 438 (1975).

1083g. P. Biswas, Metronidazole in haemorrhoids. *J. Indian Med. Assoc.* **51**, 344 (1968).

1083h. H. Eyssen and G. Parmentier, Biohydrogenation of sterols and fatty acids by the intestinal microflora. *Am. J. Clin. Nutr.* **27**, 1329 (1974).

1084. F. L. Sapico, H. Emori, L. DS. Smith, R. Bluestone, and S. M. Finegold, Absence of relationship of fecal *Clostridium perfringens* to rheumatoid arthritis and rheumatoid variants. *J. Infect. Dis.* **128**, 559 (1973).

1085. I. Mansson and H. Colldahl, The intestinal flora in patients with bronchial asthma and rheumatoid arthritis. *Acta Allergol.* **20**, 94 (1965).

1086. L. E. Bartholomew and F. R. Nelson, *Corynebacterium acnes* in rheumatoid arthritis. II. Identification of antigen in synovial fluid leucocytes. *Ann. Rheum. Dis.* **31**, 28 (1972).

1087.   J. J. R. Duthie, S. M. Stewart, W. R. M. Alexander, and R. E. Dayhoff, Isolation of diphtheroid organisms from rheumatoid synovial membrane and fluid. *Lancet* **1**, 142 (1967).

1088.   H. A. L. Clasener and P. J. Biersteker, Significance of diphtheroids isolated from synovial membranes of patients with rheumatoid arthritis. *Lancet* **2**, 1031 (1969).

1089.   J. W. Berg, M. A. Howell, and S. J. Silverman, Dietary hypotheses and diet-related research in the etiology of colon cancer. *Health Serv. Rep.* **88**, 915 (1973).

1090.   M. J. Hill, B. S. Drasar, V. Aries, J. S. Crowther, G. Hawksworth, and R. E. O. Williams, Bacteria and aetiology of cancer of large bowel. *Lancet* **1**, 95 (1971).

1091.   B. S. Drasar and M. J. Hill, Intestinal bacteria and cancer. *Am. J. Clin. Nutr.* **25**, 1399 (1972).

1092.   J. W. Cole, Carcinogens and carcinogenesis in the colon. *Hosp. Pract.* **8**, 123 (1973).

1093.   V. C. Aries, P. Goddard, and M. J. Hill, Degradation of steroids by intestinal bacteria. III. 3-oxo-5-β-steroid-Δ^1-dehydrogenase and 3-oxo-5-β-steroid-Δ^4-dehydrogenase. *Biochim. Biophys. Acta* **248**, 482 (1971).

1094.   P. Goddard and M. J. Hill, Degradation of steroids by intestinal bacteria. IV. The aromatisation of ring A. *Biochim. Biophys. Acta* **280**, 336 (1972).

1094a.   P. Goddard, F. Fernandez, B. West, M. J. Hill, and P. Barnes, The nuclear dehydrogenation of steroids by intestinal bacteria. *Med. Microbiol.* **8**, 429 (1975).

1095.   D. P. Burkitt, An epidemiological approach to gastrointestinal cancer. (Editor interviews Dr. Burkitt.) *CA* **20**, 147 (1970).

1096.   F. P. Herter and C. A. Slanetz, Jr., Preoperative intestinal preparation in relation to the subsequent development of cancer at the suture line. *Surg., Gynecol. Obstet.* **127**, 49 (1968).

1097.   A. R. Prévot, B. Halpern, G. Biozzi, C. Stiffel, D. Mouton, J.-C. Morard, Y. Bouthillier, and C. Decreusefond, Médecine expérimentale—stimulation du système reticuloéndothelial (S.R.E.) par les corps microbiens tués de *Corynebacterium parvum*. *C. R. Hebd. Seances Acad. Sci.* **257**, 13 (1963).

1097a.   C. S. Cummins and J. L. Johnson: *Corynebacterium parvum*: a synonym for *Propionibacterium acnes*? *J. Gen. Microbiol.* **80**, 433 (1974).

1098.   A. R. Prévot and J. Tran Van Phi, Étude comparative de la stimulation du système reticuloéndothelial par différentes souches de corynebactéries anaérobies et d'espèces voisines. *C. R. Hebd. Seances Acad. Sci.* **258**, 4619 (1964).

1098a.   W. H. McBride, J. Dawes, N. Dunbar, A. Ghaffar, and M. F. A. Woodruff, A comparative study of anaerobic coryneforms. *Immunology* **28**, 49 (1975).

1099.   B. N. Halpern, G. Biozzi, C. Stiffel, and D. Mouton, Inhibition of tumour growth by administration of killed *Corynebacterium parvum*. *Nature (London)* **212**, 853 (1966).

1100.   M. F. A. Woodruff and J. L. Boak, Inhibitory effect of injection of *Corynebacterium parvum* on the growth of tumour transplants in isogenic hosts. *Br. J. Cancer* **20**, 345, (1966).

1101.   J. R. Möse, *Clostridium* strain M 55 and its effect on malignant tumours. *In* "The Anaerobic Bacteria" (V. Fredette, ed.), p. 229. Inst. Microbiol. Hyg., Montreal University, Laval-des-Rapides, Quebec, Canada 1968.

1102.   D. Kayser, The use of spores of anaerobic bacteria for special investigations in cancer research. *In* "The Anaerobic Bacteria" (V. Fredette, ed.), p. 249. Inst. Microbiol. Hyg, Montreal University, Laval-des-Rapides, P.Q., Canada, 1968.

1103.   H. Kretschmer, Erste Erfahrungen bei der Behandlung maligner Hirngeschwulste mit Sporen des Stammes M 55. *Dtsch. Gesundheitswes.* **26**, 1704 (1971).

1104.   M. A. Peppercorn and P. Goldman, The role of intestinal bacteria in the metabolism of salicylazosulfapyridine. *J. Pharmacol. Exp. Ther.* **181**, 555 (1972).

1105.   G. Hawksworth, B. S. Drasar, and M. J. Hill, Intestinal bacteria and the hydrolysis of glycosidic bonds. *J. Med. Microbiol.* **4**, 451 (1971).

1105a. H. A. Soleim and R. R. Scheline, Metabolism of xenobiotics by strains of intestinal bacteria. *Acta Pharmacol. Toxicol.* **31**, 471 (1972).

1105b. V. D. Bokkenheuser, J. B. Suzuki, S. B. Polovsky, J. Winter, and W. G. Kelly, Metabolism of deoxycorticosterone by human fecal flora. *Appl. Microbiol.* **30**, 82 (1975).

1105c. K.-T. Chung, G. M. Anderson, and G. E. Fulk, Formation of indoleacetic acid by intestinal anaerobes. *J. Bacteriol.* **124**, 573 (1975).

1105d. G. Inglis, G. W. G. Bird, A. A. B. Mitchell, G. R. Milne, and J. Wingham, Erythrocyte polyagglutination showing properties of both T and Tk, probably induced by *Bacteroides fragilis* infection. *Vox Sang.* **28**, 314 (1975).

1105e. G. Inglis, G. W. G. Bird, A. A. B. Mitchell, G. R. Milne, and J. Wingham, Effect of *Bacteroides fragilis* on the human erythrocyte membrane: pathogenesis of Tk polyagglutination. *J. Clin. Pathol.* **28**, 964 (1975).

1105f. G. W. G. Bird, J. Wingham, G. Inglis, and A. A. B. Mitchell, Tk-polyagglutination in *Bacteroides fragilis* septicaemia. *Lancet* **1**, 286 (1975).

1105g. E. Ratner, P. Person, and D. J. Kleinman, Oral pathology and trigeminal neuralgia. I. Clinical experiences. *IADR Abst. J. Dent. Res.* **55B**, B299 (1976).

1105h. S. S. Socransky, C. Stone, E. Ratner, and P. Person: Oral pathology and trigeminal neuralgia. III. Microbiologic examination. *IADR Abst. J. Dent. Res.* **55B**, B300 (1976).

1106. W. E. C. Moore, E. P. Cato, and L. V. Holdeman, Anaerobic bacteria of the gastrointestinal flora and their occurrence in clinical infections. *J. Infect. Dis.* **119**, 641 (1969).

1107. R. J. Zabransky, Isolation of anaerobic bacteria from clinical specimens. *Mayo Clin. Proc.* **45**, 256 (1970).

1108. J. A. Washington II, Bacteremia due to anaerobic, unusual and fastidious bacteria. *In* "Bacteremia. Laboratory and Clinical Aspects" (A. C. Sonnenwirth, ed.), p. 47. Thomas, Springfield, Illinois, 1973.

1109. M. N. Swartz and A. W. Karchmer, Infections of the central nervous system. *In* "Anaerobic Bacteria: Role in Disease" (A. Balows *et al.*, eds.), Chapter XXIV. Thomas, Springfield, Illinois, 1974.

1110. J. Frederick and A. I. Braude, Anaerobic infection of the paranasal sinuses. *N. Engl. J. Med.* **290**, 135 (1974).

1111. W. J. Loesche, Dental infections. *In* "Anaerobic Bacteria: Role in Disease" (A. Balows *et al.*, eds.), Chapter XXXIII. Thomas, Springfield, Illinois, 1974.

1112. W. A. Altemeier, Liver abscess: the etiologic role of anaerobic bacteria. *In* "Anaerobic Bacteria: Role in Disease" (A. Balows *et al.*, eds.), Chapter XXXI. Thomas, Springfield, Illinois, 1974.

1113. R. M. Swenson, T. C. Michaelson, M. J. Daly, and E. H. Spaulding, Anaerobic bacterial infections of the female genital tract. *Obstet. Gynecol.* **42**, 538 (1973).

1114. H. Thadepalli, S. L. Gorbach, and L. Keith, Anaerobic infections of the female genital tract: Bacteriologic and therapeutic aspects. *Am. J. Obstet. Gynecol.* **117**, 1034 (1973).

1115. K. Sommer, Beitrage zur Bakteriologie der infizierten, nekrotischen Pulpa, mit besonderer Berucksichtigung der anaeroben Bakterien bei Gangran. *Dtsch. Monatsschr.-Zahnheilkd.* **33**, 297 (1915).

1116. A. Kamer, Anaerobe Mikroorganismen in der gangranosen Zahnpulpa. *Schweiz. Monatsschr. Zahnheilkd.* **39**, 211 (1929).

1117. H. Andre, Die obligaten Anaerobier im gangranosen Zahnwurzelkanal und auf der Zahnoberflache. *Z. Hyg. Infektionskr.* **114**, 397 (1932).

1118. O. Helm, Anaerobier im Wurzelkanal bei Gangran der Zahnpulpa. Inaugural Dissertation, Marburg University (1934).

1119. F. J. Gruchalla and C. B. Hamann, Root surgery. *J. Missouri State Dent. Assoc.* **27**, 229 (1947).

1120. J. B. MacDonald, G. C. Hare, and A. W. S. Wood, The bacteriologic status of the pulp chambers in intact teeth found to be nonvital following trauma. *Oral Surg., Oral*

Med. Oral Pathol. **10**, 318 (1957).

1121. J. M. Leavitt, I. J. Naidorf, and P. Shugaevsky, The bacterial flora of root canals as disclosed by a culture medium for endodontics. *Oral Surg., Oral Med. Oral Pathol.* **11**, 302 (1958).

1122. M. Goldman and A. H. Pearson, Postdebridement bacterial flora and antibiotic sensitivity. *Oral Surg., Oral Med. Oral Pathol.* **28**, 897 (1969).

1123. J. Mizuno and K. Tamai, Study of the oral anaerobes. VIII. Isolation of the anaerobes in the infectious root canal and drug resistance of isolated anaerobes. *In* "Second Symposium on Anaerobic Bacteria and Their Infectious Diseases" (S. Ishiyama *et al.*, eds.), p. 32. Eisai, Tokyo, Japan, 1972.

1123a. J.-O. Berg and C.-E. Nord, A method for isolation of anaerobic bacteria from endodontic specimens. *Scand. J. Dent. Res.* **81**, 163 (1973).

1123b. W. E. Kantz and C. A. Henry, Isolation and classification of anaerobic bacteria from intact pulp chambers of nonvital teeth in man. *Arch. Oral Biol.* **19**, 91 (1974).

1124. M. Heyde, Ueber Infektionen mit anaeroben Bakterien. Ein Beitrag zur Kenntnis anaerober Staphylokokken und des *Bacillus funduliformis*. *Beitr. Klin. Chir.* **68**, 642 (1910).

1125. G. Idman, Bakteriologische Untersuchungen von im Anschluss an Pulpitis purulenta und Gangraena Pulpae auftretenden periostalen Abszessen mit besonderer Berucksichtigung der obligat anaeroben Mikroorganismen. *Arb. Pathol. Inst. Univ. Helsinfors* **1**, 191 (1913).

1126. K. H. Thoma, Oral abscesses. *J. Allied Dent. Soc.* **11**, 95 (1916).

1127. J. Head and C. Roos, On the bacteriology of apical abscesses. *J. Dent. Res.* **1**, 13 (1919).

1128. A. L. Smith and R. W. Ludwick, Bacterial findings in 107 cases of abscessed teeth in children. *Nebr. State Med. J.* **4**, 131 (1919).

1129. F. S. Balyeat, Carious tooth causing staphylococcic pyemia and death—report of case. *Pac. Dent. Gaz.* **38**, 432 (1930).

1130. H. Hornung, Zur Kenntnis unklarer Sepsisfalle. (Bacillus Budaysepsis.) *Med. Welt* **14**, 1278 (1940).

1131. E. R. Brygoo and N. Aladame, Étude d'une espèce nouvelle anaérobie stricte du genre *Eubacterium: E. crispatum* n. sp. *Ann. Inst. Pasteur, Paris* **84**, 640 (1953).

1132. A. Durante, Su due casi di lesioni inflammatorie a tipo necrotico, dovute a *Spirochaeta vincent* con *Bacillus fusiformis. Acta Med. Ital. Mal. Infet. Parassit.* **8**, 326 (1953).

1133. H. Beerens, A. Gérard, and J. Guillaume; Étude de 30 souches de *Bifidibacterium bifidum* (*Lactobacillus bifidus*). Caractérisation d'une variéte buccale. Comparaison avec les souches d'origine fecale. *Ann. Inst. Pasteur, Lille* **9**, 77 (1957).

1134. J. Fukuda and K. Tamai, Study of the oral anaerobic bacteria. VII. Isolation rate of anaerobes in gingival abscess and the drug resistance of isolated anaerobes. *In* "Second Symposium on Anaerobic Bacteria and Their Infectious Diseases" (S. Ishiyama, *et al.*, eds.), p. 28. Eisai, Tokyo, Japan, 1972.

1134a. News and Notes: Epidemiology. Actinomycosis. *Brit. Med. J.* **1**, 365 (1973).

1134b. J. E. Turner, D. W. Moore, and B. S. Shaw, Prevalence and antibiotic susceptibility of organisms isolated from acute soft-tissue abscesses secondary to dental caries. *Oral Surg.* **39**, 848 (1975).

1134c. C. B. Sabiston and W. A. Gold, Anaerobic bacteria in oral infections. *Oral Surg., Oral Med. Oral Pathol.* **38**, 187 (1974).

1135. M. Hausmann, Von der Bedeutung der Anaerobier in der inneran Medizin. *Schweiz. Med. Wochenschr.* **58**, 305 (1928).

1136. I. J. Greenblatt and A. P. Greenblatt, Human infection with *Bacterium necrophorum*. *Am. J. Med. Sci.* **210**, 596 (1945).

1137. A. R. Prévot, H. Beerens, and J. Zimmes-Chaverou, Étude des caractères morphologiques, physiologiques et biochimiques de *Spherophorus varius. Ann. Inst. Pasteur, Paris* **73**, 390 (1947).

1138. M. H. Nathan, W. P. Radman, and H. L. Barton, Osseous actinomycosis of the head and neck. *Am. J. Roentgenol. Radium Ther. Nucl. Med.* **87**, 1048 (1962).

1139. F. M. S. Ameriso and G. Caffarena, Sinusopatia por estreptococo anaerobio. *Seman. Med.* **123**, 1355 (1963).

1140. L. K. Georg, G. W. Robertstad, and S. A. Brinkman, Identification of species of *Actinomyces. J. Bacteriol.* **88**, 477 (1964).

1141. M. B. Stanton, Actinomycosis of the maxillary sinus. *In* "Actinomycosis" (M. Bronner and M. Bronner, eds), p. 167. Williams & Wilkins, Baltimore, Maryland, 1969.

1142. H. Gujer, H. P. Hartmann, and E. Wiesmann, Gasbrandsepsis mit ungewöhnlicher Streuquelle. *Beitr. Gerichtl. Med.* **27**, 284 (1970).

1142a. A. Axelsson and J. E. Brorson, Bacteriological findings in acute maxillary sinusitis. *Otorhinolaryngology* **34**, 1 (1972).

1142b. J. H. Per-Lee, A. A. Clairmont, J. C. Hoffman, A. S. McKinney, and S. W. Schwarzmann, Actinomycosis masquerading as depression headache: Case report—management review of sinus actinomycosis. *Laryngoscope* **84**, 1149 (1974).

1142c. F. O. Evans, Jr., J. B. Sydnor, W. E. C. Moore, G. R. Moore, J. L. Manwaring, A. H. Brill, R. T. Jackson, S. Hanna, J. S. Skaar, L. V. Holdeman, G. S. Fitz-Hugh, M A. Sande, and J. M. Gwaltney, Jr., Sinusitis of the maxillary antrum. *N. Engl. J. Med.* **293**, 735 (1975).

1143. E. Fraenkel, Über postanginose Pyamie. *Virchows Arch. Pathol. Anat. Physiol.* **254**, 639 (1925).

1144. Hegler and Jacobsthal, Arztlicher Verein in Hamburg. Biologische Abteilung. *Klin. Wochenschr.* **4**, 1940 (1925).

1145. K. Kissling, Ueber postanginose Sepsis. *Muench. Med. Wochenschr.* **28**, 1163 (1929).

1146. A. Lemierre, A. Guy, and M. Rudolf, Septicemie à "*Bacillus fragilis.*" *Presse Med.* **37**, 1669 (1929).

1147. A. Lemierre, J. Reilly, F. Layani, and E. Friedman, Septicémie primitive dué au "*Bacillus funduliformis.*" *Bull. Mem. Soc. Med. Hôp. Paris* [3] **49**, 165 (1933).

1148. J. Cathala, J. Bourgeois, and P. Gabriel, Deux cas de septicémies primitives dués au Bacille "*funduliformis.*" *Bull. Mem. Soc. Med. Hôp. Paris* [3] **49**, 96 (1933).

1149. J. Cathala, J. Bourgeois, and P. Gabriel, Étude anatomique, bactériologique et expérimentale de deux cas de septicémies à Bacilles "*funduliformis.*" *Bull. Mem. Soc. Med. Hôp. Paris* [3] **49**, 100 (1933).

1150. L. Richon, P. Kissel, and F. Lepoire, Septicémie mortelle à "*Bacillus fragilis*" consécutive à une angine phlégmoneuse. *Rev. Med. Est.* **62**, 289 (1934).

1151. M. P. de Font-Réaulx, Un nouveau cas de septicémie primitive à "*Bacillus funduliformis.*" *Bull. Mem. Soc. Med. Hôp. Paris* [3] **51**, 72 (1935).

1152. A. Lemierre, Chap. XXII. Sur un cas de septico-pyohémie à *Bacillus funduliformis. In* "Maladies Infectieuses," p. 366. Masson, Paris, 1935.

1153. A. Lemierre and A. Meyer, Deux cas de septico-pyohémie à "*Bacillus funduliformis*" dont un termine par la guérison spontanée. *Bull. Mem. Soc. Med. Hôp. Paris* [3] **51**, 712 (1935).

1154. M. E. Donzelot, A. Meyer, and J. Olivier, Deux nouvelles observations de septicémie à *Bacillus funduliformis. Bull. Mem. Soc. Med. Hôp. Paris* [3] **52**, 743 (1936).

1155. A. Lemierre, A. Laporte, and H. Bloch-Michel, Phlegmon gazeux cervical à *Bacillus funduliformis. Bull. Mem. Soc. Med. Hôp. Paris* [3] **52**, 1558 (1936).

1156. A. Lemierre and R. Moreau, Un cas de septico-pyohémie post-angineuse à "*Bacillus funduliformis*" terminé par la guérison (1). *Bull. Mem. Soc. Med. Hôp. Paris* [3] **52**, 912 (1936).

1157. A. Lemierre, J. Reilly, Meyer-Heine, and J. Hamburger, Un cas de septico-pyohémie post-angineuse à "*Bacillus funduliformis*" à évolution rapidement mortelle. Ligature de la veine jugulaire interne. Minime abces amygdalien et thrombophlebite peri-amygdalienne. *Bull. Mem. Soc. Med. Hôp. Paris* [3] **52**, 919 (1936).

1158. M. Ternois, Les septicémies à *Bacillus fragilis*. *Ann. Med. (Paris)* **44**, 201 (1938).

1159. J. H. Lemierre, Les formes curables et les formes frustes des septicopyohémies à *Bacillus funduliformis*. Thesis, Faculty of Medicine, University of Paris, Paris (1939).

1160. A. Lemierre, A. P. Guimaraes, and J. Lemierre, Les formes curables et frustes des septico-pyohémies post-angineuses à *Bacillus funduliformis*. *Presse Med.* **48**, 97 (1940).

1161. M. Naville, M. Pictet, and A. Gampert, Septicémie à *Bacillus funduliformis*. *Rev. Med. Suisse Romande* **60**, 171 (1940).

1162. V. Siegler, Sur quelques cas de septico-pyohémie post-angineuse à *Bacillus funduliformis* évoluant spontanément vers la guérison. Thesis, Faculty of Medicine, University of Paris, Paris (1940).

1163. J. Ramadier and P. Mollaret, De la resection amygdalienne et jugulaire dans les septico-pyohémies post-angineuses; à-propos d'un cas à *Bacillus funduliformis*; opère et guéri. *Presse Med.* **49**, 898 (1941).

1164. P. P. Ravault, M. Girard, and J. Viallier, La septicémie à *Bacillus funduliformis*. *J. Med. Lyon* **22**, 21 (1941).

1165. B. Kemkes, Ueber eine Postanginöse Funduliformis-Infektion. *Zentralbl. Bakteriol., Parasitenkd., Infektionskr. Hyg. Abt. 1: Orig.* **151**, 68 (1943).

1166. W. E. Smith, M5. *Bacteroides* infection. *J. Bacteriol.* **45**, 54 (1943).

1167. J. J. Waring, J. G. Rayn, R. Thompson, and F. R. Spencer, *Bacteroides* septicemia. Report of a case with recovery. *Laryngoscope* **53**, 717 (1943).

1168. M. Aussannaire, Le syndrôme angine-infarctus pulmonaire dans les infections à *Bacillus funduliformis*. Thesis, Faculty of Medicine, University of Paris, Paris (1943).

1169. A. M. Smit, Geval van postangineuse Anaërobensepsis. *Ned. Tijdschr. Geneesk.* **88**, 739 (1944).

1170. L. Rüedi, Ueber Fortschritte in der Behandlung der Sepsis nach Angina. *Schweiz. Med. Wochenschr.* **74**, 545 (1944).

1171. Desbuquois and Iselin: Septicémie à *Bacillus fundibuliformis* post-angineuse, guérie par la pénicilline. *Bull. Mem. Soc. Med. Hôp. Paris* [3] **61**, 313 (1945).

1172. A. Lemierre, M. Morin, and M. Rathery, Quatre cas d'infection à *Bacillus funduliformis* guéris après traitement par la pénicilline. *Bull. Mem. Soc. Med. Hôp. Paris* [3] **61**, 335 (1945).

1173. J. D. Reid, G. E. Snider, E. C. Toone, and J. S. Howe, Anaerobic septicemia. Report of six cases with clinical, bacteriologic and pathologic studies. *Am. J. Med. Sci.* **209**, 296 (1945).

1174. A. Lemierre, J. Reilly, M. Morin, and M. Rathery, Action de la pénicilline dans quelques septicémies à microbes anaérobies. *Presse Med.* **54**, 49 (1946).

1175. A. C. Ruys, *Bacteroides funduliformis* (*Fusiformis necrophorus*). *J. Pathol. Bacteriol.* **59**, 313 (1947).

1176. P. Tardieux and A. Nabonne, Nouveau cas de septicémie à *Spherophorus pseudone-crophorus*. *Ann. Inst. Pasteur, Paris* **76**, 181 (1949).

1177. O. Ernst and H. Hartl, Über Funduliformis-Septikämien. *Muench. Med. Wochenschr.* **93**, 1213 (1951).

1178. G. Le Sueur, Contribution à l'étude des septico-pyohémies à *Bacillus funduliformis*. *Gaz. Med. Fr.* **58**, 1203 (1951).

1178a. E. Rubinstein, A. B. Onderdonk, and J. J. Rahal, Jr., Peritonsillar infection and bac-teremia caused by *Fusobacterium gonidiaformans*. *J. Pediat.* **85**, 673 (1974).

1179. R. L. McLaurin, Subdural infection. *In* "Cranial and Intracranial Suppuration" (E. S. Gurdjian, ed.), Chapter IV, p. 73. Thomas, Springfield, Illinois, 1969.

1180. H. T. Ballantine and C. N. Shealy, The role of radical surgery in the treatment of abscess of the brain. *Surg., Gynecol. Obstet.* **109**, 370 (1959).

1181. J. Baltus and J. Noterman, Bacteriologic study of 25 cases of brain abscess. *Acta Neurol. Belg.* **68**, 447 (1968).

1182. W. A. Buchheit, M. L. Ronis, and E. Liebman, Brain abscesses complicating head and neck infections. *Trans. Am. Acad. Ophthalmol. Otolaryngol.* **74**, 548 (1970).

1183. C. V. Fog, Brain abscess. *Dan. Med. Bull.* **5**, 260 (1958).

1184. D. H. Gregory, R. Messner, and H. H. Zinneman, Metastatic brain abscesses. A retrospective appraisal of 29 patients. *Arch. Intern. Med.* **119**, 25 (1967).

1185. E. R. Heinz and R. D. Cooper, Several early angiographic findings in brain abscess including "the ripple sign." *Radiology* **90**, 735 (1968).

1186. S. Langie, J. Hardy, and C. Bertrand, Les abcès cérébraux étude anatomo-clinique et statistique. *Union Med. Can.* **98**, 1089 (1969).

1187. E. Loeser and L. Scheinberg, Brain abscesses. A review of ninety-nine cases. *Neurology* **7**, 601 (1957).

1188. A. M. McFarlan, The bacteriology of brain abscess. *Br. Med. J.* **2**, 643 (1943).

1189. D. A. McGreal, Cerebral abscesses in children. *Can. Med. Assoc. J.* **86**, 261 (1962).

1190. W. Schiefer and S. Kunze, Stellt der Hirnabszess auch heute noch ein diagnostisches und therapeutisches Problem dar? *Beitr. Neurochir.* **15**, 272 (1968).

1191. M. P. Sperl, Jr., C. S. MacCarty, and W. E. Wellman, Observations on current therapy of abscess of the brain. *AMA Arch. Neurol. Psychiatry* **81**, 439 (1959).

1192. J. Tamalet, J. Bonnal, P. Descuns, and J. Duplay, Étude bactériologique de 372 cas d'abcès du cerveau à l'époque des antibiotiques. *Pathol. Biol.* **8**, 2053 (1960).

1193. G. Marwedel, Einige Betrachtungen über die Wundinfektionen des jetzigen Krieges. *Muench. Med. Wochenschr.* **63**, 982 (1916).

1194. S. Langie and J. Rougerie, Considérations sur le diagnostic, le pronostic et la thérapeutique des abcès cérébraux. *Union Med. Can.* **98**, 2072 (1969).

1195. M. Gernez, La centre de pénicillinothérapie du nord. *Presse Med.* **54**, 58 (1946).

1196. A. W. Miglets, Jr. and J. W. Harrington, Complications of chronic mastoiditis. *Ohio State Med. J.* **65**, 1219 (1969).

1197. A. Sevin and H. Beerens, Contribution à l'étude des Spherophoraceae (Prévot 1938). 1. Conditions et caractères de culture. *Ann. Inst. Pasteur, Lille* **2**, 108 (1949).

1198. W. C. Stevens, Taxonomic studies on the genus *Bacteroides* and similar forms. Ph.D. Thesis, Vanderbilt University, Nashville, Tennessee (1956).

1199. A. R. Prévot, Les anaérobies pathogènes en pédiatrie. *Osaka City Med. J.* **9**, 153 (1963).

1200. A. R. Prévot, L'anaérobie cet inconnu. *Union Med. Can.* **77**, 258 (1948).

1200a. G. Martin: Non-otogenic cerebral abscess. *J. Neurol., Neurosurg. Psychol.* **36**, 607 (1973).

1200b. A. J. Beller, A. Sahar, and I. Praiss. Brain abscess. Review of 89 cases over a period of 30 years. *J. Neurol., Neurosurg. Psychol.* **36**, 757 (1973).

1200c. C. A. Fischbein, A. Rosenthal, E. G. Fischer, A. S. Nadas, and K. Welch: Risk factors for brain abscess in patients with congenital heart disease. *Am. J. Cardiol.* **34**, 97 (1974).

1200d. N. S. Brewer, C. S. MacCarty, and W. E. Wellman, Brain abscess: A review of recent experience. *Ann. Intern. Med.* **82**, 571 (1975).

1200e. F. Meisel-Mikolajczyk and A. Dworczynski, Chemical composition of endotoxins of *Eggerthella convexa* (*Bacteroides fragilis*) strains. *Bull. Acad. Polonaise Sci.* **21**, 193 (1973).

1200f. E. Liske and N. J. Weikers, Changing aspects of brain abscesses. *Neurology* **14**, 294 (1965).

1200g. M. Tarpay, T. Rubio, and H. D. Riley, Jr., Intracranial suppurative disease in children. *Clin. Res.* **22**, 89A (1974).

1200h. N. L. Chernik, D. Armstrong, and J. B. Posner, Central nervous system infections in patients with cancer. *Medicine (Baltimore)* **52**, 563 (1973).

1200i. J. P. Houtteville and R. van Effenterre, Les abcès de la fosse postérieure. *Nouv. Presse Med.* **2**, 1061 (1973).

1200j. J. A. Koepke, Meningitis due to *Streptococcus anginosus* (Lancefield Group F). *J. Am. Med. Assoc.* **193**, 739 (1965).

1201. A. E. Walker and H. C. Johnson, "Penicillin in Neurology." Thomas, Springfield, Illinois, 1946.

1202. W. D. Johnson and D. Kaye, Serious infections caused by diphtheroids. *Ann. N.Y. Acad. Sci.* **174**, 568 (1970).

1203. H. Beerens, Contribution à l'étude de quelques bactéries anaérobies non sporulées. *Ristella insolita, Ristella pseudoinsolita* (nov. sp.), *Zuberella praeacuta, Vibrio crassus, Spherophorus varius, Corynebacterium anaerobium. Ann. Inst. Pasteur, Lille* **2**, 1 (1949).

1204. J. Bφe, Experimental infection of a human with *Fusobacterium* from a brain abscess. *Acta Pathol. Microbiol. Scand.* **19**, 591 (1942).

1205. H. Brocard and S. Daum, Les abcès cérébraux d'origine pulmonaire à *Bacilles fusiformes. Paris Med.* **31**, 373 (1941).

1206. A. Schoolman, C. Liu, and C. Rodecker, Brain abscess caused by *Bacteroides* infection. *Arch. Intern. Med.* **118**, 150 (1966).

1207. S. N. Chou, J. L. Story, L. A. French, and H. O. Peterson, Some angiographic features of brain abscess. *J. Neurosurg.* **24**, 693 (1966).

1208. G. Ehni and E. L. Crain, "Paradoxical" brain abscess. Report of a unique case in association with Lutembacher's syndrome. *J. Am. Med. Assoc.* **150**, 1298 (1952).

1209. C. Coronini and A. Priesel, Zur Kenntnis der *Bacillus-fusiformis*-Pyamien, zugleich ein Beitrag zur "Pseudoaktinomykose." *Frankf. Z. Pathol.* **23**, 191 (1920).

1210. G. F. Dick and L. A. Emge, Brain abscess caused by fusiform bacilli. *J. Am. Med. Assoc.* **62**, 446 (1914).

1211. A. Ghon and V. Mucha, Beitrage zur Kenntnis der anaeroben Bakterien des Menschen. VIII. Zur Aetiologie der pyamischen Prozesse. *Zentralbl. Bakteriol., Parasitenkd., Infektionskr. Hyg., Abt. 1: Orig.* **49**, 493 (1909).

1212. P. Castaigne and M. Goury-Laffont, Pleuresie purulente à "*Bacillus ramosus*" compliquée après un intervalle libre de huit mois d'abcès du cerveau. *Bull. Mem. Soc. Med. Hôp. Paris* [3] **26**, 365 (1946).

1213. H. Werner, F. Neuhaus, and C. Reichertz, Otogener Hirnabszess durch die darmbewohnende Bacteroides-Art *B. fragilis. Dtsch. Med. Wochenschr.* **95**, 343 (1970).

1214. M. Heyde, Zur Kenntnis der Gasgangran und über einen Fall von Hirnabscess, ausschliesslich bedingt durch anaerobe Bakterien. *Beitr. Klin. Chir.* **61**, 50 (1909).

1215. E. Hibler, Zur Kenntnis der pathogenen Anaeroben. ein Kleinhirnabszess, bedingt durch einen anaeroben Spaltpilz, bei chronischer eiterig-jauchiger Otitis, Sinusthrombose und Careinomentwickelung im rechten Felsenbein. *Zentralbl. Bakteriol., Parasitenkd. Infektionskr. 1: Abt. Orig.* **68**, 257 (1913).

1216. J. A. Washington, II, W. J. Martin, and R. E. Spiekerman, Brain abscess with *Corynebacterium hemolyticum*: Report of a case. *Am. J. Clin. Pathol.* **56**, 212 (1971).

1217. H. B. Marsden and W. A. Hyde, Isolation of *Bacteroides corrodens* from infections in children. *J. Clin. Pathol.* **24**, 117 (1971).

1218. W. T. Howard, Jr., The origin of gas and gas cysts of the central nervous system. I.

The occurrence of gas in the cerebral vessels without gas cysts of the brain. *J. Med. Res.* **6**, 105 (1901).

1219. U. Kurimoto, S. Suzuki, and K. Ueno, Significance of isolation and identification of strict anaerobes in the various febrile diseases. *Nihonijishimpo* No. 1939, p. 56 (1961).

1220. A. B. King, S. D. Conklin, and T. S. Collette, *Bacteroides* infections: Report of two cases unsuccessfully treated with antibiotics. *Ann. Intern. Med.* **37**, 761 (1952).

1221. R. Klinger, Ueber einen neuen pathogenen Anaeroben aus menschlichem Eiter (*Coccobacterium mucosum anaerobicum* n. sp.). *Zentralbl. Bakteriol., Parasitenkd., Infektionskr. Hyg., Abt. 1: Orig.* **62**, 186 (1912).

1222. S. Lussana, Batteriemia da *Bacillo fusiforme* con ascesso cerebrale da fissazione. *Pathologica* **14**, 73 (1922).

1223. S. Costa, Le bacille fusiforme et le spirille de Vincent, en association avec d'autres germes, dans un cas de necropyohémie. *C. R. Seances Soc. Biol. Ses Fil.* **67**, 317 (1909).

1224. R. Maresch, Zur Kenntnis der durch fusiforme bacillen bedingten pyamischen Prozesse. *Zentralbl. Bakteriol., Parasitenkd., Infektionskr., Abt. 1: Orig.* **77**, 130 (1915).

1225. N. P. Markham and C. Kershaw, *Bacteroides* infection. A report of two cases. *N. Z. Med. J.* **55**, 293 (1956).

1226. H. Newhart, Report of unusual bacterial findings in a fatal case of chronic otitis media with complications. *Laryngoscope* **49**, 1024 (1939).

1227. B. S. Salibi, *Bacteroides* infection of the brain. *Arch. Neurol (Chicago)* **10**, 629 (1964).

1228. G. D. Owen and M. S. Spink, Necrobacillosis. *Lancet* **1**, 402 (1948).

1229. J. Pecker and J. C. Clement, Abcès aigu du cerveau révélateur d'une bronchectasie, du à l'association de *Fusiformis fusiformis* et *Corynebacterium parvum.* Guérison par traitement combine chirurgical et anti-infectieux. *Bull. Mem. Soc. Med. Hôp. Paris* [3] **68**, 153 (1952).

1230. A. R. Prévot and M. Raynaud, Étude d'une variété de *Dialister pneumosintes* isolée d'une septicémie mortelle avec abcès du poumon et du cerveau. *Ann. Inst. Pasteur, Paris* **73**, 67 (1947).

1231. P. Sédallian, P. Wertheimer, R. Mansuy, P. Monnet, and J. Moinecourt, Septicopyohémie à *Bacille 'funduliformis'* avec abcès du cerveau stérilisé par la penicilline. *Bull. Mem. Soc. Med. Hôp. Paris* [3] **64**, 351 (1948).

1232. W. Silberschmidt, Ueber den Befund von spiessformigen Bacillen (*Bac. fusiforme* Vincent) und von Spirillen in einem Oberschenkelabscess beim Menschen. *Zentralbl. Bakteriol., Parasitenkd. Infektionskr., Abt. 1* **30**, 159 (1901).

1233. W. E. Smith, R. E. McCall, and T. J. Blake, *Bacteroides* infections of the central nervous system. *Ann. Intern. Med.* **20**, 920 (1944).

1234. E. Steen and T. Thjotta, Studies on *Bacteroides.* V. Cerebral abscess caused by *Bacteroides serpens. Acta Pathol. Microbiol. Scand.* **27**, 851 (1950).

1235. R. Vinzent and J. Linhard, Recherches sur *Ristella glutinosa* (Guillemot et Hallé). P. 1938. *Ann. Inst. Pasteur, Paris* **76**, 545 (1949).

1236. A. Wallgren, Über anaerobe Bakterien und ihr Vorkommen bei fotiden Eiterungen. *Zentralbl. Gynaekol.* **26**, 1095 (1902).

1237. H. Werner, Otogener Hirnabszess und Meningitis durch *Eggerthella convexa. Zentralbl. Bakteriol., Parasitenkd., Infektionskr, Hyg., Abt. 1: Orig.* **194**, 203 (1964).

1238. M. Abbott and W. E. Stern, Intracerebral hemorrhage associated with brain abscess. *J. Am. Med. Assoc.* **207**, 1111 (1969).

1239. P. R. R. Clarke, Gas gangrene abscess of the brain. *J. Neurol., Neurosurg. Psychiatry.* **31**, 391 (1968).

1240. F. H. DeLand and H. N. Wagner, "Atlas of Nuclear Medicine," Vol. 1, p. 45. Saunders, Philadelphia, Pennsylvania, 1969.

1240a. Y. Schaffner, Les bacilles anaérobies non sporules à Gram-negatif favorises par la bile. Thesis, Faculty of Medicine, University of Lille, 1963.

1240b. J. M. Kaplan, G. H. McCracken, Jr., and J. D. Nelson, Infections in children caused by the HB group of bacteria. *J. Pediat.* **82**, 398 (1973).

1240c. H. R. Ingham, J. B. Selkon, S. C. So, and R. Weiser, Brain abscess. *Brit. Med. J.* **4**, 39 (1975).

1240d. J. B. Tingelstad, H. F. Young, and R. B. David, Brain abscess in an infant with cyanotic congenital heart disease. *Pediatrics* **54**, 113 (1974).

1240e. D. Fritsche and G. Pulverer, Zur pathogenetischen Bedeutung der anaeroben Genera *Bacteroides* und *Sphaerophorus*. *Dtsch. Med. Wochenschr.* **98**, 1 (1973).

1240f. N. Khuri-Bulos, K. McIntosh, and J. Ehret, *Bacteroides* brain abscess treated with clindamycin. *Am. J. Dis. Child.* **126**, 96 (1973).

1240g. J. A. Russell and J. C. Taylor, Circumscribed gas-gangrene abscess of the brain. *Br. J. Surg.* **50**, 434 (1963).

1240h. R. Lafon, R. Labauge, J. Cadilhac, B. Vlaovitch, and A. Suire, Abces cerebraux à *Corynebacterium liquefaciens*, complication d'une pentalogie de Fallot, guéris par interventions neurochirurgicales successives. *Rev. Neurol. Paris* **94**, 386 (1956)

1240i. W. T. Howard, Jr., Acute fibrino-purulent cerebro-spinal meningitis, ependymitis, abscesses of the cerebrum, gas-cysts of the cerebrum, cerebro-spinal exudation, and of the liver, due to the *Bacillus aerogenes capsulatus*. *Johns Hopkins Hosp. Bull.* **10**, 66 (1899).

1240j. E. S. Gurdjian and J. E. Webster, Experiences in the surgical management of intracranial suppuration *Surg., Gynecol. Obstet.* **104**, 205 (1957).

1240k. A. I. Gilbert, R. S. Tolmach, and J. J. Farrell: Gas gangrene of the brain. *Am. J. Surg.* **101**, 366 (1961).

1240l. E. Rychlik, Gasabszess des Gehirns. *Muench. Med. Wochenschr.* **48**, 1713 (1916).

1240m. P. Teng: Actinomycotic cerebral abscess. A report of two cases with recovery. *J. Am. Med. Assoc.* **175**, 807 (1961).

1240n. E. F. Crocker, A. F. McLaughlin, J. G. Morris, R. Benn, J. G. McLeod, and J. L. Allsop, Technetium brain scanning in the diagnosis and management of cerebral abscess. *Am. J. Med.* **56**, 192 (1974).

1240o. P. I. Lerner: Meningitis caused by *Streptococcus* in adults. *J. Infect. Dis. Suppl.* **131**, S9 (1975).

1241. U. Gupta and V. N. Bhatia, Isolation of anaerobic bacteria from clinical material. *Indian J. Pathol. Bacteriol.* **13**, 55 (1970).

1242. W. L. White, F. D. Murphy, J. S. Lockwood, and H. F. Flippin, Penicillin in the treatment of pneumococcal, meningococcal, streptococcal and staphylococcal meningitis. *Am. J. Med. Sci.* **210**, 1 (1945).

1243. H. Braunstein, E. Tucker, and B. C. Gibson, Infections caused by unusual beta hemolytic streptococci. *Am. J. Clin. Pathol.* **55**, 424 (1971).

1244. E. G. Jackson, 28. A pleomorphic anaerobe: The etiological organism in a fatal meningitis. *J. Bacteriol.* **15**, 18 (1928).

1244a. E. D. Everett, A spectrum of anaerobic infections. *Rocky M. Med. J.* **70**, 50 (1973).

1244b. L. Eyckmans, Considerations au sujet des cas de meningite purulente observes chez l'adulte de 1966 à 1971. *Lyon Med.* **228**, 527 (1972).

1244c. D. W. Fraser, C. E. Henke, and R. A. Feldman, Changing patterns of bacterial meningitis in Olmsted County, Minnesota, 1935–1970. *J. Infect. Dis.* **128**, 300 (1973).

1244d. J.-C. Burdin, J. Schmitt, M. Weber, and B. Arnoux, Étude bactériologique et épidémiologique de 427 cas de meningites suppurées identifiées au C.H.R. de Nancy au cours des onze dernières années. *Sem. Hôp.* **49**, 1365 (1973).

1244e   C. D. Graber, L. S. Higgins, and J. S. Davis, Seldom-encountered agents of bacterial meningitis. *J. Am. Med. Assoc.* **192**, 956 (1965).

1245.   J. J. Ballenger, L. A. Schall, and W. E. Smith, Bacteroides meningitis—report of a case with recovery. *Ann. Otol., Rhinol., & Laryngol.* **52**, 895 (1943).

1246.   H. Beerens, J. J. Piquet, L. Leroy, and P. Ghestem, Un cas de meningite aigue à "*Ristelle variabilis.*" *Lille Med.* [3] **5**, 406 (1960).

1247.   F. A. Chandler and V. M. Breaks, Osteomyelitis of the femoral neck and head caused by *Bacterium necrophorum* (*Bacillus funduliformis*) *J. Am. Med. Assoc.* **116**, 2390 (1941).

1248.   M. Fontan, H. Beerens, and P. Warot, Un cas de meningite aigue à *Ristella convexa. Rev. Neurol.* **101**, 672 (1959).

1249.   F. Freund, Ueber eine durch ein anaerobes Bakterium hervorgerufene Meningitis. *Zentralbl. Bakteriol., Parasitenkd., Infektionskr. Hyg., Abt. 1: Orig.* **88**, 9 (1922).

1250.   S. Suzuki and K. Ueno, Isolation of *Fusiformis fusiformis* and *Eubacterium tortuosum* in a case of purulent meningitis and liver abscess. *Acta Sch. Med. Gifu* **8**, 1 (1960).

1251.   Grimaud, Wayoff, and Losson, Meningite otitique à *Ristella convexa.* Guérison par l'association antibiotiques—sulfamides—chirurgie. *Rev. Med. Nancy* **82**, 174 (1957).

1252.   A. Ghon, V. Mucha, and R. Muller, Beitrage zur Kenntnis der anaeroben Bakterien des Menschen. IV. Zur Aetiologie der akuten Meningitis. *Zentralbl. Bakteriol., Parasitenkd., Infektionskr. Hyg., Abt. 1: Orig.* **41**, 1 (1906).

1253.   H. Heubach, *Bacterium pneumosintes* als Erreger einer posttraumatischen Meningitis. *Klin. Wochenschr.* **17**, 271 (1938).

1254.   J. A. Intile and J. H. Richert, Cervicofacial actinomycosis complicated by meningitis. *In* "Actinomycosis" (M. Bronner and M. Bronner, eds.), p. 219, Williams & Wilkins, Baltimore, Maryland, 1969.

1255.   M. Mizushima, K. Ninomiya, A. Ojima, K. Ueno, and S. Yamada, A case history. Neonatal anaerobic meningitis confirmed by autopsy. *Jap. J. Clin. Med.* **29**, 127 (1971).

1256.   J. B. Sikorski, J. Gilroy, and J. S. Meyer, *Clostridium perfringens* (gas bacillus) septicemia and acute purulent meningitis. *Harper Hosp., Bull.* **21**, 38 (1963).

1257.   K. Ueno and S. Suzuki, Infections due to anaerobes. *Media Circle* **97**, 481 (1967).

1258.   M. Lamy and P. de Font-Réaulx, Septico-pyohémie dué au "*Bacillus funduliformis*" avec thrombo-phlebite des sinus caverneux. *Bull. Mem. Soc. Med. Hôp. Paris* [3] **52**, 979 (1936).

1259.   F. Lifschitz, C. Liu, and A. N. Thurn, *Bacteroides* meningitis. *Am. J. Dis. Child* **105**, 487 (1963).

1260.   R. Pierret, Gernez, Sevin, and J. Pierret, Meningite à "*Bacillus funduliformis*" resistante aux sulfamides. Guérison par association penicilline-sulfamides. *Presse Med.* **54**, 58 (1946).

1261.   M. I. Ganchrow and D. K. Brief, Meningitis complicating perforating wounds of the colon and rectum in combat casualties. *Dis. Colon Rectum* **13**, 297 (1970)

1262.   C. Edwards, W. A. Elliott, and K. J. Randall, Spinal meningitis due to *Actinomyces bovis* treated with penicillin and streptomycin. *J. Neurol., Neurosurg. Psychiatry* **14**, 134 (1951).

1263.   Reye, Zur Klinik und Atiologie der postanginosen septischen Erkrankungen. *Virchows Arch. Pathol. Anat. Physiol.* **246**, 22 (1923).

1264.   N. Sezen and K. Ozsan, Akciger lokalizssyonundan sonra menenjit komp-likasyonu da gosteren bir *Spherophorus pseudonecrophorus* sepsisi vak'asi. *Turk Ij. Tecr. Biyol. Derg.* **21**, 100 (1961).

1265.   P. Tardieux, A. R. Prévot, and R. Rozansky, Contribution à l'étude des Sphérophoroses: Pouvoir pathogenè de *Spherophorus gulosus. Ann. Inst. Pasteur, Paris* **88**, 124 (1955).

1266.   G. Willich and W. Muller, Die Bogdanbuday'sche Krankheit. *Med. Welt* **20**, 1409 (1951).

1267. W. S. Wood, G. P. Kipnis, H. W. Spies, H. F. Dowling, M. H. Lepper, and G. G. Jackson, Tetracycline therapy. *AMA Arch. Intern. Med.* **94**, 351 (1954).

1267a. B. Muller, P. Raoul-Duval, J. Bayle, R. Bouvet, and G. Naudin, Un cas de meningite à "*Actinomyces israeli*" guérison par traitement antibiotique. *Bull. Mem. Soc. Med. Hôp. Paris.* **70**, 277 (1954).

1267b. K. Watanabe, T. Miwa, H. Imamura, S. Kobata, I. Mochizuki, K. Ninomiya, K. Ueno, and S. Suzuki, A case of Hodgkin's disease associated with cryptococcal meningitis and bacterial meningitis due to *Clostridium sordellii. J. Infect. Dis.* **49**, 404 (1975).

1267c. R. J. Mangi, R. Quintiliani, and V. T. Andriole, Gram-negative bacillary meningitis. *Am. J. Med.* **59**, 829 (1975).

1267d. I. Brook, N. Johnson, G. Overturf, and J. Wilkins, Mixed bacterial meningitis as a complication of neurological shunts. *Clin. Res.* **24**, 112A (1976).

1267e. M. Adler and A. von Graevenitz, Mixed clostridial meningitis with septicemia. *In* "Pathogenic Microorganisms from Atypical Clinical Sources," (A. von Graevenitz and T. Sall, eds.). Dekker, New York, 1975.

1267f. D. B. Morrison, A. A. Humphrey, and J. E. Bailey, Actinomycotic meningitis with a primary focus in the finger. Report of a case diagnosed during life. *J. Am. Med. Assoc.* **110**, 1552 (1938).

1267g. R. Sartory and P. Verdure, Un cas d'actinomycose à foyers multiples (osteomyélite du tibia, noma jugual, osteomyélite à foyers multiples) et à terminaison meningée et septicemique. *Bull. Acad. Natl. Med. Paris* **128**, 488 (1944).

1268. J. D. Coonrod and P. E. Dans, Subdural empyema. *Am. J. Med.* **53**, 85 (1972).

1268a. D. I. Anagnostopoulos and P. Gortavi, Intracranial subdural abscess. *Br. J. Surg.* **60**, 50 (1973).

1268b. A. G. Beeden, C. D. Marsden, J. C. Meadows, and W. F. Michael, Intracranial complications of middle ear disease and mastoid surgery. *J. Neurol. Sci.* **9**, 261 (1969).

1268c. Y. S. Bhandari and N. B. S. Sarkari, Subdural empyema. A review of 37 cases. *J. Neurosurg.* **32**, 35 (1970).

1268d. S. A. Hollin, H. Hayashi, and S. W. Gross, Intracranial abscesses of odontogenic origin. *Oral Surg., Oral Med. Oral Pathol.* 23, 277 (1967).

1268e. W. S. Keith, Subdural empyema. *J. Neurosurg.* **6**, 127 (1949).

1268f. B. S. Ray and H. Parsons, Subdural abscess complicating frontal sinusitis. *Arch. Otolaryngol.* 37, 536 (1943).

1269. J. P. Murphy and J. D. Wilkes, Subdural abscess diagnosed by brain scanning. *South. Med. J.* **61**, 564 (1968).

1270. R. H. Wilkins and J. A. Goree, Interhemispheric subdural empyema. Angiographic appearance. *J. Neurosurg.* **32**, 459 (1970).

1271. T. Yamada, K. Inatomi, K. Watanabe, H. Ikemoto, A. Ishii, H. Inetani, T. Shimoji, N. Kozakai, and T. Oguri, A brain abscess due to microaerophilic *Streptococcus. In* "Second Symposium on Anaerobic Bacteria and Their Infectious Diseases" (S. Ishiyama *et al.*, eds.), p. 83. Eisai, Tokyo, Japan, 1972.

1271a. A. S. Baker, R. G. Ojemann, M. N. Swartz, and E. P. Richardson, Jr., Spinal epidural abscess. *N. Engl. J. Med.* **293**, 463 (1975).

1271b. N. Krumdieck and L. Stevenson, Spinal epidural abscess associated with actinomycosis. *Arch. Pathol.* **30**, 1223 (1940).

1272. S. M. Kompanejetz, Funf Falle von Siebbeinlabyrintheiterungen mit Durchbruch gegen die Orbita. *Monatsschr. Ohrenheilkd. Laryngo-Rhinol.* **57**, 261 (1923).

1273. Y. Lee and R. B. Berg, Cephalhematoma infected with *Bacteroides. Am. J. Dis. Child.* **121**, 77 (1971).

1273a. D. G. Schwartz and N. Christoff, Actinomycosis with cerebral and probable endocardial involvement. *J. Mt. Sinai Hosp. New York* **27**, 23 (1960).

1274. L. D. Stacy, Jr., J. A. Mertz, and M. Pearlman, *Catenabacterium filamentosum* isolated from a cystic tumor of the pons. *Am. J. Clin. Pathol.* **51**, 390 (1969).

1275. B. S. Tynes and J. P. Utz, *Fusobacterium* septicemia. *Am. J. Med.* **29**, 879 (1960).

1276. E. C. Rosenow, Jr., and A. E. Brown, Septicemia: A review of cases, 1934–1936 inclusive. *Proc. Staff Meet. Mayo Clin.* **13**, 89 (1938).

1277. P. Kotin, Techniques and interpretation of routine blood cultures. Observations in five thousand consecutive patients. *J. Am. Med. Assoc.* **149**, 1273 (1952).

1278. D. S. Munroe and W. H. Cockcroft, Septicaemia due to gram-negative bacilli. *Can. Med. Assoc. J.* **72**, 586 (1955).

1279. F. W. Ames and M. J. Fischer, Treatment of septic shock. A progress report. *Ohio State Med. J.* **62**, 329 (1966).

1280. H. P. Dalton and M. J. Allison, Etiology of bacteremia. *Appl. Microbiol.* **15**, 808 (1967).

1281. J. H. McCutchan and J. S. Pagano, Experience with gram-negative bacteremia. *Clin. Res.* **15**, 41 (1967).

1282. W. Brumfitt and D. A. Leigh, Gram-negative septicaemia. *Proc. Roy. Soc. Med.* **62**, 1239 (1969).

1283. M. Ridley, Bacteriological diagnosis and control of antibiotic treatment in bacterial endocarditis. *In* "Bacterial Endocarditis," Proc. Natl. Symp., (P. B. Beeson and M. Ridley, eds.), p. 49. Royal College of Physicians, London, 1969.

1284. N. Crowley, Some bacteraemias encountered in hospital practice. *J. Clin. Pathol.* **23**, 166 (1970).

1285. R. Rosner, Comparison of a blood culture system containing Liquoid and sucrose with systems containing either reagent alone. *Appl. Microbiol.* **19**, 281 (1970).

1286. R. L. Myerowitz, A. A. Medeiros, and T. F. O'Brien, Recent experience with bacillemia due to gram-negative organisms. *J. Infect. Dis.* **124**, 239 (1971).

1287. R. Rosner, A quantitative evaluation of three blood culture systems. *Am. J. Clin. Pathol.* **57**, 220 (1972).

1288. J. A. Washington, II, Evaluation of two commercially available media for detection of bacteremia. *Appl. Microbiol.* **23**, 956 (1972).

1289. D. C. Shanson, *Bacteroides* in the blood. *Lancet* **1**, 147 (1973).

1289a. E. Schoutens, M. Labbé, and E. Yourassowsky, "*Bacteroides fragilis*" septicemias. Incidence and sensitivity of the strains to antibiotics. *Pathol. Biol.* **21**, 349 (1973).

1289b. M. C. McHenry, and W. A. Hawk, Bacteremia caused by gram-negative bacilli. *Med. Clin. N. Amer.* **58**, 623 (1974).

1289c. W. R. Wilson, P. M. Jaumin, G. K. Danielson, E. R. Giuliani, J. A. Washington II, and J. E. Geraci, Prosthetic valve endocarditis. *Ann. Intern. Med.* **82**, 751 (1975).

1290. E. Arjona, J. Rof, and J. Perianes, Algunas comunicaciones sobre los temas del symposium. Endocarditis por anaerobios. *Rev. Clin. Esp.* **22**, 483 (1946).

1291. N. L. Cressy, W. J. Lahey, and P. Kunkel, Streptomycin in the treatment of bacterial endocarditis. *N. Engl. J. Med.* **239**, 497 (1948).

1292. P. A. Tumulty and A. M. Harvey, Experiences in the management of subacute bacterial endocarditis treated with penicillin. *Am. J. Med.* **4**, 37 (1948).

1293. R. Wallach and N. Pomerantz, Streptomycin in the treatment of subacute bacterial endocarditis. *N. Engl. J. Med.* **241**, 690 (1949).

1294. R. M. deHay, J. L. Bell, and J. M. Beebe, *Bacteroides* bacterial endocarditis. *Hawaii Med. J.* **9**, 235 (1950).

1295. L. A. Rantz, *Bacteroides* endocarditis. Report of a case successfully treated with terramycin. *J. Am. Med. Assoc.* **147**, 124 (1951).

1296. P. E. Hermans, J. K. Martin, Jr., G. M. Needham, and D. R. Nichols, Laboratory and clinical evaluation of cephaloridine, a cephalosporin derivative. *Antimicrob. Ag. Chemother.* p. 879 (1966).

1296a. Case records of the Massachusetts General Hospital. Presentation of Case—Case 30–1974. *N. Engl. J. Med.* **291**, 242 (1974).

1297. N. G. Lupu, S. Litarczek, G. T. Dinischiotu, and V. Lazarescu, Septico-pyohémie à *Bacillus funduliformis*. *Bull. Mem. Soc. Med. Hôp. Bucarest* **19**, 47 (1937).

1298. J. M. Alés, E. Arjona, and C. J. Díaz, Observations on the aetiology of subacute bacterial endocarditis. *Bull. Inst. Med. Res., Univ. Madrid* **5**, 147 (1952).

1299. F. H. Caselitz, D. Krebs, and W. Raabe, Serologische Studien bei *Sphaerophorus*-Infektionen. *Zentralbl. Bakteriol., Parasitenkd., Infektionskr. Hyg., Abt. 1: Orig.* **208**, 338 (1968).

1300. F. Magrassi, Studi sulle endocarditi lente non streptococciche: Un caso di endocardite lenta da cocco-bacillo di difficile classificazione affine ai germi appartenenti al genere *Dialister* (Bergey). *Boll. Inst. Sieroter. Milan.* **23**, 185 (1944).

1301. A. M. Fisher and V. A. McKusick, *Bacteroides* infections. Clinical, bacteriological and therapeutic features of fourteen cases. *Am. J. Med. Sci.* **225**, 253 (1953).

1302. M. Finland and M. W. Barnes, Changing etiology of bacterial endocarditis in the antibacterial era. Experiences at Boston City Hospital 1933–1965. *Ann. Intern. Med.* **72**, 341 (1970).

1303. R. H. More, Bacterial endocarditis due to *Clostridium welchii*. *Am. J. Pathol.* **19**, 413 (1943).

1304. M. Labraque-Bordenave, L'hémoculture systématique en anaérobiose. Thesis, Faculty of Medicine, University of Montpellier, Montpellier (1935).

1305. M. J. Robinson, J. J. Greenberg, M. Korn, and A. M. Rywlin, Infective endocarditis at autopsy: 1965–1969. *Am. J. Med.* **52**, 492 (1972).

1306. Y. Watanabe and K. Ueno, Subacute bacterial endocarditis due to *Eubacterium ventriosum*: Report of a case. *Bull. Jpn. Soc. Infect. Dis.* **42**, 78 (1968).

1306a. M. D. Sans and J. G. Crowder, Subacute bacterial endocarditis caused by *Eubacterium aerofaciens*. *Am. J. Clin. Pathol.* **59**, 576 (1973).

1307. C. H. Zierdt and P. T. Wertlake, Transitional forms of *Corynebacterium acnes* in disease. *J. Bacteriol.* **97**, 799 (1969).

1308. F. M. Griffin, G. Jones, and C. G. Cobbs, Aortic insufficiency in bacterial endocarditis. *Ann. Intern. Med.* **76**, 23 (1972).

1309. R. G. Wittler, W. F. Malizia, P. E. Kramer, J. D. Tuckett, H. N. Pritchard, and H. J. Baker, Isolation of a *Corynebacterium* and its transitional forms from a case of subacute bacterial endocarditis treated with antibiotics. *J. Gen. Microbiol.* **23**, 315 (1960).

1310. A. Davis, M. J. Binder, J. T. Burroughs, A. B. Miller, and S. M. Finegold, Diphtheroid endocarditis after cardiopulmonary bypass surgery for the repair of cardiac valvular defects. *Antimicrob. Ag. Chemother.* p. 643 (1964).

1311. E. Biocca and D. Reitano, Endocardite mortal no homem produzida por um lactobacilo. *Arq. Biol.* **27**, 114 (1943).

1312. E. Biocca, Novo caso de endocardite mortal por *Lactobacilo*. *Arq. Biol.* **28**, 18 (1944).

1313. F. Marschall, Der Doderleinsche Bacillus vaginalis als Endokarditiserreger. *Zentralbl. Bakteriol., Parasitenkd., Infektionskr. Abt. 1: Orig.* **141**, 153 (1938).

1313a. M. J. Tenenbaum and J. F. Warner, *Lactobacillus casei* endocarditis. *Ann. Intern. Med.* **82**, 539 (1975).

1314. J. Mathieu and M. Lefebvre, L'endocardite bactérienne à streptocoque anaérobic. *Union Med. Can.* **93**, 652 (1964).

1315. L. Loewe, P. Rosenblatt, and E. Alture-Werber, A refractory case of subacute bacterial endocarditis due to *Veillonella gazogenes* clinically arrested by a combination of penicillin, sodium paraaminohippurate, and heparin. *Am. Heart J.* **32**, 327 (1946).

1316. W. W. Spink, K. Osterberg, and J. Finstad, Human endocarditis due to a strain of

CO_2-dependent penicillin-resistant *Staphylococcus* producing dwarf colonies. *J. Lab. Clin. Med.* **59**, 613 (1962).

1317. A. Hollander and E. Landsberg, Acute endocarditis due to an anaerobic pneumococcus. *J. Lab. Clin. Med.* **26**, 307 (1940).

1318. A. Lemierre, J. Reilly, and P. de Font-Reaulx, "Streptococcus viridans" atypique dans une endocardite lente. *Soc. Med. Hôp. Paris* **51**, 128 (1935).

1319. M. Lefebvre and G. Côté, Rôle possible du *Streptococcus evolutus* dans un cas d'endocardite. *Union Med. Can.* **87**, 694 (1958).

1320. J. C. Harvey, G. S. Mirick, and I. G. Schaub, Clinical experience with aureomycin. *J. Clin. Invest.* **28**, 987 (1949).

1321. W. R. Vogler, E. R. Dorney, and H. A. Bridges, Bacterial endocarditis. A review of 148 cases. *Am. J. Med.* **32**, 910 (1962).

1322. J. S. Carey and R. K. Hughes, Cardiac valve replacement for the narcotic addict. *J. Thorac. Cardiovasc. Surg.* **53**, 663 (1967).

1323. T. Laxdal, R. P. Messner, R. C. Williams, Jr., and P. G. Quie, Opsonic, agglutinating and complement-fixing antibodies in patients with subacute bacterial endocarditis. *J. Lab. Clin. Med.* **71**, 638 (1968).

1324. J. J. Rahal, Jr., B. R. Meyers, and L. Weinstein, Treatment of bacterial endocarditis with cephalothin. *N. Engl. J. Med.* **279**, 1305 (1968).

1325. G. W. Hayward, R. A. Shooter, E. A. Shinebourne, and C. M. Cripps, Bacterial endocarditis 1956–1965: Analysis of clinical features and treatment in relation to prognosis and mortality. *In* "Bacterial Endocarditis," Proc. Natl. Symp., (P. B. Beeson and M. Ridley, eds.), p. 18. Royal College of Physicians, London, 1969.

1326. E. Jawetz, Antibiotics—blessings and curses. *Calif. Med.* **112**, 35 (1970).

1327. M. E. Levison, D. Kaye, G. L. Mandell, and E. W. Hook, Characteristics of patients with multiple episodes of bacterial endocarditis. *J. Am. Med. Assoc.* **211**, 1355 (1970).

1328. C. Lamanna, A non-life cycle explanation of the diphtheroid streptococcus from endocarditis. *J. Bacteriol.* **47**, 327 (1944).

1329. H. F. Wood, K. Jacobs, and M. McCarty, *Streptococcus lactis* isolated from a patient with subacute bacterial endocarditis. *Am. J. Med.* **18**, 345 (1955).

1330. L. Colebrook, Infection by anaerobic streptococci in puerperal fever. *Br. Med. J.* **2**, 134 (1930).

1331. A. M. Fisher and T. J. Abernethy, Putrid empyema with special reference to anaerobic streptococci. *Arch. Intern. Med.* **54**, 552 (1934).

1332. J. E. Geraci, Further experiences with short-term (2 weeks) combined penicillin-streptomycin therapy for bacterial endocarditis caused by penicillin-sensitive streptococci. *Proc. Staff Meet. Mayo Clin.* **30**, 192 (1955).

1333. W. B. Stason, R. W. DeSanctis, A. N. Weinberg, and W. G. Austen, Cardiac surgery in bacterial endocarditis. *Circulation* **38**, 514 (1968).

1334. A. Leonard, C. M. Comty, F. L. Shapiro, and L. Raiji, Osteomyelitis in hemodialysis patients. *Ann. Intern. Med.* **78**, 651 (1973).

1335. M. Nakamura, M. Akashi, and K. Kodama, Three cases of anaerobic bacterial infectious diseases. *In* "Second Symposium on Anaerobic Bacteria and Their Infectious Diseases" (S. Ishiyama *et al.*, eds.), p. 67. Eisai, Tokyo, Japan, 1972.

1336. H. Ikemoto, K. Watanabe, S. Nakazawa, T. Hagitani, H. Suzuki, O. Hisauchi, T. Fujii, T. Miyazaki, K. Inatomi, K. Mori, K. Sagawa, G. Nakata, N. Kozakai, and T. Oguri, Infectious diseases due to non-spore-forming anaerobes. *In* "First Symposium on Anaerobic Bacteria and Their Infectious Diseases" (S. Ishiyama *et al.*, eds.), p. 22. Eisai, Tokyo, Japan, 1971.

1337. P. de la Barreda and D. Centenera, Endocarditis lenta por un germen anaerobio. *Rev. Clin. Esp.* **32**, 270 (1949).

1337a. S. Milovanov, T. Bunovic, and M. Malenkovic, Subacute bacterial endocarditis caused by anaerobic *Peptostreptococcus* with brief review of general clinical picture. *Medicinski Pregled.* **22**, 267 (1969).

1338. C. E. Cherubin and H. C. Neu, Infective endocarditis at the Presbyterian Hospital in New York City from 1938–1967. *Am. J. Med.* **51**, 83 (1971).

1339. D. Kaye, R. C. McCormack, and E. W. Hook, Bacterial endocarditis: The changing pattern since the introduction of penicillin therapy. *Antimicrob. Ag. Chemother.* p. 37 (1962).

1340. F. Kaspar and W. Kern, Beitrage zur Kenntnis der anaeroben Bakterien des Menschen. IX. Weitere Beitrage zur Aetiologie der pyamischen Prozesse. *Zentralbl. Bakteriol., Parasitenkd., Infektionskr. Hyg., Abt. 1: Orig.* **55**, 97 (1910).

1341. A. Gilbert and A. Lippmann, Nôte sur la bactériologie des abcès tropicaux du foie. *C. R. Seances Soc. Biol. Ses Fil.* **63**, 565 (1907).

1342. C. G. McKenzie, Pyogenic infection of liver secondary to infection in the portal drainage area. *Br. Med. J.* **2**, 1558 (1964).

1343. V. B. Buhler, C. W. Seely, and D. D. Dixon, *Bacterium necrophorus* septicemia in man. *Am. J. Clin. Pathol.* **12**, 380 (1942).

1344. Case Records of the Massachusetts General Hospital, Case 28–1969. *N. Engl. J. Med.* **281**, 93 (1969).

1345. E. M. Fuss and M. Fuhrman, Multiple gaseous liver abscesses due to anaerobic *Streptococcus viridans* with recovery. *N.Y. State J. Med.* **50**, 1142 (1950).

1346. S. D. Henriksen, The bacterial flora of fetid infections. *Acta Med. Scand.* **129**, 352 (1947).

1347. A. J. C. M. Jacquot, Les septicémies et bactériémies à microbes anaérobies d'origine intestinale. Thesis, Faculty of Medicine, University of Paris, Paris (1942).

1348. F. B. St. John, E. J. Pulaski, and J. M. Ferrer, Primary abscess of the liver due to anaerobic nonhemolytic streptococcus. *Ann. Surg.* **116**, 217 (1942).

1349. M. L. Michel and W. R. Wirth, Multiple pyogenic abscesses of the liver. Cure by penicillin in case due to anaerobic streptococci. *J. Am. Med. Assoc.* **133**, 395 (1947).

1350. G. O. Wellman, Solitary pyogenic abscess of liver. *Ill. Med. J.* **93**, 327 (1948).

1351. T. J. Butler and C. F. McCarthy, Pyogenic liver abscess. *Gut* **10**, 389 (1969).

1352. J. F. Stokes, Cryptogenic liver abscess. *Lancet* **1**, 355 (1960).

1353. D. W. Todd, Pyogenic liver abscess: A case report. *Mil. Med.* **136**, 154 (1971).

1353a. H. A. Pitt and G. D. Zuidema, Factors influencing mortality in the treatment of pyogenic hepatic abscess. *Surg., Gynecol. Obstet.* **140**, 228 (1975).

1353b. S. B. Novy, S. Wallace, A. M. Goldman, and Y. Ben-Menachem, Pyogenic liver abscess, angiographic diagnosis and treatment by closed aspiration. *Am. J. Roentgenol., Rad. Ther. Nucl. Med.* **121**, 388 (1974).

1353c. J. H. C. Ranson, M. A. Madayag, S. A. Localio, and F. C. Spencer, New diagnostic and therapeutic techniques in the management of pyogenic liver abscesses. *Ann Surg.* **181**, 508 (1975).

1353d. R. H. Rubin, M. N. Swartz, and R. Malt, Hepatic abscess: Changes in clinical, bacteriologic and therapeutic aspects. *Am. J. Med.* **57**, 601 (1974).

1353e. J. Lazarchick, N. A. de Souza e Silva, D. R. Nichols, and J. A. Washington II, Pyogenic liver abscess. *Mayo Clin. Proc.* **48**, 349 (1973).

1354. J. D. Sherman and S. L. Robbins, Changing trends in the casuistics of hepatic abscess. *Am. J. Med.* **28**, 943 (1960).

1355. J. E. Flynn, Pyogenic liver abscess. Review of the literature and report of a case successfully treated by operation and penicillin. *N. Engl. J. Med.* **234**, 403 (1946).

1356. S. P. Grayer, Non-perforated appendicitis presenting as shock and liver abscess. *Am. Surg.* **35**, 461 (1969).

1357. S. I. Schwartz, "Surgical Diseases of the Liver," p. 145. McGraw-Hill, New York, 1964.

1358. R. P. May, J. D. Lehmann, and J. P. Sanford, Difficulties in differentiating amebic from pyogenic liver abscess. *Arch. Intern. Med.* **119**, 69 (1967).

1358a. H. Ikemoto, K. Watanabe, T. Miyazaki, T. Mori, A. Izumi, T. Yasa, N. Kosakai, and T. Oguri, Cases of nonsporeforming anaerobic infection.: 2 cases of liver abscess and a case of pyothorax. *In* "Third Symposium on Anaerobic Bacteria and Their Infectious Diseases." (S. Ishiyama *et al.*, eds.), p. 64. Eisai, Tokyo, Japan, 1973.

1358b. C. Futch, *Bacteroides* liver abscess. *Rev. Surg.* **30**, 300 (1973).

1358c. R. O. Friday, P. Barriga, and A. B. Crummy, Detection and localization of intra-abdominal abscesses by diagnostic ultrasound. *Arch. Surg.* **110**, 335 (1975).

1358d. G. B. Hopkins and C. W. Mende, Gallium-67 for the diagnosis and localization of subphrenic abscesses. *West. J. Med.* **122**, 281 (1975).

1358e. N. T. Bateman, S. J. Eykyn, and I. Phillips: Pyogenic liver abscess caused by *Streptococcus milleri*. *Lancet* **1**, 657 (1975).

1358f. P. C. T. Dickinson and P. Saphyakhajon, Treatment of *Bacteroides* infection with clindamycin-2-phosphate. *Can. Med. Assoc. J.* **111**, 945 (1974).

1358g. L. M. de la Maza, N. Faramarz, and L. D. Berman. The changing etiology of liver abscess; further observations. *J. Am. Med. Assoc.* **227**, 161 (1974).

1359. J. C. Henthorne, Gram-negative bacilli of the genus *Bacteroides*. M.Sc. Thesis, University of Minnesota, Rochester, Minnesota (1935).

1360. C. Futch, B. A. Zikria, and H. C. Neu, *Bacteroides* liver abscess. *Surgery* **73**, 59 (1973).

1361. Picard, Perrin, and Franck, Septicémie à "*Bacillus fragilis*" avec hemorragie intestinale et abcès du foie. *Bull. Mem. Soc. Med. Hôp. Paris* **52**, 1239 (1936).

1361a. P. McLaughlin, S. Meban, and W. G. Thompson, Anaerobic liver abscess complicating radiation enteritis. *Can. Med. Assoc. J.* **108**, 353 (1973).

1362. S. Remond, Un cas de septicémie à *Bacillus serpens*. Thesis, Faculty of Medicine, University of Paris, Paris (1941).

1363. M. Lemierre and M. Ternois, Septicémie à "*Bacillus serpens*." *Bull. Acad. Med.*, **123**, 552 (1940).

1364. B. Dandurand, M. Saint-Martin, and L. Beaudoin, Abcès microbiens du foie. A-propos de 16 observations nouvelles. *Union Med. Can.* **98**, 55 (1969).

1365. M. A. Block, B. M. Schuman, W. R. Eyler, J. P. Truant, and L. A. DuSault, Surgery of liver abscesses. *Arch. Surg.* (*Chicago*) **88**, 602 (1964).

1366. J. Berke and C. Pecora, Diagnostic problems of pyogenic hepatic abscess. *Am. J. Surg.* **111**, 678 (1966).

1367. N. M. Harris, *Bacillus mortiferus* (nov. spec.). *J. Exp. Med.* **6**, 519 (1901).

1367a Case Records of the Massachusetts General Hospital. Case 17–1975. *N. Engl. J. Med.* **292**, 963 (1975).

1368. P. Teissier, J. Reilly, E. Rivalier, and F. Layani, Contribution à l'étude des septicémies primitives à microbes anaérobies. *Paris Med.* **71**, 297 (1929).

1369. G. E. Foley, *In vitro* resistance of the genus *Bacteroides* to streptomycin. *Science* **106**, 423 (1947).

1370. A. Lemierre, Les septico-pyohémies à *Bacillus funduliformis* d'origine intestinale. *Bull. Med.* **64**, 405 (1950).

1371. D. C. Beaver, J. C. Henthorne, and J. W. Macy, Abscesses of the liver caused by *Bacteroides funduliformis*. Report of two cases. *Arch. Pathol.* **17**, 493 (1934).

1372. C. F. Goodnough, Liver abscess caused by *Bacteroides funduliformis*. *Am. J. Surg.* **53**, 506 (1941).

1373. H. Werner, "Die Gramnegativen Anaeroben Sporenlosen Stabchen des Menschen." Fischer, Jena, 1968.

1374.   K. Gusnar and H. Globig, Über eine besondere Form der Sepsis. *Dtsch. Z. Chir.* **221**, 263 (1929).

1375.   C. Hegler and H. Nathan, Über Buday-und Friedlander-Bacillen-Sepsis. *Klin. Wochenschr.* **11**, 1900 (1932).

1376.   S. Rusznyak, Sepsisfalle verursacht durch den anaeroben Bacillus von Buday. *Berl. Klin. Wochenschr.* **55**, 234 (1918).

1377.   B. Johan, Beitrage zur Biologie des *Bacillus pyogenes anaerobius*. *Zentralbl. Bakteriol., Parasitenkd., Infektionskr. Hyg., Abt. 1: Orig.* **87**, 290 (1922).

1378.   J. L. Tullis and O. E. Mordvin, Human necrobacillosis. *Am. J. Clin. Pathol.* **16**, 395 (1946).

1379.   K. Buday, Endemisch auftretende Leberabszesse bei Verwundeten, verursacht durch einen anaeroben Bacillus. *Zentralbl. Bakteriol., Parasitenkd. Infektionskr., Abt. I: Orig.* **77**, 453 (1916).

1380.   E. Nedelmann, *Bacillus symbiophiles* (Schottmüller) als alleiniger Erreger einer Pylephlebitis nach Appendizitis. *Dtsch. Arch. Klin. Med.* **160**, 40 (1928).

1381.   R. Knowles and J. A. Rinaldo, Pyogenic liver abscess probably secondary to sigmoid diverticulitis. Report of two cases. *Gastroenterology* **38**, 262 (1960).

1382.   J. M. Holt and C. J. F. Spry, Solitary pyogenic liver abscess in patients with diabetes mellitus. *Lancet* **2**, 198 (1966).

1383.   Staff Conference, St. Thomas Hospital, Gas gangrene cholecystitis. *J. Tenn. Med. Assoc.* **62**, 418 (1969).

1384.   W. M. Rambo and H. C. Black, Intrahepatic abscess. *Am. Surg.* **35**, 144 (1969).

1385.   L. G. Thorley, L. S. Figiel, S. J. Figiel, and D. K. Rush, Roentgenographic findings in accidental ligation of the hepatic artery. *Radiology* **85**, 56 (1965).

1386.   R. M. Kivel, A. Kessler, and D. J. Cameron, Liver abscess due to *Clostridium perfringens*. *Ann. Int. Med.* **49**, 672 (1958).

1387.   G. Quattrocchi, Rarissimo caso di ascesso piogassoso del fegato da *Clostridium sporogenes–Bacterium pyocianeum*. *Riforma Med.* **77**, 288 (1963).

1388.   M. J. Monod, Association bactérienne d'aérobies et d'anaérobies; gangrène du foie. *C. R. Seances Soc. Biol. Ses Fil.* **47**, 354 (1895).

1389.   A. G. MacKay, E. J. Caldwell, and R. K. Hindawi, Massive gas-bacillus infection of the liver after surgical exploration of a stenotic biliary anastomosis. *N. Engl. J. Med.* **268**, 534 (1963).

1390.   D. J. B. Ashley, Two cases of clostridial hepatitis. *J. Clin. Pathol.* **18**, 170 (1965).

1391.   N. C. Plimpton, *Clostridium perfringens* infection. *Arch. Surg.* (*Chicago*) **89**, 499 (1964).

1391a. A. Van Beek, E. Zook, P. Yaw, R. Gardner, R. Smith, and J. L. Glover, Nonclostridial gas-forming infections. *Arch. Surg.* **108**, 552 (1974).

1392.   W. Ostermiller, Jr., and R. Carter, Hepatic abscess. Current concepts in diagnosis and management. *Arch. Surg.* (*Chicago*) **94**, 353 (1967).

1393.   L. P. Dehner and J. M. Kissane, Pyogenic hepatic abscesses in infancy and childhood. *J. Pediatr.* **74**, 763 (1969).

1394.   L. D. Wruble, S. D. Mitchell, and C. F. Tate, Jr., Actinomycotic liver abscess. Report of a case. *Am. J. Dig. Dis.* **7**, 331 (1962).

1395.   C. S. Keefer, Liver abscess: A review of eighty-five cases. *N. Engl. J. Med.* **211**, 21 (1934).

1396.   R. Schwandt and P. Kunze, Zur Differentialdiagnose und Klinik der abdominalen Aktinomykose. *Dtsch. Z. Verdau.- Stoffwechselkr.* **28**, 31 (1968).

1396a. P. G. Prioleau and F. L. Brochu, Penicillin and epluchage treatment of hepatic actinomycosis. *Arch. Surg.* **109**, 426 (1974).

1396b. S. K. C. Chandarlapaty, M. Dusol, Jr., R. Edwards, R. Pereiras, Jr., R. Clark, and E. R. Schiff: [67]Gallium accumulation in hepatic actinomycosis. *Gastroenterology* **69**, 752 (1975).

1397.   C. A. Elsberg, Solitary abscess of the liver. *Ann. Surg.* **44**, 217 (1906).

1398.   M. J. Monod, Association bactérienne d'aérobies et anaérobies; gangrene du foie. *C. R. Seances Soc. Biol. Ses Fil.* **47**, 354 (1895).

1399.   A. Gilbert and A. Lippmann, Recherches bactériologiques sur les cholecystites. *C. R. Seances Soc. Biol. Ses Fil.* **54**, 989 (1902).

1400.   A. Gilbert and A. Lippmann, Bactériologie des cholecystites. *C. R. Seances Soc. Biol. Ses Fil.* **54**, 1189 (1902).

1401.   E. C. Rosenow, The etiology of cholecystitis and gallstones and their production by the intravenous injection of bacteria. *J. Infect. Dis.* **19**, 527 (1916).

1402.   A. Edinburgh and A. Geffen, Acute emphysematous cholecystitis. A case report and review of the world literature. *Am. J. Surg.* **96**, 66 (1958).

1402a. A. Posselt, Beziehungen zwischen Leber, Gallenwegen und Infektionskrankheiten. *Ergeb. Allg. Pathol. Pathol. Anat.* **28**, 115 (1934).

1403.   H. Lodenkämper, Über die Bakteriologie der Gallenwege. *Dtsch. Ges. Verdau. Stoffwechselkr. Verh.* **17**, 250 (1953).

1404.   A. J. H. Rains, G. J. Barson, N. Crawford, and J. F. D. Shrewsbury, A chemical and bacteriological study of gallstones. *Lancet* **2**, 614 (1960).

1404a. L. Paredes and M. Fernández, Estudio bacteriologico y antibiograma actual de las infecciones por bacterias anaerobias. *Rev. Med. Chile Ano.* **95**, 751 (1967).

1405.   B. Czarnecki and H. Kolsut, Mycosis of the biliary tract. *Pol. Tyg. Lek.* **24**, 1645 (1969).

1406.   M. R. B. Keighley and N. G. Graham, Infective complications of choledochotomy with T-tube drainage. *Br. J. Surg.* **58**, 764 (1971).

1406a. M. R. B. Keighley, R. B. Drysdale, A. H. Quoraishi, D. W. Burdon, and J. Alexander-Williams, Antibiotic treatment of biliary sepsis. *Surg. Clin. N. Amer.* **55**, 1379 (1975).

1407.   A. Zuber and M. P. Lereboullet, Cholecystite calculeuse. Perforation. Peritonite localisée toxique à pus fétide. Présence de microbes anaérobies dans le pus. *Gaz. Hebd. Med. Chir.* **101**, 601 (1898).

1408.   G. Potez and A. Compagnon, Sur la bacille anaérobie isolé d'une cholecystite suppurés chez l'homme: *Bacillus trichoides. C. R. Seances Soc. Biol. Ses Fil.* **87**, 339 (1922).

1409.   W. A. Wilson, Acute cholecystitis due to gas-producing organisms. *Br. J. Surg.* **45**, 333 (1958).

1410.   W. H. Cole, Suppurative cholangitis. *Surg. Clin. N. Amer.* **27**, 23 (1947).

1411.   C. J. Heifetz and H. R. Senturia, Acute pneumocholecystitis. *Surg., Gynecol. Obstet.* **86**, 424 (1948).

1412.   S. H. Rubin, P. K. Bornstein, C. Perrine, A. D. Rubin, and D. Schwimmer, Aureomycin treatment of recurrent cholangitis with liver abscesses due to *Bacteroides funduliformis. Ann. Intern. Med.* **35**, 468 (1951).

1413.   B. Tedesco, Pyo-pneumo-cholécyste. Contribution à l'étude de la cholécystite gangrénogazeuse. *Arch. Mal. Appar. Dig. Mal. Nutr.* **40**, 906 (1951).

1414.   H. Beerens, Procède de différenciation entre *Spherophorus necrophorus* (Schmorl 1891) et *Spherophorus funduliformis* (Hallé 1898). *Ann. Inst. Pasteur, Paris* **86**, 384 (1954).

1415.   F. C. Bigler, Acute gaseous cholecystitis. *Am. J. Med.* **29**, 181 (1960).

1416.   L. Parsons, Jr., Pneumocholecystitis. *Am. J. Surg.* **106**, 544 (1963).

1417.   J. M. Bennett and P. J. M. Healey, Spherocytic hemolytic anemia and acute cholecystitis caused by *Clostridium welchii. N. Engl. J. Med.* **268**, 1070 (1963).

1418.   R. B. Sawyer, T. J. Kennedy, J. E. List, and K. C. Sawyer, Acute emphysematous cholecystitis. *Arch. Surg. (Chicago)* **86**, 484 (1963).

1419.   J. S. Aldrete and E. S. Judd, Gas gangrene. *Arch. Surg. (Chicago)* **90**, 745 (1965).

1419a. R. V. Sarmiento, Emphysematous cholecystitis. *Arch. Surg.* **93**, 1009 (1966).

1420.   W. J. Boerema and R. A. McWilliam, Emphysematous cholecystitis: An unusual form of presentation. *Aust. N. Z. J. Surg.* **39**, 258 (1970).

1420a. M. Davies and N. C. Keddie, Abdominal actinomycosis. *Br. J. Surg.* **60**, 18 (1973).

1420b. F. H. Fukunaga, Gallbladder bacteriology, histology, and gallstones. *Arch. Surg.* **106**, 169 (1973).

1420c. K. Shimada, E. Yamazaki, K. Adachi, M. Koyasa, and K. Sawada, Interesting cases of *Bacteroides fragilis* infection. *In* "Third Symposium on Anaerobic Bacteria and Their Infectious Diseases" (S. Ishiyama *et al.*, eds.), p. 55. Eisai, Tokyo, Japan, 1973.

1420d. D. D. Von Hoff and R. S. Imball, *Eikenella corrodens* infection. *Ann. Intern. Med.* **81**, 273 (1974).

1420e. A. W. Chow, J. Z. Montgomerie, and L. B. Guze, Parenteral clindamycin therapy for severe anaerobic infections. *Arch. Intern. Med.* **134**, 78 (1974).

1420f. D. A. Leigh, Clinical importance of infections due to *Bacteroides fragilis* and role of antibiotic therapy. *Br. Med. J.* **2**, 225 (1974).

1420g. J. L. Davis, F. D. Milligan, and J. L. Cameron, Septic complications following endoscopic retrograde cholangiopancreatography. *Surg., Gynecol. Obstet.* **140**, 365 (1975).

1421. F. Patocka and J. Laplanche, Étude des caractères biochimiques de deux espèces de *Spherophorus*: *Sph. gulosus* et *Sph. mortiferus. Ann. Inst. Pasteur, Paris* **73**, 1206 (1947).

1422. L. Wechsler, Forschungsergebnisse aus Medizin und Naturwissenschaft. *Med. Klin. (Munich)* **31**, 1110 (1935).

1423. G. Podgorny, Splenic abscess causing obstruction of the large intestine: First reported case. *Am. Surg.* **37**, 269 (1971).

1423a. A. L. Rosenblum, H. Bonner, Jr., M. S. Milder, V. J. Brenner, M. A. Weinstein, A. J. Cook, and P. P. Carbone, Cavitating splenic infarction. *Am. J. Med.* **56**, 720 (1974).

1423b. T. Gadacz, L. W. Way, and J. E. Dunphy, Changing clinical spectrum of splenic abscess. *Am. J. Surg.* **128**, 182 (1974).

1424. J. Rosenthal and M. J. Tobias, Acute phlegmenous gastritis with multiple perforations. *Am. J. Surg.* **59**, 117 (1943).

1425. A. Behrend, A. B. Katz, and J. W. Robertson, Acute necrotizing gastritis. *AMA Arch. Surg.* **69**, 18 (1954).

1426. L. L. Gonzalez, C. Schowengerdt, H. H. Skinner, and P. Lynch, Emphysematous gastritis. *Surg., Gynecol. Obstet.* **116**, 79 (1963).

1427. T. J. Smith, Emphysematous gastritis associated with adenocarcinoma of the stomach. *J. Lipid Res.* **7**, 327 (1966).

1428. L. F. Urdaneta, R. P. Belin, J. Cueto, and R. C. Doberneck, Intramural gastric actinomycosis. *Surgery* **62**, 431 (1967).

1429. M. D. Lagios and M. J. Suydam, Emphysematous gastritis with perforation complicating phytobezoar. *Am. J. Dis. Child.* **116**, 202 (1968).

1430. W. Ringler and S. Tarbiat, Interessante Einzelbeobachtungen. Gasbrand des Magens. *Zentralbl. Chir.* **94**, 103 (1969).

1430a. G. E. Smith, Subacute phlegmonous gastritis simulating intramural neoplasm: Case report and review. *Gastrointest. Endosc.* **19**, 23 (1972).

1431. V. K. Russ, Ueber ein Influenzabacillen-ahnliches anaerobes Stabchen. *Zentralbl. Bakteriol., Parasitenkd., Infektionskr. Hyg., Abt. 1: Orig.* **39**, 357 (1905).

1432. H. Lodenkämper, Beitrag zur Ursache von Gangranosen bzw. fotiden Eiterungen. *Z. Hyg.* **130**, 260 (1949).

1433. V. Bokkenheuser, Étude d'une nouvelle espéce anaérobie du genre *Pasteurella*: *P. serophila. Ann. Inst. Pasteur, Paris* **80**, 548 (1951).

1434. M. Pereira, Contribution to the study of non-sporing anaerobes with particular reference to the selective cultivation of the genus *Fusiformis*. M.Sc. Thesis, Laval University, Quebec, Canada (1958).

1435. D. Danielsson, D. W. Lambe, Jr., and S. Persson, The immune response in a patient to an infection with *Bacteroides fragilis* ss. *fragilis* and *Clostridium difficile*. *Acta Pathol. Microbiol. Scand., Sect. B* **80**, 709 (1972).

1435a. G. Marks, W. V. Chase, and T. B. Mervine, The fatal potential of fistula-in-ano with abscess: Analysis of 11 deaths. *Dis. Colon Rectum* **16**, 224 (1973).

1435b. A. Larcan, M. C. Laprevote-Heully, G. Fieve, and J. F. Haemmerle, La gangrene gazeuse à propos de 24 observations intérêt de l'association chirurgie-oxygenotherapie hyperbare. *Ann. Chir.* **28**, 445 (1974).

1435c. T. D. Wilkins, W. E. C. Moore, S. E. H. West, and L. V. Holdeman, *Peptococcus niger* (Hall) Kluyver and van Niel 1936: Emendation of description and designation of neotype strain. *Int. J. Syst. Bacteriol.* **25**, 47 (1975).

1436. J. Albarran and J. Cottet, Note sur le rôle des microbes anaérobies dans les infections urinaires. *Sess. Assoc. Fr. Urol., 3rd, 1898*, p. 83 (1898).

1437. J. Albarran and J. Cottet, Des infections urinaires anaérobies. *Trans. Cong. Int. Med. Paris, C. R. 1900, Sect. Chir. Urinaire* **11**, 281 (1900).

1438. O. Wyss, Ueber einen neuen anaeroben pathogenen Bacillus. Beitrag zur Aetiologie der akuten Osteomyelitis. *Mitt. Grenzgeb. Med. Chir.* **13**, 199 (1904).

1439. M. Jungano, Caratteri biologici e culturali dei piu frequenti anaerobii delle affezioni urinarie. *Tommasi* **2**, 289 (1907).

1440. M. Jungano, Sur un cas d'infection renale, d'origine sanguine, due à certains microbes, dont un anaérobie strict (nouvelle espèce). *C. R. Seances Soc. Biol. Ses Fil.* **63**, 302 (1907).

1441. M. Jungano, "La flore de l'appareil urinaire normal et pathologique." Jacques, Paris, 1908.

1442. M. Jungano, Un caso di uretrite acuta non gonococcica. *Assoc. Ital. Urol. Roma* **1**, 378 (1908).

1443. V. Babes and A. Babes, Note sur un cas de phlegmon emphysemateux et sur son microbe. *C. R. Seances Soc. Biol. Ses Fil.* **66**, 324 (1909).

1444. E. Rach and A. von Reuss, Zur Aetiologie der Cystitis im Sauglingsalter (*Bacillus bifidus communis* und ein Paracoli-bacillus). *Zentralbl. Bakteriol., Parasitenkd., Infektionskr. Hyg., Abt. 1: Orig.* **50**, 169 (1909).

1445. R. R. Mellon, A contribution to the bacteriology of a fuso-spirillary organism, with special reference to its life history. *J. Bacteriol.* **4**, 505 (1919).

1446. W. W. Oliver and W. B. Wherry, Notes on some bacterial parasites of the human mucous membranes. *J. Infect. Dis.* **28**, 341 (1921).

1447. I. C. Hall, *Micrococcus niger*. A new pigment-forming anaerobic coccus recovered from urine in a case of general arteriosclerosis *J. Bacteriol.* **20**, 407 (1930).

1448. L. Thompson and D. C. Beaver, Bacteremia due to anaerobic gram-negative organisms of the genus *Bacteroides*. *Med. Clin. N. Amer.* **15**, 1611 (1932).

1449. J. B. Lee, Cystitis emphysematosa. *Arch Intern. Med.* **105**, 618 (1960).

1450. T. L. Schulte, *Bacteroides* and anaerobic streptococci in infection of the urinary tract: Report of case. *Proc. Staff Meet. Mayo Clinic* **14**, 536 (1939).

1451. W. Schultz, Über Infektionen des Menschen mit streng anaeroben, gramnegativen Stabchen, insbesondere mit dem *Bacterium symbiophiles*. *Zentralbl. Gynaekol.* **63**, 2458 (1939).

1452. B. B. Madison, Perinephric abscess. Review of twenty-one cases, with special reference to an anaerobic infection. *Wis. Med. J.* **39**, 932 (1940).

1453. A. T. Willis, "Clostridia of Wound Infection." Butterworths, London, 1969.

1454. M. Galinovic-Weisglass, Old and new about actinomycosis. *Lijec. Vjesn.* **89**, 179 (1967).

1455. A. R. Prévot, E. Pouliquen, and P. Tardieux, Contribution à l'étude des spherophoroses: Aspect chirurgical du pouvoir pathogène de "*Spherophorus freundi*." *Bull. Acad. Natl. Med., Paris* **138**, 308 (1954).

1456. L. Persky, G. Austen, Jr., and W. E. Schatten, Recent experiences with prostatic abscess. *Surg., Gynecol. Obstet.* **101**, 629 (1955).

1457. H. Lodenkämper, M. Fischer, and H. Nickel, Über die klinische Bedeutung und Differentialdiagnose der anaeroben *Corynebakterien. Z. Hyg.* **143**, 467 (1957).

1458. A. D. Amar and R. K. Ratliff, Gas gangrene of urinary bladder and abdominal wall following catheterization. *J. Urol.* **80**, 130 (1958).

1459. G. A. Jutzler, W. Leppla, and H.-J. Brunck, Primare doppelseitige Nierenaktinomykose mit Papillennekrose als Ursache eines Nierenversagens. *Arzneim. Forsch.* **15**, 340 (1961).

1460. L. H. Stahlgren and G. Thabit, Jr., An important sign of intraabdominal abscess. *Ann. Surg.* **153**, 126 (1961).

1461. M. C. McHenry, W. J. Martin, M. M. Hargraves, and A. H. Baggenstoss, Bacteremia due to *Clostridium perfringens* complicating leukemia: Report of a case with associated clostridial pyelonephritis. *Proc. Staff Meet. Mayo Clin.* **38**, 23 (1963).

1462. E. G. Lufkin, M. Silverman, J. J. Callaway, and H. Glenchur, Mixed septicemias and gastrointestinal disease. *Am. J. Dig. Dis.* **11**, 930 (1966).

1463. M. A. Feldman, R. E. Cotton, and W. M. Gray, Carcinoma of the colon presenting as left perinephric abscess. *Br. J. Surg.* **55**, 21 (1968).

1464. T. Nishiura, M. Tamura, K. Ueno, and K. Ninomiya, Spontaneous cure of acute cystitis and judgment of drug effect: Clinical trial of cephalexin. *Acta Urol. Jpn.* **16**, 185 (1970).

1465. H. E. Pearson and G. V. Anderson, 659. Genital bacteroidal abscesses in women. *Am. J. Obstet. Gynecol.* **107**, 1264 (1970).

1466. F. H. Caselitz, V. Freitag, and F. Lantzius, Zur Kasuistik der Infektion mit anaeroben Staphylokokken. *Z. Med. Mikrobiol. Immunol.* **156**, 259 (1971).

1466a. W. A. Altemeier, W. R. Culbertson, and W. D. Fullen, Intraabdominal sepsis. *Advan. Surg.* **5**, 281 (1971).

1466b. Y. Shimizu and Z. Mo, Anaerobic organisms isolated from urine specimens (clinical study). *In* "Third Symposium on Anaerobic Bacteria and Their Infectious Diseases" (S. Ishiyama *et al.*, eds), p. 92. Eisai, Tokyo, Japan, 1973.

1466c. H. Werner, G. Rintelen and C. Lohner, Identifizierung und medizinische Bedeutung von 8 aus pathologischem Material isolierten Clostridium-Arten. *Zbl. Bakt. Hyg. I. Abt. Orig. A.* **224**, 220 (1973).

1467. J. P. Marcel, F. N. Guillermet, D. Hautier, and P. Vincent, Septicémies à *Ristella insolita* et *pseudo-insolita. Lyon Med.* **224**, 835 (1970).

1468. N. M. Sullivan, V. L. Sutter, M. M. Mims, V. H. Marsh, and S. M. Finegold, Bacteremia following genito-urinary tract manipulation. Clinical Aspects. *J. Infect. Dis.* **127**, 49 (1973).

1469. J. A. Guilbeau, E. B. Schoenbach, I. G. Schaub, and D. V. Latham, Aureomycin in obstetrics: Therapy and prophylaxis. *J. Am. Med. Assoc.* **143**, 520 (1950).

1470. J. W. Harris and J. H. Brown, A clinical and bacteriological study of 113 cases of streptococcic puerperal infection. *Bull. Johns Hopkins Hosp.* **44**, 1 (1929).

1471. O. Schwarz and W. J. Dieckmann, Anaerobic streptococci: Their role in puerperal infection. *South. Med. J.* **19**, 470 (1926).

1471a. D. F. Hawkins, L. H. Sevitt, P. F. Fairbrother, and A. U. Tothill: Management of septic abortion with renal failure. *N. Engl. J. Med.* **292**, 722 (1975).

1472. T. K. Brown, The incidence of puerperal infection due to anaerobic streptococci. *Am. J. Obstet. Gynecol.* **20**, 300 (1930).

1473. O. H. Schwarz and W. J. Dieckmann, Puerperal infection due to anaerobic streptococci. *Am. J. Obstet. Gynecol.* **13**, 467 (1927).

1474. W. J. Ledger, A. M. Reite, and J. T. Headington, A system for infectious disease surveillance on an obstetric service. *Obstet. Gynecol.* **37**, 769 (1971).

1475. S. R. Steinhorn, The possible role of bacterial synergism in puerperal infections due to anaerobic streptococci. *Am. J. Obstet. Gynecol.* **50**, 63 (1945).

1475a. L. T. Hibbard, E. N. Snyder, and R. M. McVann, Subgluteal and retropsoal infection in obstetric practice. *Obstet. Gynecol.* **39**, 137 (1972).

1475b. D. R. Wenger and R. G. Gitchell, Severe infections following pudendal block anesthesia: Need for orthopaedic awareness. *J. Bone Jt. Surg.* **55A**, 202 (1973).

1475c. R. L. Sweet and W. J. Ledger, Puerperal infectious morbidity. *Am. J. Obstet. Gynecol.* **117**, 1093 (1973).

1476. L. Boez and R. Keller, Septicémie anaérobie à *"Micrococcus foetidus"* de Veillon diagnostic par la méthode d'hémoculture anaérobie en milieu solide (1). *Bull. Mem. Soc. Med. Hôp. Paris* [3] **48**, 1757 (1924).

1477. L. Nash, N. A. Janovski, and S. M. Bysshe, Localized clostridial chorioamnionitis. *Obstet. Gynecol.* **21**, 481 (1963).

1477a. J. MacVicar: Chorioamnionitis. *In*: "Obstetric and Perinatal Infections," (D. Charles and M. Finland, eds.). Lea & Febiger, Philadelphia, 1973.

1478. B. Carter, C. P. Jones, R. A. Ross, and W. L. Thomas, A bacteriologic and clinical study of pyometra. *Am. J. Obstet. Gynecol.* **62**, 793 (1951).

1479. J. R. Willson and J. R. Black, III, Ovarian abscess. *Am. J. Obstet. Gynecol.* **90**, 34 (1964).

1480. W. J. Ledger, C. Campbell, D. Taylor, and J. R. Willson, Adnexal abscess as a late complication of pelvic operations. *Surg., Gynecol. Obstet.* **129**, 973 (1969).

1481. A. Mickal and A. H. Sellmann, Management of tubo-ovarian abscess. *Clin. Obstet. Gynecol.* **12**, 252 (1969).

1482. S. Friedman and M. L. Bobrow, Pelvic inflammatory disease in pregnancy. *Obstet. Gynecol.* **14**, 417 (1959).

1483. A. G. Dudley, F. Lee, and D. Barclay, Ovarian and tubo-ovarian abscess in pregnancy: Report of a case and a review of the literature. *Mil. Med.* **135**, 403 (1970).

1484. J. Lip and X. Burgoyne, Cervical and peritoneal bacterial flora associated with salpingitis. *Obstet. Gynecol.* **28**, 561 (1966).

1485. C. A. D. Ringrose, Clinical, etiological, and economic aspects of salpingitis. *Can. Med. Assoc. J.* **83**, 53 (1960).

1486. W. A. Nebel and W. E. Lucas, Management of tubo-ovarian abscess. *Obstet. Gynecol.* **32**, 382 (1968).

1486a. A. W. Chow, K. L. Malkasian, J. R. Marshall, and L. B. Guze, Acute pelvic inflammatory disease and clinical response to parenteral doxycycline. *Antimicrob. Ag. Chemother.* **7**, 133 (1975).

1486b. A. W. Chow, K. L. Malkasian, J. R. Marshall, and L. B. Guze. The bacteriology of acute pelvic inflammatory disease. *Am. J. Obstet. Gynecol.* **122**, 876 (1975).

1486c. D. A. Eschenbach, T. M. Buchanan, H. M. Pollock, P. S. Forsyth, E. R. Alexander, J.-S. Lin, S.-P. Wang, B. B. Wentworth, W. M. McCormack, and K. K. Holmes, Polymicrobial etiology of acute pelvic inflammatory disease. *N. Engl. J. Med.* **293**, 166 (1975).

1487. W. L. Hall, I. Sobel, C. P. Jones, and R. T. Parker, Anaerobic postoperative pelvic infections. *Obstet. Gynecol.* **30**, 1 (1967).

1488. W. J. Ledger, C. Campbell, and J. R. Willson, Postoperative adnexal infections. *Obstet. Gynecol.* **31**, 83 (1968).

1489. E. E. Lee and M. Turko, A survey of postoperative infection on a gynecological service. *Can. Med. Assoc. J.* **84**, 1302 (1961).

1490. W. J. Ledger, R. L. Sweet, and J. T. Headington, Prophylactic cephaloridine in the prevention of postoperative pelvic infections in premenopausal women undergoing vaginal hysterectomy. *Am. J. Obstet. Gynecol.* **115**, 766 (1973).

1491. H. Kamiya, K. Ninomiya, T. Ushigima, K. Ueno, and S. Suzuki, Six cases of anaerobic bacterial infections after gynecological operations. *In* "Second Symposium on Anaerobic Bacteria and Their Infectious Diseases" (S. Ishiyama *et al.*, eds.), p. 113. Eisai, Tokyo, Japan, 1972.

1492. H. P. Jones, Septicaemia due to *Bact. necrophorum* and an anaerobic streptococcus. *Lancet* **2**, 824 (1944).

1493. C. S. Stevenson and C.-C. Yang, Septic abortion with shock. *Am. J. Obstet. Gynecol.* **83**, 1229 (1962).

1494. R. H. Adams and J. A. Pritchard, Bacterial shock in obstetrics and gynecology. *Obstet. Gynecol.* **16**, 387 (1960).

1495. M. Spring and S. Kahn, Nonclostridial gas infection in the diabetic. *Arch. Intern. Med.* **88**, 373 (1951).

1496. G. Qvist, Anaerobic cellulitis and gas gangrene. *Br. Med. J.* **2**, 217 (1941).

1497. M. R. Wills and M. W. Reece, Non-clostridial gas infections in diabetes mellitus. *Br. Med. J.* **2**, 566 (1960).

1498. J. T. Flynn and C. M. Karpas (discussed by D. B. Louria), Clinicopathologic Conference—cirrhosis, septicemia, and subcutaneous gas formation. *N.Y. State J. Med.* **67**, 2852 (1967).

1499. J. W. Meade and C. B. Mueller, Necrotizing infections of subcutaneous tissue and fascia. *Ann. Surg.* **168**, 274 (1968).

1499a. R. L. Nichols and J. W. Smith, Gas in the wound: What does it mean? *Surg. Clin. N. Amer.* **55**, 1289 (1975).

1499b. D. B. Roberts and L. L. Hester, Jr., Progressive synergistic bacterial gangrene arising from abscesses of the vulva and Bartholin's gland duct. *Am. J. Obstet. Gynecol.* **114**, 285 (1972).

1499c. R. Rudolph, M. Soloway, R. G. DePalma, and L. Persky, Fournier's syndrome: Synergistic gangrene of the scrotum. *Am. J. Surg.* **129**, 591 (1975).

1500. P. H. Hennessy and W. Fletcher, Infection with the organisms of Vincent's angina following man-bite. *Lancet* **2**, 127 (1920).

1501. I. Pilot and K. A. Meyer, Fusiform bacilli and spirochetes. XII. Occurrence in Gangrenous lesions of fingers: Report of a case. *Arch. Dermatol. Syph.* **12**, 837 (1925).

1502. G. C. Bower and H. B. Lang, A case report of finger infection due to fusi-spirochetal organisms. *N. Y. State J. Med.* **30**, 975 (1930).

1503. C. E. Welch, Human bite infections of the hand. *N. Engl. J. Med.* **215**, 901 (1936).

1504. R. M. Smith and W. F. Manges, Roentgen treatment of infection from human bite. *Am. J. Roentgenol. Radium Ther.* [ns] **38**, 720 (1937).

1505. R. Cohn, Infections of the hand following human bites. *Surgery* **7**, 546 (1940).

1506. E. C. G. Butler, Anaerobic infection of the finger. Systemic penicillin. *Proc. Roy. Soc. Med.* **38**, 69 (1944).

1507. A. T. Andreasen, Bone abscess from human bite. Report of a case. *Br. J. Surg.* **34**, 411 (1947).

1508. P. Tardieux, Étude de deux espèces nouvelles du genre *Spherophorus*. *Ann. Inst. Pasteur, Paris* **80**, 275 (1951).

1509. E. C. Rosenow and R. Tunnicliff, Pyemia due to an anaerobic polymorphic bacillus, probably *Bacillus fusiformis*. *J. Infect. Dis.* **10**, 1 (1912).

1510. L. Colebrook, A report upon 25 cases of actinomycosis, with especial reference to vaccine therapy. *Lancet* **1**, 893 (1921).

1511. V. Z. Cope, Actinomycosis of bone with special reference to infection of the vertebral column. *J. Bone Jt. Surg., Br. Vol.* **33**, 205 (1951).

1512. K. E. Fritzell, Infections of the hand due to human mouth organisms. *J.—Lancet* **60**, 135 (1940).

1513. H. J. Burrows, Actinomycosis from punch injuries with a report of a case affecting a metacarpal bone. *Br. J. Surg.* **32**, 506 (1945).

1514. M. G. Henry, Conservative treatment of human bite infections. Report of two cases. *Mil. Surg.* **97**, 122 (1945).

1515. L. J. McCormack, J. A. Dickson, and A. R. Reich, Actinomycosis of the humerus. *J. Bone Jt. Surg., Am. Vol.* **36**, 1255 (1954).

1516. W. M. Wearne, Actinomycosis of the finger. *Proc. Roy. Soc. Med.* **53**, 884 (1960).

1517. N. Statman and S. Spitzer, *Bacteroides* septicemia and osteomyelitis. *Ohio State Med. J.* **58**, 1374 (1962).

1518. R. W. McIntire, Actinomycotic osteomyelitis of the fibula. *Clin. Orthop.* **27**, 206 (1963).

1519. P. J. Kelly, W. J. Martin, and M. B. Coventry, Bacterial arthritis of the hip in the adult. *J. Bone Jt. Surg., Am. Vol.* **47**, 1005 (1965).

1520. B. G. Mendelsohn, Actinomycosis of a metacarpal bone. *J. Bone Jt. Surg., Br. Vol.* **47**, 739 (1965).

1521. M. R. Himalstein, Osteomyelitis of the maxilla due to an anaerobic organism. Case report and management. *Laryngoscope* **77**, 559 (1967).

1522. P. J. Kelly, W. J. Martin, and M. B. Coventry, Chronic osteomyelitis. II. Treatment with closed irrigation and suction. *J. Am. Med. Assoc.* **213**, 1843 (1970).

1523. W. M. Kirsch and J. C. Stears, Actinomycotic osteomyelitis of the skull and epidural space. *J. Neurosurg.* **33**, 347 (1970).

1524. E. B. D. Neuhauser, Actinomycosis and botryomycosis. *Postgrad. Med.* **48**, 59 (1970).

1525. F. A. Waldvogel, G. Medoff, and M. N. Swartz, Osteomyelitis: A review of clinical features, therapeutic considerations and unusual aspects. *N. Engl. J. Med.* **282**, 198, 260, and 316 (1970).

1525a. W. A. Limongelli, B. Connaughton, and A. C. Williams, Suppurative osteomyelitis of the mandible secondary to fracture. *Oral Surg., Oral Med. Oral Pathol.* **38**, 850 (1974).

1525b. P. M. Sharp, R. C. Meador, and R. R. Martin, A case of mixed anaerobic infection of the jaw. *J. Oral Surg.* **32**, 457 (1974).

1525c. H. D. Isenberg, L. S. Lavine, B. G. Painter, W. H. Rubins, and J. I. Berkman, Primary osteomyelitis due to an anaerobic microorganism. *Am. J. Clin. Pathol.* **64**, 385 (1975).

1525d. M. Silbermann, F. J. Chiminello, H. C. Doku, and P. L. Maloney, Mandibular actinomycosis: Report of case. *J. Am. Dent. Assoc.* **90**, 162 (1975).

1525e. P. J. Kelly, C. J. Wilkowske and J. A. Washington, II: Comparison of gram-negative bacillary and staphylococcal osteomyelitis of the femur and tibia. *Clin. Orthop. Related Res.* **96**, 70 (1973).

1525f. M. C. McHenry, R. J. Alfidi, A. H. Wilde, and W. A. Hawk, Hematogenous osteomyelitis. A changing disease. *Cleveland Clin. Quart.* **42**, 125 (1975).

1525g. M. E. Levison, J. L. Bran, and K. Ries, Treatment of anaerobic bacterial infections with clindamycin-2-phosphate. *Antimicrob. Ag. Chemother.* **5**, 276 (1974).

1525h. P. J. Kelly, C. J. Wilkowske, and J. A. Washington, II, Musculoskeletal infections due to *Serratia marcescens*. *Clin. Orthop. Related Res.* **96**, 76 (1973).

1525i. J. H. Newman and R. G. Mitchell, Diphtheroid infection of the cervical spine. *Acta Orthop. Scand.* **46**, 67 (1975).

1525j. L. Gold and E. E. Doyne, Actinomycosis with osteomyelitis of the alveolar process. *Oral Surg.* **5**, 1056 (1952).

1525k. D. Stenhouse and D. G. MacDonald, Low grade osteomyelitis of the jaws with actinomycosis. *Int. J. Oral Surg.* **3**, 60 (1974).

1525l. J. H. P. Main and I. T. MacPhee, Actinomycosis of the maxilla in relation to a periodontal abscess. *Oral Surg.* **17**, 299 (1964).

1526. J. Ernst and E. Ratjen, Actinomycosis of the spine. *Acta Orthop. Scand.* **42**, 35 (1971).

1527. P. Godeau, J. Hewitt, A. Meyer, M. Sebald, and D. Sicard, Actinomycose disséminée avec foyers cutanes et osseux. *Ann. Med. Intern.* **122**, 1129 (1971).

1528. R. Lewis and S. L. Gorbach, *Actinomyces viscosus* in man. *Lancet* **1**, 641 (1972).

1529. Editorial, *Bacteroides* bacteraemia. *Br. Med. J.* **1**, 686 (1973).

1530. L. Albright, S. Toczek, V. J. Brenner, and A. K. Ommaya, Osteomyelitis and epidural abscess caused by *Arachnia propionica*. Case Report. *J. Neurosurg.* **40**, 115 (1974).

1531. G. F. Brooks, J. M. O'Donoghue, J. P. Rissing, K. Soapes, and J. W. Smith, *Eikenella corrodens*, a recently recognized pathogen. *Medicine* (*Baltimore*) **53**, 325 (1974).

1532. A. A. Quayle, *Bacteroides* infections in oral surgery. *J. Oral Surg.* **32**, 91 (1974).

1533. G. A. L. Inge and F. L. Liebolt, The treatment of acute suppurative arthritis: Report of thirty-six cases treated by operation. *Surg., Gynecol. Obstet.* **60**, 86 (1935).

1534. J. A. Heberling, A review of two hundred and one cases of suppurative arthritis. *J. Bone Jt. Surg.* **23**, 917 (1941).

1535. W. A. Altemeier and T. Largen, Antibiotic and chemotherapeutic agents in infections of the skeletal system. *J. Am. Med. Assoc.* **150**, 1462 (1952).

1536. M. B. Watkins, R. L. Samilson, and D. M. Winters, Acute suppurative arthritis. *J. Bone Jt. Surg., Am. Vol.* **38**, 1313 (1956).

1537. J. H. Kellgren, J. Ball, R. W. Fairbrother, and K. L. Barnes, Suppurative arthritis complicating rheumatoid arthritis. *Br. Med. J.* **1**, 1193 (1958).

1538. Y. Chartier, W. J. Martin, and P. J. Kelly, Bacterial arthritis: Experiences in the treatment of 77 patients. *Ann. Intern. Med.* **50**, 1462 (1959).

1539. J. Ward, A. S. Cohen, and W. Bauer, The diagnosis and therapy of acute suppurative arthritis. *Arthritis Rheum.* **3**, 522 (1960).

1540. R. F. Willkens, L. A. Healy, and J. L. Decker, Acute infectious arthritis in the aged and chronically ill. *Arch. Intern. Med.* **106**, 354 (1960).

1541. A. C. Ortiz and W. E. Miller, Treatment of a septic joint. *South. Med. J.* **54**, 594 (1961).

1542. A. Baitch, Recent observations of acute suppurative arthritis. *Clin. Orthop.* **22**, 157 (1962).

1543. L. Borella, J. E. Goobar, R. L. Summitt, and G. M. Clark, Septic arthritis in childhood. *J. Pediatr.* **62**, 742 (1962).

1544. R. J. Argen, C. H. Wilson, Jr., and P. Wood, Suppurative arthritis: Clinical features of 42 cases. *Arch. Intern. Med.* **117**, 661 (1966).

1545. J. H. Bulmer, Septic arthritis of the hip in adults. *J. Bone Jt. Surg., Br. Vol.* **48**, 289 (1966).

1546. P. J. Kelly, W. J. Martin, and M. B. Coventry, Bacterial (suppurative) arthritis in the adult. *J. Bone Jt. Surg., Am. Vol.* **52**, 1595 (1970).

1547. T. M. Kolawole and S. P. Bohrer, Acute septic arthritis in Nigeria: A review of 65 cases involving the hip and shoulder joints. *Trop. Geogr. Med.* **24**, 327 (1972).

1548. J. D. Nelson, The bacterial etiology and antibiotic management of septic arthritis in infants and children. *Pediatrics* **50**, 437 (1972).

1548a. R. D. Feigin, L. K. Pickering, D. Anderson, R. E. Keeney and P. G. Shackleford: Clindamycin treatment of osteomyelitis and septic arthritis in children. *Pediatrics* **55**, 213 (1975).

1549. A. S. Russell and B. M. Ansell, Septic arthritis. *Ann. Rheum. Dis.* **31**, 40 (1972).

1550. L. Lidgren and L. Lindberg, Twenty-nine cases of bacterial arthritis. *Acta Orthop. Scand.* **44**, 263 (1973).

1551. J. M. Singh and A. N. Bhattacharyya, Septic arthritis in children. *Indian J. Pediatr.* **40**, 129 (1973).

1552. A. Bogdan, Eine bisher unbekannte Infektionskrankheit bei Verwundeten. *Med. Klin.* (*Munich*) **12**, 383 (1916).

1553. G. E. Rockwell, The influence of carbon dioxide on the growth of bacteria. *J. Infect. Dis.* **32**, 98 (1923).

1554. P. Teissier, J. Reilly, E. Rivalier, and V. Stefanesco, Les septicémies primitives dués au *Bacillus funduliformis. Ann. Med. (Paris)* **30**, 97 (1931).

1555. A. Lemierre, R. Grégoire, A. Laporte, and R. Couvelaire, Les aspects chirurgicaux des infections à "*Bacillus funduliformis.*" *Bull. Acad. Med.*, **119**, 352 (1938).

1556. R. H. Williams, Fusospirochetosis: Recovery of the causative organisms from the blood, with report of two cases. *Arch. Intern. Med.* **68**, 80 (1941).

1557. L. Dienes, The isolation of L type cultures from *Bacteroides* with the aid of penicillin and their reversion into the usual bacilli. *J. Bacteriol.* **56**, 445 (1948).

1558. H. Harders and H. Hornbostel, Infektion mit *Bact. necroseos* beim Menschen. *Zentralb. Bakteriol. Parasitenkd. Infektionskr. Hyg., Abt. 1: Orig.* **156**, 582 (1951).

1559. J. M. Miller and R. L. Engle, Jr., Metastatic suppurative arthritis with subcutaneous emphysema caused by *Escherichia coli. Am. J. Med.* **10**, 241 (1951).

1560. E. Jansson, O. Wager, M. Miettinen, and S. Ignatius, *Bacteroides* Markaisen Niveltulehduksen Aiheuttajana. *Duodecim* **81**, 587 (1965).

1561. J. McNae, An unusual case of *Clostridium welchii* infection. *J. Bone Jt. Surg., Br. Vol.* **48**, 512 (1966).

1562. J. D. Nelson and W. C. Koontz, Septic arthritis in infants and children: A review of 117 cases. *Pediatrics* **38**, 966 (1966).

1563. W. S. Smith and R. M. Ward, Septic arthritis of the hip complicating perforation of abdominal organs. *J. Am. Med. Assoc.* **195**, 1148 (1966).

1564. M. E. Ament and S. A. Gaal, *Bacteroides* arthritis. *Am. J. Dis. Child.* **114**, 427 (1967).

1565. J. J. Comtet, G. Henry, L. Fischer, and Y. Auffray, Aspect actual des arthrites aigues à germes non spécifiqués. *Lyon Med.* **29**, 727 (1968).

1566. J. S. Torg and T. R. Lammot, III, Septic arthritis of the knee due to *Clostridium welchii.* Report of two cases. *J. Bone Jt. Surg., Am. Vol.* **50**, 1233 (1968).

1567. I. Karten, Septic arthritis complicating rheumatoid arthritis. *Ann. Intern. Med.* **70**, 1147 (1969).

1568. I. Ziment, A. Davis, and S. M. Finegold, Joint infection by anaerobic bacteria: A case report and review of the literature. *Arthritis Rheum.* **12**, 627 (1969).

1569. P. Bradley, Actinomycosis of the temporomandibular joint. *Br. J. Oral Surg.* **9**, 54 (1971).

1570. R. G. Aptekar and A. D. Steinberg, *Peptococcus* septic arthritis after surgery in rheumatoid arthritis. *Clin. Orthop.* **88**, 92 (1972).

1571. B. Nolan, W. D. Leers, and J. Schatzker, Septic arthritis of the knee due to *Clostridium bifermentans. J. Bone Jt. Surg., Am. Vol.* **54**, 1275 (1972).

1572. J. D. Schlenker, G. Vega, and K. G. Heiple, *Clostridium* pyoarthritis of the shoulder associated with multiple myeloma. *Clin. Orthop. Relat. Res.* **88**, 89 (1972).

1573. A. W. Chow and L. B. Guze, Bacteroidaceae bacteremia: Clinical experience with 112 patients. *Medicine (Baltimore)* **53**, 93 (1974).

1574. A. L. Esposito and R. A. Gleckman, Acute polymicrobic septic arthritis in the adult: Case report and literature review. *Am. J. Med. Sci.* **267**, 251 (1974).

1575. P. B. Sherer and J. Dobbins, Actinomycosis arthritis: A case report. *Med. Ann. D.C.* **43**, 66 (1974).

1576. B. Albrecht, Infektionen durch Nekrosebacillen. *In* "Handbuch der Pathologischen Mikroorganism" (W. Kolle, R. Kraus, and P. Uhlenhuth, eds.), Vol. 6, p. 673. Fischer, Jena, 1929.

1577. A. Athanasiu, Angine ulcero-membraneuse aigue à *Bacilles fusiformes* de Vincent et spirilles, Thesis, Faculty of Medicine, University of Paris, Paris (1900).

1578   A. A. Gunn, *Bacteroides* infection in the surgery of childhood. *Arch. Dis. Child.* **32**, 523 (1957).

1579.   G. Kapsenberg and A. E. Beute, Otitis media, Gepaard met septische Toestanden, veroorzaakt door *Bacteroides fundiliformis*. *Ned. Tijdschr. Geneesk.* **90**, 890 (1946).

1580.   D. T. Smith, The diagnosis and treatment of pulmonary abscess in children. *J. Am. Med. Assoc.* **103**, 971 (1934).

1581.   J. Sobel and C. Herrman, Ulceromembranous angina associated with the fusiform bacillus (Vincent): A report of twelve cases in children. *N. Y. Med. J.* **74**, 1037 (1901).

1582.   B. M. King, B. A. Ranck, F. D. Daugherty, and C. A. Rau, *Clostridium tertium* septicemia. *N. Engl. J. Med.* **269**, 467 (1963).

1583.   C. I. Allen and J. F. Blackman, Treatment of lung abscess with report of 100 consecutive cases. *J. Thorac. Surg.* **6**, 156 (1936).

1584.   T. C. Moore and J. S. Battersby, Pulmonary abscess in infancy and childhood. *Ann. Surg.* **151**, 496 (1960).

1585.   C. M. Nice, Abdominal abscesses in children. *South. Med. J.* **52**, 659 (1959).

1586.   C. E. Lucas and A. J. Walt, Acute gangrenous acalculous cholecystitis in infancy: Report of a case. *Surgery* **64**, 847 (1968).

1587.   J. F. Hultgen, Partial gangrene of the left index-finger caused by the symbiosis of the fusiform bacillus and the *Spirochaeta denticola*. *J. Am. Med. Assoc.* **55**, 857 (1910).

1588.   R. C. Brock, "Lung Abscess." Thomas, Springfield, Illinois, 1952.

1589.   E. J. Pulaski, "Common Bacterial Infections." Saunders, Philadelphia, Pennsylvania, 1964.

1590.   J. J. Byrne, "The Hand: Its Anatomy and Diseases." Thomas, Springfield, Illinois, 1959.

1591.   C. G. Dunaif, Fournier's gangrene. Report of a case and review of the literature. *Plast. Reconstr. Surg.* **33**, 84 (1964).

1592.   L. A. French and S. N. Chou, *In* "Cranial and Intracranial Suppuration" (E. S. Gurdjian, ed.), Chapter III. Thomas, Springfield, Illinois, 1969.

Subject Index